T0418600

Perception-Action Cycle

Springer Series in Cognitive and Neural Systems

Volume 1

Series Editors

John G. Taylor
King's College, London, UK

Vassilis Cutsuridis
Boston University, Boston, MA, USA

For further volumes
http://www.springer.com/series/8572

Vassilis Cutsuridis • Amir Hussain
John G. Taylor

Editors

Perception-Action Cycle

Models, Architectures, and Hardware

 Springer

Editors

Vassilis Cutsuridis
Department of Psychology
Center for Memory and Brain
Boston University
Boston, MA 02215
USA
vcut@bu.edu

John G. Taylor
Department of Mathematics
King's College
London
UK
john.g.taylor@kcl.ac.uk

Amir Hussain
Department of Computing Science and
Mathematics
University of Stirling
Stirling FK9 4LA
UK
ahu@cs.stir.ac.uk

ISBN 978-1-4419-1451-4   e-ISBN 978-1-4419-1452-1
DOI 10.1007/978-1-4419-1452-1
Springer New York Dordrecht Heidelberg London

Library of Congress Control Number: 2011920805

Printed on acid-free paper

Springer is part of Springer Science+Business Media (www.springer.com)

Preface

The perception–action cycle has been described by the eminent neuroscientist JM Fuster as the circular flow of information that takes place between the organism and its environment in the course of a sensory-guided sequence of behaviour towards a goal. Each action in the sequence causes certain changes in the environment that are analysed bottom-up through the perceptual hierarchy and leads to the processing of further action, top-down through the executive hierarchy, towards motor effectors. These cause new changes that are analysed and lead to new action, and so on.

This book provides a snapshot and a resumé of the current state-of-the-art of the ongoing research avenues concerning the perception–action cycle. The central aim of the volume is to be an informational resource and a methodology for anyone interested in constructing and developing models, algorithms and hardware implementations of autonomous machines empowered with cognitive capabilities.

The book is divided into three thematic areas: (1) computational neuroscience models, (2) cognitive architectures and (3) hardware implementations. In the first thematic area, leading computational neuroscientists present brain-inspired models of perception, attention, cognitive control, decision making, conflict resolution and monitoring, knowledge representation and reasoning, learning and memory, planning and action, and consciousness grounded on experimental data. In the second thematic area, architectures, algorithms and systems with cognitive capabilities and minimal guidance from the brain are discussed. These architectures, algorithms and systems are inspired from the areas of cognitive science, computer vision, robotics, information theory, machine learning, computer agents and artificial intelligence. In the third thematic area, the analysis, design and implementation of hardware systems with robust cognitive abilities from the areas of mechatronics, sensing technology, sensor fusion, smart sensor networks, control rules, controllability, stability, model/knowledge representation and reasoning are discussed.

This engaging volume will be invaluable to computational neuroscientists, cognitive scientists, robotists, electrical engineers, physicists, mathematicians and others interested in developing cognitive models, algorithms and systems of the perception–action cycle. Graduate level students and trainees in all of these fields will find this book a significant source of information.

Finally, there are many people whom we would like to thank for making this book possible. This includes all the contributing authors who did a great job. We would like to thank Ann H. Avouris, our Springer senior editor, and members of the production team, for their consistent help and support. We dedicate this work to our families.

Boston, USA   *Vassilis Cutsuridis*
Stirling, UK   *Amir Hussain*
London, UK   *John G. Taylor*

Contents

Contributors

William H. Alexander Department of Psychological and Brain Sciences, Indiana University, 1101 E. Tenth Street, Bloomington, IN 47405, USA, wialexan@indiana.edu

Panagiotis K. Artemiadis PostDoctoral Associate, Massachusetts Institute of Technology, 77 Massachusetts Avenue, Cambridge, MA 02139, USA, partem@mit.edu

Yannis Avrithis Image, Video and Multimedia Systems Laboratory, Computer Science Division, School of Electrical and Computer Engineering, National Technical University of Athens, Iroon Polytexneiou 9, 15780 Zografou, Greece, iavr@image.ntua.gr

Joshua W. Brown Department of Psychological and Brain Sciences, Indiana University, 1101 E. Tenth Street, Bloomington, IN 47405, USA, jwmbrown@indiana.edu

Antonio Chella Department of Computer Engineering, University of Palermo, Viale delle Scienze – Building 6, 90128 Palermo, Italy, chella@unipa.it

José L. Contreras-Vidal School of Public Health, Department of Kinesiology, University of Maryland, College Park, MD 20742, USA
and
Graduate Program in Neuroscience and Cognitive Science (NACS), University of Maryland, College Park, MD 20742, USA
and
Graduate Program in Bioengineering, University of Maryland, School of Public Health, College Park, MD 20742, USA, pepeum@umd.edu

L. Andrew Coward College of Engineering and Computer Science, Australian National University, Canberra, ACT 0200, Australia, andrew.coward@anu.edu.au

Cristiano Cuppini Department of Electronics, Computer Science and Systems, University of Bologna, Viale Risorgimento 2, I-40136, Bologna, Italy, cristiano.cuppini@unibo.it

Artur S. d'Avila Garcez Department of Computing, City University, London EC1V 0HB, UK, aag@soi.city.ac.uk

Rodolphe Gentili Department of Kinesiology, University of Maryland School of Public Health, College Park, MD 20742, USA
and
Graduate Program in Neuroscience and Cognitive Science (NACS), University of Maryland School of Public Health, College Park, MD 20742, USA, rodolphe@umd.edu

John N. Karigiannis Division of Signals, Control and Robotics, School of Electrical and Computer Engineering, National Technical University of Athens, Zographou Campus, Athens 15773, Greece, john@fhw.gr

Stathis Kasderidis Computational Vision and Robotics Lab, FORTH, Heraklion, Greece, stathis@ics.forth.gr

Stefanos Kolias National Technical University of Athens, School of Electrical and Computer Engineering, Division of Computer Science, Iroon Polytechniou 9, Athens 15780, Greece, stefanos@cs.ntua.gr

Kostas J. Kyriakopoulos Control Systems Laboratory, Mechanical Engineering, National Technical University of Athens, 9 Heroon Polytechniou Street, Zografou, Athens 15700, Greece, kkyria@central.ntua.gr

Luis C. Lamb Departamento de Informática Teórica, Instituto de Informática, UFRGS, Porto Alegre, RS, Brazil, LuisLamb@acm.org

Daniel S. Levine Department of Psychology, University of Texas at Arlington, Arlington, TX 76019–0528, USA, levine@uta.edu

Alvin S. Lim Department of Computer Sciences and Engineering, Auburn University, Auburn, AL 36849, USA, lim@eng.auburn.edu

Pedro U. Lima Institute for Systems and Robotics, Instituto Superior Técnico, Av. Rovisco Pais, 1 1049-001 Lisboa, Portugal, pal@isr.ist.utl.pt

Elisa Magosso Department of Electronics, Computer Science and Systems, University of Bologna, Viale Risorgimento 2, I-40136, Bologna, Italy, elisa.magosso@unibo.it

Riccardo Manzotti Institute of Communication and Behaviour, IULM University, Via Carlo Bo, 8, 20143 Milan, Italy, riccardo.manzotti@iulm.it

Giorgio Metta Cognitive Humanoids Lab, Robotics Brain and Cognitive Sciences Department, Italian Institute of Technology, Genoa, Italy, pasa@liralab.it

Vishwanathan Mohan Cognitive Humanoids Lab, Robotics Brain and Cognitive Sciences Department, Italian Institute of Technology, Genoa, Italy, vishwanathan.mohan@iit.it

Javier Molina Department of Systems Engineering and Automation, Technical University of Cartagena, C/Dr Fleming S/N. 30202, Cartagena, Spain, javi.molina@upct.es

Pietro Morasso Cognitive Humanoids Lab, Robotics Brain and Cognitive Sciences Department, Italian Institute of Technology, Genoa, Italy and Department of Communication Computer and System Sciences, University of Genova, Italy, morasso@dist.unige.it

Alan F. Murray School of Engineering, The University of Edinburgh, King's Buildings, Mayfield Road, Edinburgh, EH9 3JL, UK, A.F.Murray@ed.ac.uk

Ryunosuke Nishimoto RIKEN Brain Science Institute, 2-1 Hirosawa, Wako-shi, Saitama 351-0198, Japan, ryu@brain.riken.jp

Hyuk Oh Graduate Program in Neuroscience and Cognitive Science (NACS), University of Maryland School of Public Health, College Park, MD 20742, USA, hyukoh@umd.edu

Stavros J. Perantonis Institute of Informatics and Telecommunications, NCSR "Demokritos", Patriarchou Grigoriou and Neapoleos St. GR-15310, Aghia Paraskevi, Attiki, Greece, sper@iit.demokritos.gr

Sergios Petridis Institute of Informatics and Telecommunications, NCSR "Demokritos", Patriarchou Grigoriou and Neapoleos St. GR-15310, Aghia Paraskevi, Attiki, Greece, petridis@iit.demokritos.gr

Daniel Polani Adaptive Systems Research Group, School of Computer Science, University of Hertfordshire, Hatfield, AL10 9AB Hertfordshire, UK, d.polani@herts.ac.uk

Konstantinos Rapantzikos Image, Video and Multimedia Systems Laboratory, Computer Science Division, School of Electrical and Computer Engineering, National Technical University of Athens, Iroon Polytexneiou 9, 15780 Zografou, Greece, rap@image.ece.ntua.gr

Theodoros Rekatsinas Division of Signals, Control and Robotics, School of Electrical and Computer Engineering, National Technical University of Athens, Zographou Campus, Athens 15773, Greece, rekatsinas@gmail.com

Edmund T. Rolls Oxford Centre for Computational Neuroscience, Oxford, UK, Edmund.Rolls@oxcns.org

Albert L. Rothenstein Department of Computer Science and Engineering, Centre for Vision Research, York University, Toronto, ON, Canada, albertir@cs.yorku.ca

Ron Sun Cognitive Science Department, Rensselaer Polytechnic Institute, 110 Eighth Street, Carnegie 302A, Troy, NY 12180, USA, rsun@rpi.edu

Tong Boon Tang School of Engineering, The University of Edinburgh, King's Buildings, Mayfield Road, Edinburgh EH9 3JL, UK, Tong-Boon.Tang@ed.ac.uk

Jun Tani Laboratory for Behavior and Dynamic Cognition, RIKEN Brain Science Institute, 2-1 Hirosawa Wako-shi, Saitama, 351-0198 Japan, tani@brain.riken.jp

John G. Taylor Department of Mathematics, King's College, London, UK, john.g.taylor@kcl.ac.uk

Naftali Tishby School of Engineering and Computer Science, Interdisciplinary Center for Neural Computation, The Suadrsky Center for Computational Biology, Hebrew University Jerusalem, Jerusalem, Israel, tishby@cs.huji.ac.il

John K. Tsotsos Department of Computer Science and Engineering, Centre for Vision Research, York University, Toronto, Ontario, Canada, tsotsos@cse.yorku.ca

Costas S. Tzafestas National Technical University of Athens, School of Electrical and Computer Engineering, Division of Signals, Control and Robotics, Zographou Campus, Athens 15773, Greece, ktzaf@softlab.ntua.gr

Mauro Ursino Department of Electronics, Computer Science and Systems, University of Bologna, Viale Risorgimento 2, I-40136 Bologna, Italy, mauro.ursino@unibo.it

Nick Wilson Cognitive Science Department, Rensselaer Polytechnic Institute, 110 Eighth Street, Carnegie 302A, Troy, NY 12180, USA, wilson3@rpi.edu

Qi Zhang Sensor System, 8406 Blackwolf Drive, Madison, WI, USA, qizhang_sensor@yahoo.com

Part I
Computational Neuroscience Models

Vassilis Cutsuridis, Amir Hussain, and John G. Taylor

In this part, leading computational neuroscientists present neural network models of the various components of the perception–action cycle, namely perception, attention, cognitive control, decision making, conflict resolution and monitoring, knowledge representation and reasoning, learning and memory, planning and action, and consciousness at various levels of detail. The architectures of these models are *heavily* guided by knowledge of the human and animal brain. The models then allow the synthesis of experimental data from different levels of complexity into a coherent picture of the system under study.

In the chapter entitled "The role of attention in shaping visual perceptual processes", Tsotsos and Rothenstein argue that an optimal solution to the generic problem of visual search, which is robust enough to apply to *any* possible image or target, is unattainable because the problem of visual search has been proven intractable. The brain, however, is able to solve this problem effortlessly and hence that poses a mystery. To solve this mystery, Tsotsos and Rothenstein argue that the brain is not solving that same generic visual search problem *every time*. Instead, the nature of the problem solved by the brain is fundamentally different from the generic one. They describe a biologically plausible and computationally well-founded account of how the brain might deal with these differences and how the attentional brain mechanisms dynamically shape the visual perceptual processes of humans and animals.

In the next chapter entitled "Sensory fusion", Ursino, Magosso and Cuppini present two computational models of multisensory integration, inspired by real neurophysiological systems. The first model considers the integration of visual and auditory stimuli in the superior colliculus, whereas the second one considers the integration of tactile stimuli and visual stimuli close to the body to form the perception of the peripersonal space. Although both models attack different problems, the mechanisms delineated in the models (lateral inhibition and excitation, non-linear neuron characteristics, recurrent connections, competition) may govern more generally the fusion of senses in the brain. The models, besides improving our comprehension of brain function, drive future neurophysiological experiments and provide valuable ideas to build artificial systems devoted to sensory fusion.

In the chapter entitled "Modeling memory and learning consistently from psychology to physiology", Coward describe the Recommendation Cognitive Architecture that maps the information model for each major anatomical brain structure into more detailed models for its substructures, and so on all the way down to neuron physiology. The architecture explains how the more detailed models are implemented physiologically, and how the detailed models interact to support higher level models, up to descriptions of memory and learning on a psychological level.

In the chapter entitled "Value maps, drives and emotions", Levine discusses value maps, drives and emotions through the modelling of decision making, judgment and choice. He presents the Distributed Emotional Connections Influencing Decisions and Engaging Rules (DECIDER) model, which is based on interactions among a large number of brain regions such as loci for emotions (amygdala and orbital prefrontal cortex), rule encoding and executive function (orbital and dorsolateral prefrontal and anterior cingulate), and behavioural control (striatum, thalamus and premotor cortex). In order to incorporate the capacities for using either heuristic or deliberative decision rules in real time in a changing and uncertain environment, the model addresses a variety of questions concerning the interface between emotional valuation and numerical calculation.

In the chapter entitled "Computational neuroscience models: Error monitoring, conflict resolution and decision making", Brown and Alexander discuss computational neuroscience models of how performance monitoring and cognitive control can monitor the outcomes of decisions and actions and depending on the success of these actions in achieving the desired goals can implement corrective actions as quickly as possible.

In the chapter entitled "Neural network models for reaching and dexterous manipulation in humans and anthropomorphic robotic systems", Gentili, Oh, Molina and Contreras-Vidal present a modular neural network model able to learn the inverse kinematics of an anthropomorphic arm [7 degrees of freedom (DOF)] and three fingers (4 DOFs) and perform accurate and realistic arm reaching and finger grasping movements. They present the model's architecture in detail including both the model's components and parameters. They show simulation results of the proposed model and discuss them in terms of their biological plausibility and model assumptions. Finally, they conclude with their model's limitations along with future solutions that could challenge it.

In the chapter entitled "Schemata learning", Nishimoto and Tani describe a dynamic neural network model that accounts for the neuronal mechanism of schemata learning. Schemata learning refers to sequences of actions on objects learnt as "schemata" by humans, so as to be able to function in fast, even an automatic, manner in many well-known situations (e.g. going to a restaurant, going to bed, etc). Their model shows that the functional hierarchical structures that emerge through the stages of development assist through their fast and slow dynamics behaviour primitives and motor imagery to be generated in earlier stages and compositional sequences of achieving goals to appear in later stages. They conclude with a discussion on how schemata of goal-directed actions could be acquired with gradual development of the internal image and compositionality for the actions.

In the chapter entitled "The perception-conceptualisation-knowledge representation-reasoning representation-action cycle: The view from the brain", Taylor considers new and important aspects of brain processing related to perception, attention, reward, working memory, long-term memory, spatial and object recognition, conceptualization and action, and how they can be melded together in a coherent manner. His approach is based mainly on work done in the EU GNOSYS project to create a reasoning robot using brain guidance, starting with the learning of object representations and associated concepts (as long-term memory), with the inclusion of attention, action, internal simulation and creativity.

In the chapter entitled "Consciousness, decision making and neural computation", Rolls describes a computational theory of consciousness named higher order syntactic thought (HOST). He argues that the adaptive value of higher order thoughts is to solve the credit assignment problem that arises if a multi-step syntactic plan needs to be corrected. He suggests qualia arise secondarily to higher order thoughts and sensations and emotions are there because it is unparsimonious for the organism *not* to be able to feel something. Brain-inspired models of decision making are described based on noise-driven and probabilistic integrate-and-fire attractor neural networks, and it is proposed that networks of this type are involved when decisions are made between the explicit and implicit routes to action. Rolls argues that the confidence one has in one's decisions provides an objective measure of awareness, but it is shown that two coupled attractor networks can account for decisions based on confidence estimates from previous decisions. On the implementation of consciousness, Rolls shows that the threshold for access to the consciousness system is higher than that for producing behavioural responses. He argues that the adaptive value of this may be that the systems in the brain that implement the type of information processing involved in conscious thoughts are not interrupted by small signals that could be noise in sensory pathways. He concludes that oscillations are not a necessary part of the implementation of consciousness in the brain.

In the final chapter entitled "Review of models of consciousness", Taylor reviews the main computational models of consciousness and develops various tests to assess them. Although many models are successful in passing some of these tests under certain conditions, only one model (the CODAM model) is able to pass all of them.

Chapter 1
The Role of Attention in Shaping Visual Perceptual Processes

John K. Tsotsos and Albert L. Rothenstein

Abstract It has been known now for over 20 years that an optimal solution to a basic vision problem such as visual search, which is robust enough to apply to any possible image or target, is unattainable because the problem of visual search is provably intractable ("Tsotsos, The complexity of perceptual search tasks, Proceedings of the International Joint Conference on Artificial Intelligence, 1989," "Rensink, A new proof of the NP-completeness of visual match, Technical Report 89–22, University of British Columbia, 1989"). That the brain seems to solve it in an apparently effortless manner then poses a mystery. Either the brain is performing in a manner that cannot be captured computationally, or it is not solving that same generic visual search problem. The first option has been shown to not be the case ("Tsotsos and Bruce, Scholarpedia, 3(12), 6545, 2008"). As a result, this chapter will focus on the second possibility. There are two elements required to deal with this. The first is to show how the nature of the problem solved by the brain is fundamentally different from the generic one, and second to show how the brain might deal with those differences. The result is a biologically plausible and computationally well-founded account of how attentional mechanisms dynamically shape perceptual processes to achieve this seemingly effortless capacity that humans – and perhaps most seeing animals – possess.

1.1 Introduction

The computational foundations for attentive processes have been recently detailed in Tsotsos and Bruce (2008). There, it was argued that the visual computation processes of the brain can indeed be modeled computationally and that the generic problem of visual search is intractable. What this means is that the brain cannot be

J.K. Tsotsos (✉)
Department of Computer Science and Engineering, Centre for Vision Research,
York University, Toronto, Ontario, Canada
e-mail: tsotsos@cse.yorku.ca

V. Cutsuridis et al. (eds.), *Perception-Action Cycle: Models, Architectures, and Hardware*, Springer Series in Cognitive and Neural Systems 1, DOI 10.1007/978-1-4419-1452-1_1, © Springer Science+Business Media, LLC 2011

solving the generic problem – that is, it cannot be providing optimal solutions for all possible image and target pairs. This further implies that our illusion of perfect vision is just that, an illusion. The impression we have that we see everything in just a glance is false. Nevertheless, the brain is doing something to perpetuate that illusion, and is doing it very well indeed. Tsotsos and Bruce continue by summarizing ways, first presented in Tsotsos (1987) by which the problem can be reshaped so that the reshaped version is quickly solvable. Interestingly, the difference between the generic and the reshaped problems is the set of visual perception tasks that require more than just a glance. Computational vision has been making strong progress on those problems that need just a glance, and an excellent recent volume provides an up-to-date snapshot of that progress (Dickinson et al. 2009). This chapter will focus on those problems that fall in the difference class.

First, it is important to summarize how the problem is reshaped as mentioned above, and we start by providing an answer to the question "what is the generic vision problem?" One possible definition follows. Given a time-varying sequence of images, projections of a dynamic scene for each pixel determine whether it belongs to some particular object or other spatial construct, localize all those objects in space, detect and localize all events in time, determine the identity of all the objects and events in the sequence, determine relationships among objects, and relate all objects and events to the available world knowledge. The visual search problem is a subset of this and requires that one determine whether or not a particular target appears in a given image. It has been shown that the visual search problem, if task guidance using the target image or other knowledge is prohibited, is provably intractable (it is NP-Complete; Tsotsos 1989). If a vision system needs to search through the set of all possible image locations (pixels) and compare them to each element of memory, then without any task guidance or knowledge of the characteristics of the subset it seeks, it cannot know which subset may be more likely than another. As a result, it is the powerset of all locations that gives the number of image subsets to examine, an exponential function.

Human vision certainly solves visual search, even in conditions of casual, undirected, viewing. How can one deal with an intractable problem in practice? NP-completeness means that a completely optimal and general algorithm is not possible, but Garey and Johnson (1979) provide a number of guidelines for how to proceed in practice. Here we use the problem's natural parameters to guide the search for approximation algorithms, algorithms that solve the problem but not optimally, providing a solution only to within some specified error tolerance. One way of doing this is to reduce the exponential effect of the largest valued parameters. The process was presented in Tsotsos (1987, 1988, 1990, 1991a).

The parameters of the visual search computation are N (number of items in memory), P (image size in pixels), and M (number of features computed at each pixel), and its worst-case time complexity is $O(N 2^{PM})$. N is a large number but any reduction leads to linear improvements. P is also a large number but reduction in P can lead to exponential improvement, as does reduction in M. However, M is not so large a number. We can conclude that the best improvement would come from reductions in P. What would such changes look like? Here are several that are straightforward yet sufficient (but not necessary):

1. Hierarchical organization takes search of model space from $O(N)$ to $O(\log_2 N)$.
2. Search within a pyramidal representation of the image (a layered representation, each layer with decreasing spatial resolution and with bidirectional connections between adjacent layers) operates in a top-down fashion, beginning with a smaller more abstract image, and is then refined locally thus reducing P.
3. Spatiotemporally localized receptive fields reduce number of possible receptive fields from 2^P to $O(P^{1.5})$ (this assumes contiguous receptive fields of all possible sizes centered at all locations in the image array).
4. Spatial selectivity can further reduce the $P^{1.5}$ term if one selects the receptive field that is to be processed. This is not a selection of location only, but rather a selection of a local region and its size at a particular location.
5. Feature selectivity can further reduce the M term, that is, which subset of all possible features actually is represented in the image or is important for the task at hand.
6. Object selectivity can further reduce the N term, reflecting task-specific information.

After applying the first three constraints, $O(P^{1.5}2^M\log_2 N)$ is the worst-case time complexity. The next three reduce P, N, and M all to 1 to bring the expression down to perhaps its lowest value and are all manifestations of attentive processing.

But how do these actions affect the generic perception problem described above? Hierarchical organization does not affect the nature of the vision problem. However, the other mechanisms have the following effects:

- Pyramidal abstraction affects the problem through the loss of location information and signal combination.
- Spatiotemporally localized receptive fields force the system to look at features across a receptive field instead of finer grain combinations, and thus arbitrary combinations of locations are disallowed.
- Attentional selection further limits what is processed in the location, feature, and object domains.

As a result of these, the generic problem as defined earlier has been altered. Unfortunately, it is not easy to formally characterize this altered problem. It is, however, possible to say something about the difference between instances of visual search that can be solved at a glance and those that cannot. First, it is important to abandon use of the casual phrase "at a glance" and replace it with something better defined. Instances of visual search, or of the generic vision problem, that can be solved using only a single feed-forward pass through the visual processing machinery of the brain are those that correspond to the "at a glance" problems (call this set of problems the AG set). This implies that any visual quantities needed are available at the end of that pass – or within 150 ms or so of stimulus onset (consistent with Marr's (1982) theory, with Bullier 2001, Lamme and Roelfsema 2000, and with Thorpe et al. 1996, among many others). The only action required is to select from those quantities the subset that may satisfy the task at hand and to verify that they in fact do satisfy the task. It is implied by the time course that the selection is clear, that is, there are no other potential subsets that may compete for task satisfaction; so the first selected is the correct one. It is important to note that the vast majority of modern computer vision limits consideration to exactly this scenario (e.g., see Dickinson et al. 2009).

What remains of the generic vision problem? There are an enormous number of instances of visual search and other vision tasks that fall outside the above limitation; we may call them the "more than a glance" problems (MG). The way to specify them can be guided by the shortcomings of the overall architecture described earlier. A partial specification would include scenes that contain more than one copy of a given feature each at different locations, contain more than one object/event each at different locations, or contain objects/events that are composed of multiple features and share at least one feature type. The tasks must also be specified and would include those where simple detection or naming of a single item in a scene do not suffice. For example, time-varying scenes require more than a glance. The questions Yarbus asked subjects about image contents require more than a glance (Yarbus 1967). Non-pop-out visual search requires more than a glance (Treisman and Gelade 1980; Wolfe 1998). Tasks where a behavior is required such as an eye movement or a manipulation require more than a glance. There are many more and it is not too much of a stretch to suggest that the majority of vision tasks we face in our everyday lives are not of the single glance variety. Such tasks will be further examined below.

1.2 Connecting Attention, Recognition, and Binding

Recognition, binding, and attention are rarely considered together. For example, although the Dickinson et al. (2009) collection of papers on object categorization is excellent, attention does not appear in any of the computational models. Others have explicitly written that binding may not be required for recognition (Riesenhuber and Pogio 1999; Ghose and Maunsell 1999). Marr (1982) had no mention of either attention or binding in his classic work. Why then would it be justified to think of attention, recognition, and binding as being connected? In order to address this question, it is important to be clear about what each of these three terms – attention, binding, and recognition – means. In other words, the usual, almost casual use of these terms will be rejected and new viewpoints presented.

Considering attention first, Tsotsos and Bruce (2008) provide a list of mechanisms, all attributed to attentional processes. This list is expanded here:

- Selection of

  ○ Spatiotemporal region of interest
  ○ Features of interest
  ○ World, task, object, or event model
  ○ Gaze and viewpoint
  ○ Best interpretation or response

- Restriction to
  - Task relevant search space
  - Location cues
  - Fixation points
  - Task-related search depth
  - Task appropriate neural tuning profiles

- Suppression of
  - Spatial and feature local context (surround inhibition)
  - Previously processed items (inhibition of return)
  - Task irrelevant computations

These are all attributed to attention because they each contribute to the key need for attention, namely, reduction in the amount of information that the visual system must process. Recall how attentional processing arose during the earlier discussion on how to lower the complexity of visual search. This list shows that attention cannot be regarded as a single, monolithic process. It is a set of mechanisms, perhaps not all applied together and with differing parameters, that is dynamically deployed depending on the stimulus and task at hand. For some stimuli and tasks – those that can indeed be perceived at a glance – the use of these mechanisms may be minimal. For more difficult stimuli and tasks, their use can be quite extensive. The balance of this chapter will argue for a particular arrangement of these mechanisms for specific kinds of perceptual scenarios.

Attention, then, is a set of mechanisms that dynamically control the processes of vision so that they perform as well as possible for the visual scene and task at hand. Recognition, on the other hand, differs because there are many variations in tasks and the term recognition is only a "catch-all". Macmillan and Creelman (2005) provide good definitions for many types of recognition. For them, the variations are all tied to the kind of experimental paradigm that is used to explore them. One-interval experimental design involves a single stimulus presented on each trial. *Discrimination* is the ability to tell two stimuli apart. The simplest example is a *correspondence* experiment in which the stimulus is drawn from one of two stimulus classes, and the observer has to say from which class it is drawn. This is perhaps the closest to the way much of modern computer vision currently operates. A *detection* task is where one of the two stimulus classes is null (noise) and the subject needs to choose between noise or noise $+$ signal and the subject responds if he or she sees the signal. In a *recognition* task neither stimulus is noise. More complex versions have more responses and stimuli. If the requirement is to assign a different response to each stimulus, the task is *identification*. If the stimuli are to be sorted into a smaller number of classes – say, M responses to sort N stimuli into categories – it is a *classification* task. The *categorization* task requires the subject to connect each stimulus to a prototype, or class of similar stimuli (cars with cars, houses with houses). The *within-category identification* task has the requirement that a stimulus is associated with a particular subcategory from a class (e.g., bungalows, split-level, and other such house types). There are also N-interval tasks where more than one stimulus is presented per trial. In the *same–different* task, a pair of stimuli is presented on each trial, and the observer must decide if its two elements are the same or different. For the *match-to-sample* task, three stimuli are shown in sequence and the observer must decide which if the first two is matched by the third. *Odd-man-out* is a task where the subject must locate the odd stimulus from a set where all stimuli are somehow similar while one is not. For ease of reference, we will term these *extended discrimination tasks*. They include two-or-more interval designs such as visual search, odd-man-out, resolving illusory conjunctions, determining transparency, recognizing objects in cluttered scenes, any task requiring sequences of saccades, or pursuit eye movements.

In each experiment, subjects are required to produce a response for each trial. Responses can vary. They may be verbal, eye movement to target, the press of a particular button, pointing to the target, and more. The choice of response method can change the processing needs and overall response time. The need for a subject to respond leads us to define a new task that is not explicitly mentioned in Macmillan and Creelman, the *localization* task where the subject is required to extract some level of stimulus location information to produce the response requested by the experimenter. In fact, this may be considered as an implicit subtask for any of the standard tasks if they also require location information to formulate a response.

There are many more such tasks in Macmillan and Creelman for the interested reader. It should be clear even from this abbreviated listing that recognition is not a single task, that it has many variants, well-defined and understood variants, and this decomposition is important. It provides a finer scale level of study and each subclass points to the need for specific solution mechanisms.

Before performing any of the above tasks, subjects are provided with knowledge of the experiment, what to expect in terms of stimuli, what responses are required, and so on. In other words, subjects are "primed" in advance for their task (Posner et al. 1978). Thus, in any model of vision, the first set of computations to be performed is priming the hierarchy of processing areas. Task knowledge, such as fixation point, target/cue location, task success criteria, and so on must somehow be integrated into the overall processing; they *tune* the hierarchy. In the terminology of the attentional mechanisms listed above, they restrict processing to the relevant computations in one or more of several ways. It has been shown that such task guidance must be applied 80 to 300 ms before stimulus onset to be effective (Müller and Rabbitt 1989). This informs us that significant processing time is required for this step alone, a sufficient amount of time to complete a top-down traversal of the hierarchy before any stimulus is shown. Even for casual, non-task-specific viewing, our lifelong experience plays a role in priming our vision. When walking along the street on a beautiful sunny day, one is never surprised by a group of other people walking by. But if a group of camels were to walk by, this is not in the frame of our current expectations, and our reaction is exactly the same as for the invalid cue in the Posner paradigm (Posner et al. 1978). This does not mean that the camels cannot be processed; it simply means that it will take just a bit longer to have the system respond to the unexpected (see Zucker et al. 1975).

There is no reason to believe that the same kind of action that priming might perform in advance of a stimulus cannot also be performed during the processing of a stimulus. Indeed, if the priming is incorrect for any reason, Posner and many others have shown that response times are greater than if no priming at all was given. This implies there is some level of additional processing taking place. One possibility could be that the overall system has realized that a readjustment is needed, explicitly changing the priming to permit a different analysis to proceed.

The finer scale decomposition of both attention and recognition reveals that their components exhibit natural connections. Table 1.1 shows some potential linkages. The table gives a list of recognition tasks and the corresponding attentional processes that would be needed for their completion. It demonstrates that indeed attention and recognition have a real relationship to one another, and the final column provides the crucial link – timing. It is well evidenced that different kinds of visual recognition tasks require different amounts of time for human subjects to respond correctly. It was documented earlier that a single feed-forward pass through the visual cortical areas takes about 150 ms. The discrimination task (and its variants correspondence, detection, recognition, categorization, classification), as long as no location information is required for a response, seems to take about 150 ms (Thorpe et al. 1996). This kind of "yes–no" response can be called "pop-out" in visual search, with the added condition that the speed of response is the same regardless of number of distractors (Wolfe 1998). The categorization task also seems to take the same amount of time (Grill-Spector and Kanwisher 2005). The only required attentional mechanism here is response selection. The other tasks described above seem to take more time.

Table 1.1 This table shows how recognition tasks may be related to the attentional processes required for their solution. The organization is by the amount of processing time required for the task. Support for the choices of timing intervals is given in the text

Recognition	Attention	Timing
Priming	Suppression of task irrelevant features, stimuli or locations; restrict to location, fixation; selection of task success criteria	-300 to $-80\,\mathrm{ms}$
Stimulus onset		$0\,\mathrm{ms}$
Discrimination	Response selection	$\approx 150\,\mathrm{ms}$
Within-category-identification	Response selection; top-down feature selection	$\approx 215\,\mathrm{ms}$
Localization	Response selection; top-down feature and location selection	$\approx 250\,\mathrm{ms}$
Extended discrimination	Sequences of discrimination, localization and/or identification tasks, perhaps with task priming specific to each	$>250\,\mathrm{ms}$

There are then two options to consider. For those tasks that require longer than 150 ms, does the processing occur outside the visual cortex or within it? If outside, then where? If within, then how? Several investigators have demonstrated that single neuron responses in monkey show attentional effects well beyond 150 ms for a given visual stimulus (e.g., Mehta et al. 2000). Hence, at least we may conclude that processing includes the visual cortex beyond 150 ms. An obvious – but perhaps not the only – conclusion then is that recurrence likely plays a role. Here, it is proposed that this recurrent action is a top-down attentional feedback mechanism (shown to be exactly the case by Boehler et al. 2009). Top-down feature or location selection would involve restriction and selection processes and would occur after a top-level attentional selection is made.

To provide more detail about a stimulus, such as for a within-category identification task, additional processing time, 65 ms or so, is needed (Grill-Spector and Kanwisher 2005; Evans and Treisman 2005). If the first feed-forward pass can provide the basic category of the stimulus, such as "house," where are the details that allow one to determine the type of house? The sort of detail required would be size, shape, structure, and so forth. These are clearly features of lower abstraction, and thus they can only be found in earlier levels of the visual hierarchy. They can be accessed by looking at which feature neurons feed into those category neurons.

One way to achieve this is to traverse the hierarchy downward, beginning with the category neuron and moving downward through the needed feature maps. This downward traversal is what requires the additional time observed. The extent of downward traversal is determined by the task, that is, the aspects of identification that are required.

If localization is required for description or a motor task (pointing, grasping, etc.), then how much time is needed for this? A lever press response seems to need 250–450 ms in monkey (Mehta et al. 2000). During this task, the temporal pattern of attention modulation shows a distinct top-down pattern over a period of 35–350 ms poststimulus. The "attention dwell time" needed for relevant objects to become available to influence behavior seems to be about 250 ms (Duncan et al. 1994). Pointing to a target in humans seems to need anywhere from 230 to 360 ms (Gueye et al. 2002; Lünenburger and Hoffman 2003). None of these experiments cleanly separate visual processing time from motor processing time; as a result, these results can only provide an encouraging guide and further experimental work is needed. Still, it seems that behavior, i.e., an action relevant to the stimulus, requires some degree of localization. The location details are available only in the early layers of the visual processing hierarchy because that is where the finer spatial resolutions of neural representation can be found.

The extended discrimination task includes two-or-more interval designs, visual search, odd-man-out, resolving illusory conjunctions, determining transparency, recognizing objects in cluttered scenes, any task requiring sequences of saccades or pursuit eye movements, and more. These tasks include changes to the input – the input is time-varying because the scene is time-varying, because the eyes move, and so on. In reality, the input to the retina is continuous and one would expect that the feed-forward path also continuously reflects changes in the input. Our apparent ability to respond to such a continuous change seems limited. VanRullen et al. (2007) showed that we can process only at a rate of about seven items per second.

Extended discrimination tasks would then require sequences of feed-forward and top-down processing, and speed of response is limited by these actions, consistent with Van Rullen et al.

Finally, we come to the binding problem. Binding is usually thought of as taking one kind of visual feature, such as a color, and associating it with another feature, such as location, to provide a united representation of an object (Rosenblatt 1961). Such an association is important when more than one visual object is present, to avoid incorrect combinations of features. The binding literature is large; no attempt is made here to review it due to space limitations (see Roskies 1999).

Classical demonstrations of binding in vision seem to rely on two things: the existence of representations in the brain that have no location information and representations of pure location for all stimuli. It is not clear whether these assumptions are indeed valid. However, it is clear that location is *partially* abstracted away within pyramidal representation as part of the solution to complexity. A single neuron receives converging inputs from many neurons and each provides input for many neurons. Precise location is lost in such a network of diverging feed-forward paths yet increasing convergence onto single neurons. The challenge then is to specify how location may be recovered.

Since time course was effective in helping to organize attentional and recognition elements, let us continue to use it here for binding. What occurs during the first feed-forward pass? In a feed-forward process, we see neural convergence layer by layer in the processing network from lower to higher order representations. In order to make a decision at the highest layer, the attention process involved is selection of maximum response. This action can be termed *convergence binding*; it achieves the *discrimination task* for AG scenes. This is consistent with previous views on this problem (Treisman 1999; Reynolds and Desimone 1999). This type of binding will suffice only when stimulus elements that fall within the larger receptive fields are not too similar or otherwise interfere with the response of the neuron to its ideal tuning properties.

For the other tasks, a top-down process seems to be required. During the attentive first recurrent pass following the convergence binding process, attended stimuli can be localized in each layer of the hierarchical representation. Recurrent traversals through the visual processing hierarchy "trace" the pathways of neural activity that lead to the strongest responding neurons at the top of the hierarchy. This will be termed *full recurrence binding* and it achieves the *localization task*. It is needed for any of the recognition tasks where localization is needed (for description or behavior for example, regardless of whether or not the scene is an AG or a MG). Depending on task needs or allowable processing time, the full recurrence binding process might not complete. This is called *partial recurrence binding*. Partial recurrence binding can find the additional information needed to solve the *Identification Task* if it is represented in intermediate layers of the processing hierarchy. Also, coarse localization tasks can be solved (such as "in which quadrant is the stimulus?"). If this is not deployed directly due to task needs but is due to interruption, then this may result in illusory conjunctions. The attention process involved is top-down feature search guided by local max selection. A variety of different effects may be observed depending on when the top-down traversal process is interrupted.

Many tasks require time longer than the approximately 250 ms required for full recurrence binding to complete. *Iterative recurrence binding* is needed for visual search, stimulus sequences for any of the N-interval experiments where $N \geq 2$, and other more complex scenarios, i.e., extended discrimination tasks. Iterative recurrence binding is defined as one of more convergence binding-full recurrence binding cycles. The processing hierarchy may be primed for the task before each traversal as appropriate. The iteration terminates when the task is satisfied. The attention mechanisms include sequences of convergence and recurrence binding, perhaps with task priming specific to each pass.

There are at least two types of iterative recurrence binding. The first, Type I, deals with tasks where multiple fixations are required. Visual search is such a task. The second, Type II, permits different pathways to be invoked. Consider a motion stimulus, motion-defined form where a square of random elements rotates in a background of similar random elements. A rotating square is perceived even though there is no edge information present in the stimulus. After one cycle of full recurrence binding, the motion can be localized and the surround suppressed. The suppression changes the intermediate representation of the stimulus so that any edge detecting neurons in the system now see edges, edges that were not apparent because they were hidden in the noise. As a result, the motion is recognized, and with an additional processing cycle the edges can be detected and bound with the motion. Table 1.2 summarizes the preceding paragraphs, extending Table 1.1 to include the four binding processes described.

One key aspect is how is the decision to use one or the other type of binding made? Figure 1.1 gives the algorithm in its simplest form. For this algorithm, it is assumed that the kind of visual task is known in advance. Three kinds of

Table 1.2 The Table 1.1 is extended to include the four binding processes described in the text and their connections recognition tasks and attentional mechanisms

Recognition	Attention	Binding	Timing
Priming	Suppression of task irrelevant features, stimuli or locations; restrict to location, fixation; selection of task success criteria	N/A	-300 to -80 ms
Stimulus onset			0 ms
Discrimination	Response selection	Convergence binding	≈ 150 ms
Within-category-identification	Response selection; top-down feature selection	Partial recurrence binding	≈ 215 ms
Localization	Response selection; top-down feature and location selection	Full recurrence binding	≈ 215 ms
Extended discrimination	Sequences of discrimination, localization and/or identification tasks, perhaps with task priming specific to each	Iterative recurrence binding	>250 ms

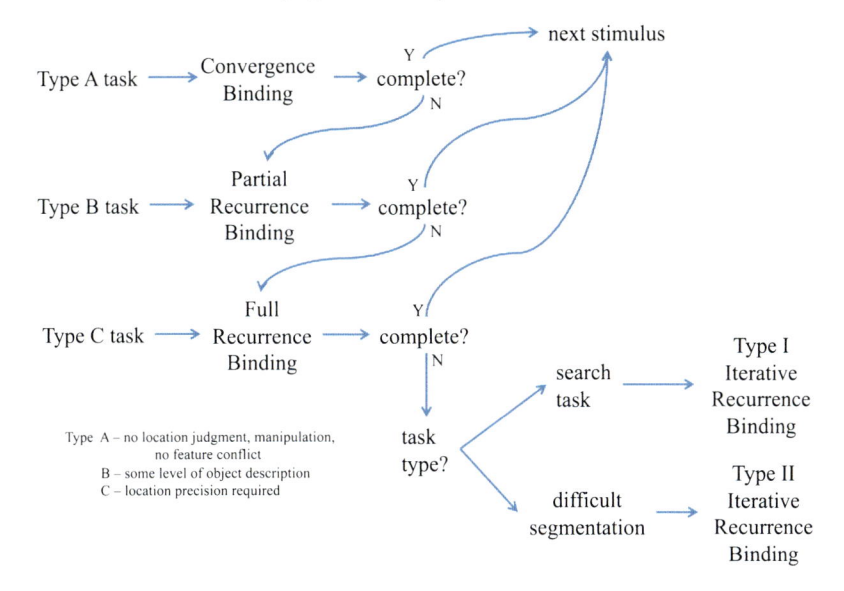

Fig. 1.1 The algorithm for deciding on the type of binding process to apply for a given task and input is summarized

tasks are included. Task A is the class where no location judgment or object manipulations are needed (i.e., no visually activated or guided behavior) and no conflicts in the images are expected. In Task B, a bit more is needed. Here, there is some visually guided behavior, but not much detail is needed for it. Tasks in C class require detailed observations to support visual-guided behavior. For classes A and B, if the processing thought sufficient by the task description is found to not be satisfactory for the task, then processing moves to the next higher class. For example, if for a particular image one expects convergence binding to suffice but it does not, partial recurrence binding is tried. As in Posner's cueing paradigm,when the cue is misleading, processing takes longer, similarly, here, the expectation does not match the stimulus, processing is revised until it does. In the case of free viewing, with no particular task, one may assume a Type A task, that is, one where no specific judgments of the perceived world are needed. If something is seen were this does not suffice, the system automatically moves to a perhaps more appropriate strategy.

1.3 Finding the Right Subset of Neural Pathways on a Recurrent Pass

A key element of the above connections among attention, recognition, and binding has been the top-down tracing of neural connections. How is the right set of pathways through the visual processing network selected and bound together to represent an object? The entire scheme hinges on this.

This idea of top-down attentional processing seems to have appeared first in Milner (1974), and it was used in the attentive NeoCognitron of Fukushima (1986). He explicitly described it as a backward tracing of connections and provided an algorithm that was based on choosing maximum responses and then suppressing all afferents that did not lead to this maximum. Milner provided only discussion, while Fukushima provided a simulation as evidence of the algorithm's operation. Tsotsos (1990) also describes the idea but with the added twist of coupling it with a suppressive surround realizing a Branch-and-Bound search strategy. In comparison with Fukushima, Tsotsos limited the suppression of afferents to those within receptive fields, while Fukushima extended it to the full visual field. The two max-finding algorithms also differed. Implementations appeared in Culhane and Tsotsos (1992) and Tsotsos (1993), and proofs of its properties in Tsotsos et al. (1995). Backward tracing is also featured in the Reverse Hierarchy Model of Ahissar and Hochstein (1997) but in a descriptive fashion only as it appears more recently in Lamme and Roelfsema (2000) and DiLollo (2010). Here, the Selective Tuning model version of Tsotsos and colleagues is used.

Figure 1.2 illustrates the connections among the various processes of attention and binding for tasks organized along a latency time line, just like in the previous figures. What is added in this figure is the action of feed-forward and recurrent processes. The first stage, the leftmost element of the figure, shows the priming stage. The attention processes involved include: suppression of task irrelevant features, stimuli or locations, and imposing the selectivity of a location cue or of a fixation point. The selection of task model and success criteria must also be completed. Then, the stimulus can be presented.

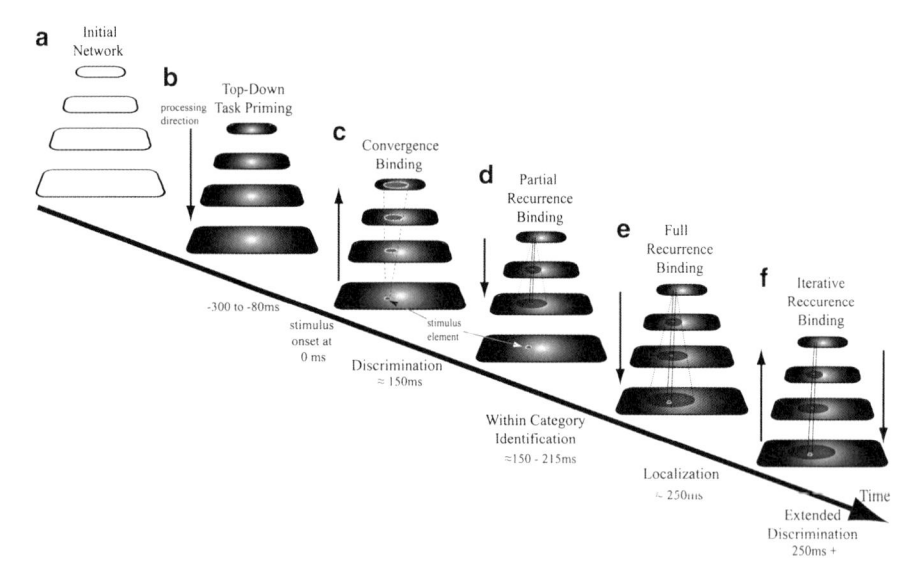

Fig. 1.2 The aspects of recognition, attention, and binding described in the text are now all connected along a time line of processing latency. The flow of processing is shown for each of the hypothetical pyramid representations. Also, the attentive suppressive surround is clearly shown for each layer of representation

The third element of the figure represents convergence binding for the solution of the one-interval discrimination tasks without localization. The fourth from the left element of the figure portrays the partial recurrence binding process that is claimed to provide for within-category identification and coarse localization tasks. This first shows the major difference between the Fukushima and the Selective Tuning versions of recurrence tracing. On the figures, areas of neural suppression or inhibition surrounding the attended stimulus appear as gray regions. This area is defined by the afferents to the chosen neuron at the top. Inputs corresponding to the stimulus most closely matching the tuning characteristics of the neuron form the signal, while the remainder of the input within that receptive field is the noise. Any lateral connections are also considered as noise for this purpose. Thus, if it can be determined what those signal elements are, the remainder of the receptive field is suppressed, enhancing the overall signal-to-noise ratio of processing for that neuron. This was first described in Tsotsos (1990), the method for achieving it first described in Tsotsos (1991b), and fully detailed together with proofs of convergence and other properties in Tsotsos et al. (2005). Supporting evidence now abounds for this attentive suppressive surround (see Tsotsos et al. 2008).

The next element of the figure, the second from the right, depicts a full recurrence binding where the top-down tracing has completed its journey from the highest to the lowest levels of the visual processing pyramid. At this point sufficient information has been fond to solve localization tasks. The final stage in time is iterative recurrence binding corresponding to the extended discrimination tasks.

Each of these mechanisms and binding processes has been detailed elsewhere, and many examples of performance have been shown. They form the Selective Tuning model of attention and the interested reader is referred to Tsotsos et al. (2008, 2005, 1995) and Tsotsos (1990). The structure of the model, although complex, has made many predictions for human vision that now have significant experimental support. Tsotsos (2011), Rothenstein et al. (2008) and Tsotsos et al. (2008) provide details.

Type I iterative recurrence binding is needed for a visual search task, of the classic kind that Treisman and Gelade (1980) or Wolfe (1998) have described. Several examples of how the above process operates for visual search tasks using both shape and motion stimuli have been presented in Rodriguez-Sanchez et al. (2007). Here, a simple example is included to illustrate Type II iterative recurrence binding. The recovery of shape from motion is a particularly good illustration of the Type II case, especially for cases where the shape information is not even present in the input image, and motion detection is a clear prerequisite. The overall strategy is presented schematically in Fig. 1.3 and has as basic elements a motion-processing hierarchy and a shape-processing hierarchy. The motion hierarchy detects translation motion (12 different directions, 3 different speeds) in a V1-like area, spatial derivatives of velocity (12 directions, 3 speeds for each of 12 local motion directions) in an MT-like area, and spiral motions patterns (rotation, expansion, contraction, and their combinations) in an MST-like area. The 12 basic motion directions and three speeds are illustrated with the spoked-wheel design in the figure. In addition, at both MT and MST levels there is a second set of detectors for translational motion, just like in V1, but at progressively coarser scales. This motion hierarchy is detailed in Tsotsos

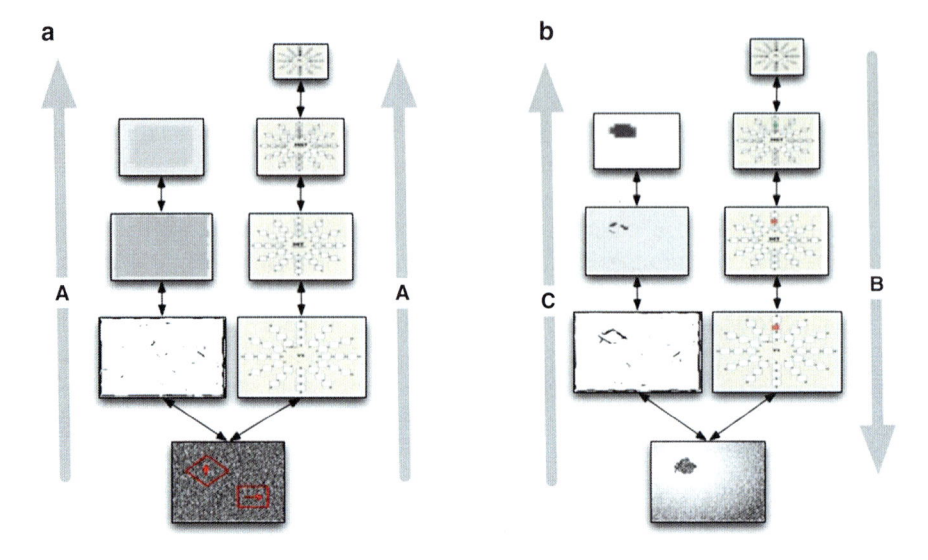

Fig. 1.3 An example of Type II iterative recurrence binding operating on a Form-from-Motion stimulus input. (**a**) Two forms – *diamond* and *rectangle* – are moving as indicated by the *red outlines* and *arrows*, defined by coherently moving dots on a field of randomly moving dots, *lowest panel*. There is no luminance variation across the image that would sufficient for an edge detection process to find a luminance discontinuity. The processing pathways are for motion – areas V1, MT, and MST (*on the right*) and for shapes V1, V4, and IT (*on the left*) and are described in the text. The arrow A shows flow of processing. Note how the shape pathway "sees" nothing. (**b**) The same pathways but after the first feed-forward pass. The arrow B depicts first recurrent pass that localizes the selected motion stimulus (in *red* for the motion representations) and imposes a suppressive surround. The arrow C represents the second feed-forward pass following the recurrent pass. Note how the lowest processing level of the shape pathway is able to localize the rectangular shape once the suppressive surround removes the surrounding motion

et al. (2005). The shape hierarchy detects basic geometric shapes and consists of edge detectors (four directions, one scale), and two layers of shape detectors, loosely corresponding to visual areas V1, V4 (local shape), and IT (global shape). The system processes the image in parallel, along both of these independent processing pathways, detecting the presence of the different shapes and motion patterns. Selective Tuning's selection process chooses the top-level representation for further analysis, i.e., localization. For a motion-defined shape – that is, shape created by a set of coherently moving dots on a background of randomly moving dots with no other information indicating a boundary – only the motion pathway will detect anything and the shape pathway will be silent. There is no luminance discontinuity for the edge detectors to respond to. In the figure, the moving forms are shown with red outlines and an arrow indicating their direction of motion. Only the translational motion units of the hierarchy are activated with these motions. The flows indicated by the arrows labeled A highlight the fact that both paths process the input simultaneously. The flow indicated by the arrow B represents Selective Tuning's recurrence process that localizes the corresponding pixels in the input image. The resulting suppressive surround also enhances the relative saliency of the attended stimulus. However, by doing so – suppressing the randomly moving dots surrounding the coherent stimulus – a luminance discontinuity is created where there was none in the original stimulus. Feed-forward signals from the stimulus are continually flowing through the system, and as the suppressive surround is imposed, the shape pathways now see these discontinuities and detect shape, the flow represented by the arrow labeled C. The second feed-forward pass through the pathways refines the representation of the selected object. The key here is that the process of localization and surround inhibition inherent in the top-down neural trace reveals image structure not previously visible to the shape and object processing hierarchy. The system can then process the revealed features normally, permitting detection of a rectangle at the top layers and, if needed, localizing the details of the rectangle's shape by another full recurrence pass. The process can be repeated until the desired target is found, implementing visual search.

1.4 Vision as Dynamic Tuning of a General Purpose Processor

The importance of generality in computer vision systems whose goal is to achieve near-human performance was emphasized early in the field's development (Zucker et al. 1975; Barrow and Tenenbaum 1981). To be specific, Zucker et al. write:

> How can computer vision systems be designed to handle unfamiliar or unexpected scenes? Many of the current systems cope quite well with limited visual worlds, by making use of specialized knowledge about these worlds. But if we want to expand these systems to handle a wide range of visual domains, it will not be enough to simply employ a large number of specialized models. It will also be important to make use of what might be called "general purpose models."

With similar thinking, Barrow and Tenenbaum say:

> Like relaxation, hierarchical synthesis is conceptually parallel, with plausible hypotheses emerging from a consensus of supporting evidence. When resources are finite (the extreme case being a conventional serial computer), control can be allocated among processes representing various levels of hypotheses in accordance with strength of support, goals, expectations, and so forth. In this way, data-driven and goal-driven operation can be integrated as suggested earlier.... the principles and techniques brought together in this paper represent a significant step beyond current special-purpose computer vision systems, towards systems approaching the generality and performance of human vision.

Yet, this generality in practice has been elusive. Here, a proposal is made in this direction. Using formal methods from complexity theory, we have shown an architecture for vision that has two main properties:

- It can solve a particular class of vision problems very quickly.
- It can be tuned dynamically to adapt its performance to the remaining subclasses of vision problems but at a cost of greater time to process.

Furthermore, we have shown that a major contributor in this dynamic tuning process is the set of mechanisms that have come to be known as attention. Importantly, the data-driven processes of Barrow and Tenenbaum are the same as the feed-forward ones described herein, but goal-driven processing is not the only top-down mechanism. The top-down and recurrent mechanisms here include both goals as well as top-down attention mechanisms not specific to tasks or goals. This is a significant difference and one that seems to be missing from current computational vision works as well. The combined application of elements from the set of attentional mechanisms listed in Sect. 1.2 provide for a means to tune the general purpose, but limited in functionality, processing network to enable a far broader set of visual tasks to be solved albeit at the expense of greater processing time.

References

Ahissar, M., Hochstein, S. (1997). Task difficulty and the specificity of perceptual learning, *Nature*, 387, 401–406.

Barrow, H., Tenenbaum, J. (1981). Computational vision, *Proc. IEEE*, 69(5), 572–595.

Boehler, C.N., Tsotsos, J.K., Schoenfeld, M., Heinze, H.-J., Hopf, J.-M. (2009). The center-surround profile of the focus of attention arises from recurrent processing in visual cortex, *Cereb. Cortex*, 19, 982–991.

Bullier, J. (2001). Integrated model of visual processing, Brain Res. Rev., 36, 96–107.

Culhane, S., Tsotsos, J.K. (1992). An attentional prototype for early vision. In: Sandini, G. (ed.), Computer Vision—ECCV'92. Second European Conference on Computer Vision Santa Margherita Ligure, Italy, May 19–22, 1992. *Lect. Notes in Comput. Sci.* 588, pp. 551–560, Springer-Verlag.

Dickinson, S., Leonardis, A., Schiele, B., Tarr, M. (2009). Object categorization, Cambridge University Press, New York.

DiLollo, V. (2010). Iterative reentrant processing: a conceptual framework for perception and cognition. In: Coltheart, V. (ed.), Tutorials in Visual Cognition, Psychology Press, New York.

Duncan, J., Ward, J., Shapiro, K. (1994). Direct measurement of attentional dwell time in human vision, *Nature*, 369, 313–315.

Evans, K., Treisman, A. (2005). Perception of objects in natural scenes: is it really attention free? *J. Exp. Psychol. Hum. Percept. Perform.*, 31–6, 1476–1492.

Fukushima, K. (1986). A neural network model for selective attention in visual pattern recognition, *Biol. Cybern.*, 55(1), 5–15.

Garey, M., Johnson, D. (1979). Computers and intractability: A guide to the theory of NP-completeness, Freeman, San Francisco.

Ghose, G., Maunsell, J. (1999). Specialized representations in visual cortex: a role for binding? *Neuron*, 24, 79–85.

Grill-Spector, K., Kanwisher, N. (2005). Visual recognition: as soon as you know it is there, you know what it is, *Psychol. Sci.*, 16, 152–160.

Gueye, L., Legalett, E., Viallet, F., Trouche, E., Farnarier, G. (2002). Spatial orienting of attention: a study of reaction time during pointing movement, *Neurophysiol. Clin.*, 32, 361–368.

Lamme, V., Roelfsema, P. (2000). The distinct modes of vision offered by feedforward and recurrent processing, *TINS*, 23(11), 571–579.

Lünenburger, L., Hoffman, K.-P. (2003). Arm movement and gap as factors influencing the reaction time of the second saccade in a double-step task, *Eur. J. Neurosci.*, 17, 2481–2491.

Macmillan, N.A., Creelman, C.D. (2005). Detection theory: a user's guide, Routledge.

Marr, D. (1982). Vision: a computational investigation into the human representation and processing of visual information. Henry Holt and Co., New York.

Milner, P.M. (1974). A model for visual shape recognition. *Psychol. Rev.*, 81–86, 521–535.

Mehta, A., Ulbert, I., Schroeder, C. (2000). Intermodal selective attention in monkeys. I: distribution and timing of effects across visual areas. *Cereb. Cortex*, 10(4), 343–358.

Müller, H., Rabbitt, P. (1989). Reflexive and voluntary orienting of visual attention: time course of activation and resistance to interruption. *J. Exp. Psychol. Hum. Percept. Perform.*, 15, 315–330.

Posner, M.I., Nissen, M., Ogden, W. (1978). Attended and unattended processing modes: the role of set for spatial locations. In: Pick Saltzmann (ed.), Modes of perceiving and processing information. Erlbaum, Hillsdale, NJ, pp. 137–158.

Rensink, R. (1989). A new proof of the NP-Completeness of Visual Match, Technical Report 89–22, Department of Computer Science, University of British Columbia.

Reynolds, J., Desimone, R. (1999). The role of neural mechanisms of attention in solving the binding problem, *Neuron*, 24, 19–29.

Riesenhuber, M., Pogio, T. (1999). Are cortical models really bound by the "Binding Problem"? *Neuron*, 24, 87–93.

Roskies, A. (1999). The binding problem–introduction, *Neuron*, 24, 7–9.

Rosenblatt, F. (1961). Principles of neurodynamics: perceptions and the theory of brain mechanisms. Spartan Books.

Rodriguez-Sanchez, A.J., Simine, E., Tsotsos, J.K. (2007). Attention and visual search, *Int. J. Neural Syst.*, 17(4), 275–88.

Rothenstein, A., Rodriguez-Sanchez, A., Simine, E., Tsotsos, J.K. (2008). Visual feature binding within the selective tuning attention framework, *Int. J. Pattern Recognit. Artif. Intell.*, Special Issue on Brain, Vision and Artificial Intelligence, 861–881.

Thorpe, S., Fize, D., Marlot, C. (1996). Speed of processing in the human visual system, *Nature*, 381, 520–522.

Treisman, A. (1999). Solutions to the binding problem: progress through controversy and convergence, *Neuron*, 24(1), 105–125.

Treisman, A.M., Gelade, G. (1980). A feature-integration theory of attention, *Cogn. Psychol.*, 12(1), 97–136.

Tsotsos, J.K., Rodriguez-Sanchez, A., Rothenstein, A., Simine, E. (2008). Different binding strategies for the different stages of visual recognition, *Brain Res.*, 1225, 119–132.

Tsotsos, J.K. (1987). A 'Complexity Level' analysis of vision, Proceedings of 1st International Conference on Computer Vision, London, UK.

Tsotsos, J.K. (1988). A 'Complexity Level' analysis of immediate vision, *Int. J. Comput. Vision*, Marr Prize Special Issue, 2(1), 303–320.

Tsotsos, J.K. (1989). The complexity of perceptual search tasks, Proceedings of the International Joint Conference on Artificial Intelligence, Detroit, pp. 1571–1577.

Tsotsos, J.K. (1990). Analyzing vision at the complexity level, *Behav. Brain Sci.*, 13–3, 423–445.

Tsotsos, J.K. (1991a). Localizing Stimuli in a Sensory Field Using an Inhibitory Attentional Beam, October 1991, RBCV-TR-91–37.

Tsotsos, J.K. (1991b). Is complexity theory appropriate for analysing biological systems? *Behav. Brain Sci.*, 14–4, 770–773.

Tsotsos, J.K. (1993). An inhibitory beam for attentional selection. In: Harris, L., Jenkin, M. (Eds.), Spatial vision in humans and robots. Cambridge University Press, pp. 313–331.

Tsotsos, J.K., Culhane, S., Wai, W., Lai, Y., Davis, N., Nuflo, F. (1995). Modeling visual attention via selective tuning, *Artif. Intell.*, 78(1–2), 507–547.

Tsotsos, J.K., Liu, Y., Martinez-Trujillo, J., Pomplun, M., Simine, E., Zhou, K. (2005). Attending to visual motion, *Comput. Vis. Image Underst.*, 100(1–2), 3–40.

Tsotsos, J.K., Bruce, N.D.B. (2008). Computational foundations for attentive processes, *Scholarpedia*, 3(12), 6545.

Tsotsos, J.K. (2011). A computational perspective on visual attention, MIT, Cambridge, MA.

VanRullen, R., Carlson, T., Cavanaugh, P. (2007). The blinking spotlight of attention, *Proc. Natl. Acad. Sci. USA*, 104–49, 19204–19209.

Wolfe, J.M. (1998). Visual search. In: Pashler, H. (Ed.), Attention. Psychology Press, Hove, UK, pp. 13–74.

Yarbus, A.L. (1967). Eye movements and vision. Plenum, New York.

Zucker, S.W., Rosenfeld, A., Davis, L.S. (1975). General-purpose models: expectations about the unexpected, Proceedings of the 4th International Joint Conference on Artificial Intelligence, Tblisi, USSR pp. 716–721.

Chapter 2
Sensory Fusion

Mauro Ursino, Elisa Magosso, and Cristiano Cuppini

Abstract Multisensory integration is known to occur in many regions of the brain, and involves several aspects of our daily life; however, the underlying neural mechanisms are still insufficiently understood. This chapter presents two mathematical models of multisensory integration, inspired by real neurophysiological systems. The first considers the integration of visual and auditory stimuli, as it occurs in the superior colliculus (a subcortical region involved in orienting eyes and head toward external events). The second model considers the integration of tactile stimuli and visual stimuli close to the body to form the perception of the peripersonal space (the space immediately around our body, within which we can interact with the external world). Although devoted to two specific problems, the mechanisms delineated in the models (lateral inhibition and excitation, nonlinear neuron characteristics, recurrent connections, and competition) may govern more generally the fusion of senses in the brain. The models, besides improving our comprehension of brain function, may drive future neurophysiological experiments and provide valuable ideas to build artificial systems devoted to sensory fusion.

2.1 Introduction

A fundamental problem in cognitive neuroscience is how we perceive and represent the external world starting from information produced by our senses. Traditional research on this topic generally adopted a "sense by sense" approach, i.e., attention was focused on the single sensory modalities to study how information is coded in each sensory channel and subsequently processed and transmitted. The motivation of this approach was to reveal the characteristics of single senses (vision, hearing, touch, etc.) to point out their specific neural substrates and correlate this information with perception. Indeed, it has been assumed for many decades that information

M. Ursino (✉)
Department of Electronics, Computer Science and Systems, University of Bologna,
Viale Risorgimento 2, I-40136 Bologna, Italy
e-mail: mauro.ursino@unibo.it

V. Cutsuridis et al. (eds.), *Perception-Action Cycle: Models, Architectures, and Hardware*, Springer Series in Cognitive and Neural Systems 1, DOI 10.1007/978-1-4419-1452-1_2, © Springer Science+Business Media, LLC 2011

from individual senses is initially coded and processed in segregate brain areas (such as the visual cortex in the occipital lobes, the auditory cortex in the temporal lobes, and the somatosensory cortex in the parietal lobes) and that only later in the processing pathways, this distributed information may be merged together. This traditional view posits that a large part of the brain can be reduced into a collection of unisensory modules that can be studied in isolation.

However, recent research challenges this traditional view, clearly demonstrating that our senses are designed to work in concert, and not as separate modules. Indeed, almost any experience in daily life consists of a multisensory stream of data that provide concurrent and complementary information about the same external event or object: the multiple sensory information must be simultaneously considered by the brain and globally processed to reach a unique coherent percept. Moreover, compelling evidence now exists that information represented by each of the primary sensory systems (visual, auditory, somatosensory, and olfactory) is highly susceptible to influences from other senses. This signifies that the theory of multisensory integration must move beyond the notion of a purely sum of individual contributions and must consider a more complex cooperation among the senses.

This sophisticate integration of multiple sensory cues provides animals and humans with enormous behavior and response flexibility. Indeed, the fundamental aim of the brain with respect to the external world is to reach a clear recognition of objects and events (especially of those essential for surviving) and to drive the suitable behavior and physical reactions. Decision on objects and events in the environment requires that information from all the senses is simultaneously exploited to maximize the probability of a correct detection and minimize the probability of errors. To reach this objective, sensory signals that originate from congruent spatial positions and in temporal proximity – hence that likely derive from the same event – must be linked together, whereas sensory signals that do not exhibit temporal and/or spatial coherence must be separated. Moreover, conflict situations must be resolved favoring one or another possibility (either sense merging or sense separation).

Multisensory processing is known to occur in many cortical and subcortical regions of the brain, and involves a lot of different aspects of our daily life and overt behavior: among the others, the rapid head and eye movements directed toward external visual and auditory cues; the merging of smell and taste during eating; the integration of tactile and visual information close to the body, to create the perception of a peripersonal space; and the merging of auditory, visual, and other senses to recognize objects (for instance, the shape of a cow and its moo). The importance of all these aspects for survival is evident.

Unfortunately, despite the enormous role that multisensory integration plays in our daily life, the neural mechanisms that subserve this property are still insufficiently understood. It is worth noting that neural circuits for multisensory integration must not merely sum the information coming from different sensory modalities, but must perform complex, nonlinear computation to emphasize stimuli coming from coherent spatial and temporal characteristics, neglect incongruent stimuli, and solve conflicts. As it is illustrated throughout this chapter, this requires the presence of complex excitatory and inhibitory mechanisms and a sophisticate topological representation of the external space.

The aim of this chapter is to present two recent mathematical models of multisensory integration, inspired by real neurophysiological systems. The first model considers the integration of visual and auditory stimuli, as it occurs in a subcortical region of the midbrain, named the superior colliculus, which is implicated in driving overt responses (such as eye movements) toward external events. The second model treats the problem of how visual stimuli close to the body (for instance, stimuli on and close to the hands) and tactile stimuli in the same body parts are integrated to form the perception of a peripersonal space (i.e., the space immediately around our body surface, from which objects can hit us and where we can manipulate objects).

As just stated above, the two models are devoted to two different and specific problems; they have been formalized separately, and of course, their development is founded on different data in the literature. The visual–auditory model is mainly based on electrophysiological recordings in superior colliculus neurons, and, to a less extent, behavioral data concerning interplay between vision and audition; the visual–tactile model has been conceptualized on the basis of single-cell recordings in multisensory regions of the associative cortex (frontal–parietal cortex) and on the basis of neuroimaging and behavioral data on both healthy subjects and brain damage patients. The so obtained models obviously differ for some aspects (as they are conceived for different aims); however, they share some important features concerning the architecture, the synaptic mechanisms, and the individual characteristics of the single neurons: in both models, these features are fundamental to reproduce and explain in vivo data. First of all, both models include upstream areas of unimodal neurons, which project feedforward convergence to a downstream area; hence, neurons in the downstream area are multisensory, responding to stimuli in more than one modality. The patterns of the feedforward synapses are fundamental to establish the shape and position of the multiple receptive fields of the multimodal neurons. Moreover, in both models, neurons in the same area are connected via lateral synapses characterized by short-range excitation and long-range inhibition; this lateral connectivity implements a *topological organization* of the single areas, realizes an efficient competitive mechanism among neurons in the same area, and influences the shape of the neurons' receptive fields. Again, both models include feedback synapses from the higher level multimodal area down to the lower level unimodal areas; accordingly, a stimulus in one modality may influence the activity of neurons coding for a different modality. This mechanism explains how the subjective experience in one modality may be dramatically affected by stimulation in another modality, even creating illusory perception. Finally, neurons in both models are characterized by a nonlinear sigmoidal input–output relationship (with a lower threshold and an upper saturation). Such nonlinearity is fundamental to reproduce a general principle of multisensory integration, named *inverse effectiveness*, according to which cross-modal enhancement is highly effective when the information provided by the unisensory channels is weak.

The important fact is that in both models these mechanisms are essential to replicate experimental data, although the two models refer to different stimuli combinations (visual and auditory on the one hand, visual and tactile on the other hand) and to different regions of multisensory integration in the brain (the superior colliculus, a subcortical structure, and the parieto–frontal associative cortex). Hence, we can postulate that the mechanisms exploited in these two models may have a larger validity, can be probably found in many other multisensory regions of the brain, and explain integration among other different modalities. In fact, as pointed out by some authors (Dalton et al. 2000), the same principles governing the fusion of some senses (for instance, integrating visual and auditory information) apply to other combinations of senses, and multisensory integration probably shares some common principles.

As we describe in detail in subsequent sections, the developed models not only enhance coherent cross-modal stimuli but also allow depression of incoherent stimuli; in certain conditions, they afford conflict resolution among stimuli and account for phenomena such as ventriloquism or phantom perception. They may be subject to learning and plasticity, thus allowing a modification of our multisensory perception driven by previous experience; finally, they may be under attention mechanisms, which modify our multisensory perception on the basis of exogenous or endogenous expectation. Finally, they may be used to simulate clinical deficits, by simply adjusting some model parameters. Of course, some of these aspects have been considered more extensively in one model than in the other and vice versa, in relation to the specific aim of each model and to data available in the literature. A crucial point is that both models allow behavioral and psychophysical results (e.g., ventriloquism in one case, extinction in the other case) to be interpreted in terms of the individual neuron properties and of the reciprocal interconnections among neurons; hence, they may give an important contribution to bridge the gap between behavioral and neuronal responses.

As a last point, we wish to stress that the present models, although inspired by neurophysiology, and especially aimed at improving our understanding of brain function, may furnish valuable ideas to build artificial systems devoted to sensory fusion. Some of the basic mechanisms presented below (mutual excitation, mutual inhibition, enhancement vs. depression, spatial arrangement of the inputs, and non-linear sensitivity to the strength of the stimuli) have a general role in perception and can represent paradigms for any advanced multisensory processing system.

2.2 Audio–Visual Integration

2.2.1 Audio–Visual Integration in the Superior Colliculus: Neurophysiological and Behavioral Evidence (Overview and Model Justification)

Let us consider the problem of integration of visual and auditory stimuli to drive overt behavior. The concepts described below refer to a particular midbrain area, the superior colliculus (SC), that has been deeply studied in the context of multisensory integration; however, they may have a more general validity and are suitable to illustrate how a biologically inspired neural network can realize multisensory integration to improve the response to external stimuli.

The role of the SC is to initiate and control overt movements in response to important stimuli from the external world, for instance, to control the shift of gaze or to orient various sensory organs to a correct direction (Stein and Meredith 1993). Various brain regions involved in auditory, somatosensory, and visual processing send inputs to the SC (Edwards et al. 1979; Huerta and Harting 1984; Stein and Meredith 1993).

The behavior of the SC neurons exhibits quite trivial characteristics in response to a single spatially localized unisensory stimulus (either auditory or visual or somatosensory) (Perrault et al. 2005). If the stimulus amplitude is too small, it is neglected and the neuron does not respond at all. If the stimulus overcomes a given threshold, it activates a response: the higher the stimulus the higher the response up to an upper saturation. This kind of behavior can be reproduced fairly well using a dynamical block (for instance, a low-pass filter), which mimics the temporal aspects of the response, followed by a sigmoidal relationship. Moreover, different neurons in the SC have different receptive fields (RFs), i.e., they respond to stimuli coming from different positions in space. These RFs are topographically organized, so that proximal neurons in the SC have RFs with proximal centers in the environment. Of course, only those neurons whose receptive field overlaps the stimulus will respond, by driving the behavior.

However, different problems complicate the previous scenario. First, while some neurons in the SC are unisensory, and they respond only to stimuli of a single modality [visual: 21% of neurons, auditory: 15% of neurons, somatosensory: 6% of neurons (Kadunce et al. 1997)], more than half are multisensory, i.e., they respond to stimuli of different sensory modalities [visual–auditory: 35%, visual–auditory–somatosensory: 8% (Kadunce et al. 1997); however, different percentage values are reported in Wallace et al. (1998): visual: 23%, auditory: 13%, somatosensory: 10%, audio–visual: 25%, trimodal: 9%]. Multisensory neurons, in general, have receptive fields for different modalities in spatial register; this means not only that a visual–auditory neuron will have two RFs (one for the auditory and one for the visual modality) but also that these RFs have a large superimposed region (Meredith and Stein 1996).

The presence of multisensory neurons, whose RFs are in spatial register, can explain a phenomenon named "multisensory enhancement": when two cross-modal stimuli (for instance, one visual and one auditory) come from proximal positions of space and in close temporal proximity, the response of the SC neuron is generally greater than each of the individual unisensory responses. This enhancement may be subadditive, additive, or superadditive depending on whether the cross-modal response is smaller, equal, or greater than the sum of the individual unisensory responses (Kadunce et al. 2001; Perrault et al. 2005). Furthermore, the response of a multisensory SC neuron follows a rule named "inverse effectiveness": the enhancement produced by two spatially aligned cross-modal stimuli is inversely related to the effectiveness of the individual unimodal stimuli (Perrault et al. 2005). Consequently, the weaker the unisensory stimuli, the greater the enhancement they produce when occurring together in spatial proximity. The practical effect of these rules on behavior is easily understandable: two-weak targets, which do not produce an appreciable response if applied alone, can induce a vigorous response when occurring together but from the same spatial source.

The complexity of the SC response, however, is much greater than that emerging from a single nonlinearity, i.e., from the behavior of a single neuron. Several other aspects related with the interactions among neurons should be considered.

First, if two within-modal stimuli (i.e., two stimuli of the same modality, for instance, both auditory or both visual) or two cross-modal stimuli (i.e., stimuli of different modalities, one auditory and the other visual) originate from disparate positions in space, the final response of the SC is reduced or eliminated compared with the response to an individual stimulus alone ("within modal and cross-modal suppression") (Kadunce et al. 1997). This behavior implicates the presence of some competitive interactions among neurons, whose RFs are located at different spatial positions.

Another important property of SC neurons, however, is that within-modal suppression is stronger and more robust than cross-modal suppression. Kadunce et al. (1997) demonstrated that, in the cat, power within-modal suppression can be present even in the absence of cross-modal suppression. However, the converse is not true, i.e., cross-modal suppression always occurs together with within-modal suppression. This result suggests the presence of at least two different mechanisms: the first, which may affect both cross-modal and within-modal suppression, can be ascribed to a competition among multimodal neurons in the SC; the second, which affects unimodal suppression only, probably reflects interactions occurring at the level of the single input channels (i.e., a competition in unimodal areas targeting to the SC).

Finally, a further mechanism should be mentioned, related to the well-known phenomenon of ventriloquism: in certain cases, in the presence of two stimuli with disparate spatial positions, a stronger stimulus may capture the position of the other one (Bermant and Welch 1976; Woods and Recanzone 2004). This occurs in case of a conflict between the position of two stimuli, and the system solves the conflict assuming a single position for both. Worth noting is that, in general, the visual stimulus predominates over the position of the auditory stimulus. We are not aware whether a ventriloquism effect actually occurs in the SC, but inclusion of this possibility makes our study of multisensory integration much more general and suitable for a variety of further applications.

A further related aspect is that perception of a unimodal stimulus (for instance, a poor auditory or visual target) can be improved, not only at the SC level (multimodal), but also at the level of unimodal areas, by a second stimulus of the other modality given at the same spatial position (Bolognini et al. 2005).

Despite the large amount of results published in the literature and qualitative explanations currently available, we were not aware of a comprehensive theoretical model able to summarize these data. Previous important models were especially focused on information theory, but were not inspired by neurobiological mechanisms. In particular, Anastasio et al. (2000), Anastasio and Patton (2003),

Patton et al. (2002), Patton and Anastasio (2003) developed some models in which neurons implement the Bayes rule to compute the probability that a target is present in their receptive field. With these models, they were able to account for the existence of both multimodal and unimodal neurons and for the existence of cross-modal enhancement as well as within-modal suppression (Anastasio and Patton 2003; Patton and Anastasio 2003). A similar approach was used by Colonius and Diederich (2004) who developed a Bayes' ratio model of multisensory enhancement.

In earlier years, we presented a model (Magosso et al. 2008; Ursino et al. 2009) that is inspired by biological mechanisms and can explain most of the results delineated above; moreover, by changing some model parameters (especially those related with the synaptic strength), the model can account for the presence of neurons with different characteristics, as experimentally observed.

In the following, the main aspects of the model are first presented and justified. Then, some simulation exempla are given and commented on the basis of the mechanisms incorporated in the model.

2.2.2 Model Components

A qualitative sketch of the model is given in Fig. 2.1. Fundamental aspects are explained below, while all equations, mathematical details, and parameter numerical values can be found in previous publications of the authors (Magosso et al. 2008; Ursino et al. 2009).

- Each neuron is described through a sigmoidal relationship (with lower threshold and upper saturation) and a first-order dynamic (with a given time constant). Neurons normally are in a silent state (or exhibit just a mild basal activity) and can be activated if stimulated by a sufficiently strong input. In vivo the sigmoidal nonlinearity can be ascribed to the typical characteristics of neurons, which need a sufficient input current to generate spikes and which saturate: this behavior may be further accentuated by nonlinearities in the receptor responses at the synapse levels (for instance, the response of NMDA receptors).
- The model is composed of three areas (Fig. 2.1). Elements of each area are organized in $N \times M$ dimension matrices, so that the structure keeps a spatial and geometrical similarity with the external world: neurons of each area respond only to stimuli coming from a limited zone of the space. Furthermore, the two upstream areas are unimodal, and respond to auditory and visual stimuli, respectively. A third downstream area represents neurons in the SC responsible for multisensory integration. These three areas have a topological organization, i.e., proximal neurons respond to stimuli in proximal position of space.
- Each element of the unisensory areas has its own receptive field (RF) that can be partially superimposed on that of the other elements of the same area. The elements of the same unisensory area interact via lateral synapses, which can be both excitatory and inhibitory. These synapses are arranged according to a Mexican hat disposition (i.e., a circular excitatory region surrounded by a larger inhibitory annulus).

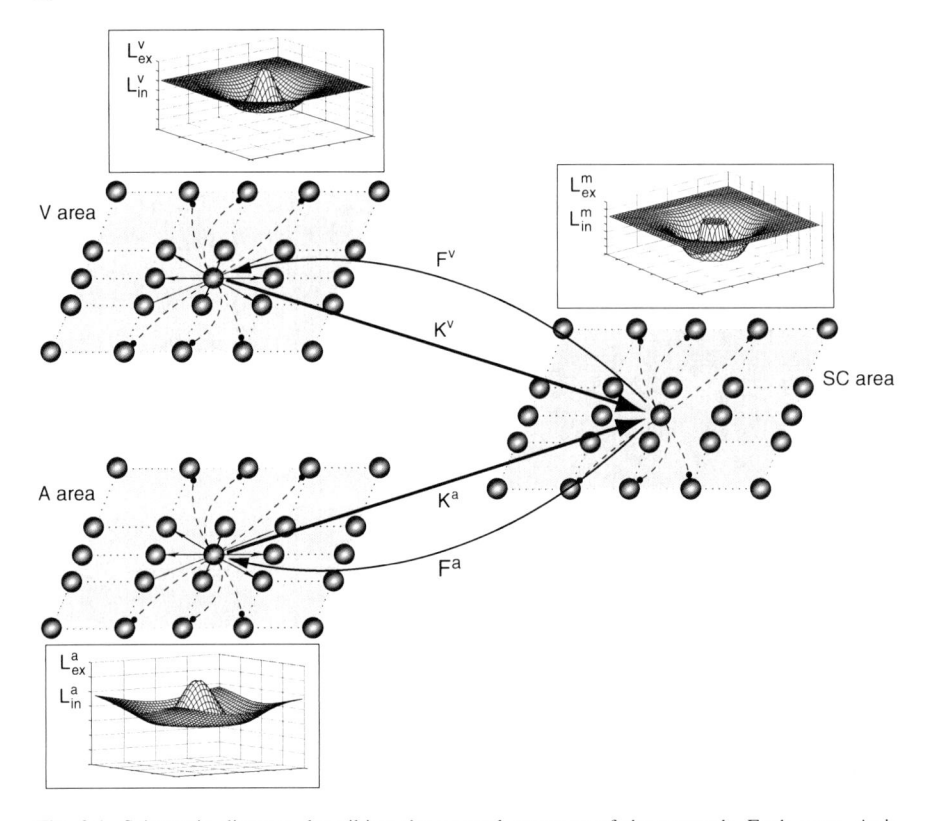

Fig. 2.1 Schematic diagram describing the general structure of the network. Each *gray circle* represents a neuron. Neurons are organized into three distinct areas of 40 × 40 elements. Each neuron of these areas (*V* visual, *A* auditory, *SC* multimodal in the superior colliculus) is connected with other elements in the same area via lateral excitatory and inhibitory intra-area synapses (*arrows* L$_{ex}$ and L$_{in}$ within the area). Neurons of the unimodal areas send feedforward excitatory interarea synapses to multimodal neurons in the superior colliculus area located at the same position (*arrows* K). Multimodal neurons, in turn, send excitatory feedback interarea connections to neurons of the unisensory areas (*arrows* F)

- The elements of the multisensory area in the superior colliculus receive inputs from the two neurons in the upstream areas (visual and auditory) whose RFs are located in the same spatial position. Moreover, elements in the SC are connected by lateral synapses, which also have a Mexican hat disposition.
- The multimodal neurons in the SC send a feedback excitatory input to the unimodal neurons whose RFs are located in the same spatial position; in this way, detection of a multimodal stimulus may help reinforcement of the unisensory stimuli in the upstream areas.

2.2.3 Results

In the following lines, we separately consider how the different aspects of the model may contribute to explain the results of multisensory integration delineated before.

2.2.3.1 Enhancement and Inverse Effectiveness

Let us first consider the problem of multisensory enhancement and the inverse effectiveness rule. To simulate this phenomenon, we imagine two cross-modal stimuli (one auditory and one visual) located at approximately the same position in space. Each of these two stimuli causes an activation bubble in the respective unisensory area, as a consequence of superimposed RFs and lateral connections among neurons; hence, some contiguous neurons are simultaneously active. Since multimodal neurons in the third area have the visual and auditory RFs in spatial register, a group of contiguous multimodal neurons receive excitation from both the visual and auditory neurons, which are active together. Finally, this input is converted into a multimodal output, which can drive overt behavior, via the monotonic sigmoidal relationship. It is worth noting that enhancement will occur only if the two cross-modal stimuli fall within the receptive field of the same multimodal neuron, i.e., they must be in close spatial register. But how can we explain the inverse effectiveness? This property simply depends on the presence of the sigmoidal characteristic, which describes neuron output, as illustrated in Fig. 2.2.

If the two cross-modal stimuli have a small intensity, each of them produces just a negligible response in the multimodal neurons when acting alone, since the neuron working point is located close to the lower threshold of the sigmoidal relationship. In this condition, a second cross-modal stimulus moves the working point of the multimodal neuron to the linear region of the sigmoid, thus causing a disproportionate increase in the final response. Hence, we may observe this quasi-paradoxical result: two stimuli, which produce no response when acting separately, may cause a strong response if they act together, provided they occur in close spatial and temporal proximity. This property is named superadditivity and is illustrated via a quantitative exemplum in Fig. 2.2 upper panel.

If the two unimodal stimuli have sufficient amplitude, so that each of them can set the working point of the multimodal neuron to the linear region, their simultaneous occurrence may cause a quasi-linear behavior, named simple additivity: this signifies that the cross-modal response is close to the sum of the individual responses (Fig. 2.2, middle panel).

Finally, in case of strong stimuli, the simultaneous occurrence of both may lead the multimodal neuron to its saturation region: as a consequence, the multimodal response is much smaller than the sum of the individual responses (subadditivity, see Fig. 2.2, bottom panel).

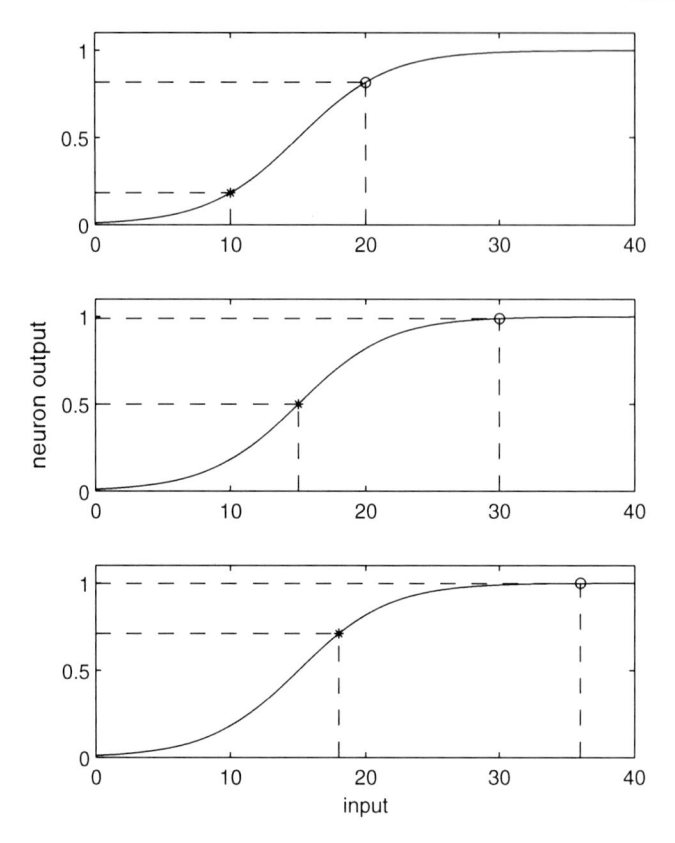

Fig. 2.2 Explanation of the inverse effectiveness. The figure represents the sigmoidal characteristic of a multimodal neuron. *Upper panel*: a small input (value $= 10$) is applied to the neuron, causing just a mild activity (*asterisk*). A second equal cross-modal stimulus moves the working point into the linear region (*open circle*). Enhancement is about 400%. *Middle panel*: a moderate input (value $= 15$) is applied to the neuron, leading the working point exactly at the mid of the linear region (*asterisk*). A second equal cross-modal stimulus moves the working point close to saturation (*open circle*). Enhancement is about 100%. *Bottom panel*: a stronger input (value $= 18$) is applied to the neuron, leading the working point proximal to saturation (*asterisk*). A second equal cross-modal stimulus moves the working point inside the saturation region (*open circle*). Enhancement is less than 50%

A summary of the results is presented in Fig. 2.3. Here, enhancement is computed through the so-called "interactive index"; this is a measure of the response increase induced by two cross-modal stimuli compared to a single stimulus, and is defined as follows:

$$\text{Interactive Index} = \left[\frac{\text{Mr} - \text{Ur}_{\max}}{\text{Ur}_{\max}} \right] \times 100, \tag{2.1}$$

where Mr (multisensory response) is the response evoked by the combined-modality stimulus and $\text{Ur}_{\max}$ (unisensory response) is the response evoked by the most effective unisensory stimulus. In these results, enhancement is presented as a function

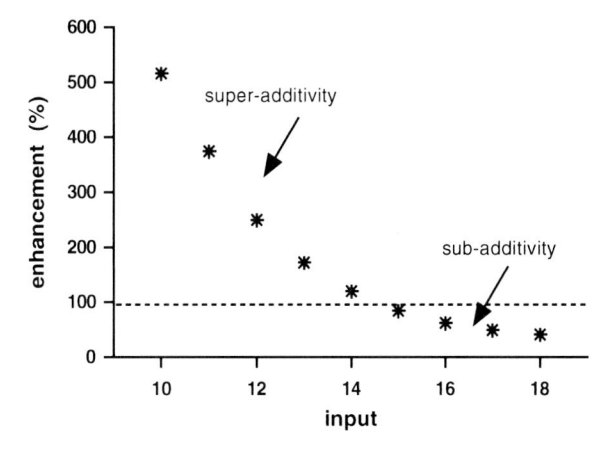

Fig. 2.3 Multisensory enhancement computed with the model in steady state conditions, in response to two superimposed cross-modal stimuli of the same intensity, placed at the center of the RF. The intensity of the stimuli is plotted in the x-axis. Hundred percent represents the threshold between subadditivity and superadditivity. Enhancement decreases with stimulus strength, according to the inverse effectiveness principle explained in Fig. 2.2

of the input intensity, assuming two cross-modal stimuli (one auditory and one visual) with identical strength (i.e., which can produce the same SC output when applied individually) and located at the same spatial position. In case of weak stimuli, enhancement can reach 500% or more (i.e., the cross-modal response is fivefold greater than the unisensory response). In case of strong stimuli, enhancement becomes close to zero. It is worth noting that, in these simulations, 100% is the threshold between subadditivity and superadditivity.

In conclusion, the previous results can be explained by the following characteristics of our model (1) the presence of two unimodal areas, with unimodal RFs; (2) the presence of a downstream multimodal area, whose neurons have auditory and visual RFs in spatial register; and (3) the presence of a sigmoidal relationship for neurons.

However, further aspects must be incorporated to explain suppression.

2.2.3.2 Cross-Modal Suppression

For the model to explain cross-modal suppression, we need the presence of lateral synapses among multimodal neurons. To this end, as anticipated above, we included lateral synapses in the SC with a Mexican hat disposition: proximal neurons in the multimodal area send reciprocal excitatory connections, but send and receive inhibitory connections to/from more distal neurons. Hence, each neuron in the multimodal area is surrounded by a small excitatory region and a wider inhibitory annulus. Indeed, the presence of this inhibitory suppressive region can explain both cross-modal and within-modal suppression. Two exempla are provided, the first

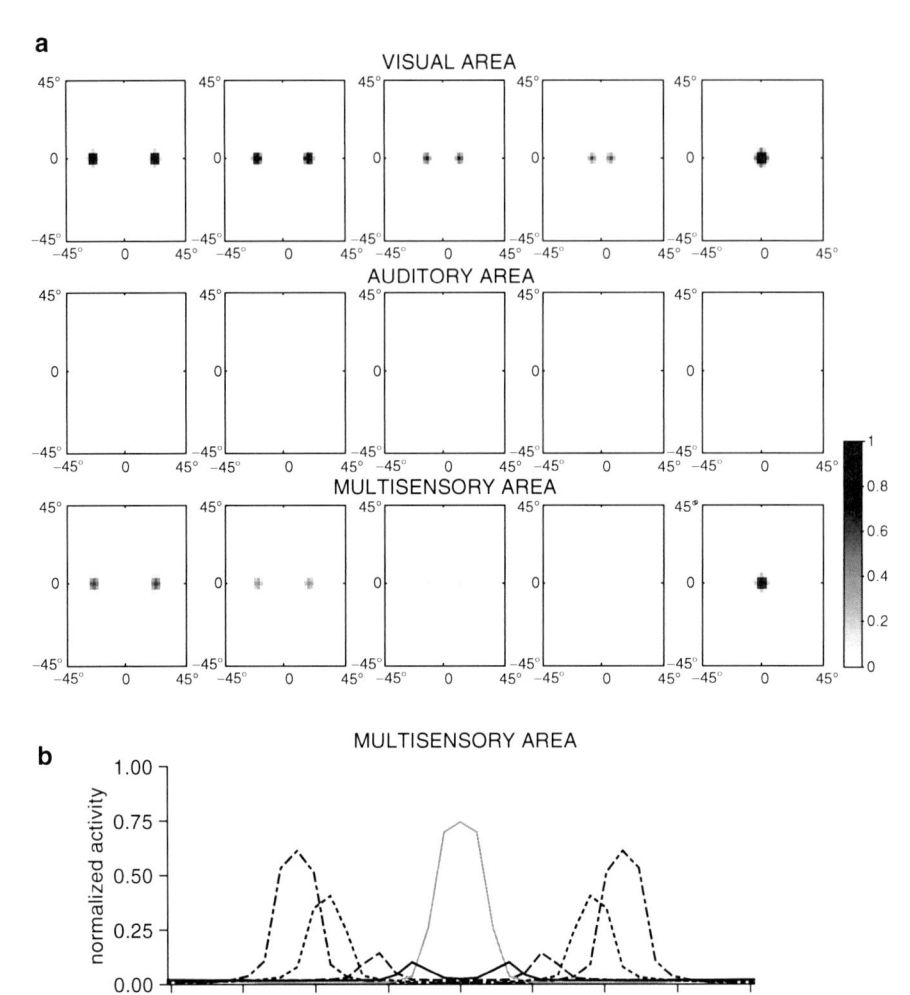

Fig. 2.4 Model response to two simultaneous visual stimuli placed at five different positions in space. (**a**) Each column depicts the activity in the three areas of the model (visual, auditory and multisensory) in steady state conditions, after application of two stimuli with decreasing spatial distance. A strong within modality suppression in the multisensory area is evident in the third and fourth columns. (**b**) Profile showing the response of neurons in the multisensory area whose RF is centered at the vertical coordinate $0°$ (i.e., positioned at the middle of the vertical field) during the five simulations depicted in (**a**). Within modality suppression is evident (greater than 80%). Within modality enhancement of two superimposed stimuli is mild

(Fig. 2.4) with reference to two within-modal stimuli located at different positions in space, the second (Fig. 2.5) with reference to two cross-modal stimuli. The simulations have been repeated by varying the distance between the two stimuli, and examining its effect on the response of the multimodal neurons. Results show that a

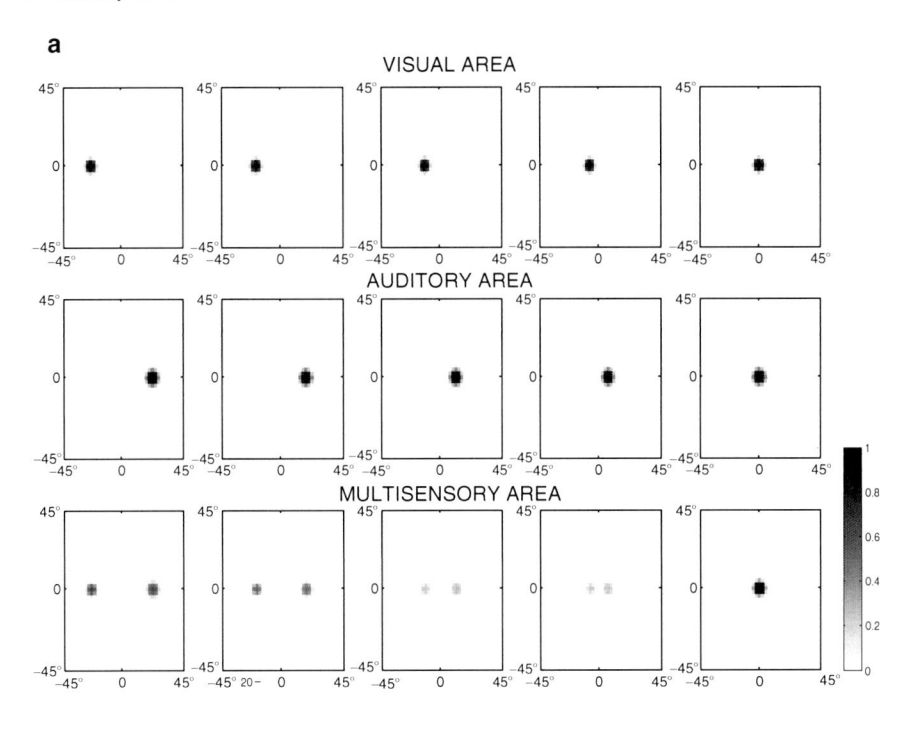

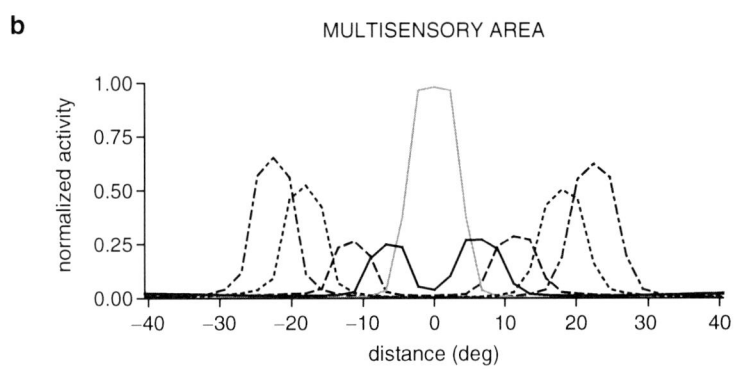

Fig. 2.5 Model response to two simultaneous cross-modal stimuli placed at five different positions in space. The meaning of panel (**a**) and panel (**b**) is the same as in Fig. 2.4. Cross modality suppression is evident (greater than 60%). Cross-modal enhancement of two superimposed stimuli is also evident (about +50%)

second stimulus of a different modality located within the receptive field causes significant cross-modal enhancement (Fig. 2.5), whereas in the case of a within-modal stimuli, the enhancement is mild (i.e., a second stimulus of the same modality located inside the RF does not evoke a significantly greater response, Fig. 2.4). If the second stimulus is moved away from the RF, one can observe significant within-modal suppression (up to 80%, Fig. 2.4) and significant cross-modal suppression (about 60%, Fig. 2.5). The suppressive regions are quite large (25–30°) in accordance with physiological data (Kadunce et al. 1997).

2.2.3.3   Within-Modal Suppression Without Cross-Modal Suppression

According to the previous simulations, a single mechanism (i.e., lateral inhibition within the multimodal area) can explain both within-modal and cross-modal suppression. However, results in the literature summarized in the introduction (Kadunce et al. 1997), suggest the presence of a further mechanism responsible for within-modal suppression. In fact, while cross-modal suppression is always accompanied by within-modal suppression, the reverse is not true. Furthermore, within-modal suppression is generally stronger than cross-modal suppression, as already evident comparing simulation results in Figs. 2.4 and 2.5. According to these results, in our model, within-modal suppression is affected not only by the presence of lateral inhibition within the multimodal area (that is the mechanism responsible for suppression in Fig. 2.5) but also by a second similar mechanism operating at the level of the unimodal areas. In fact, in our model also unimodal neurons receive and send lateral synapses arranged as a Mexican hat. If the suppressive annulus in the unimodal area is weak compared to that in the multimodal area, within-modal and cross-modal suppression will have approximately the same strength. Conversely, if we assume the existence of strong inhibitory synapses in one unimodal area, but poor inhibitory synapses in the multimodal area, we obtain strong within-modal suppression without cross-modal suppression. In this case, within-modal suppression is determined by inhibitory competition within the unimodal area, whereas two cross-modal stimuli do not compete in the multimodal area due to the absence of significant reciprocal inhibition. An example of the latter behavior, which has been experimentally observed in some SC neurons (Kadunce et al. 1997), is illustrated in Fig. 2.6.

2.2.3.4   Cross-Modal Facilitation and Ventriloquism Phenomenon

As shown above, the presence of lateral inhibitory mechanisms can explain suppression between spatially disparate stimuli and the model developed until now can account for a variety of phenomena observed in the intact superior colliculus. However, at least another important phenomenon deserves attention when discussing audio–visual integration: auditory and visual stimuli interact not only at the level of the multimodal areas (like the SC) but also at the level of the unimodal areas. By way of example in patients with unimodal visual impairments, the visual perception is improved when acoustic stimuli are given in close temporal and spatial proximity to visual stimuli (Bolognini et al. 2005). In other words, the presence of an auditory cue may improve visual perception of a poorly perceived visual cue. Another example is provided by ventriloquism tricks, in which a visual stimulus "captures"

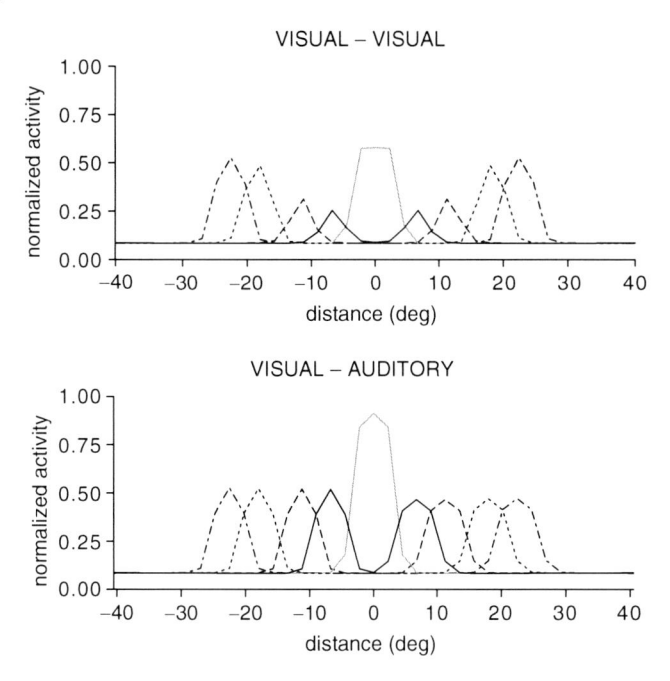

Fig. 2.6 Response of multisensory neurons with RF centered at the vertical coordinate 0° (i.e., positioned at the middle of the vertical field) during the five simulations depicted in Fig. 2.4 (visual–visual, *upper panel*), and in Fig. 2.5 (visual–auditory, *lower panel*). Results differ from those in Figs. 2.4 and 2.5, since the strength of lateral synapses in the multisensory area has been set at zero

or biases the perceived location of a spatially discordant auditory sound (Bermant and Welch 1976). Inclusion of such phenomena within a model for audio–visual integration requires that the two unimodal areas communicate, so that activity in one area may affect activity in the other. This might occur in two alternative ways: either postulating the existence of direct connections among the unimodal areas, or assuming the presence of a feedback from the multimodal neurons to the unimodal areas. In both cases, an input stimulus in one unimodal area would affect the other unimodal area. Actually, there is now some preliminary neurophysiological evidence showing that brain areas traditionally assumed as unimodal can be affected by input coming from different modalities (Ghazanfar and Schroeder 2006; Schroeder and Foxe 2005).

In the present model, we assumed the existence of a feedback link from the multimodal neuron in the SC to the neurons (visual and auditory) in the unimodal areas at the same spatial position (Fig. 2.1). In this way, activity in a unimodal area can be affected by a cross-modal stimulus via a top-down strategy (i.e., from a higher hierarchical level, represented by the multisensory area in our model). If feedback synapses are weak, this effect is almost negligible and the model replicates all simulations shown above. Conversely, if the feedback synapse gain is increased, one can observe significant interaction between the two unimodal areas.

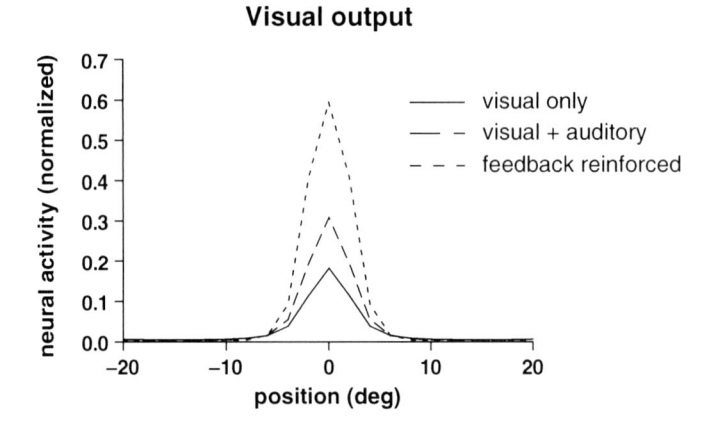

Fig. 2.7 Response of neurons in the visual unisensory area, whose RF is centered at the vertical coordinate $0°$ (i.e., positioned at the middle of the vertical field), in the steady-state condition following a poor visual stimulus, applied at the horizontal coordinate $0°$

An example, in which an auditory stimulus may help perception of a visual stimulus (cross-modal facilitation) is shown in Fig. 2.7. In this case, to simulate a patient with poor visual perception, a visual stimulus with low-amplitude was applied to the visual area. The stimulus produces just a moderate visual activity (about 10–15% of the maximal activity), which, due to intrinsic neural noise, may not be perceived in real conditions. However, let us assume that an acoustic stimulus is provided to the patient at the same spatial point. Since the acoustic area is not impaired, we assumed that this stimulus is strong enough to cause strong activation in the unisensory acoustic area (almost 100% at the corresponding position) and is able to activate the multimodal neuron. The latter, in turn, causes an increased activity in the visual area (up to 30–35%) thanks to the presence of feedback synapses from the multimodal to the unimodal areas. Hence, a visual stimulus, which is just poorly perceived, may be perceived much better, thanks to the simultaneous acoustic help. It is worth noting that, in this case, the acoustic stimulus not only evokes a response in the multimodal area, which is quite obvious, but can also help the formation of some activity in the visual area, i.e., it actively concurs to the formation of a visual target.

Let us now assume that the situation depicted above occurs frequently (the visual stimulus is continuously paired with an acoustic stimulus) and that the feedback synapses improve. In this condition, an acoustic stimulus can evoke a stronger activity (up to 60%) in the visual unimodal area (Fig. 2.7). This situation may have important potential consequences in therapeutic rehabilitation procedures.

As a last example, let us consider the case of ventriloquism. To simulate this phenomenon, we imagined the presence of a strong visual stimulus occurring together with a weaker auditory stimulus at different (but not too far) positions (Fig. 2.8). When the model are given basal parameter values (in particular, weak feedback synapses), both stimuli cause activation in the respective unimodal area (hence, the

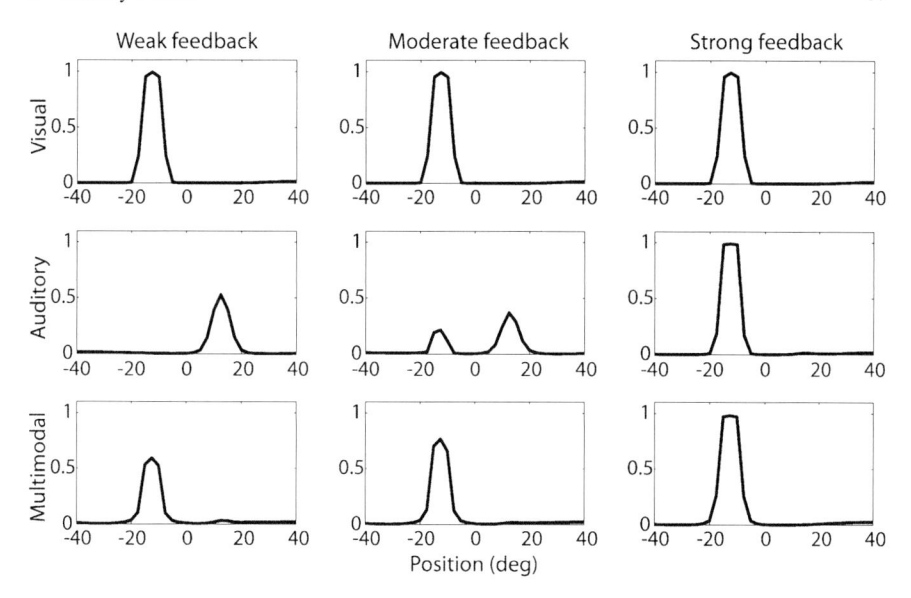

Fig. 2.8 Response of neurons, whose RF is centered at the vertical coordinate 0° (i.e., positioned at the middle of the vertical field), in the visual (*upper line*), auditory (*middle line*) and multi-modal (*bottom line*) areas, in response to two cross-modal stimuli at different horizontal positions. The visual stimulus is stronger and applied at position −15°. The auditory stimulus is weaker and applied at position +15°. The *left column* refers to the case of normal feedback synapses from the multimodal to the unisensory areas (basal synapse strength = 1). Results in the *middle column* were obtained by increasing the feedback synapse strength from the multimodal to the auditory area (synapse strength = 15) while feedback synapse to the visual area was unchanged. It is re-markable the appearance of a second "phantom" activity in the auditory area. The *right column* describes the case in which the feedback synapse strength from the multimodal to the auditory area was further increased (synapse strength = 20).The original auditory activity is suppressed and the visual stimulus captures the auditory one

subject is perceiving both a visual and an auditory stimulus, at different positions) (Fig. 2.8, left column). However, only the stronger visual stimulus can excite the multimodal neurons thus producing overt behavior. In fact, activity of multimodal neurons induced by the auditory stimulus is almost entirely suppressed in the SC as a consequence of lateral competition. Let us now consider that, to solve a conflict be-tween the auditory and visual stimuli, some mechanisms (maybe attentive) increase the feedback from the multimodal area to the auditory unimodal area. Increasing this feedback, one can first observe the occurrence of two activities at different locations in the auditory area (Fig. 2.8, middle column): the subject perceives the original activity and a second "phantom" activity at the same position as the visual input. If feedback is increased further, the original auditory activity is suppressed by competition in the unimodal area, and only the second "phantom" auditory activity keeps on (Fig. 2.8, right column). This signifies that the visual stimulus captures the auditory one at its position, suppressing the original auditory input.

2.2.4 Successes, Limitations and Future Challenges

The present model is based on the idea that multimodal neurons in the superior colliculus receive their inputs from two upstream unimodal areas, one devoted to a topological organization of visual stimuli and another devoted to a topological organization of auditory stimuli. For the sake of simplicity, in this model, somatosensory stimuli are neglected, i.e., we consider only the problem of audio–visual integration. Moreover, the exact location of these areas is not established in our model, i.e., we did not look for a definite anatomical counterpart.

Several mechanisms have been included in this simple basal circuit, each with a specific significance and a possible role in affecting final responses:

1. Nonlinearities in the activation function of single neurons (i.e., a lower threshold and upper saturation, expressed with a sigmoidal relationship). These nonlinearities are essential to understand some important properties of multisensory integration, such as the inverse effectiveness, and the possibility of superadditive, additive, or subadditive integration.
2. Lateral (mainly inhibitory) synapses among multisensory neurons. These synapses are necessary to obtain a significant cross-modal suppression between spatially separated auditory and visual stimuli, as documented in recent experiments.
3. Lateral synapses (excitatory and inhibitory) among neurons in the same unimodal area. They have been modeled with a classical "Mexican hat" disposition, i.e., a close facilitatory area surrounded by an inhibitory annulus. These synapses play a fundamental role in producing the receptive field of multimodal neurons and in producing an "activation bubble" (i.e., a group of contiguous neurons simultaneously active) in response to punctual stimuli. Moreover, they contribute to the within-modal suppression documented in many experiments (Kadunce et al. 1997, 2001) even in the absence of cross-modal suppression (Kadunce et al. 2001).
4. Feedforward connections from unimodal to multimodal neurons. The strength of these synapses affects the sensitivity of multimodal neurons and the maximum response they can attain to a single stimulus of a given modality.
5. Excitatory backward connections from multimodal neurons to unimodal neurons at the same spatial position. Inclusion of these connections considers the possibility that the response by a multimodal neuron reinforces the response at an earlier unimodal area and allows a reciprocal influence between unisensory areas.

By incorporating the previous mechanisms, the model can make several predictions, which can be compared with experimental data. In the following, the main simulation results are critically commented:

1. Inverse effectiveness – As it is evident in Fig. 2.3, the capacity of multisensory neurons to integrate cross-modal stimuli depends on the intensity of unisensory inputs. As in Perrault et al. (2005), enhancement is affected by the intensity of the unisensory inputs, and it exhibits a significant decrease if stimulus intensity is progressively raised.

2. Cross-modal vs. within-modal integration – According to the literature (Stein and Meredith 1993), in our model, a combination of two cross-modal stimuli within the RF results in significant enhancement of the SC response, but the same effect is not visible when the two stimuli are presented as a within-modal pair. A second within-modal stimulus applied within the RF causes just a mild enhancement (Fig. 2.4).

3. Spatial relationship between two (within-modal or cross-modal) stimuli – In agreement with experimental data (Kadunce et al. 1997, 2001), our model shows that, as the spatial distance between two stimuli increases, multisensory integration in SC layer shifts from enhancement to suppression. In the present model, with the basal parameter values used in Figs. 2.4 and 2.5, the suppressive effect is evident both using within-modal and cross-modal stimuli.

4. Differences between within-modal and cross-modal suppression – By increasing the strength of lateral inhibition in the unimodal areas and decreasing the strength of lateral inhibition in cross-modal areas, the model can explain the occurrence of within-modal suppression even in the absence of cross-modal suppression, a result that has been reported in the literature. Furthermore, if different lateral inhibitory synapses are used for the two unisensory areas (for instance, weak inhibition in the visual area and strong inhibition in the auditory), model can explain the occurrence of within-modal suppression in one modality only.

5. By introducing feedback synapses, the model can explain phenomena like ventriloquism, or the improvement in perception of one unisensory stimulus occurring in the presence of a cue from the other modality. However, to this end, we need to suppose that feedback synapses are stronger than normal. It is worth noting that ventriloquism was simulated assuming a huge increase in feedback synapses directed toward the auditory unisensory area only. In this way, the visual stimulus created a phantom activity in its position, which suppresses the other activity. Conversely, improvement in perception can occur even with moderate feedback synapses, and appears as a more normal behavior (i.e., one which does not require excessive parameter changes from basal).

The latter aspect opens the problem of synaptic plasticity in the model, an issue that may have the greatest importance both in clinical-therapeutic problems (for instance, for the rehabilitation of patients with neurological deficits) and for achieving a deeper understanding of adaptation to complex situations. One possibility, which should be tested in future works, is that some synapses are subjected to rapid change, in a short-time basis, to adapt model behavior to the particular conditions. Another possibility is that some model parameters are under the influence of top-down attentional mechanisms, which may work by modulating the threshold of some neurons (thus making them more or less prompt to respond to input stimuli) or may affect synaptic transmission via a gating mechanism. All these aspects may represent future applications of the model.

A further future possible application of the model concerns the study of multisensory maturation in early life. Indeed, multisensory enhancement is not present in SC at birth and develops gradually with sensory experience (Stein 2005). To study this aspect, however, a more sophisticate model is necessary, which comprehends more inputs to the SC. In fact, several recent neurophysiological works show that the SC neurons receive at least four distinct unimodal paths (two visual and two auditory), which have a different functional role in multisensory integration: descending paths from cortico-collicular regions (especially the anterior ectosylvian sulcus and the rostral lateral suprasylvian area) are responsible for multisensory integration; ascending paths coming from a variety of other subcortical sources do not produce multisensory integration (Jiang et al. 2001; Jiang and Stein 2003; Wallace et al. 1993). The present model considers only two unimodal descending inputs, i.e., those responsible for multisensory integration. It is probable that, at birth, only the ascending subcortical inputs are functionally active, whereas the descending paths maturate subsequently with sensory experience. Of course, formulation of a more complex model (with four inputs related through nonlinear multiplicative relationships and with a learning rule for the descending paths) might be the subject of future refinements and extensions.

Finally, we wish to stress that the present model circuitry, inspired by neurophysiological considerations, may not only contribute to our understanding of the neural system, but also drive the project of artificial systems for sensory fusion. In fact, most of the mechanisms adopted here can have a general validity: nonlinearity to regulate the degree of integration among different stimuli and favor enhancement in the presence of poor stimulation; the presence of lateral competition to suppress less-relevant information; the presence of feedback plastic mechanisms to realize more sophisticate top-down control strategies. Moreover, all these mechanisms are working in parallel and in an integrated fashion to achieve a highly distributed and efficient system, as that emerged after millions of years of animal and human evolution.

2.3 Visual–Tactile Integration

2.3.1 Visual–Tactile Representation of Peripersonal Space: Neurophysiological and Behavioral Evidence (Overview and Model Justification)

The crucial factor that distinguishes the space immediately surrounding our body [peripersonal space (Rizzolatti et al. 1997)] from the more distant space (extrapersonal space) is our potential ability to interact with objects located within it. Objects within peripersonal space may be reached, grasped, and manipulated; potentially harmful objects (requiring avoidance and defensive movements) are those closest to, and moving rapidly toward, our body. Hence, it makes functional sense that the brain represents peripersonal space differently from the more distant space. In particular, objects near to or in contact with the body may be perceived via a sensory system (the touch), which is not involved at all by objects located in the extrapersonal space.

In the last two decades, neurophysiological research on monkeys has yielded a large body of evidence supporting the notion that peripersonal space is represented in a multisensory fashion, by integrating visual and tactile stimuli. Multimodal neurons have been found in several brain structures (putamen, parietal, premotor areas) (Bremmer et al. 2001; Duhamel et al. 1998; Fogassi et al. 1996; Graziano et al. 1997; Rizzolatti et al. 1998), which respond both to touches delivered on a specific body part (e.g., the hand or the face) and to visual stimuli presented close to the same body part, where the tactile receptive field (RF) is located. The visual RF of these neurons has a limited extension in depth, being typically restricted to the space immediately surrounding the body part: the neuronal response to visual stimuli decreases as the distance between stimuli and the cutaneous RF increases (Duhamel et al. 1998; Fogassi et al. 1996; Graziano et al. 1994).

The main evidence for the existence of a multisensory system in humans, functionally similar to that in monkeys, comes from neuropsychological studies conducted on right brain-damaged (RBD) patients with extinction (di Pellegrino et al. 1997; Làdavas et al. 1998, 2000; Làdavas 2002). Extinction patients fail to detect a contralesional stimulus only under conditions of bilateral (ipsilateral and contralateral) simultaneous stimulation. Extinction phenomenon has been attributed to an unbalance competition between concurrent representations: the unilateral brain damage gives rise to a weaker representation of the contralateral side, which is disadvantaged in terms of competitive weights (Mattingley et al. 1997). Extinction can occur when the concurrent stimuli are in the same modality (unimodal extinction) or in different modalities (cross-modal extinction). In cross-modal extinction, presentation of a visual stimulus in the ipsilesional (right) visual field can extinguish a simultaneous tactile stimulus on the contralesional (left) hand. Crucially, tactile extinction on the contralesional hand is modulated by the spatial arrangement of the simultaneous visual stimulus with respect to the body (Làdavas and Farnè 2004; Làdavas et al. 1998, 2000). Tactile extinction is more severe when the visual stimulus is presented near ($\sim$5 cm) the ipsilesional hand, than when it is presented away ($\sim$35 cm) from the ipsilesional hand; moreover, a visual stimulus presented near the contralesional hand improves the detection of the tactile stimulus applied to the same hand. Such findings can be explained by referring to the activity of bimodal neurons, similar to those observed in monkeys, which have tactile receptive fields on the hand and corresponding visual receptive field in the space immediately adjacent to the tactile field; the activation of such bimodal neurons would activate the perceptual representation of the corresponding hand (di Pellegrino et al. 1997).

Recent functional neuroimaging studies, indeed, have identified multimodal structures in the human brain responding selectively to tactile and visual information on a single body part (the hand or head), suggesting that such areas may represent the human equivalent of macaque's areas for peripersonal space representation (Bremmer et al. 2001; Makin et al. 2007; Sereno and Huang 2006).

Peripersonal space representation both in humans and monkeys has basically a motor function (Bremmer et al. 2001; Cooke and Graziano 2004; Rizzolatti et al. 1998); spatial locations of multisensory stimuli are encoded in relationship to body parts to generate appropriate motor responses (goal-directed, defensive, or avoidance movements) (Graziano and Cooke 2006; Làdavas and Farnè 2004; Legrand et al. 2007; Rizzolatti et al. 1998). Normally, such action space is delimited by the physical length of our effectors (mainly the arms). However, we can use many different tools to extend our physical body structure, and consequently our action space. While using a tool, the tactile information felt at the hand can be related to visual information from distant objects. There are several evidences that the use of a tool, linking tactile events with far visual events, induces a plastic modification of peripersonal space, with a recoding of the far space as near space.

Resizing of peri-hand space representation following tool use was first reported in monkeys by Iriki et al. (1996). They observed that, after the animal had repeatedly used a tool to retrieve distant food, the visual RF of intraparietal bimodal neurons was elongated to include the entire length of the tool, whereas originally it was limited to the space around the hand. The emergence of novel projections from visual-related areas to bimodal intraparietal regions (Hihara et al. 2006; Ishibashi et al. 2002) has been suggested as a possible neural mechanism underlying such phenomenon. Evidence for a similar tool incorporation into peri-hand space representation has been reported in humans at behavioral level. In extinction patients, a visual stimulus located at the end of a right hand-held tool induced more severe left tactile extinction immediately after the tool use than before (Farnè and Làdavas 2000; Maravita et al. 2001), suggesting an extension of the integrative peri-hand space.

In summary, a massive amount of data contributed to describe functional and dynamical properties of peripersonal space representation. Recently, we have exploited computational modeling via artificial neural networks to shed light on the neural mechanisms and circuits underlying such properties (Magosso et al. 2010a, b). In particular, with our models, we aspire to address the following questions: What is the organization of the neural network subserving peripersonal space representation? How does it relate with the multimodal neurons identified by electrophysiological studies? How do multimodal and modality-specific areas communicate reciprocally? Which are the alterations in the neural circuitry that may explain extinction in brain-damaged patients? Which are the neural correlates of peripersonal space plasticity following tool use? How can behavioral results be related with responses of individual neurons?

Although some computational models have been proposed in the past to investigate some properties of multimodal neurons in the parietal cortex (Avillac et al. 2005; Denève et al. 2001; Pouget and Sejnowski 1997; Salinas and Abbott 1995), none of them has tackled the above questions explicitly.

2.3.2 A Neural Network Model for Peri-Hand Space Representation: Simulation of a Healthy Subject and a RBD Patient (Model Components and Results 1)

The model consists of two subnetworks, reciprocally interconnected; each subnetwork refers to the contralateral hand of a hypothetical subject (Fig. 2.9).

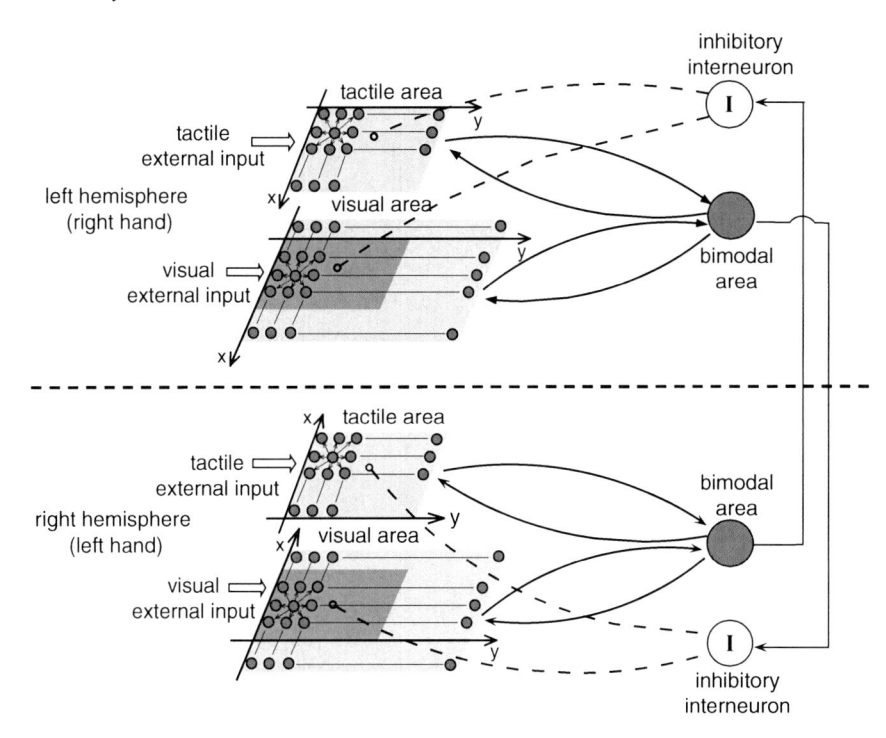

Fig. 2.9 Layout of the neural network. The model includes two subnetworks, one per hemisphere, each corresponding to the contralateral hand and surrounding space. The *gray circles* represent excitatory neurons; the *continuous arrows* linking neurons or areas of neurons denote excitatory connections, the *dashed lines* denote inhibitory connections. I indicate inhibitory interneurons

The single subnetwork embodies three areas of neurons. The two upstream areas are bidimensional matrices of unimodal neurons: neurons in one area respond to tactile stimuli on the contralateral hand (*tactile area*); neurons in the other area respond to visual stimulation on the same hand and around it (*visual area*). Each neuron has its own receptive field (RF), described by means of a Gaussian function. In both areas, the RFs are in hand-centered coordinates and are arranged at a distance of 0.5 cm, so that proximal neurons within each area respond to stimuli coming from proximal positions of the hand and space. RFs of proximal neurons are partially superimposed. The tactile area maps a surface of 10×20 cm, roughly representing the surface of the hand. The visual area covers a space of 15×100 cm, representing the visual space *on* the hand and *around* it (extending by 2.5 cm on each side and 80 cm ahead). Moreover, the units within each unimodal area interact via *lateral synapses* with a "Mexican hat" arrangement (i.e., with short-range excitation and long-range inhibition).

The unimodal neurons send *feedforward synapses* to a third downstream multimodal area devoted to visual–tactile representation of peri-hand space. The multimodal area might represent multisensory regions in the premotor or parietal cortex, which receive feedforward projections from sensory-specific areas (Duhamel et al. 1998; Graziano et al. 1997; Rizzolatti et al. 1981). For the sake of simplicity, we considered a single visual–tactile neuron, covering the entire peri-hand space. Data in the literature (Graziano et al. 1997; Rizzolatti et al. 1981) indeed, stress the existence of multimodal neurons with a RF as large as the whole hand. The tactile feedforward synapses have a uniform distribution (their strength is independent of the position of single tactile neuron's RF).

The strength of the visual feedforward synapses is constant on the hand and decreases exponentially as the distance between the neuron's RF and the hand increases. Figure 2.10a shows the pattern arrangement of the feedforward synapses

a

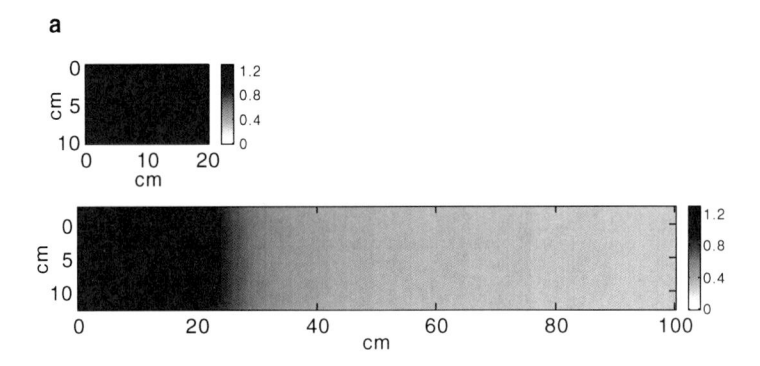

b

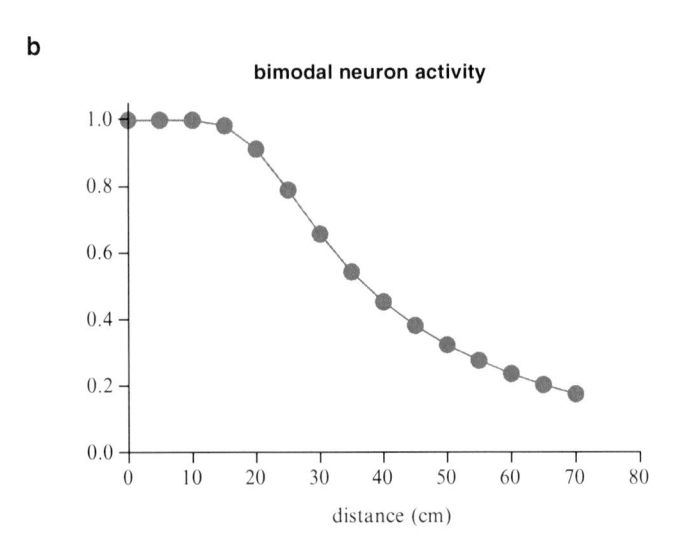

Fig. 2.10 (**a**) Pattern of the feedforward synapses from the tactile (*upper plot*) and visual (*lower plot*) area to the downstream bimodal area in the left hemisphere (for basal parameter values, i.e., healthy subject). The *x* (*vertical*) and *y* (*horizontal*) axes represent the coordinates of the RF center of the unimodal neurons; the *gray scale* indicates the strength of the synapse connection. (**b**) Response of the bimodal neuron in one hemisphere to a visual stimulus at different distances from the contralateral hand. Neuron response is normalized with respect to its maximum saturation activity (i.e., value one corresponds to the maximal neuron activation)

from the tactile and visual area. According to such synapses arrangement, the bimodal neuron has a tactile RF covering the entire hand, and a visual RF matching the tactile RF and extending some centimetres around it. Figure 2.10b displays the response of the bimodal neuron in one hemisphere to a visual stimulus located at different distances from the contralateral hand. Activity of the neuron is evaluated after the initial transient has exhausted and the network has reached a new steady state (the stimulus is maintained throughout the entire simulation). The visual-related activity of the bimodal neuron decreases as the distance between the visual stimulus and the hand increases, in agreement with neurophysiological data (Graziano et al. 1997; Rizzolatti et al. 1981). It is worth noticing that in the model, activation of the bimodal neuron in one hemisphere mimics the perceptual representation of the contralateral hand, triggered by a somatosensory stimulation or by a near visual stimulation.

The visual–tactile neuron within one hemisphere sends *feedback excitatory synapses* to the upstream unimodal areas in the same hemisphere, in agreement with recent neuroimaging data suggesting that higher level multimodal areas may send back-projections to modality-specific areas (Kennett et al. 2001; Macaluso et al. 2000; Taylor-Clarke et al. 2002). The feedback synapses have the same arrangement as the feedforward synapses, with different parameter values.

The two hemispheres interact via a competitive mechanism realized by means of inhibitory interneurons. The inhibitory interneuron in one hemisphere receives information from the bimodal neuron in the other hemisphere via a synapse, characterized by a pure delay to account for the interhemispheric transit time. Then, the interneuron sends inhibitory synapses locally to the unimodal visual and tactile neurons. The inhibitory synapses from the interneuron to the tactile and visual areas have the same spatial arrangement as the feedforward and feedback synapses, with a different set of parameters.

The input–output relationship of each neuron (both unimodal, bimodal and inhibitory) includes a first-order dynamics and a static sigmoidal relationship with a lower threshold and an upper saturation. Each neuron is normally in a silent state and can be activated if stimulated by a sufficiently high excitatory input.

Basal values for all model parameters were assigned on the basis of neurophysiological and behavioral literature to reproduce a healthy subject. In particular, in basal conditions, the two hemispheres are characterized by the same parameter values.

2.3.2.1 Simulation of the Healthy Subject

Figure 2.11 shows network activity in response to bilateral cross-modal stimulations in the healthy subject (basal parameter values). Each panel (a and b) shows the activity in the unimodal areas (represented by a gray plot) and in the bimodal area (represented by a 3D bar), in the left and right hemispheres. Network activity is displayed in the steady-state conditions reached by the network following the stimuli application. In panel a, a tactile stimulus is applied on the left hand [stimulus position: x(vertical) $= 7.5$ cm; y(horizontal) $= 4$ cm] and a simultaneous visual

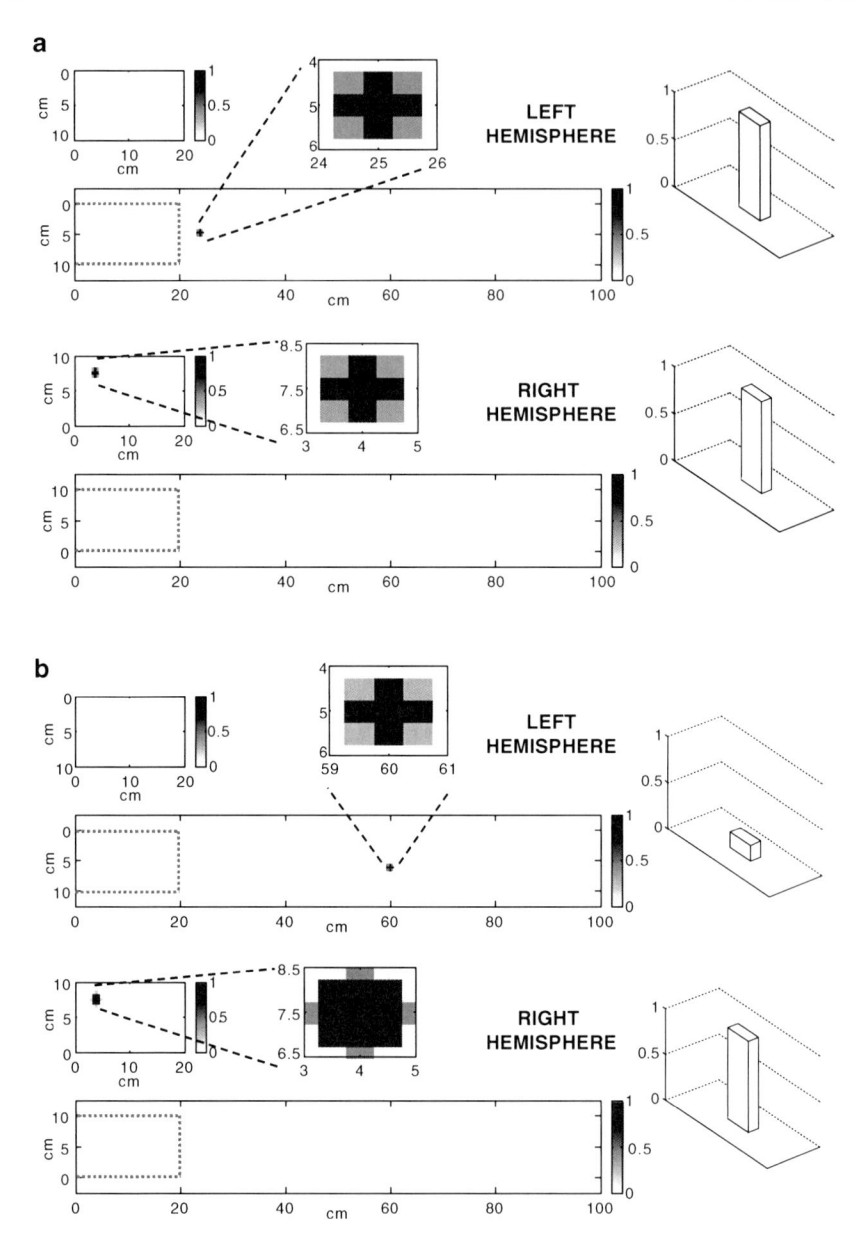

Fig. 2.11 Network activity in response to two different bilateral cross-modal stimulations in the healthy subject. Each panel shows the activity in the unimodal areas (represented by a *gray plot*, where the *x* and *y* axes represent the coordinates of the neuron's RF center, and the *gray scale* denotes the activation level of the neurons), and in the bimodal area (represented by a 3D bar), in the left and right hemispheres. The inserted plots are magnified images of the activated group of unimodal neurons. The *dotted gray line* in the visual areas borders the visual space *on* the hand. (**a**) Network response to a tactile stimulus applied on the left hand, and to a visual stimulus applied *near* the right hand (5 cm distance). (**b**) Network response to a tactile stimulus applied on the left hand, and to a visual stimulus applied *far* from the right hand

stimulus is applied *near* the right hand (stimulus position: $x = 5\,cm$; $y = 25\,cm$, at a distance of 5 cm from the hand). Each stimulus produces a significant activation in the respective unimodal area, that is, an activation bubble emerges in each area. The formation of an activation bubble is due to the partial superimposition of the RFs of adjacent neurons and to the lateral excitation that produces reciprocal reinforcement of neighboring neurons activity. Unimodal activity, in turn triggers – via feedforward synapses – the corresponding bimodal neuron; it is worth noticing that activation of the bimodal neuron further reinforces the activity in the unimodal area owing to the feedback excitatory projection on and near the hand. Moreover, the simultaneous activation of the two bimodal neurons leads to a competition between the two hemispheres. In this case (healthy subject), the competition is unbiased and the final outcome is the coexistence of activations in both hemispheres; in particular, in each hemisphere the bimodal neuron is in the *on* state. This model result agrees with in vivo data showing that in healthy subjects, the representations of both hands coexist in case of a simultaneous right- and left-hand stimulation (Hillis et al. 2006).

In Fig. 2.11b, the left tactile stimulus is applied in the same position ($x = 7.5\,cm$; $y = 4\,cm$), whereas the right visual stimulus is applied *far* from the hand (stimulus position: $x = 5\,cm$; $y = 60\,cm$, i.e., at a distance of 40 cm from the hand). The two stimuli have the same intensity. The left tactile stimulus triggers the right hemisphere bimodal neuron to its maximum saturation. The far visual stimulus still activates a group of neuron in the unimodal visual area; however, because of its distance from the hand, it produces only a slight activity of the corresponding bimodal neuron. At that position, indeed, feedforward synapses are weaker (see Fig. 2.10a). Accordingly, the far visual stimulus does not trigger the right-hand representation, and only the left hand representation is activated, boosted by the somatosensory stimulus.

2.3.2.2 Simulation of the RBD Patient with Left Tactile Extinction

In order to simulate an RBD patient suffering from left tactile extinction, we decreased the strength of all excitatory synapses (both lateral and feedforward) originating from the tactile unimodal neurons in the right hemisphere. The hypothesized reduction in synaptic strength has to be interpreted not as a real synaptic depression, but, rather, as the effect of a reduction in the number of effective excitatory units that contribute to activity in that region. Of course, the smaller the number of effective excitatory cells, the smaller the overall excitatory input emerging from that area. Under these conditions, we replicated the same bilateral stimulations as in Fig. 2.11; results are reported in Fig. 2.12a, b.

In Fig. 2.12a, the left tactile stimulus (stimulus position: $x = 7.5\,cm$; $y = 4\,cm$) is applied simultaneously with a right visual stimulus *near* the right hand (stimulus position: $x = 5\,cm$; $y = 25\,cm$, i.e., at a distance of 5 cm from the hand). The two stimuli have the same intensity. The *near* right-hand visual stimulus activates the bimodal neuron in the left hemisphere, competing with the simultaneous left tactile stimulus. In this case, the competition is unbalanced, since right hemisphere

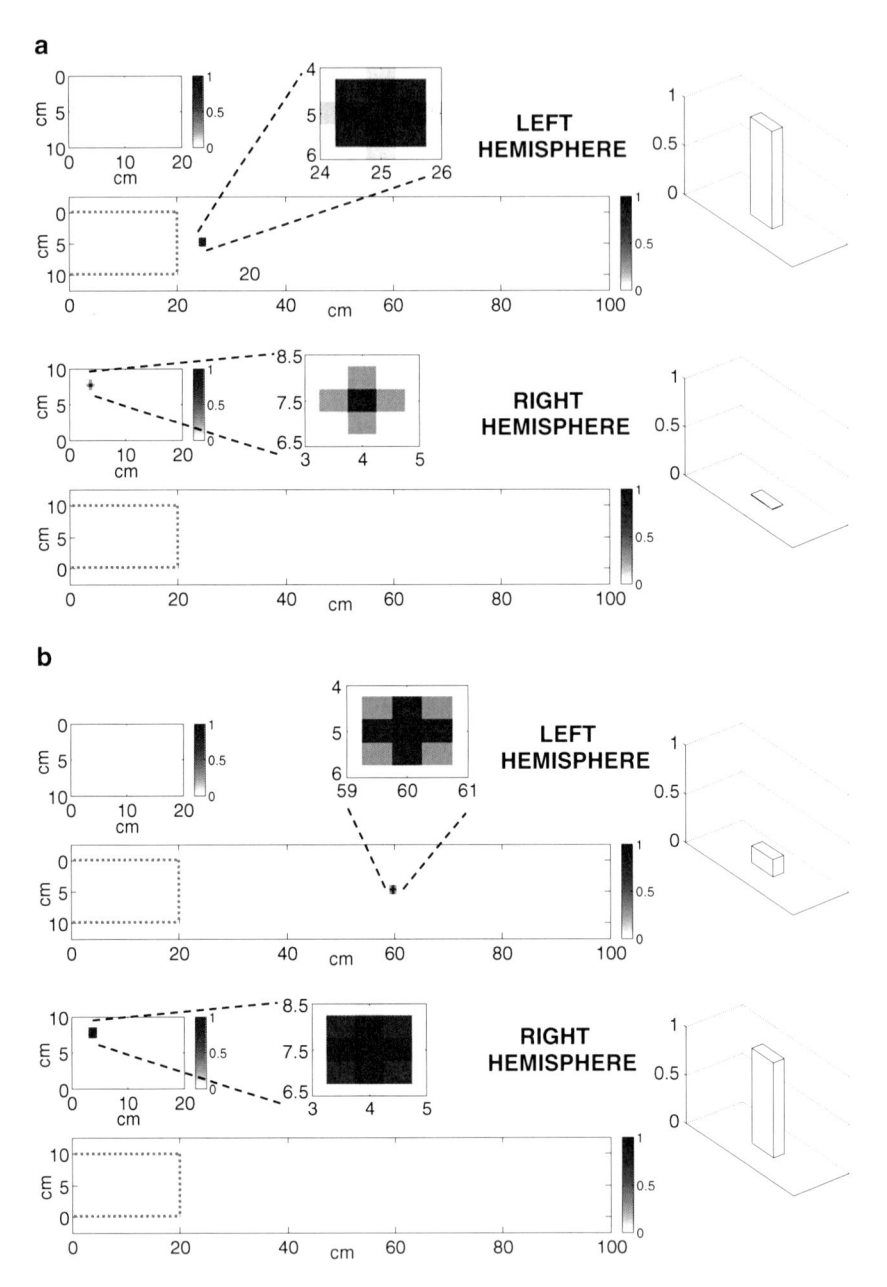

Fig. 2.12 Network activity in response to two different bilateral cross-modal stimulations in the RBD patient. The figure has the same meaning as Fig. 2.11. (**a**) Network response to a tactile stimulus applied on the left hand, and to a visual stimulus applied *near* the right hand (5 cm distance). The activity of the bimodal neuron in the right hemisphere is extinguished by the concurrent activation in the opposite hemisphere. (**b**) Network response to a tactile stimulus applied on the left hand, and to a visual stimulus applied *far* from the right hand (40 cm distance). The far visual stimulus does not extinguish the activation of the right hemisphere bimodal neuron

activation is weakened by the lesion. In particular, the reduction of lateral excitation in the right tactile area diminishes the intensity and extension of the activation bubble in response to a tactile stimulus; the reduction of feedforward synapses from the right tactile area impairs the ability for tactile unimodal neurons to trigger the bimodal neuron. Thus, the right visual stimulus has a higher competitive strength than the left tactile stimulus. The final outcome is a strong reduction of the activity in the right hemisphere tactile area and a consequent deactivation of the bimodal neuron; i.e., the left hand representation is extinguished and only the right-hand representation survives. In Fig. 2.12b, the tactile stimulus at the same position is associated with a visual stimulus *far* from the right hand (stimulus position: $x = 5$ cm; $y = 60$ cm, i.e., at a distance of 40 cm from the hand). The two stimuli have the same intensity. The right visual stimulus, being far from the hand, exerts only a weak competition with the left tactile stimulus; then, the latter is able to activate the corresponding bimodal neuron, triggering the left hand representation. Simulation results displayed in Fig. 2.12 show that the model can reproduce the near–far modulation of left tactile extinction as reported in vivo (Làdavas and Farnè 2004; Làdavas et al. 1998, 2000). It is worth noticing that extinction of the left touch can be obtained also by applying a tactile stimulus on the right hand, in agreement with experimental data (unpublished simulations) (di Pellegrino et al. 1997; Làdavas et al. 1998, 2000).

Furthermore, behavioral studies in extinction patients indicate that under conditions of bilateral stimulation, the detection of the left tactile stimulus is ameliorated by a simultaneous left visual stimulus (Làdavas et al. 1998, 2000) (cross-modal facilitation). This situation is simulated in Fig. 2.13, where a visual stimulus is applied near the right hand (position: $x = 5$ cm; $y = 25$ cm, i.e., 5 cm distance from the hand) and a double stimulation (tactile and visual) is delivered to the left hand (stimuli position: $x = 7.5$ cm, $y = 4$ cm). All stimuli have the same intensity. In this case, the left visual stimulus sustains the activation of the bimodal neuron (despite the competition with the concurrent right-hand visual stimulus); the bimodal neuron, in turn, reinforces the activity in the tactile area via the feedback projections. Consequently, the activation in the right hemisphere tactile area ameliorates significantly (compare Fig. 2.13 with Fig. 2.12a); this might correspond to an improved detection of the left tactile stimulus.

2.3.3 Modeling Peri-Hand Space Resizing: Simulation of Tool-Use Training (Model Components and Results 2)

The model can be exploited to investigate the neural mechanisms underlying plasticity of the peri-hand space representation consequent to tool use. In particular, we hypothesized a neurobiological mechanism for synapses plasticity, and assessed whether this mechanism can account for dynamic changes in peri-hand space representation after tool use.

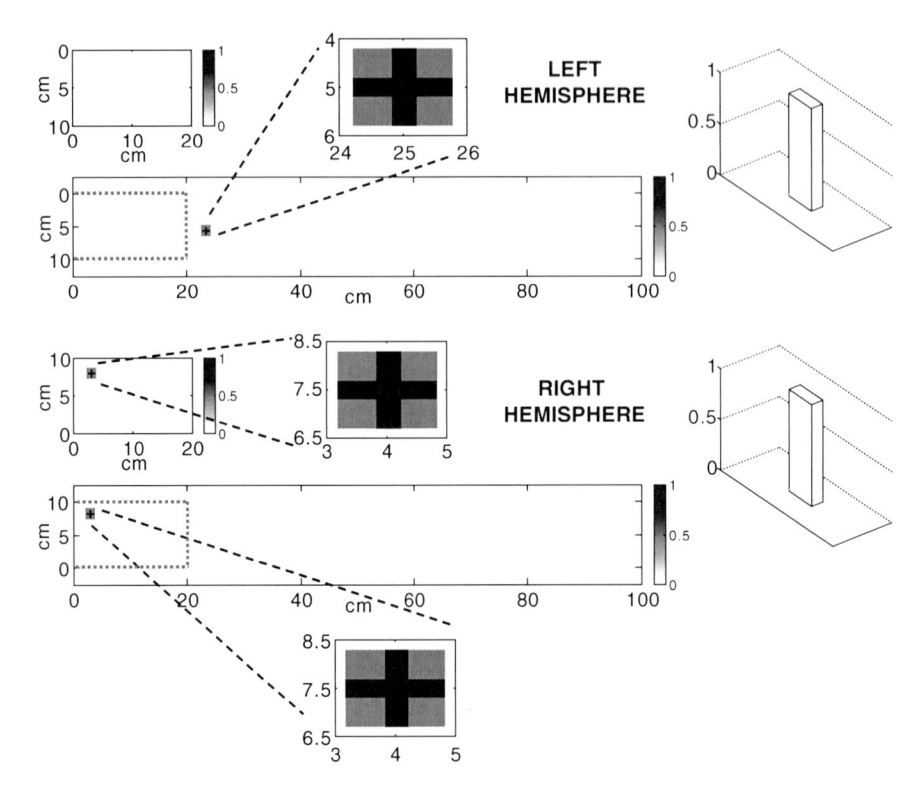

Fig. 2.13 Cross-modal facilitation in the RBD patient. Response to a bilateral stimulation: the left tactile stimulus is paired with a simultaneous visual stimulus on the left hand, and a visual stimulus is applied *near* the right hand. Bimodal neuron in the right hemisphere is not extinguished (thanks to the visual stimulus), and reinforces the activity in the right tactile area via back projections (compare with Fig. 2.12a)

To this aim, a training experiment has been simulated in which the hypothetical subject utilizes a tool with the right hand to interact with the far space. We assumed that the tip of the tool lies in position 5, 60 cm in the x, y plane (i.e., at a distance of 40 cm from the hand along the parasagittal plane). The use of the tool by the right hand has been mimicked by applying both a tactile and a visual input to the left hemisphere (Fig. 2.14a).

The tactile input represents the portion of the hand stimulated while holding the tool. The visual input represents the region of the visual space functionally relevant for the tool use, selected, for instance, by top-down attentive mechanisms. Here, we adopted an elongated visual input, centered on the tip of the tool, and spread both before and behind it: such input might simulate the case of using a rake to retrieve objects from the far space toward the body, a task adopted in several in vivo studies (Bonifazi et al. 2007; Farnè et al. 2005, 2007; Iriki et al. 1996). Indeed, retrieving objects may require attention to cover a wide portion of the visual space, including the area where the distant objects are located and the region of space between the

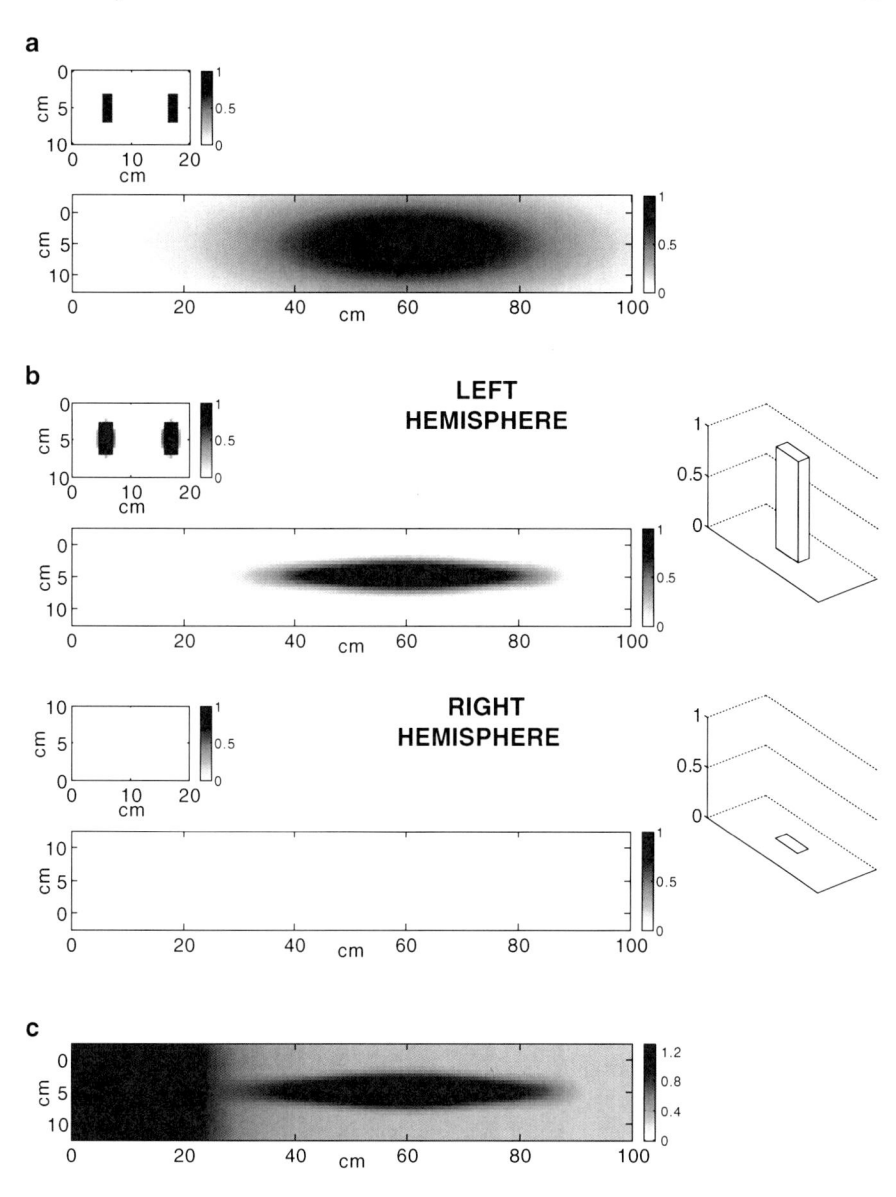

Fig. 2.14 Tool-use training. (**a**) Tactile and visual inputs used to simulate tool-use training with the model. These inputs were applied to the left hemisphere since we simulated the use of the tool with the right hand. (**b**) Network response to the previous inputs. Only the left hemisphere is active since no stimulation was applied to the right hemisphere. (**c**) Feedforward synapses from the visual area to the bimodal neuron in the left hemisphere after training, computed via the application of a Hebbian rule during stimulation of the network by the tool-related inputs (compare with Fig. 2.10a)

tip of the tool and the hand, along which the objects are dragged toward the subject body. The application of the previous inputs to the network produces the activation of the corresponding regions in the unimodal areas within the left hemisphere, as well as the activation of the bimodal neuron (Fig. 2.14b).

We assumed that during the application of these inputs, the feedforward synapses linking unimodal neurons to the bimodal neuron in the left hemisphere modify via a classical Hebbian learning rule with an upper saturation: i.e., synapses are reinforced in presence of the simultaneous activation of the presynaptic and postsynaptic neurons, until a maximal value is reached. Moreover, we assumed that synapses *on* the hand are already at their maximum value even before tool use. This assumption is reasonable since these synapses are frequently and repeatedly involved in the daily perception of the peri-hand space. Therefore, synapses from the tactile neurons and from the visual neurons coding the space on the hand do not modify; conversely, visual synapses reinforce significantly along the extended visual input highlighted during the training. Figure 2.14c shows the pattern of the visual feedforward synapses after the Hebbian learning.

All equations and parameters concerning model training and plasticity can be found in our previous paper (Magosso et al. 2010a).

To evaluate the effect of the tool-use training on the integrative visual–tactile peri-hand area, bilateral visual–tactile stimulations have been simulated in the RBD patient, with the right visual stimulus at different distances from the right hand, before and after the training. For each position, the tactile and visual stimuli have the same intensity. Results of these simulations are reported in Fig. 2.15, which merely shows the activation of the bimodal neurons in the two hemispheres as a function of the right visual stimulus location. Before the training, the bimodal neuron in the left hemisphere responds only to near visual stimuli. Consequently, left tactile extinction occurs only in case of a visual stimulus in proximity to the right hand (5 cm distance); more distant visual stimuli on the right side do not compete, or compete only slightly, with the left hand representation, which, therefore, survives (Fig. 2.15a; near–far modulation of left tactile extinction). Conversely, after the training, cross-modal extinction is no longer modulated by the distance of the right visual stimulus; the right visual stimulus located in any of the four positions activates the bimodal neuron in the left hemisphere, and left hand representation is extinguished in all four cases (Fig. 2.15b). Hence, the model predicts an extension of the visual–tactile integrative area to include the elongated visual space highlighted during tool training. These model predictions are in agreement with in vivo results. Iriki and colleagues, in their work on monkeys (Iriki et al. 1996), documented an elongation of visual RFs of bimodal neurons along the tool. Studies on extinction patients report that, after 5 min of tool use, left tactile extinction produced by visual stimuli located at the distal edge of the tool, and midway between the hand and tip, was as severe as that obtained by a visual stimulation near the ipsilesional hand (Bonifazi et al. 2007; Farnè et al. 2007). Moreover, in another study (Farnè et al. 2005), extinction was also caused by a visual stimulus placed several centimetres beyond the tip of the tool, suggesting that the elongated peri-hand area is not coincident with the tool length, but includes also space located beyond the tip (as

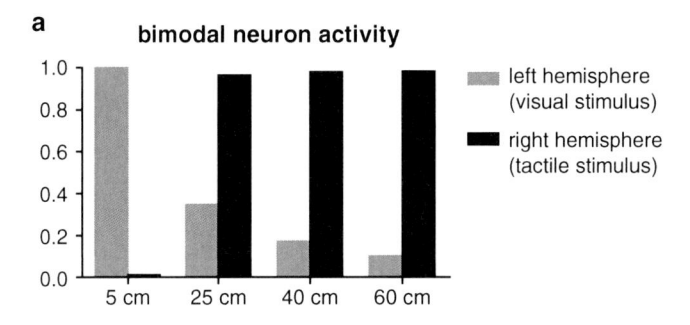

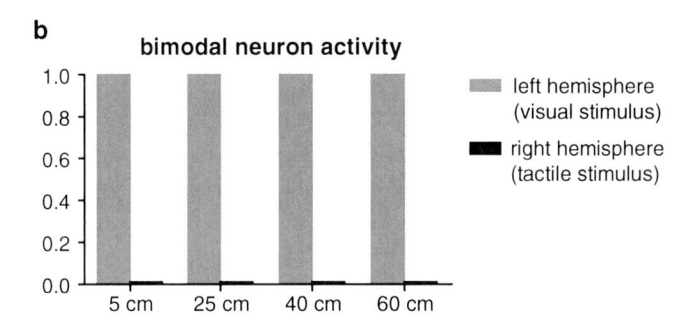

Fig. 2.15 Effects of tool-use training on visual–tactile interaction. The histogram shows the activity of the bimodal neurons in response to bilateral cross-modal stimulation (left touch and right visual stimulus) in the RBD patient as a function of the distance of the visual stimulus from the right hand. (**a**) Before training conditions. Left tactile extinction occurs only in case of a near visual stimulus. (**b**) After training conditions. Left tactile extinction occurs also for visual stimuli far from the right hand, at several positions along the tool axis

in our simulations). It is worth noticing that in all the mentioned in vivo studies, participants were required to use a rake to reach distant items, located out of the hand-reaching space, and to bring them back to the near space.

2.3.4 Successes, Limitations and Future Challenges

We proposed a simple network architecture for visual–tactile integration and peri-hand space representation, consisting of two unimodal areas and one bimodal area in each hemisphere connected via feedback and feedforward synapses, and including a competitive mechanism between the two hemispheres via inhibitory interneurons. The model does not aspire to have a definite neurophysiological and neuroanatomical counterpart, but rather to identify a plausible structure of the network and the functional links between its different parts, able to account for psychophysical and behavioral results. In particular, one of the main over-simplification in the present model is that the spatial arrangement of visual and tactile receptive fields of the unimodal neurons has been set a priori on and around the hand; that is, we avoid considering explicitly the problem of coordinate transformations between different reference frames (e.g., from eye- to hand-centered coordinates). We assumed that the computational analysis performing coordinate transformation of the visual stimuli from retinotopic to hand-centered reference frames has been carried out by other areas in upstream, not represented, levels of the neural network, using postural information (e.g., eye, head, and hand positions). The problem of coordinate transformations has been widely investigated in other studies by means of neural network models (Avillac et al. 2005; Denève et al. 2001; Pouget et al. 2002).

The model is able to reproduce a variety of results concerning peripersonal space representation and its plastic modifications; in the following, we will highlight how the model may help interpretation of in vivo data, and, on the basis of the generated predictions, can suggest new experiments to validate the involved hypotheses.

An important point is that the model may be of value to gain insights into the neural mechanisms underlying extinction in unilateral brain-damaged patients. Consistent reproduction of in vivo data in these patients has been obtained by assuming a reduction of the excitatory synapses (both lateral and feedforward) emerging from the tactile neurons in the right hemisphere. Such reduction in synaptic strength wishes to represent the effect of a reduction in the number of effective excitatory units, which contribute to activity in the right tactile region. With this modification, a left tactile stimulus is able to activate the corresponding bimodal neuron (thus boosting the representation of the left hand), in absence of a simultaneous competition with the right-hand representation (Fig. 2.12b). This model outcome reflects neuroimaging data in extinction patients showing that right parietal and frontal regions (corresponding to the downstream bimodal area in the model) are activated when the left tactile stimulus is perceived (Sarri et al. 2006). Conversely, when a competition with the right-hand representation occurs (Fig. 2.12a), a weak activity still survives in the right tactile area, but it is insufficient to excite the bimodal neuron. This might corresponds to the lack of activation of the right parietal–frontal cortex, despite activation of sensory cortex, reported by fMRI studies in case of extinguished left touches (Sarri et al. 2006). Hence, the model identifies potential functional alterations in the neural circuitry able to explain extinction and relating cortical phenomena; in particular, the model suggests that several mechanisms (reduction of the overall excitation emerging from the right tactile area and the presence of a competition between two simultaneous spatial representations) are concurrently involved in such pathological sign.

The scenario provided by the model is also able to reproduce the phenomenon of cross-modal visual–tactile facilitation in the pathological subject (see Fig. 2.13), thanks to the back-projection from the bimodal area to the unimodal areas. Several recent studies have evidenced the adaptive advantage of multisensory integration in producing response enhancement when the information from one modality is weak (Calvert et al. 2004; Kennett et al. 2001; Press et al. 2004; Tipper et al. 1998), and have suggested the exploitation of brain multisensory capabilities to recovery sensory or spatial deficits after damage (see Làdavas 2008 for a review). In perspective, the model could provide important contribution in this field, by shedding light on the neural correlates of rehabilitation procedures and suggesting new strategies of rehabilitation. For example, according to the model, systematic visuo-tactile stimulation of the pathological side in extinction patients may promote a Hebbian reinforcement of the feedback and feedforward synapses in the damage hemisphere that could be effective to reequilibrate – in a long-lasting way – the competition between the two hemispheres.

A neurobiologically plausible hypothesis has been generated by the model, concerning the resizing of visual–tactile integrative area following tool use. The model assumes that the modification of peri-hand space arises from a Hebbian growing of visual synapses converging into the multimodal area, which extends the visual RF of the peripersonal bimodal neurons. Crucially, in the model, the change in visual RF strictly depends upon the visual input used during the learning phase (see Fig. 2.14); the latter might be selected – during the use of the tool – by top-down, probably attentive, mechanisms that identify the region of the visual space of interest for the task. Two important predictions descend from the previous model hypotheses that may be validated experimentally:

1. The hypothesis that peripersonal space modification depends on reinforcement of visual synapses converging into the bimodal area has the following implication: once the effect of tool use has been achieved via the training, the recoding of a far visual stimulus as a near one should be independent of the presence of the tool (at least for a certain period immediately after the training). Thus, after tool use, an extension of the visuo-tactile peri-hand area should be observed even in absence of any physical connection between the patient's hand and the far visual stimulus. At best of our knowledge, no in vivo study in the literature has evaluated visual–tactile interaction after tool use in conditions of tool absence. Hence, ad hoc behavioral experiments may be conceived to confirm (or refute) model prediction, thus supporting (or rejecting) the validity of the hypothesized mechanism of plasticity. We obtained a preliminary validation of model hypothesis and prediction in our recent work (Magosso et al. 2010a) via a behavioral experiment on one RBD patient with extinction: the same extension of peri-hand space was measured after tool use both when the patient held the tool with her right hand, and when the tool was removed from the patient's hand.

2. The model predicts that different tool use-mediated tasks (e.g., retrieving objects, pressing buttons with the tip, sorting objects in the far space, etc.), requiring direction of movements and attention toward different regions of the visual space, may produce different resizing of the peri-hand visual–tactile space (e.g., the formation of a novel integrative peri-hand area at the tip of the tool rather than an elongation along the tool axis). These model predictions might be tested in vivo, for example, by exposing the participants to different kinds of tool-mediated tasks.

In conclusions, the present model of visual–tactile interaction exemplarily illustrates the importance of integrating experimental research with theoretical and computational studies. On the one hand, empirical results are fundamental to build the mathematical model, identifying model components and parameters. On the other hand, the model is fundamental to synthesize the data into a coherent theoretical framework, helping interpretation of behavioral results in terms of neuron responses and interconnections; moreover and of great importance, simulations results can generate new predictions and inspire new related experiments, which may further support, in a feedback fashion, the validity of the model.

2.4 Conclusions

In conclusion of this chapter, we wish to underline some basic ideas and fundamental mechanisms that emerge from the previous two models: knowledge of these aspects may drive the implementation of future neurocomputational models for multisensory integration in the brain, and perhaps the design of innovative devices for sensory fusion.

The reader can certainly recognize that the two models presented above, although devoted to different problems and simulating different brain regions (the superior colliculus in the first model, a parietal cortex association area in the second), share some common mechanisms. They are:

1. *A topological organization of the input space.* A topological organization (i.e., similar stimuli are coded by proximal neurons) can be encountered everywhere in the cortex and allows the implementation of a very efficient competitive mechanism directly in the single areas. This has several advantages: (a) *robustness*: a stimulus is coded by a group of mutually excited units, not by a single cell; (b) *similarity*: similar stimuli are coded by proximal units and, due to the superimposition of their receptive fields, a same unit responds to various similar stimuli; (c) *suppression*: an incongruent stimulus can be depressed or even eliminated by a proximal stronger stimulus.
2. *A nonlinear (sigmoid-like) input–output response.* This kind of response offers several major advantages, both in case of within-modal and cross-modal stimulation. (a) *thresholding*: all stimuli below a given threshold are neglected, in accordance with a parsimony requirement; (b) *saturation*: strongest stimuli do not produce an excessive (and often deleterious) response; (c) *inverse effectiveness*: the benefit of multisensory integration is higher in case of lower stimuli, i.e., in conditions where the individual stimuli carry uncertain information and may produce an inaccurate response, hence the need for sensory integration is higher.
3. *A feedback from multisensory to unisensory areas.* Our models assume that the multisensory representation sends a feedback to the primary sensory areas

(see also Driver and Spence 2000). In view of this feedback, the unisensory areas can be affected by the other sensory modality, i.e., the unisensory representation can change as a consequence of a change in the other unisensory representation, with the occurrence of interesting cross-talk effects. This is an essential aspect to implement a reciprocal influence between the two unisensory representations (for instance, to mimic ventriloquism). It is worth noting that our models suggest that the strength of this feedback mechanism should be reinforced in situations of conflict (to favor the prevalence of the stronger stimulus), but reduced in normal conditions (to avoid illusory experience). This consideration opens the problem of how the system can be actively controlled by external inputs. Such a problem, in turn, may deal with the role of attention in neurocognitive science, and with the existence of higher hierarchical centers, which can plan and implement more sophisticate high-level strategies.

4. *Parameter changes*. Several parameters in the model can be modified to simulate individual variability and/or pathological conditions. This aspect opens the possibility to build a family of models, which share the same theoretical structure and make use of the same mechanisms, but exhibit different behavior in response to the same stimuli. While the potentialities of this approach are evident for what concerns the study of neuroclinical problems and the simulation of procedures for neurorehabilitation, they may also be exploited to build flexible artificial systems.

5. *Synaptic plasticity*. Certainly the most intriguing aspects of neurocomputational models consist in the possibility to learn from the external environment and to adapt behavior on the basis of previous experience. The model of visuo-tactile integration presented above is an excellent example of these possibilities, in that it can mimic how the peripersonal space representation may be plastic and modified by practice. Presently, we are working on an extension of the audio–visual integration model in the superior colliculus, assuming that multisensory enhancement is not an innate property of the system, but one which develops from experience in a multisensory environment [this hypothesis agrees with physiological experiments, see (Stein 2005)]. The reader can find preliminary results on this extended model in Cuppini et al. (2008). We claim that inclusion of synaptic plasticity may open enormous possibilities to any multisensory system, providing it with the capability to track the statistical changes in its environment, and to adapt its behavior to maximize specific goals.

Although we are aware that inclusion of all these aspects in real devices is still at a pioneering stage, we hope that at least some of these ideas may inspire new artificial systems devoted to sensory fusion in a not-too-distant future. Regardless of practical applications, however, neurocomputational models are invaluable to improve our comprehension of how the brain works, and to summarize the plethora of existing data on sensory merging into a coherent theoretical structure. The latter may be a repository of our knowledge and may drive future neurophysiologic experimentation, as well as may inspire new ideas for future research.

References

Anastasio TJ, Patton PE (2003) A two-stage unsupervised learning algorithm reproduces multisensory enhancement in a neural network model of the corticotectal system. J Neurosci 23:6713–6727

Anastasio TJ, Patton PE, Belkacem-Boussaid K (2000) Using bayes rule to model multisensory enhancement in the superior colliculus. Neural Comput 12:1165–1187

Avillac M, Denève S, Olivier E et al (2005) Reference frames for representing visual and tactile locations in parietal cortex. Nat Neurosci 8:941–949

Bermant RI, Welch RB (1976) Effect of degree of separation of visual-auditory stimulus and eye position upon spatial interaction of vision and audition. Percept Mot Skills 43:487–493

Bolognini N, Rasi F, Coccia M et al (2005) Visual search improvement in hemianopic patients after audio-visual stimulation. Brain 128:2830–2842

Bonifazi S, Farnè A, Rinaldesi L et al (2007) Dynamic size-change of peri-hand space through tool-use: spatial extension or shift of the multi-sensory area. J Neuropsychol 1:101–114

Bremmer F, Schlack A, Shah NJ et al (2001) Polymodal motion processing in posterior parietal and premotor cortex: a human fMRI study strongly implies equivalencies between humans and monkeys. Neuron 29:287–296

Calvert GA, Spence C, Stein BE (2004) The handbook of multisensory processes. MIT, Cambridge

Colonius H, Diederich A (2004) Why aren't all deep superior colliculus neurons multisensory? A Bayes' ratio analysis. Cogn Affect Behav Neurosci 4:344–353

Cooke DF, Graziano MS (2004) Sensorimotor integration in the precentral gyrus: polysensory neurons and defensive movements. J Neurophysiol 91:1648–1660

Cuppini C, Ursino M, Magosso E et al (2008) A neural network model of multisensory maturation in superior colliculus neurons. Program No. 457.5. 2008 Neuroscience Meeting Planner. Washington, DC: Society for Neuroscience, 2008. Online

Dalton P, Doolittle N, Nagata H et al (2000) The merging of the senses: integration of subthreshold taste and smell. Nat Neurosci 3:431–432

Denève S, Latham PE, Pouget A (2001) Efficient computation and cue integration with noisy population codes. Nat Neurosci 4:826–831

di Pellegrino G, Làdavas E, Farnè A (1997) Seeing where your hands are. Nature 388:730

Driver J, Spence C (2000) Multisensory perception: beyond modularity and convergence. Curr Biol 10:R731-R735

Duhamel JR, Colby CL, Goldberg ME (1998) Ventral intraparietal area of the macaque: congruent visual and somatic response properties. J Neurophysiol 79:126–136

Edwards SB, Ginsburgh CL, Henkel CK et al (1979) Sources of subcortical projections to the superior colliculus in the cat. J Comp Neurol 184:309–330

Farnè A, Làdavas E (2000) Dynamic size-change of hand peripersonal space following tool use. Neuroreport 11:1645–1649

Farnè A, Iriki A, Làdavas E (2005) Shaping multisensory action-space with tools: evidence from patients with cross-modal extinction. Neuropsychologia 43:238–248

Farnè A, Serino A, Làdavas E (2007) Dynamic size-change of peri-hand space following tool-use: determinants and spatial characteristics revealed through cross-modal extinction. Cortex 43:436–443

Fogassi L, Gallese V, Fadiga L et al (1996) Coding of peripersonal space in inferior premotor cortex (area F4). J Neurophysiol 76:141–157

Ghazanfar AA, Schroeder CE (2006) Is neocortex essentially multisensory? Trends Cogn Sci 10:278–285

Graziano MS, Cooke DF (2006) Parieto-frontal interactions, personal space, and defensive behavior. Neuropsychologia 44:2621–2635

Graziano MS, Yap GS, Gross CG (1994) Coding of visual space by premotor neurons. Science 266:1054–1057

Graziano MS, Hu XT, Gross CG (1997) Visuospatial properties of ventral premotor cortex. J Neurophysiol 77:2268–2292

Hihara S, Notoya T, Tanaka M et al (2006) Extension of corticocortical afferents into the anterior bank of the intraparietal sulcus by tool-use training in adult monkeys. Neuropsychologia 44:2636–2646

Hillis AE, Chang S, Heidler-Gary J et al (2006) Neural correlates of modality-specific spatial extinction. J Cogn Neurosci 18:1889–1898

Huerta MF, Harting JK (1984) The mammalian superior colliculus: studies of its morphology and connections. In: Vanegas H (ed) Comparative neurology of the optic tectum. Plenum, New York

Iriki A, Tanaka M, Iwamura Y (1996) Coding of modified body schema during tool use by macaque postcentral neurones. Neuroreport 7:2325–2330

Ishibashi H, Hihara S, Takahashi M et al (2002) Tool-use learning selectively induces expression of brain-derived neurotrophic factor, its receptor *trk*B, and neurotrophin 3 in the intraparietal mutlisensorycortex of monkeys. Brain Res Cogn Brain Res 14:3–9

Jiang W, Stein BE (2003) Cortex controls multisensory depression in superior colliculus. J Neurophysiol 90:2123–2135

Jiang W, Wallace MT, Jiang H et al (2001) Two cortical areas mediate multisensory integration in superior colliculus neurons. J Neurophysiol 85:506–522

Kadunce DC, Vaughan JW, Wallace MT et al (1997) Mechanisms of within- and cross-modality suppression in the superior colliculus. J Neurophysiol 78:2834–2847

Kadunce DC, Vaughan JW, Wallace MT et al (2001) The influence of visual and auditory receptive field organization on multisensory integration in the superior colliculus. Exp Brain Res 139:303–310

Kennett S, Taylor-Clarke M, Haggard P (2001) Noninformative vision improves the spatial resolution of touch in humans. Curr Biol 11:1188–1191

Làdavas E (2002) Functional and dynamic properties of visual peripersonal space. Trends Cogn Sci 6:17–22

Làdavas E (2008) Multisensory-based approach to the recovery of unisensory deficit. Ann N Y Acad Sci 1124:98–110

Làdavas E, Farnè A (2004) Visuo-tactile representation of near-the-body space. J Physiol Paris 98:161–170

Làdavas E, di Pellegrino G, Farnè A et al (1998) Neuropsychological evidence of an integrated visuotactile representation of peripersonal space in humans. J Cogn Neurosci 10:581–589

Làdavas E, Farnè A, Zeloni G et al (2000) Seeing or not seeing where your hands are. Exp Brain Res 131:458–467

Legrand D, Brozzoli C, Rossetti Y et al (2007) Close to me: multisensory space representations for action and pre-reflexive consciousness of oneself-in-the-world. Conscious Cogn 16:687–699

Macaluso E, Frith CD, Driver J (2000) Modulation of human visual cortex by crossmodal spatial attention. Science 289:1206–1208

Magosso E, Cuppini C, Serino A et al (2008) A theoretical study of multisensory integration in the superior colliculus by a neural network model. Neural Netw 21:817–829

Magosso E, Ursino M, di Pellegrino G et al (2010a) Neural bases of peri-hand space plasticity through tool-use: insights from a combined computational-experimental approach. Neuropsychologia 48:812–830

Magosso E, Zavaglia M, Serino A et al (2010b) Visuotactile representation of peripersonal space: a neural network study. Neural Comput 22:190–243

Makin TR, Holmes NP, Zohary E (2007) Is that near my hand? Multisensory representation of peripersonal space in human intraparietal sulcus. J Neurosci 27:731–740

Maravita A, Husain M, Clarke K et al (2001) Reaching with a tool extends visual-tactile interactions into far space: evidence from cross-modal extinction. Neuropsychologia 39:580–585

Mattingley JB, Driver J, Beschin N et al (1997) Attentional competition between modalities: extinction between touch and vision after right hemisphere damage. Neuropsychologia 35:867–880

Meredith MA, Stein BE (1996) Spatial determinants of multisensory integration in cat superior colliculus neurons. J Neurophysiol 75:1843–1857

Patton PE, Anastasio TJ (2003) Modelling cross-modal enhancement and modality-specific suppression in multisensory neurons. Neural Comput 15:783–810

Patton PE, Belkacem-Boussaid K, Anastasio TJ (2002) Multimodality in the superior colliculus: an information theoretic analysis. Brain Res Cogn Brain Res 14:10–19

Perrault TJ Jr, Vaughan JW, Stein BE et al (2005) Superior colliculus neurons use distinct operational modes in the integration of multisensory stimuli. J Neurophysiol 93:2575–2586

Pouget A, Sejnowski JT (1997) Spatial transformations in the parietal cortex using basis functions. J Cogn Neurosci 9:222–237

Pouget A, Deneve S, Duhamel JR (2002) A computational perspective on the neural basis of multisensory spatial representations. Nat Rev Neurosci 3:741–747

Press C, Taylor-Clarke M, Kennett S et al (2004) Visual enhancement of touch in spatial body representation. Exp Brain Res 154:238–245

Rizzolatti G, Scandolara C, Matelli M et al (1981) Afferent properties of periarcuate neurons in macaque monkeys. II. Visual responses. Behav Brain Res 2:147–163

Rizzolatti G, Fadiga L, Fogassi L et al (1997) The space around us. Science 277:190–191

Rizzolatti G, Luppino G, Matelli M (1998) The organization of the cortical motor system: new concepts. Electroencephalogr Clin Neurophysiol 106:283–296

Salinas E, Abbott LF (1995) Transfer of coded information from sensory to motor networks. J Neurosci 15:6461–6474

Sarri M, Blankenburg F, Driver J (2006) Neural correlates of crossmodal visual-tactile extinction and of tactile awareness revealed by fMRI in a right-hemisphere stroke patient. Neuropsychologia 44:2398–2410

Schroeder CE, Foxe J (2005) Multisensory contributions to low-level, 'unisensory' processing. Curr Opin Neurobiol 15:454–458

Sereno MI, Huang RS (2006) A human parietal face area contains aligned head-centered visual and tactile maps. Nat Neurosci 9:1337–1343

Stein BE (2005) The development of a dialogue between cortex and midbrain to integrate multisensory information. Exp Brain Res 166:305–315

Stein BE, Meredith MA (1993) The merging of the senses. MIT, Cambridge

Taylor-Clarke M, Kennett S, Haggard P (2002) Vision modulates somatosensory cortical processing. Curr Biol 12:233–236

Tipper SP, Lloyd D, Shorland B et al (1998) Vision influences tactile perception without proprioceptive orienting. Neuroreport 9:1741–1744

Ursino M, Cuppini C, Magosso E et al (2009) Multisensory integration in the superior colliculus: a neural network model. J Comput Neurosci 26:55–73

Wallace MT, Meredith MA, Stein BE (1993) Converging influences from visual, auditory, and somatosensory cortices onto output neurons of the superior colliculus. J Neurophysiol 69:1797–1809

Wallace MT, Meredith MA, Stein BE (1998) Multisensory integration in the Superior Colliculus of the alert cat. J Neurophysiol 80:1006–1010

Woods TM, Recanzone GH (2004) Visually induced plasticity of auditory spatial perception in macaques. Curr Biol 14:1559–1564

Chapter 3
Modelling Memory and Learning Consistently from Psychology to Physiology

L. Andrew Coward

Abstract Natural selection pressures have resulted in the physical resources of the brain being organized into modules that perform different general types of information processes. Each module is made up of submodules performing different information processes of the general type, and each submodule is made up of yet more detailed modules. At the highest level, modules correspond with major anatomical structures like the cortex, hippocampus, basal ganglia, cerebellum etc. In the cortex, for example, the more detailed modules include areas, columns, neurons, and a series of neuron substructures down to molecules. Any one memory or learning phenomenon requires many information processes performed by many different anatomical structures. However, the modular structure makes it possible to describe a memory phenomenon at a high (psychological) level in terms of the information processes performed by the major anatomical structures. The same phenomenon can be described at each level in a hierarchy of more detailed descriptions, in terms of the information processes performed by anatomical substructures. At higher levels, descriptions are approximate but can be mapped to more detailed, more precise descriptions as required down to neuron levels and below. The total information content of a high level description is small enough that it can be fully understood. Small parts of a high level phenomenon, when described at a more detailed level, also have a small enough information content to be understood. The information processes and resultant hierarchy of descriptions therefore make it possible to understand cognitive phenomena like episodic, semantic or working memory in terms of neuron processes via a series of intermediate levels of description.

L.A. Coward (✉)
College of Engineering and Computer Science, Australian National University,
Canberra, ACT 0200, Australia
e-mail: andrew.coward@anu.edu.au

V. Cutsuridis et al. (eds.), *Perception-Action Cycle: Models, Architectures, and Hardware*, Springer Series in Cognitive and Neural Systems 1,
DOI 10.1007/978-1-4419-1452-1_3, © Springer Science+Business Media, LLC 2011

3.1   Introduction

Understanding how the human brain supports higher cognitive phenomena like memory and learning cannot depend on models at the psychological level and other models at the physiological level with no clear connection between them. A hierarchy of descriptions of the same phenomenon from psychological to physiological is needed, with clear mapping between the levels (Coward and Sun 2004; Sun et al. 2005). At any one level, it must be possible to describe how at each point in time the observed situation causes the situation at the next point in time, in other words, descriptions must be causal.

The mapping between levels must be well understood. As in the physical sciences, the higher levels will be more approximate, but there must be clear understanding of when a more detailed level is necessary to achieve a given degree of quantitative accuracy (Coward and Sun 2007).

Such a hierarchy of causal descriptions requires consistent information models for components at each level of description. The information models make it possible to map precisely from a causal description on one level of detail to descriptions on other levels of detail.

For memory and learning, at the highest level a causal description in psychological terms is required. At a more detailed level, a causal description in terms of major anatomical structures, such as the cortex, thalamus, basal ganglia, cerebellum, etc., is needed that precisely maps into the psychological description. At an even more detailed level, a description in terms of components of the major anatomical structures (such as cortical columns and subcortical nuclei) must map into the higher level descriptions. At an even more detailed level, a causal description in terms of neuron algorithms must map into higher anatomical descriptions.

Experience with the design of very complex electronic real-time systems indicates that a number of practical considerations place severe constraints on the architectures of such systems (Coward 2001). These practical considerations include resource limitations and the need to make changes and additions to some features without undesirable side effects on other features. "System architecture" means the way in which the information handling resources required to support system features are separated into subsystems. Each subsystem is individually customized for efficient performance of a different type of information recording, processing and/or communication, and there are limits on the type and degree of information exchange between subsystems. The separation between memory and processing and the sequential execution of information processes, often known as the von Neumann architecture, are important aspects of these architectural constraints.

Although there are minimal direct resemblances between such electronic systems and brains, natural selection results in a number of practical considerations that influence brain architectures in analogous ways (Coward 2001, 2005). If the brains of two species can learn the same set of behaviours, but the brain architecture of one of these species requires fewer resources, then the species with the more efficient architecture will have a significant natural selection advantage. If one species can learn new behaviours with less interference to previously learned behaviours than another species, then again the more effective species will have natural selection advantages.

It can be demonstrated theoretically (Coward 2001) that these and other practical considerations tend to constrain the architecture of any sufficiently complex learning system into some specific architectural forms, analogous with but qualitatively different from the von Neumann architecture. There is considerable evidence (Coward 2005) that the mammal brain has been constrained into these forms, known as the recommendation architecture.

As a result of these architectural forms, different general information models can be assigned to different major anatomical structures including the cortex, hippocampal system, thalamus, basal ganglia, amygdala, hypothalamus and cerebellum. More specific information models that support the general models can be assigned to substructures of these structures, such as areas and columns in the cortex; CA fields, dentate gyrus and associated cortices in the hippocampal system; and striatum, substantia nigra, globus pallidus, nucleus accumbens, etc. in the basal ganglia. Yet more specific supporting information models can be assigned to neurons, synapses and ion channels.

This hierarchy of consistent information models is the foundation for modelling memory phenomena consistently on a psychological level, on the level of major anatomical structures, on the level of more detailed structures and on the level of neuron physiology. Cognitive phenomena such as retrieval of episodic memories can be mapped into sequences of activities in major anatomical structures that result in the memory retrieval, an activity within one such structure (such as the cortex) can be mapped into sequences of activities in substructures of that structure (such as cortical columns), and an activity in a substructure can be mapped into a sequence of activities at a neuron level. If necessary, a neuron activity can be mapped into a sequence of chemical activities at the synapse level, etc. The end result is understanding of the psychological phenomenon in terms of neurons.

3.2 The Recommendation Architecture Model

The information models for the major subsystems in the recommendation architecture are illustrated in Fig. 3.1, along with the anatomical structures of the mammal brain that correspond with each subsystem (Coward 1990, 2001, 2005, 2009a; Coward and Gedeon 2009). The primary separation is between a subsystem called clustering (corresponding with the cortex) that organizes the system resources used to define and detect conditions within the information available to the system and a subsystem called competition (corresponding with the basal ganglia and thalamus) that interprets each current condition detection as a set of recommendations in favour of many different behaviours, each with an individual weight, and implements the behaviour with the largest total weight across all current

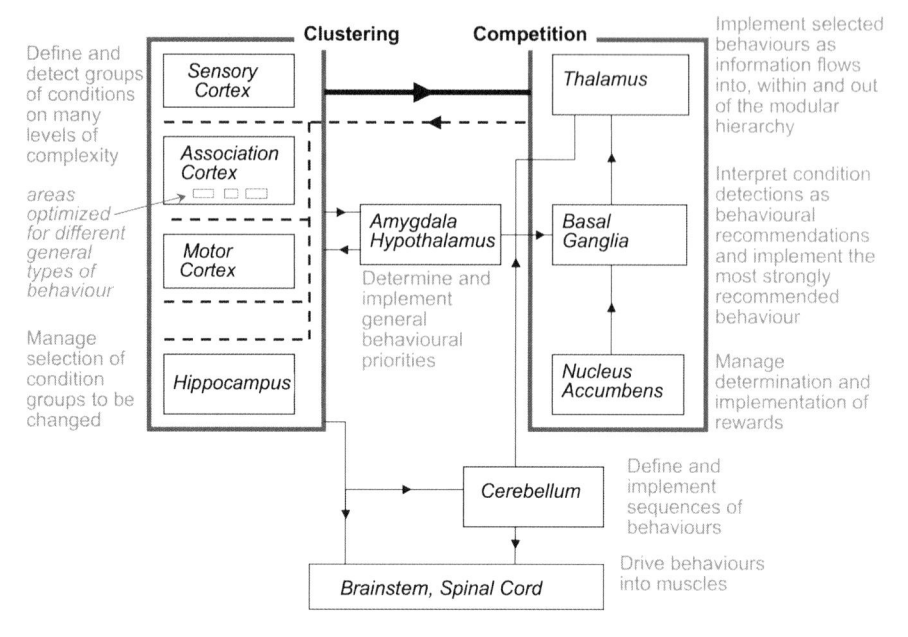

Fig. 3.1 The recommendation architecture mapped into mammal brain anatomy. The primary separation is between a modular hierarchy (called clustering and corresponding with the cortex) that defines and detects conditions in the information available to the brain, and a component hierarchy (called competition and corresponding with the thalamus and basal ganglia) that at each point in time receives some of the conditions detected by clustering, interprets each condition as a set of recommendations in favour of a range of different behaviours, each with an individual weight, and implements the most strongly recommended behaviour. Most conditions must be defined heuristically, and within clustering there is a subsystem (corresponding with the hippocampal system) that determines when and where new conditions will be recorded. Positive or negative reward feedback following a behaviour can adjust the recently active weights that recommended the behaviour, but cannot change condition definitions without severe interference with all the other behaviours recommended by the same conditions. Detections of conditions indicating that a reward is appropriate are received by the nucleus accumbens, which applies changes to weights in the basal ganglia. Most behaviours are implemented by release of information flows into, out of or within clustering, and a separate subsystem (corresponding with the thalamus) is required to efficiently manage such information releases. Conditions indicating the appropriateness of general types of behaviours (e.g. aggressive, fearful, food seeking, etc.) are provided to structures (corresponding with the amygdala and hypothalamus) that modulate the relative probability of such behaviour types. Frequently required sequences of behaviours that need to be executed rapidly and accurately are recorded and executed by the cerebellum

condition detections. Reward feedback results in adjustments to the weights that recommended recently implemented behaviours. However, such reward feedback cannot change condition definitions because such changes would interfere with all the other behaviours recommended by the same condition.

Clustering is organized as a modular hierarchy, with more detailed modules detecting sets of very similar conditions and higher level modules made up of groups of more detailed modules detecting larger sets of somewhat less similar conditions.

The primary driving force that generates this hierarchy is the need to share the resources used to detect similar conditions. Individual modules detect conditions relevant to many different behaviours, and module outputs therefore have very complex behavioural meanings.

Competition is organized as a component hierarchy. The use of reward feedback within competition means that component outputs can only have very simple behavioural meanings, and components must correspond with one individual behaviour or with one general type of behaviour.

The information available to the cortex includes sensory inputs and inputs indicating internal activity of the brain (including the cortex). A condition is defined by a set of inputs and a specified state for each input. A condition is detected if a high proportion of the inputs that define it are in their specified state.

Two conditions are similar if there is significant overlap in the information defining them and/or they often tend to be present at the same time (i.e. in the same system input states). This definition of similarity implies that two similar conditions will tend to have similar behavioural implications. Rather than connecting every individual condition detection to the component hierarchy, resource economy can be achieved by organizing conditions into groups on the basis of similarity. The conditions making up the group are recorded in a module, and the module generates an output to the component hierarchy only if a significant subset of the conditions in the group is present. If necessary, sufficiently different subsets can be indicated by different outputs. The group of similar conditions defines the receptive field of the module.

Conditions (and receptive fields) can be defined on many different levels of complexity, where the complexity of a condition is the total number of raw system inputs (including all duplicates) that contribute to the condition, either directly or via intermediate conditions. Receptive fields on different levels of complexity will tend to be effective in discriminating between different types of circumstances with different types of behavioural implications. For example, relatively simple receptive fields will be able to discriminate between different visual features (e.g. tail, wing and tooth). More complex receptive fields will be able to discriminate between different categories of visual object (e.g. cat, dog and bird). Yet more complex receptive fields will be able to discriminate between different types of groups of visual objects (e.g. cat chased by dog, cat confronting dog and cat avoiding dog). Simple receptive fields will be elements in the definitions of more complex receptive fields.

In a learning system, most conditions and receptive fields must be defined heuristically from system experience. Hence the information model for the cortex as a whole includes definition and detection of receptive fields at different levels of complexity within the information available to the brain.

In the basal ganglia, high-level components correspond with general types of behaviour, more detailed components within a high level component correspond with more specific behaviours of the general type. A receptive field detection by the cortex is communicated to a large number of components in the basal ganglia. Each component interprets the detection as a recommendation in favour of its corresponding behaviour, with a weight that is different for each component. The information model for the basal ganglia is therefore interpretation of cortical receptive field detections as behavioural recommendations and implementation of the most strongly recommended behaviours.

In general, a behaviour will be implemented by an information flow into, within or out of the cortex. There is therefore a requirement for a subsystem that provides coordinated management of all such information flows. This subsystem requires information on current cortical activity and information on currently selected behaviours. In the mammal brain, the thalamus corresponds with this subsystem. Thus the thalamus receives information from the cortex and gates information flows into the cortex (attention behaviours), out of the cortex (e.g. from the motor cortex to drive motor behaviours) and within the cortex (e.g. various cognitive behaviours).

If a behaviour is implemented, it will have consequences. If these consequences are good, the probability of the behaviour being selected in the future can be increased, if the consequences are bad, the probability should be decreased. A reward subsystem is therefore required that receives indications of positive and negative consequences of behaviours and changes recently active weights that led to the selection of recent behaviours. The nucleus accumbens and associated structures within the ventral basal ganglia corresponds with this subsystem. Because components in the component hierarchy correspond with behaviours, appropriate targetting of reward feedback is straightforward.

Indications of the positive or negative consequences of behaviours will often be receptive field detections by the cortex. For example, the anterior cingulate cortex detects receptive fields that correlate with the presence of error conditions (Kiehl et al. 2000). The anterior cingulate cortex projects to the nucleus accumbens (Devinsky et al. 1995), where such receptive field detections are interpreted as recommendations in favour of negative rewards. If the recommendations are accepted (i.e. if there is adequate total weight into the nucleus accumbens), then the recently accepted behavioural weights elsewhere in the basal ganglia will be weakened.

Any one receptive field detection recommends a wide range of different behaviours, and reward feedback follows an individual behaviour. Hence changes to receptive field definitions as a result of reward feedback would damage the integrity of the recommendation weights in favour of all the other behaviours. In general, any major change to a receptive field will damage the integrity of its associated behavioural meanings. As a result, receptive field changes must be strongly constrained. To a first approximation, a receptive field can be expanded slightly by addition of similar conditions, but previously added conditions cannot be changed or deleted. One exception to this rule is that if a condition is added and does not occur again within a significant period of time, it can probably be removed again. Another exception is that if a receptive field ceases to occur for some long period of time (perhaps because the source of its inputs has been damaged), it can probably be removed and its resources reassigned to another receptive field.

An implication of this constraint on receptive field changes is that individual fields cannot be guided to correspond with unambiguous cognitive circumstances such as object categories. Any one receptive field may be detected within some instances of many different such object categories and will therefore have recommendation strengths in favour of behaviours appropriate to all of the categories (such as naming them). However, an array of such fields can be evolved so that it discriminates between such categories. In other words, the predominant recommendation strength across all the receptive fields detected within an instance of any category will be in favour of behaviours appropriate for that category, even though no one receptive field corresponds with any one category (Coward 2001, 2005).

There is another important implication of the stability of receptive fields and their associated behavioural meanings. In response to an input state, a number of receptive fields will be detected, contributing their behavioural recommendation strengths. However, there are some potentially relevant recommendation strengths that are associated with receptive fields not actually being detected. For example, suppose there is a receptive field that is not currently being detected, but that has often been detected in the past when many of the currently detected receptive fields were also detected. Given this past consistency, there could be behavioural value in generating receptive field detections on the basis of frequent past simultaneous activity with currently detected receptive fields. Such value could be accessed by each receptive field also having recommendation strengths in favour of indirect activation of any receptive fields that have often been active in the past at the same time. Behavioural value could also exist for receptive field activations on the basis of recent simultaneous activity and also on the basis of simultaneous receptive field expansion. Furthermore, there could also be behavioural value in receptive field activations on the basis of past activity just after the activity of currently detected receptive fields, where the temporally correlated activity could be frequent, recent or receptive field expansion based.

If unrestricted, such indirect activations would generate a huge amount of activity, and they must therefore be behaviours that compete with other behaviours for acceptance. These indirect activation mechanisms are the information basis for a wide range of memory phenomena.

In order to achieve a high integrity behaviour, there must be an adequate range of recommendations available. To achieve an adequate range, there must be at least a minimum number of receptive field detections in response to every overall system input state. If this minimum is not reached, some receptive fields must expand in order to extend the range of available recommendations. However, to minimize changes to receptive fields, those requiring the least degree of expansion must be identified.

There is therefore a requirement for a further major subsystem that determines whether receptive field expansions are required, and if so identifies the modules in which the least such expansion will be required and drives expansion in those modules. In the mammal brain, the hippocampal system corresponds with this resource management subsystem. The information model for the hippocampal system is therefore determination of when and where in the cortex receptive field expansions are appropriate and driving the required expansions.

There is an additional requirement to record frequently used sequences of behaviours to ensure their rapid and accurate execution whenever required. In the mammal brain, the cerebellum corresponds with this subsystem.

There is a requirement for a subsystem that modulates the relative probabilities of different general types of behaviours. Such general types could include aggressive, fearful, food-seeking, etc. There are two ways in which relative probabilities could be affected. One is to change the relative arousal of components in the component hierarchy corresponding with the behaviour types. The other is to change the probability of recommendations of the general type being generated. To economize on resources, most receptive field detections must be able to recommend any type of behaviour. However, if there are substantial behavioural advantages in separate receptive field definitions at some levels of complexity for different behavioural types, the advantages might outweigh the resource costs. Hence there could be some modules within the cortex that tend to recommend one general type of behaviour, and temporarily broadening their receptive fields would increase the probability of such behaviours being selected.

Finally, there are two dynamic considerations. Efficient use of receptive field detection resources requires that the same set of resources must be able to detect receptive fields at one level of complexity simultaneously within multiple different sensory circumstances, keeping the different detections separate until it is appropriate to combine them in a controlled fashion. For example, if two dogs are chasing a squirrel, receptive fields must be detected within each dog and also the squirrel. There must be a set of receptive fields at a level of complexity that is effective for discriminating between objects. Resource economy dictates that each object will result in receptive field detections which are subsets of the same set, generally with some overlap between the subsets. Once receptive fields have been detected within all the individual objects, the detections must be combined to detect more complex receptive fields that can discriminate between different types of groups of objects. The problem is that the sets of simpler receptive fields must be kept completely separate (e.g. one activated set of receptive fields must not correspond with the head of a dog on the body of a squirrel) but must all be active at the same time for them to be combined to detect the more complex receptive fields (two dogs chasing a squirrel). As will be discussed in more detail later, one role of the gamma band frequency in the EEG is to maintain a separation between receptive fields detected within different objects but using the same neural resources.

The second dynamic consideration is that because there is behavioural value in indirect activation of receptive fields on the basis of activity shortly after currently active fields, there is an analogous requirement for simultaneous activity in the same resources corresponding with a sequence of points in time. This simultaneous activity is required so that the appropriate links supporting future indirect activations can be established. As discussed later, one role of the theta band frequency in the EEG is to maintain separation between receptive field detections corresponding with different points in time.

3.3   Review of Experimental Data Literature

Tulving (1984, 1985) and later Schacter and Tulving (1994) developed an approach to classifying memory and learning phenomena. They suggested there were three criteria that could be used to identify a memory system in the brain, which was separate from other memory systems. These criteria are the existence of a group of memory tasks which have some common characteristics, the existence of a list of features different from the list for any other system and the existence of multiple dissociations between any two systems. A dissociation is a way in which a similar manipulation of tasks performed by different systems produces different effects on the performance. An important category of dissociations is observations of patients with brain damage that affects one type of memory and not another.

On the basis of a wide range of evidence, Schacter and Tulving (1994) proposed that there are five independent memory systems: semantic memory; episodic memory; priming memory; procedural memory and working memory. Semantic memory is the memory for facts and the meaning of words, without recall of the context in which those facts or words were learned. Episodic memory is the memory for events, including autobiographical memory for events with personal involvement. Semantic and episodic memories are together known as declarative memory because they are consciously accessible and can be described verbally. Priming memory is the ability to make use of experiences in the recent past to enhance behaviour, even when there is no conscious awareness or memory of those experiences. Procedural memory is the ability to learn skills, including motor skills. Working memory is the ability to maintain direct access to multiple objects, so that information derived from the objects is immediately available to influence behaviour.

3.3.1   Semantic Memory

Semantic memory is defined as the ability to recall a wide range of organized information including facts and word meanings (Tulving 1972). The typical experimental test of semantic memory is category verification, where a subject is presented with paired category names and category (e.g. mammal, monkey or mammal and pigeon) and asked to identify if the pairing is correct. Identification speed is slightly slower for non-typical instances (e.g. mammal-bat) than for typical or incorrect pairings (Rips et al. 1973).

Functional neuroimaging indicates that semantic knowledge is encoded in many different cortical areas, especially the posterior temporal and frontal cortices, with the areas activated during semantic memory tasks generally being those also active during sensory or motor processing (Martin 2007). However, damage to the most anterior portions of the temporal cortices results in general loss of semantic memory capabilities, although this area does not show strong activation during semantic memory tasks (Rogers et al. 2006). There appears to be no consistent evidence for cortical area specialization for semantic domain (e.g. natural or man-made objects) or category (e.g. animals, fruit, tools and vehicles) although there is animal specific activity in the left anterio-medial temporal pole and tool specific activity in the left posterior middle temporal gyrus which appears in a subset of experiments with lower statistical confidence (Devlin et al. 2002).

3.3.2 Episodic Memory

In the laboratory, episodic memory is measured by both recognition and recall experiments. In recognition experiments, subjects are shown a set of novel objects and later shown a mixture of further novel objects and objects from the earlier set and asked to identify objects seen before. With photographs, subjects have a remarkable high capability to identify previously perceived objects (Standing et al. 1970). In recall experiments, subjects are asked to describe one past event. To trigger the recall, subjects are given a word (Robinson 1976) or group of words (Crovitz and Schiffman 1974). Such experiments are often used to measure the past time period for which episodic memory retrieval has been degraded and the degree of degradation (Kensinger et al. 2001).

Observation of the severity of retrograde amnesia in amnesic patients with damage to their hippocampal system indicates graduations between semantic and episodic memory (Nadel and Moscovitch 1997). In such patients, the most severe amnesia is for personal autobiographic memories. Amnesia for personal information, public events and persons is less severe, and amnesia for words and general facts is often minimal.

Functional neuroimaging of the brain during episodic memory recall indicates that there is strong activity in the prefrontal cortex during episodic recall (Fletcher et al. 1997) and also in visual and visual association areas (Addis et al. 2007). There is somewhat weaker activity in the hippocampal system (Fletcher et al. 1997). Strong cerebellar activity is observed during episodic memory retrieval (Fliessbach et al. 2007).

3.3.3 Procedural Memory

Experimental tests of procedural memory are generally confined to simple learning tasks and are focussed on clarifying the distinction between procedural and other types of memory, in particular declarative memory (meaning both semantic and episodic). Typical investigations test the ability of patients who have lost the ability to create new semantic and episodic memories to learn simple procedural skills. For example, a wide range of such amnesic patients showed the ability to learn to read words reflected in a mirror at the same rate as normal controls and retain the skill for at least 3 months, despite failing to recall any familiarity with the task at the start of each session (Cohen and Squire 1981).

For more complex skills, declarative knowledge speeds up the learning process (Mathews et al. 1989; Sun et al. 1996), and declarative memory is required for high levels of procedural skill performance (Mathews et al. 1989). However, there can be inconsistencies between procedural and declarative knowledge. When highly skilled subjects describe their skill, the descriptions often correspond with beginner methods rather than actual methods, and generating a verbal description can result in reversion to the less effective beginner method (Bainbridge 1977; Berry 1987).

There is a range of evidence derived from the cognitive deficits associated with degeneration of the basal ganglia indicating that this structure plays an important role in procedural learning. The symptoms of Parkinson's disease include difficulty with voluntary movement and with initiation of movement and in general slowness of movement. The observed physical deficit (Jankovic 2008) is degeneration of dopaminergic neurons in the substantia nigra compacta (SNc) nucleus of the basal ganglia. The major symptom of Huntingdon's disease is the intrusion of irregular, unpredictable, purposeless, rapid movements that flow randomly from one body part to another (Berardelli et al. 1999). The observed physical deficit is loss of striatal cells that project into the indirect pathway (Starr et al. 2008).

In addition, the cerebellum plays an important role in procedural learning (Torriero et al. 2007), although it also plays a role in a wide range of higher cognitive processes (Leiner et al. 1993) including semantic (Devlin et al. 2002), episodic and working memory (Cabeza et al. 2002).

3.3.4 Working Memory

Working memory refers to the number of different objects that can be maintained active in the brain at the same time. A typical working memory experiment is list recall. Subjects are shown a sequence of objects and immediately afterwards asked to list all the objects in any order. Normal subjects can fully recall sequences of seven (plus or minus two) random digits, but only four or five random words or letters (McCarthy and Warrington 1990). Miller (1956) suggested that the limit of seven plus or minus two is a fundamental information processing limit. Cowan (2000) argued that seven is an overestimate because subjects were able to rehearse and/or chunk items and proposed that the true information limit is close to four items, based on a wide range of observations. One key observation is that there are performance discontinuities around the number four, with errorless performance below four and sharp increases in errors above four (Mandler and Shebo 1982).

In a variation of the list recall test, the number of items that subjects can report on longer lists is measured. Typically, there is enhanced recall for the first few items on the list and enhanced recall for the last few items (Baddeley 2000). A brief delay occupied with another task eliminates the recency effect but has much less influence on the primacy effect (Glanzer 1972). List recall capability is greater if there is a semantic connection between objects on the list. Recall of meaningful sentences can extend to 16 words or more, but recall for random words is limited to four or five (Baddeley 1987).

Neuroimaging indicates considerable overlap in the cortical areas active during working memory and declarative memory tasks. For example, Cabeza et al. (2002) used fMRI to compare brain activity during an episodic memory task (recalling if a word was on a list of 40 words studied much earlier) with a working memory task (recalling if a word was in a list of four words presented 15 s earlier). They found that the cerebellum and left dorsolateral cortex areas were active during both tasks, bilateral anterior and ventrolateral cortex areas were more active during episodic retrieval and Broca's area and bilateral posterior/dorsal areas were more active during working memory retrieval. A patient with damage to the left parietal lobe showed a deficit in working memory but unaffected declarative memory (Warrington and Shallice 1969; Shallice and Warrington 1970).

3.3.5   Priming Memory

Priming memory is a short-term effect of exposure to a stimulus on the response to a similar later stimulus. Priming memory decays rapidly over a period of minutes, then more slowly over periods of hours and days. One experimental test of priming is word stem completion, in which subjects are asked to complete each of a list of three letter word stems with the first English word that comes to mind. A stem is the first three letters of a word, and in such experiments there are typically about ten possible completions. Previous study of the word increases the probability of the word being generated, provided the study is less than about 2 h prior to test (Graf et al. 1984). Amnesic patients with no ability to create new declarative memories have priming memory that is the same as and decays at the same rate as in normal subjects (Graf et al. 1984).

Another important priming memory experiment is tachistoscopic image recognition, when a subject is shown a sequence of brief (<100 ms) presentations of images of objects, each image separated by a masking pattern to prevent retinal, etc. afterimages. In this situation, few images can be accurately identified. However, if later there is a repeat exposure to the same image, identification accuracy increases considerably (Bar and Biederman 1998; Badgaiyan 2000).

It has been argued that unconscious priming involves different mechanisms from when the earlier stimulus is consciously used. However, McBride et al. (2001) presented evidence that similar mechanisms operate in both cases. Their experiments used word fragments, which were words from which two to four letters had been replaced by spaces. Subjects were given word lists to study. After a measured period of time, they were given word fragments and asked to complete the word. Two types of experiment were performed. In one, the subjects were asked to complete the fragment with a word studied earlier (i.e. a conscious approach). In the other, they were asked to complete the fragment with the first word that came to mind (i.e. an unconscious approach). The proportion of completions with studied words declined with time, rapidly in the first 10 min, but both performance and rate of decline were the same with both conscious and unconscious instructions.

3.3.6 Dissociations Indicating Separate Memory Systems

The evidence for the different memory systems is extensively discussed in Schacter and Tulving (1994). Any theory of memory and learning at anatomical and physiological levels must provide an account for this evidence.

Some of the most striking evidence for the separation of semantic and episodic memory systems from each other and from other memory types comes from observations of patients with damage to their hippocampal systems. In the 1950s, a number of patients had experimental surgery to treat intractable epilepsy. The surgery involved sectioning significant parts of their hippocampal systems, and although successful in reducing the frequency of epileptic seizures, it had some drastic side effects on their memory capabilities (Scoville and Milner 1957). One of these patients, HM, was extensively studied until his death in 2008.

HM lost all capability to acquire new semantic or episodic memories. He could learn no new facts or words and could recall no events after his surgery (Scoville and Milner 1957; Corkin 2002). However, he retained normal working memory (Wickelgren 1968) demonstrating a dissociation between working memory and declarative memory. He retained skills learned prior to surgery, including speech and reasoning skills (Scoville and Milner 1957). He could still acquire simple motor skills such as learning mirror writing, showing steady improvement over a number of sessions, even though at each session he had no memory of attempting the task before (Gabrieli et al. 1993), demonstrating a dissociation between declarative and procedural memory. His priming memory was retained (Gabrieli et al. 1990). In addition, his ability to access episodic memories for 11 years prior to his surgery was also impaired (Sagar et al. 1985), but his semantic memory for word meanings learned in the same 11-year period was retained (Kensinger et al. 2001), demonstrating a dissociation between episodic and semantic memory.

Evidence for the separation between priming and procedural memories comes from the study of Huntington's Syndrome patients. Such patients are characterized by damage to the basal ganglia and exhibit severe deficits in motor skill learning, but their priming memory appears intact (Heindel et al. 1989).

As mentioned earlier, patients have been observed to exhibit deficits in working memory, with no apparent deficit in declarative memory (Warrington et al. 1971). In these patients, performance in an immediate memory span test in which they were presented with strings of one to four digits, letters or words revealed good recall for one item strings but well below normal recall for strings with more than one item. However, performance in recall of a short story showed performance slightly better than for normal subjects.

3.4 Other Modelling Approaches

There is a very large number of attempts to model memory phenomena, ranging from attempts to prove that synaptic weight changes are always present during learning (Martin and Morris 2002) to high-level psychological models such as Baddeley's

(1986) working memory model that uses subsystems like phonological memory and central executive with no attempt to map into physiology. In general, previous models focus on one or two levels of description and do not present a consistent hierarchy of descriptions with information models on each level which can be mapped between all levels from psychology to physiology.

The five system memory model proposed by Schacter and Tulving (1994) and described earlier is in fact a psychological level model with some implications for high level anatomy. As discussed earlier, there have been successful attempts to map the model into major anatomical structures by the evidence from deficits resulting from local brain damage (Schacter and Tulving 1994) and various imaging techniques including fMRI and PET. For example, Devlin et al. (2002) and Rogers et al. (2006) analyze the brain regions active during semantic memory processes. Kassubek et al. (2001) investigated brain regions active during procedural memory processes. Fletcher et al. (1997) and Addis et al. (2007) have investigated the brain regions active during episodic memory processes. These investigations demonstrate differences in the cortical areas active during different types of memory and are valuable high level descriptions, but do not provide information models for the processes that can be mapped into deeper level descriptions.

Another type of approach has been the development of phenomenological models like Baddeley's working memory model (1986) and the spreading activation model for semantic memory (Collins and Loftus 1975). These models attempt information models for the phenomena at high level but do not offer any mapping to more detailed levels. The ACT-R model (Anderson 1996) offers a detailed information model which can, for example, model working memory in a fair amount of quantitative detail (Lovett et al. 1999). However, although there have been attempts to map ACT-R to fMRI imaging, the ACT-R information models do not provide any mapping into information models for more detailed anatomical or physiological structures.

Another extensive modelling approach to memory is that of Hasselmo and his collaborators. For example, they have proposed a model for the operation of working memory and episodic memory that uses reinforcement learning and attempts to account for a range of observations on rats (Zilli and Hasselmo 2007). This model postulates the existence of a number of buffers which can be in states reflecting current sensory inputs or a range of past sensory inputs that have been recorded. A key aspect of the model is the concept of actions which can be taken on the memory systems themselves in addition to actions on the external environment. There are some analogies between these proposed self actions and the indirect activation of receptive field information in the recommendation architecture. An implementation of their model has been described (Zilli and Hasselmo 2007), but the implementation does not provide mapping into plausible physiology. Furthermore, the model does not contain the recommendation architecture separation between clustering (i.e. condition definition and detection) and competition (i.e. reinforcement learning-based interpretation of conditions into behavioural recommendations). As a result, the model would have problems scaling up to learn complex combinations of behaviours.

At the neurophysiological level there have been numerous proposals that synaptic plasticity supports memory (e.g. Martin and Morris 2002). However, although these proposals are based on experimental evidence that long-term changes to synaptic weights are associated with learning, they do not offer neuron level information models that can be mapped (through intermediate anatomical levels) into, for example, semantic and episodic memory.

Another important set of models are those which attempt to understand the role of different EEG frequencies in memory. The beta frequency (12–30 Hz), gamma frequency (30–80 Hz) and the theta frequency (5–12 Hz) occur throughout the neocortex and hippocampus. These frequencies appear as modulations placed upon the firing of pyramidal neurons and are probably managed by interneuron activity (Whittington and Traub 2003). A number of proposals have been made that these frequencies play various roles in memory.

For example, Hasselmo et al. (2002) proposed that different functions are supported in different phases of the theta frequency in the hippocampus. In one part of a theta cycle, associations between sensory events are learned. In the other part of the cycle, previously learned associations are retrieved. As a result, the theta rhythm plays an important role in the reversal of previously learned associations (e.g. when the physical location of a reward changes). This model has been simulated with a considerable degree of physiological detail (Cutsuridis et al. 2008, 2010; Cutsuridis and Wenneckers 2009). However, the model focusses on the hippocampus and does not address the role of the cortex in these memory functions, other than the general view that it is the long term storage location for declarative memory, with the hippocampus providing intermediate term storage. How declarative memories are transferred from hippocampus to cortex is not addressed.

A number of workers have suggested that the gamma frequency binds together the activity of neurons representing the features of the stimulus that is the focus of attention (e.g. Engel and Singer 2001). To this proposal has been added the idea that the gamma frequency is important for the formation of declarative memories (Axmacher et al. 2006).

There is a relationship between these proposals and the recommendation architecture requirement for separation between different populations of receptive field detections corresponding with different cognitive stimuli but within the same neural resources. This relationship will be discussed below. However, the various dynamic models do not provide causal descriptions of higher level cognitive phenomena which can be mapped into the dynamic models.

Previously proposed memory models can model memory phenomena at one or two levels of detail, but do not provide a hierarchy of information models which can map consistently from psychological phenomena to neuron physiology.

The primary focus of this chapter is on memory and learning phenomena. However, the recommendation architecture is a general cognitive architecture that can provide descriptions of a wide range of higher cognitive phenomena in terms of brain anatomy and physiology (Coward 2005). A comparison between the recommendation architecture approach and a wide range of alternative cognitive architectures including Haikonen's neural architecture, Baar's global workspace, virtual machine models, simulation models, kernel architectures and forward models has been performed (Coward and Gedeon 2009). This study concluded that the alternative approaches do not take adequate account of natural selection pressures on brain resources, and the mapping between higher cognition and detailed anatomy and physiology is more plausible for the recommendation architecture.

3.5 Brain Anatomy and the Recommendation Architecture Model

The RA cognitive architecture maps the information model for each major anatomical structure into more detailed models for its substructures and so on all the way down to neuron physiology. The architecture explains how the more detailed models are implemented physiologically, and how the detailed models interact to support higher level models, up to descriptions of memory on a psychological level.

3.5.1 Cortical Structure

The cortex is a 3–4 mm thick sheet of tissue, with an area in adult humans of about $2,600 \, cm^2$. The cortical sheet is organized into six layers, with the layers differing in cell type, size and density and in intralayer and interlayer connectivity. Six layers are generally prominent, but sublayers are often visible, and sometimes a major layer can be absent. The cortex is organized perpendicular to the sheet into groups of about 100 cells linked across the layers. These groups, called minicolumns, are produced by the cortical growth process, each minicolumn from a small set of progenitor cells. Cortical columns are much larger vertical structures, perhaps formed by binding together a number of minicolumns. Cortical columns vary from 300 to 600 μm in diameter and are distinguished by similarity of the receptive fields of all their principal neurons and by common short-range horizontal connections (Mountcastle 1997).

The cortical sheet is separated into at least 50 areas (Brodmann 1908; Petrides and Pandya 1999, 2002; Morasan et al. 2001). These areas differ in the cell types, sizes and densities observed in each layer and sublayer, in the concentration of various neurochemicals such as the neurofilament protein that influences neuron size and shape, in the relative degree of myelinization of each layer and sublayer, in the interconnectivity with adjacent layers and in the interconnectivity with other areas.

3.5.2 Cortical Information Models

As discussed earlier, the information model for the cortex is definition and detection of receptive fields on many different levels of complexity within the information available to the brain. The number of detections must reach at least a minimum but not an excessive level in response to every overall brain input state. A receptive field is defined by a set of similar information conditions, with the receptive field being detected if a significant subset of the conditions is detected.

3.5.2.1 Information Model for a Cortical Area

The information model for an area is definition of receptive fields within one range of complexity and detection of the most significant receptive fields in each input state. These receptive fields must adequately address five requirements. The *first requirement* is that, in order to preserve previously acquired behavioural meanings, receptive fields can expand to detect additional, similar conditions but with some tightly restricted exceptions cannot change or discard previously added conditions. The *second requirement* is that, in response to any brain input state that results in inputs to the area, the number of receptive field detections must reach at least a minimum level. This second requirement is to ensure that, as discussed earlier, enough alternative behavioural recommendations are available to generate a high integrity behavioural selection. If detections are below the minimum, some receptive fields will expand until the minimum is reached. The *third requirement* is that, in response to any brain input state that results in inputs to the area, the number of receptive field detections does not reach an excessive level. The *fourth requirement* is that, to conserve resources, two receptive fields should not be detected consistently in the same input states. In other words, receptive fields must be as orthogonal as possible, or as statistically independent as possible. The *fifth requirement* is that the set of receptive fields programmed in one area must be able to discriminate between different perceptual circumstances, whenever such a difference implies that a different behavioural response is appropriate. The type of perceptual circumstance is different for different areas, the receptive fields in one area discriminate between visual features, in another between object categories, in yet another between different types of groups of objects, etc. as discussed earlier.

The conceptual process for definition of receptive fields can be understood by consideration of Figs. 3.2 and 3.3. In figures, there is a visual input domain which is an array of pixels. Visual inputs appear within the domain. A condition is defined by a set of elements (in this case pixels) and a state for each element, the condition is detected if a high proportion of the elements are in their defined state. As illustrated in Fig. 3.2, two conditions are similar if their elements are the same and/or often occur in the same input states. The modules illustrated in Fig. 3.3 have receptive fields defined by groups of similar conditions, with the receptive field being detected if a high proportion of the conditions is detected. An array of modules such as the one illustrated in Fig. 3.3 detects different receptive fields, all in one range of complexity. Such an array is required to detect at least a minimum number of receptive fields in every input state. If the minimum is not achieved, modules that are detecting a significant number of conditions but less than the proportion for receptive field detection add conditions until the proportion for receptive field

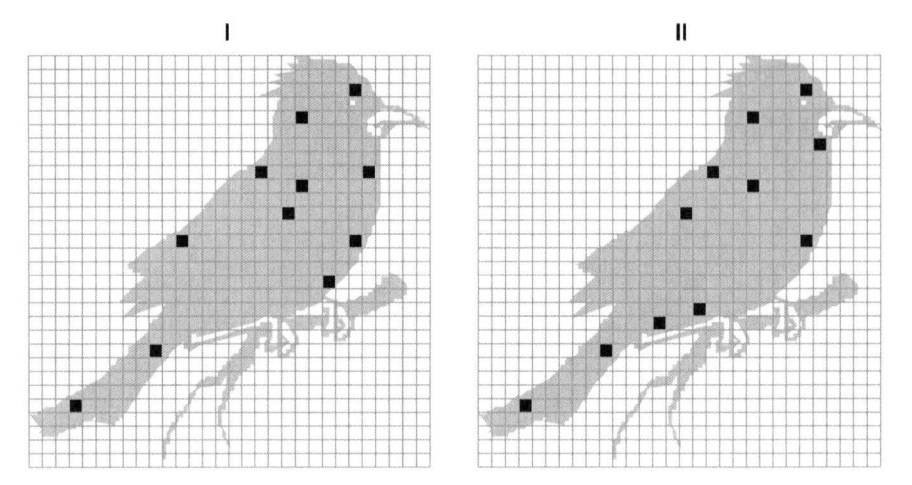

Fig. 3.2 The concept of conditions and condition similarity. A visual domain of 30 × 30 pixels is defined. A visual object is present in the domain. A condition is defined by a set of elements (in this case pixels), with a state specified for each element of the set. Such conditions are illustrated in a and b. If a visual input results in a high proportion of the pixels in the set being in their specified state, the condition is present. Two conditions are similar if a high proportion of the elements and states that define them are the same and/or often occur in the same input states. Conditions **I** and **II** are very similar by this definition, since seven of their elements are the same, and all of the elements occur in the same input state. With this definition of similarity, similar conditions will tend to have similar behavioural implications (such as recommending saying "that is a bird")

detection is reached. These conditions are selected (by a biased random process to be discussed later) from the large number of conditions that are present in the current input state. Receptive field expansions therefore occur in those modules for which the smallest expansion is required, causing the least damage to existing behavioural meanings of the receptive fields.

Arrays of modules are arranged in a sequence, with the conditions detected in one array being combinations of the conditions detected in the preceding array. Hence there is a steady increase in receptive field complexity down the sequence, where complexity can be defined as the number of raw sensory inputs (in the conceptual example pixels) that must be present in their appropriate state for the receptive field to be detected.

It may be possible to describe receptive fields in areas close to sensory input in sensory terms [such as the class of visual shapes that contain the receptive field, see Tanaka (2003)]. However, for higher areas, receptive fields can best be understood simply as groups of receptive fields in other areas that have tended to be detected at similar times in the past. The receptive field in one area receives inputs from the group of receptive fields in other areas and is detected if a high proportion of the group is present.

The mechanism for definition of receptive fields means that in general one receptive field will not correspond with one unambiguous cognitive circumstance such as an object category. In other words, no single receptive field will always

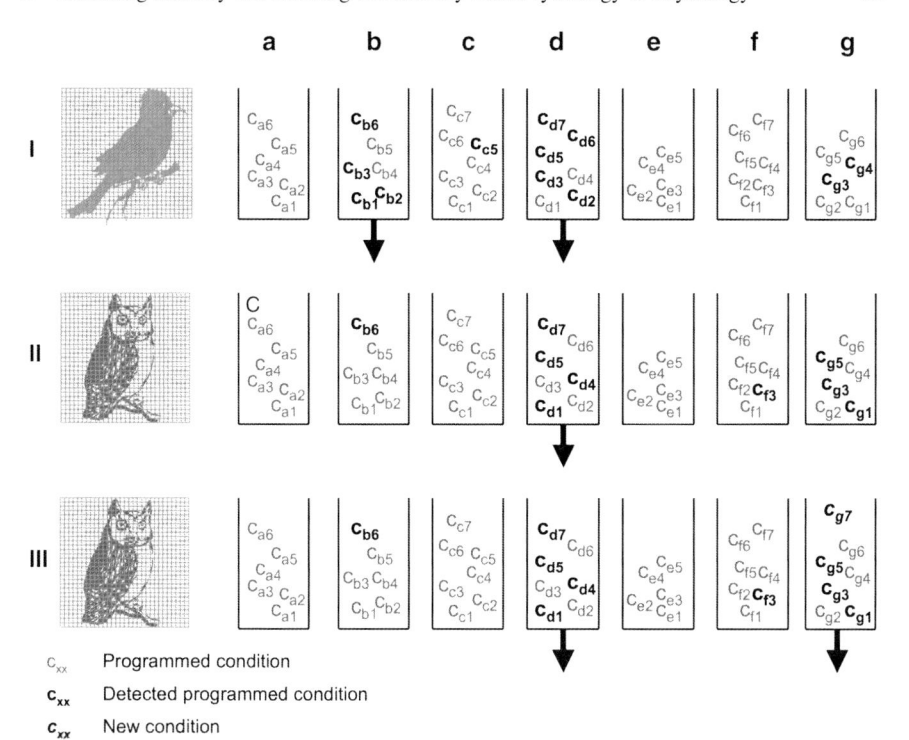

c_{xx} Programmed condition

$\mathbf{c_{xx}}$ Detected programmed condition

$\boldsymbol{c_{xx}}$ New condition

Fig. 3.3 The concept of receptive field expansion management. In the illustration there are six modules, each module having a receptive field defined by a set of similar conditions as described in Fig. 3.2. If a significant proportion of the conditions programmed in a module are detected, the module receptive field is detected and the module produces an output. The array of modules is required to detect at least two receptive fields in every input state. In scenario **I**, the visual input state results in detection of a significant proportion of conditions in modules b and d, corresponding with receptive field detection and resulting in outputs from those modules. The level of condition detection in other modules is relatively low. In scenario **II**, the visual input state is less familiar, and initially only module d has enough condition detection for receptive field detection. Because this is less than the minimum, the module with the highest degree of condition detection below the proportion for receptive field detection (i.e. module g) is selected, and additional conditions actually present in the current input are added to the module. Such additions expand the receptive field of the module and result in enough condition detections for receptive field detection and module output (scenario **III**). Selection of the module without receptive field detection but with the highest degree of condition detection is equivalent to selecting the module requiring the least receptive field expansion

be detected in all instances of one object category and in no instances of any other category. However, an array can be evolved to discriminate between types of cognitive circumstance, with arrays with different receptive field complexities being effective for discriminating between different types as illustrated in Fig. 3.4.

The meaning of discrimination can be understood from Fig. 3.5. The modules in the array illustrated in the figure detect receptive fields within a range of complexity appropriate for discriminating between visual objects. Each receptive field

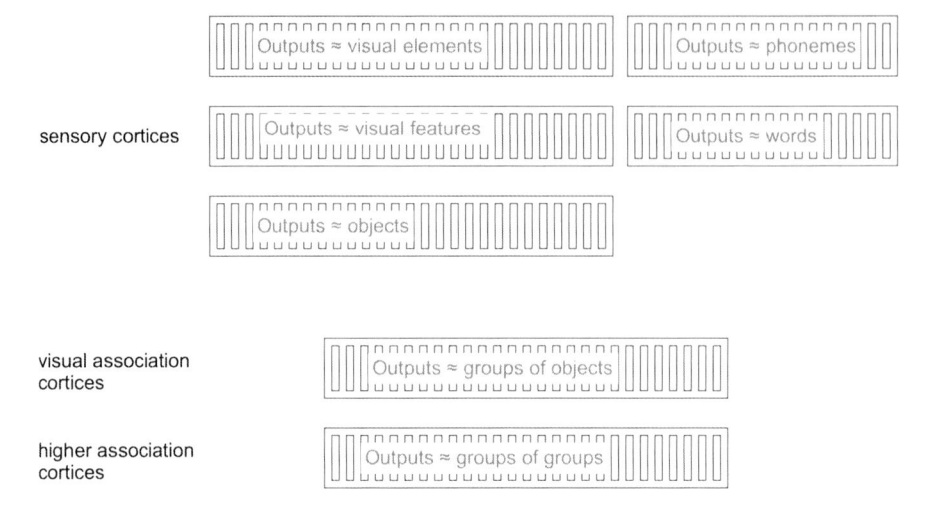

Fig. 3.4 Sequence of arrays of modules in which the conditions detected in one array are combinations of receptive field detections by modules in the preceding array. Few receptive fields will correspond with cognitively unambiguous circumstances, but at the appropriate level of receptive field complexity, the sets of receptive fields detected within instances of one category will be sufficiently different from the sets detected within instances of any other category that the currently detected set can effectively guide selection of a behaviour appropriate to current category. In other words, the array of columns will be able to discriminate between different categories. Receptive fields on different levels of complexity will be able to discriminate effectively between different types of cognitive categories (features, objects, groups of objects, etc.). Multiple levels of receptive field complexity may be relevant to, for example, discrimination between different types of groups of objects

corresponds with some visual circumstance which may occur in some instances of many different object categories. Thus, for example, one module may detect its receptive field in some instances of the category DOG, but also in some instances of the category CAT. In general, one receptive field will be detected in some instances of many different categories. There are no modules that detect their receptive fields in all DOG instances and in no instances of other categories, and individual modules are therefore cognitively ambiguous. However, the set of modules that detect their receptive fields in response to an actual CAT instance will have a predominant recommendation strength in favour of CAT-appropriate behaviours.

Module receptive field definition is not directly guided by the existence of cognitive categories, and an array might lack discrimination in some cases. A discrimination problem and the mechanism for resolving it are illustrated in Fig. 3.6. In the figure, because of a visual similarity an instance of a dog and an instance of a cat activate the same receptive fields. In other words, the array of receptive fields cannot discriminate between the two instances. The effect will be that the same behaviours will be recommended for both objects. Suppose that the dog was seen first, a DOG-appropriate behaviour was performed and rewarded, and the set of columns now have a strong overall recommendation strength in favour of repeating that behaviour

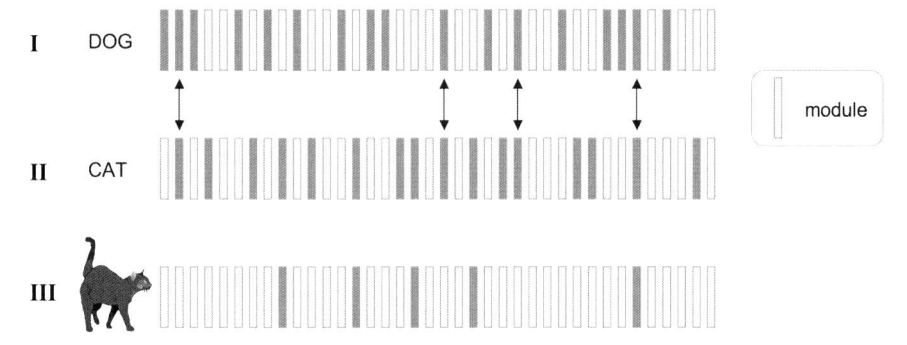

Fig. 3.5 An array of modules detecting receptive fields at a level of complexity effective for discriminating between different visual object categories like DOG and CAT. In **I**, the modules that detect their receptive fields in some instances of DOG are shaded. A different subset of these shaded modules will detect its receptive field in each actual instance of DOG, and each shaded module will have some recommendation strengths (in subcortical structures) in favour of DOG-appropriate behaviours such as saying "that is a dog". In **II**, the modules in the same array that detect their receptive fields in some instances of CAT are shaded. There is some overlap between the CAT and DOG modules illustrated in **I** and **II**, reflecting the existence of some visual similarities between cats and dogs. Receptive fields detected in some instances of both CAT and DOG will have recommendation strengths in favour of both DOG- and CAT-appropriate behaviours. All the modules will also have recommendation strengths in favour of behaviours appropriate to many other categories of object. In **III**, the modules in the same array that detect their receptive fields in one actual CAT instance are shaded. Although some modules with DOG-appropriate recommendation strengths are active, the predominant recommendation strength across all receptive fields will be CAT-appropriate. The module array can thus discriminate between dogs and cats

in the future. When the cat is seen, there is strong recommendation strength in favour of a DOG-appropriate behaviour, but the behaviour is punished because it is incorrect. This strong recommendation followed by negative reward triggers expansion of receptive fields in modules with strong internal activity, even if the number of active modules already reaches the minimum required level. There is a reasonable probability that these modules will not detect their receptive fields in a repetition of the DOG instance, making it possible to discriminate between the instances in the future.

Because receptive fields expand over time, there is a requirement to limit the total number that is activated in response to one input state. This limitation is achieved by inhibition between receptive fields so that only the most strongly present are detected. The means by which this is achieved will be discussed after the information model for a pyramidal neuron has been described (Fig. 3.9).

3.5.2.2 Information Model for a Cortical Column

The information model for a cortical column is definition and detection of a receptive field, including identification of the circumstances in which expansion of its receptive field could be appropriate. This information model can be understood

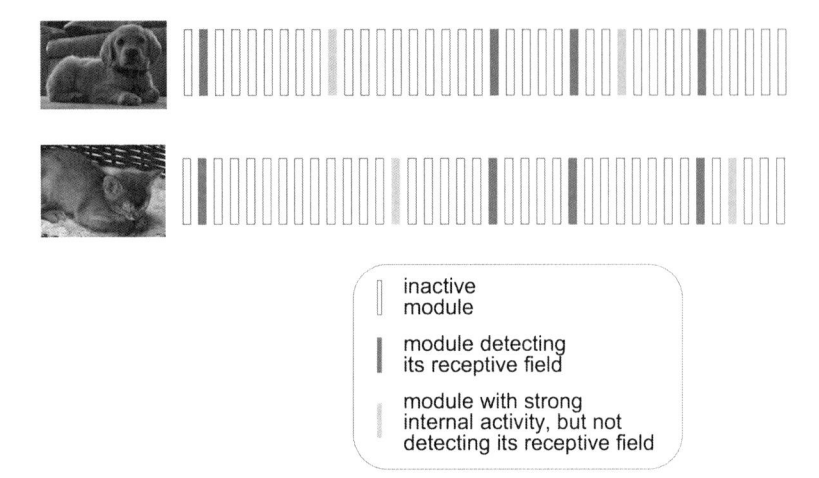

inactive
module

module detecting
its receptive field

module with strong
internal activity, but not
detecting its receptive field

Fig. 3.6 Managing discrimination problems. Initially, the visual similarities between the two pictures result in activation of the same sets of columns at the level of receptive field complexity that discriminates between object categories. Hence the same behaviour will be recommended in response to the two pictures (such as saying "cat") even though one is a cat and the other is a dog. The contradictory consequence feedback in response to the same behaviour following activation of the same columns results in forced receptive field expansions which lead to different column sets in response to the two objects

by consideration of Fig. 3.7. The cortical column information model also includes indirect activation of a receptive field under special circumstances in which the field is not actually present in the current sensory input state. Such indirect activations are behaviours that must be adequately recommended by currently active receptive fields.

In Fig. 3.7, there are three layers of principal neurons (generally pyramidal). The inputs that define the conditions detected in the top layer come from other areas of the cortex. The inputs that define the conditions detected in the middle layer come from the top layer, and the inputs that define conditions detected in the bottom layer come from the middle layer. The bottom layer provides outputs to other cortical areas and to subcortical structures where they are interpreted as behavioural recommendations. Note that this is a conceptual model, there could be more layers for various functional reasons (Coward 2005), and connectivity need not necessarily be sequential from top to bottom layer. From the point of view of the information model, the key point is that the conditions detected in the layer that provides outputs are less complex than the conditions detected in earlier layers. A situation could therefore arise in which there in no output from the column as a whole, but significant condition detection in earlier layers. The implication of such a situation is that the current input state contains many of the conditions that contribute to the definition of the column receptive field but not sufficient to trigger receptive field detection. Hence the degree of activity of neurons in the middle layer is a good indicator of the degree to which expansion of the column receptive field would be appropriate if required in response to the current input state.

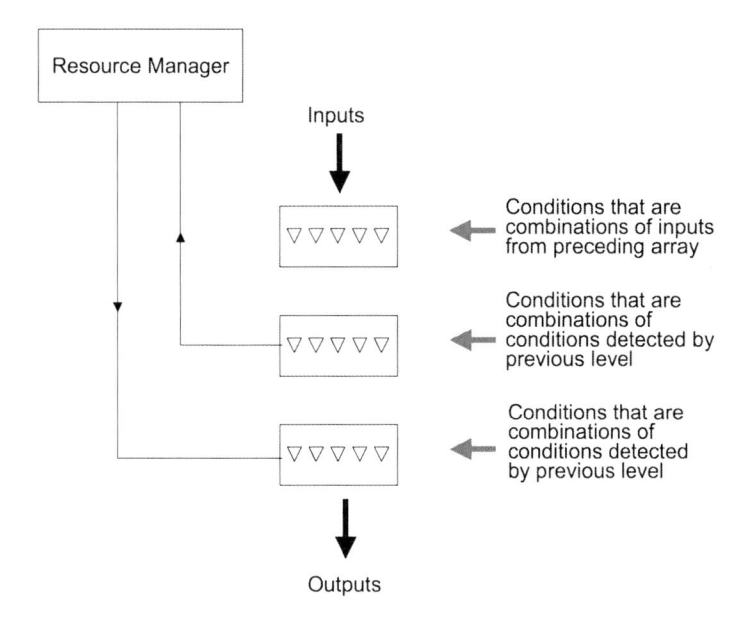

Fig. 3.7 Information model for a cortical column. A simple cortical column model made up of three layers of principal neurons is illustrated. The conditions detected by neurons in the top layer are combinations of receptive field detections in the areas that provide inputs to the column. The conditions detected by neurons in the second layer are combinations of receptive field detections by neurons in the top layer. The conditions detected by neurons in the bottom layer are combinations of receptive field detections by neurons in the middle layer. Outputs from the bottom layer are the column receptive field detections and go to other cortical areas and also to subcortical structures where they are interpreted as behavioural recommendations. Receptive field complexities therefore increase somewhat from the top to the bottom layer. If in response to some input state there is no activity in the bottom layer but significant activity in the middle layer, this indicates that there is some similarity between the current state and other states that have contained the column receptive field. Such middle layer activity therefore indicated that if the receptive field of the column were expanded slightly, it would be detected in the current state. The strength of middle layer activity therefore indicates the degree to which receptive field expansion would be appropriate if required in response to the current input state. Outputs from the middle layer go to the resource manager, where they are used to determine the set of columns with the strongest degree of internal activity. If additional receptive field detections are required, then set of columns receive inputs from the resource manager that drive receptive field expansion

The requirement is to determine if receptive field expansions are required (based on the overall degree of activity of columns across the area) and if so to determine the columns with the highest degree of internal activity and trigger receptive field expansions in those columns. Although in principle this requirement could be met by all-to-all connectivity between columns, it is more efficient and effective to use a resource manager (Coward 1990, 2005, 2009a). In Fig. 3.7, the output from the middle layer of the column to the resource manager carries information on the internal activity of the layer, and the inputs from the resource manager drives receptive field expansions as described below.

3.5.2.3   Information Model for a Pyramidal Neuron

The information model for a pyramidal neuron is definition and detection of a receptive field within a column. There are a number of aspects to this information model. *First*, separate information conditions that constitute the receptive field must be defined by groups of elements, with a condition being detected if enough of the elements are in the appropriate state. *Second*, the receptive field of the neuron must be detected and an output generated if a significant proportion of the conditions are detected. *Third*, it must be possible to expand the receptive field in appropriate circumstances by adding appropriate conditions. *Fourth*, it must be possible to limit the total activity across an array of columns, ultimately by limiting neuron activity in an appropriate fashion. *Fifth*, it must be possible to activate the neuron in some circumstances in the absence of its receptive field.

The pyramidal neuron information model is conceptually illustrated in Fig. 3.8. In this information model, conditions correspond with groups of synapses on one branch of the apical dendrite. These synapses are connection points for axons coming from other pyramidal neurons. An incoming action potential causes a synapse to inject voltage potential into its branch. The magnitude of the injected potential is proportional to the weight of the synapse. In the conceptual figure, synaptic weights are in arbitrary units (w). As described in the next section, an injected potential increases rapidly after the arrival of the action potential and then decays more slowly. A significant proportion of these synapses must receive incoming action potentials for the total potential in the branch to reach a threshold enabling a potential contribution deeper into the dendrite, and thus raise the chance of the soma producing an output action potential indicating detection of its receptive field. Because of the decay in potentials, the inputs to one branch must occur within a relatively short time so that the potentials resulting from action potentials at different synapses add to each other. Synapses on one branch thus correspond with the elements defining one information condition, which is detected if a high proportion of the inputs are active.

The neuron will only produce an output if a number of branches inject potential deeper into the dendrite within a relatively short period of time. The group of conditions corresponding with branches thus define the receptive field of the neuron, which is detected if a significant proportion of the conditions are detected.

A neuron can expand its receptive field by addition of new conditions. Such addition is achieved by means of provisional conditions such as the one illustrated in the lower left of Fig. 3.8. The total weight of all the synapses that define the elements of the condition is not sufficient for the branch to reach the threshold for potential injection deeper into the dendrite. However, there are additional synapses on the branch that come from the resource manager. If within a relatively short period of time, a high proportion of the condition defining elements are active and the inputs from the resource manager are also active, there will be enough potential in the branch to inject potential deeper into the dendrite. If a number of regular conditions are also being detected, then there will be enough potential to trigger firing of the neuron. If the neuron fires, a backpropagating acting potential into the branch increases the weights of any synapses that have recently received an action potential

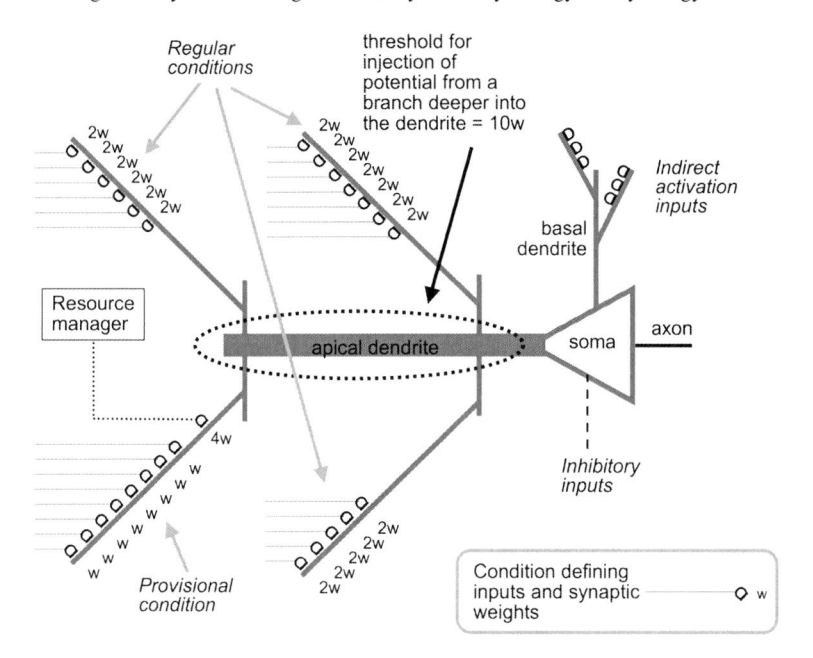

Fig. 3.8 Information model for a pyramidal neuron. The neuron has a body (or soma) and two sets of dendritic trees. Inputs that define the conditions that in turn define the receptive field of the neuron synapse on different branches of the apical dendrite. If an action potential arrives at a synapse, it injects a voltage potential into the branch that is proportional to the weight of the synapse. If the total weight injected by action potentials arriving at all the synapses on a branch exceeds a threshold, the branch injects potential deeper into the dendrite. The group of synapses on a branch thus define an information condition. If the total potential injected into the dendrite reaches a high enough level (i.e. enough conditions are detected), the dendrite will inject potential into the soma. If the potential injected into the soma exceeds a threshold, the soma will generate an output action potential. The group of conditions defined on apical branches thus define the receptive field of the neuron. Inhibitory signals from local interneurons target the soma (or the proximal dendrites) and reduce the chance of the neuron producing an output. The basal dendrite can also inject potential to trigger an output action potential; the inputs to the basal dendrite come from sources that can generate such an output in the absence of the neuron receptive field

from their source neuron. This is the long-term potentiation mechanism observed by Bi and Poo (1998). The effect is that the weights of the condition defining inputs are increased to the point that they can contribute to the firing of the neuron independent of the state of the inputs from the resource manager. Effectively, a new condition has been recorded.

A key requirement is that new conditions on a neuron must be similar to previously recorded conditions on the neuron or on other neurons in the same layer of the same column. As discussed earlier, similarity means that the elements defining the condition are the same as and/or often active in the past at the same time as the elements defining other conditions. There are a number of factors that ensure this similarity. First, for a new condition to be recorded, conditions already programmed

inputs

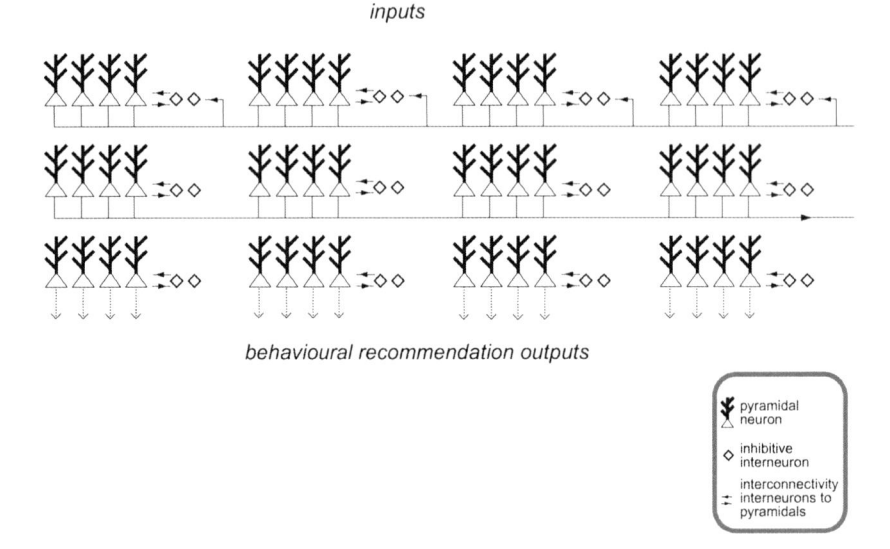

behavioural recommendation outputs

¥	pyramidal neuron
◇	inhibitive interneuron
‡	interconnectivity interneurons to pyramidals

Fig. 3.9 Interneuron connectivity to limit column activity. Interneuron outputs inhibit their targets. Interneurons target pyramidal neurons in the same layer of the same column. Interneurons can receive inputs from pyramidals in the same layer of the same column or from the same layer of columns in the same area. Connectivity between columns can be economized by biasing it in favour of connectivity between interneurons and pyramidals often active at the same time. Connectivity within a column limits overall activity within the column. Connectivity between columns effectively results in a competition between columns, reducing the total number of active columns

on the same neuron must be detected. Second, the inputs to a provisional condition are selected from those available in the neighbourhood. Third, the inputs to a provisional condition are selected from those often active in the past when the neuron is also active. This final factor is achieved utilizing REM sleep (Coward 1990 etc.). In REM sleep, there is a rerun of a selection of past experience. Provisional connectivity is established between neurons often active at the same time during this rerun.

Limiting activity within a column and also across an array of columns is achieved using inhibitory interneurons with the connectivity illustrated in Fig. 3.9.

As discussed earlier, there is behavioural value in the ability to indirectly activate columns on the basis of different types of temporally correlated past activity. Inputs to indirectly activate neurons in a column must be segregated from inputs defining receptive fields. Hence these inputs target a different dendrite system, the basal dendrites.

The staged integration model described in this section is generally consistent with the ideas of Hausser and Mel (2003). However, the details of the staging could be different; the key requirement is to achieve appropriate receptive field expansions.

3.5.2.4 Pyramidal Neuron Dynamics

As described earlier, a number of frequencies can be observed in the EEG which reflects the firing patterns of pyramidal neurons. In the recommendation architecture model, these frequencies reflect different information processing functions by networks of pyramidal neurons. In general terms, one action potential generated by a neuron indicates a detection of its receptive field at one point in time. The degree to which the receptive field is present is reflected in the rate at which action potentials are generated, and such rates averaged across many neurons result in the beta band frequencies. As discussed earlier, effective use of cortical resources requires the ability to maintain separate but simultaneously active populations of receptive field detections within the same resources. The gamma band reflects an information process that maintains a separation between simultaneous receptive field detections within different sensory objects in the same neuron network. The theta band reflects an information process that maintains a separation between groups of receptive fields that are detected at a sequence of different points in time.

The information model for the response of a pyramidal neuron to a sequence of action potentials is illustrated in Fig. 3.10. For simplicity, dendritic structure and staged integration is omitted for the figure, and it is assumed that the synaptic strengths are all the same. The potential injected by one action potential rises

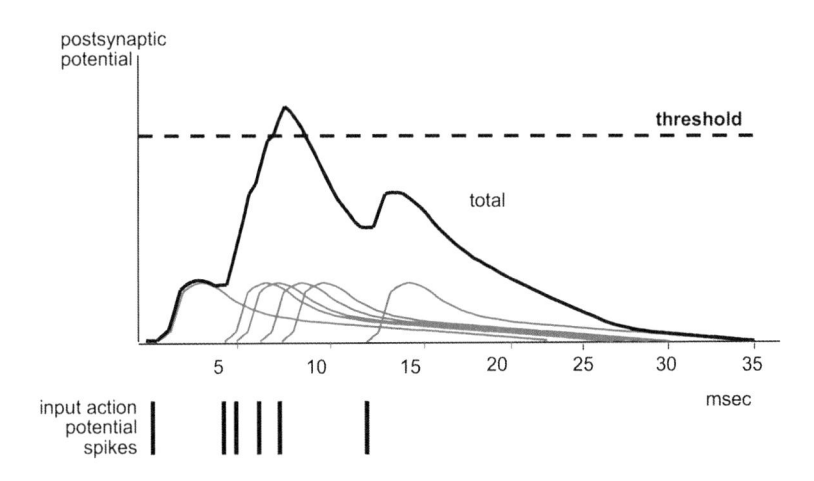

Fig. 3.10 Leaky integration and integration windows in a pyramidal neuron. For simplicity, staged integration is omitted. Action potential spikes arriving at different synapses each result in the injection of potential into the neuron. This injected potential rises over a period of a couple of milliseconds, then decays again with a decay constant of the order of 10 ms. It is assumed that the synaptic strengths are all the same. The postsynaptic potential resulting from different spikes adds linearly, and the total potential must exceed a threshold to result in (or contribute to) firing of the neuron. Because of the decay constant, two spikes must arrive close together in time to reinforce each other. Unless a number of spikes arrive within a fairly short period of time (of the order of the decay constant), the threshold will not be reached. This situation thus supports the concept of an integration window or period within which a minimum number of spikes must arrive for there to be a contribution to target neuron firing

relatively rapidly after the arrival of the input action potential, peaks after 2–3 ms and then decays with a time constant of the order of 8 ms. This time constant implies that action potentials arriving significantly more than 8 ms apart do not reinforce each other. Hence if the threshold for a group of synapses to contribute to the firing of the neuron is several times the maximum potential injected by one spike, at least a minimum number of spikes must arrive within a period of the order of the time constant for the group of spikes to contribute to the firing of the neuron. The concept of an integration window is therefore useful, where at least a minimum number of spikes must arrive within the integration window to contribute to neuron firing. The integration window is not completely fixed, it depends both on the ratio of leaky integration peak to threshold and on how close together the spikes arrive, but simulations demonstrate that it is a viable concept (Coward 2004).

An important concept that makes the integration window functionally useful is frequency modulation as illustrated in Fig. 3.11. A frequency modulation signal applied to a neuron shifts each of its output spikes towards the nearest peak in the frequency modulation signal. Such a frequency modulation could be imposed by a regular sequence of spikes at the modulation frequency. Interneuron inputs generally inhibit the firing of their targets, but if an inhibitory spike arrives more than about 5 milliseconds before an excitatory spike, it adds to the excitatory effect (Gulledge and Stuart 2003). A regular stream of interneuron spikes will therefore tend to impose a frequency modulation on the outputs of pyramidal neurons.

One functional value of frequency modulation can be understood from Fig. 3.12. If unmodulated inputs from some visual domain are provided to an array of columns, then the degree of receptive field detection will be much lower than if the inputs are modulated. Hence a domain in the visual field can be selected by placing a frequency modulation on all the inputs from that domain, and the effect will be preferential receptive field detection (and therefore behavioural recommendation generation) with respect to that domain. This is the information model for the attention function.

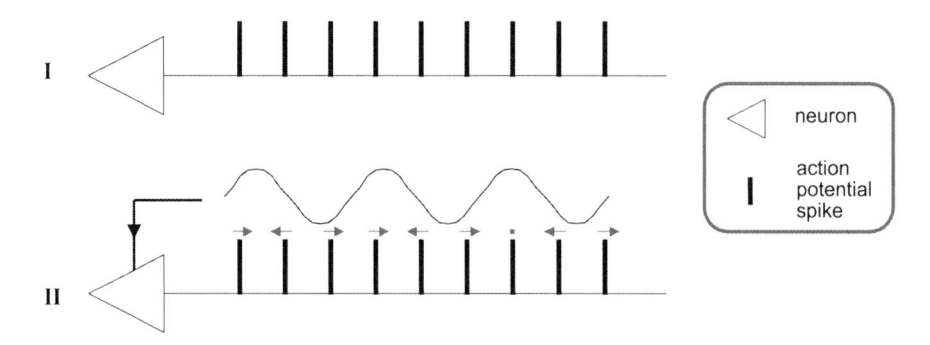

Fig. 3.11 Concept of frequency modulation. In **I**, a neuron produces an output made up of a sequence of action potential spikes. For ease of explanation, the illustrated sequence is regular; the actual sequences produced by neurons are irregular but the concept applies in the same way. In **II**, a frequency modulation signal is applied to the neuron. The effect of the signal is to shift output spikes towards the nearest peak in the modulation signal

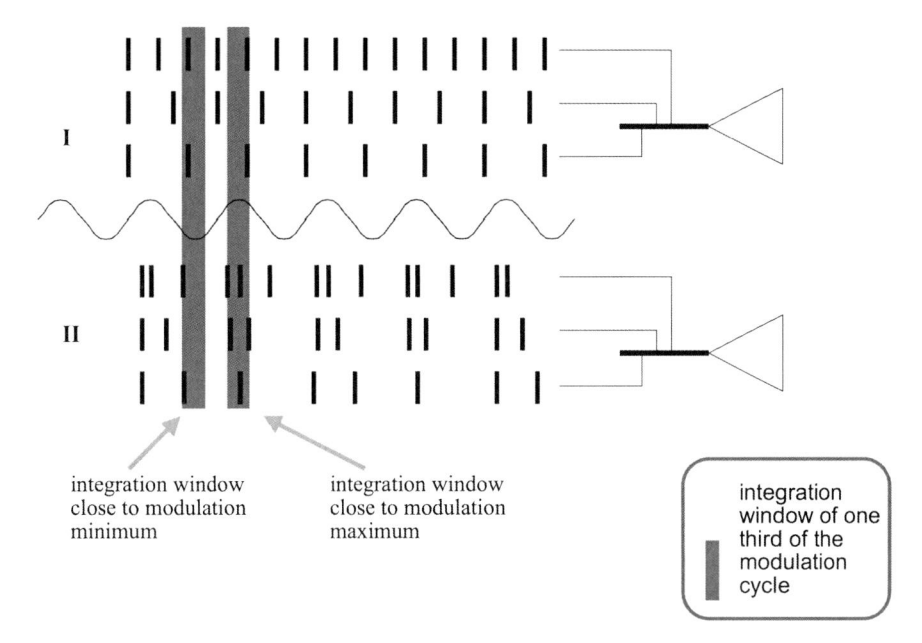

Fig. 3.12 Frequency modulation of inputs from a source. In **I**, three unmodulated input sources target a neuron. Within any integration window, there is only a total of about two spikes. If at least 4 spikes must be present to generate an output, there will be no outputs. In **II**, the inputs are modulated. This results in the same or fewer spikes in the integration window around the modulation minimum, but more around the maximum. There are sufficient around the maximum to fire the target neuron. Hence if a group of inputs from one source (such as a domain in the visual field) are frequency modulated, receptive fields will be detected only within the inputs from that domain

Note that if the modulation is placed on inputs, the outputs of the neurons will also be frequency modulated, and the modulation will propagate through following layers of neurons.

Another functional value can be seen from Fig. 3.13. If the interval between peaks in the modulation frequency is much larger than the integration window, then several integration windows can fit within one modulation cycle. If inputs from several different sources are modulated with the same frequency but at different phases, the spikes from the different sources will tend to arrive in different integration windows. Hence neuron receptive fields will be detected separately and independently within the different input sources. In other words, receptive fields within several different input sources (such as visual objects) can be detected simultaneously without crosstalk in the same neural resources.

The principle of maintaining a separation between different populations of receptive field detections can be extended to make use of multiple different frequencies. For example, in order to separate detection groups of receptive fields corresponding with sensory circumstances at a sequence of different points in time, a different frequency could be used.

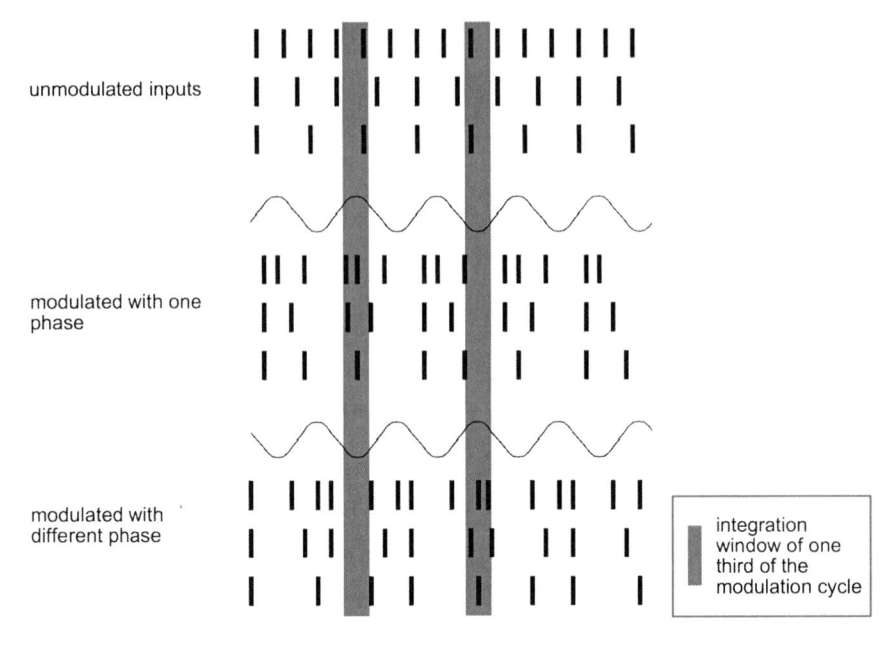

Fig. 3.13 Frequency modulation enabling simultaneous receptive field detections in multiple objects without interference. Inputs from different sources are given different phases of the same frequency modulation. As a result, spikes from the different sources tend to arrive in different integration windows, meaning that receptive fields present within different input sources can be detected without interference in the same neural resources, even in the same neuron

3.5.3   Structure of the Basal Ganglia and Thalamus

The basal ganglia and thalamus are organized into nuclei: clusters of neurons separated by regions containing mainly axons. The major nuclei and connectivity of the basal ganglia and thalamus can be understood by consideration of Figs. 3.14–3.17.

The thalamus is made up of a number of major nuclei as illustrated in Fig. 3.14. Each nucleus has strong reciprocal excitatory connectivity with a different cortical area, plus inputs from some other areas. Between each nucleus and the cortex, there is a nucleus called the thalamic reticular nucleus (TRN). As illustrated in Fig. 3.15, all axons between a thalamic nucleus and its associated cortices pass through a sector of the TRN, which regulates the information flow (Guillery et al. 1998).

There are a number of separate nuclei that make up the basal ganglia. These nuclei and the major connectivity between them and with the thalamus and cortex are illustrated in Fig. 3.16.

There are in fact several parallel paths from the cortex through the striatum, GPi/SNr and thalamus and back to the cortex (Alexander et al. 1986). These paths start and end in a cortical area different for each path, and each path goes through different subnuclei of the striatum, GPi, SNr and thalamus. In each path, the striatum also receives cortical inputs from some additional areas, but there is no return connectivity from the path to those areas.

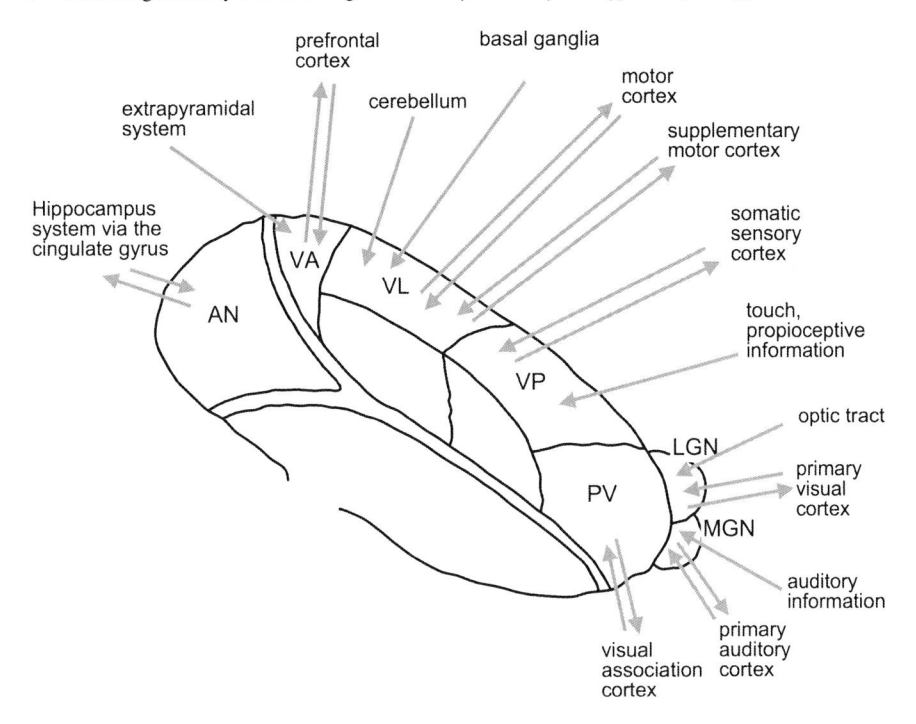

Fig. 3.14 Organization of the thalamus into nuclei with reciprocal connectivity between a nucleus and specific cortical areas. In addition, different additional cortical areas provide inputs to each nucleus but do not receive inputs in return. All connectivity between the thalamus and the cortex are excitatory

Individual neurons in the striatum and in the globus pallidus correspond with specific aspects motor activity such as the direction of arm movement but not to the underlying pattern of muscle activity associated with that movement (Crutcher and DeLong 1984; Mitchell et al. 1987).

The nucleus accumbens is sometimes regarded as part of the basal ganglia. As illustrated in Fig. 3.18, it gets substantial inputs from the amygdala and from the orbitofrontal cortex, and its outputs target the GPi and SNr nuclei of the basal ganglia.

3.5.4 Information Models for the Thalamus and Basal Ganglia

The general information model for these structures is interpretation of cortical receptive field detections as behavioural recommendations and determining and implementing the most strongly recommended behaviour. Reward feedback following a behaviour adjusts recently utilized recommendation weights. There are five components to this information model. The first is determination of raw total recommendation weights for each behaviour. The second is a competition between

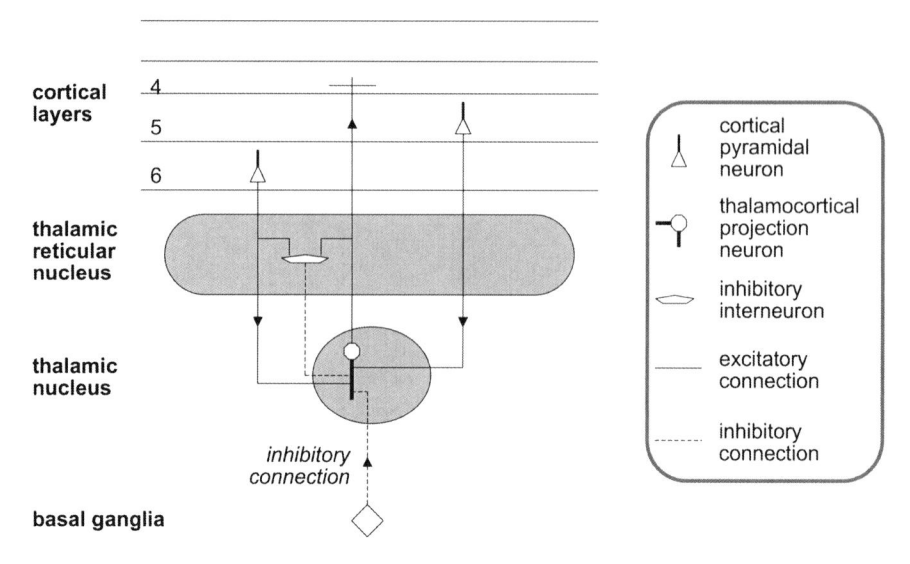

Fig. 3.15 All connectivity between a thalamic nucleus and the cortex passes through a sector of the TRN. The primary connection paths for one type of thalamic nucleus are illustrated (Guillery et al. 1998). The principal neurons of the thalamic nucleus, thalamocortical projection neurons, send excitatory connections to layer 4 of the cortex. Thalamocortical neurons receive excitatory inputs from pyramidal neurons in layers 5 and 6. Axons from layer 6 pyramidals and from thalamocortical neurons branch in the TRN to form synapses on the inhibitory interneurons that are the most common cells in the TRN, but the axons from layer 5 do not form such synapses. These inhibitory interneurons project to thalamocortical neurons. In addition, thalamocortical neurons receive inhibitory inputs from neurons in the basal ganglia

different behaviours to determine the most appropriate. The third is modulation to ensure that one and only one behaviour is selected. The fourth is implementation of the one selected behaviour. The fifth is utilizing reward feedback to modulate the probability of the same behaviour being selected in similar circumstances in the future.

3.5.4.1 Information Model for the Thalamus

The thalamus implements selected behaviours. Such an implementation is generally the release of information into the cortex, between cortical areas, or out of the cortex. Principal cells in the thalamus excite different groups of cortical columns and correspond with the releasing of information from those groups. Thalamic principal cells are excited by a range of columns in their groups and elsewhere. This connectivity can be viewed as the cortical columns recommending the release of outputs from the group. However, the basal ganglia tonically inhibits the thalamic cells, and release will not occur unless the tonic inhibition is reduced. The release of information includes imposing frequency modulation with an appropriate phase.

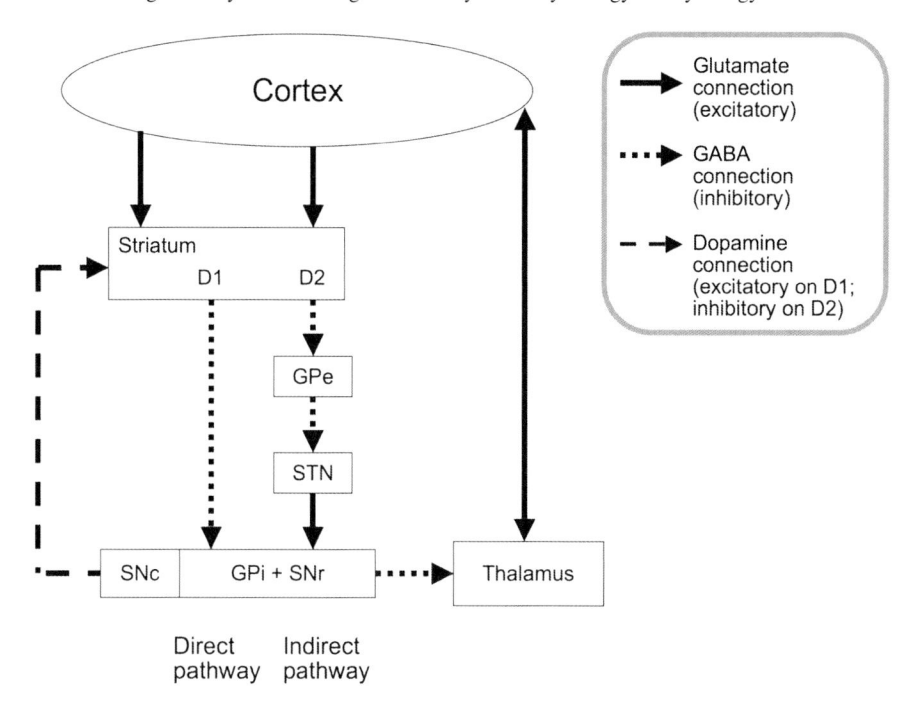

Fig. 3.16   The major nuclei and connectivity of the basal ganglia. Excitatory inputs from the cortex arrive at spiny projection neurons in the striatum. Each such neuron receives one or two inputs from a very large number of different cortical pyramidals. Striatal spiny projection neurons inhibit their targets, and there are two populations of these neurons. D1 population neurons project directly to the globus pallidus internal segment (GPi) and substantia nigra pars reticula (SNr) and are excited by dopaminergic inputs from the substantia nigra pars compacta (SNc). D2 neurons project indirectly to the GPi and SNr via two additional nuclei, the globus pallidus external segment (GPe) and the subthalamic nucleus (STN). As a result, the D2 population neurons ultimately excite the GPi and SNr. D2 population neurons are inhibited by dopaminergic inputs from the SNc. The GPi and SNr generate constant (tonal) inhibitory outputs to the thalamus. The direct path therefore tends to reduce inhibition of the thalamus while the indirect path tends to maintain this inhibition. As described earlier, the thalamus has reciprocal excitatory connectivity with the cortex. There is no direct return path from the basal ganglia to the cortex

This frequency modulation could be the role of the TRN, with TRN interneurons inhibiting thalamocortical neuron activity out of phase with the targetted modulation peaks.

3.5.4.2   Information Model for the Striatum

The information model for the striatum is that different striatal projection neurons correspond with different behaviours or types of behaviour. Each neuron receives inputs from many cortical pyramidals, which can be interpreted as raw recommendation weights in favour of the behaviour corresponding with the striatal neuron.

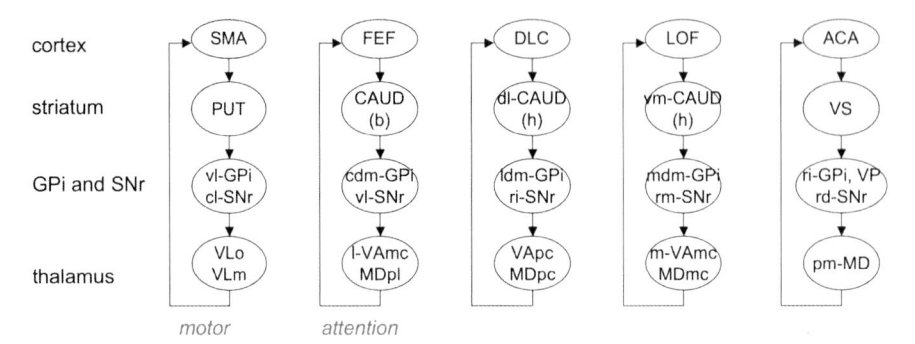

Fig. 3.17 Parallel paths linking basal ganglia, thalamus and cortex. In each path, one striatal subnucleus receives inputs from one primary area, sends outputs to one pair of GPi and SNr subnuclei, which send outputs to one thalamic nucleus that connects reciprocally with the primary cortical area. In each path, a different group of cortical areas also provide inputs to the striatal subnucleus but does connect reciprocally with the corresponding thalamic nucleus. Abbreviations follow Alexander et al. (1986)

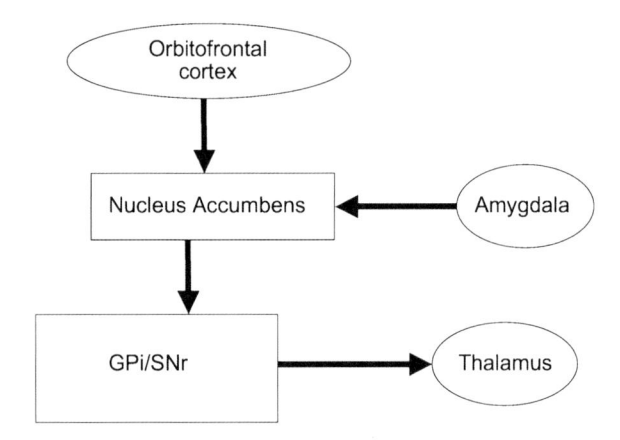

Fig. 3.18 The primary connectivity of the nucleus accumbens. Following a behaviour, both the orbitofrontal cortex and amygdala provide information indicating the detection of circumstances in which reward feedback is appropriate. The nucleus accumbens interprets these inputs as recommendations in favour of adjustments of the weights in the GPi and SNr that resulted in the recent behaviour and applies the most strongly recommended adjustments. The weight adjustments therefore affect the probability of acceptance of similar behaviours in similar circumstances in the future

For reasons discussed below, there is a pair of striatal projection neurons corresponding with each behaviour, with similar inputs, but one population D1 and the other population D2.

Competition within the striatum is implemented by the extensive local axon collaterals of striatal projection neurons, which mainly target other striatal projection neurons (Somogyi et al. 1981). In addition, there are populations of striatal interneurons that receive inputs from the cortex and target striatal projection neurons (Tepper and Bolam 2004).

3.5.4.3 Information Model for the GPi and SNr

The information model for the GPi and SNr is that individual principal neurons correspond with behaviours or types of behaviour. A principal neuron produces steady (tonic) inhibitory output to thalamic principal neurons corresponding with the release of cortical outputs that implement the behaviour corresponding with that principal neuron. With this connectivity arrangement, if a principal neuron in the GPi and SNr is inhibited, the result will be implementation of its corresponding behaviour.

A GPi and SNr principal neuron corresponding with a particular type of behaviour is targetted over the direct path (Fig. 3.16) by the D1 population striatal neuron corresponding with the same behaviour. This connectivity is inhibitory and therefore tends to encourage the implementation of the behaviour. The GPi and SNr neuron is targetted over the indirect path by D2 population neurons corresponding with different behaviours. This connectivity is excitatory and therefore tends to discourage the behaviour. GPi and SNr therefore supplement the competition between alternative currently recommended behaviours.

3.5.4.4 Information Model for the Nucleus Accumbens

The nucleus accumbens is observed to be associated with rewards (e.g. Ritz 1999). It targets the area of the basal ganglia in which the competition between different alternative behaviours occurs. The information model is therefore that following a positive reward it increases the weights that favoured recently accepted behaviours and decreases the weights that opposed them and vice versa for a negative reward.

3.5.5 Structure of the Hippocampal System

The hippocampus proper is made up of the CA fields and the dentate gyrus. The hippocampal system is made up of the hippocampus proper, some associated cortices and some associated subcortical structures. There is extensive connectivity between these structures as illustrated in Fig. 3.19.

All cortical areas except primary sensory project to the perirhinal or the parahippocampal cortices. These cortices project to the entorhinal cortex, which in turn projects to all the components of the hippocampus proper. CA1 generates outputs that go back through the entorhinal, perirhinal and parahippocampal cortices to the cortical areas from which inputs are derived.

The CA fields are cortex like, but with just one layer of pyramidal neurons. Within the hippocampus proper, there are two positive feedback loops: dentate gyrus granule cells excite mossy cells and CA3 pyramidals excite large numbers of other CA3 pyramidals. Each granule cells excites a small number of CA3 pyramidals and (via CA3 interneurons) inhibits a much larger number of CA3 pyramidals.

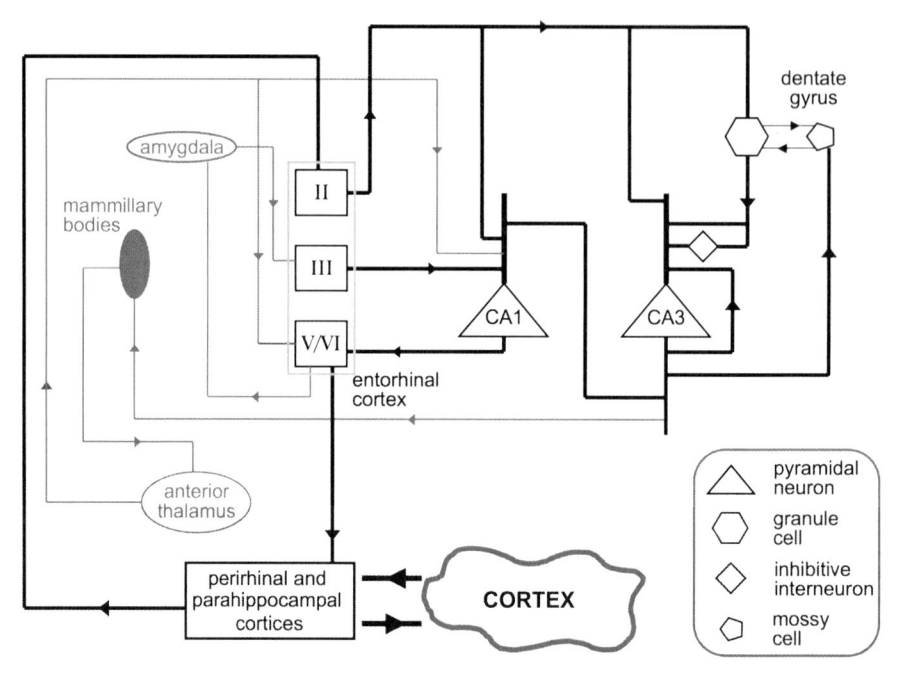

Fig. 3.19 The structure of the hippocampal system. Outputs from all cortical areas except primary sensory areas go through the perirhinal and parahippocampal cortices to the entorhinal cortex and from there to all regions of the hippocampus proper. There is massive connectivity between the various CA fields and the dentate gyrus of the hippocampus proper. Outputs from CA1 return via the subicular complex, and the entorhinal, perirhinal and parahippocampal cortices to the cortex. In addition, there are several connectivity loops involving subcortical structures. There is connectivity from the CA fields to the mammillary bodies, connectivity from the mammillary bodies to the anterior thalamus and so back to CA1. The subicular complex and entorhinal cortex project to the amygdala, which it turn projects back to those same structures

3.5.6 Information Model for the Hippocampal System

The information model for the hippocampal system is definition of which cortical columns will expand their receptive fields at each point in time. Identification of the most appropriate columns for such expansion uses two sources of information. One is information on the current internal activity of all cortical columns. The other is information on groups of columns that have tended to expand their receptive fields at similar times in the past.

The information model for the associated cortices is illustrated in Fig. 3.20. Columns in the PHC and PRC have receptive fields that are internal activity by groups of cortical columns that have tended to expand their receptive fields at similar times in the past, where internal activity is indicated by outputs from column middle layers. Bottom layer outputs from a PHC or PRC column target the cortical columns that make up their receptive fields. These bottom layer outputs are not generated unless there is both receptive field detection and input from the ERC. Such

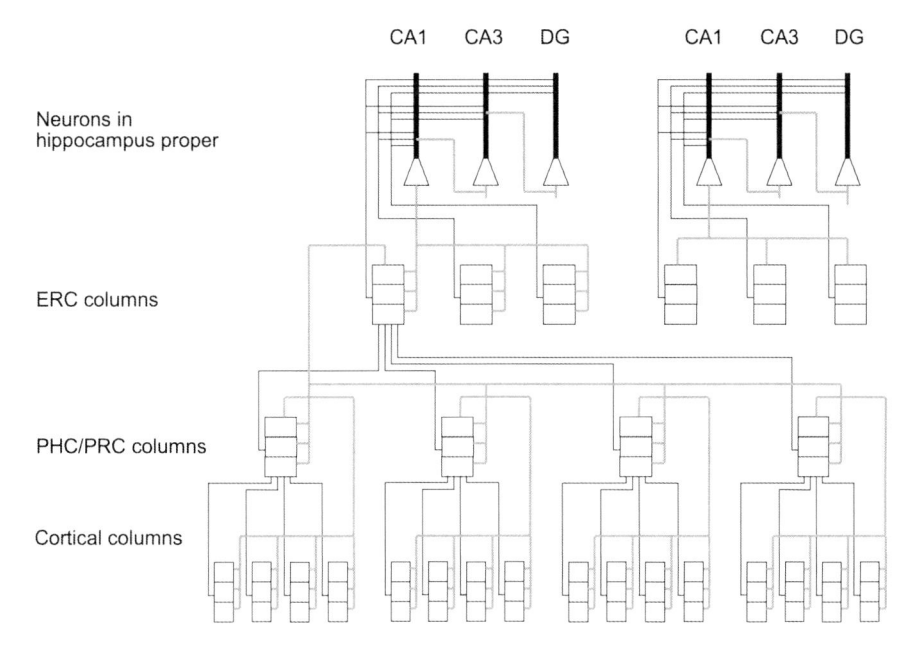

Fig. 3.20 Information model for the hippocampal cortices. Columns in the PHC and PRC have receptive fields which are internal activity of groups of cortical columns that have tended to expand their receptive fields at the same time in the past. One cortical column may appear in the receptive fields of a number of PHC or PRC columns. An output from the bottom layer of one of these PHC and PRC columns encourages receptive field expansion in the cortical columns that form its receptive field. However, to produce a bottom layer output, a PHC or PRC column requires both strong receptive field detection and an input from ERC columns. Such an ERC input encourages PHC or PRC output, with receptive field expansion if necessary. Similarly, columns in the ERC have receptive fields which are internal activity of groups of PHC and PRC columns that have tended to expand their receptive fields at the same time in the past and require both strong receptive field detection and input from CA1 pyramidals to generate an output back to the PHC and PRC columns that form their receptive fields. CA1 pyramidals, CA3 pyramidals and granule cells all have receptive fields which are internal activity of groups of EC columns that have tended to expand their receptive fields at the same time in the past

an input from the ERC triggers a PHC or PRC column output, with receptive field expansion of the PHC or PRC column if required. Such a receptive field expansion would therefore add a new group of cortical columns about to expand their receptive fields at the same time.

Columns in the ERC have receptive fields that are internal activity by groups of PHC and PRC columns that have tended to expand their receptive fields at similar times in the past. Bottom layer outputs from an ERC column target the PHC and PRC columns that make up their receptive fields. Such bottom layer outputs are only generated if there is both receptive field detection and inputs from CA1 pyramidals. Inputs from CA1 pyramidals trigger bottom layer outputs with receptive field expansions as required. Again, such receptive field expansions add groups of PHC and PRC columns about to expand their receptive fields at the same time.

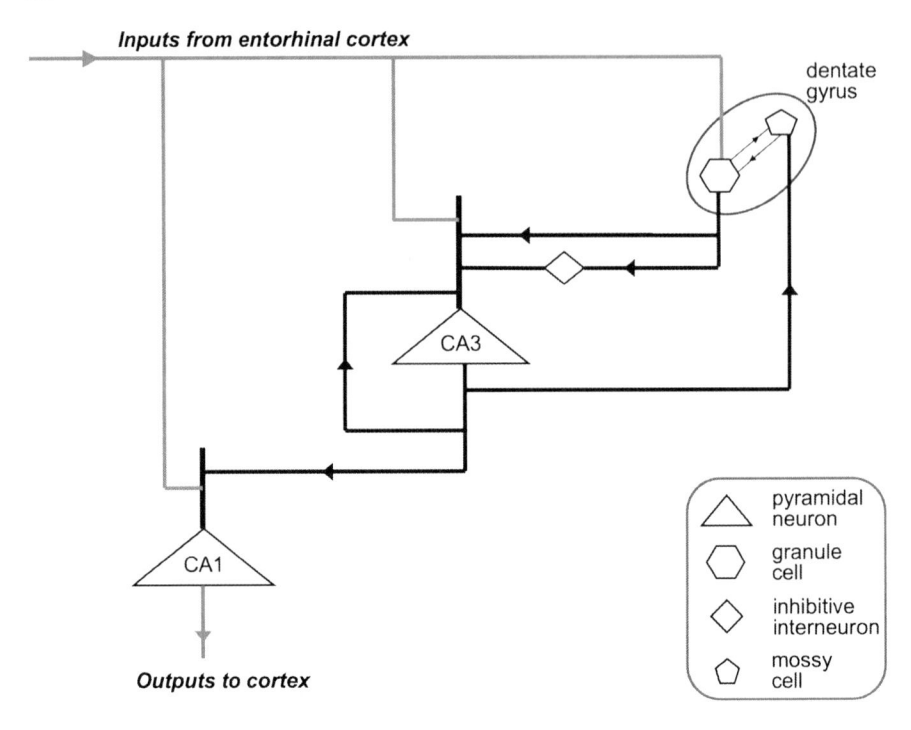

Fig. 3.21 Connectivity within the hippocampus proper. CA1 pyramidals, CA3 pyramidals and dentate gyrus granule cells all have receptive fields that are groups of ERC columns that tend to expand their receptive fields at the same time. However, the CA1 pyramidals have sharply focussed receptive fields, CA3 pyramidals less sharply focussed, and granule cells relatively poorly focussed. There are two positive feedback loops: Each CA3 pyramidal excites a large number of other CA3 pyramidals, granule cells excite mossy cells and mossy cells excite granule cells. The two positive feedback loops are linked: CA3 pyramidals exciting mossy cells; granule cells excite a small number of CA3 pyramidals and (via CA3 interneurons) inhibit a much wider range of CA3 pyramidals. CA3 pyramidals target CA1 pyramidals

The connectivity within the hippocampus proper is illustrated in Fig. 3.21. CA1 pyramidals and CA3 pyramidals all have receptive fields which are groups of ERC columns that tend to have expanded their receptive fields at the same time in the past. Granule cells have similar receptive fields, but poorly focussed so that they correspond more with groups of ERC columns that tend to be active at similar times in the past. The degree of input activity from the ERC indicates the degree of familiarity in the current sensory input state to the cortex. If there is a high level of such input activity across all cortical areas, there is a high degree of granule cell activity. This high level of granule cell activity means that inhibition of CA3 pyramidals is predominant and there is no CA3 activity and therefore no CA1 activity and no cortical receptive field expansions. If there is a degree of novelty in the current sensory input state to the cortex, some of the ERC input will be lower. Lower granule cell activity allows development of some CA3 pyramidal activity. The internal feedback within CA3 means that activity tends to develop in a group of CA3 pyramidals corresponding with large groups of cortical columns that have strong internal activity and also have tended to expand their receptive fields at the same time in the past. Direct inputs from granule cells encourage CA3 pyramidal receptive field expansion if required.

The projections from CA3 pyramidals to mossy cells mean that as CA3 pyramidal activity increases, granule cell activity will increase. Increasing granule cell activity will eventually reach a level where CA3 interneuron activity will cut off further increases in CA3 pyramidal activity. The lower the level of input from the ERC to granule cells, the higher the eventual level of activity in CA3 pyramidals. In other words, the degree of CA3 pyramidal activity will be proportional to the degree of novelty in the current sensory input state to the cortex.

The outputs of CA3 pyramidals target CA1 pyramidals with similar receptive fields. These outputs encourage CA1 pyramidal outputs, with receptive field expansion if required. CA3 activity is also communicated to the mammillary bodies. The mammillary bodies indicate to the anterior thalamus when the feedback competition within CA3 has settled down, and the anterior thalamus encourages CA1 outputs. To produce an output, a CA1 pyramidal therefore requires inputs from the ERC indicating receptive field detection, inputs from CA1 pyramidals indicating that a similar receptive field has won the competition for receptive field expansion and inputs from the anterior thalamus indicating that CA1 pyramidal activity has stabilized.

An output from a CA1 pyramidal triggers a cascade of activity and receptive field expansion through the ERC, the PHC and PRC to the cortical columns appropriate for receptive field expansion. These receptive field expansions bring the overall level of cortical activity up to the minimum level required to generate a wide enough range of behavioural recommendations to support a high integrity behaviour selection.

Receptive field expansions by granule cells are unguided. Hence they are poorly focussed on groups of columns that expanded their receptive fields at similar times in the past. However, these granule cells provide receptive field expansion guidance to CA3 pyramidals, which therefore have more sharply focussed receptive fields. CA1 pyramidals have receptive field expansions guided by CA3 pyramidals and therefore have the most sharply focussed receptive fields, appropriate for guiding the cortex.

The amygdala plays a role in emotion (Anderson and Phelps 1997; Whalen 1998). The information model for the role of the amygdala within the hippocampal system is that in emotional situations it increases the level of signals driving receptive field expansion in certain cortical areas, in particular those areas detecting receptive fields that discriminate between different general situations.

At a particular location, there will be a relatively constant surrounding sensory environment. As a result, the large group of cortical columns activated in response to this environment will have tended to expand their receptive fields at the same time in the past. Many of the receptive fields in the hippocampal system will therefore correspond with particular locations, leading to the existence of the observed place fields in the higher levels of the hippocampal system (Fyhn et al. 2004; Leutgeb et al. 2004).

In order to support indirect activation on the basis of receptive field expansion shortly after currently active columns, the simultaneous activity of hippocampal system neurons with receptive fields corresponding with different large groups of cortical columns that have expanded their receptive fields at a sequence of points in time will be necessary. For example, a sequence of place fields must be maintained active during physical movement. These different hippocampal receptive field detections must only interact in a controlled fashion, and their separation is managed by the theta frequency in the EEG.

This use of the theta frequency to prevent interaction between different neural populations has some similarities with the proposal of Hasselmo et al. (2002) that different functions (learning and retrieval of earlier learning) are supported by activity in different phases of the theta cycle. However, the recommendation architecture information model is somewhat more consistent with the observations that the phase of the theta modulation shifts in a regular fashion across place fields (O'Keefe and Recce 1993).

3.5.7 Structure of the Cerebellum

The cerebellum has an outer sheet of tissue surrounding a body of white matter (i.e. axons) and a core of nuclei. The sheet has three major layers. The innermost layer is made up of very large numbers of granule cells. The middle layer is made up of Purkinje cells. The outer layer is made up of the dendritic trees of the purkinje cells, penetrated by the axons of the granule cells.

One major source of excitatory inputs to the cerebellum is inputs containing motor information derived from the motor cortex and the spinal cord, via a nucleus called the inferior olive. These inputs target both the cerebellar nuclei and Purkinje cells. One input from the inferior olive targets a small set of Purkinje cells, and one Purkinje cell receives inputs from only one inferior olive input. A second source of excitatory inputs to the cerebellum is inputs containing sensory information derived from the cortex, via a structure called the pons. These inputs target both the cerebellar nuclei and granule cells.

Purkinje cells receive very large numbers of excitatory inputs from granule cells, each individual input is very weak. The single axon from the inferior olive that targets a Purkije cell makes multiple synapses and strongly excites the cell. Purkinje cells make inhibitory connections on to cells in the cerebellar nuclei.

Excitatory outputs from the cerebellar nuclei target both the spinal cord through the brain stem and the thalamus.

3.5.8 Information Model for the Cerebellum

The information model for the cerebellum is rapid and accurate implementation of a frequently used sequence of behaviours. The way this information model is

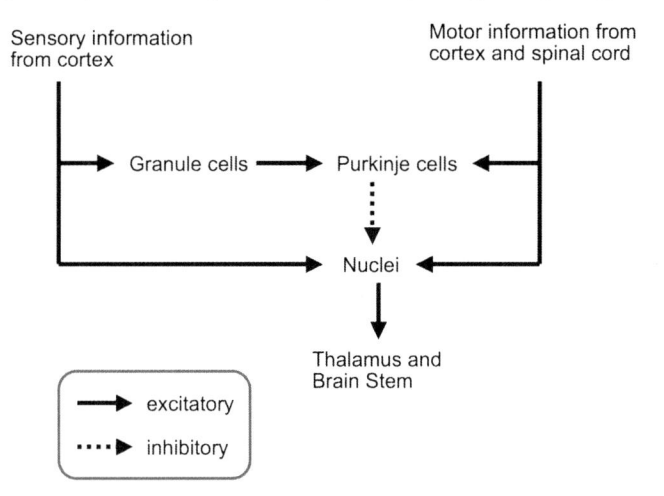

Fig. 3.22 Information model for the cerebellum. A purkinje cell corresponds with a "next behaviour". It receives an input containing motor information indicating that its preceding behaviour has taken place. It also receives inputs containing sensory information. Sensory information corresponding with completion of its "next behaviour" is weak compared with any other sensory input. Hence a Purkinje cell will be active from when its preceding behaviour is indicated until the completion of its own behaviour is indicated. Cerebellar nuclei cells correspond with different behaviours, and a Purkinje cell inhibits all except its corresponding next behaviour

partitioned between different parts of the cerebellum is illustrated in Fig. 3.22. The next step in a sequence is instantiated by a cerebellar nucleus cell and by a Purkinje cell. The Purkinje cell receives an excitatory input from the inferior olive indicating that the behaviour preceding its corresponding behaviour has occurred and excitatory inputs indicating the presence of a wide range of sensory circumstances. However, the weights of inputs indicating the sensory circumstances that follow the performance of its corresponding behaviour are weak. The Purkinje cell inhibits all cerebellar nuclei cells except the one corresponding with its own behaviour. Hence when the programmed preceding behaviour occurs and the sensory inputs are not consistent with completion of its programmed behaviour, the Purkinje inhibits all except its programmed behaviour. Once the sensory inputs are consistent with completion of its behaviour, the Purkinje cell outputs cease. At the start of learning, a Purkinje has inputs from a very wide range of sensory circumstances, but during learning the inputs that are present soon after it produces an output are weakened by the long-term depression (LTD) mechanism (Linden 2003).

Outputs from the cerebellar nuclei drive the next behaviour. For sequences of motor behaviours, these outputs can go directly to the spinal cord via the brain stem. For cognitive behaviours only involving releases of information within the cortex, outputs go to the thalamus.

Note that a behaviour sequence programmed in the cerebellum biases behaviour acceptance in favour of a series of types of behaviour, but the actual behaviour at each point will be the recommendation of the type that is most strongly recommended across the currently active cortical column population. At each point in the sequence, a different column population will be active as a result of the previous step in the sequence. Hence a sequence could proceed even with damage to the cerebellum because all the recommendation strengths will be in place, but it will be slower and more prone to errors. This is consistent with observations that cerebellar damage results in a generalized tendency towards poorer cognitive performance rather than severe specific deficits (Bracke-Tolkmitt et al. 1989).

3.6 Modelling of Memory and Learning Phenomena

There are a number of types of information recorded in different parts of the brain that are relevant to memory and learning. These information types are summarized in Table 3.1, along with the ways in which the information changes and the physiological instantiation of the information and the change mechanisms.

3.6.1 Receptive Fields Stability and Memory

In general the receptive field of a column can expand slightly with learning but does not contract or make major shifts. This relative stability of receptive fields and their sets of associated behavioural recommendations is the critical factor supporting many observed memory phenomena. If a receptive field is detected within the current sensory input, its associated behavioural recommendations have a high probability of being currently relevant. However, because of receptive field stability, some additional receptive fields that are not detected but can be identified also have a reasonable probability of having relevant behavioural recommendations.

For example, suppose that the receptive field of a column is not being detected and the column is therefore inactive, but the column was recently active at the same time as a number of the columns that are currently active. There is a reasonable probability that the recommendations associated with that column could be relevant. Similarly, if a column is inactive, but has often been active in the past at the same time as a number of the columns that are currently active, again its recommendations could be relevant. Finally, if a column is inactive, but it expanded its receptive field at the same time in the past as a number of currently active columns, its recommendations could also be relevant. Furthermore, the behaviourally relevant past activity of the inactive column could be at the same time as just before or just after the past activity of currently active columns. There is therefore potential behavioural value in indirect activation of columns on the basis of various types of past activity that is temporally correlated activity with the past activity of currently active columns. Such indirectly activated columns could in turn indirectly activate further columns on the basis of past temporally correlated activity and so on. An indirect activation behaviour applied to an already active column would have the effect of prolonging its activity.

Table 3.1 Different types of information that support memory and learning in the brain. The major types are receptive fields in the cortex that record similarity circumstances, recommendation weights in the basal ganglia that associate receptive field detections with behaviours, and behaviour sequences in the cerebellum that associate behaviours with frequent next behaviours. Many of the memory relevant behaviours and next behaviours recorded in the cerebellum are indirect activation behaviours acting on cortical columns

Information type	Changes with time	Physiological instantiation of information	Physiological instantiation of changes
Receptive fields			
Receptive field detecting presence of a type of sensory circumstance	Slightly expands IF only a slight expansion is needed to result in current detection AND *either* the total number of receptive field detections is low *or* contradictory reward feedback has occurred in the past following activation of currently active group of columns	Columns in sensory cortices	LTP mechanism operating on pyramidal neurons within the changing column to add conditions to neuron receptive field definition
Receptive field detecting activity of a group of cortical columns with frequent past temporally correlated activity		Columns in anterior temporal cortices	
Receptive field detecting activity of a group of cortical columns with past temporally correlated receptive field expansion		Columns in cortices associated with the hippocampal system	
Recommendation weights			
Indirect activation of receptive fields on the basis of past temporally correlated activation	Starts off high, initially (minutes) decays rapidly, then (hours and days) more slowly. Decays more slowly if repeated or if an indirect activation is followed by a reward	Weights of connections from columns in the anterior temporal cortex into the basal ganglia	Inputs from nucleus accumbens drive synaptic weight changes in the basal ganglia

(continued)

Table 3.1 (continued)

Information type	Changes with time	Physiological instantiation of information	Physiological instantiation of changes
Indirect activation of receptive fields on the basis of past temporally correlated receptive field expansion	Starts off high, initially (hours and days) decays slowly, then (weeks and months) more slowly. Decays more slowly or increases if an indirect activation is followed by a reward	Weights of connections from columns in the cortices associated with the hippocampal system into the basal ganglia	
Behaviour sequences			
Bias in favour of one behaviour type whenever a behaviour of another type has just been completed	Increases with repetition of sequence if followed by positive reward	Cerebellar purkinje neuron corresponding with the first behaviour that inhibits all except second behaviour	LTD mechanism reducing synaptic weights of sensory circumstances corresponding with completion of first behaviour

However, if such indirect activations took place without any limitations, there would be large numbers of columns indirectly activated in all situations. In some situations, only the directly activated columns are necessary to select appropriate behaviour and indirectly activated columns could reduce the appropriateness of the selection. Hence indirect activations must themselves be behaviours that are recommended by currently active columns, with competitions determining whether such recommendations are accepted.

3.6.2 Development and Evolution of Indirect Activation Recommendation Strengths

If two columns are active at the same time, there is a chance that an indirect activation strength would be useful in the future, but the highest probability of being useful is soon after the simultaneous activity. However, if the simultaneous activity occurs repeatedly, there is a higher chance that an indirect activation would be useful further in the future. If an indirect activation strength is utilized and is followed by a reward, there is an even higher chance of future value.

Hence if two columns are active at the same time, they will immediately acquire recommendation strengths in favour of activating each other in the future, but this strength will decay rapidly with time. If the simultaneous activity often occurs, the recommendation strength will decay more slowly. If the recommendation strength results in an indirect activation and the activation is followed by a positive reward, the weight will be increased and decay will be much slower.

If two columns expand their receptive fields at the same time, there is a much higher probability of future behavioural relevance than for simple simultaneous activity. The initial indirect activation recommendation strengths will therefore decay much more slowly. Reward feedback following use of a recommendation strength will again increase and stabilize the weight. Frequent use of a recommendation weight in favour of activation on the basis of simultaneous receptive field expansion would mean that the columns are often active at similar times, and indirect activation weight on that basis would develop which could even become stronger than the original weight based on receptive field expansion.

3.6.3 Semantic Memory

The primary information mechanism supporting semantic memory is indirect activation of cortical columns on the basis of frequent past simultaneous activity. This will be illustrated by the example of learning the meaning of the word "bird". Visual experiences of different birds result in receptive field detections in column arrays that discriminate between visual elements, between visual features and between visual objects (as in Fig. 3.4). Auditory experiences of the spoken word "bird" result in receptive field detections in column arrays that discriminate between phonemes and between words (as in Fig. 3.4). Because there are some similarities between different birds, there are some columns in the $\approx$visual objects array in Fig. 3.4 that tend to detect their receptive fields relatively frequently in different bird instances. There will be less consistency in the $\approx$visual features array and even less in the $\approx$visual elements array. Similarly, there will be some columns in the $\approx$word array that frequently tend to detect their receptive fields in different auditory experiences of the word "bird".

In Fig. 3.23, the column activations in arrays $\approx$visual objects and $\approx$words are illustrated for a number of experiences a bird was perceived visually at the same time as the word "bird" was heard. In each visual experience, a different group of columns was activated, but because of the similarities between birds, there are some columns that tend to detect their receptive fields relatively frequently in different instances. Similarly, although a different group of auditory columns detect their receptive fields in each "bird" experience, some columns are active relatively frequently. There is therefore a set of $\approx$visual object columns that are frequently active at the same time as a set of $\approx$word columns. These two sets of columns will acquire recommendation strengths in favour of indirectly activating each other.

In any future experience of the word "bird", a significant subset of the auditory set will be activated, and this subset will have a strong total recommendation strength in favour of indirect activation of the visual set. This indirect activation will be experienced as if a bird that was an average over past experiences were experienced. Because there is less consistency in activation at the $\approx$visual features and $\approx$visual elements levels, the experience will not be a visual hallucination; it will be confined

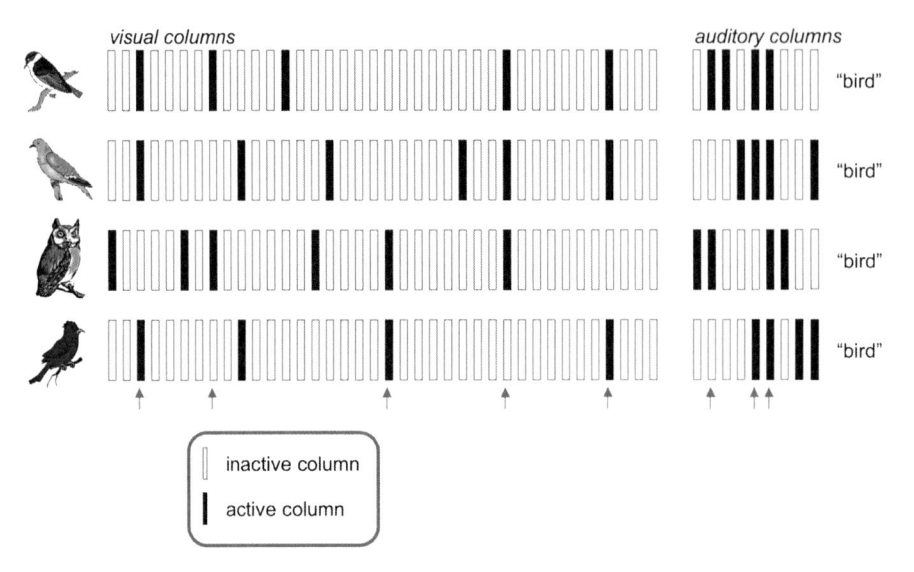

Fig. 3.23 Information model for semantic memory. The column arrays that discriminate between visual objects and that discriminate between auditory words (as shown in Fig. 3.4) are illustrated for four situations. In each situation, a visual instance of a bird and auditory instance of the word "bird" are experienced simultaneously, resulting in activation of the columns that detect their receptive fields. Although the set of columns activated is different for each visual and each auditory experience, the similarity between the experiences means that some receptive fields tend to be detected relatively frequently in many visual bird instances and in many auditory "bird" instances, as indicated by the *grey arrows*. As a result, there is a set of visual columns that are often active at the same time as a set of auditory columns. The set of auditory columns therefore tend to acquire recommendation strengths in favour of indirect activation of the set of visual columns and vice versa. An instance of the word "bird" will contain a significant proportion of the auditory set, and there will therefore be significant total recommendation strength in favour of activating the visual set. The experience of activation of this visual set will be as if a visual bird instance that is an average of past experience were perceived, except that the columns close to visual input (i.e. arrays ≈visual elements and ≈visual features in Fig. 3.4) will not be activated, and the experience will therefore not be a visual hallucination.

to higher receptive field complexities. The activated ≈visual object columns will have all their recommendation strengths, including relatively strong bird appropriate total recommendation strengths.

At a deeper level of description, the information model for this type of indirect activation is illustrated in Fig. 3.24. The indirect activation behaviour must win a competition for selection in the basal ganglia, and the behaviour must be implemented by the thalamus. Columns with receptive fields corresponding with groups of ≈words auditory columns that are often active at the same time are required. These are the linking columns in Fig. 3.24. These linking columns target visual columns that are often active at the same time, but by connections on to basal dendrites of pyramidal neurons in those columns rather than on to the apical dendrites that define receptive fields.

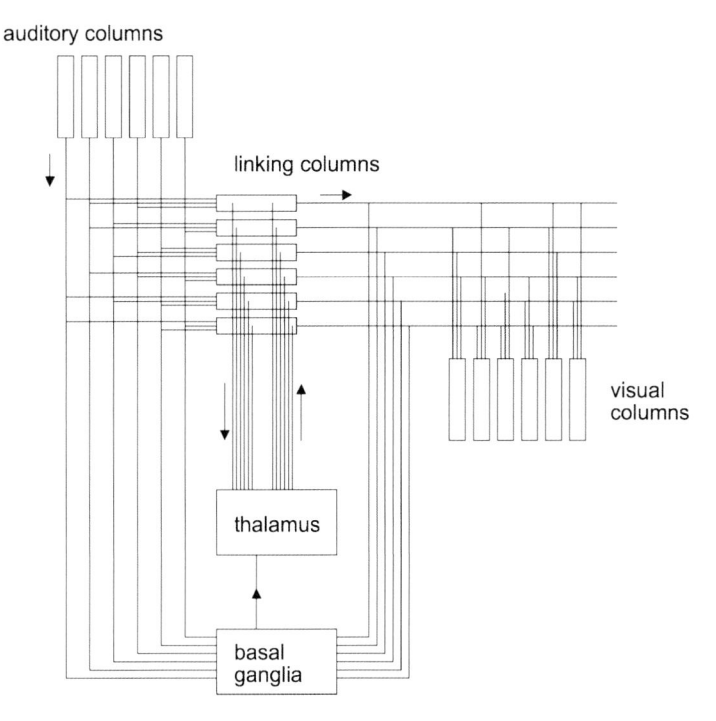

Fig. 3.24 Semantic memory for word meanings supported by management of indirect activation on the basis of frequent past simultaneous activation. The linking columns have receptive fields corresponding with groups of auditory columns that are often active at the same time. The linking columns excite visual columns that are often active at the same time but not as elements in their regular receptive fields. Auditory columns and other columns including the linking columns target the basal ganglia components corresponding with the behaviour of indirect activation on the basis of frequent past simultaneous activity. These connections into these components have different synaptic weights which correspond with recommendation strengths in favour of the indirect activation behaviour. If the behaviour is accepted by the basal ganglia, the outputs from the linking columns are released by the thalamus to drive visual column activations. Note that the outputs from the linking columns recommending the indirect activation behaviour to the basal ganglia are from a different column layer to the outputs to the visual columns

The receptive fields activated in the experience of a semantic memory will be those active during sensory or motor processing, and one receptive field will be present in instances of many different categories of object. Neuroimaging is consistent with this picture (e.g. Martin 2007). However, the activation of a semantic memory requires a cortical area in which columns have receptive fields corresponding with groups of columns elsewhere that are often active at the same time. Damage to this area would result in general loss of semantic memory capability. Damage to the most anterior portions of the temporal cortices results in this type of general loss of semantic memory capabilities (Rogers et al. 2006).

3.6.4 Working Memory

In the recommendation architecture model, the working memory of an object is simple the activation of the columns corresponding with the semantic memory of the object. The primary information mechanism supporting working memory is frequency modulation of receptive field detections within different sources. The cognitive significance of the mechanism can be understood by reference to Fig. 3.25. Suppose there are three objects in the visual field, a dog, a cat and a tree, with the cat up the tree and the dog barking at the cat. The appropriate behaviour in response to the group (e.g. chasing the dog away) needs information about each of the objects that must be derived from receptive fields detected within those objects. For example, relevant information could include whether the cat is my pet, whether the dog looks fierce and the height of the tree. Receptive fields are detected within the three objects in the ≈objects array in Fig. 3.25, but detections must be kept separate in this array (e.g. must not have information indicating a barking cat, or a fierce tree, etc.). However, the information derived from the three objects must be integrated in the ≈groups array to activate columns with appropriate recommendation strengths.

The frequency modulation mechanism makes it possible to place a different phase of modulation on the receptive fields detected within the dog, cat and tree. As a result, the receptive field detections are kept independent, even though they are all active in the same cortical array, potentially in some of the same columns or even some of the same pyramidal neurons. When all three populations are active, their outputs can be brought into the same phase and released to the ≈groups array where receptive fields within the group are detected.

This type of mechanism supports the observed working memory limits on the number of different objects that can be retained in working memory at the same time. Cowan (2000) suggested a limit of approximately four items, based on

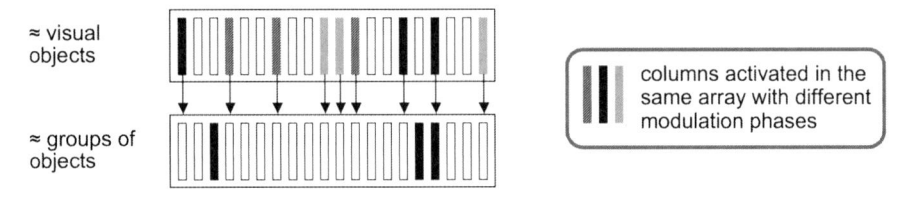

Fig. 3.25 Cognitive significance of frequency modulation in maintaining independent populations of active cortical columns. Two arrays of visual columns are illustrated, one at a level of receptive field complexity effective for discriminating between objects, the other groups of objects. The receptive fields that discriminate between groups of objects need inputs from receptive fields that discriminate between objects. The modulation mechanism makes it possible for attention to be shifted to three different objects with column populations in response to the different objects being maintained active simultaneously but without interaction in the ≈objects array until all the members of the group have been attended. The outputs from the three populations are then brought into the same phase and released to the ≈groups array, where columns with recommendation strengths appropriate for responding to the group of objects then detect their receptive fields

performance discontinuities such as errorless performance in immediate recall when the number of items is less than four and sharp increases in errors for larger numbers. In information model terms, the primary limit is defined by the number of active column populations which can be maintained independently. This number is approximately the ratio of the period of the modulation frequency to the postsynaptic potential decay constant. If the gamma frequency in the EEG is interpreted at the modulation signal on the basis of its association with attention (Muller et al. 2000), a 40 Hz signal has a period of 25 ms. With a postsynaptic decay constant of 10 ms (Fan et al. 2005), this would allow of the order of 2–3 independent column populations. The actual number would depend upon the gamma band frequency (which can vary from 20 to 50 Hz) and the shape of the postsynaptic potential decay curve.

The content of an item in working memory is defined by a group of columnar receptive fields in a cortical area that are active at the same phase of frequency modulation. The column receptive fields are the same fields that record the information which makes up the content of semantic and episodic memories. This is consistent with the observations that there is considerable overlap in the cortical areas active during working memory and declarative memory tasks (Cabeza et al. 2002).

At a psychological level, the requirement is to keep information relevant to different objects active simultaneously, without interference. In particular, it is sometimes important to allow processing of a new sensory object while preserving information about an earlier object. However, it will not in general be necessary to keep information on the earlier object active at all levels of receptive field complexity, only at the levels with relevant behavioural recommendation strengths. In addition, there is a biological cost to maintaining information active, both the direct cost of the activation and the indirect cost of occupying resources that therefore cannot be used for other purposes. Hence maintaining information active in a specific cortical area must be a behaviour that will compete with other alternative behaviours for acceptance.

This competition for acceptance will depend upon what else is already active in different cortical areas. There is therefore a requirement for a cortical area with receptive fields corresponding with, for example, the number of different groups of columns in another area that are active with different phases of frequency modulation. Damage to this area would result in general deterioration of working memory capabilities. Neuroimaging indicates higher activity in Brodmann's area 40 in the left parietal cortex during working memory tasks (Cabeza et al. 2002), and a patient with damage to the left parietal lobe showed a deficit in working memory but not declarative memory (Warrington and Shallice 1969; Shallice and Warrington 1970).

If we are asked to imagine a number of different objects, keeping the images separate, the maximum number is 3–4, consistent with Cowan's (2000) limit. However, objects can be remembered as groups, and it is possible to remember objects as both visual and verbal terms. Cycling between these representations makes it possible to increase the apparent size of working memory, hence the "magic number" of seven (Miller 1956).

This model has some analogies with the proposal that different working memories are encoded as a different high frequency (40 Hz) subcycle of a low frequency

(10 Hz) oscillation (Lisman and Idiart 1995). However, their model leads to a working memory content of seven, which is significantly higher that the actual experimental number.

3.6.5 Episodic Memory

The primary information mechanism supporting episodic memory is indirect activation of cortical columns on the basis of past simultaneous receptive field expansion. This will be illustrated by the example of recalling the news of the terrorist attack exploding a bomb in a bar in Bali. The state of column activation at some point while television news of the bombing was being viewed is shown in Fig. 3.26.

In the figure, attention is paid to a sequence of visual objects on the television screen. Different sets of cortical columns in the ≈objects array detect their receptive

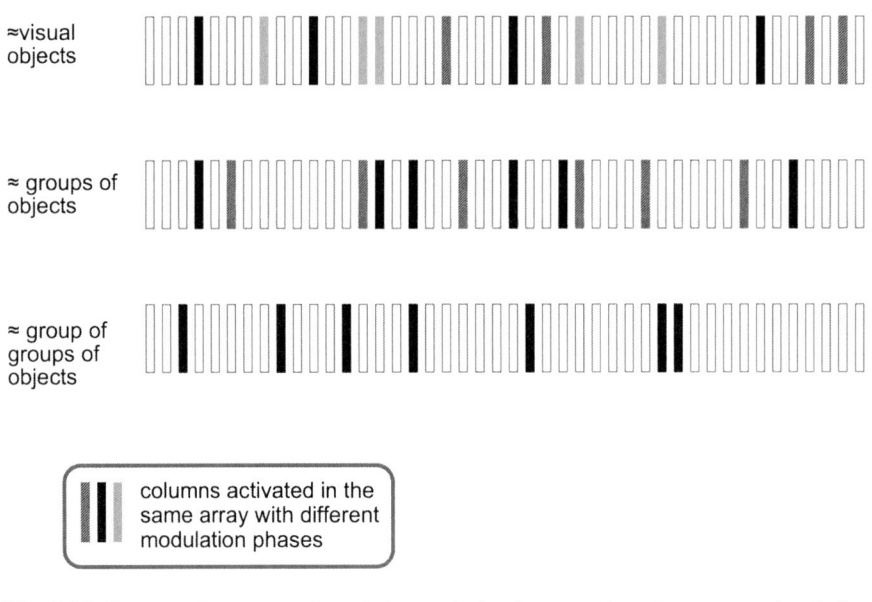

Fig. 3.26 Conceptual representation of the cortical columns activated at some point during viewing of television news of the first Bali bombing. The pictured activation was built up by a series of steps. Several objects on the television screen were sequentially the focus of attention, resulting in separate activations in the ≈objects array, at different phases of frequency modulation. Outputs from these separate activations were synchronized to the same phase and released to the ≈groups array, an active column population was generated in that array at a specific phase of frequency modulation, and the populations in the ≈objects array were extinguished. More visual objects generated activations in the ≈objects array, their outputs were again synchronized, and a second independent population generated in the ≈groups array at a different modulation phase from the first. In due course, the outputs from several independent populations in the ≈groups array were synchronized to the same phase and released to the ≈groups of groups array where a column population was activated

fields within each object, and several sets corresponding with different objects are maintained active simultaneously at different phases of frequency modulation. Outputs from the different sets are then synchronized and released to the $\approx$groups array where a set of columns detect receptive fields. The outputs from several sets of columns in the $\approx$groups array are in due course synchronized and released to the $\approx$groups of groups array.

There is relatively little novelty to the individual visual objects, which might include objects typical of Bali, of bars and of explosion aftermath, all of which are separately familiar. Hence little receptive field expansion is required in the $\approx$objects array to achieve the minimum required number of column receptive field detections. However, at the $\approx$groups level there will be rather more novelty, and hence a significant degree of receptive field expansion will be required, while at the $\approx$groups of groups level there will be significant novelty, and substantial numbers of receptive field expansions will be required.

Now suppose that later the words "Bali" and "bombing" are heard. By the semantic memory mechanism, sets of columns will be indirectly activated in the $\approx$visual objects array corresponding with the visual experiences that have often occurred at the same time as the two words. The outputs from these sets would be synchronized and released to the $\approx$groups array where they would generate a set of active columns. However, the overlap between the resultant column populations and the columns active in Fig. 3.26 will be slight.

If now the recommendations of the currently active columns in favour of indirect activation of other columns on the basis of past simultaneous receptive field expansion are accepted, a secondary population will be activated. This secondary population will contain columns that have expanded their receptive fields at the same time in the past as a significant subset of the columns activated in response to the words. A tertiary population could then be generated on the same basis from the secondary population. This sequence of indirect activations will tend to result in a population in which all columns tend to have expanded their receptive fields at the same time in the past. If the initial words are well chosen the final population will be a good approximation to the population active at the time of the original experience, but without the activations in the earlier cortical arrays ($\approx$visual features and $\approx$visual elements) in which the degree of receptive field expansion will be minimal. Nevertheless, this reconstructed population will have appropriate recommendation strengths in favour of describing what was seen. In other words, an episodic memory of the original experience has been constructed.

Column recommendation strengths in favour of activating other columns that expanded their receptive fields just before or just after expansion in the column provide the ability to move through the sequence of experiences to reconstruct an episode.

As discussed earlier, cortical columns in the parahippocampal, perirhinal and entorhinal cortices and pyramidal neurons in CA1 and CA3 define receptive fields corresponding with groups of cortical columns that have tended to expand their receptive fields at the same time. These receptive field definitions result from the primary resource management role of the hippocampal system. Hence these hippocampal system structures can be used to reconstruct episodic memories, with the restriction that outputs that drive receptive field expansions cannot be used without generating unnecessary and undesirable receptive field changes. The outputs to drive episodic memory recall must therefore come from different layers in the PHC, PRC and EC columns from those layers driving receptive field changes.

Note that if an episode is often recalled, there may be sufficient recommendation strengths acquired on the basis of frequent past simultaneous activity for that mechanism to take over from the receptive field expansion-based mechanism. This accounts for the observations that hippocampal damage has the largest effect on autobiographical memories and less effect on memories of personal information, notable public personalities and notable public events that may have been frequently recalled (Nadel and Moscovitch 1997).

This model of episodic memory recall involves a specific sequence of cortical behaviours: generation of activations in response to words; a series of indirect activations on the basis of past simultaneous receptive field expansion; evolution of the end population on the basis of slightly later receptive field expansion; and generation of speech driven by the final population. This sequence must be learned and later performed rapidly and effectively. This sequence will therefore be instantiated in the cerebellum.

Episodic memory thus requires activity in cortical areas with receptive fields that can discriminate between different types of groups and different types of groups of sensory objects. These types of receptive fields are polymodal and therefore located in the frontal cortex. Episodic memory will also require activity in the higher visual areas to provide receptive fields corresponding with individual objects and in the hippocampal system to manage the activation of the appropriate receptive fields in all the other cortical areas. The hippocampal activity would be expected to be targetted at specific groups of cortical columns and therefore smaller. Because episodic memory retrieval requires a specific sequence of indirect activation behaviours, cerebellar activity is therefore to be expected. Functional neuroimaging of the brain during episodic memory recall is consistent with this picture, with strong activity in the prefrontal cortex (Fletcher et al. 1997) and in visual and visual association areas (Addis et al. 2007); somewhat weaker activity in the hippocampal system (Fletcher et al. 1997); and strong cerebellar activity (Fliessbach et al. 2007).

3.6.6 Priming Memory

The primary information mechanism supporting priming memory is indirect activation of cortical columns on the basis of recent simultaneous receptive field detection. This will be illustrated by the example of identification of pictures following subliminal presentation. In one experiment (Badgaiyan 2000), subjects are shown a number of line drawings, studying each for 3 s. A few minutes after completing this study phase, the subjects are shown a series of pictures, each for 16 ms, and asked to identify them. The subjects had an 82% success rate for studied pictures, but only a 5% success rate for identifying new unstudied pictures.

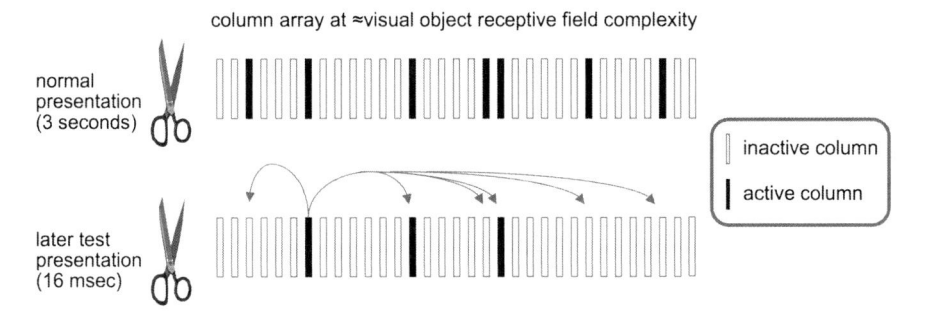

Fig. 3.27 Column activations during study and test phases of a priming experiment. The study phase presentation results in a substantial set of column activations. A later test presentation is very brief, and only a small subset of the columns activated at study is activated. However, these columns have significant recommendation strengths in favour of activation of the other columns in the full set, on the basis of recent simultaneous activity

As illustrated conceptually in Fig. 3.27, the study presentation results in activation of a normal sized set of columns. The very brief test presentation results on activation of a small subset of the original set that in general does not have enough total recommendation strength in favour of naming the picture. However, if the picture has been recently studied, the columns in the subset all have strong recommendation strengths in favour of activation of the other columns in the full set on the basis of recent simultaneous activity. The resultant larger set of columns often has enough total recommendation strength to identify the picture.

Indirect activations based on recent simultaneous activity decay fairly rapidly unless repeated or used in a situation that result in reward feedback. Priming memory is the situation in which neither of these factors is present. Semantic memory can start from the same initial weights, but repetition and reward feedback strengthens and stabilizes the weights long term.

3.6.7 Procedural Memory

The primary information mechanism supporting procedural memory is adjustment of the behavioural recommendation weights of receptive field detections on the basis of reward feedback. In learning a new skill, if previously existing receptive fields have enough discrimination to distinguish between situations in which different skilled behaviours are appropriate, the skill could be learned by adjustment to recommendation weights in the basal ganglia without changes to cortical receptive fields. If the sensory environment in which the skill is practiced is novel, some receptive field changes will be required. The model thus accounts for the observations that patients who have lost the ability to create new declarative memories can nevertheless acquire simple mechanical skills (Corkin 1968).

The symptoms of a number of disorders associated with the basal ganglia, such as Parkinson's and Huntingdon's syndromes, can be understood in terms of the information model for the basal ganglia shown in Fig. 3.16.

The symptoms of Parkinson's disease include difficulty with voluntary movement and with initiation of movement and in general slowness of movement. The observed physical deficit (Jankovic 2008) is degeneration of dopaminergic neurons in the SNc. In the information model for the basal ganglia, this will mean that the loop that ensures that one and only one behaviour is selected in response to each sensory input state is unbalanced. The lack of dopamine will result in increased activity in the indirect pathway (Fig. 3.16), leading to no behaviour being selected.

The major symptom of Huntingdon's disease is the intrusion of irregular, unpredictable, purposeless, rapid movements that flow randomly from one body part to another (Berardelli et al. 1999). The observed physical deficit is loss of striatal cells that project into the indirect pathway (Starr et al. 2008). In the information model for the basal ganglia, this can be understood as reducing the inhibition of all behaviours. Any individual receptive field detection by the cortex will recommend a wide range of behaviours. The weakening of the selection management leads to the selection of multiple behaviours. The motor system can only implement one or a consistent set of behaviours, but the ultimate selection will be fairly random.

3.7   Mapping Between Different Levels of Description

As an illustration of how the recommendation architecture model makes it possible to describe the same phenomenon consistently on different levels of detail from psychology to physiology, consider again an example of episodic memory.

At the highest level, the causal description begins with a subject with a range of sensory inputs including the verbal input "What do you remember about your first day at your current job?" In the situation of the subject, the verbal input causes the subject to give priority to answering the enquiry over other possible demands. This priority causes attention to be paid to the verbal input. The verbal input causes a sequence of internal brain activities. The end point of these brain activities causes a verbal response describing the suggested event.

At a more detailed level, a range of sensory input is reaching the subject, including auditory input of the spoken words. A range of receptive fields are detected by the cortex within the words and other sensory inputs, each with a range of behavioural recommendation weights into the basal ganglia. There is substantial total recommendation weight in favour of paying attention to the auditory inputs, and as a result, the auditory information is released by the thalamus for more detailed receptive field detections by the cortex. More detailed cortical receptive fields detected within the word "remember" and receptive fields indirectly activated as a result have substantial recommendation weight in favour of an episodic memory behaviour. This weight is the current largest, and therefore a sequence of behaviours supporting episodic memory is activated in the cerebellum. This sequence drives indirect activation behaviours in the cortex, using information recorded in the hippocampal system, and favours acceptance of speech recommendation weights of the final population.

At the next level of detail, the description of the complete end to end psychological process would be very lengthy. As in the physical sciences, understanding is based on establishing more detailed descriptions of key small segments of the overall high level process.

One segment could be a description of the handling of the sequence of words making up the question. Receptive fields detected within the auditory input recommend a standard speech processing behavioural sequence recorded in the cerebellum. Such a behavioural sequence generates bias in favour of the weights of currently detected receptive fields in favour of the next behaviour in the sequence. The sequence of behaviours would be indirect activation of visual and associative receptive fields by the auditory receptive fields detected within a word, prolonging the indirectly activated population with a specific phase of frequency modulation, generating an indirectly activated population in response to a second word and prolonging it with a different phase of frequency modulation, and once several words have corresponding indirectly activated populations, bringing their outputs into the same modulation phase and releasing them to a cortical level detecting a higher level of receptive field complexity.

A second segment would be to take the population indirectly activated at the higher level of receptive field complexity by the combination of words and drive a series of further indirect activations on the basis of simultaneous past receptive field expansion, using information recorded in the hippocampal system. Again, this segment would be a sequence of behaviours instantiated in the cerebellum. The segment would include releasing the outputs of the population to the hippocampal system, where columns would detect receptive fields if a significant proportion of the columns making up their receptive fields were active. The active hippocampal columns would then encourage activation of all the cortical columns defining their receptive field. Some cortical columns could appear in multiple activated hippocampal receptive fields, and such cortical columns would be most likely to be activated. All the releases of column outputs would be managed by the basal ganglia and thalamus on the basis of active column recommendation weights, modulated by the bias placed on weights in favour of certain behavioural types by the behaviour sequence instantiated in the cerebellum.

At an even deeper level, there could be descriptions of the processes for column receptive field detection, for behaviour selection through the basal ganglia and thalamus, for management of behaviour sequence through the cerebellum and for hippocampal system management of receptive field expansions.

At a yet deeper level, descriptions would be in terms of neuron receptive field definition using LTP type mechanisms, neuron receptive field detection using leaky integration and frequency modulation.

An even deeper level could describe synapses, neurotransmitters and ion channels supporting leaky integration at a higher level, and the changes to ion channels supporting LTP mechanisms.

The model thus has the capability to map consistently all the way from psychology to the chemical processes underlying physiology. This consistent mapping means that it is possible to have confidence in the intermediate level descriptions (e.g. in terms of cortical columns) which are most critical for understanding the psychological phenomena.

3.8 More Complex Cognitive Processes

The focus of this chapter has been on memory and learning phenomena. However, in this section an outline will be provided of how the mechanisms discussed for memory and learning relate to major cognitive management systems such as attention and emotion, and how the mechanisms support more complex cognitive phenomena including imagination and creativity.

3.8.1 Attention

The attention function on a psychological level selects a subset of the currently available sensory information and favours that subset in the determination of behaviour. Unmodulated sensory input from the whole visual field enters the cortex via the thalamus. Because the inputs are unmodulated, they do not penetrate deeply into the visual cortex. Receptive field detection is mainly in V1, where receptive fields correspond with boundary elements in different boundaries within the visual field. Each such receptive field recommends focussing attention on the retinal area in a band perpendicular to the boundary element and on both of its sides as illustrated in Fig. 3.28.

For a closed boundary, the recommendation strengths within the boundary from all the surrounding boundary elements will reinforce each other, and there will be a strong total recommendation strength in favour of the area within the boundary, in other words, an object in the visual field.

The strongest such recommendation, totalled across all currently detected receptive fields, is accepted. This acceptance is implemented by placing a modulation on the sensory inputs within the corresponding closed boundary. The effect of the modulation is that receptive fields are detected deep into the visual cortex and beyond. These receptive field detections recommend behaviours appropriate to the visual object within the selected closed boundary.

Receptive field detections within multiple objects can be retained active without interference with each other by modulating the detections at different phases of modulation as illustrated in Fig. 3.13. This support of multiple sets of receptive field detections supports working memory as discussed earlier.

Analogous mechanisms support attention in auditory and propioceptic sensory processing.

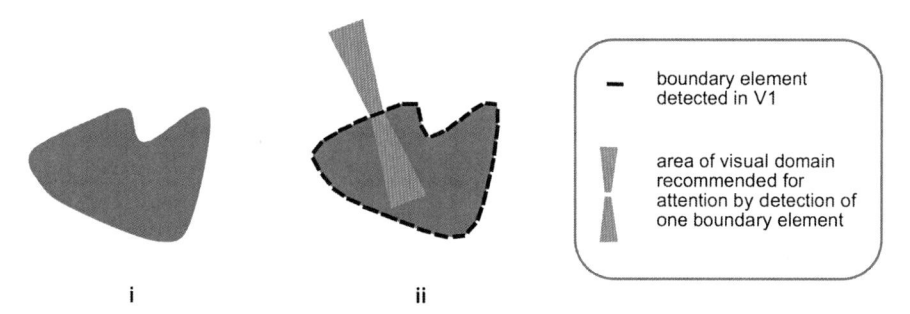

Fig. 3.28 Management of attention by boundary element detection in V1. In (i), a visual field is illustrated with one object. In (ii), some of the boundary element receptive fields detected in visual area V1 within the object illustrated in (i) are illustrated, along with the retinal area recommended for attention by one receptive field detection. Within a closed boundary, attention recommendations will reinforce each other, resulting in a strong overall recommendation in favour of attention on the area within the closed boundary

3.8.2 Emotion and Reward

Different cortical areas detect receptive fields within different ranges of complexity. Because of the need to economize on resources, receptive field detections in any area recommend many different behaviours. However, as discussed earlier, there may be behavioural advantages in having areas that specialize in receptive fields effective in discriminating between circumstances in which different behaviours of a particular general type (e.g. food seeking, aggressive, avoidance, etc.) are appropriate. Sometimes the behavioural advantage of such areas may outweigh the extra resource cost. Such areas have the additional advantage that the relative probability of their behaviour type can be modulated.

Emotions correspond with different general types of behaviour, and the presence of the emotion encourages its corresponding type. Anger encourages aggressive behaviour, fear encourages avoidance (i.e. fearful) behaviour, disgust encourages rejection behaviour, surprise encourages recording information and delaying behaviour, sadness encourages avoidance of behaviour because of radical change in circumstances, happiness encourages repetition of recent behaviours and hunger encourages food seeking behaviour, etc.

Some cortical receptive field detections recommend such general types of behaviour. These recommendations are instantiated by connections to the amygdala and/or hypothalamus, with different weights. If the total recommendation weight in favour of a behaviour type is high enough, the amygdala and/or hypothalamus will be activated to release a neurotransmitter to targets in the cortex within areas that generate more specific behavioural recommendations of the general type. Such a neurotransmitter release corresponds with an emotional state and will have the effect of modulating the thresholds of pyramidal neurons in the targetted areas. This modulation will result in much more activity recommending specific behaviours of the general type.

Emotion signals generated by the amygdala and hypothalamus also target the striatum, where they influence the relative probability of a behaviour of their corresponding type being accepted. Thus, for example, the basolateral complex of the amygdala, which is associated with fear, projects strongly to both the striatum and the prefrontal cortex (Sah et al. 2003).

Note that if a set of columns recommending an emotion are active at the same time as a set of cortical columns elsewhere in the cortex, the two sets are active at the same time and can therefore acquire recommendation strengths in favour of activating each other. At a psychological level, this means that, for example, the memory of an event may generate an emotion.

Note also that the generation of a reward that modulates recent recommendation strengths is itself a behaviour that is in general recommended by cortical receptive field detections. If a set of columns are active at the same time as some reward recommending columns, the set may acquire recommendation strengths in favour of indirect activation of the reward columns. The circumstances in which a particular type of behaviour is appropriate will be a very complex combination of sensory inputs able to discriminate between subtle social differences. Reward discrimination in many cases also requires subtle discrimination between different social circumstances. The receptive field complexity for the two types of discrimination may be comparable, resulting in the same cortical area being associated with both types of behavioural recommendation. The orbitofrontal cortex does appear to be involved in both emotion and rewards (O'Doherty et al. 1991).

3.8.3 Sleep

A major role of sleep in the recommendation architecture model is the configuration of neural resources that are as appropriate as possible for learning in subsequent waking periods (Coward 1990, 2001). Recording of information in the brain occurs by creation of new synapses and by changes to synaptic weights. If new synaptic connections are needed, it would not be practical to create those synapses at the instant they were required. It is therefore necessary to create "provisional" synapses in advance, which can be utilized if appropriate during subsequent experience. Such provisional synapses could be between randomly selected pre- and postsynaptic neurons, but this would result in considerable resource expenditure on the creation of synapses which are not useful and which in some circumstances could reduce behavioural effectiveness (Coward 2001).

One way to improve the probable effectiveness of provisional synapses is to utilize past experience. If two neurons have never in the past been active at the same time, the probability that a connection between them will be useful is low. Conversely, if two unconnected neurons have often been active at the same time in the past, the probability is higher. If the simultaneous activity is recent, the probability is even higher. The strategy is therefore to bias the creation of provisional synapses in favour of connections between neurons that have tended to be active at similar times in the past, with the recent past having the strongest weight. Provisional connectivity will also be required to support indirect activation on the basis of past temporally correlated activity, such as activity of one neuron at the same time as or shortly after another neuron.

Connectivity is required between cortical neurons and between cortical and hippocampal neurons to support declarative memory. Connectivity is also required between cortical neurons and basal ganglia and thalamic neurons, and within the basal ganglia and thalamus, to support procedural memory. Different stages of sleep may support different aspects of provisional connectivity creation, including extension of axon segments and building of synapses and connectivity in different brain structures. Partial reruns of past experience will be required to identify the best candidates for provisional connectivity.

Consistent with the model, reruns of past patterns of neuron activity have been observed in both REM sleep (Pavlides and Winson 1989) and in slow wave sleep (SWS) (Lee and Wilson 2002). There is some evidence that reruns of recent experience occur in SWS, with reruns of more remote experience in REM sleep (Hoffman and McNaughton 2002).

This model is radically different from consolidation models (Squire and Alvarez 1995) in which memory information is initially recorded in the hippocampus and gradually over time transferred to the cortex. Such information transfers would be very complex, and modelling efforts have generally been limited to just the hippocampus (e.g. Redish and Touretzky, 1998). The resource configuration model does not require these complex information transfers, memories are fully defined in cortical information terms at the end of the few hundred millisecond period of the remembered experience. There may be chemical processes required to consolidate the memory (Tronson and Taylor 2007), but the information content of the memory is not significantly changed or relocated. The hippocampus is only required in the future to identify the sets of cortical information associated with individual episodic memories. Furthermore, the rerun process is not indispensible for declarative memories, since random connectivity can support such memories with reduced resource effectiveness. The resource configuration model is therefore more consistent with the observation that substantial or complete suppression of REM sleep by various antidepressant drugs or by bilateral damage to the pons has no apparent effect on declarative memory (Vertes and Eastman 2000).

3.8.4 Mental Image Manipulation

Consider how a brain might be able to imagine the appearance of a currently viewed object as it would appear if viewed from a different angle. As an example, suppose that a subject is presented with an image of an object viewed from an unfamiliar angle and is asked to identify the object. The requirement is to be able to mentally rotate the current visual image to generate a mental image that is more recognizable.

In the subject's past experience, many different objects have been examined from many different angles. During these examinations, the cortical columns activated in response to viewing from one angle have been active shortly before or after the columns activated in response to viewing from a different angle. As a result, columns will have acquired recommendation strengths in favour of activating other columns on the basis of frequent past activity just before and just after the other columns.

The behaviour of "mentally rotating" is therefore one of imposing a bias on the behavioural selection process in favour of column recommendations of indirect activation on the basis of frequent past activity just before and just after the other columns. Any one column will have been activated in the past in response to many different objects. There will be a population of columns currently active in response to the visual image. The effect of this recommendation type will be activation of a population of columns, each of which has often been active in the past just before or after a number of these currently active columns. This indirectly activated (or secondary) population will be maintained active at a different phase of frequency modulation from the directly activated columns, to avoid meaningless mixing of information.

The secondary population will therefore correspond with an estimate of what the object would look like from a different angle based on an average of many past experiences of viewing rotations of different objects. The recommendation strengths of this population in favour of naming an object will be accepted. This process is of course not guaranteed to work, since it depends upon the existence of reasonably frequent, relevant past experiences.

A key point is that the brain does not "know about" rotations, but can implement processes based on indirect column activations. Certain processes generate results that are rewarded. For example, when initially a child is asked to imagine an object as it would be if rotated, various indirect activation processes may be tried randomly. Once the appropriate indirect activation process is tried, the result will be rewarded, and the process will tend to be used again in the future if the word "rotated" is heard.

3.8.5 Self-Awareness

Consider a child who is learning his name, and suppose that name is Michael. The name is used on many occasions when his attention is being directed towards himself. When his attention is directed towards himself, cortical columns detect their receptive fields within visual information derived from looking at himself and within proprioceptic information derived from his own body ("that's Michael's foot"). Other columns may be active recommending various emotions. In addition, the word "Michael" activates auditory column receptive fields.

Over a period of time, there will be some columns (visual, proprioceptic, polymodal, emotional, etc.) that are often active at the same time as a group of auditory columns often active when the word "Michael" is spoken. Hence the auditory columns will acquire recommendation strengths in favour of indirect activation of the visual, propioceptic, polymodal and emotional columns. The result is that the auditory columns activate an internal state that is a kind of average of Michael's experiences at the times when his name has been used.

Michael has often heard the word "boy" when his visual attention has been directed towards various boys, and frequent simultaneous column activity means that the columns activated in response to hearing the word will tend to activate a set of visual columns that are an average of the boys Michael was looking at when he heard the word.

Furthermore, Michael sometimes hears the words "Michael is a boy". At this point, the visual columns indirectly activated by hearing "boy" are active at the same time as the auditory columns within the word "Michael". Hence those auditory columns will acquire recommendation strengths in favour of activating "boy" visual columns. Finally, Michael sometimes looks in a mirror and is told "that's Michael", and again the auditory receptive fields detected within the word are active at the same time as the visual receptive fields detected within the reflection of Michael.

The result is that hearing the word "Michael" can result in activation of a population of columns that are a kind of average of Michael's emotions, visual inputs and body awareness when his attention was on himself viewed from the "inside", plus a population of columns that are a kind of average of Michael's sensory experiences of other boys and himself viewed from the "outside". These two self images will generally be activated at different phases of frequency modulation from each other and from current sensory inputs, etc. so that they will not be confused. The "inside" and "outside" images correspond with the "me" and the "I" perspectives (Jaynes 1976), the associations with the "me" and "I" words being again created by frequent simultaneous activity of columns.

3.8.6 Imagination

Suppose that someone is asked "imagine that you are at a party with Bill Clinton". The word "you" drives activation of a self column population as described in the previous section. This activation could be at many different levels of column receptive field complexity. The words "Bill Clinton" drive activation of a column population on the basis of frequent past simultaneous activity with the auditory columns, corresponding with mental images that are an average of past visual observations of Bill Clinton on television, columns activated in response to stories about Bill Clinton, etc. The word "party" will similarly activate a population of columns including, for example, columns active during past parties actually attended.

There will be strong active column populations in the $\approx visual\ objects$ and $\approx groups\ of\ objects$ cortical areas, because self is familiar, Bill Clinton and some actions of Bill Clinton are familiar, and parties are familiar. However, the populations in the $\approx groups\ of\ objects$ and especially in the $\approx groups\ of\ groups$, etc. areas will have some weaknesses, because no party including self and Bill Clinton has been attended. These weaknesses will drive receptive field expansions, which because they are adding conditions containing information derived from self, parties and Bill Clinton will be the type of expansions that would occur if a party with Bill Clinton present was attended. However, there will be little related activity in the cortical areas very close to sensory inputs, so the imagining will not be a visual hallucination. The result will therefore be an experience as if recalling a party with Bill Clinton.

Because there has been receptive field expansion, it will be possible to recall imagining the event. However, in such a recall the level of activity close to sensory inputs will be even less than in recall of a real experience, so in general it will not be confused with a real memory.

Initially, the imagining will be at a relatively high level of receptive field complexity, and therefore experienced as a relatively abstract experience. However, the active receptive fields will have recommendation strengths to activate other receptive fields on the basis of frequent past simultaneous activity. Some simpler receptive fields closer to sensory input will have been active in the past at the same time as the higher level receptive fields and will therefore be activated. If the degree of activity at the simpler level is small, there could be receptive field expansion at this simpler level. Hence a less abstract experience can be generated. It is possible that such a process could lead to an imagined experience that in retrospect was indistinguishable from a real experience, in other words a false memory as observed by Loftus and Pickrell (1995).

This model is consistent with the observations that the brain regions activated in response to recalling a real event are very similar to the regions activated in response to imagining an event that has never happened (Addis et al. 2007).

The capability to imagine events well outside of actual past experience at anything other than a very abstract level is one aspect of what is called creativity. This capability will depend on whether some required receptive field expansions are possible.

The column activations that support detailed imagination are indirect activations of simpler receptive fields on the basis of past simultaneous activity with more complex receptive fields that are already active.

For a simpler receptive field to be activated, it must have been active in the past at the same time as a significant number of the more complex receptive fields. For an imagined experience that is radically different from any past experience, there will be relatively few simpler receptive fields that have the required level of temporally correlated past activity. Hence the total activity in the cortical area within which the columns with the simpler receptive fields are located will be lower than the minimum required level. The inputs driving activation in this area are not regular sensory but indirect activation inputs. Receptive field expansions could occur to increase the degree of area activity, but these expansions would be new groupings of currently active columns at the more complex level. In information terms, this introduces a kind of semantic memory based not on actual experience but on imagined experience.

Receptive field expansions can only occur if at the pyramidal neuron level there are adequate provisional conditions. In this case, such provisional conditions would be combinations of higher complexity columns supporting indirect activation. Such provisional conditions would need to be created during sleep processing.

There are behavioural benefits and costs to the creation of such provisional conditions, which are essentially opposite sides of the same coin. The benefit is the ability to imagine circumstances well outside past experience, and therefore, for example, the ability to plan for such unusual circumstances. The cost is the introduction of indirect activation capabilities that are not supported by actual experience and may be misleading for future planning. The degree to which a brain supports such provisional condition configuration in the hippocampal system, the anterior temporal cortex and the monomodal sensory and polymodal cortices is therefore a compromise which may be different for different brains. This compromise defines the degree of creativity which the brain will support.

3.8.7 Planning

To illustrate planning, suppose that someone is asked "What will you do after the concert?" The words "you" and "concert" activate a primary population of auditory cortical columns, which in turn indirectly activate a secondary population of visual and polymodal columns on many levels of receptive field complexity. Hearing the words "what", "do" and "after" encourages favouring a behaviour sequence recorded in the cerebellum. This sequence biases behaviour acceptance in the basal ganglia and thalamus in favour of a sequence of different indirect activation behaviours.

The first step in the sequence biases acceptance of recommendation strengths in the secondary population of columns in favour of activation of other columns often active in the past after the columns active in the secondary population were active. The effect will be to activate a tertiary population made up of fragments of populations that were active during different past "after concert" experiences.

The next step in the sequence biases acceptance of recommendation strengths in the tertiary population in favour of activation of other columns that expanded their receptive fields in the past at the same time as the columns active in the tertiary population. This step also preserves the activity of the tertiary population, at a different modulation phase from the developing quaternary population. The effect is to establish a quaternary population approximating to the population active during the event corresponding with the fragment with the strongest total recommendation strengths in the tertiary population. In other words, this quaternary population corresponds with the past "self after concert" experience with the largest active column representation in the tertiary population.

The next step in the sequence biases acceptance of recommendation strengths in the quaternary population in favour of activation of other columns that were often active in the past after the columns in the quaternary population. The effect is to establish a quinary population approximating to the population active a little later than the population active during the "self after concert" past experience. If this population contains a lot of columns active in the past at the same time as columns recommending a positive reward, the quaternary population becomes the "plan" for after concert.

If there is little positive reward recommendation in the quinary population, the brain can go back to the tertiary population (made up of fragments of different "after concert" experiences) and generate another population by activation of other columns that expanded their receptive fields in the past at the same time as the columns active in the tertiary population. However, in this case, there is a bias placed against recently active columns, resulting in the activation of a population corresponding with the past "self after concert" experience with the second largest active column representation in the tertiary population. A further population is indirectly activated to test reward level after the new quaternary population.

This tertiary to quaternary to quinary process can be repeated until a high reward level is found in a quinary population, at which point the final quaternary population becomes the "plan". The final step is to repeat the evolution from the "self after concert" population to the final quaternary population.

Note that for simplicity, more complex sequences involving imaginary scenarios developed in a similar way to that described in the previous section could also be developed and evaluated for overall final positive reward recommendation strengths. Also, in reality there will be intermediate populations approximating to different stages in past experiences. Such intermediate populations can be added to the above description without difficulty.

At the end of the concert, a population with a fair amount of overlap with the tertiary population generated earlier will develop. Because the populations corresponding with the plan will have recently been active shortly after that population, acceptance of recommendation strengths in favour of activation on the basis of recent activity shortly after the currently active columns will result in development of the column populations corresponding with the plan.

All active column populations will also have recommendation strengths in favour of motor behaviours to get to the next stage in the experience. However, during the concert, these recommendation strengths will not be the strongest in total. In the new sensory circumstances corresponding with the end of the concert, these motor behaviours will have the predominant total recommendation strengths.

3.8.8 *Stream of Consciousness*

Perhaps the classical definition of human consciousness is the experience of a stream of mental images relatively unrelated to current sensory inputs. According to James, the experience is of a stage with relatively consistent mental images, separated by periods of vague evolution (James 1892).

In the recommendation architecture, the starting point for this process is some population of columns directly activated by the presence of their receptive fields within current sensory inputs. Each activated column has a set of recommendation strengths in favour of activating other columns on the basis of past temporally correlated activity which could be recent activity at the same time, before or after, frequent past activity at the same time, before or after, or past receptive field expansion at the same time, before or after.

If such indirect expansions are encouraged, in the absence of any strong total recommendation in favour of an externally directed behaviour, a vague secondary population will develop. Evolution could then continue through tertiary, quaternary, etc. populations, but because of the huge range of indirect activation recommendation strengths the cognitive content of the population could become extremely vague.

However, there is a way to focus a population at any point in time to make it potentially more cognitively useful (e.g., for developing a plan as discussed earlier). Any population will also have recommendation strengths in favour of speech. If the strongest such recommendation is accepted, but only to activate auditory columns often active in the past before the columns in the population were active, the experience will be of pseudohearing of words that are the closest representation of the population. If now the population is replaced with another population generated on the basis of frequent past activity at the same time as the auditory columns, the effect will be a population with a more sharply focussed cognitive meaning.

Evolution of the new population could then occur, eventually followed by another focussing stage. This model thus accounts for the stream of consciousness experience as described by James. For more detail see Coward and Gedeon (2009).

3.9 Electronic Implementations

The recommendation architecture memory model is dependent upon a number of information mechanisms. One of the most critical is unsupervised organization of experience into column condition groups, in such a way that a set of columns can discriminate between different circumstances if the difference is behaviourally significant. A second mechanism is association of different groups of columns with different appropriate behaviours using reward feedback, in such a way that interference between new and prior learning is minimized. A third mechanism is support for different independent populations of active columns within the same resources, using different phases of frequency modulation. A fourth is indirect activation of columns on the basis of past temporally correlated activity. A fifth is management of the configuration of provisional connectivity for future receptive field expansions using past experience.

There have been various electronic implementations of these information mechanisms that confirm their capabilities. These implementations have in general used three layer columns as illustrated in Fig. 3.2, pyramidal neuron receptive fields that can expand slightly but not contract or change qualitatively, and have used software (Smalltalk and C++) models for physiological structures. Early implementations (e.g. Gedeon et al. 1999; Coward 2001; Ratnayake and Gedeon 2003; Coward et al. 2004) have used a relatively simple pyramidal neuron model in which all synapses have the same weight and inputs from one sensory state arrive synchronously. Later neuron models (e.g. Coward 2004, 2009b) are dynamic, with staged leaky integration across their dendritic trees, and use an LTP algorithm for learning.

These electronic simulations have demonstrated that experience can be organized into column modules, where each column detects a gradually expanding similarity space that is relatively orthogonal to the spaces detected by other columns, in such a way that the column ensemble can discriminate between circumstances with behaviourally different implications (Gedeon et al. 1999; Coward 2001, 2009b; Ratnayake and Gedeon 2003). The ability to associate partially ambiguous columns with behaviours using reward feedback and the capability of imitation to improve the efficiency of rewardbased learning (Coward 2005) have been demonstrated, including the management of behavioural selection by competition between components corresponding with the different behaviours (Coward et al. 2004). It has also been demonstrated that the gradual expansion of column portfolios means that the architecture does not experience catastrophic interference (Coward et al. 2004). Behaviours have included appropriate responses to objects and groups of objects. Simulations have also demonstrated the effectiveness of indirect activation mechanisms in supporting activation of pseudovisual images in response to verbal inputs and supporting activation of pseudovisual images of objects often present in the past at the same time as currently perceived objects (Coward 2001). The capability of the frequency modulation mechanism to implement attention functions at the physiological level has also been demonstrated (Coward 2004). The use of a sleep-like process for configuration of provisional conditions has also been implemented, with the expected improvement to the behavioural effectiveness of recorded conditions (Coward 2001).

3.10 Conclusions

The recommendation architecturebased cognitive model establishes the consistent causal descriptions of memory and learning phenomena on different levels of detail that are the essential core of any scientific understanding. Causal descriptions of memory and learning at the psychological level can be precisely mapped into causal descriptions of the same phenomena at the level of anatomical structures. The anatomical descriptions can be precisely mapped into causal descriptions at the level of neuron algorithms, and the neuron level descriptions can be mapped into known physiology. Alternative theories do not demonstrate the same degree of consistent multilevel modelling.

A critical intermediate level of description is at the level of cortical columns, because the most complex types of information processing occur in the cortex. However, units of cortical information processing cannot be mapped simply into units of cognitive processing. The units of information processing in the cortex are detections of similarity circumstances (or receptive fields) and expansion of receptive fields when necessary to reach a minimum required level of detection. One such unit of information processing is shared by many different cognitive processes. All cognitive processing can be described in terms of direct detection of receptive fields, indirect activation of receptive fields on the basis of past temporally correlated activity with currently active receptive fields and interpretation of receptive field detections as behavioural recommendations, including indirect activation recommendations.

The brain has no a priori definitions of cognitive processes like episodic memory, imagining, planning, etc. All the brain has is a set of information processes (or internal behaviours) for activating and evolving populations of receptive field detections. At each stage the currently active population is interpreted into a predominant behaviour recommendation which is then implemented. Learning uses reward feedback both to discover sequences of information processes that are behaviourally valuable and to associate different circumstances with invocation of different sequences. Different sequences supporting different cognitive processes such as episodic memory, semantic memory and imagination or planning often use overlapping cortical resources.

Electronic modelling of the recommendation architecture demonstrates that the information models that form the foundation of the description hierarchy support observed memory and learning phenomena.

References

Addis DA, Wong AT, Schacter DL (2007) Remembering the past and imagining the future: Common and distinct neural substrates during event construction and elaboration. *Neuropsychologia* 45:1363–1377

Alexander GE, DeLong MR, Strick PL (1986) Parallel organization of functionally segregated circuits linking basal ganglia and cortex. *Annual Reviews of Neuroscience* 9:357–381

Anderson AK, Phelps EA (1997) Emotional memory: what does the amygdala do? *Current Biology* 7:R311–R314

Anderson JR (1996) ACT: A simple theory of complex cognition. *American Psychologist* 51:355–365

Axmacher N, Morman F, Fernandez G, Elger C, Fell J (2006) Memory formation by neuronal synchronization. *Brain Research Reviews* 52:170–182

Baddeley AD (1986) *Working Memory*. Oxford, Oxford University Press

Baddeley AD, Vallar G, Wilson BA (1987) Sentence comprehension and phonological memory: some neuropsychological evidence. In: *Attention and Performance XII: The psychology of reading*. London: Erlbaum

Baddeley AD (2000) Short-term and working memory. In: *The Oxford Handbook of Memory*, Tulving E, Craik FIM (eds). Oxford: Oxford University Press

Badgaiyan RD (2000) Neuroanatomical organization of perceptual memory: An fMRI study of picture priming. *Human Brain Mapping* 10:197–203

Bainbridge L (1977) Verbal reports as evidence of the process operator's knowledge International. *Journal of Man-Machine Studies* 11:411–436

Bar M, Biederman I (1998) Subliminal visual priming. *Psychological Science* 9:464–469

Berardelli A, Noth J, Thompson PD et al. (1999) Pathophysiology of chorea and bradykinesia in Huntington's disease. *Movement Disorders* 14:398–403

Berry DC (1987) The problem of implicit knowledge. *Expert Systems* 4:144–151

Bi G-Q, Poo M-M (1998) Synaptic modifications in cultured hippocampal neurons: Dependence on spike timing, synaptic strength, and postsynaptic cell type. *Journal of Neuroscience* 18:10464–10472

Bracke-Tolkmitt R, Linden A, Canavan AGM et al. (1989) The cerebellum contributes to mental skills. *Behavioral Neuroscience* 103:442–446

Brodmann K (1908) Beitraege zur histologischen Lokalisation der Grosshirnrinde. VI. Mitteilung: Die Cortexgliederung des Menschen. *Journal of Psychology and Neurology (Lzp)* 10:231–246T

Cabeza R, Dolcos F, Graham R et al. (2002) Similarities and differences in the neural correlates of episodic memory retrieval and working memory. *Neuroimage* 16:317–30

Cohen JJ, Squire LR (1981) Preserved learning and retention of patternanalysing skill in amnesia: Dissociation of knowing how and knowing that. *Science* 210:207–210

Collins A, Loftus E (1975) A spreading-activation theory of semantic processing. *Psychological Review* 82:407–428

Corkin S (2002) What's new with the amnesic patient H.M.? *Nature Reviews Neuroscience* 3:153–160

Corkin S (1968) Acquisition of motor skill after bilateral medial temporal-lobe excision. *Neuropsychologia* 6:225–264

Cowan N (2000) The magical number 4 in short-term memory: a reconsideration of mental storage capacity. *The Behavioural and Brain Sciences* 24:87–185

Coward LA (1990) *Pattern Thinking*. New York, Praeger

Coward LA (2001) The Recommendation Architecture: lessons from the design of large scale electronic systems for cognitive science. *Journal of Cognitive Systems Research* 2:111–156

Coward LA (2004) Simulation of a proposed binding model. In: *Brain Inspired Cognitive Systems 2004*, Smith LS, Hussain A, Aleksander I (eds). Stirling, University of Stirling

Coward LA, Sun R (2004) Some criteria for an effective scientific theory of consciousness and examples of preliminary attempts at such a theory. *Consciousness and Cognition* 13:268–301

Coward LA, Gedeon TD, Ratanayake U (2004) Managing interference between prior and later learning. ICONIP 2004, Calcutta. *Lecture Notes in Computer Science* 3316:458–464

Coward LA (2005) *A System Architecture Approach to the Brain: from Neurons to Consciousness*. New York, Nova Science Publishers

Coward LA, Sun R (2007) Hierarchical approaches to understanding consciousness. *Neural Networks* 20:947–954

Coward LA (2009a) The Hippocampal system as the cortical resource manager: A model connecting psychology, anatomy and physiology. In: *Brain Inspired Cognitive Systems*. Cutsuridis V, Hussain A, Barros AK, Aleksander I, Smith L, Chrisley R. (eds). Berlin, Springer

Coward LA (2009b) The Hippocampal system as the manager of neocortical declarative memory resources. In: *Connectionist Models of Behaviour and Cognition II*, Mayor J, Ruh N, Plunkett K (eds). London, World Scientific

Coward LA, Gedeon TO (2009) Implications of resource limitations for a conscious machine. *Neurocomputing* 72:767–788

Crovitz HF, Schiffman H (1974) Frequency of episodic memories as a function of their age. *Neuropsychologia* 21:213–234

Crutcher MD, DeLong MR (1984) Single cell studies of the primate putamen II. Relations to direction of movement and pattern of muscular activity. *Experimental Brain Research* 53:244–258

Cutsuridis V, Cobb S, Graham BP (2010) Encoding and retrieval in the hippocampal CA1 microcircuit model. *Hippocampus* 20(3):423–446

Cutsuridis V, Wenneckers T (2009) Hippocampus, microcircuits and associative memory. *Neural Networks* 22(8):1120–1128

Cutsuridis V, Cobb S, Graham BP (2008) Encoding and retrieval in a CA1 microcircuit model of the Hippocampus. In: *Lecture Notes in Computer Science 5164*, Kurkova V, et al. (eds). Berlin, Springer, 238–247

Devinsky O, Morrell MJ, Vogt BA (1995) Contributions of anterior cingulate cortex to behaviour. *Brain* 118:279–306

Devlin JT, Russell RP, Davis MH et al. (2002) Is there an anatomical basis for category-specificity? Semantic memory studies in PET and fMRI. *Neuropsychologia* 40:54–75

Engel AK, Singer W (2001) Temporal binding and the neural correlates of sensory awareness. *Trends in Cognitive Sciences* 5:16–25

Fan Y, Zou B, Ruan Y et al. (2005) In vivo demonstration of a late polarizing postsynaptic potential in CA1 pyramidal neurons. *Journal of Neurophysiology* 93:1326–1335

Fliessbach K, Trautner P, Quesada CM et al. (2007) Cerebellar contributions to episodic memory encoding as revealed by fMRI. *NeuroImage* 35:1330–1337

Fletcher PC, Frith CD, Rugg MD (1997) The functional neuroanatomy of episodic memory. *Trends in Neurosciences* 20:213–218

Fyhn M, Molden S, Witter MP, Moser EI, Moser M-B (2004) Spatial representation in the entorhinal cortex. *Science* 305:1258–1264

Gabrieli JDE, Milberg W, Keane MM et al. (1990) Intact priming of patterns despite impaired memory. *Neuropsychologia* 28:417–427

Gabrieli JDE, Corkin S, Mickel SF et al. (1993) Intact acquisition and long-term retention of mirror tracing skill in Alzheimer's disease and in global amnesia. *Behavioral Neuroscience* 107:899–910

Gedeon T, Coward LA, Bailing Z (1999) Results of Simulations of a System with the Recommendation Architecture. *Proceedings of the 6th International Conference on Neural Information Processing* I:78–84

Graf P, Squire LR, Mandler G (1984) The information that amnesic patients do not forget. *Journal of Experimental Psychology: Learning, Memory and Cognition* 10:164–178

Guillery RW, Feig SL, Lozsadi DA (1998) Paying attention to the thalamic reticular nucleus. *Trends in Neuroscience* 21:28–32

Gulledge AT, Stuart GJ (2003) Excitatory actions of GABA in the cortex. *Neuron* 37:299–309

Hasselmo M, Bodelon C, Wyble B (2002) A proposed function of the hippocampal theta rhythm: Separate phases of encoding and retrieval of prior learning. *Neural Computing* 14:793–817

Hausser M, Mel B (2003) Dendrites: bug or feature? *Current Opinion in Neurobiology* 13:372–383

Heindel WC, Salmon DP, Shults CW et al. (1989) Neuropsychological evidence for multiple implicit memory systems: A comparison of Alzheimer's, Huntington's, and Parkinson's disease patients. *The Journal of Neuroscience* 9:582–587

Hoffman KL, McNaughton BL (2002) Sleep on it: cortical reorganization after the fact. *Trends in Neuroscience* 25(1):1–2

James W (1892) The stream of consciousness. *Psychology* ChapterXI. World Publishing Company, Cleveland and New York /http://psychclassics.yorku.ca/ James/jimmy11.htm

Jankovic J (2008) Parkinson's disease: clinical features and diagnosis. *Journal of Neurology, Neurosurgery and Psychiatry* 79:368–376

Jaynes J (1976) *The Origin of Consciousness in the Breakdown of the Bicameral Mind*. Boston, Harvard

Kassubek J, Schmidtke K, Kimmig H et al. (2001) Changes in cortical activation during mirror reading before and after training: an fMRI study of procedural learning. *Cognitive Brain Research* 10:207–217

Kensinger EA, Ullman MT, Corkin S (2001) Bilateral medial temporal lobe damage does not affect lexical or grammatical processing: Evidence from amnesic patient H.M. *Hippocampus* 11:347–360

Kiehl KA, Liddle PF, Hopfinger JB (2000) *Psychophysiology* 37:216–223

Lee AK, Wilson MA (2002) Memory of sequential experience in the hippocampus during slow wave sleep. *Neuron* 36:1183–1194

Leiner HC, Leiner AL, Dow RS (1993) Cognitive and language functions of the human cerebellum. *Trends in Neurosciences* 16:444–447

Leutgeb S, Leutgeb JK, Treves A, Moser M-B, Moser EI (2004) Distinct ensemble codes in Hippocampal areas CA3 and CA1. *Science* 305:1295–1298

Linden DJ (2003) From molecules to memory in the Cerebellum. *Science* 301:1682–1685

Lisman JE, Idiart MAP (1995) Storage of 7 ± 2 short term memories in oscillatory subcycles. *Science* 267:1512–1515

Loftus EF, Pickrell JE (1995) The formation of false memories. *Psychiatric Annals* 25:720–725

Lovett MC, Reder LM, Lebiere C (1999) Modeling working memory in a unified architecture: An ACT-R perspective. In: *Models of Working Memory*, Miyake A, Shah P (eds). Cambridge, Cambridge MA, 135–182

Mandler G, Shebo BJ (1982) Subitizing: An analysis of its component processes. *Journal of Experimental Psychology: General* 111:1–22

Martin A (2007) The representation of object concepts in the brain. *Annual Review of Psychology* 58:25–45

Martin SJ, Morris RGM (2002) New life in an old idea: the synaptic plasticity and memory hypothesis revisited. *Hippocampus* 12:609–636

Mathews R, Buss R, Stanley W et al. (1989) Role of implicit and explicit processes in learning from examples: a synergistic effect. *Journal of Experimental Psychology: Learning, Memory and Cognition* 15:1083–1100

McBride DM, Dosher BA, Gage NM (2001) A comparison of forgetting for conscious and automatic memory processes in word fragment completion tasks. *Journal of Memory and Language* 45:585–615

Miller GA (1956) The magical number seven, plus or minus two: Some limits on our capacity for processing information. *Psychological Review* 63:81–97

Mitchell SJ, Richardson RT, Baker FH et al. (1987) The primate globus pallidus: neuronal activity related to direction of movement. *Experimental Brain Research* 68:491–505

Morasan P, Rademacher J, Schleicher A et al. (2001) Human primary auditory cortex: Cytoarchitectonic subdivisions and mapping into a spatial reference system. *NeuroImage* 13:684–701

Mountcastle VH (1997) The columnar organization of the neocortex. *Brain* 120:701–722

Muller MM, Gruber T, Keil A (2000) Modulation of induced gamma band activity in the human EEG by attention and visual information processing. *International Journal of Psychophysiology* 38:283–299

Nadel L, Moscovitch M (1997) Memory consolidation, retrograde amnesia and the hippocampal complex. *Current Opinion in Neurobiology* 7:217–227

O'Doherty J, Kringelbach ML, Rolls ET et al. (1991) Abstract reward and punishment representations in the human orbitofrontal cortex. *Nature Neuroscience* 4:95–102

O'Keefe J, Recce ML (1993) Phase relationship between hippocampal place units and the EEG theta rhythm. *Hippocampus* 3:317–330

Pavlides C, Winson J (1989). Influences of hippocampal place cell firing in the awake state on the activity of these cells during subsequent sleep episodes. *Journal of Neuroscience* 9:2907–2918

Petrides M, Pandya DN (1999) Dorsolateral prefrontal cortex: comparative cytoarchitectonic analysis in the human and the macaque ventrolateral prefrontal cortex and corticocortical connection patterns. *European Journal of Neuroscience* 11:1011–1036

Petrides M, Pandya DN (2002) Comparative cytoarchitectonic analysis in the human and the macaque brain and corticocortical connection patterns in the monkey. *European Journal of Neuroscience* 16:291–310

Ratnayake U, Gedeon TD (2003) Extending the recommendation architecture model for text mining. *International Journal of Knowledge-Based Intelligent Engineering Systems* 7:139–148

Redish AD, Touretzky DS (1998) The role of the hippocampus in solving the Morris water maze. *Neural Computation* 10:73–111

Rips L, Shoben J, Smith E (1973) Semantic distance and verification of semantic relations. *Journal of Verbal Learning and Verbal Behaviour* 12:1–20

Ritz M (1999) Chapters 5 and 6. In: *Drugs of Abuse and Addiction: Neurobehavioral Toxicology*, Niesink R, Jaspers RMA, Hollinger MA et al. (eds). Boca Raton, CRC

Robinson JA (1976) Sampling autobiographical memory. *Cognitive Psychology* 8:578–595

Rogers TT, Hocking J, Noppency U et al. (2006) Anterior temporal cortex and semantic memory: reconciling findings from nuropsychology and functional imaging. *Cognitive, Affective and Behavioural Neuroscience* 6:201–213

Sagar JH, Cohen NJ, Corkin, S et al. (1985) Dissociations among processes in remote memory. *Annals of the New York Academy of Science* 444:533–535

Sah P, Faber ESL, De Armentia ML et al. (2003) The amygdaloid complex: Anatomy and physiology. *Physiological Review* 83:803–834

Schacter JC, Tulving E (1994) What are the memory systems of 1994? In: *Memory Systems 1994*, Schacter JC, Tulving E (eds). Cambridge, MA, MIT

Scoville WB, Milner B (1957) Loss of recent memory after bilateral hippocampal lesions. *Journal of Neurology, Neurosurgery, and Psychiatry* 20:11–21

Shallice T, Warrington EK (1970) Independent functioning of verbal memory stores: a neuropsychological study. *Quarterly Journal of Experimental Psychology* 22:261–273

Somogyi P, Bolam JP, Smith AD (1981) Monosynaptic cortical input and local axon collaterals of identified striatonigral neurons. A light and electron microscopic study using the Golgiperoxidase transport-degeneration procedure. *Journal of Comparative Neurology* 195:567–584

Squire LR, Alvarez P (1995) Retrograde amnesia and memory consolidation: a neurobiological perspective. *Current Opinion in Neurobiology* 5:169–177

Standing L, Conexio J, Haber RN (1970) Perception and memory for pictures: single-trial learning of 2500 visual stimuli. *Psychonomic Science* 19:73–74

Starr PA, Kang GA, Heath S et al. (2008) Pallidal neuronal discharge in Huntington's disease: support for selective loss of striatal cells originating the indirect pathway. *Experimental Neurology* 211:227–33

Sun R, Peterson T, Merrill E (1996) Bottom-up skill learning in reactive sequential decision tasks. *Proceedings of 18th Cognitive Science Society Conference*. Lawrence Erlbaum Associates, Hillsdale, NJ

Sun R, Coward LA, Zenzen MJ (2005) On levels of cognitive modeling. *Philosophical Psychology* 18:613–637

Tanaka K. (2003) Columns for complex visual object features in the inferotemporal cortex: clustering of cells with similar but slightly different stimulus selectivities. *Cerebral Cortex* 13:90–99

Tepper JM, Bolam JP (2004) Functional diversity and specificity of neostriatal interneurons. *Current Opinion in Neurobiology* 14:685–692

Torriero S, Oliveri M, Koch G et al. (2007) Cortical networks of procedural learning: Evidence from cerebellar damage. *Neuropsychologia* 45:1208–1214

Tronson NC, Taylor JR (2007) Molecular mechanisms of memory reconsolidation. *Nature Reviews Neuroscience* 8:262–275

Tulving E (1972) Episodic and semantic memory. In: *Organization of Memory*, Tulving E, Donaldson EW (eds). New York, Academic

Tulving E (1984) Multiple learning and memory systems. In: *Psychology in the 1990's*, Lagerspetz KMJ, Niemi P (eds). Holland, Elsevier

Tulving E (1985) How many memory systems are there? *American Psychologist* 40:385–398

Vertes RP, Eastman KE (2000) The case against memory consolidation in REM sleep. *Behavioral and Brain Sciences* 23:867–876

Warrington K, Shallice T (1969) The selective impairment of auditory verbal short-term memory. *Brain* 92:885–896

Warrington EK, Logue V, Pratt RTC (1971) The anatomical localisation of selective impairment of auditory verbal short-term memory. *Neuropsychologia* 9:377–387

Wickelgren WA (1968) Sparing of short-term memory in an amnesic patient: Implications for strength theory of memory. *Neuropsychologia* 6:235–244

Whalen PJ (1998) Fear, vigilance, and ambiguity: Initial neuroimaging studies of the human amygdala. *Current Directions in Psychological Science* 7:177–188

Whittington M, Traub R (2003) Inhibitory interneurons and network oscillations in vitro. *Trends in Neuroscience* 26:676–682

Zilli EA, Hasselmo ME (2007) Modeling the role of working memory and episodic memory in behavioral tasks. *Hippocampus* 18:193–209

Chapter 4
Value Maps, Drives, and Emotions

Daniel S. Levine

Abstract This chapter discusses value maps, drives, and emotions through the modeling of decision making, judgment, and choice. Ever the since the seminal work of Amos Tversky and Nobel Laureate Daniel Kahneman (Tversky and Kahneman 1974, 1981), it has been known that decision models based on rational maximization of expected utility do not capture the typical choices that people or nonhuman animals make in risky situations, even when extensive numerical or probabilistic information is available. Moreover, many of those choices tend to be strongly influenced by emotions and values in ways that are predictable, repeatable, and therefore amenable to theory development.

4.1 Overview of the DECIDER Model

Decision making under risk has been an active focus of cognitive modeling for over 30 years. Tversky and Kahneman's own Prospect Theory (Kahneman and Tversky 1990) and several other theories to be discussed in later sections (Decision Field Theory, Decision Affect Theory, and Affective Balance Theory) have provided both qualitative and quantitative fits to a large amount of experimental data on risky decision making. Yet none of those theories so far has been fit to a biologically plausible underlying mechanism. This chapter represents a partial effort to synthesize the major strengths of these theories with relevant recent results from cognitive neuroscience.

An extensive body of data on choices between risky alternatives shows that heuristics derived from emotion and experience frequently, but not always, trump accurate mathematical calculation. Yet some choices by some decision makers do in fact conform to what would be predicted from mathematical laws such as Bayes' rule. Hence, both experimenters and theorists are interested in differences of both context and personality that mediate the use of an automatic heuristic rule

D.S. Levine (✉)
Department of Psychology, University of Texas at Arlington, Arlington, TX 76019–0528, USA
e-mail: levine@uta.edu

V. Cutsuridis et al. (eds.), *Perception-Action Cycle: Models, Architectures, and Hardware*, Springer Series in Cognitive and Neural Systems 1, DOI 10.1007/978-1-4419-1452-1_4, © Springer Science+Business Media, LLC 2011

versus a deliberative numerical rule for choice. The model discussed in this chapter is designed to incorporate variability due to both situational and dispositional differences.

The DECIDER (Distributed Emotional Connections Influencing Decisions and Engaging Rules) model discussed here is an extension of networks developed in previous articles (Levine 2007, 2009a; the name "DECIDER" first appears in the 2009 article) and is based on the interactions among a large number of brain regions. These regions include loci for emotions (amygdala and orbital prefrontal cortex), rule encoding and executive function (orbital and dorsolateral prefrontal and anterior cingulate), and behavioral control (striatum, thalamus, and premotor cortex). Interactions among all these regions are mainly governed by shunting nonlinear differential equations (see, e.g., Grossberg 1988). The shunting type equations are based on realistic neurophysiological membrane equations and have been shown to be efficacious for a large variety of important effects in cognitive information processing; notably, short-term memory storage with contrast enhancement and noise suppression, and enhancement of matched patterns combined with inhibition of mismatched patterns.

In order to incorporate the capacities for utilizing either heuristic or deliberative decision rules in real time in a changing and uncertain environment, the model must address a variety of questions about the interface between emotional valuation and numerical calculation. Among these questions are the following:

> How are probabilities of events in the outside world encoded in the brain?

> How do rewards and punishments change the likelihoods of particular future behaviors?

> Under what conditions is a more risky or a less risky alternative preferred when the two alternatives are approximately equal in expected net reward?

> What differences in choice behavior arise from learning probabilities explicitly via language versus learning the same probabilities implicitly via feedback?

> How much are choices determined by relative versus absolute desirability of alternatives? How do choices between two desirable alternatives differ from choices between two undesirable alternatives, both behaviorally and in the brain?

> How are choices among gambles that include probabilities of obtaining or of losing a particular resource influenced by the affective value attached to the resource in question?

Most of this chapter will be devoted to discussing the data we seek to explain and the qualitative brain network answers we propose for the above questions. Toward the end, we will present some simulations of specific data sets based on simplified partial versions of the larger model.

4.2 Review of Experimental Data

4.2.1 Behavioral Data on Risky Decision Making

The behavioral data we review have to do with choices among two or more alternatives under risk. Decision psychologists distinguish *risk*, which means that the outcomes may be uncertain but the probabilities of the possible outcomes are known, from *uncertainty*, which means that the probabilities are unknown. Even though uncertainty is more the rule in real-life decisions such as those involving jobs, marriage, stocks, and so forth, risk is more often utilized in experimental psychology laboratories as a simplified paradigm for deconstructing the effects of anticipated outcomes on actions.

Before the work of Tversky and Kahneman, and previously of Allais (1953), it was generally believed that decision makers (DMs) possessed a constant "utility function" for amounts of money, lives saved, or other resources. It was also believed that in choices between alternatives that could be risky or certain, the DM would calculate the expected (in a mathematical sense) utility of each alternative by multiplying the utilities of all possible outcomes by their probabilities of occurrence and summing these products, then choosing the alternative with maximum expected utility.

This maximum utility formulation of decision making has remained influential in economics, despite mounting evidence against it from experimental psychology. For example, Allais (1953) gave some of his experimental participants the following choice:

A: Certainty of getting 100 million francs
B: 10% chance of getting 500 million francs, 89% chance of getting 100 million francs, 1% chance of getting nothing

He gave other participants the following choice:

A′: 11% chance of getting 100 million francs, 89% chance of getting nothing
B′: 10% chance of getting 500 million francs, 90% chance of getting nothing

The results were that the vast majority of those asked to choose between A and B chose A, the certain 100 million francs, despite its lower expected value. Yet the vast majority of those asked to choose between *A′* and *B′* chose *B′*, because the difference in amount of money (500 million versus 100 million) loomed larger than the difference in probability of obtaining that money (11% versus 10%). Simple algebra shows that the combination of both those choices is incompatible with Expected Utility Theory because A′ results from removing an 89% chance of 100 million francs from A, whereas B′ results from removing the same 89% chance of 100 million francs from B.

Tversky and Kahneman (1981) found numerous other violations of the predictions of Expected Utility Theory. One of their best-known violations occurred when they asked their participants to make *concurrent* choices between the following pairs of two alternatives:

A: Certainty of getting $240
B: 25% chance of getting $1,000, 75% chance of getting $0

and

C: Certainty of losing $750
D: 75% chance of losing $1,000, 25% chance of losing $0

Despite the instructions to choose concurrently, participants tended to make each choice separately without regard for the other choice. Most of them chose A over B (certainty compensating for slightly lower expected value) and chose D over C (avoiding certain loss when expected values were equal). Yet if those choices are combined, A and D together is clearly worse than B and C together:

A plus D: 75% chance of losing $760, 25% chance of getting $240
B plus C: 75% chance of losing $750, 25% chance of getting $250

Both of these results illustrate the principle that certainty looms large in people's emotional evaluation of possible outcomes. Hence, the psychological distance between 0 and 1% probabilities, or between 99 and 100% probabilities, is much larger than, say, the psychological distance between 10 and 11% probabilities, or between 50 and 51% probabilities.

Another principle that Tversky and Kahneman's (1981) results illustrate is the importance of framing and reference points. In perhaps their most famous experiment (known to decision psychologists as the "Asian disease problem"), they asked one group of participants the following:

Imagine that the US is preparing for the outbreak of an unusual Asian disease, which is expected to kill 600 people. Two alternative programs to combat the disease have been proposed. Assume that the exact scientific estimate of the consequences of the programs are as follows:

If Program A is adopted, 200 people will be saved.
If Program B is adopted, there is 1/3 probability that 600 people will be saved, and 2/3 probability that no people will be saved. (p. 453).

Which of the two programs would you favor?

A second group of participants was given the same Asian disease story with a different choice of programs:

If Program C is adopted 400 people will die.
If Program D is adopted there is 1/3 probability that nobody will die, and 2/3 probability that 600 people will die.

Which of the two programs would you favor?

The participants strongly favored Program A over B and Program D over C, despite the fact that the effects of A and C are identical and the effects of B and D are identical. The difference was that the first set of participants thought in terms of lives saved (i.e., were anchored on the reference point of all 600 dying), whereas the second set thought in terms of lives lost (i.e., were anchored on the reference point of all 600 surviving). This illustrated the dependency of DMs' choices on reference points, combined with risk aversion for gains and risk seeking for losses.

Many of these results could be explained by *Prospect Theory* (Kahneman and Tversky 1990), which posits a nonlinear weighting function for probabilities such as shown in Fig. 4.1. This means that low nonzero probabilities of gains are valued more than would be predicted from their probabilities, which explains the appeal of lotteries. Likewise, low nonzero probabilities of losses are dreaded more than

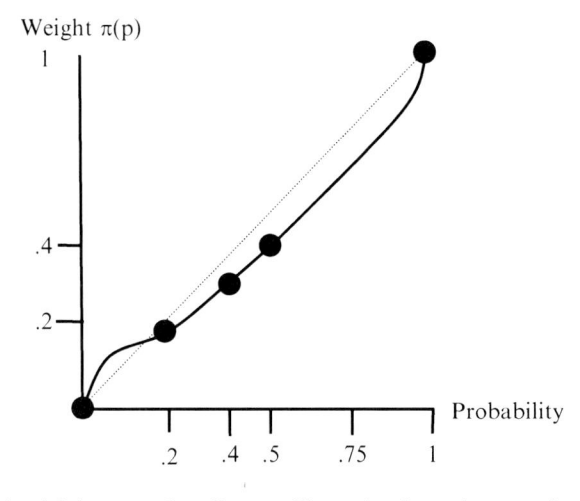

Fig. 4.1 Typical weighting curve from Prospect Theory (made continuous at 0 and 1)

would be predicted from their probabilities, which explains the appeal of insurance policies. Conversely, high, non-unity probabilities of both gains and losses are underweighted. Kahneman and Tversky, however, provided no explanation for the underlying cognitive processes that generated these nonlinear weightings. In later sections, we describe a possible neurocognitive mechanism for generating a curve like that of Fig. 4.1. However, we note experimental findings that suggest that the curvature of the Prospect Theory weighting function can be influenced by emotional coloration. Rottenstreich and Hsee (2001) found examples where the certainty of an affect-poor resource such as a tuition break was preferred to the certainty of an affect-rich resource such as a trip to a foreign country or the kiss of a movie star, whereas a low probability of the affect-rich resource was preferred to the same low probability of the affect-poor resource. The authors' interpretation of the results was that the S-shaped Prospect Theory curve of Fig. 4.1 had a sharper curvature (exaggerated weighting of low probabilities) in the affect-rich case as compared with the affect-poor case.

Other investigators have found that the tendency to overweight low probabilities and underweight high probabilities also depends on the method by which alternatives are presented. Several recent articles have investigated the gap between *decision from description* and *decision from experience*. Decision from description is the case Tversky and Kahneman studied: probabilities of alternatives are explicitly stated and the DM has to make a one-shot choice between them. Decision from experience means that probabilities are not explicitly stated but the DM learns those probabilities over time from the consequences of his or her choices. Barron and Erev (2003) and Hertwig et al. (2004) found that several of Tversky and Kahneman's effects, including risk seeking for losses and risk aversion for gains, were mitigated or reversed if the DM learned the probabilities from experience (in anywhere from 20 to several hundred trials). Barron and Erev (2003) fit their data fairly well using a reinforcement learning model of the same type as models previously used to simulate effects in the time evolution of single-attribute decisions (Busemeyer and Townsend 1993) and multiattribute decisions (Roe et al. 2001).

Decisions based on probabilities can also be inaccurate or inconsistent in cases when there is confusion because the alternative with the larger probability of a gain or loss also has the smaller numerical frequency. Several experimental psychologists have studied the phenomenon of *ratio bias*, meaning that the same small probability is perceived as larger when it is the ratio of two larger numbers; for example, 10 out of 100 is perceived as larger than 1 out of 10. A consequence of the ratio bias is that, in some cases, the smaller of two probabilities of an emotionally positive outcome is chosen over the larger probability. For example, Denes-Raj and Epstein (1994) showed their participants two bowls containing red and white jellybeans, told them they would win a certain amount of money if they randomly selected a red jellybean, and instructed them to choose which bowl gave them the best chance of winning money. In one of the bowls, there were a total of 10 jellybeans out of which 1 was red. In the other bowl, there were a total of 100 jellybeans out of which some number greater than 1 but less than 10 were red. Hence, choice of the bowl with a larger frequency of red jellybeans was always nonoptimal, because the probability of drawing red from that bowl was less than 1/10. The majority of participants made the nonoptimal choice when the choice was 9 out of 100 versus 1 out of 10, about a quarter still chose 5 out of 100 over 1 out of 10, but no participant chose 2 out of 100 over 1 out of 10.

4.2.2   Data on Neural Bases of Cognitive-Emotional Decision Making

Bechara et al. (1994, 2000) developed the *Iowa Gambling Task* (IGT) as a means of illuminating the roles of emotionally related brain regions [specifically, the amygdala and orbitofrontal cortex (OFC)] in decision making. Bechara and his colleagues gave participants a sequence of 100 choices among four decks of cards that provided different gains and losses of play money with different probabilities. In their original protocol, two of these decks (Decks 1 and 2) yielded higher short-term payoffs ($100 per card as opposed to $50 for Decks 3 and 4; Table 4.1). However, in addition to these payoffs, the decks also yielded losses of money on every second or every tenth draw. The randomization of these losses was structured so that the two decks with higher short-term gains also yielded long-term expected losses, whereas the two decks with lower short-term gains yielded long-term expected gains. Bechara and his colleagues tested normal participants on the IGT along with patients with damage to either the OFC or amygdala. They found that normals began favoring the disadvantageous decks (Decks 1 and 2) because of the higher payoffs, but after

Table 4.1 Decks used in the Iowa Gambling Task and their expected earnings (reprinted from Levine et al. 2005, with the permission of the IEEE)

Deck 1 (bad): $100 gain/card, $p\ (-\$1{,}250) = 0.1 \rightarrow$ EV/10 cards $= -\$250$
Deck 2 (bad): $100 gain/card, $p\ (-\$250) = 0.5 \rightarrow$ EV/10 cards $= -\$250$
Deck 3 (good): $50 gain/card, $p\ (-\$250) = 0.1 \rightarrow$ EV/10 cards $= \$250$
Deck 4 (good): $50 gain/card, $p\ (-50) = 0.5 \rightarrow$ EV/10 cards $= \$250$

experiencing losses, they gradually shifted to the advantageous decks (Decks 3 and 4). Patients with damage to either the OFC or amygdala never learned the advantageous strategy, but continued to selectively choose from the disadvantageous decks.

Bechara and his colleagues further differentiated the roles of amygdala versus OFC in decision making. They discovered that people who lost money on a deck selection experienced a skin conductance response (SCR), denoting emotional arousal. Patients with amygdalar damage did not experience the SCR after a loss, whereas patients with OFC damage did experience that SCR after a loss. However, normal participants with no brain damage also experienced an anticipatory SCR at times where they made a selection from a disadvantageous deck even after learning it would be likely to lead to a loss. Patients with OFC damage, like those with amygdalar damage, did not experience this anticipatory SCR. This led Bechara and his colleagues to conclude that the OFC is important for storing the memory of previous gains and losses.

The OFC is one of the key executive areas of the brain: other prefrontal areas whose executive functions are now well established include the anterior cingulate cortex (ACC) and dorsolateral prefrontal cortex (DLPFC). Recent functional magnetic resonance imaging (fMRI) studies have helped to parcel out dissociable functions for each of these three areas in high-level emotionally influenced decision making.

An fMRI study by DeMartino et al. (2006) showed that individuals who were not susceptible to framing effects on a monetary decision task exhibited different brain activation patterns than those who were susceptible to those effects. In their study, participants had to choose between a sure option and a gamble option, where the sure option was expressed either in terms of gains (e.g., keep £20 out of the £50 they initially received) or in terms of losses (e.g., lose £30 out of the initial £50). As in the Asian disease problem of Tversky and Kahneman (1981), the majority of participants chose the sure option with a gain frame and the gamble option with a loss frame. Yet significant minorities of participants chose the gamble with a gain frame or the sure option with a loss frame, in violation of the usual heuristics. fMRI measurements showed that heuristics-violators exhibited more activation than heuristics-followers in both the OFC and the ACC. Conversely, heuristics-followers exhibited more activation than heuristics-violators in the amygdala, the subcortical area most involved with primary emotional experience.

DeNeys et al. (2008) gave their fMRI participants a task that manifests *base rate neglect* (Kahneman and Tversky 1973); that is, judging the probability of a conditional event based purely on description without regard to prior probabilities.

In this task, the participants were told that a certain group of people consisted of specific numbers of lawyers and engineers: the numbers could either be balanced (500 lawyers and 500 engineers) or imbalanced (995 lawyers and 5 engineers, or the reverse). A hypothetical person was identified as a member of that group and given a description fitting a stereotype of one of the two professions; then the participant was asked to determine the probability of that person being a lawyer or an engineer. Bayes' rule indicates that the probability of being a lawyer or engineer should partly depend on the distribution of lawyers and engineers in the sample (*base rates*). De Neys et al. found that those participants (the minority) who took the base rate into account showed greater activation of the right DLPFC than those who neglected the base rate (i.e., judged solely on the stereotype). Also, for both types of participants, the ACC was most activated when there was a conflict between possible choice criteria (e.g., when a stereotypical description of an engineer was combined with a statement that the group had 995 lawyers and 5 engineers).

The results of these decision studies are consistent with other studies that have pointed to overarching executive functions for the three areas of OFC, ACC, and DLPFC. OFC damage has been observed to lead to poor decision making and inappropriate social behavior in many patients including the famous nineteenth century patient Phineas Gage (Damasio 1994). These clinical observations, combined with some animal lesion studies, indicate that OFC forms and sustains mental linkages between specific sensory events in the environment and positive or negative affective states. This region creates those linkages via connections between neural activity patterns in the sensory cortex that reflect past sensory events, and other neural activity patterns in subcortical regions (the amygdala and hypothalamus) that reflect emotional states. Long-term storage of affective valences is likely to take place at connections from OFC to amygdala (Rolls 2000).

The DLPFC is a working memory region, and is involved with information processing at a higher level of abstraction than the OFC. For example, in monkeys, whereas OFC lesions impair learning of changes in reward value within a stimulus dimension, DLPFC lesions impair learning of changes in which dimension is relevant (Dias et al. 1996). Evidence from a variety of human fMRI studies (Bunge 2004; Huettel and Misiurek 2004) indicates that DLPFC is involved in accurate, rule-based stimulus-response contingencies, particularly when there is a need to override a prevailing response.

The ACC is involved in detection of either potential response error or conflict among signals promoting competing responses (Botvinick et al. 2001; Brown and Braver 2005). Bush et al. (2000) found that the ACC is activated when a subject must select or switch among different interpretations or aspects of a stimulus. Also, Barch et al. (2001) review fMRI studies of the Stroop task, whereby participants are instructed to respond based on a color-naming rule that must compete with a prepotent word-reading rule. The ACC was consistently found to be more active when the two rules conflict (a word for one color is written in ink of another color) than when the two rules agree.

The results of DeNeys et al. (2008) confirm this task division between ACC and DLFC, with ACC involved in noting the presence of a potential conflict of rules and

DLPFC in choosing the right rule and the right actions that conform to that rule. The task of DeMartino et al. (2006) did not appreciably involve the DLPFC because it was less cognitively complex, involving the OFC in making an informed selection based on memory and the ACC again in detecting a conflict between such a decision and the framing heuristics.

Yet the behavioral decision results we have reviewed indicate that the roles of these prefrontal areas (particularly the OFC and the ACC) depend not only on the cognitive information rules being employed but also on the positive or negative affective value attached to that information. The role of OFC in affective preference was clarified in a seminal article of Tremblay and Schultz (1999) on single-cell recording in monkeys. The monkeys had been previously shown to exhibit clear preferences among foods, preferring raisin to apple and apple to cereal. On a delayed response task, pictures suggesting two of these foods were presented in a given trial block, and the animals had to choose to move its arm in one direction or another; it would obtain one of those foods as a reward depending on where it moved. The same OFC neurons that fired in anticipation of the raisin when it chose between raisin and apple also fired in anticipation of the apple when it chose between apple and cereal. In other words, these OFC neurons coded *relative* reward preference.

A similar pattern of OFC activity correlated with relative preference occurred in human fMRI studies of monetary reward tasks by Blair et al. (2006) and Elliott et al. (2008). Elliott et al. (2008) found that medial OFC response to a particular stimulus depended on whether the stimulus predicted the more desirable or the less desirable one of two rewards. Yet Blair et al. (2006) found that the response to alternatives was influenced by their absolute as well as their relative desirability. In a gambling task, they compared choosing which of two monetary rewards was greater with choosing which of two monetary punishments was lesser, and discovered greater OFC activity in the reward/reward choice and greater ACC activity in the punishment/punishment case. These results are broadly consistent with the conception of OFC as a keeper of information about reward contingencies and of ACC as a detector of potential conflict, error, or need to somehow reset the executive system. Another fMRI study by Breiter et al. (2001) indicates that subcortical emotional regions (the "extended amygdala") are also sensitive to relative and absolute monetary values.

Thus far, there have been few neural network models addressing the roles of specific brain regions in decision making. Most of them so far have been models of particular decision tasks, notably the IGT due to Bechara and Damasio (Frank and Claus 2006; Levine et al. 2005; Wagar and Thagard 2004; Table 4.1). However, there have been numerous influential cognitive models of decision making under risk, many of which have captured important principles governing real-time choices and judgments. These models, and the few actual neural network models of these processes, are reviewed in the following section. Then we discuss our efforts to fit the best features of all these cognitive models into brain-based neural network structures.

4.3   Review of Previous Decision Models

4.3.1   Psychological Models Without Explicit Brain Components

The most influential and widely utilized decision theory is Tversky and Kahneman's own Prospect Theory (PT; Kahneman and Tversky 1990; Tversky and Kahneman 2004). The original formulation of PT was designed to make the smallest change in Expected Utility Theory that would account for such phenomena as the Allais paradox, certainty effects, and framing data. Hence, PT posited a utility (value) function that is concave for gains and convex for losses, with the curvature being steeper for losses, combined with the probability weighting function of Fig. 4.1. In this formulation, as in Expected Utility Theory, the DM calculates the value of each "prospect" or anticipated alternative (certain or risky) then always chooses the one with the highest value.

Yet other decision theorists noted that the same DM does not always make the same decision between a given two alternatives, but rather will exhibit some random variation in his or her choices. Moreover, the DM's preference may depend on the amount of time he or she has to make the decision. This led to the development of decision models with explicit probabilistic components, most notably the *Decision Field Theory* (DFT) of Busemeyer and Townsend (1993). DFT combines a random version of Expected Utility Theory with reinforcement learning, approach and avoidance gradients, and sequential sampling of information over time. The original version of the theory could account for various results including changes in decision criteria, and therefore in choices, as the amount of time available for decision increased. Later, DFT was extended to multiattribute decision making and incorporated into a connectionist framework (Roe et al. 2001).

Prospect Theory has been extended from one-shot to real-time decision making with the development of *Cumulative Prospect Theory* (Tversky and Kahneman 2004). In the cumulative theory, the DM still makes the same choice the same two alternatives in all instances, assuming no changes in framing. Rieskamp (2008) developed a probabilistic version of Cumulative Prospect Theory and tested its performance on a large number of gamble choices in comparison with DFT, along with another theory based on limiting the attributes under consideration. Rieskamp found that of the three theories, the predictions of the DFT had the best fit with the decision makers' observed choices.

Yet to date, neither PT nor DFT has been solidly grounded in an underlying brain-based neural network structure, though there have been steps in that direction with both theories (for PT, Trepel et al. 2005; for DFT, Busemeyer et al. 2006). In addition to probabilistic information processing, such a neural network structure would need to encompass what is known about emotional influences on decision making. Two significant quantitative theories that have explicitly considered emotions are *Affective Balance Theory* (ABT) (Grossberg and Gutowski 1987) and *Decision Affect Theory* (DAT) (Heyman et al. 2004; Mellers 2000).

ABT was designed to explain the basic findings of Tversky and Kahneman (1974, 1981) in a real-time connectionist framework. Preferences among gambles were

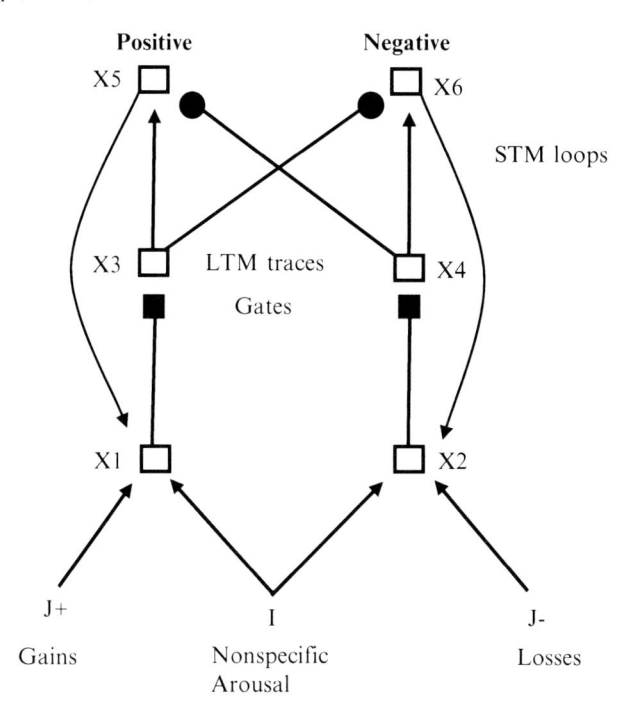

Fig. 4.2 Generic gated dipole network as in Grossberg (1972a, b). Arrows denote excitation, filled circles inhibition, filled squares transmitter depletion. Left and right pathways denote inputs of opposite affective sign, in this case, gains and losses. If $J+$ is larger than $J-$, indicating inputs of net positive affective valence, the left pathway also undergoes more depletion at the corresponding filled square synapse. If the inputs are shut off, due to the nonspecific arousal both pathways are still active and $X4$ is transiently larger than $X3$. By feedforward inhibition this means $X6 > 0$, that is, there is a negative affective rebound. By analogy, shutting off an input with net negative affective valence leads to a positive affective rebound at $X5$

explained as steady states of a *gated dipole* (Grossberg 1972a, b), a type of network previously designed to explain both relief and frustration effects in conditioning. The gated dipole network comprises two pathways of opposite emotional valence and utilizes neurotransmitter depletion to make one pathway transiently active when there is a decrease of activity in the opposite pathway (Fig. 4.2). Variants of ABT have been applied to simulation of consumer preference behavior (Leven and Levine 1996) and animal foraging under predation risk (Coleman et al. 2005). ABT is restricted in its explanatory range; for example, in processing alternative prospects it lumps together payoffs and probabilities as affectively positive or negative inputs of certain sizes. Yet the theory's nonlinear interactions capture some of the principles required for a larger, more comprehensive network theory, such as emotional reactions to contrasts between obtained positive or negative resources and other possible outcomes of a choice (cf. the neural data of Breiter et al. 2001 and Tremblay and Schultz 1999).

DAT was designed to account for effects of the DM's anticipated pleasure or displeasure about the consequences of his or her choice. This affective reaction was based partly on the actual outcome of a choice and partly on counterfactual comparison both with other possible outcomes of the same choice (because it is probabilistic) and with possible consequences of choices not made. Heyman et al. (2004) developed this principle quantitatively using a multiple linear regression model of gamble choices, with three independent variables consisting of actual money obtained, money obtained minus what would have been obtained if the gamble had come out differently, and money obtained minus what would have been obtained if the other alternative had been chosen, all of them weighted by the appropriate probabilities of occurrence.

Is it possible to synthesize the best features of PT, DFT, ABT, and DAT into a comprehensive neural network model that also specifies realistic neural processes for generating the required network interactions? There is some doubt among decision psychologists as to whether a comprehensive descriptive theory of human decision making is now possible (see Johnson et al. 2008). The variability of human choices, not only with framing but also with mood, context, and personality traits, seems to preclude the validity of any "supertheory," at least one as simple as Expected Utility Theory or even as the original version of Prospect Theory. Yet with the extensive knowledge, we now possess of the cognitive neuroscience of emotionally influenced decision making, it has become possible to develop a more realistic account of the underlying processes involved in decision, judgment, and choice. Debates among competing theoretical approaches become sterile in the absence of an intuitively understandable mechanistic account of these processes. Hence, at the current stage of knowledge, it is important to judge decision theories not only by how much of the behavioral data they can reproduce but also by how compatible they are with process theories that are realistic on both the neural and cognitive levels.

4.3.2 Models of Brain Area Involvement in Cognitive-Emotional Decision Making

The IGT model of Wagar and Thagard (2004) specifically focuses on the production of covert emotional reactions. Their model particularly simulated the "somatic marker" finding of Bechara et al. (1994, 2000), whereby participants with intact OFC and amygdala developed an anticipatory SCR to disadvantageous decks of cards before being cognitively aware of the decks being disadvantageous. In order to model this behavior, they embedded the OFC and amygdala in a network that also included the ventral striatum as a behavioral gate. However, Maia and McClelland (2004) developed a more sensitive test showing the cognitive awareness appears earlier in the deck selection trials than Bechara et al. had found. Also, Wagar and Thagard lumped the two good decks (Decks 1 and 2) together despite their significant differences in variability, and likewise lumped the two bad decks (Decks 3 and 4) together. Hence, their model does not capture the more detailed probabilistic learning required of real-life IGT participants.

The IGT models of Frank and Claus (2006) and Levine et al. (2005) are also embedded in large networks that included parts of prefrontal cortex, amygdala, and basal ganglia. The Frank and Claus model is perhaps closer to the goal of a comprehensive model: it can also be applied to data on OFC involvement in reversal learning and distinguishes subsystems that learn magnitudes from those than learn probabilities or frequencies. Yet their model, like that of Wagar and Thagard, does not distinguish bad decks or good decks of different payoff variability. The Levine et al. model includes payoff variability and the conflict detection function of the anterior cingulate, but does not make the probability/magnitude distinction as clearly as does the Frank and Claus model. Both models deal only with probabilities learned by experience and do not reproduce data such as that of Tversky and Kahneman (1981) on one-shot learning from description; therefore, neither model deals with the description–experience gap described by Hertwig et al. (2004). Also, neither model deals with differences among resources of different affective richness (Rottenstreich and Hsee 2001).

Hence, the work of building a model that encompasses all these decision effects is still ahead of us. This chapter represents a start, explicitly specifying some parts of the model and laying out the principles required for other parts.

4.4 Organization of the Model

4.4.1 Fuzzy Emotional Traces and Adaptive Resonance

We will start the discussion of a proposed model with the first question we asked: how are probabilities represented in the brain? Since there is considerable behavioral evidence in support of an S-shaped probability weighting function such as the one shown in Fig. 4.1, we seek a plausible neurobiological substrate for that type of function. The dynamics of this neural substrate must be compatible with results suggesting that the curvature of the function is different for experience- and description-based decisions, and with other results suggesting that the curvature is different for choices involving affect-rich and affect-poor resources. Finally, the brain model that emerges needs to be based in a network structure that allows for the use of either deliberative, long-term utility maximizing decision rules or automatic, short-term heuristic decision rules.

One of the clues to understand the nonlinear probability weights, and the effects of language on those weights, arises from *fuzzy trace theory* (Reyna et al. 2003). Fuzzy trace theory posits the coexistence and interaction of two distinct systems for encoding information: literal or *verbatim* encoding and intuitive or *gist* encoding. Verbatim encoding means literal storage of facts, whereas gist encoding means storing the essential intuitive meaning or "gist" of a situation.

As Reyna et al. (2003) note, gist encoding of probabilities tends toward all-or-none representations of risk. For example, the gist encoding of an 80% probability of

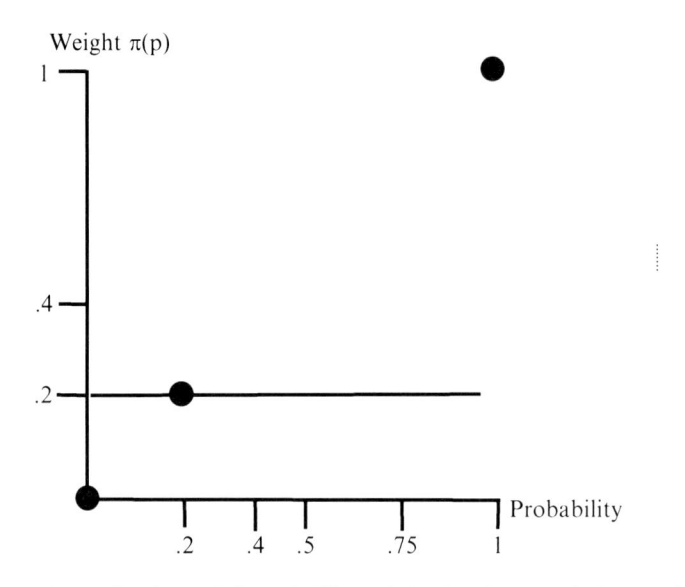

Fig. 4.3 Double step function ("gist") probability weights. Break-even point is set at 0.2, a round number within the plausible range suggested by data. Value is 0 at $p = 0$, 1 at $p = 1$, 0.2 at all other probabilities ("possible yes or no")

losing \$4,000 with a 20% probability of losing nothing is simply "some risk of losing \$4,000," whereas the gist encoding of a sure loss of \$3,000 is "certainty of losing \$3,000." Hence, gist encoding tends to reduce the relative attractiveness of sure losses and enhance the relative attractiveness of sure gains in comparison with risky alternatives, and deemphasizes the precise amounts gained or lost. The S-shaped function of Fig. 4.1 can be interpreted as some kind of nonlinear weighted average of an all-or-none step function arising from gist encoding (Fig. 4.3) and a linear function arising from verbatim encoding (Fig. 4.4).

The results of Rottenstreich and Hsee (2001) indicate that the S curve has sharper turns when the choice involves an affect-rich resource than when it involves an affect-poor resource. This suggests that the stronger the affective associations of the resource in question, the greater is the relative influence of the "gist encoding" indicated by Fig. 4.3 compared to the "verbatim encoding" of Fig. 4.4.

We add the caveat that the gist–verbatim distinction is *not* equivalent to the emotion–reason distinction that has been central to some other decision theories (e.g., Pacini and Epstein 1999). Reyna and Brainerd (2008) point to the gradual transformation from verbatim to gist encoding as the significance of items and their relationship to other items is learned during development, whether those items are emotional or not. Yet those authors also acknowledge that affect "impinges on every aspect of judgment and decision making, from the nature of gist representations … to the role of emotions as retrieval cues for knowledge and experience" (p. 97). Emotions play a particularly strong role in decisions that involve a preference between two or more alternatives, which are the major focus of this chapter.

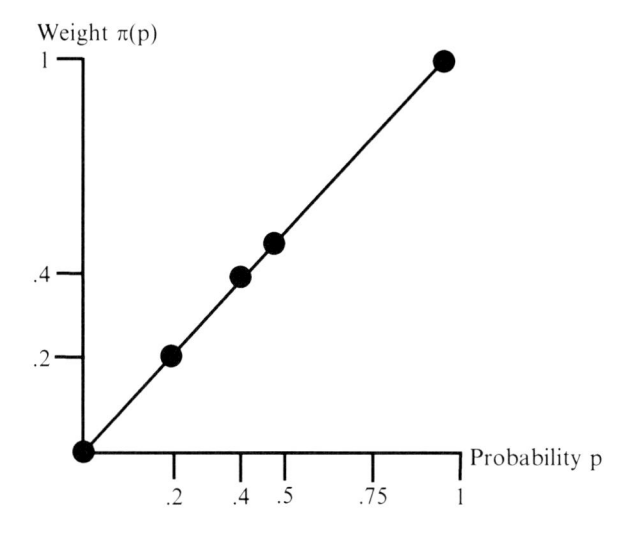

Fig. 4.4 Linear ("verbatim") probability weights

Hence, the gist processing that influences preferences on decision tasks seems likely to engage at least two brain regions that play major roles in emotional encoding: the amygdala and OFC. How might these two regions interact in the processing of emotional significance of sensory stimuli or of anticipated actions?

The amygdala and OFC are connected by extensive reciprocal pathways (see, e.g., Schoenbaum et al. 2003), and evidence points to the importance of those pathways for the consciousness of emotional experience (Barbas and Zikopoulos 2006). A wealth of data indicates that the OFC's representations of emotional values of stimuli are more sophisticated and abstract than the amygdala's. For example, stimulus-reinforcement associations can be more rapidly learned and reversed by OFC neurons than by amygdala neurons (Rolls 2006). Also, as discussed earlier, Bechara et al. (1994) showed that OFC damage, but not amygdalar damage, abolishes anticipatory autonomic responses to aversive events before they occur.

The results of Rolls, Bechara, Schoenbaum, and others suggest a hierarchical relationship between amygdala and OFC in the processing of emotional stimuli. Hierarchical connections between levels of processing have been modeled successfully by a variety of neural network architectures since the 1970s (see Levine 2000, Chap. 6 for a partial review). Hence, it seems plausible to model the amygdala–OFC relationship using one of those architectures. The model of that relationship discussed here is based on *adaptive resonance theory* or *ART* (Carpenter and Grossberg 1987; Carpenter et al. 1991a, b).

In its simplest form (*ART1*; Fig. 4.5), the ART network consists of two interconnected layers of nodes, called F_1 and F_2. F_1 consists of nodes that respond to input features and F_2 consists of nodes that respond to categories of F_1 node activity patterns. Interlayer synaptic connections are modifiable in both directions. F_2 nodes compete in a recurrent on-center off-surround network. Inhibition from F_2 to F_1

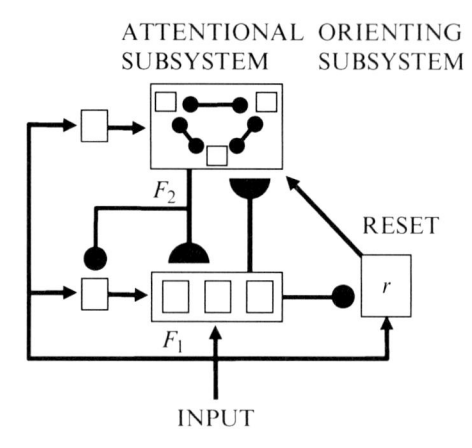

Fig. 4.5 ART 1 architecture. Short-term memory is encoded at the feature level F_1 and category level F_2, and learning at interlevel synapses. The orienting system generates F_2 reset when bottom-up and top-down patterns mismatch at F_1, that is, when some function representing match between those two patterns is less than the *vigilance r*. Arrows denote excitation, filled circles inhibition, and semicircles learning (adapted from Carpenter and Grossberg 1987, with the permission of Academic Press)

shuts off most F_1 activity if the input mismatches the active category's prototype. If the match is close, most of the F_1 nodes excited by the input are also excited by the active F_2 category node, making total F_1 activity large enough to overcome non-specific inhibition from F_2 and thereby inhibit the reset node. If mismatch occurs, by contrast, the reset node is activated and shuts off the active category node as long as the current input is present. Matching is said to occur when some function representing correspondence between top-down and bottom-up patterns is greater than a positive constant r called *vigilance*.

Carpenter and Grossberg (1987) proposed that adaptive resonance modules similar to the one shown in Fig. 4.5 exist in multiple brain regions. For our decision model, we interpret the level F_1 in Fig. 4.5 as a set of amygdalar encodings of positively and negatively valenced simple events, and F_2 as a set of OFC categories of compound events represented by F_1 activity patterns. We interpret the reset node as corresponding to some part of ACC, based on that brain region's role in conflict or error detection. Then how might the ART interactions explain the fMRI result of DeMartino et al. (2006) that OFC and ACC activity is greater for participants not subject to the framing effect while amygdalar activity is greater for participants subject to the framing effect?

Recall that the gist representations (Reyna and Brainerd 2008) of prospects involving probabilities of gains and losses tend to be simplified to the four categories of "sure gain," "risk of gain or no gain," "sure loss," or "risk of loss or no loss." These gist representations favor choices of sure gains over risky gains and choices of risky losses over sure losses, in accordance with the framing heuristic. Consider a network corresponding to a participant who is subjected to the framing effect, his or her decisions are presumably dominated by these simple all-or-none gist representations. Those four gain/loss categories could be encoded at the OFC, and the actual alternatives (e.g., gain \$400 with probability 80%) at the amygdala. There is resonant feedback between the amygdala and OFC, indicating that the emotional input is characterized stably as being in one of those four categories, and the resulting increased F_1 activity inhibits reset (i.e., ACC) activity. Both amygdala (F_1) and OFC (F_2) are activated by the input and the interlevel feedback, but the only F_2 nodes activated are the two corresponding to the "sure gain" and "risky gain" categories.

Now consider a network corresponding to a second participant not subject to the framing effect, for whom all-or-none gist representations are assumed to be weaker, and either verbatim or more nuanced gist representations stronger, than in the first participant. The network corresponding to that participant will have higher vigilance and may experience mismatches between those inputs and the corresponding simple categories, based on sensitivity to probability and/or magnitude information about the potential gains or losses. That mismatch means that F_1 activity no longer inhibits reset (i.e., ACC) activity. Also, there is greater activity at F_2 (i.e., OFC) than for the first participant because the all-or-none interpretation of the input is challenged and other (existing or novel) categories of gain-loss-probability configurations at that level are considered. The enhanced F_2 activity, in turn, nonspecifically inhibits F_1 (i.e., amygdalar) activity. This interpretation of the heuristics is consistent with the suggestion of Trepel et al. (2005) that the Prospect Theory probability weighting function of Fig. 4.1 is related to impulsiveness. It is also consistent with the network analysis of Frank and Claus (2006) suggesting that the OFC is required for precise discrimination of magnitudes of reward or punishment.

4.4.2 Effects of Learning

ART1 for binary inputs and its extension to continuous inputs, *fuzzy ART* (Carpenter et al. 1991a) deal with self-organization of input patterns from the environment. The supervised version of ART, called *ARTMAP*, was introduced in Carpenter et al. (1991a). The learning that takes place during development, or during the course of a decision task, is probably closer to the ARTMAP version because it involves learning the emotional significance of patterns even when that significance is not immediately apparent from their sensory properties. ARTMAP involves learnable interconnections between two ART modules that each can create stable recognition categories in different domains. Such an operation is important for learning that a stimulus previously regarded as neutral predicts consequences with positive or negative emotional significance. Hence, the OFC categorization layer of Fig. 4.1, which represents categories of amygdalar activation patterns, could be connected with another layer of nodes that represents contexts or environmental stimuli. In Frank and Claus (2006), such a layer is located in another part of the OFC that has stronger connections with other parts of association cortex representing semantic information.

For simplicity, we do not include another ART module or cortical semantic stores in the current model, but telescope the learnable interactions into connections between prefrontal cortex and amygdala that are modifiable by the experience of reward or punishment, as will be described in the next section. In Levine et al. (2005), these fronto-amygdalar connections enabled the network to learn "good" and "bad" characterizations of the card decks in the IGT, which were neutral stimuli when the task began, by a process such as occurs in classical conditioning. We assume the "good" and "bad" stimuli get classified in categories at the F_2 level of an ART network.

The difference between the resonant categorizations of learned and unlearned stimuli can help illuminate the description–experience gap in decision making (Barron and Erev 2003; Hertwig et al. 2004). Consider, for example, the following two scenarios: (a) being asked explicitly, and in less than a minute, to choose between the certainty of gaining \$30 and an 80% probability of gaining \$40; (b) choosing repeatedly between card decks A and B over 15 minutes to an hour, and learning from experience that A will yield \$30 each time, whereas B will yield \$40 approximately 80% of the time and nothing the other 20%. Scenario (a) involves fitting the choices given by the experimenter into one of four well-established gist categories: (1) sure gain, (2) risk of gain or no gain, (3) sure loss, and (4) risk of loss or no loss. Scenario (b) involves continual updating of the magnitude of affective evaluation of two card decks, a process for which the OFC is known to be crucial (Bechara et al. 1994; Frank and Claus 2006; Rolls 2006). Hence, in scenario (b), the OFC gist categorizes into which the alternatives would be more nuanced than (1) through (4). This explains why choices in scenario (b) would be more likely than choices in scenario (a) to violate the Tversky–Kahneman heuristics of risk aversion with gains and risk seeking with losses. Also, this leads me to predict that if the fMRI experiment of DeMartino et al. (2006) was repeated, but with the gambles or sure alternatives learned from experience instead of description, the differences in OFC, amygdala, and ACC activation between participants whose choices conform to the Tversky–Kahneman heuristics (who might no longer be a majority) and those whose choices violate the heuristics should be much reduced, if not disappear altogether.

4.4.3 Adaptive Resonance and Its Discontents: Relative Versus Absolute Emotional Values

How does "reset" take place in an adaptive resonance network? First a tentative choice is made of the category at the upper level F_2 in which to classify the input pattern, and that category is the main or the only one activated at that level (Carpenter and Grossberg 1987). Then, if the input is found to sufficiently mismatch the previously stored prototype of that category, activation of that category is disabled. At that time, either another category is tried or an uncommitted F_2 node is assigned to a new category for the input pattern.

In the original computational algorithm of Carpenter and Grossberg (1987), the deactivation of the mismatched active category node was simply accomplished through an instruction to set its activity to 0. The authors hinted that this deactivation could be accomplished within a real-time network dynamical system if F_2 were a *dipole field* whereby each node was replaced by "on" and "off" channels such as in a gated dipole (Grossberg 1972a, b; Fig. 4.2), which encodes pairs of (emotional, sensory, or motor) opposites such that one side of the pair is transiently activated by the offset of the other side. The dipole field was not explicitly contained in the algorithm of the original article, but was studied mathematically and computationally in a later article by Raijmakers and Molenaar (1997).

We add dipole fields to both the OFC and the amygdalar parts of the ART network of Fig. 4.5. The dipole field at the OFC level could be operating through the medial and lateral parts of the OFC itself, which have been identified as responding to inputs of positive and negative affective valence, respectively (Elliott et al. 2008). Alternatively, this dipole field could be operating through the OFC's connections with mutually inhibitory neural populations in the ventral striatum of the basal ganglia. The existence of parallel loops from prefrontal areas to basal ganglia to thalamus and then back to cortex, and their role in cognitively and emotionally influenced behavioral control, is well established (Alexander et al. 1990). Extensive interactions among the loops from OFC, ACC, and DLPFC in decision making have been identified (Haber et al. 2006). A dipole-like structure arises in the basal ganglia through the *direct pathway*, which is excitatory in its effects on the cortex, and the *indirect pathway*, which is inhibitory in its effects on the cortex. In some recent models of decision-making tasks (Frank and Claus 2006; Levine et al. 2005), positive emotional inputs activate the direct pathway and acquire positive incentive motivational properties, whereas negative emotional inputs activate the indirect pathway and acquire negative incentive properties. In another network model that deals with creative brainstorming (Iyer et al. 2009), the basal ganglia are interpreted as playing the role of "critic" that decides whether to accept or reject ideas that other parts of the network generate.

Gated dipole theory yields some clues about how to understand expectation and comparison effects at the neural level. Recall that both ABT (Grossberg and Gutowski 1987), which is based on the gated dipole, and DAT (Heyman et al. 2004; Mellers 2000) involve computation of the desirability of outcomes by comparisons with other possible outcomes, either from the choice the DM made or the choice the DM rejected. The dipole theory was developed as a neural instantiation of the psychological principle of opponent processing (e.g., Solomon and Corbit 1974). The direct and indirect basal ganglia pathways represent a type of opponent organization that is functionally, if not structurally, similar to a gated dipole. There appears to be another opponent processing system at the amygdala. It has long been known that a pathway located in the basolateral amygdala is potentiated during classical fear conditioning. Royer and Paré (2002) found neurophysiological evidence that there is a parallel pathway, located in intercalated amygdalar neurons, that is potentiated during extinction of conditioned fear.

The ART network of Fig. 4.5 with F_1 as amygdala and F_2 as OFC simply classifies potential alternatives in terms of emotionally desirable and undesirable properties. For these classifications to be translated into the behavior of choosing one alternative over another, that network needs to be connected to another network that actually decides on actions (either expressing a preference, as in the paradigm of Tversky and Kahneman 1981, or selecting one of two or more inputs, as in the paradigms of Barron and Erev 2003 and Bechara et al. 1994). The network for translating preferences into actions needs to include "dipoles" in the striatum or OFC and in the amygdala, as well as the connections of those areas to ACC and thalamus. We propose a synthesis of the ART network with a modification of the network developed by Levine et al. (2005) to simulate the IGT.

The network of Levine et al. (2005) includes separate modules for response planning and response execution. In that network, both response modules are placed in ACC because that region is part of a loop involving the ventral striatum that is interpreted as an action gate. However, it is more realistic that the actual response execution take place in premotor cortex, so we are modifying the network to place response execution in premotor with response planning still in ACC. Both layers of OFC encode positive and negative affective representations of all alternatives from which the choice is being made. The superficial layer of OFC sends conditioned reinforcer signals to the amygdala, and projects to a deep layer of OFC that also receives incentive motivational signals from medial prefrontal areas (Öngur and Price 2000). In this chapter, we treat both layers as stations along dipole pathways in the OFC. In addition to ACC and amygdala, the network includes representations of alternatives at the basal ganglia and thalamus for action gating.

The decision process begins by considering potential courses of action. Simultaneous activity at multiple action representations excites the conflict monitor, which is assumed to be part of the ACC. This conflict signal in turn activates medial prefrontal cortex, which suppresses amygdala-to-striatum bias signals. The alternative action whose activity is strongest is selected for further evaluation. The emergence of one predominating action representation in ACC leads to a decrease in conflict monitor activity that disinhibits the amygdala, allowing conditioned reinforcer signals to pass through amygdala to the action gate in the ventral striatum. Appetitive signals facilitate the direct pathway; aversive signals facilitate the indirect pathway. The gate opens (a thalamic node fires) when direct pathway input sufficiently counteracts indirect pathway input.

Figure 4.6 shows the network for affectively based decisions, which synthesizes ART, gated dipoles, and the IGT network of Levine et al. (2005). Modifiable connections from amygdala to OFC enable computation of expected reward or punishment values, in a similar manner to other network models of those regions that reproduce reversal learning and other conditioning data (Dranias et al. 2008; Frank and Claus 2006). Then the dipole field at the amygdalar level enables computation of experienced or anticipated emotions to be based partly on comparison of expected outcomes with other possible outcomes, from the choice made and from possible competing choices. The dipole field at the OFC level facilitates changes in

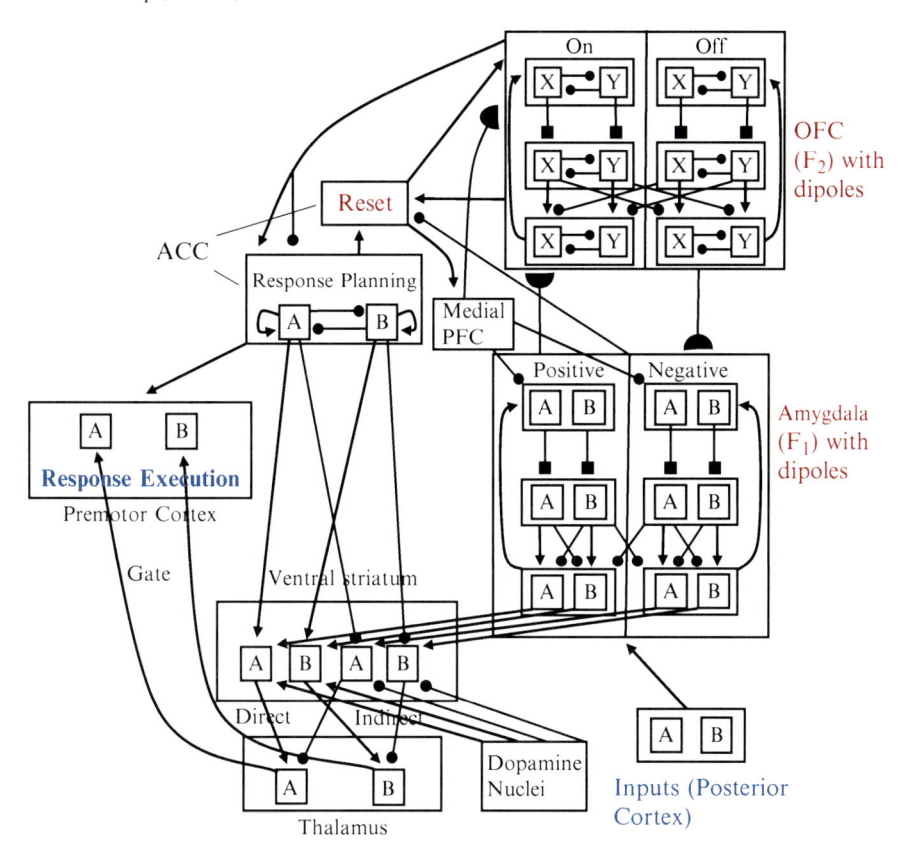

Fig. 4.6 Proposed neural network for making choices between risky alternatives. A and B represent (risky or sure) alternative stimuli or the responses to those stimuli. X and Y represent competing categories. The network combines several features of previous models: (**a**) An adaptive resonance module connects amygdala (affective input processing); OFC (categorization); and ACC (reset). (**b**) The amygdalar dipole field allows positive or negative affective values for each alternative to be compared with corresponding values for competing alternatives. Positive representations of each alternative compete both with negative representations of the same alternative and with positive representations of different alternatives. (**c**) The OFC dipole field allows removal from consideration of categories determined to be mismatched. (**d**) Positive amygdalar representations excite the striatal direct pathway which is part of the fronto-striato-thalamic loop for approach. Negative amygdalar representations excite the striatal indirect pathway which is part of the fronto-striato-thalamic loop for inhibiting approach

categorizations of inputs based on the outcomes of choosing those inputs. Also, top-down connections from OFC to amygdala provide a mechanism for the influence of cognitive appraisal on emotional reactions (Lazarus 1982).

The network model of Fig. 4.6 is compatible with data showing that the brain executive regions are sensitive to *both* relative and absolute desirability of alternatives (Blair et al. 2006; Elliott et al. 2008; Tremblay and Schultz 1999). The gated dipole fields allow for the influence of transmitter depletion from the presentation of other alternatives on the calculated desirability of a particular alternative, in accordance with the main ideas of both ABT and DAT. Yet relative preference is not the only influence on the executive system's response to the alternatives; otherwise, choices between a greater and lesser reward would be treated in the same way as choices between a lesser and a greater punishment. By contrast, the results of Blair et al. (2006) and others indicate that reward/reward choices activate different neural patterns than punishment/punishment choices. These investigators found that ACC activity was larger in choices between two punishments than in choices between two rewards. This suggests that, in addition to its role in monitoring conflict, the ACC might respond to a "discontent" signal indicating that none of the choices are desirable. As in the ART categorization network of Fig. 4.5, high ACC activity indicates that the network is open to different possible choices and can be influenced by semantic information from the environment. Preliminary behavioral results from my laboratory suggest that choices between gambles that involve losses or infrequent gains are more labile, that is, more subject to changes in mode of preference elicitation, than choices between gambles that involve frequent gains.

Such discontent signals are expected to be larger in those participants who are less subjected to framing effects, and who, in general, use more deliberative rather than heuristic strategies on judgment and decision tasks. The framing task of DeMartino et al. (2006) is not the most sensitive for making that distinction, and therefore engages little activity from the DLPFC. Yet the DLPFC is highly active in more complex decision tasks that engage its function of choosing the most appropriate actions based on current rules (Bunge 2004).

Now we add another layer to the network that enables coexistence of both deliberative and heuristic rules, with competition between the rules whose outcome is subjected to wide individual differences. In our theory, this requires an additional adaptive resonance module linking OFC with DLPFC.

4.4.4 Higher Level and Deliberative Rules

The results of DeMartino et al. (2006) suggest personality differences in susceptibility to framing effects in decision making. One personality factor that has been identified is *need for cognition* (NFC) (Cacioppo and Petty 1982; Cacioppo et al. 1996). NFC, typically measured by a questionnaire widely used in psychology laboratories, is defined as the tendency to enjoy effortful cognitive activity and the motivation to analyze carefully the source of arguments for particular viewpoints. High-NFC individuals have been found to be more rational and less subject to framing effects than low-NFC individuals on the Tversky–Kahneman type of decision tasks (Curşeu 2006).

Differences in NFC particularly influence performance on tasks that rely on accurate numerical calculation, whether or not the calculations are connected with emotional preferences. Examples of operations used in such tasks include updating prior probabilities in the face of new information and deciding which of two ratios is larger. There is much evidence that the DLPFC is required for performing those operations accurately. The evidence includes the fMRI results of DeNeys et al. (2008) showing that the DLPFC is most active in participants who do not neglect base rates in judging the likelihood of a person being a member of a particular class. An fMRI study is in progress (Krawczyk et al. 2008) to test the hypothesis that the DLPFC should also be most active in participants who are most logically accurate on a ratio bias task similar to that of Denes-Raj and Epstein (1994).

As the DLPFC codes properties at a higher level of abstraction than the OFC, including reward or punishment values of those abstract properties (Dias et al. 1996), I have conjectured elsewhere (Levine 2007, 2009a) that DLPFC bears roughly the same relationship to OFC as OFC does to amygdala. This conjecture suggests the operation of a three-layer hierarchical ART network, such as developed in Carpenter and Grossberg (1990). Figure 4.7 shows a three-layer ART network combining amygdala, OFC, ACC, and DLPFC.

Note that in an ART network, the reset node is influenced by both layers. The greater the influence of the upper level on the reset node, the more the network's categorizations or discussion will be influenced by higher level rules. Applying that reasoning to the upper ART module of Fig. 4.7 leads to the prediction that individuals high in NFC should have stronger synaptic connections from DLPFC to ACC than individuals low in NFC. Hypotheses dealing with connections rather than single brain regions have only begun to be tested in an fMRI setting but advances have been made in such neural pattern studies (O'Toole et al. 2007).

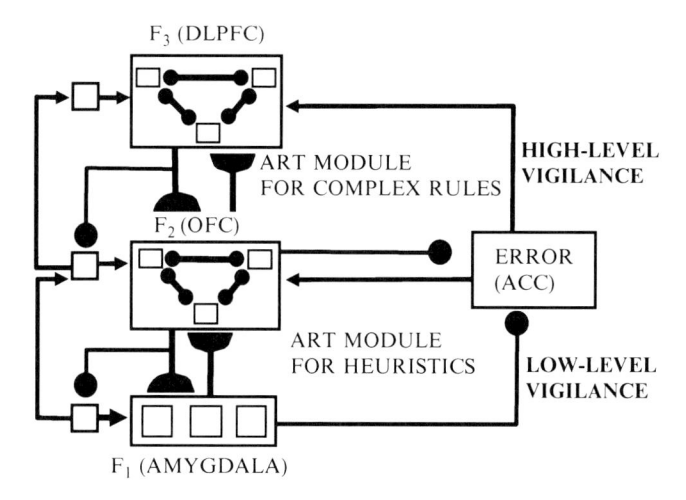

Fig. 4.7 Network that encodes both heuristic and deliberative rules. With low vigilance, the ART module that combines F_1 and F_2 (amygdala and OFC) makes decisions based on simple heuristics. With high vigilance, discontent with outcomes of simple decisions generates activity in the orienting (error) module (ACC). ACC activity in turn may generate a search for more complex decision rules at F_3 (DLPFC) (adapted from Levine 2007, copyright © IEEE, by permission of the publishers)

4.5 A Simplified Simulation

The article of Levine and Perlovsky (2008a) dealt with a simulation of a ratio bias task using simultaneous encoding of a heuristic and a deliberative rule. The network model utilized in this simulation was a very much simplified version of the DECIDER model, which includes parameters that mimic the overall functions we have ascribed in this chapter to ACC and DLPFC. Those "ACC" and "DLPFC" parameters help to determine which rule (heuristic or deliberative) is used, and how accurately, in a given context.

Participants in the ratio bias experiment of Denes-Raj and Epstein (1994) were assigned randomly either to a *win condition* or a *loss condition*. In the win condition, they were shown two bowls containing red and white jellybeans, told they would win a certain amount of money if they randomly selected a red jellybean, and instructed to choose which bowl gave them the best chance of winning money. In one of the bowls there were always 10 jellybeans, out of which 1 was red. In the other bowl there were 100 jellybeans, out of which some number greater than 1 but less than 10 were red. Hence, choice of the bowl with a larger frequency of red jellybeans was always nonoptimal, because the probability of drawing red from that bowl was less than 1/10. The loss condition used the same two bowls, but the participants were told they would lose a certain amount of money if they selected a red jellybean, so the bowl with more jellybeans was the optimal choice.

We review here only the simulation of the win condition; the loss condition data were also simulated by the same network with different parameter values. Figure 4.8 shows percentages of nonoptimal responses in the win condition. "Nonoptimal response size" in that graph means the difference between the chosen option and 10 out of 100, which was equivalent to 1 out of 10; that is, 1 represents the choice of 9 out of 100 over 1 out of 10, 2 represents the choice of 8 out of 100, and so on. Note that the majority of participants chose 9 out of 100 (nonoptimal response size 1) over 1 out of 10, and about a quarter still chose 5 out of 100 (nonoptimal response size 5) over 1 out of 10.

The decision between the two alternative gambles is based on either one of two rules, a *heuristic rule* based on frequencies and a *ratio rule* based on probabilities.

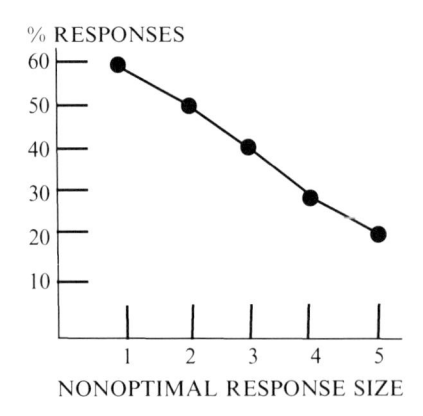

Fig. 4.8 Experimental results on percentages of time the higher-frequency, lower-probability alternative was chosen on win trials. See the text for explanation of nonoptimal response size (adapted from Denes-Raj and Epstein 1994, with permission of the second author)

The "ACC" parameter, called α, determines the likelihood of choosing the ratio rule for a given pair of gambles. If the ratio rule is chosen, the "DLPFC" parameter, called δ, determines the probability that the optimal response is made.

The heuristic rule is defined by the fuzzy concept (Zadeh 1965) of "much larger than 1." It is assumed that in the absence of sufficient ACC activity, decision is controlled by the amygdala using a rule "choose k out of 100 over 1 out of 10 if the numerator k is much larger than one." The denominator may affect the values of the fuzzy set that the rule generates but is then ignored. The fuzzy membership function of k in the "much larger" category, called $\psi(k)$, is set to be a ramp function that is linear between the values 0 at $k = 3$ and 1 at $k = 13$, hence,

$$\psi(k) = \begin{cases} 0, k < 3 \\ 0.1(k-3), \quad 3 \le k \le 13. \\ 1, k > 13 \end{cases} \tag{4.1}$$

The ACC parameter α, across all choices made by all participants in the experiment, varies uniformly over the interval $[0, 1]$. If the function $\psi(k)$ of (4.1) is less than or equal to α, the heuristic "much larger" rule is chosen. Otherwise, a rule of "largest ratio of numerator to denominator" is chosen, that is,

$$\begin{cases} \text{heuristic chosen if } \alpha \le \psi(k) \\ \text{ratio rule chosen if } \alpha > \psi(k) \end{cases}. \tag{4.2}$$

But that ratio rule, while more likely to lead to the choice of the higher probability alternative than the heuristic rule, does not guarantee the higher probability alternative (in this case, 1 out of 10) will be chosen. This is because of the imprecision of numerosity detectors in the parietal cortex (Piazza et al. 2004), which have been suggested as a neural substrate for imprecise numerical gists (Reyna and Brainerd 2008). We assume that the numerators and denominators of both alternatives (k, 100, 1, and 10), each activate a Gaussian distribution of parietal numerosity detectors. Hence, before the ratios are computed and compared, each of those numbers is multiplied by a normally distributed quantity with mean 1. Based on the DLPFC's working memory functions, we assume that DLPFC inputs to parietal cortex sharpen the tuning of these numerosity detectors. Hence, higher DLPFC activity should lead to a smaller standard deviation and thereby greater accuracy of relative probability estimations. Specifically, the standard deviation of each normal quantity is proportional to $1 - \delta$, with δ being the DLPFC parameter.

Hence, if the ratio rule is chosen, the nonoptimal choice of k out of 100 over 1 out of 10 is made if the *perceived* ratio of red jellybeans to total jellybeans is higher in the first alternative than in the second alternative. Based on the Gaussian perturbations of numerators and denominators described earlier, this means that a nonoptimal choice is made if and only if

$$\begin{cases} \dfrac{k(1+\varphi \, r_1)}{100(1+\varphi \, r_2)} > \dfrac{(1+\varphi \, r_3)}{10(1+\varphi \, r_4)} \end{cases}. \tag{4.3}$$

where $\varphi = (1 - \delta); 0, i = 1, 2, 3, 4$ are unit normals.

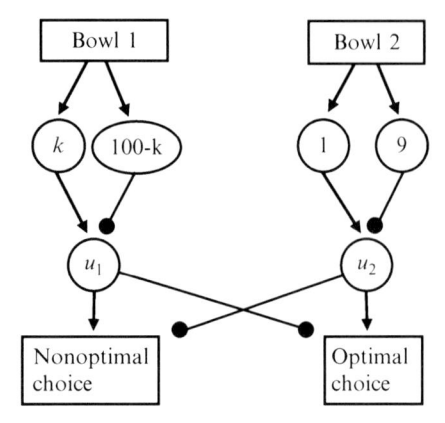

Fig. 4.9 Network representing choice between k-out-of-100 and 1-out-of-10 assuming use of the ratio rule. k out of 100 is interpreted as k good and 100–k bad; similarly, 1 out of 10 is 1 good and 9 bad. Probabilities of drawing red in each bowl are steady state values of (4.4) for node activity at u_1 and its analog at u_2, representing "utilities" of the two bowls (reprinted from Levine and Perlovsky 2008a, copyright © IEEE, by permission of the publishers)

Thus far we have described the simulation algorithm as a mathematical process without reference to a neural network diagram. However, ratios such as shown in (4.3) can be interpreted as steady states of a shunting on-center off-surround network, as follows. Present the two alternatives as inputs to the network shown in Fig. 4.9. Assuming perfect accuracy of numerical perceptions (otherwise the values k, 100, 10, and 1 in the circles of that figure are replaced by their normally perturbed values), the activity of the node u_1, representing the utility (i.e., reward value) of the bowl with k red out of 100, can be described by a nonlinear shunting differential equation with excitatory input k and inhibitory input $100 - k$:

$$\frac{du_1}{dt} = -\lambda u_1 + k(1 - u_1) - (100 - k)\,u_1, \tag{4.4}$$

Setting the left hand side of (4.4) to 0, we find that the steady state value of u_1 is $k/100$, which is exactly the probability of drawing a red jellybean from bowl 1. Similarly, the steady state value of u_2 is 1/10, the probability of drawing a red jellybean from bowl 2. The mutual nonrecurrent inhibition between those nodes leads to the choice of whichever bowl has the larger u_i value.

Returning to the algorithm, by (4.2), since α is uniformly distributed across [0, 1], the probability of the heuristic rule being chosen for a given value of k is $1 - \psi(k)$ as defined by (4.1). Assuming that the heuristic rule does not engage the ACC and thereby always leads to a nonoptimal choice, this means the probability of a nonoptimal choice is

$$\psi(k) + (1 - \psi(k))r(k), \tag{4.5}$$

where $r(k)$ is the probability that the inequality (4.3) holds, that is, the probability of a nonoptimal choice if the ratio rule is chosen. Hence, (4.5) was graphed as a

Fig. 4.10 Results of a neural network simulation of the Denes-Raj and Epstein (1994) experiment. The DLPFC parameter δ was normally distributed with mean 0.5 and standard deviation 0.25. The standard deviation of the normal variables by which ratios were multiplied was set to $0.1(1 - \delta)$ (reprinted from Levine and Perlovsky 2008a, copyright © IEEE, by permission of the publishers)

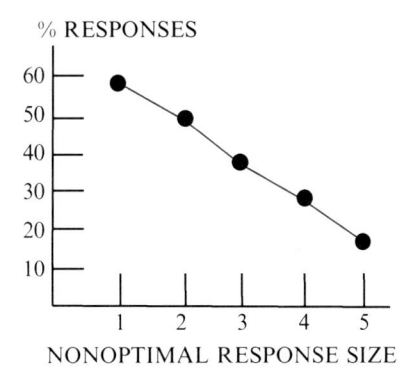

function of nonoptimal response size (which is equal to $10 - k$) in order to simulate the data shown in Fig. 4.8. This was done via Monte Carlo simulations in MATLAB R2006a, the program being run 1000 times with δ allowed to vary normally about a mean of 0.5 with standard deviation 0.25.

Figure 4.10 shows the results of this simulation of the win condition in the experiment of Denes-Raj and Epstein (1994). The simulation fits the data of Fig. 4.8 fairly closely, going from a maximum of over 60% nonoptimal responses for $k = 1$ to slightly above 20% nonoptimal responses for $k = 5$.

The model used in these simulations is much simpler than the full DECIDER model described in earlier sections, but captures some of the main principles of the larger model. It encodes the ability to switch between a heuristic and a deliberative rule and some functions of brain executive regions that mediate potential rule switches. The same model is likely to be generalizable to base rate neglect data (Krynski and Tenenbaum 2007) and to the development of the Prospect Theory curve of Fig. 4.1.

4.6 Concluding Remarks

4.6.1 Predictions and Syntheses

The neural network theory of emotionally influenced decision making based on Figs. 4.6 and 4.7 of this chapter leads to various predictions about the relationships between executive regions of prefrontal cortex and other key brain regions such as amygdala, striatum, and thalamus. Some of these predictions (notably about fMRI data) are scattered through this chapter's text, whereas others are implicit in the diagrams themselves.

As far as behavioral results on decision making, the model herein does not generate new predictions so much as confirm the overall insights (and thereby some of the predictions) of other modelers whose structural modules did not include explicit brain region assignments. The key features of decision-making models by Tversky and Kahneman, Barron and Erev, Busemeyer, Grossberg, Mellers, Reyna, and Brainerd, and their colleagues are fit into a common framework that is compatible with the known results on the cognitive-affective neuroscience of several brain regions.

Our theory also suggests the efficacy of paying more attention to some lines of behavioral research that have so far been under-represented in the decision-making literature. When we look at decision phenomena from the viewpoint of brain region involvement, various differences among tasks and paradigms emerge as important. For example, domains that differ in affective evocativeness, such as human lives versus money, are likely to elicit differential patterns of amygdalar–OFC interaction, which in turn influence tendencies toward risk aversion or risk seeking. Yet since the groundbreaking study of Rottenstreich and Hsee (2001), there have been few other studies that compared risky choice patterns across different types of resources. Weber et al. (2004), in a meta-analysis of dozens of decision studies (each involving a single resource domain), found that choices involving lives tended to be more risk seeking for losses and less risk averse for gains than comparable choices involving money. This pattern corresponds to Rottenstreich and Hsee's (2001) finding of a sharper low-probability portion of the S curve for emotionally evocative resources, but is opposite to Rottenstreich and Hsee's findings of a sharper high-probability portion of those same S curves. Yet the latter investigators noted that their findings concentrated most heavily on the low-probability choices, so the two studies may or may not actually be in conflict. Moreover, the Weber et al. (2004) analysis is between participants, and the network theory developed herein would be better tested by more *within-participant* studies along the Rottenstreich–Hsee lines, that is, involving comparable choices in multiple domains. A study is underway in my laboratory whereby the same participants make analogous choices involving gains and losses in two sessions separated in time, one involving an affect-rich resource (life of a pet), the other an affect-poor resource (life of a business; Ramirez and Levine (2009a). Tests of the theory should include animal as well as human studies: for example, we predict that monkeys will have a greater tendency to overweight small probabilities and underweight large probabilities when their choices involve a preferred food (e.g., banana) compared with a less-preferred food (e.g., lettuce).

We note that the gambling tasks that provide the bulk of our model's data base are a somewhat artificial decision environment, in that they involve objects (amounts of money) whose values are already well understood and known or learnable probabilities of obtaining or losing those objects. Yet the brain processes required to effectively perform those tasks include, and even go slightly beyond, conditioned learning of predicted rewards or punishments. For example, the MOTIVATOR model of Dranias et al. (2008) reproduces data on effects of the OFC and the amygdalar lesions in various animal classical conditioning paradigms, and in their model the connections from amygdala to OFC are modifiable but unidirectional. Yet the rule-driven decision-making paradigms discussed here require utilizing previously stored information for affective evaluation, a process that Bechara et al. (1994) found to be impaired by OFC damage. Hence we propose that such memory-driven appraisal is instantiated by modifiable OFC-to-amygdala connections, which have previously been implicated in cognitive control of emotional reactions (cf. Barbas and Zikopoulos 2006). The conditioning paradigms discussed in Dranias et al. (2008) can be explained without those reciprocal connections because those tasks do not require cognitive control.

At the higher level of the hierarchy, we have noted that the DLPFC is most involved in decision tasks that require significant deliberation. The fMRI study by DeMartino et al. (2006) of framing biases engaged some but relatively little DLPFC activity, whereas the study by DeNeys et al. (2008) of base rate neglect engaged substantial DLPFC activity that correlated with the sophistication of the participant's probability judgments. This observation suggests a research agenda of fMRI experiments on a large number of judgment and decision tasks differing both in their complexity and whether they possess definite correct or incorrect responses (as, e.g., the ratio bias task does but the framing task does not). The results of such imaging studies will enable more precise specification of the DLPFC's cognitive role, including the dynamics of that region's connections with OFC, ACC, striatum[1], and thalamus. Future development of the DECIDER model will incorporate those dynamics into a synthesis of Figs. 4.6 and 4.7 that can utilize either heuristic or deliberative rules on a wide range of decision tasks.

4.6.2   Extension to a Multi-Drive Multiattribute Decision Model

The parts of the model presented so far in this chapter assume that there is a single source of positively or negatively valenced emotion or value. However, many decision conflicts are based on multiple emotions or sources of value, and the multiplicity of drives in humans and other animals. These include, at least in humans, not just the standard physiological drives (e.g., hunger, thirst, sex, safety) but also such psychological drives as curiosity, need for stimulation, need for knowledge, and self-actualization. Levine (2009a) proposes an integration of the models presented in Figs. 4.4–4.6 with a competitive-cooperative network of drive representations, assumed to be located in the hypothalamus. The parts of the DECIDER network that include the hypothalamus and its decision-relevant connections with other brain areas are not shown here for space reasons but are depicted in Levine (2009a). The ensuing network is likely to lead to a brain-based refinement of previous network models of multiattribute decision making (e.g., Leven and Levine 1996; Roe et al. 2001).

---

[1] The DLPFC's direct connections with the striatum are predominantly with the dorsal portion of that region (caudate), not the ventral portion shown in Fig. 4.6. Yet there is also mutual influence between different cortico-striato-thalamic loops (Haber et al. 2006), so the DLPFC can indirectly influence the ventral striatum via the dorsal striatum.

4.6.3   The Larger Human Picture

My original attraction to the neural network field as a graduate student (in the late 1960s, before neural networks were widely known) was motivated by a desire to understand systematically the bases for real-life human choices. First of all, I asked what psychological motivations underlie societal choices leading to outcomes that are largely detrimental to our welfare, such as war, widespread poverty, and environmental degradation. Second, I discovered that at least one of my own significant choices (between several comparably attractive and prestigious graduate programs) had been based on a nonrational criterion that had not occurred to me when the school application process had begun. Yet human choices are far from being either uniformly harmful or uniformly irrational. Despite the fact that we create human miseries we are also motivated to reduce them, and despite the fact that we often decide based on hunches or short-term emotions, we can also often be systematic and deliberate. Also, the interplay between emotion and rationality enables a wider range of choices that is ultimately beneficial to our needs.

Since human decisions, judgments, and choices arise out of the workings of a natural system, it seemed likely to me that scientific study of this system would yield insights into the origins of decisions. This in turn would lead to some practical suggestions for structuring social institutions and human interactions in a manner that optimizes human welfare. Practical suggestions were not forthcoming in the late 1960s, but have been to emerge from the exponential growth in our knowledge of cognitive and affective neuroscience since then. A few articles have in fact suggested scientifically founded hypotheses that are potentially relevant to the structuring of societal institutions and customs (e.g., Eisler and Levine 2002; Levine 2009b; Levine and Perlovsky 2008b).

Yet the enormous complexity of the networks shown in Figs. 4.6 and 4.7, and other figures from the parts of the DECIDER model not shown here, illustrate the major challenges in understanding, and ultimately trying to influence, the process of decision making and mapping emotional values. The complex neural value and decision system depicted in these figures arose from conflicting evolutionary pressures. One of these pressures was to understand a complex environment as closely and accurately as possible, which Perlovsky (2006) has termed the *knowledge instinct*. Another pressure was to be able to respond quickly and reliability to routine situations that typically do not require the maximum use of our cognitive capacities. The combined need for knowledge maximization and effort minimization, in different circumstances, also required the development of an executive system for what Reyna and Brainerd (2008) term *task calibration*: appraising the requirements of the current situation to decide whether knowledge maximization or effort minimization is most appropriate. As Levine and Perlovsky (2008b) discuss, the imprecision of the executive task calibration system is one source of the difficulty of the human condition. Understanding how the executive system directs the interplay between emotion and cognition (Pessoa 2008), both in humans and in intelligent machines, is a challenge that computational cognitive neuroscience has just begun to meet.

References

Alexander G. E., Crutcher, M. D., and DeLong, M. R. (1990). Basal ganglia-thalamocortical circuits: parallel substrates for motor, oculomotor, "prefrontal" and "limbic" functions. *Progress in Brain Research*, *85*, 119–146.

Allais, P. M. (1953). Le comportement de l'homme rationnel devant le risqué: Critique des postulats et axiomes de l'école américaine. *Econometrica*, *21*, 503–546.

Barbas, H., and Zikopoulos, B. (2006). Sequential and parallel circuits for emotional processing in primate orbitofrontal cortex. In D. Zald and S. L. Rauch (Eds.), *The orbitofrontal cortex* (pp. 57–80). Oxford: Oxford University Press.

Barch, D. M., Braver, T. S., Akbudak, E., Conturo, T., Ollinger, J., and Snyder, A. (2001). Anterior cingulate cortex and response conflict: Effects of response modality and processing domain. *Cerebral Cortex*, *11*, 837–848.

Barron, G., and Erev, I. (2003). Small feedback-based decisions and their limited correspondence to description-based decisions. *Journal of Behavioral Decision Making*, *16*, 215–233.

Bechara, A., Damasio, A. R., Damasio, H., and Anderson, S. W. (1994). Insensitivity to future consequences following damage to human prefrontal cortex. *Cognition*, *50*, 7–15.

Bechara, A., Damasio, H., and Damasio, A. R. (2000). Emotion, decision making, and the orbitofrontal cortex. *Cerebral Cortex*, *10*, 295–307.

Blair, K., Marsh, A. A., Morton, J., Vythilingam, M., Jones, M., Mondillo, K., Pine, D. C., Drevets, W. C., and Blair, J. R. (2006). Choosing the lesser of two evils, the better of two goods: Specifying the roles of ventromedial prefrontal cortex and dorsal anterior cingulate in object choice. *Journal of Neuroscience*, *26*, 11379–11386.

Botvinick, M. M., Braver, T. S., Barch, D. M., Carter, C. S., and Cohen, J. D. (2001). Conflict monitoring and cognitive control. *Psychological Review*, *108*, 624–652.

Breiter, H. C., Aharon, I., Kahneman, D., Dale, A., and Shizgal, P. (2001). Functional imaging of neural responses to expectancy and experience of monetary gains and losses. *Neuron*, *30*, 619–639.

Brown, J. W., and Braver, T. S. (2005). Learned predictions of error likelihood in the anterior cingulate cortex. *Science*, *307*, 1118–1121.

Bunge, S. A. (2004). How we use rules to select actions: A review of evidence from cognitive neuroscience. *Cognitive, Affective, & Behavioral Neuroscience*, *4*, 564–579.

Busemeyer, J. R., Jessup, R. K., Johnson, J. G., and Townsend, J. T. (2006). Building bridges between neural models and complex decision making behavior. *Neural Networks*, *19*, 1047–1058.

Busemeyer, J. R., and Townsend, J. T. (1993). Decision field theory: A dynamic-cognitive approach to decision making in an uncertain environment. *Psychological Review*, *100*, 432–459.

Bush, G., Luu, P., and Posner, M. I. (2000). Cognitive and emotional influences in anterior cingulate cortex. *Trends in Cognitive Science*, *4*, 215–222.

Cacioppo, J. T., and Petty, R. E. (1982). The need for cognition. *Journal of Personality and Social Psychology*, *42*, 116–131.

Cacioppo, J. T., Petty, R. E., Feinstein, J. A., and Jarvis, W. B. G. (1996). Dispositional differences in cognitive motivation: The life and times of individuals varying in need for cognition. *Psychological Bulletin*, *119*, 197–253.

Carpenter, G. A., and Grossberg, S. (1987). A massively parallel architecture for a self-organizing neural pattern recognition machine. *Computer Vision, Graphics, and Image Processing*, *37*, 54–115.

Carpenter, G. A., Grossberg, S., and Reynolds, J. H. (1991a). ARTMAP: Supervised real-time learning and classification of nonstationary data by a self-organizing neural network. *Neural Networks*, *4*, 565–588.

Carpenter, G. A., Grossberg, S., and Rosen, D. B. (1991b). Fuzzy ART: Fast stable learning and categorization of analog patterns by an adaptive resonance system. *Neural Networks*, *4*, 759–771.

Coleman, S., Brown, V. R., Levine, D. S., and Mellgren, R. L. (2005). A neural network model of foraging decisions made under predation risk. *Cognitive, Affective, and Behavioral Neuroscience, 5*, 434–451.

Curşeu, P. L. (2006). Need for cognition and rationality in decision-making. *Studia Psychologica, 48*, 141–156.

Damasio, A. R. (1994). Descartes' error: Emotion, reason, and the human brain. New York: Grosset/Putnam.

DeMartino, B., Kumaran, D., Seymour, B., and Dolan, R. (2006). Frames, biases, and rational decision-making in the human brain. *Science, 313*, 684–687.

Denes-Raj, V., and Epstein, S. (1994). Conflict between intuitive and rational processing: When people behave against their better judgment. *Journal of Personality and Social Psychology, 66*, 819–829.

DeNeys, W., Vartanian, O., and Goel, V. (2008). Smarter than we think: When our brain detects we're biased. *Psychological Science, 19*, 483–489.

Dias, R., Robbins, T. W., and Roberts, A. C. (1996). Dissociation in prefrontal cortex of affective and attentional shifts. *Nature, 380*, 69–72.

Dranias, M., Grossberg, S., and Bullock, D. (2008). Dopaminergic and non-dopaminergic value systems in conditioning and outcome-specific revaluation. *Brain Research, 1238*, 239–287.

Eisler, R., and Levine, D. S. (2002). Nurture, nature, and caring: We are not prisoners of our genes. *Brain and Mind, 3*, 9–52.

Elliott, R., Agnew, Z., and Deakin, J. F. W. (2008). Medial orbitofrontal cortex codes relative rather than absolute value of financial rewards in humans. *European Journal of Neuroscience, 27*, 2213–2218.

Frank, M. J., and Claus, E. D. (2006). Anatomy of a decision: Striato-orbitofrontal interactions in reinforcement learning, decision making, and reversal. *Psychological Review, 113*, 300–326.

Grossberg, S. (1972a). A neural theory of punishment and avoidance. I. Qualitative theory. *Mathematical Biosciences, 15*, 39–67.

Grossberg, S. (1972b). A neural theory of punishment and avoidance. II. Quantitative theory. *Mathematical Biosciences, 15*, 253–285.

Grossberg, S. (1988). Nonlinear neural networks: Principles, mechanisms, and architectures. *Neural Networks, 1*, 17–61.

Grossberg, S., and Gutowski, W. (1987). Neural dynamics of decision making under risk: Affective balance and cognitive-emotional interactions. *Psychological Review, 94*, 300–318.

Haber, S. N., Kim, K.-S., Mailly, P., and Calzavara, R. (2006). Reward-related cortical inputs define a large striatal region in primates that interface with associative cortical connections, providing a substrate for incentive-based learning. *Journal of Neuroscience, 26*, 8368–8376.

Hertwig, R., Barron, G., Weber, E. U., and Erev, I. (2004). Decisions from experience and the effect of rare events in risky choice. *Psychological Science, 15*, 534–539.

Heyman, J., Mellers, B. A., Tishcenko, S., and Schwartz, A. (2004). I was pleased a moment ago: How pleasure varies with background and foreground reference points. *Motivation and Emotion, 28*, 65–83.

Huettel, S. A., and Misiurek, J. (2004). Modulation of prefrontal cortex activity by information toward a decision rule. *Neuroreport, 15*, 1883–1886.

Iyer, L. R., Doboli, S., Minai, A. A., Brown, V. R., Levine, D. S., and Paulus, P. B. (2009). Neural dynamics of idea generation and the effects of priming. *Neural Networks, 22*, 674–686.

Johnson, E. J., Schulte-Mecklenbeck, M., and Willemsen, M. C. (2008). Process models deserve process data: Comment on Brandstätter, Gigerenzer, and Hertwig (2006). *Psychological Review, 115*, 263–273.

Kahneman, D., and Tversky, A. (1973). On the psychology of prediction. *Psychological Review, 80*, 237–251.

Kahneman, D., and Tversky, A. (1990). Prospect theory: An analysis of decision under risk. In P. K. Moser (Ed.), *Rationality in action: Contemporary approaches* (pp. 140–170). New York: Cambridge University Press. (Originally published in Econometrica, 47, 263–291, 1979).

Krawczyk, D., Levine, D. S., Ramirez, P. A., Robinson, R., and Togun, I. (2008). *fMRI study of rational versus irrational choices on a ratio bias task*. Poster at the annual meeting of the Society for Judgment and Decision Making, Chicago.

Krynski, T. R., and Tenenbaum, J. B. (2007). The role of causality in judgment under uncertainty. *Journal of Experimental Psychology, General, 136*, 430–450.

Lazarus, R. S. (1982). Thoughts on the relations between emotion and cognition. *American Psychologist, 37*, 1019–1024.

Leven, S. J., and Levine, D. S. (1996). Multiattribute decision making in context: a dynamical neural network methodology. *Cognitive Science, 20*, 271–299.

Levine, D. S. (2000). *Introduction to neural and cognitive modeling* (2nd ed.) Mahwah, N. J.: Lawrence Erlbaum Associates.

Levine, D. S. (2007). *Seek simplicity and distrust it: Knowledge maximization versus effort minimization*. Proceedings of KIMAS 2007.

Levine, D. S. (2009a). Brain pathways for cognitive-emotional decision making in the human animal. *Neural Networks, 22*, 286–293.

Levine, D. S. (2009b). Where is utopia in the brain? *Utopian Studies, 20*, 249–274.

Levine, D. S., Mills, B. A., and Estrada, S. (2005). Modeling emotional influences on human decision making under risk. *Proceedings of International Joint Conference on Neural Networks*, 1657–1662.

Levine, D. S., and Perlovsky, L. I. (2008a). A network model of rational versus irrational choices on a probability maximization task. IEEE: *Proceedings of WCCI* 2008.

Levine, D. S., and Perlovsky, L. I. (2008b). Simplifying heuristics versus careful thinking: scientific analysis of millennial spiritual issues. *Zygon, 43*, 797–821.

Maia, T. V., and McClelland, J. L. (2004). A reexamination of the evidence for the somatic marker hypothesis: What participants really know in the Iowa gambling task. *Proceedings of the National Academy of Sciences, 101*, 16075–16080.

Mellers, B. (2000). Choice and the relative pleasure of consequences. *Psychological Bulletin, 126*, 910–924.

Öngur, D., and Price, J. L. (2000). The organization of networks within the orbital and medial prefrontal cortex of rats, monkeys and humans. *Cerebral Cortex, 10*, 206–219.

O'Toole, A. J., Jiang, F., Abdi, H., Pénard, N., Dunlop, J. P., and Parent, M. A. (2007). Theoretical, statistical, and practical perspectives on pattern-based classification approaches to the analysis of functional neuroimaging data. *Journal of Cognitive Neuroscience, 19*, 1735–1752.

Pacini, R., and Epstein, S. (1999). The relation of rational and experiential information processing styles to personality, basic beliefs, and the ratio-bias phenomenon. *Journal of Personality and Social Psychology, 76*, 972–987.

Perlovsky, L. I. (2006). Toward physics of the mind: Concepts, emotions, consciousness, and symbols. *Physics of Life Reviews, 3*, 23–55.

Pessoa, L. (2008). On the relationship between emotion and cognition. *Nature Reviews in Neuroscience, 9*, 148–158.

Piazza, M., Izard, V., Pinel, P., Le Bihan, D., and Dehaene, S. (2004). Tuning curves for approximate numerosity in the human intraparietal sulcus. *Neuron, 44*, 547–555.

Raijmakers, M. E. J., and Molenaar, P. C. M. (1997). Exact ART: A complete implementation of an ART network. *Neural Networks, 10*, 649–669.

Ramirez, P. A., and Levine, D. S. (2009). Expanding beyond the foundations of decision making: Perceived differences in the value of resources. Poster at the annual meeting of the Society for Judgment and Decision Making, Boston.

Reyna, V. F., and Brainerd, C. J. (2008). Numeracy, ratio bias, and denominator neglect in judgments of risk and probability. *Learning and Individual Differences, 18*, 89–107.

Reyna, V. F., Lloyd, F. J., and Brainerd, C. J. (2003). Memory, development, and rationality: An integrative theory of judgment and decision making. In S. Schneider and J. Shanteau (Eds.), *Emerging perspectives on judgment and decision making* (pp. 201–245). New York: Cambridge University Press.

Rieskamp, J. (2008). The probabilistic nature of preferential choice. *Journal of Experimental Psychology: Learning, Memory, and Cognition, 34*, 1446–1465.

Roe, R. M., Busemeyer, J. R., and Townsend, J. T. (2001). Multialternative decision field theory: A dynamic connectionist model of decision making. *Psychological Review, 108*, 370–392.

Rolls, E. T. (2000). The orbitofrontal cortex and reward. *Cerebral Cortex, 10*, 284–294.

Rolls, E. T. (2006). The neurophysiology and functions of the orbitofrontal cortex. In D. Zald and S. L. Rauch (Eds.), *The orbitofrontal cortex* (pp. 95–124). Oxford: Oxford University Press.

Rottenstreich, Y., and Hsee, C. (2001). Money, kisses, and electric shocks: On the affective psychology of risk. *Psychological Science, 12*, 185–190.

Royer, S., and Paré, D. (2002). Bidirectional synaptic plasticity in intercalated amygdala neurons and the extinction of conditioned fear responses. *Neuroscience, 115*, 455–462.

Schoenbaum, S., Setlow, B., Saddoris, M., and Gallagher, M. (2003). Encoding predicted outcome and acquired value in orbitofrontal cortex during cue sampling depends upon input from basolateral amygdala. *Neuron, 39*, 855–867.

Solomon, R. L., and Corbit, J. D. (1974). An opponent-process theory of motivation: I. Temporal dynamics of affect. *Psychological Review, 81*, 119–145.

Tremblay, L., and Schultz, W. (1999). Relative reward preference in primate orbitofrontal cortex. *Nature, 398*, 704–707.

Trepel, C., Fox, C. R., and Poldrack, R. A. (2005). Prospect theory on the brain? Toward a cognitive neuroscience of decision under risk. *Cognitive Brain Research, 23*, 34–50.

Tversky, A., and Kahneman, D. (1974). Judgment under uncertainty: Heuristics and biases. *Science, 185*, 1124–1131.

Tversky, A., and Kahneman, D. (1981). The framing of decisions and the rationality of choice. *Science, 211*, 453–458.

Tversky, A., and Kahneman, D. (2004). Advances in prospect theory: Cumulative representation of uncertainty. In E. Shafir, *Preference, belief, and similarity: Selected writings by Amos Tversky* (pp. 673–702). Cambridge, MA: MIT. This was originally published in Journal of Risk and Uncertainty, 5, 297–323, 1992.

Wagar, B., and Thagard, P. (2004). Spiking Phineas Gage. *Psychological Review, 111*, 67–79.

Weber, E. U., Shafir, S., and Blais, A.-R. (2004). Predicting risk sensitivity in humans and lower animssals: Risk as variance or coefficient of variation. *Psychological Review, 111*, 430–445.

Zadeh, L. (1965). Fuzzy sets. *Information and Control, 8*, 338–353.

Chapter 5
Computational Neuroscience Models: Error Monitoring, Conflict Resolution, and Decision Making

Joshua W. Brown and William H. Alexander

Abstract A critical part of the perception–reason–action cycle is performance monitoring, in which the outcomes of decisions and actions are monitored with respect to how well the actions are (or are not) achieving the desired goals. If the current behavior is not achieving the desired goals, then the problems must be detected as early as possible, and corrective action must be implemented as quickly as possible. The cognitive process of *performance monitoring* refers to the ability to detect such problems, and *cognitive control* (or executive control) refers to the processes of directing actions toward their intended goals and correcting failures to do so.

In this chapter, the basic neural mechanisms of goal-directed behavior are re-viewed through the lens of computational neural models on the topic, and the neural mechanisms of decision-making are reviewed briefly. We then go on to discuss how top-down goal representations are chosen, instantiated, strengthened, and when necessary, changed.

Progress on computational neuroscience models of cognitive control began with earlier, more abstract qualitative models. Norman and Shallice (1986) delineated a hierarchy of control with several components. In their framework, *schema* formed the basic building blocks that linked stimuli to corresponding responses. In some cases, schema might conflict with each other, and this required a *contention scheduling* mechanism to resolve the conflicts. This was conceived of as operating by lateral inhibition and competition between schemas, so that the dominant schema gains exclusive control of the output. In some cases of especially complex or novel situations, an appropriate schema might not be readily available, and so a *supervisory attentional system* must be invoked. This system redistributes attention to the most relevant stimuli, which then leads to the activation of an appropriate schema. This seminal paper by Norman and Shallice delineated qualitative roles for the interaction of bottom-up stimuli in driving responses (via schema and contention scheduling),

J.W. Brown (✉)
Department of Psychological and Brain Sciences, Indiana University, 1101 E. Tenth Street, Bloomington, IN 47405, USA
e-mail: jwmbrown@indiana.edu

V. Cutsuridis et al. (eds.), *Perception-Action Cycle: Models, Architectures, and Hardware*, Springer Series in Cognitive and Neural Systems 1, DOI 10.1007/978-1-4419-1452-1_5, © Springer Science+Business Media, LLC 2011

vs. top-down cognitive control (via the supervisory attentional system). A line of cognitive models called *production systems* has continued to develop in this computational cognitive tradition (Newell 1991; Meyer and Kieras 1994; Anderson et al. 2004). The main underlying theme of these models is the existence of a set of behavioral goals, which in turn motivate a set of rules that define how environmental states are mapped to actions. A set of *control* processes and mechanisms serve to instantiate these goals, and these mechanisms correspond essentially to the supervisory attentional system in the Norman and Shallice framework.

A cognitive system may have a variety of goals ranging from simple behaviors such as pressing a button to complex goals such as finding a mate. Goals may involve attaining a reward or avoiding an aversive event. A general requirement for adaptive behavior is that an agent should be capable of modifying its behavior moment-by-moment in order to achieve some goal while simultaneously adjusting for errors and changing environmental conditions. The ability to direct ongoing behavior toward specific desirable goals is seemingly a simple observation dating back to Thorndike's law of effect (Thorndike 1911). Nonetheless, the existence of more complex goal-directed behavior spawns a number of additional related questions. Once a goal is selected, how is progress toward the goal evaluated? Relatedly, what events, internal or external, underlie performance monitoring? In the course of ongoing behavior, when is cognitive control exerted? How does an organism learn what indicates the need for cognitive control?

These questions are related in that an answer to one implies answers to the others. In a simple instrumental conditioning task, for instance, changes in behavioral strategies may be indicated by the failure to receive a reward after generating a previously rewarded response, even without an explicit cue regarding what the correct response should be. Subsequently, a single error of this sort may drive changes in the responses the organism generates (Grant and Berg 1948; Rabbitt 1967; Laming 1968). In this case, cognitive control is exerted only after the erroneous response, and the need to learn the association between environmental cues and outcomes is minimal. This kind of cognitive control may be termed *reactive* (Braver et al. 2007).

On the other hand, what if we instead suppose that cognitive control is deployed as means to avoid an error in the first place rather than after the fact? This kind of control may be termed *proactive* (Braver et al. 2007). Often it is advantageous for the organism not only to adjust behavior in response to an error but also to learn what environmental factors are associated with the *possibility* of an error. This in turn generates additional questions: what factors in the environment does the organism associate with potential error? How is this association learned? The remainder of the chapter describes recent progress on computational models of performance monitoring and cognitive control.

5.1 Models of Cognitive Control

5.1.1 Biased Competition Model

Models of cognitive control attempt to identify how the interaction of top-down control systems and bottom-up perceptual systems interact to produce observed behavior of an animal. Such models may be purely algorithmic, without regard for the particular brain systems underlying behavior. Neurocomputational models, in addition to describing the computation underlying control, attempt to identify how such computation is implemented in the brain. Perhaps the most fundamental neural model of goal-directed cognition is the *biased competition model* (Miller and Cohen 2001), shown in Fig. 5.1 (Cohen et al. 1990). In this model, goal representations maintained in working memory by lateral prefrontal cortex (PFC) bias the competition between responses that may be made to the current bottom-up sensory stimuli. In that regard, the model follows closely the rubric established by Norman and Shallice (1986). The interaction between bottom-up and top-down influences can be seen in the classic example of the Stroop task (Stroop 1935). In this task, subjects may be asked to name the ink color of a color word, for example, to respond "green" when shown the word "RED" written in green ink. To perform correctly, subjects must suppress the strong tendency to read the word and instead substitute the correct response of naming the ink color. In the biased competition model, a representation of the goal of color naming in the model PFC increases the activation of units that name the ink color. The Stroop task exemplifies a fundamental decision between a prepotent stimulus that is inconsistent with a goal vs. a less potent option that is more compatible with a goal.

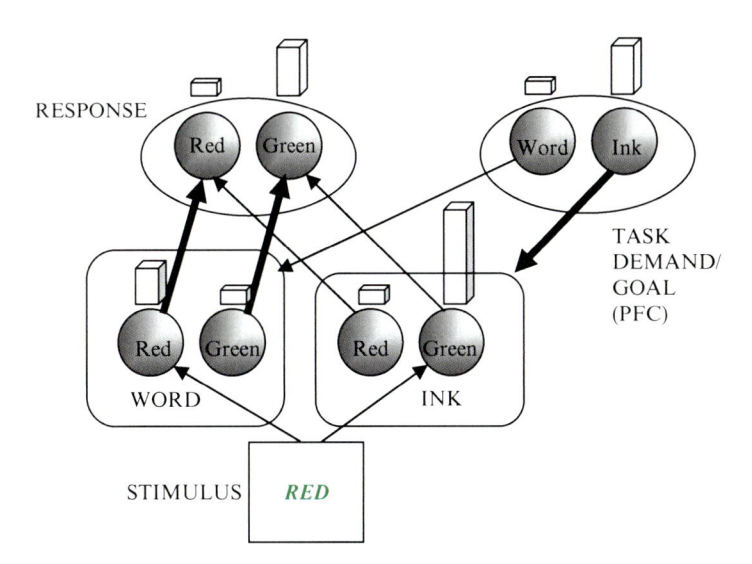

Fig. 5.1 The biased competition model (Miller and Cohen 2001). The appropriate but less prepotent response is generated due to increased goal representations of ink color naming in the PFC

5.1.2 Neural Models of Decision-Making

The biased competition model of decision-making in the Stroop task incorporates a specific model of decision making, namely an accumulator model with a fixed threshold. In these models, the firing rate of individual neurons representing various possible decisions increases over time, and the first representation to reach a threshold becomes the chosen action. At the neural level, it has been shown that action is initiated when neural activity reaches a fixed threshold, and the variability in decisions and reaction times is due to variability in the rate of rise of neural activity toward the fixed threshold (Hanes and Schall 1996; Schall and Boucher 2007). A number of models of the neural processes of decision-making have been proposed (Reddi and Carpenter 2000; Ratcliff and Smith 2004; Cutsuridis et al. 2007), and many of these different models have been shown to be mathematically equivalent as members of a class of drift-diffusion models (Bogacz et al. 2006). Other effects related to nonrational decision-making, such as preference reversals, have been modeled elsewhere (Busemeyer and Townsend 1993) but are beyond the scope of this chapter.

5.2 Medial Prefrontal Cortex and Performance Monitoring

With a basic framework for how goals are represented and subsequently influence decision-making, the remainder of this chapter will focus on how those goal representations are instantiated, strengthened, and changed by performance monitoring processes. The medial prefrontal cortex (mPFC), and especially the anterior cingulate cortex (ACC), has been a target of keen interest in the study of cognitive control and error processing. ACC is a target of projections from a diverse range of brain areas, including amygdala, involved in processing of emotional or affective information, prefrontal cortex, implicated in executive function, and midbrain structures underlying reward processing (Barbas and Pandya 1989). In turn, ACC innervates targets in dorsolateral prefrontal cortex (DLPFC), which has been identified as a brain region that implements control, as well as areas related to reward processing (nucleus accumbens and substantia nigra, via descending striatal projections) and structures involved in reward-related action selection (basal ganglia) (Eblen and Graybiel 1995).

Research on mPFC has suggested that ACC acts as a primary source of signals necessary for cognitive control and decision making. The strongest of these effects is the response to errors. Errors are a strong signal that the level of cognitive control must be increased, which may include an increased effort to implement the current strategy, or alternatively, a change of strategies. The error-related negativity (ERN) is an error-related activation that has been observed in human EEG studies (Hohnsbein et al. 1989; Gehring et al. 1990) and localized to a region in or near the ACC (Dehaene et al. 1994). Error signals have also been found in the ACC of monkeys (Gemba et al. 1986; Ito et al. 2003; Emeric et al. 2008).

The occurrence of an error can be considered a major cue for the need for increased cognitive control. Errors in an ongoing behavioral task lead to increased reaction times on subsequent trials, as well as decreased error rates (Rabbitt 1966; Laming 1979), both hallmarks of cognitive control manifested as increased caution (but see Mayr et al. 2003). While increased caution with the same strategy is appropriate in some cases, an error may also signal that the current strategy is not working satisfactorily and that a new approach should be tried (Shima and Tanji 1998; Bush et al. 2002). Conversely, lesions of ACC in monkeys lead to deficits in changing strategies in response to an error (Shima and Tanji 1998) and problems with error correction in humans (Modirrousta and Fellows 2008).

While the occurrence of an error indicates the need to reactively increase cognitive control, it would be even better if the potential for an error could be detected proactively so that errors could be pre-empted before they occur. In the late 1990s, a series of high-profile papers showed that ACC activity increased not only when an error was committed but also when there was a state of conflict between two mutually incompatible response processes (Carter et al. 1998), even when no error was committed. Further work showed that the ACC was specifically involved in performance monitoring rather than control (Botvinick et al. 1999). Cognitive control was more directly implemented by DLPFC, once activated by performance monitoring signals from ACC (MacDonald et al. 2000; Kerns et al. 2004).

5.2.1 Models of Performance Monitoring

5.2.1.1 Comparator Model

The early findings of error effects led to a straightforward interpretation that ACC computes discrepancies between actual and intended actions (Scheffers and Coles 2000) or outcomes (Holroyd and Coles 2002). Computationally, this can be implemented as a *subtractive* operation between a representation of the actual outcome vs. a representation of the desired goal. In this kind of model, the intended outcome is represented by neural activity. Then if the intended outcome actually occurs, the representations of the actual outcome inhibit the representation of the intended outcome. Conversely, if the intended outcome does not actually occur, then the representation of the intended outcome is not inhibited by the actual outcome, and this unopposed activity representing the intended outcome stands as an error signal.

5.2.1.2 Conflict Monitoring Model

The findings of response conflict effects in ACC led to the development of a computational model of conflict detection in ACC and how conflict detection leads to cognitive control signals (Botvinick et al. 2001). In the conflict model (Fig. 5.2), a signal of response conflict is computed as the Hopfield energy or

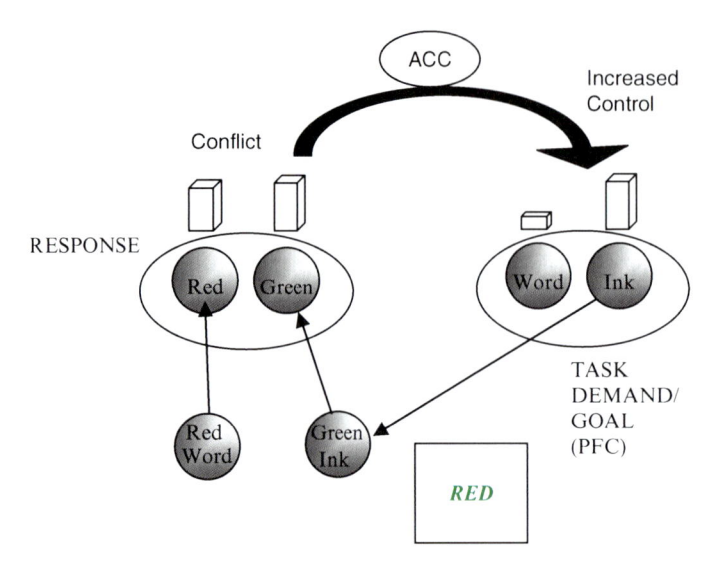

Fig. 5.2 In the Stroop task, response conflict is detected by ACC and in turn drives an increased level of cognitive control in the biased competition model. The effect of cognitive control here is to increase the processing of the task-relevant components of the stimulus

multiplicative product of the activations of two mutually incompatible response processes (Botvinick et al. 2001). This conflict model, when combined with the biased competition model, was able to capture a variety of putative control effects on behavior such as post-trial slowing and increases in reaction time (Botvinick et al. 2001), as well as attentional focusing (Posner and DiGirolamo 1998). Another instantiation of the conflict model was able to account for effects in ACC in sequences of trials, as measured with fMRI (Jones et al. 2002).

In the conflict model of ACC, decisions about which response to choose of several are biased by top-down attentional processes that reflect the amount of conflict present on previous trials. Conflict in the model reflects the level of coactivation of response output units over the course of a trial (Botvinick et al. 2001). On subsequent trials, the computed level of conflict acts as a control signal that biases attention to task stimuli. When only a single response is prepared, the computed level of conflict remains low. However, for trials in which multiple outcomes are indicated, calculated conflict rises as a result of two simultaneously active response processes that compete for access to model output.

Subsequent work with the conflict model argued that a single conflict signal may be sufficient to account for both error and conflict effects (Yeung et al. 2004). The reasoning was that in two-alternative forced choice (2AFC) tasks in which an error is made, the correct response representation is also likely to be highly active (just below the decision threshold), which leads to a state of response conflict. Conversely, when a correct response is made, the incorrect response is likely to be relatively inactive, resulting in less response conflict. Therefore, the authors argued that error effects in the ERN may be due to the presence of greater response conflict on error than correct trials (Yeung et al. 2004). This proposal may account for the ERN during error commission, although it is controversial (Burle et al. 2008). Nonetheless, it is unclear how response conflict may account for error effects when the error cannot be determined by the response itself, but must await subsequent feedback (the feedback ERN) (Holroyd and Coles 2002).

Overall, a successfully behaving agent should not only have the ability to select from a range of possible responses but also the ability to learn when a given action is likely to result in an undesirable result. In this interpretation, cognitive control is not merely enforced in order to avoid further errors after the fact, but rather control is enforced in order to avoid errors prior to their commission. In this case, the need for control must be detected by a combination of internal and environmental cues. In order to exert proactive control, ACC should show increased activation prior to feedback, reflecting the probability of committing an error in a given context.

5.2.1.3 Action Selection Model

Looking toward more proactive control, Holroyd and Coles (2002) proposed a model in which ACC learns to associate errors with the actions that lead to error via a temporal difference learning algorithm. In the model, the ACC is the recipient of ascending dopamine projections originating from the midbrain as well as inputs from brain areas associated with motor control. Transient depression in baseline activity of dopamine neurons due to an error (Schultz 1998) disinhibits the ACC component of the model. In that model, the effect is to drive a change in strategies, updating the values of motor controllers associated with commission of the error and allowing a new motor controller to become active and instantiate a different strategy. In this model, the ACC effectively acts as a control filter, determining which of perhaps several possible responses receives access to effectors.

5.3 Error Likelihood Model

Following Holroyd and Coles (2002), Brown and Braver (2005) proposed another model of proactive control, where error signals might not only activate ACC when an error occurs but also train it to respond more strongly to the conditions that preceded an error. This hypothesis was distilled as the error likelihood hypothesis: *the ACC activation will be proportional to the perceived likelihood of an error*. This hypothesis was then developed into the error Likelihood computational (EL) model (Brown and Braver 2005), as shown in Fig. 5.3. The EL model is an adaptive model of ACC. Each time an error occurs, cells in the model ACC are recruited to represent the conditions that preceded the error. Over time, the model ACC cells become allocated to represent task cues in proportion to how often those cues precede an error (Fig. 5.1). The core component of the Error Likelihood model is a self-organizing map (SOM) (Kohonen 1982), which learns representations of task stimuli using a

Fig. 5.3 Error likelihood model. Adapted with permission from (Brown and Braver 2005) © 2005 AAAS

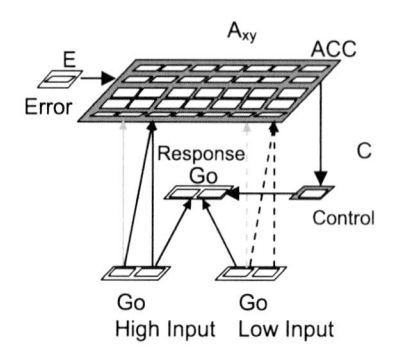

dopamine-gated learning signal. Following Holroyd and Coles (2002), the learning signal is the result of depression below baseline firing rates of dopamine neurons, which disinhibits units in the SOM. Stochastic sampling of units in the SOM combined with postsynaptic activity determines which units are eligible for learning on a given iteration of the model. On the occurrence of an error, weights between task-relevant stimuli and units in the SOM that have been randomly sampled or exceed a threshold of activity are incrementally updated. Over the course of training, activity of the adaptive ACC increases in response to the appearance of stimuli that are more predictive of error.

A second component of the model simulates task behavior. Input units to the model representing task-relevant cues project to both the adaptive SOM as well as to Response units. The Response units represent the continuous output of the network in a behavioral task, corresponding with behavioral options available. When a Response unit reaches a predefined threshold as described above, an action is generated and the network receives feedback (correct/error).

The adaptive SOM interacts with the behavior component both through feedforward connections from Input units as well as a feedback control loop via a Control unit. In general, a signal of high error likelihood might lead to a variety of possible control effects as appropriate to a given task. In the current computational model, activity in the Control unit suppresses activity in the Response layer, essentially slowing down the responses to minimize the chance of a premature response error. Over the course of learning, the adaptive SOM learns to associate stimuli and the likelihood with which those stimuli predict the occurrence of an error. As the adaptive SOM learns which stimuli are more indicative of future errors, its activity increases in response to those stimuli, resulting in activation of the Control unit and subsequent inhibition of responses.

5.3.1 Testing the Error Likelihood Model

The error likelihood model was first applied to a change signal task (CST) (Brown and Braver 2005). In the CST, a participant is presented with an arrow cue indicating

Fig. 5.4 Change signal task.
Adapted with permission
from (Brown and
Braver 2005) © 2005 AAAS

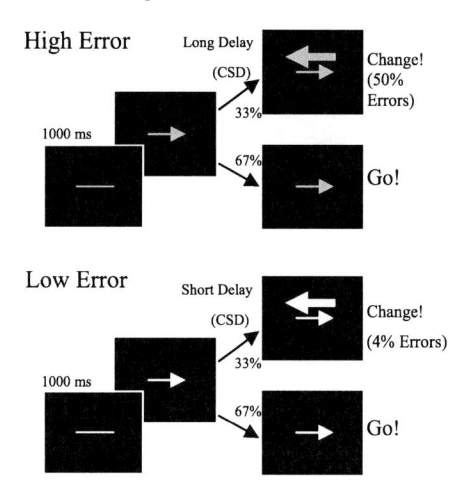

which of two responses should be made (Fig. 5.4). On a proportion of the trials, a
second cue (the change signal) is presented at a delay after the first cue. The time
interval between the onset of the first and second cues is the change signal delay
(CSD). The second cue indicates that the participant should, if possible, withhold
the response to the first cue and instead substitute a response to the second cue. The
longer the delay between the first and second cue, the more likely it is that the subject
will mistakenly respond to the first cue. There were two error likelihood conditions.
In the high error likelihood condition, a target error rate of 50% was maintained with
a relatively long CSD, while in the low error likelihood condition, error rates were
maintained at 5% with a relatively short CSD. The error likelihood model predicts
that activity in ACC should be greater in the high error likelihood condition than in
the low error likelihood condition. In comparison, the conflict model predicts that
ACC activity for both conditions should be the same.

An fMRI study with the CST confirmed the predictions of the Error Likelihood
model (Brown and Braver 2005). In addition to showing the increased activity in
ACC in response to error commission and effects of conflict during trials on which
a change signal was presented, ACC activity was observed to be higher for trials in
which no signal was presented in the high error likelihood condition, as compared
to the same trial type in the low error likelihood condition. The error likelihood
model is able to capture this pattern of activity, including both conflict and error
likelihood effects. The adaptive component of the model learns representations of
task stimuli in proportion to the number of times with which they have been paired
with an error. Task stimuli in the high error likelihood condition are therefore represented to a greater extent than for the low error likelihood condition. ACC activity in
the conflict model, in contrast, reflects only the presence of mutually incompatible
response cues.

While it is of use to construct a model that accounts for the results from a single
study, a key goal of computational neural models is to predict further results based
on the additional examination of the model. The error likelihood model, in addition

to accounting for the error likelihood effect described above, is also capable of generating novel, empirically testable predictions about ACC activity as well as human behavior.

In the error likelihood computational model, further simulations showed that the strength of the model ACC response depends both on the frequency with which a task cue is associated with error as well as the magnitude of the error signal. Therefore, a task condition in which an error has more severe consequences will be represented to a greater extent than a task condition in which the consequences of an error are relatively benign, even when the frequency of error commission is similar for both conditions. This *a priori* prediction suggests that ACC should show differential activity not only for different levels of error likelihood but also for different levels of expected severity of an error, even when no error actually occurs. Taken together, these predictions suggest that ACC responds to cues predicting both a greater probability of an error and greater severity of an error, the combination of which constitutes a measure of the expected risk of an action. Here, we define risk as the product of the probability of a loss and the potential magnitude of a loss, following the definitions used in operations research (Saaty 1987).

5.3.2 Risk

An fMRI study using a variant of the CST, the incentive change signal task (ICST) (Brown and Braver 2007), was conducted to test the model prediction of expected risk. The CST was expanded to include two factors, i.e., two levels of error likelihood (as in the previous study) and two levels of error consequences, for a total of four task conditions. As before, effects of error likelihood were observed in dorsal ACC. In addition, the same region showed effects of expected risk magnitude, as suggested by the error likelihood model. Interestingly, these effects were observed to be stronger in individuals who had scored low on a measurement of propensity to engage in financial risk taking (Weber et al. 2002), which suggests that greater ACC sensitivity to anticipated risk correlates with greater avoidance of risky behavior, consistent with previous findings (Fishbein et al. 2005; Magno et al. 2006).

This latter finding suggested a new hypothesis: perhaps some individuals are more risk averse because their ACC is more efficient at learning to signal risk based on the experience of past mistake in similar circumstances. To explore this, we conducted a new set of simulations of the error likelihood model in which the model learning rate parameter was manipulated to reflect differing levels of learning due to errors (Brown and Braver 2008). A high learning rate parameter leads to a better-learned representation of error likelihood in the model ACC, and it corresponds with risk-averse personality traits. Conversely, a low learning rate corresponds with risk-insensitive or risk-seeking personality traits. Simulations with the high learning rate yielded the expected effect of error likelihood. However, when the learning rate was reduced, error likelihood effects in the model were severely attenuated. Similarly, the effects of error magnitude were likewise reduced for the low learning rate simulations, as predicted.

Strikingly, manipulation of the model learning rate not only affected the error likelihood and error magnitude effects but also modulated the size of conflict effects within the error likelihood model. The conflict effect in the model was found to be larger for lower learning rates (i.e., decreased sensitivity to risk) than for higher learning rates. This agrees with fMRI data (Brown and Braver 2007), which show stronger conflict effects in risk-seeking individuals.

5.3.3 Multiple Response Effects

Further simulations of the error likelihood computational model predict that response conflict effects do not depend on the cued responses being mutually incompatible. This suggests that the well-known conflict effect may, in fact, reflect the consideration of multiple action plans, regardless of whether they are mutually incompatible. In the CST, a subject must generate one of two mutually incompatible responses on trials in which a change signal is presented. But what if the subject were instead instructed to generate *all* cued responses (i.e., both the originally cued "go" response as well as the response cued by the "change" signal)? The adaptive component of the ACC associates task-relevant stimuli with the commission of an error. In the CST, both change and go cues are associated with error, and so the error likelihood computational model is maximally active when both go and change responses are cued. Conversely, when only the go cue is present, activity in ACC is proportionately less.

This *multiple response hypothesis* of ACC was investigated in an fMRI study using two tasks, which were modified versions of the CST and the Eriksen Flanker task (Brown 2009b). The tasks were modified such that for half the trials, the subjects were instructed to generate all cued responses. This means that in Change trials, subjects were instructed to generate both left and right responses simultaneously. For the other half of the trials, they were instructed to generate only one of the cued responses. The trials were further divided such that half of all trials cued two (incongruent) responses, while the other half only cued one of two potential responses. For conditions in which the subject was instructed to generate one response and when two incongruent responses were cued, the typical conflict effect was observed in ACC. However, for conditions in which the subject was instructed to generate all responses, the same region likewise showed increased activation, despite the fact that the two responses were not mutually exclusive. This finding provides support for the multiple response hypothesis suggested by the error likelihood model, and it suggests a reinterpretation of response conflict effects.

A primary assumption of the error likelihood model is that the ACC learns to associate task-relevant stimuli with the occurrence of an error through a dopaminergic reinforcement signal. Specifically, errors in a reinforcement learning task are associated with transient depressions below baseline activity in midbrain dopamine neurons (Schultz 1998). Pauses in dopamine firing may result in disinhibition of neurons in the ACC (Holroyd and Coles 2002), allowing associations to be formed at the time of an error. While it is known that ACC is innervated by ascending projections from midbrain reward structures (Williams and Goldman-Rakic 1998), decisive evidence that dopamine actively inhibits ACC has not been found. An alternative hypothesis is that pauses in dopamine neurons are the result of inhibition of dopamine neurons by descending projections from ACC via striatum (Eblen and Graybiel 1995; Frank et al. 2007). Further work is needed to distinguish the precise mechanisms by which ACC processes and learns to predict error.

5.3.3.1 Cognitive Control Effects Driven by Error Likelihood, Conflict, and Errors

In the error likelihood model, cognitive control is implemented as a general inhibition of prepotent, cued responses (Brown and Braver 2005). Control is exerted proportionally to the activity of the adaptive ACC component of the model. In the context of the CST, this is reasonable because it slows responses, and delaying responses may better allow them to be countermanded later, thus avoiding an error. Nonetheless, other task situations may exist in which errors occur if a response is too slow instead of premature. Other models of cognitive control driven by conflict include both general slowing and attentional focusing (Botvinick et al. 2001). To test for the possibility of multiple simultaneous control effects, we conducted a behavioral study in which different error likelihood cues signaled the greater likelihood of premature vs. tardy errors. We found that the likelihood of each kind of error would lead to an appropriate control response of slowing down vs. speeding up, respectively (Brown 2009a). This suggests that ACC may compute the likelihood of multiple different kinds of errors and drive corresponding appropriate control effects. Similarly, activity in mPFC has been found in ERN studies to correlate with increased response times and lower error commission (Gehring et al. 1993), consistent with the notion that activity in ACC suppresses more automatic response processes in favor of more deliberative, top-down control. Evidence for this assumption is mixed, however. In some cases, mPFC activity (especially in the more dorsal pre-SMA has been observed to correlate with decreased response times (Forstmann et al. 2008), while other studies have observed that reaction times after surprising correct outcomes increase on subsequent trials (Notebaert et al. 2009). These findings suggest a more nuanced view of cognitive control instead of a single, scalar control signal. Likewise, we have found behavioral evidence for dissociation between control signals driven by response conflict vs. error likelihood (Brown 2009a). Increases in reaction times after surprising events, whether positive or negative, may also reflect additional processing time needed for the ACC to incorporate new information regarding environmental contingencies (Notebaert et al. 2009). Besides modulating performance, cognitive control signals may adaptively tune learning rates based on the anticipated level of uncertainty in the environment (Yu and Dayan 2005; Behrens et al. 2007).

5.4   Future Challenges

5.4.1   Reward as well as Error Likelihood?

The error likelihood model in its present form only simulates ACC neurons that are sensitive to error and error-related task cues. This simplification, while corresponding with the vast amount of literature on human ACC, is at odds with single-unit recording studies performed in monkey (Shima and Tanji 1998; Shidara and Richmond 2002; Ito et al. 2003; Matsumoto et al. 2003; Amiez et al. 2005, 2006; Nakamura et al. 2005). In these monkey studies, ACC and related areas are found to respond to other kinds of signals, including both errors and rewards, and units that reflect error and reward predictions. Recently, fMRI studies with humans have suggested that in addition to ACC activity reflecting error likelihood and risk magnitude, ACC also responds differentially to levels of expected value (Alexander and Brown, submitted).

If human ACC learns information regarding both error and reward (as is the case for other primates), this would present a significant challenge to the error likelihood model. Since the model, as originally formulated, has no mechanism by which the model ACC can learn information related to reward, a likely extension of the error likelihood model would involve implementing such a mechanism. In the present model, the ACC learns representations of error through depression of dopamine firing. One possible extension of the model is that unexpected rewards associated with phasic increases in dopamine activity may be used to train complementary neurons in the ACC to respond to anticipated reward. Alternatively, other neuromodulators may be involved in reward learning. Serotonin has been proposed as a neuromodulatory complement of the dopamine system (Daw et al. 2002); transient depressions in serotonergic neurons in response to reward may conceivably disinhibit ACC neurons to allow learning of reward-related information. Another possibility is that acetylcholine or noradrenaline, which have been suggested as mediating information regarding uncertainty (Yu and Dayan 2005), may be recruited in an extended model to allow a simulated ACC to process reward.

5.5   Toward a More Comprehensive Model
of Performance Monitoring

Looking forward, models of mPFC are faced with several open questions First, how can the proactive and reactive components of performance monitoring be reconciled? More specifically, how can mechanisms that yield predictive signals such as conflict and error likelihood be reconciled with reactive signals such as the detection of an error? Second, what is the functional role of both reward and error signals in performance monitoring and how can they be reconciled in a single model? To address these questions, we have begun working on a new computational neural model of medial PFC, the *predicted response-outcome model*. In this model, *mPFC first predicts the various potential outcomes of a planned action, including positive and negative outcomes, and then compares the predicted vs. the actual outcomes.* The new model casts the mPFC as predicting both what outcomes might occur as a result of the current actions (Walton et al. 2004; Rudebeck et al. 2008) and when they might occur, so that the comparisons of actual vs. predicted outcomes are performed in real time. The two components are mutually interdependent. The prediction mechanism provides the best estimate of the probability of each of the possible outcomes, and the discrepancies between actual and predicted outcomes provide a training signal that updates the predictions.

The new predicted response–outcome model has the potential to account for a variety of effects. First, error effects are seen as an actual outcome that was not predicted to occur with high probability. ACC is known to respond to unlikely events (Braver et al. 2001), and recent ERP studies suggest that the error effect in mPFC can reverse if errors are more likely than correct outcomes (Oliveira et al. 2007). Our own fMRI results suggest that ACC error signals are reduced as errors are more likely (Brown and Braver 2005). Second, conflict signals may be due to a prediction of multiple possible outcomes of a task rather than response conflict per se; this result is now suggested by recent findings that apparent response conflict effects persist even when task rules are changed to allow multiple conflicting responses to be generated simultaneously (Brown 2009b). Third, error likelihood effects may be seen as a special case of a more general mechanism that learns to predict a variety of possible outcomes, including correct or rewarding outcomes as well as errors. Fourth, studies of the ERN show that errors can be detected even before environmental feedback signals an error, meaning that the knowledge of a task and what response was made is sufficient to generate an error. The new model incorporates real-time predictions of outcomes in terms of both the movements made (i.e., an ERN) and the feedback received (i.e., the feedback ERN) (Holroyd and Coles 2002), so that the earliest reliable predictor of an unexpected outcome can be detected.

5.6   Concluding Remarks

The Error Likelihood model described in this chapter is a neurocomputational model, which suggests that a specific brain region, ACC, shows effects of errors, response conflict, and error likelihood, and in turn exerts control over behavior. While the ACC is considered a key locus of cognitive control, performance monitoring, and executive function, it is unlikely that this region is the sole area of the brain involved in these processes. Computational processes involved in executive control and decision making are likely conducted in a distributed fashion throughout cortex (Miller and Cohen 2001), with individual brain regions supplying information relevant to the generation of adaptive behaviors.

Acknowledgments Supported in part by AFOSR FA9550-07-1-0454, A NARSAD Young Investigator Award, the Sidney R. Baer, Jr. Foundation, R03 DA023462, and R01 DA026457.

References

Amiez C, Joseph JP, Procyk E (2005) Anterior cingulate error-related activity is modulated by predicted reward. Eur J Neurosci 21:3447–3452.

Amiez C, Joseph JP, Procyk E (2006) Reward encoding in the monkey anterior cingulate cortex. Cereb Cortex 16:1040–1055.

Anderson JR, Bothell D, Byrne MD, Douglass S, Lebiere C, Qin Y (2004) An integrated theory of the mind. Psychol Rev 111:1036–1060.

Barbas H, Pandya DN (1989) Architecture of intrinsic connections of the prefrontal cortex in the Rhesus monkey. J Comp Neurol 286:353–375.

Behrens TE, Woolrich MW, Walton ME, Rushworth MF (2007) Learning the value of information in an uncertain world. Nat Neurosci 10:1214–1221.

Bogacz R, Brown E, Moehlis J, Holmes P, Cohen JD (2006) The physics of optimal decision making: a formal analysis of models of performance in two-alternative forced-choice tasks. Psychol Rev 113:700–765.

Botvinick MM, Nystrom L, Fissel K, Carter CS, Cohen JD (1999) Conflict monitoring versus selection-for-action in anterior cingulate cortex. Nature 402:179–181.

Botvinick MM, Braver TS, Barch DM, Carter CS, Cohen JC (2001) Conflict monitoring and cognitive control. Psychol Rev 108:624–652.

Braver TS, Barch DM, Gray JR, Molfese DL, Snyder A (2001) Anterior cingulate cortex and response conflict: effects of frequency, inhibition, and errors. Cereb Cortex 11:825–836.

Braver TS, Gray JR, Burgess GC (2007) Explaining the many varieties of working memory variation: dual mechanisms of cognitive control. In: Variation of working memory (A. Conway CJ, M. Kane, A. Miyake, & J. Towse, eds). Oxford: Oxford University Press.

Brown J, Braver TS (2007) Risk prediction and aversion by anterior cingulate cortex. Cogn Affect Behav Neurosci 7:266–277.

Brown JW (2009a) Multiple cognitive control effects of error likelihood and conflict. Psychol Res 73:744–750.

Brown JW (2009b) Conflict effects without conflict in anterior cingulate cortex: multiple response effects and context specific representations. Neuroimage 47:334–341.

Brown JW, Braver TS (2005) Learned predictions of error likelihood in the anterior cingulate cortex. Science 307:1118–1121.

Brown JW, Braver TS (2008) A computational model of risk, conflict, and individual difference effects in the anterior cingulate cortex. Brain Res 1202:99–108.

Burle B, Roger C, Allain S, Vidal F, Hasbroucq T (2008) Error negativity does not reflect conflict: a reappraisal of conflict monitoring and anterior cingulate cortex activity. J Cogn Neurosci 20:1637–1655.

Busemeyer JR, Townsend JT (1993) Decision field theory: a dynamic-cognitive approach to decision making in an uncertain environment. Psychol Rev 100:432–459.

Bush G, Vogt BA, Holmes J, Dale AM, Greve D, Jenike MA (2002) Dorsal anterior cingulate cortex: a role in reward-based decision making. Proc Natl Acad Sci U S A 99:507–512.

Carter CS, Braver TS, Barch DM, Botvinick MM, Noll DC, Cohen JD (1998) Anterior cingulate cortex, error detection, and the online monitoring of performance. Science 280:747–749.

Cohen JD, Dunbar K, McClelland JL (1990) On the control of automatic processes: a parallel distributed processing account of the stroop effect. Psychol Rev 97:332–361.

Cutsuridis V, Smyrnis N, Evdokimidis I, Perantonis S (2007) A neural model of decision-making by the superior colliculus in an antisaccade task. Neural Netw 20:690–704.

Daw ND, Kakade S, Dayan P (2002) Opponent interactions between serotonin and dopamine. Neural Netw 15:603–616.

Dehaene S, Posner MI, Tucker DM (1994) Localization of a neural system for error detection and compensation. Psychol Sci 5:303–306.

Eblen F, Graybiel AM (1995) Highly restricted origin of prefrontal cortical inputs to striosomes in the macaque monkey. J Neurosci 15:5999–6013.

Emeric EE, Brown JW, Leslie M, Pouget P, Stuphorn V, Schall JD (2008) Performance monitoring local field potentials in the medial frontal cortex of primates: anterior cingulate cortex. J Neurophysiol 99:759–772.

Fishbein DH, Eldreth DL, Hyde C, Matochik JA, London ED, Contoreggi C, Kurian V, Kimes AS, Breeden A, Grant S (2005) Risky decision making and the anterior cingulate cortex in abstinent drug abusers and nonusers. Brain Res Cogn Brain Res 23:119–136.

Forstmann BU, Dutilh G, Brown S, Neumann J, von Cramon DY, Ridderinkhof KR, Wagenmakers EJ (2008) Striatum and pre-SMA facilitate decision-making under time pressure. Proc Natl Acad Sci U S A 105:17538–17542.

Frank MJ, D'Lauro C, Curran T (2007) Cross-task individual differences in error processing: neural, electrophysiological, and genetic components. Cogn Affect Behav Neurosci 7:297–308.

Gehring WJ, Coles MGH, Meyer DE, Donchin E (1990) The error-related negativity: an event-related potential accompanying errors. Psychophysiology 27:S34.

Gehring WJ, Goss B, Coles MGH, Meyer DE, Donchin E (1993) A neural system for error detection and compensation. Psychol Sci 4:385–390.

Gemba H, Sasaki K, Brooks VB (1986) 'Error' potentials in limbic cortex (anterior cingulate area 24) of monkeys during motor learning. Neurosci Lett 70:223–227.

Grant DA, Berg EA (1948) A behavioral analysis of degree of reinforcement and ease of shifting to new responses in a Weigl type card sorting problem. J Exp Psychol 38:404–411.

Hanes DP, Schall JD (1996) Neural control of voluntary movement initiation. Science 274:427–430.

Hohnsbein J, Falkenstein M, Hoorman J (1989) Error processing in visual and auditory choice reaction tasks. J Psychophysiol 3:32.

Holroyd CB, Coles MG (2002) The neural basis of human error processing: reinforcement learning, dopamine, and the error-related negativity. Psychol Rev 109:679–709.

Ito S, Stuphorn V, Brown J, Schall JD (2003) Performance monitoring by anterior cingulate cortex during saccade countermanding. Science 302:120–122.

Jones AD, Cho R, Nystrom LE, Cohen JD, Braver TS (2002) A computational model of anterior cingulate function in speeded response tasks: effects of frequency, sequence, and conflict. Cogn Affect Behav Neurosci 2:300–317.

Kerns JG, Cohen JD, MacDonald AW, 3rd, Cho RY, Stenger VA, Carter CS (2004) Anterior cingulate conflict monitoring and adjustments in control. Science 303:1023–1026.

Kohonen T (1982) Self-organized formation of topologically correct feature maps. Biol Cybern 43:59–69.

Laming D (1979) Choice reaction performance following an error. Acta Psychol 43:199–224.

Laming DRJ (1968) Information theory of choice reaction times. London: Academic.

MacDonald AW, Cohen JD, Stenger VA, Carter CS (2000) Dissociating the role of the dorsolateral prefrontal cortex and anterior cingulate cortex in cognitive control. Science 288:1835–1838.

Magno E, Foxe JJ, Molholm S, Robertson IH, Garavan H (2006) The anterior cingulate and error avoidance. J Neurosci 26:4769–4773.

Matsumoto K, Suzuki W, Tanaka K (2003) Neuronal correlates of goal-based motor selection in the prefrontal cortex. Science 301:229–232.

Mayr U, Awh E, Laurey P (2003) Conflict adaptation effects in the absence of executive control. Nat Neurosci 6:450–452.

Meyer DE, Kieras DE (1994) EPIC computational models of psychological refractory-period effects in human multiple-task performance. In. Ann Arbor: University of Michigan.

Miller EK, Cohen JD (2001) An integrative theory of prefrontal cortex function. Annu Rev Neurosci 21:167–202.

Modirrousta M, Fellows LK (2008) Dorsal medial prefrontal cortex plays a necessary role in rapid e rror prediction in humans. J Neurosci 28:14000–14005.

Nakamura K, Roesch MR, Olson CR (2005) Neuronal activity in macaque SEF and ACC during performance of tasks involving conflict. J Neurophysiol 93:884–908.

Newell A (1991) Unified theories of cognition. Cambridge: Cambridge University Press.

Norman D, Shallice T (1986) Attention to action: willed and automatic control of behavior. In: Consciousness and self regulation: advances in research and theory (Davidson R, Schwartz G, Shapiro D, eds). New York: Plenum.

Notebaert W, Houtman F, Opstal FV, Gevers W, Fias W, Verguts T (2009) Post-error slowing: an orienting account. Cognition 111:275–279.

Oliveira FT, McDonald JJ, Goodman D (2007) Performance monitoring in the anterior cingulate is not all error related: expectancy deviation and the representation of action-outcome associations. J Cogn Neurosci 19:1994–2004

Posner MI, DiGirolamo GJ (1998) Executive attention: conflict, target detection and cognitive control. In: The attentive brain (Parasuraman R, ed), pp 401–423. Cambridge: MIT.

Rabbitt P (1967) Time to detect errors as a function of factors affecting choice-response time. Acta Psychol (Amst) 27:131–142.

Rabbitt PMA (1966) Errors and error correction in choice-response tasks. J Exp Psychol 71:264–272.

Ratcliff R, Smith PL (2004) A comparison of sequential sampling models for two-choice reaction time. Psychol Rev 111:333–367.

Reddi BA, Carpenter RH (2000) The influence of urgency on decision time. Nat Neurosci 3:827–830.

Rudebeck PH, Behrens TE, Kennerley SW, Baxter MG, Buckley MJ, Walton ME, Rushworth MF (2008) Frontal cortex subregions play distinct roles in choices between actions and stimuli. J Neurosci 28:13775–13785.

Saaty TL (1987) Risk – its priority and probability: the analytic hierarchy process. Risk Anal 7:159–172.

Schall JD, Boucher L (2007) Executive control of gaze by the frontal lobes. Cogn Affect Behav Neurosci 7:396–412.

Scheffers MK, Coles MG (2000) Performance monitoring in a confusing world: error-related brain activity, judgments of response accuracy, and types of errors. J Exp Psychol Hum Percept Perform 26:141–151.

Schultz W (1998) Predictive reward signal of dopamine neurons. J Neurophysiol 80:1–27.

Shidara M, Richmond BJ (2002) Anterior cingulate: single neuronal signals related to degree of reward expectancy. Science 296:1709–1711.

Shima K, Tanji J (1998) Role of cingulate motor area cells in voluntary movement selection based on reward. Science 282:1335–1338.

Stroop JR (1935) Studies of interference in serial verbal reactions. J Exp Psychol 18:643–662.

Thorndike EL (1911) Animal intelligence: experimental studies. New York: MacMillan.

Walton ME, Devlin JT, Rushworth MF (2004) Interactions between decision making and performance monitoring within prefrontal cortex. Nat Neurosci 7:1259–1265.

Weber E, Blais A, Betz N (2002) A domain-specific risk-attitude scale: measuring risk perceptions and risk behaviors. J Behav Decis Mak 15:263–290.

Williams SM, Goldman-Rakic PS (1998) Widespread origin of the primate mesofrontal dopamine system. Cereb Cortex 8:321–345.

Yeung N, Cohen JD, Botvinick MM (2004) The neural basis of error detection: conflict monitoring and the error-related negativity. Psychol Rev 111:931–959.

Yu AJ, Dayan P (2005) Uncertainty, neuromodulation, and attention. Neuron 46:681–692.

Chapter 6
Neural Network Models for Reaching and Dexterous Manipulation in Humans and Anthropomorphic Robotic Systems

Rodolphe J. Gentili, Hyuk Oh, Javier Molina, and José L. Contreras-Vidal

Abstract One fundamental problem for the developing brain as well as for any artificial system aiming to control a complex kinematic mechanism, such as a redundant anthropomorphic limb or finger, is to learn internal models of sensorimotor transformations for reaching and grasping. This is a complex problem since the mapping between sensory and motor spaces is generally highly nonlinear and depends of the constraints imposed by the changing physical attributes of the limb and hand and the changes in the developing brain. Previous computational models suggested that the development of visuomotor behavior requires a certain amount of simultaneous exposure to patterned proprioceptive and visual stimulation under conditions of self-produced movement—referred to as 'motor babbling.' However, the anthropomorphic geometrical constraints specific to the human arm and finger have not been incorporated in these models for performance in 3D.

Here we propose a large scale neural network model composed of two modular components. The first module learns multiple internal inverse models of the kinematic features of an anthropomorphic arm and fingers having seven and four degree of freedom, respectively. Once the 3D inverse kinematics of the limb/finger are learned, the second module learns a simplified control strategy for the whole hand shaping during grasping tasks that provides a realistic coordination among fingers. These two bio-inspired neural models functionally mimic specific cortical features and are able to reproduce reaching and grasping human movements. The high modularity of this neural model makes it well suited as a high-level neuro-controller for planning and control of grasp motions in actual anthropomorphic robotic system.

J.L. Contreras-Vidal (✉)
School of Public Health, Department of Kinesiology, University of Maryland, College Park, MD 20742, USA
and
Graduate Program in Neuroscience and Cognitive Science (NACS), University of Maryland, College Park, MD 20742, USA
and
Graduate Program in Bioengineering, University of Maryland, School of Public Health, College Park, MD 20742, USA
e-mail: pepeum@umd.edu

V. Cutsuridis et al. (eds.), *Perception-Action Cycle: Models, Architectures, and Hardware*, Springer Series in Cognitive and Neural Systems 1, DOI 10.1007/978-1-4419-1452-1_6, © Springer Science+Business Media, LLC 2011

6.1 Introduction

One fundamental problem for the developing brain as well as for any robotic controller aiming to command a complex kinematic mechanism, such as a redundant, multijointed, anthropomorphic limb, is to learn internal models of forward and inverse sensorimotor transformations for reaching and grasping. This is a complex problem since the mapping between sensory and motor spaces is generally highly nonlinear and depends on the constraints imposed by the changes in the physical attributes of the human or robotic limb/hand as well as by changes in the environment. Moreover, the definition of a grasping posture associated with an object's intrinsic properties has been one of the most challenging problems since it implies the satisfaction of a large number of constraints related not only to the hand's and the object's structure but also to the requirement of the task and the state of the environment. Therefore, considering the numerous degrees of freedom (DOF) involved in the control of dexterous hands and robotic limbs, it is not surprising that both neuroscientists and roboticists have struggled to develop adaptive robot controllers for systems working in unknown environments (Conforto et al. 2009). More specifically, several questions arise: how the human brain coordinates and controls multijointed limbs taking into account task demands (e.g., "as fast as possible," "as accurate as possible," "with the least effort," etc.) or changing environmental conditions (e.g., force fields and other external forces affecting the limb's intended movement)? How is sensory information integrated with movement? How are objects perceived? And how do planning and movement control occur? To address these questions, motor neuroscientists and modelers have studied adaptive motor control using experimental and computational neuroscience approaches. Among these approaches, biologically plausible neural network modeling (e.g., Bullock et al. 1993; Grosse-Wentrup and Contreras-Vidal 2007; Guenther and Micci-Barreca 1997; Contreras-Vidal et al. 1997; Gentili et al. 2009a) and optimal feedback control theory (Todorov 2004) have attracted considerable interest.

Here we build on the neural network framework and present a large-scale (modular) neural network model that is able to learn the inverse kinematics of a 7 DOF anthropomorphic arm and 4 DOFs fingers, and perform accurate and realistic arm reaching as well as finger grasping movements. First, the neural network model will be introduced with a focus on what are the questions addressed and the level of detail of the model. Second, the relevant experimental and computational neuroscience literature will be reviewed in order to provide both experimental and theoretical arguments supporting the design of this neural model. Third, the architecture of the neural network model will be presented in detail including both its model components and parameters. In a fourth and last section, the simulations of the proposed neural network model will be presented and discussed in terms of its biological plausibility, the model assumptions and limitations along with future solutions that could challenge them.

6.2   Overview

6.2.1   Overview of the Neural Network Model for Arm Reaching and Grasping

The neural network model is based on the DIRECT (DIrection-to-Rotation Effector Control Transform) model of redundant reaching (Bullock et al. 1993; Guenther and Micci-Barreca 1997) that functionally reproduces the population vector coding process evidenced in the motor and premotor cortices (Georgopoulos et al. 1984, 1986). The model presented here is a neural network architecture that learns neural representations encoding the inverse kinematics of the arm/finger as well as grasp manipulation of different objects.[1] More precisely, this model learns inverse kinematics computations for the control of an anthropomorphic human arm and fingers with multiple DOFs. The acquisition of these neural representations is based on a strategy involving two learning stages (inverse kinematics and object manipulation stages) that uses previous learned representations for the encoding of subsequent knowledge. Thus, this neural model provides a substrate for a progressive (i.e., developmental) learning of grasping and manipulation tasks that can be achieved by means of selective reweighting of synaptic connections during the learning process without overriding established connections and thus, without "forgetting" what has been previously learned. Once the learning process is completed, the neuronal populations of the cortical network architecture reveal a comparable activity to that observed in populations of biological neurons. The high modularity of this neural model makes it well suited as a high-level neuro-controller for planning and control of grasp motions in actual anthropomorphic robotic system.

6.2.2   Modular Multinetwork Architecture for Learning Reaching, and Grasping Tasks

The model is composed of four modules corresponding to distinct neural networks that generally mimic the function of different biological functional circuits including the primary motor (M1), the dorsal and ventral premotor (PMd, PMv), the posterior parietal (PP), the posterior temporal (PT) cortices as well as subcortical areas such as cerebellum involved in the sensorimotor error computation (Albus1971; Doya 1999; Ito 1984; Marr 1969). The proposed neural model generates five major types of sensorimotor information that are involved in the control of visually

---

[1] The models described in this chapter focus on the inverse kinematic problems, ignoring the effects of inertial and interaction forces with external loads on planned prehension movements. Although the limb's dynamic computations are very important for the execution of planned kinematic trajectories, their study is outside the scope of this chapter. The reader is referred to Contreras-Vidal et al. (1997) and Bullock et al. (1998) for models of voluntary arm movements under variable force conditions that are compatible with the proposed model in this chapter.

guided grasps: (1) the neural drive conveying information about motor command for actual performance, (2) the proprioceptive information providing the current state of the limb/finger (e.g., angular position, velocity) resulting from the sensory consequences (i.e., feedback) of the motor commands, (3) the visual information related to the localization of the targets in the extra-personal space, the body segments and object properties, (4) the task and goal related information involved in motor planning, and finally, (5) the motor errors computed at subcortical centers that are employed in the model to tune the various local networks throughout the successive learning steps. By combining these five major types of sensorimotor information this neural architecture is able to perform adaptive reaching and grasping, as well as object manipulation.

Learning within each of the four modules comprising a distributed parieto–frontal network (Burnod et al. 1999; Contreras-Vidal et al. 1997; Grosse-Wentrup and Contreras-Vidal 2007; Molina-Vilaplana et al. 2009) follows a stage-wise learning approach. Such an approach can be described by considering two learning modules. The first learning module (LM1) learns the inverse kinematics of the arm and the fingers. Four different networks (one for the arm and three for each finger: index, middle, and thumb) learn separately the correspondence between visual, proprioceptive and motor command signals of the moving arm/finger. Upon learning, motor commands can be generated to perform arm/finger movements with various orientations to reach targets and grasp objects by contacting fingertips on pre-specified contact points on the object. This learning stage illustrates the acquisition of multimodal internal representations of the inverse kinematic for the arm and fingers. The second learning module (LM2) in turn learns the intrinsic properties of objects by employing adequate grasping postures. The relationship between visual features of an object and sensorimotor features of the hand are acquired by a local neural network (named GRASP). Once the learning is complete, hand motor commands are generated to achieve an appropriate grasp on the object.

6.3 Experimental and Computational Neurosciences Background

6.3.1 Review of Experimental Data Literature

The model presented here is based on the assumption that the connective architecture of cortex defines a learning architecture that provides the substrate for various stages of learning enabling a progressive learning of reaching and grasping. Two evolutionary developments appear to be crucial for the human cortical control of arm/hand movements and grasps. First, the increasingly dominant role played by the motor cortex and cortico-spinal tract (Porter and Lemon 1993), and second, the development of stereoscopic vision. In addition to these two developments, the motor cortex and its linkages to visual, proprioceptive, and somatosensory areas evolved into the major descending pathway for the visuo-motor control of human dexterity. Indeed, there is strong evidence that proprioception influences the control of ongoing movement (Ghez et al. 1994) as well as accounts for correction due to any disturbances encountered throughout the movement (Hulliger 1984). Moreover, while visual information is also essential for accurate reaching and pre-shaping of the grasp as well as for object manipulation (Jeannerod 1991), the parietal (Jeannerod 1994; Johnson et al. 1997; Pause et al. 1989; Sakata et al. 1995) and premotor (Rizzolatti et al. 1990, 1988) areas are also implicated in visually guided reaching and grasping. Specifically, the dorsal visual stream (Goodale and Milner 1992; Ungerleider and Mishkin 1982) projecting to the posterior parietal cortex carries critical information for visual control of the movement (Caminiti et al. 1996; Ferraina et al. 1997a, b). These areas are strongly and reciprocally linked to premotor areas (Wise et al. 1997). It must be noted that the analysis of the ontogenetic development of reaching and grasping suggest a progressive improvement in neural control (Wimmers et al. 1998). Namely, this suggests that beyond the emergence of new motor skills (Contreras-Vidal et al. 2005), there is also a progressive improvement that takes place on a longer time scale (Contreras-Vidal 2006). Also, from a behavioral point of view it is well established that 2D or 3D pointing movements produce sigmoid-shape displacements, velocity profiles that are single-peaked and bell-shaped as well as trajectories relatively straight (Abend et al. 1982; Morasso 1981; Gordon et al. 1994). It must be noted that endpoint trajectories were found not completely straight but instead gently curved in many parts of the workspace, particularly in the sagittal plane (e.g., Atkeson and Hollerbach 1985; Hollerbach et al. 1986). Also, the analysis of finger movements lead to very similar kinematics profiles (e.g., bell-shaped and single peaked velocity profiles) when investigated during human precision-grip movements (Grinyagin et al. 2005). Thus, the aim of this work is to extract principles from experimental results that guide learning of reaching and grasping by a neural model, which can provide a mechanistic account of neural circuits engaged in reaching and grasping, as well as a controller for anthropomorphic robots.

6.3.2 Review of Previous Modeling Attempts

Several computational research efforts that aimed to simulate arm kinematics have been proposed. A first approach was to employ minimization techniques to reproduce straight trajectories and bell-shaped velocity profiles. Such an approach aims to minimize some cost functions based on specific criterion (e.g., minimum jerk, Flash and Hogan 1985; minimum torque change, Uno et al. 1989). More recently the feedback optimal control theory based on the minimal intervention principle where task irrelevant deviations from the average behavior are left uncorrected to maximize performance (Todorov and Jordan 2002). Although interesting these approaches do not include learning and also are unrelated to any neural substrate. Alternatively, other researchers proposed neural models to learn the inverse kinematics with redundant manipulator (Bullock et al. 1993; Fiala 1994; Guenther and Micci-Barreca 1997; Molina-Vilaplana and López-Coronado 2006; Molina-Vilaplana et al. 2007).

These models were based on the computation of the movement direction using a process compatible with primate single cells properties in motor and premotor cortices whose activity level depend on movement direction (e.g., Caminiti et al. 1990; Georgopoulos et al. 1984, 1986; Kalaska et al. 1990). More recently, this type of model was implemented in robotic platforms to learn the control of a redundant robotic arm (Molina-Vilaplana et al. 2004; Pedreño-Molina et al. 2005). It must be noted that many different methods are available to solve the inverse kinematic problem when applied to redundant manipulators; however, most of these approaches do not take into account any neural mechanisms (e.g., Gan et al. 2005; Sheng et al. 2006). Therefore, here, we extended previous models (Guenther and Micci-Barreca 1997; Molina-Vilaplana et al. 2007) to include a neural component able to learn the inverse kinematics of a 7 DOFs redundant anthropomorphic human arm to perform accurate arm reaching movements in a 3D workspace.

In a previous work, grasping was commonly approached either by employing optimization processes or control composition (Gorce and Fontaine 1996; Kerr and Roth 1986; Pons et al. 1999; Woelfl and Pfeiffer 1995). Different human grasping classifications (Iberall 1997; Napier 1956) were used to design solutions to the grasp shape planning problem in robotics. Engineers, mathematicians, computer and cognitive scientists have designed various forms of computational architectures that carry out the mapping between input information and the observed output prehensile behavior. These systems were designed using artificial intelligence programming techniques and expert systems (Cutkosky 1989; Cutkosky and Howe 1990; Iberall 1987; Iberall et al. 1988; Kang and Ikeuchi 1997). Recently, an emerging approach based on neural networks has been employed to define grasping configurations or to learn the relationship between object shapes and hand configuration (called grasp choice; Guigon et al. 1994; Kuperstein 1991; Uno et al. 1995; Taha et al. 1997). These studies emphasized the correspondence between an object and a hand shape. However, it is important to note that the same grasping posture can be employed to grasp objects with various shapes. Also, instead of considering the general shape of an object, the recognition of its graspable parts may be an important feature that can affect the grasping (i.e., the choice of the graspable parts with suitable dimensions allowing the grasp). Thus, the graspable parts of an object have been called graspable features (Moussa and Kamel 1998) or grasp affordances (Fagg and Arbib 1998). If a neural model can acquire a representation of these graspable features or affordances, such a model can deal with additional situations and thus allows an enhanced flexibility for planning and execution of grasp. Moreover, it is expected to integrate task requirements to choose the appropriate graspable features depending on the context. For instance, if we want to drink a glass of water, instead of grasping it from the top part, considering its circumference being as a grasp feature, we will grasp it from the lateral side, which is another grasp feature more appropriate to drink the contain of this glass. Thus, by following such an approach, Moussa and Kamel (1998) proposed a computational architecture that would enable to learn grasping rules called "generic grasping functions." Such general grasping rules could be used for gross planning while a more local grasping strategy (related to a final grasp synthesis) would be used once the object surface has been touched.

Molina-Vilaplana and López-Coronado (2006) proposed a simplified control strategy for the whole hand shaping during grasping tasks that provides a realistic coordination among fingers. This strategy is based on the increasing body of empirical evidence that supports the view of a synergistic control of the whole fingers during prehension (Braido and Zhang 2004; Mason et al. 2001; Santello et al. 2002). By employing such an approach, only two parameters are needed to define the evolution of hand shape during the task performance. This scheme constitutes a powerful dimensionality reduction stage that could simplify the learning of mapping previously mentioned.

6.4 The Neural Network Model Architecture

6.4.1 Model Components

6.4.1.1 The Basic Module

Sensorimotor transformations or multimodal internal representations of objects and effectors are implemented by the interaction of various cortical modules. Notably, computation, learning, and memorization of such transformations are performed by similar neuronal substrates (Pouget and Snyder 2000). Such neuronal substrate includes neurons that have sigmoidal or Gaussian firing curves in response to postural and visual stimuli. These neurons provide what Regularization theory (Tikhonov and Arsenin 1977) describes as a set of basis functions (BF) that provide the possibility to approximate through learning any nonlinear and multivariate mapping by means of simple linear combination of these BF. Moreover, it was demonstrated that the regularized solution to a problem of approximating a nonlinear multivariate function can be expressed in terms of multilayer neural networks called regularization networks or HYPerBasis Functions networks (HYPBF; Poggio and Girosi 1989). These networks are able to implement the approximation (denoted f*(x)) of the multivariate function (denoted f(x)) by employing the following equation:

$$f(x) \approx f^{*}(x) = \sum_{\alpha=1}^{n} c_{\alpha} h_{\alpha} = \sum_{\alpha=1}^{n} c_{\alpha} G\left(\|x - t_{\alpha}\|_{W}^{2}\right), \qquad (6.1)$$

where the parameters t_{α} (called "centers") define the response of the neuron in the hidden layer $\left(h_{\alpha} = G\left(\|x - t_{\alpha}\|_{W}^{2}\right)\right)$ and coefficients c_{α} are the synaptic weights between the neurons located in the hidden and output layers. G and x are the radial function and input vector, respectively.

The norm in (6.1) is a weighted norm

$$\|x - t_{\alpha}\|_{W}^{2} = (x - t_{\alpha})^{\mathrm{T}} W^{\mathrm{T}} W (x - t_{\alpha}), \qquad (6.2)$$

where W is a square matrix unknown and ^{T} denotes the transpose operator. The elements of the diagonal w_i in the matrix W correspond to the weight for each coordinate of the x input vector. Thus, the neural network can be described as follows: Once the learning (e.g., using gradient descent) is complete, the centers t_α, matrix W and coefficients c_α are tuned.

The centers of neurons in the hidden layer or BF correspond to prototype points in the multidimensional input space. Each neuron of the hidden layer computes a weighted distance (with the matrix W) between the input vector x and its associated centre. When considering Gaussian functions as BF, neurons in the hidden layer respond with a maximal activity when the input vector provided to the network perfectly matches the center of these neurons. Conversely, if the matching is reduced, this will lead to a reduction of the activity of the neurons located in the hidden layer depending on the receptive field (or width of the BFs) employed to model the neurons. Thus, beyond the weight associated with the neuron, two additional parameters can be adapted through learning and play an important role in the approximation of the nonlinear multidimensional mapping: (1) the position of the centers of the neurons and (2) the width of the receptive field of the neurons. Thus, the positions of the centers and the widths of the receptive fields for each neuron can also be adapted during learning. This corresponds to a process of input data clustering and the search for optimum values of W that corresponds to reduce the dimensionality of the problem (Poggio and Girosi 1989). The linear combination of the weighted activity of all neurons in the hidden layer will provide the outputs of this neural network.

6.4.1.2 Learning the Inverse Kinematics of the Arm and Fingers: LM1

This section describes the learning of the inverse kinematics of the arm and the three fingers (index, middle, and thumb) that constitute the virtual arm and hand by means of four HYPBF neural networks. The arm and each of the fingers has 7 and 4 DOFs, respectively, and each HYPBF neural network learns the inverse kinematics mapping. Learning is induced by action-perception cycles (Fig. 6.1).

The modeling of the geometrical models included an actual human redundant arm/fingers including 7 DOFs and 4 DOFs, respectively. The forward models for the geometry of the arm/fingers were obtained by employing the Denavit–Hartenberg (HartenBerg and Denavit, 1964) parameterization (Fig. 6.2).

The learning scheme showed in Fig. 6.1 allowed the acquisition of the correspondence between spatial displacements of the arm/fingertip and the motor commands that generate these displacements at each joints. Such a correspondence is modulated by the actual configuration of the arm/finger. For the arm and each finger, the implementation used here is an extension of a previously proposed approach for a 3 DOFs redundant manipulators (Guenther and Micci-Barreca 1997). Each HYPBF neural network was trained to learn its associated inverse kinematics. The spatial directions Δx in the 3D Cartesian space are mapped into joint increments $\Delta \theta$ for each arm and finger joint by means of the following equation:

$$\Delta \theta = A(\theta).\Delta x, \qquad (6.3)$$

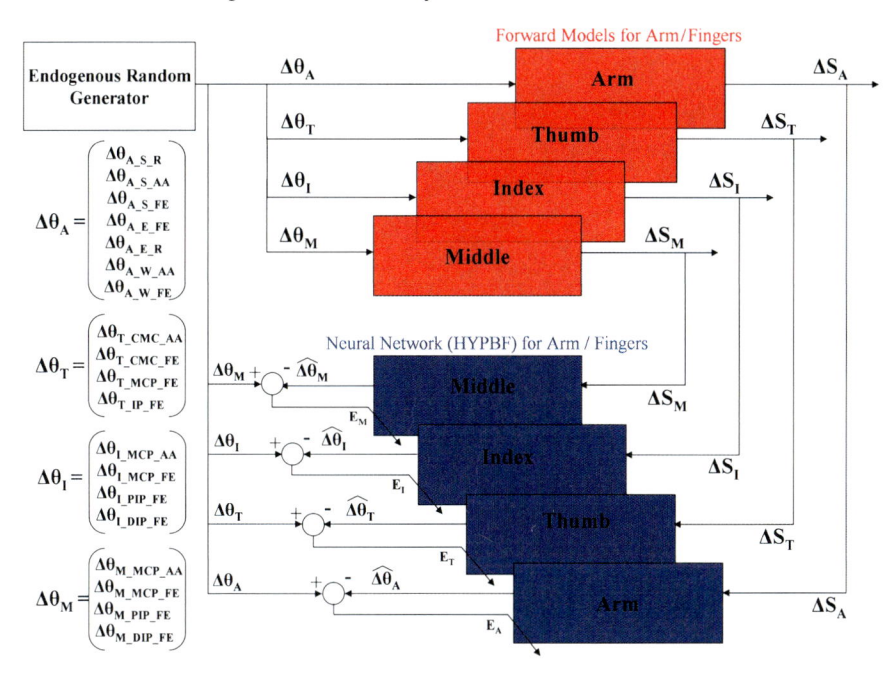

Fig. 6.1 Learning scheme for LM1 employed to acquire internal models (*blue boxes*) of the inverse kinematic for the arm and the three fingers (*index, middle, and thumb*). The forward (or direct) kinematic (*red boxes*) of the arm/fingers is employed to mimic the role of an active vision system that would compute the spatial displacements of fingertips throughout learning. *A* arm, *S* shoulder, *E* elbow, *W* wrist, *AA* abduction-adduction, *FE* flexion-extension, *R* rotation, *MCP* metacarpophaleangeal, *PIP* proximalinterphalangeal, *DIP* distalinterphalangeal, *CMC* carpometacarpal, *IP* interphalangeal. $\Delta\theta$: small angular displacement. ΔS: small spatial displacement. The hat symbol refers to the estimated values

where $A(\theta)$ is a (7×3 for the arm and 4×3 for each finger) matrix and the elements $a_{ij}(\theta)$ of the matrix $A(\theta)$ are the outputs of the HYPBF network. Index i and j refers to the arm/finger configuration space and to the 3D workspace, respectively. These elements are computed with the following equations:

$$a_{ij}(\theta) = \sum_k \left(\frac{g_{ijk}(\theta)}{\sum_l g_{ijk}(\theta)} \right) \left(w_{ijk} + \sum_m c_{ijkm} z_{ijkm} \right), \tag{6.4}$$

$$c_{ijkm}(\theta) = \frac{\theta_m - \mu_{ijkm}}{\sigma_{ijkm}}, \tag{6.5}$$

$$g_{ijk}(\theta) = \exp\left(-\sum_m c_{ijkm}^2 \right), \tag{6.6}$$

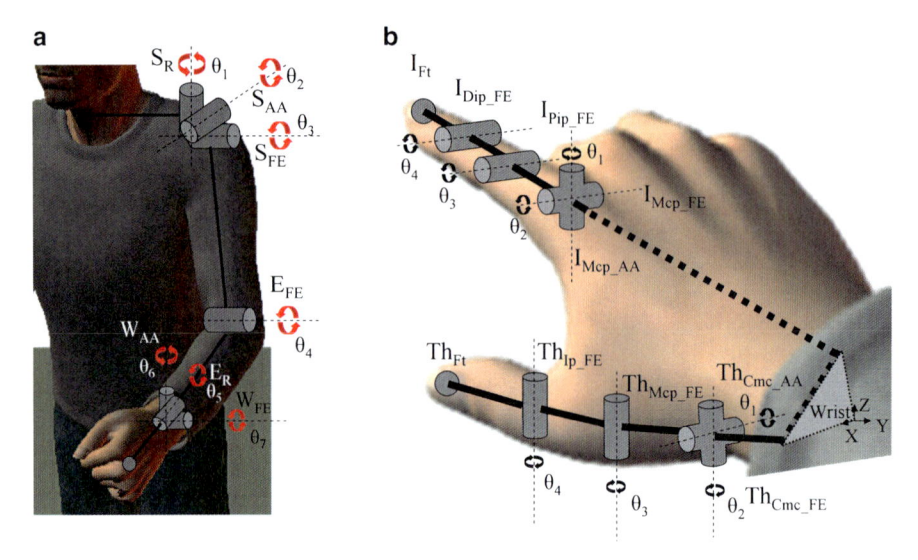

Fig. 6.2 (**a**) The glenohumeral joint is modeled with three revolute joints including rotation (θ_1), adduction-abduction (θ_2) and flexion-extension (θ_3), respectively. The elbow joint is modeled with two revolute joints (Morrey and Chao 1976) allowing flexion-extension (θ_4) and pronation-supination (θ_5). The hand is connected to the arm by the wrist articulation which controls the hand orientation allowing the adduction-abduction (θ_6) and flexion-extension (θ_7). (**b**) For the index (and middle) finger, the MCP has two revolute joints allowing adduction-abduction (θ_1) and flexion-extension (θ_2) while the DIP and PIP has one revolute joint allowing flexion-extension (θ_3 for PIP and θ_4 for DIP). For the thumb, the CMC has two revolute joints allowing adduction-abduction (θ_1 or palmar abduction) and flexion-extension (θ_2 or radial abduction) while the MCP and IP joints were modeled with one revolute joint allowing flexion-extension (θ_3 for MCP and θ_4 for IP). *MCP* metacarpophaleangeal, *PIP* proximal interphalangeal, *DIP* distal interphalangeal, *CMC* carpometacarpal, *IP* interphalangeal

where the $g_{ijk}(\theta)$ are the BF of the network (Gaussian functions) while $\mu_{ijkm}(\theta)$ and $\sigma_{ijkm}(\theta)$ are parameters corresponding to the mean and variance of the kth BF along dimension m of the input space (joint space), respectively. For each hyperplane a BF ($g_{ijk}(\theta)$) is associated to a scalar weight "w_{ijk}," related to the magnitude of the data "under its receptive field." The set of weights "z_{ijkm}" allow to locally and linearly approximate the slope of the data "under its receptive field." An action–perception cycle is generated by random joint velocity vectors $\Delta\theta^R$ (where R denotes random movements) at the arm/finger joints level. These joint rotations are performed from a given joint configurations denoted by θ, that are provided as inputs to the HYPBF network. Random joint rotations induce spatial displacements (denoted Δx) of the arm/fingers by employing their forward kinematics. Then, the HYPBF network computes an estimation ($\Delta\theta$) of the random joint movements ($\Delta\theta^R$). The difference between this estimation and the corresponding random joint movement permits to compute the cost function H used to derive the gradient descent method providing the learning rules associated to the network parameters w_{ijk}, z_{ijkm}, μ_{ijkm}, σ_{ijkm} (Molina-Vilaplana et al. 2007). Thus, the network learns to compute a locally linear approximation of the Jacobian pseudo inverse at each joint configuration, θ. The functional of error that has been used in training the neural networks implementing LM1 is:

$$H = \sum_i (\Delta\theta_i^R - \Delta\theta_i)^2, \tag{6.7}$$

where $\Delta\theta_i$ are the joint rotation vector components that are computed through the HYPBF network outputs as noted in (6.3). And the network parameters are updated according to the following equations:

$$\Delta w_{ijk} = -\alpha \left(\frac{\partial H}{\partial w_{ijk}} \right) = -2\alpha(\varepsilon_i)(\Delta x_j)(h_{ijk}(\theta)), \tag{6.8}$$

$$\Delta z_{ijkm} = -\alpha \left(\frac{\partial H}{\partial z_{ijkm}} \right) = -2\alpha(\varepsilon_i)(\Delta x_j)\left(h_{ijk}c_{ijkm}\right) \tag{6.9}$$

$$\Delta\mu_{ijkm} = -\alpha \left(\frac{\partial H}{\partial \mu_{ijkm}} \right) = -2\alpha(\varepsilon_i)(\Delta x_j)\left(\frac{1}{\sigma_{ijkm}}\right)$$
$$\times \left[\psi_{ijk}\left(w_{ijk} + \sum_m c_{ijkm}z_{ijkm}\right) - z_{ijkm} \right], \tag{6.10}$$

$$\Delta\sigma_{ijkm} = -\alpha \left(\frac{\partial H}{\partial \sigma_{ijkm}} \right) = -2\alpha(\varepsilon_i)(\Delta x_j)\left(\frac{c_{ijkm}}{\sigma_{ijkm}}\right)$$
$$\times \left[\psi_{ijk}\left(w_{ijk} + \sum_m c_{ijkm}z_{ijkm}\right) - z_{ijkm} \right] \tag{6.11}$$

where ψ_{ijk} and h_{ijk} are defined as:

$$\psi_{ijk} = \|c_{ijk}\| h_{ijk}\left(1 - h_{ijk}\right), \tag{6.12}$$

$$h_{ijk}(\theta) = \frac{g_{ijk}(\theta)}{\sum_k g_{ijk}(\theta)}. \tag{6.13}$$

Both voluntary control of movement duration and generation of realistic velocity profiles were obtained by employing a movement-gating $GO(t)$ signal (Bullock et al. 1993). This signal multiplies the joint rotation commands computed by the system prior to their integration at the joint angle command stage. Even if the $GO(t)$ signal is zero, a spatial Δx is measured and desired joint rotations $\Delta\theta$ are computed according to (6.3); however, no movement occurs. When the $GO(t)$ signal becomes positive, the arm/fingers movement rate is proportional to its magnitude multiplied by $\Delta\theta$. In order to modulate the arm/finger joint movement velocity, the following $GO(t)$ signal has been used:

$$GO(t) = g_0 \frac{t^2}{(\gamma + t^2)}. \tag{6.14}$$

Additionally, it must be noted that this model was implemented using two similar but slightly different versions of these four neural networks. Both were based on the same theory and both provided very similar results; however, compared to the first version, the second was faster and more parsimonious since it employed a number of neurons in the hidden layer significantly smaller (approximately ten times less; Gentili et al. 2008, Gentili et al. 2009b) which makes it particularly well suited for embedded control system and real-time application with actual anthropomorphic robotic system.

6.4.1.3 Learning to Associate Object's Intrinsic Properties and Grasping Postures: LM2

Once the learning of the inverse kinematics of the arm and fingers is complete, the subsequent learning of the correct grasping postures for a set of selected objects in the GRASP module is performed. This stage is presented through five steps.

First, the input vector to the GRASP system needs to be defined and codified. This input vector (X) consists in a mixture of multimodal information related to the perception (i.e., visual aspects) of the object. Namely, the type and dimensions of the object as well as the aspects related to tasks execution such as the number of fingers needed to perform a particular grasp. This input vector has seven components. The first three components are used to codify the object type.

Sub-vector encoded as [1 0 0], [0 1 0], and [0 0 1] denotes that the object has to be grasped as cube, a sphere, and cylinder, respectively. Another 3-components sub-vector is used to define object dimensions (Fig. 6.3). The sub-vector [Sx, Sy, Sz] encodes, with scalar within 0 and 1, the different dimensions of the object. In the case of spheres $Sx = Sy = Sz = R$, where R is the radius of the sphere. The maximum and minimum dimensions of the objects in our simulations are about 6 and 2 cm, respectively.

Also, a neuron in the input vector encodes the intention to make the grasp with two (value 0) or with three (value 1) fingers.

Second, a heuristic selection of the contact points over the object and of the positioning of the center of the hand (the palm) in relation with the center of mass of the object to be grasped need to be defined. Here, the term "heuristic" does not refer to any automatic/autonomous search of contact points but to a software-based selection of contact points driven by a human expert with natural skills to identify and choose stable contact points attending only to geometrical aspects of the object.

Third, once the selection of the contact points for the object is done, the module that encodes (by learning, see previous section) the inverse kinematics generates the movements from initial finger configurations to final configurations that allow the contact of the fingertips with the selected points on the object. The facts that these contact points had to belong to the object's surface and had to be coplanar were the unique restrictions imposed on the model (Fig. 6.3). If the fingers do not collide during movement and if the final posture is qualitatively evaluated as a comfortable posture, a 12D vector $\theta_0 = [\theta_P \theta_M \theta_I]$ that encodes final joint

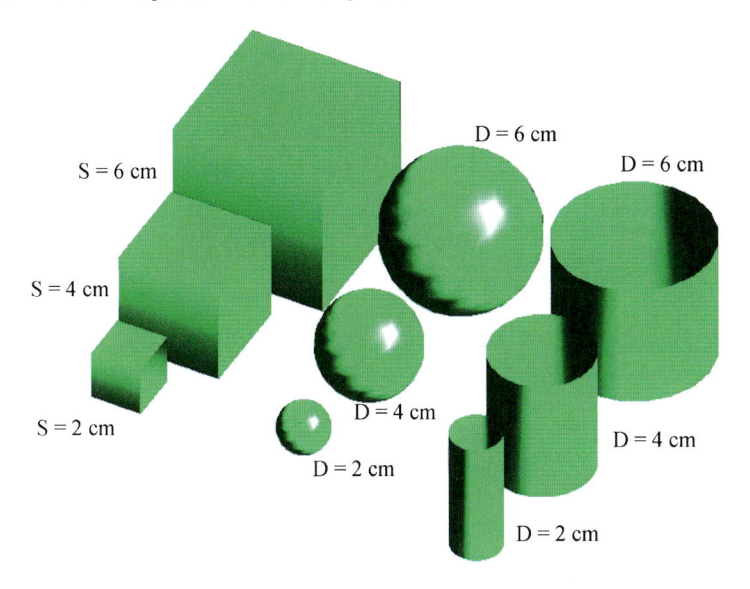

Fig. 6.3 Different types of object employed in simulation. Cubes are defined by their width (Sx), height (Sz), and depth (Sy). Spheres are described by their associated radius whereas cylinders are defined by the radius of the two bases (Sy and Sz) and by their height (Sx). The selection of contact points on the object surfaces has a unique restriction: all contact points have to lie on a plane parallel to the "floor," described by the distance h. Training set for different objects; three cubes (side: 2, 4, and 6 cm), three spheres (diameter: 2, 4, and 6 cm), and three cylinders (diameter: of bases 2, 4, and 6 cm). (Adapted with permission from Molina-Vilaplana et al. 2007)

configurations is created and stored once the fingers grasping movement is completed. A comfortable posture (i.e., a posture close to a natural relaxed one) is defined as a posture in which all final joint values are sufficiently closed from the center of its natural range.

Fourth, the input vector (X) propagates through successive layers of the GRASP HYPBF network (Fig. 6.4). The output of this network is a 12D vector θ. This vector is an estimation of vector θ_0. The processing in the hidden layer of the GRASP neural network is described by the following equations:

$$c_{kl}(\theta) = \frac{X_l - \mu_{kl}}{\sigma_{kl}}, \tag{6.15}$$

$$g_k(\theta) = \exp\left(-\frac{1}{2}\sum_m c_{kl}^2\right), \tag{6.16}$$

where g_k corresponds to activity of the kth Gaussian neuron of the hidden layer and c_k is a 7D vector that "measures" the matching between input vector and the center of the kth Gaussian BF. The activity of the output layer is an adaptive weighted combination of the activity in the hidden layer and is described by the following equation:

$$\theta_i(\theta) = \sum_{k=1}^{K} g_k w_{ki}, \tag{6.17}$$

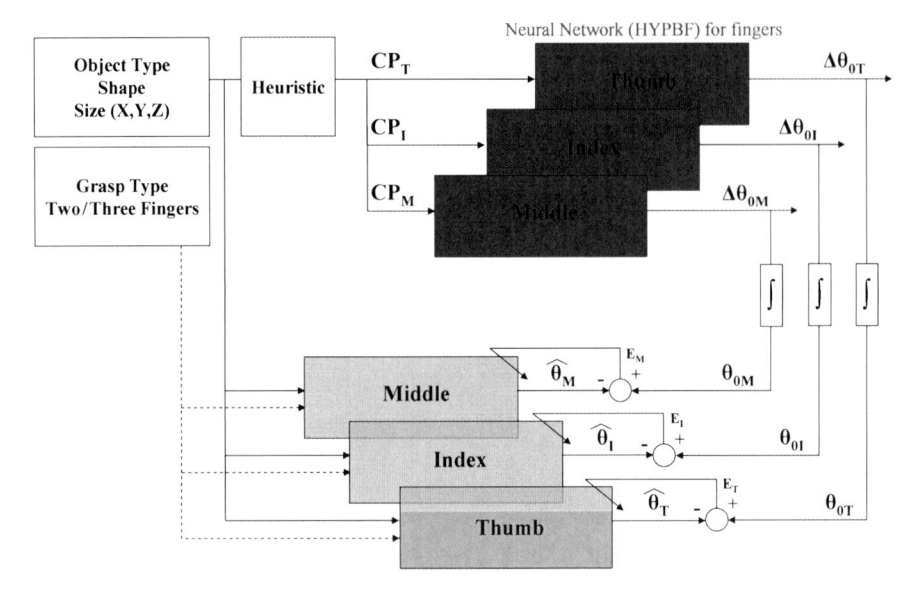

Fig. 6.4 Learning scheme for LM2. Previously acquired knowledge is used at this stage to learn correct mappings between objects properties and suitable grasping postures. GRASP neural network computes an estimation for a grasping posture related with a concrete object (θ), and this estimation is compared with and stored hand configuration (θ_0) related with a grasping movement implementation from an initial open-hand state to a final contact configuration defined by an heuristical selection of 3D contact points. The mismatch between θ and θ_0 is used as a learning signal that drives the learning of network parameters. *CP* contact point, *T* thumb, *I* index, *m* middle finger, *E* error. The hat symbol refers to the estimated values

where w_{ki} is the adaptive weight between k^{th} hidden layer neuron and output layer neuron i. The activity of this neuron (θ_i) is an estimation of the ith hand-joint value, stored in the vector (θ_0).

Finally, a functional of error $H = \sum_i (\Delta\theta_{0i} - \Delta\theta_i)^2$ is defined. This cost function is employed in order to derive learning equations for adaptive parameters w_{ki}, μ_{kl}, and σ_{kl} of the network by using gradient descent method until $H < \phi$ (fixed before learning occurs). In each learning iteration, parameters w_{ki}, μ_{kl}, and σ_{kl} are updated using the following equations:

$$\Delta w_{ki}(\theta) = -2\alpha(\varepsilon_i A_k), \tag{6.18}$$

$$\Delta \mu_{kl}(\theta) = -2\alpha \left(\frac{c_{kl}}{\sigma_{kl}}\right) A_k \sum_i \varepsilon_i w_{ki}, \tag{6.19}$$

$$\Delta \sigma_{kl}(\theta) = -2\alpha \left(\frac{1}{\sigma_{kl}}\right) A_k \sum_i \varepsilon_i w_{ki}, \tag{6.20}$$

where $\varepsilon_i = \theta_i^0 - \theta_i$.

6.5 Simulations Results, Limitations, and Future Challenges

6.5.1 Simulation Results

6.5.1.1 Generation of Reaching and Grasping Trajectories

First, we assessed the capability of module LM1 to perform 3D center-out movements toward multiple targets (Figs. 6.5 and 6.6) placed in the Cartesian workspace once the inverse kinematic was learned for both the arm and the fingers.

For the reaching component, the computer simulations revealed that, generally, the angular and linear displacements were sigmoid-shaped and that the velocity profiles were single-peaked and bell-shaped (not shown). Globally, the paths were slightly curved and the targets accurately reached (Fig. 6.5). For the fingers, the findings revealed that all the targets were accurately reached with paths slightly curved

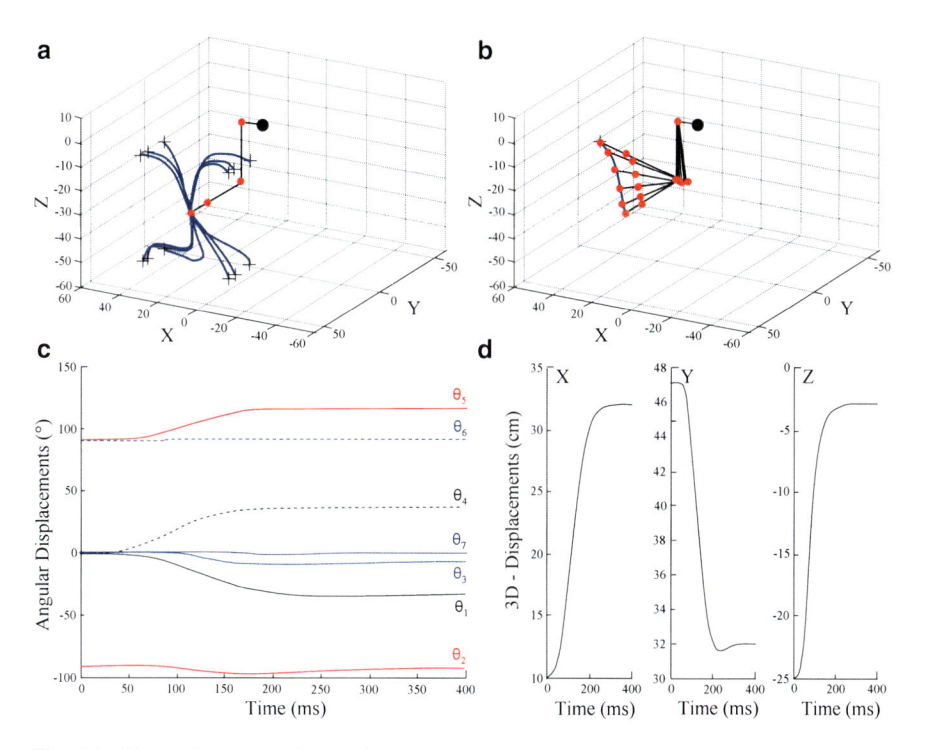

Fig. 6.5 Kinematics results for the 3D center-out arm reaching movements with seven DOFs. (**a**) Dispositions of the 12 targets (*black crosses*) with the trajectory (*blue line*) performed for each reaching movement. (**b**) Stick diagram and path (*blue line*) for the upper right target (*rear one*). (**c**) Angular displacement for the seven angles to reach the upper right target (*rear one*). (**d**) Corresponding Cartesian displacements in the 3D workspace. These results were obtained using the faster and more parsimonious (i.e., second) model; however, very similar results were obtained using the first implementation (not shown)

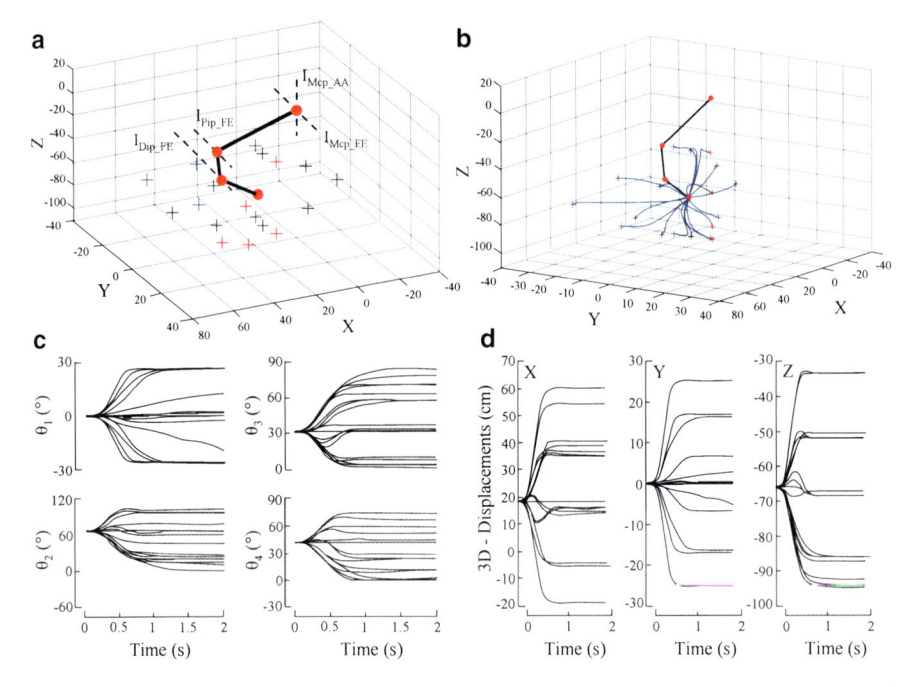

Fig. 6.6 Kinematics results for the 3D center-out finger reaching movements. (**a**) Dispositions of the 18 targets to reach by the finger to test LM1. Target to reach in the back plane ($n=5$), in the middle plane ($n=8$), and in the front plane ($n=5$) are indicated in *blue, grey,* and *red,* respectively. (**b**) Trajectory performed for each reaching movement. (**c**) Angular displacement for the four angles to reach all the targets. (**d**) Corresponding Cartesian displacements in the 3D workspace. These results were obtained using the faster and more parsimonious (i.e., second) model; however, very similar results were obtained using the first implementation (not shown)

(Fig. 6.6a, b). Generally, angular and linear displacements (sigmoid-shaped), as well as velocity profiles (single peaked and bell-shaped, not shown), were comparable to those found in human studies (Grinyagin et al. 2005) (Fig. 6.6c, d).

For the simulation of LM1, four HYPBF neural networks (Fig. 6.1) have been trained to learn the inverse kinematics of the arm and fingers. The training process consisted in generating action–perception cycles (10,000 and 1,000 for the arm and fingers, respectively) in which correlations between arm/finger motor commands and their sensory consequences were established. After training, the contact points were heuristically chosen on different objects. From an initial hand configuration (characterized by an extended finger posture), the HYPBF neural networks was employed to generate spatial trajectories of the fingertips from their initial position to the selected contact points.

Three specific snapshots of these movements are showed in Fig. 6.7a. In order to use the results of these simulations for the subsequent training of LM2, we carried out simulations in which, once the movement of the fingers to the contact points selected on the object was performed, the actual configuration of these fingers was

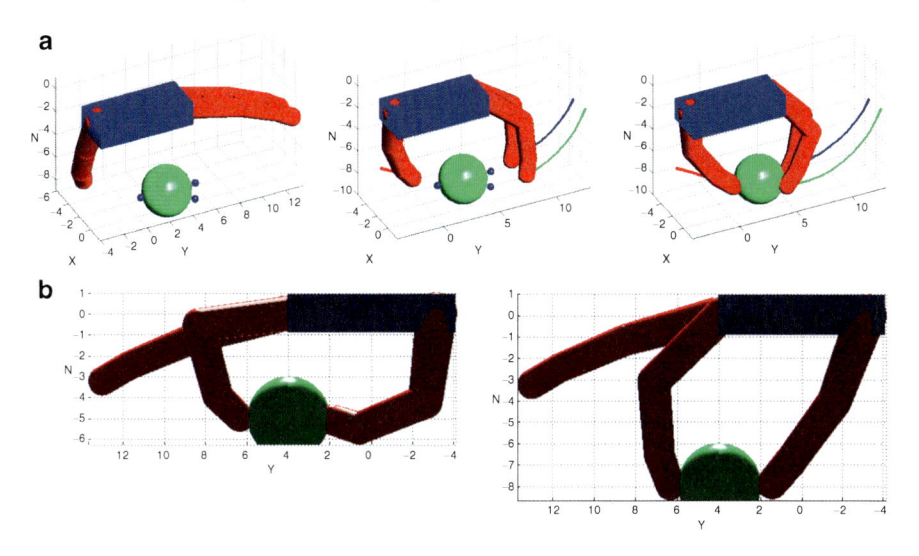

Fig. 6.7 (**a**) Three snapshots of a grasping movement. From an initial open finger configuration (*left*), LM1 controls the three fingers movements toward the selected contact points (*small points*) on the contact surface until the fingertips match their location (*right*). (**b**) Final finger posture depends on initial relative positioning of palm to the center of gravity of the object. For training LM2, grasping movements that finishes with a comfortable final finger configuration (*right*) was chosen as opposed to those that requires a final finger configuration in which some joints have extreme values within their range (*left*). (Adapted with permission from Molina-Vilaplana et al. 2007)

stored as a vector θ_0 associated to a suitable grasping posture to be learned by LM2. The movements of the three fingers were coordinated by using a common gating modulatory GO(t) signal ((6.8), $g_0 = 100$, $\gamma = 1$) that allows temporal equifinality of the three movements (Bullock and Grossberg 1988; Ulloa and Bullock 2003). This configuration θ_0 was stored when the final posture of the hand satisfies some degree of comfort (Fig. 6.7b). We performed the simulation for nine objects (Three cubes, three spheres, and three cylinders, see Fig. 6.3). The stored patterns of hand configurations achieved after completion of successful grasp movements have been employed for the training of LM2 (Fig. 6.3).

6.5.1.2   Learning Capabilities, Training, and Generalization Errors of the GRASP Module

An HYPBF neural network (GRASP module) for learning the association of the intrinsic properties of some simple objects (Fig. 6.3) with the adequate grasping postures was trained. During training, the objects with dimensions equal or smaller than 4 cm had to be grasped with two fingers while objects with higher dimensions had to be grasped with three fingers. As far as we know, this has to be considered as an ad hoc experimental protocol imposed on the training procedure without no relation or support from experimental data. The GRASP neural network includes 30 hidden neurons ($K = 30$ in (6.17)–(6.20)). The system was trained until the functional $H = \sum_i (\Delta\theta_{0i} - \Delta\theta_i)^2$ reaches a value smaller than 0.01.

The learning curve is depicted in Fig. 6.8a while Fig. 6.8b illustrates the errors that GRASP module produces when, after training, a suitable grasping posture for a set of objects is estimated.

Some objects of this set were included in the training set (plain solid color histograms) while the others were not used during the training process (stripped histograms). Figures 6.9 and 6.10 show qualitative (graphics) and quantitative (error

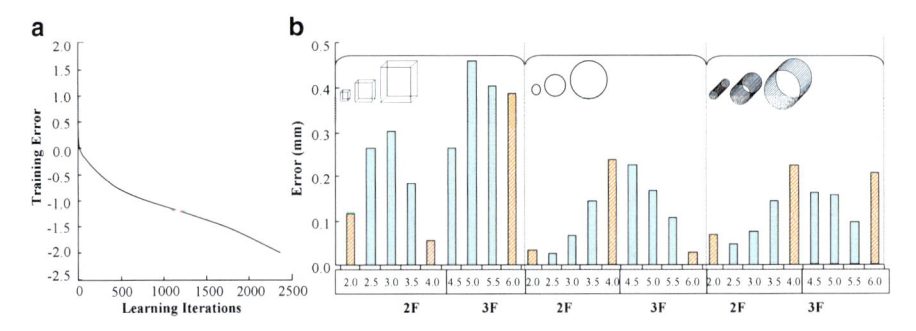

Fig. 6.8 (a) Learning curve ($\log_{10}$) for LM2. (b) Error produced by the GRASP neural module when performing grasping postures for objects included (*stripped histograms*) and not included (*solid color histograms*) in the training set. Absolute values of errors were more pronounced in the case of cubes. It was important to note that generalization errors (*solid color histograms*) were smaller than training errors (*stripped histograms*) in the case of spheres and cylinders while greater in the case of cubes. (Adapted with permission from Molina-Vilaplana et al. 2007)

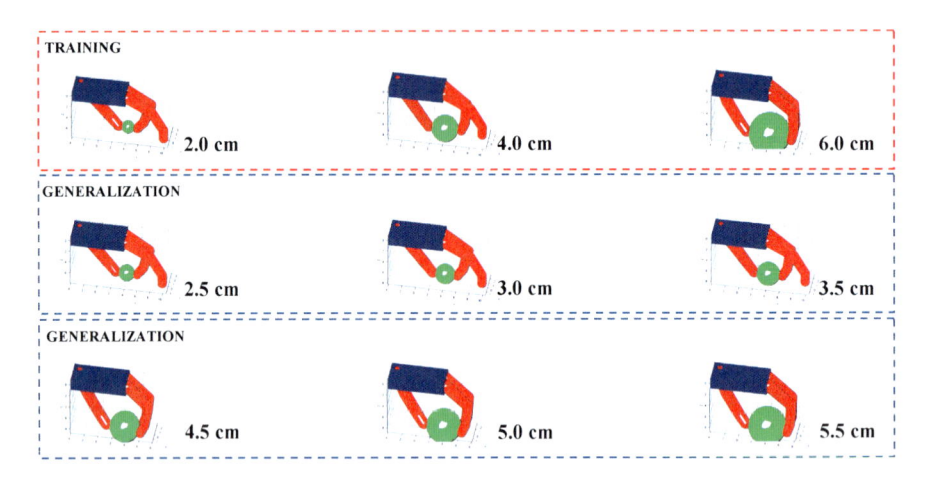

Fig. 6.9 Illustration of training and generalization performance of the GRASP module in the case of the spheres. The red and blue dashed boxes represent the training and generalization stages, respectively. (Adapted with permission from Molina-Vilaplana et al. 2007)

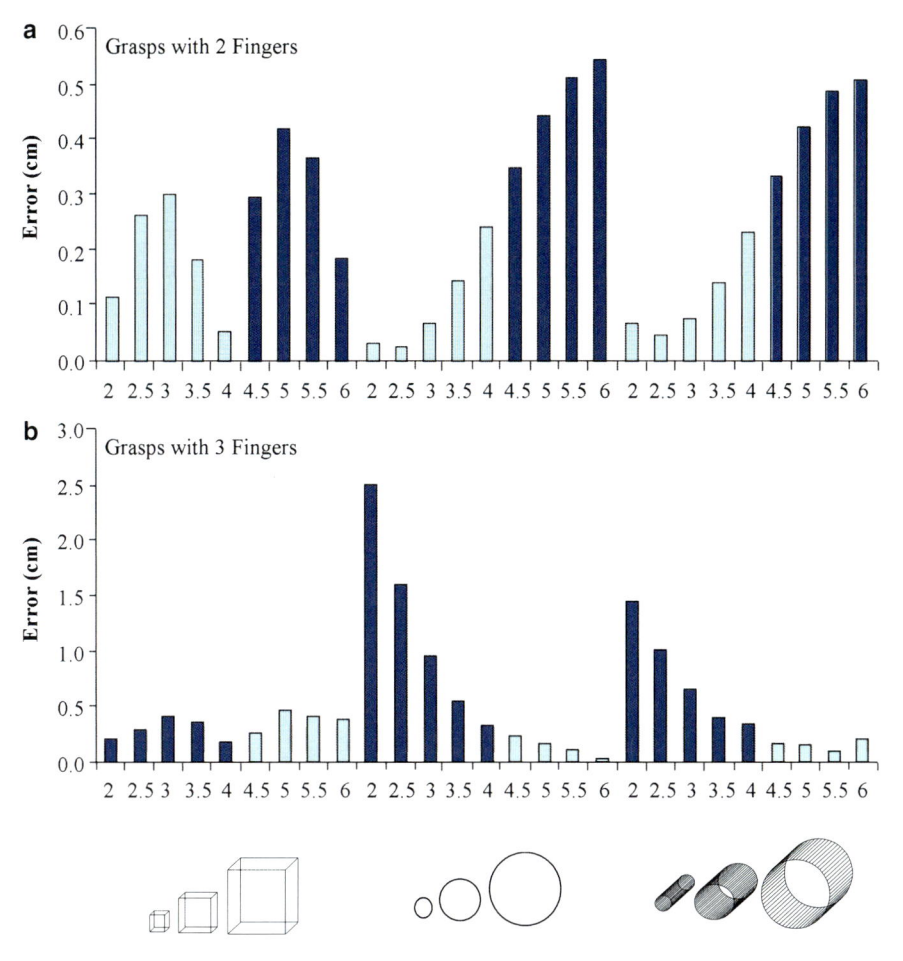

Fig. 6.10 (**a**) Errors produced by the GRASP network when it performed grasping postures with two fingers in all cases although for some objects the training of GRASP module were biased toward a grasp with three fingers (*dark histograms*). (**b**) Errors produced by the GRASP network when it performed grasping with three fingers for all cases although for some objects the training of the GRASP was biased toward a grasp with two fingers (*dark histograms*)

diagrams) measurements of the interpolative capabilities of the GRASP network. Specifically, Fig. 6.8 illustrates the type of object, its dimensions, and the number of fingers involved in the task, whereas Fig. 6.7 shows the graphical results of simulations with spheres.

In order to measure the capabilities of the GRASP module to extrapolate the knowledge acquired during learning, simulations in which all objects were grasped with two fingers (Fig. 6.10a) or three fingers (Fig. 6.10b) were performed. This figure illustrates that GRASP module is still able to predict suitable grasp postures when a grasp configuration with two (or three fingers) is desired though the model has not been necessarily trained for every grasp configurations. Thus, the GRASP

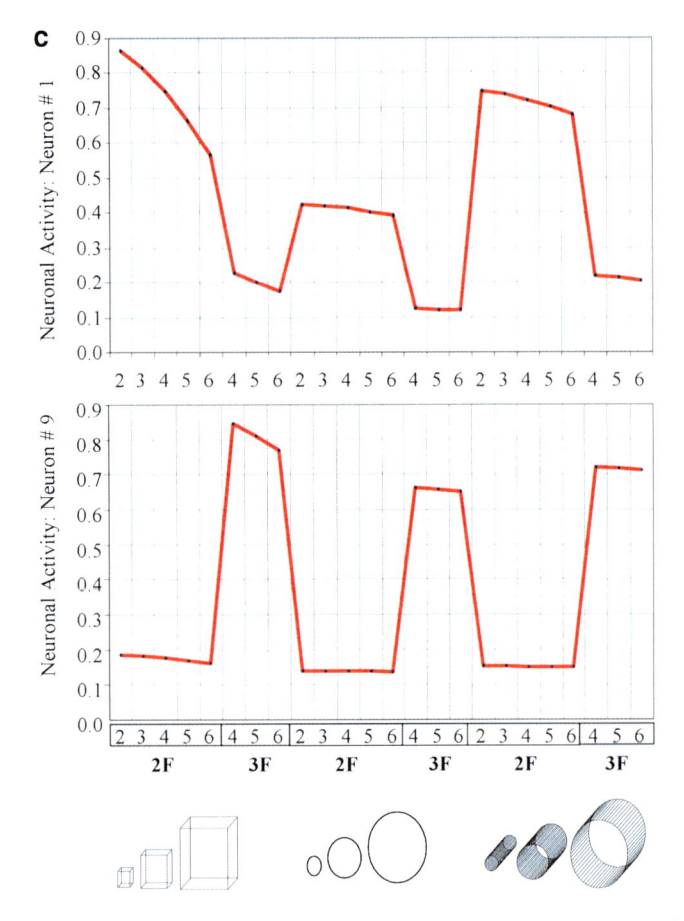

Fig. 6.10 (continued) (**c**) Neural firing properties of hidden neurons with an associated motor-dominant neurons behavior. While activity modulations were present, neuron #1 fired specifically when a grasping posture involved two fingers independently of the object type. The motor-dominant neurons (neuron #9) fired when a grasping posture was performed with three fingers independently of the type of object grasped. (Adapted with permission from Molina-Vilaplana et al. 2007)

network generated suitable grasping postures in these extrapolative conditions although errors were more pronounced than during performance under interpolative conditions.

6.5.1.3   Analysis of the Neural Activity of the GRASP Module During Performance

The neural activity patterns of the hidden layer of the GRASP module during different grasping tasks were analyzed. Similar neural activity was recorded in areas F5, as well as in the anterior intraparietal area (AIP) of monkey premotor, and in the posterior parietal cortex corresponding to the three following different types of neural firing "behavior": (1) motor-dominant neurons, (2) visual-dominant neurons, and (3) visual and motor neurons, respectively. Thus, a neuron in the hidden layer of the GRASP HYPBF neural network was labeled as a motor-dominant neurons, visual-dominant neurons, and visual and motor neurons when its neural activity revealed a certain selectivity to (1) the motor aspects of the task such as the number of fingers involved independently of the visual properties of the object, (2) the visual aspects of the object to be grasped (object's type and size), (3) both visual and motor aspects of the grasping task, respectively. Namely, Fig. 6.10c shows the pattern of neural activity of two neurons of the GRASP module located in the hidden layer labeled as motor-dominant neurons. Neuron #9 fired selectively when grasping tasks was executed with three fingers irrespective of object type while neuron #1 preferentially fired for grasping tasks with two fingers. Figure 6.11a shows the neural activity of two neurons that was identified as visual-dominant neurons (e.g., neuron #20) that fired when grasping task was applied on cubes. This neural activity

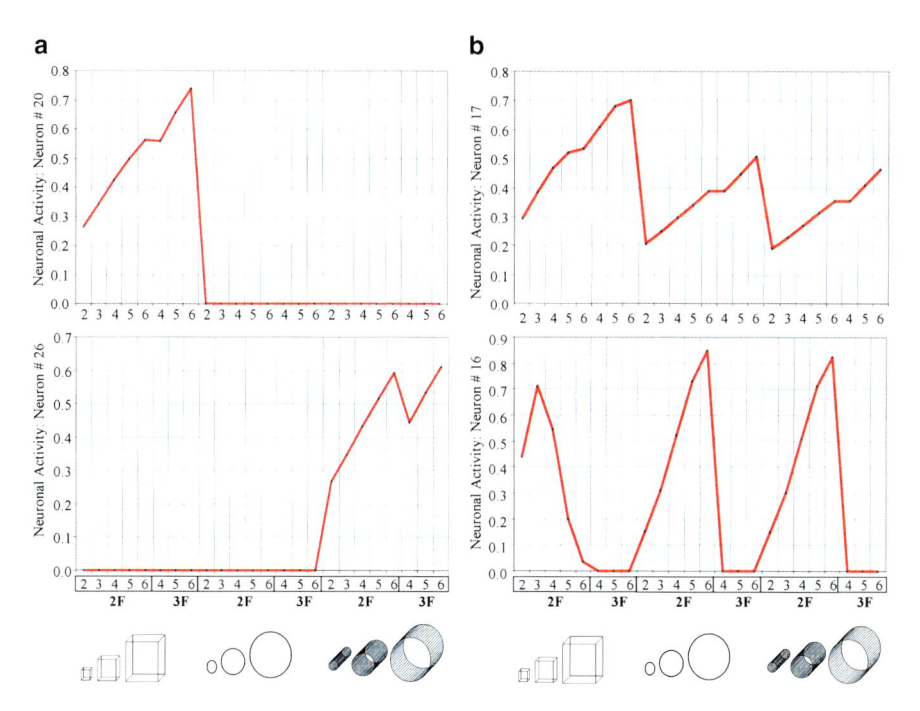

Fig. 6.11 (**a**) Simulated neural firing properties of hidden neurons of GRASP module with an associated visual-dominant neurons behavior. Neurons #20 and #26 fire specifically to perception of cubes and cylinders, respectively. No modulation of activity induced by another factors affects the firing profile of neuron #26 while in neuron #20 a small modulation of activity induced by the type of grasp intended with the cubes (two or three fingers) was detected. (**b**) Neural firing properties of hidden neurons of the GRASP module with an associated visual and motor neurons behavior. Neurons #16 and #17 fire when grasp are performed with two and three fingers, respectively; however, this activation was strongly modulated by the size of the *cubes, spheres*, and *cylinders*. (Adapted with permission from Molina-Vilaplana et al. 2007)

was strongly modulated by cube's size and slightly modulated by the number of fingers involved in task completion while a visual-dominant neurons behavior was detected in neuron #26 revealing a more pronounced activity. The neural activity of this neuron was selective to cylinder perception with a significant modulation induced by cylinder size. Finally, Fig. 6.11b depicts the neural activity of two neurons of the GRASP module in the hidden layer identified as visual and motor neurons.

Specifically, neuron #16 fired preferentially to grasping tasks with two fingers (motor aspect of the grasp) and its activity was strongly modulated by object's size (visual aspect of the grasp) irrespective of the object type, whereas neuron #17 had a similar behavior but the main activity was related to three fingered grasping tasks.

6.5.2 Discussion

6.5.2.1 Reaching and Grasping Performance

Here we report a multimodular neural model that is able to compute the inverse kinematics for an anthropomorphic arm (7 DOFs) and three fingers (4 DOFs) to reach and manipulate objects with various physical features. The first neural module (LM1) aims, in a 3D workspace, to bring the hand to the target and to perform finger motion to contact the object surface. The results revealed that this neural module was able to learn the computation of the inverse kinematics of the arm and of the three fingers to reach targets with a good accuracy. Moreover, this module could generate comparable displacement (e.g., sigmoid-shaped) and velocity profiles (e.g., unimodal and bell-shaped) to those found in human studies (Abend et al. 1982; Atkeson and Hollerbach 1985; Grinyagin et al. 2005; Hollerbach et al. 1986; Morasso 1981). The neural model showed robust control of the effectors when the limb/fingers were close from singularities, that is, effector configurations (e.g., alignment of the links) that can produce misbehavior of the manipulator (e.g., in the neighborhood of a singularity small velocity in the operational space may cause high velocity in the joint space). Also, the second version of this model, which is faster and more parsimonious, included a limited number of neurons (approximately ten times less, Gentili et al. 2008, Gentili et al. 2009b), and thus, is particularly well-suited for embedded control system and real-time applications with actual anthropomorphic robotic system (Molina-Vilaplana et al. 2004; Pedreño-Molina et al. 2005).

Once the training of LM1 and LM2 was complete, the GRASP module was able to estimate a suitable grasping posture for a set of objects albeit some error (Fig. 6.8b). While some objects were included in the training set, the performance of the neural model was tested with those not previously presented to the neural network. These simulations represent an interpolation condition for the knowledge acquired by the GRASP system throughout training. The postures predicted by the GRASP module for this interpolative condition (in the case of spheres) are illustrated in Fig. 6.8. It clearly appears that the GRASP network achieves a good performance in predicting suitable grasp postures for similar objects to those presented during training when these objects have similar dimensions. Furthermore, when the GRASP network has to perform all the grasps with two or three fingers, it appears that the GRASP module is still able to predict suitable (i.e., with acceptable errors) grasp postures in such extrapolative conditions. Nevertheless, as expected, compared to the interpolative condition, the performance was accompanied with greater errors as previously mentioned. Although, here only changes in input parameters related to size of the objects or to the number of fingers involved in the task were considered, performance including conditions with a higher degree of extrapolation could be used to test the performance of the GRASP system. For instance, this could be done by employing simple objects similar to those used in the training set with very different dimensions such as spheres with diameter of 8 cm. Even higher degree of extrapolation condition could be considered. Under such conditions, it should be necessary to test the system for grasping objects that are similar but different enough from those used during training. For example, it would be the case when grasping a cup of tea with a truncated conic shape similar to but different from the cylinder used here. Another example would be when grasping a truncated pyramid similar but different enough from a cube. Such extrapolative simulations could be employed in future studies in order to test the sensitivity of the network related to the input that encodes the shape of objects (considering that the new shapes should differ from the original shapes within a reasonable range).

6.5.2.2 Neural Activity of the GRASP Module

The study of neural activity in hidden layer of the GRASP module presented in this paper allowed the identification of a variety of neural firing behaviors that can be, to some extent, related to neural firing properties of neurons located in the AIP and F5 areas of the distributed parietal–frontal biological neural network involved in visual guidance of grasping (Fogassi et al. 2001; Gallese et al. 1997; Luppino et al. 1999; Murata et al. 1996, 1997, 2000; Sakata et al. 1995).

Thus, we report an artificial neural network that processes multisensory information through a hidden layer of radial BFs that is able to acquire throughout learning an abstract grasping-object encoding. Such encoding can be used to identify (using neural firing patterns) neurons in parietal and premotor areas of monkeys when they perform visually guided grasping behaviors. This result constitutes an emergent property of the model since such feature was not explicitly included in the design of the GRASP system. This appears to result from particular characteristics of the neural computation based on the BFs. As previously mentioned, during learning, the tuning process of centers in the HYPBF implies a dimensionality reduction process as well as an input data clustering. So, it is possible to hypothesize that parietal computation of signals involved in visually guided grasping, integrates the motor-dominant neurons, visual-dominant 451 neurons and visual and motor neurons codification of the task reflecting a process of data clustering and a reduction of the problem dimensionality. However, this property is not enough to establish a direct relationship between the different parts of the GRASP module with different

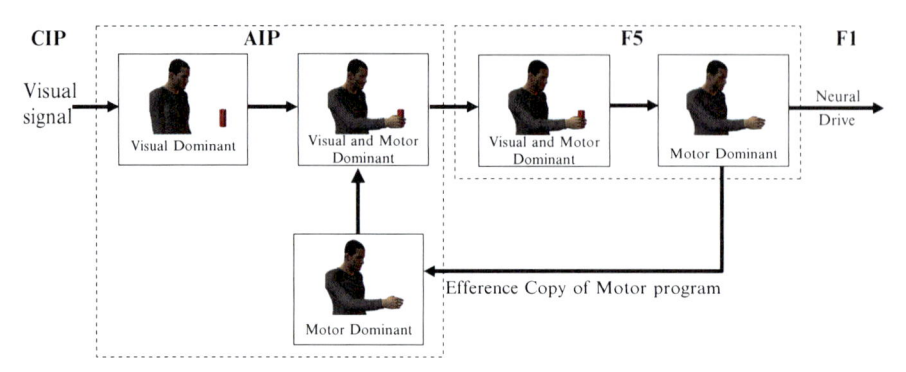

Fig. 6.12 A diagram of the model for the cortical network for visually guided grasping. (Adapted from Murata et al. 1997 and Molina-Vilaplana et al. 2007)

sub-networks of the biological circuit since this neural model (in its present form) does not achieve the conceptual model for visual guidance of grasping proposed in Murata et al. 1997; Molina-Vilaplana et al. 2007 (Fig. 6.12). However, several suggestions can be proposed from this model. First, this model suggests that the visual signals related to 3D shape and orientation of the objects are processed at the caudal intraparietal and that the resultant outputs are sent to the visual-dominant neurons in the AIP area. Second, this model proposes that the information related to the grasping objective is conveyed to the visual- and motor-dominant neurons in the AIP area. At this stage, the AIP motor-dominant neurons also send an efference copy of actual motor commands to the AIP visual- and motor-dominant neurons.

Thus, the visual- and motor-dominant neurons of AIP continuously compute a sort of "visuomotor error" based on the comparison between the internal representation of the target state and the internal representation of the actual motor effector state (Cohen and Andersen 2002; Desmurget et al. 1999). This visuomotor error computed at the level of visual and motor-dominant neurons of AIP is then sent via cortico-cortical pathways to visual and motor neurons in F5. Third, this model also suggests that the processed information in F5 visual and motor neurons is sent to motor-dominant neurons in F5. These neurons would send the final motor information to the primary motor cortex (F1) for final execution of the movement. Finally this model proposes that an efference copy of the activity in F5 motor neurons is sent to AIP motor-dominant neurons closing, thus, the circuit. This closed circuit constitutes a feedback circuit that generates proper signals sent from AIP to F5 by continuously monitoring some type of parietal visuomotor error in order to perform the movement. Thus, a model for grasp learning and execution that includes elements of sensorimotor information processing previously mentioned could be developed according to the new perspectives proposed by Burnod and collaborators (Baraduc et al. 2001; Baraduc and Guigon 2002). These new perspectives are based on the hypothesis that parietal cortex areas carry out multimodal sensory correlations from different neural representations employing a common representation frame (Cohen and Andersen 2002) through a process that Burnod and collaborators has called a "matching" process. This process is performed by different "matching units." A matching unit can be interpreted as a set of two neural networks, each of them processing separately two different sensory modalities or sensory stimuli coupled to motor effectors activity. In a matching unit, these two neural networks perform a transformation between frames of reference, that is, the outputs of these networks are established in a common neural representation which allows the subsequent comparison and computation of the "mismatch" between these outputs. If such a mismatch exists, it is used as a learning signal to update the internal parameters of both networks in order to achieve a final "match" between the outputs of the two networks. The process implies the correlation of two different sensory modalities or between a sensory modality related to the ongoing motor command. In this framework, the AIP area could be considered as one of these matching units. The neural processing in one neural network could be related to visual-dominant neurons of AIP (the input of this network is related to object's visual properties) while the second neural network would be involved in the processing that takes place in motor-dominant neurons of the AIP (the input of this network is the copy of the ongoing motor command from F5). The dynamic computation of the mismatch between these two different neural representations is related to the type of computation that visual- and motor-dominant neurons of AIP should perform. Thus, it is possible to consider that the use of the matching unit concept can improve the neural plausibility of cortical functions during the acquisition of grasping skills for a model with similar architectural concepts to the one presented in this chapter. The matching unit concept previously discussed could be extended by combining a matching unit and a neural network able to generate appropriate motor commands (related with motor processing and execution in F5 and F1) reducing, thus, the initial mismatch generated when a graspable object is detected.

6.5.2.3 Model Assumptions, Limitations, and Possible Solutions to Challenge Them

The proposed model considers three important assumptions. The first one, as previously mentioned, is that the brain employs numerous cortical modules for performing sensorimotor transformations or multimodal internal representations of objects and effectors (i.e., internal models). This assumption is relatively common in modeling and is supported by experimental data (e.g., Shadmehr and Wise 2005). The second important assumption is the a priori existence of a forward model of the biomechanical system that enables to learn the inverse kinematic. However, it would be possible to learn the forward model during a babbling period with a faster dynamics than the inverse model. Under such conditions, even an imperfect (i.e., under development) forward model could be used to learn the inverse model as it has been previously suggested (Jordan and Rumelhart 1992). The third and last important assumption is that this neural network model controls a biomechanical system that is considered as purely kinematic. In other words, the forward and inverse model of the fingers include only geometrical features without incorporating any dynamics components related to the mass or inertia of the fingers as well as the associated gravitational, inertial, and interaction torques. This could be realized by a third learning step were internal representations of these dynamics could be performed by modeling more explicitly specific structures that have been considered to encode this type of learning such as the cerebellum (Bullock et al. 1998; Contreras-Vidal et al. 1997; Gentili et al. 2006, 2007, 2009b).

Furthermore, it must be noted that the model presented here does not include two important neural mechanisms. First, this model focuses on how the brain could learn multiple inverse kinematic mappings for reaching and grasp coordination; thus, motor attention is outside the scope of this model. Such a "motor attention" component could be included in a possible extension of this neural model integrating additional parietal subsections (e.g., intraparietal sulcus) (Snyder et al. 1997). Second, this model does not explicitly address the neural mechanisms related to the supercession of control to shift from feedback (attended) to feedforward (unattended) mode during sensorimotor learning. However, our model is compatible with such mechanism since our neural architecture has been defined within the internal model framework. Therefore, throughout learning as the internal model is progressively acquired, movement becomes progressively more automatized with less reliance on feedback shifting gradually toward a feedforward control of the effectors. Interestingly, considering this internal model framework, a recent experiment combined a graphomotor learning task with fMRI (functional Magnetic Resonance Imagery) and revealed a shift from (1) a first dominant visuomotor stage involved in the formation of an internal representation of the motor sequence where cortical processes (e.g., dorsal and ventral visual streams) were involved to (2) a second, slower and prolonged learning stage during which this internal representation is progressively refined (i.e., automatization) and encoded sub-cortically (e.g., basal ganglia, cerebellum) as performance improves (Swett et al. 2010). Specifically, the first learning period was associated with activation of the dorsal (processing of the visual object to draw) and ventral (visually guided drawing movements) visual streams and the dorsal premotor cortex (e.g., Burnod et al. 1999; Caminiti et al. 1996) while the second revealed less active visual regions and additional involvement of subcortical brain regions associated to movement refinement/automatization processes. Therefore, beyond the compatibility of our model with the supercession mechanism previously mentioned in the computational framework, our neural model incorporates neural structures (e.g., dorsal, ventral premotor areas) that appear to be involved differently from the early (relying mainly on feedback control) to the late (relying mainly on feedforward control) learning phase. Thus, a possible extension of this model would explicitly account for this supercession mechanism in order to model and understand the shift from feedback to feedforward control and the role of "motor attention" in visuomotor learning.

6.6 Conclusion

We presented a modular neural network model able to learn throughout a stepwise approach the inverse kinematics to perform accurate arm and finger motion for reaching and grasping tasks with objects having various properties. This model exhibited features similar to those recorded in populations of real neurons in the distributed parietal–frontal biological neural network involved in visual guidance for reaching and grasping. Future works related to this neural model include the modeling of the acquisition of the internal forward and inverse models throughout the developmental period, the learning of the dynamics, and robotic applications using anthropomorphic robotic arm/fingers.

Acknowledgments This work was supported in part by the Office of Naval Research (N000140910126) and the National Institutes of Health (PO1HD064653). Rodolphe J. Gentili would like to sincerely thank *La Fondation Motrice*, Paris, France, for the continued support of his research.

References

Abend W, Bizzi E, Morasso P. (1982). Human arm trajectory formation. *Brain*, 705:331–348

Albus JS. (1971). A theory of cerebellar function. *Math. Biosci.*, 10:25–61

Atkeson CG, Hollerbach JM. (1985). Kinematic features of unrestrained vertical arm movements. *J. Neurosci.*, 5(9):2318–2330

Baraduc P, Guigon E, Burnod Y. (2001). Recoding arm position to learn visuomotor transformations. *Cereb. Cortex*, 11(10):906–917

Baraduc P, Guigon E. (2002). Population computation of vectorial transformations. *Neural Comput.*, 14(4):845–871

Braido P, Zhang X. (2004). Quantitative analysis of finger motion coordination in hand manipulative and gestic acts. *Hum. Mov. Sci.*, 22:661–678

Bullock D, Grossberg S. (1988). Neural dynamics of planned arm movements: emergent invariants and speed-accuracy trade-offs during trajectory formation. *Psychol. Rev.*, 95(1):49–90

Bullock D, Grossberg S, Guenther FH. (1993). A self organizing neural model for motor equivalent reaching and tool use by a multijoint arm. *J. Cogn. Neurosci.*, 5(4):408–435

Bullock D, Cisek P, Grossberg S. (1998). Cortical networks for control of voluntary arm movements under variable force conditions. *Cereb. Cortex*, 8(1):48–62

Burnod Y, Baraduc P, Battaglia-Mayer A, Guigon E, Koechlin E, Ferraina S, Lacquaniti F, Caminiti R. (1999). Parieto-frontal coding of reaching: an integrated framework. *Exp. Brain Res.*, 129:325–346

Caminiti R, Johnson PB, Urbano A. (1990). Making arm movements within different parts of space: dynamic aspects in the primate motor cortex. *J Neurosci.*, 10:2039–2058

Caminiti R, Ferraina S, Johnson PB. (1996). The sources of visual input to the primate frontal lobe: a novel role for the superior parietal lobule. *Cereb. Cortex*, 6:102–119

Cohen YE, Andersen R. (2002). A common reference frame for movement plans in posterior parietal cortex. *Nat. Rev. Neurosci.*, 3:553–562

Conforto S, Bernabucci I, Severini G, Schmid M, D'Alessio T. (2009). Biologically inspired modelling for the control of upper limb movements: from concept studies to future applications. *Front. Neurorobotics*, 3(3):1–5

Contreras-Vidal JL, Grossberg S, Bullock D. (1997). A neural model of cerebellar learning for arm movement control: cortico-spino-cerebellar dynamics. *Learn. Mem.*, 3(6):475–502

Contreras-Vidal JL, Bo J, Boudreau JP, Clark JE. (2005). Development of visuomotor representations for hand movement in young children. *Exp. Brain Res.*, 162(2):155–164

Contreras-Vidal JL. (2006). Development of forward models for hand localization and movement control in 6- to 10-year-old children. *Hum. Mov. Sci.*, 25(4–5):634–645

Cutkosky MR. (1989). On grasp choice, grasp models and the design of hands for manufacturing tasks. *IEEE Trans. Rob. Autom.*, 5(3):269–279

Cutkosky MR, Howe RD. (1990). Human grasp choice and robotic grasp analysis. In ST Venkatara-man and T Iberall (Eds), *Dextrous robot hands*. New York: Springer, pp. 5–31

Desmurget M, Epstein CM, Turner RS, Prablanc C, Alexander GE, Grafton ST. (1999). Role of the posterior parietal cortex in updating reaching movements to a visual target. *Nat. Neurosci.,* 2(6):563–567

Doya K. (1999). What are the computations of cerebellum, the basal ganglia, and the cerebral cortex? *Neural Netw.,* 12:961–974

Fagg AH, Arbib MA. (1998). Modeling parietal-premotor interactions in primate control of grasp-ing. *Neural Netw.,* 11(7–8):1277–1303

Ferraina S, Garasto MR, Battaglia-Mayer A, Ferraresi P, Johnson PB, Lacquaniti F, Caminiti R. (1997a). Visual control of hand-reaching movement: activity in parietal area7m. *Eur. J. Neu-rosci.,* 9:1090–1095

Ferraina S, Johnson PB, Garasto MR, Battaglia-Mayer A, Ercolani L, Bianchi L, Lacquaniti F, Caminiti R. (1997b). Combination of hand and gaze signals during reaching: activity in parietal area 7m of the monkey. J. Neurophysiol., 77:1034–1038

Fiala JC. (1994). A network for learning kinematics with application to human reaching models. *Proc. IEEE Int. Conf. Neural Netw.,* 5:2759–2764

Flash T, Hogan N. (1985). The coordination of arm movements: an experimentally confirmed math-ematical model. *J. Neurosci.,* 5(7):1688–1703

Fogassi L, Gallese V, Buccino G, Craghiero L, Fadiga L, Rizzolatti G. (2001). Cortical mechanism for the visual guidance of hand grasping movements in the monkey: a reversible inactivation study. *Brain,* 124:571–586

Gallese V, Fadiga L, Fogassi L, Luppino G, Murata A. (1997). A parietal-frontal circuit for hand grasping movements in the monkey: evidence from reversible inactivation experiments. In P Thier and HO Karnath (Eds), *Parietal lobe contributions to orientation in 3D space*. Berlin: Springer, pp. 255–270

Gan JQ, Oyama E, Rosales EM, Hu H. (2005). A complete analytical solution to the inverse kine-matics of the pioneer 2 robotic arm. *Robotica,* 23(1):123–129

Gentili RJ, Papaxanthis C, Ebadzadeh M, Ouanezar S, Eskiizmirliler S, Darlot C, Maier M. (2006). Internal representation of gravitational forces in cerebellar pathways allows for the dynamic inverse computation of vertical pointing movements of a robot arm. Program No. 57.7. 2006 Neuroscience Meeting Planner. Atlanta, GA: Society for Neuroscience, 2006. Online.

Gentili R, Papaxanthis C, Ebadzadeh M, Ouanezar S, Eskiizmirliler S, Darlot C, Maier MA. (2007). Sensorimotor predictions in cerebellar pathways allows inverse dynamic computation of the gravitational forces on vertical pointing movement of a robot arm. *NEASB Conference,* University of Maryland, College Park, USA

Gentili RJ, Contreras-Vidal JL. (2008). A neural model of cortico-spino-cerebellar learning for force computation during precision grip. Program No. 77.9/NN31. 2008 Neuroscience Meeting Planner. Washington, DC: Society for Neuroscience, 2008. Online

Gentili RJ, Charalambos P, Ebadzadeh M, Eskiizmirliler S, Ouanezar S, Darlot C. (2009a). Integra-tion of gravitational torques in cerebellar pathways allows for the dynamic inverse computation of vertical pointing movements of a robot arm. PLoS One, 4(4):e5176

Gentili RJ, Oh H, Contreras-Vidal JL. (2009b) A cortical neural model for inverse kinematics com-putation of an anthropomorphic robot finger. Program No. 862.15. 2009 Neuroscience Meeting Planner. Chicago, IL: Society for Neuroscience, 2009. Online.

Georgopoulos AP, Kalaska JF, Curtcher MD, Caminiti R, Massey JT. (1984). The representation of movement direction in the motor cortex: single cell and population studies. In GM Edelman, WE Gall and WM Cowan (Eds), *Dynamic aspects of cortical function*. New York: Wiley, pp. 501–524

Georgopoulos AP, Schwartz AN, Kettner RE. (1986). Neuronal population coding of movement direction. *Science,* 233:1416–1419

Ghez C, Gordon J, Ghilardi MF, Sainburg R. (1994). Contributions of vision and perception to accuracy in limb movements. In MS Gazzaniga (Eds), *The cognitive neurosciences*. Cambridge: MIT, pp. 549–564

Goodale MS, Milner AD. (1992). Separate visual pathways for perception and action. *Trends Neurosci.*, 15:20–25

Gorce P, Fontaine JG. (1996). Design methodology for flexible grippers. *J. Intell. Robot. Syst.*, 15(3):307–328

Gordon J, Ghiraldi MF, Ghez C. (1994). Accuracy of planar reaching movements. I. independence of direction and extent variability, *Exp. Brain Res.*, 99:97–111

Grinyagin IV, Biryukova EV, Maier MA. (2005). Kinematic and dynamic synergies of human precision-grip movements. *J. Neurophysiol.*, 94(4):2284–2294

Grosse-Wentrup M, Contreras-Vidal JL. (2007). The role of the striatum in adaptation learning: a computational model. *Biol. Cybern.*, 96(4):377–388

Guenther FH, Micci-Barreca D. (1997). Neural models for flexible control of redundant systems. In PG Morasso and V Sanguinetti (Eds), *Self-organization, computational maps and motor control*. North-Holland Psychology series, Elsevier, pp. 383–421

Guigon E, Grandguillaume P, Otto I, Boutkhil L, Burnod Y. (1994). Neural network models of cortical functions based on the computational properties of the cerebral cortex. *J. Physiol.*, 88:291–308

HartenBerg RS, Denavit J. (1964). *Kinematic synthesis of linkages.* McGraw-Hill, New York

Hollerbach JM, Moore SP, Atkeson CG. (1986). Workspace effect in arm movement kinematics derived by joint interpolation. In G Ganchev, B Dimitrov and P Patev (Eds), *Motor control.* New York: Plenum, pp. 197–208

Hulliger M. (1984). The mammalian muscle spindle and its central control. *Rev. Physiol. Biochem. Pharmacol.*, 101:1–110

Iberall T. (1987). Grasp planning for human prehension. *Proceedings of the 10th International Conference on Artificial Intelligence,* Milan, Italy, pp. 1153–1156

Iberall T, Jackson J, Labbe L, Zamprano R. (1988). Knowledge based prehension: capturing human dexterity. *Proceedings of the IEEE International Conference on Robotics and Automation,* Philadelphia, PA, USA, pp. 82–87

Iberall T. (1997). Human prehension and dexterous robot hands. *Int. J. Rob. Res.*, 16(3):285–299

Ito M. (1984). *The cerebellum and neural control.* Raven, New York

Jeannerod M. (1991). The interaction of visual and proprioceptive cues in controlling reaching movements. In DR Humphrey and HJ Freund (Eds), *Motor control: concepts and issues.* New York: Wiley, pp. 277–291

Jeannerod M. (1994). The hand and the object: the role of posterior parietal cortex in forming motor representations. *Can. J. Physiol. Pharmacol.*, 72:535–541

Johnson PB, Ferraina S, Garasto MR, Battaglia-Mayer A, Ercolani L, Burnod Y, Caminiti R. (1997). From vision to movement: cortico-cortical connections and combinatorial properties of reaching related neurons in parietal areas V6 and V6A. In P Their and O Karnath (Eds), *Parietal lobe contributions to orientation in 3D space.* Heidelberg: Springer, pp. 221–236

Jordan MI, Rumelhart DE. (1992). Forward models: supervised learning with a distal teacher. *Cogn. Sci.*, 16:307–354

Kalaska JF, Cohen DAD, Prud'homme M, Hyde ML. (1990). Parietal area 5 neuronal activity encodes movement kinematics, not movement dynamics. *Exp. Brain Res.*, 80:351–364

Kang SB, Ikeuchi K. (1997). Toward automatic robot instruction from perception: mapping human grasps to manipulator grasps. *IEEE Trans. Rob. Autom.*, 13(1):81–95

Kerr J, Roth R. (1986). Analysis of multifingered hands. *Int. J. Rob. Res.*, 4:3–17

Kuperstein M. (1991). Infant neural controller for adaptive sensory-motor coordination. *Neural Netw.*, 4(2):131–146

Luppino G, Murata A, Govoni P, Matelli M. (1999). Largely segregated parietofrontal connections linking rostral intraparietal cortex (areas AIP and VIP) and ventral premotor cortex (areas F5 and F4). *Exp. Brain Res.*, 128:181–187

Marr D. (1969). A theory of cerebellar cortex. *J. Physiol.*, 202:437–470

Mason CR, Gomez JE, Ebner TJ. (2001). Hand synergies during reach-to-grasp. *J. Neurophysiol.*, 86(6):2896–2910

Molina-Vilaplana J, Pedreno-Molina JL, Lopez-Coronado J. (2004). Hyper RBF model for accurate reaching in redundant robotic systems. *Neurocomputing,* 61:495–501

Molina-Vilaplana J, López-Coronado J. (2006). A neural network model for coordination of hand gesture during reach to grasp. *Neural Netw.,* 19:12–30

Molina-Vilaplana J, Feliu-Batlle J, Lopez-Coronado J. (2007). A modular neural network architecture for step-wise learning of grasping tasks. *Neural Netw.,* 20(5):631–645

Molina-Vilaplana J, Contreras-Vidal JL, Herrero-Ezquerro MT, López Coronado J. (2009). A model for altered neural network dynamics related to prehension movements in parkinson disease. *Biol. Cybern.,* 100(4):271–287

Morasso P. (1981). Spatial control of arm movements. *Exp. Brain Res.,* 42(2):223–227

Moussa MA, Kamel MS. (1998). An experimental approach to robotic grasping using a connectionist architecture and generic grasping functions. *IEEE Trans. Syst. Man Cybern. C Appl Rev.,* 28(2):239–253

Murata A, Gallese V, Kaseda K, Sakata H. (1996). Parietal neurons related to memory guided hand manipulation. *J. Neurophysiol.,* 75:2180–2186

Murata A, Fadiga L, Fogassi L, Gallese V, Raos V, Rizzolatti G. (1997). Object representations in the ventral premotor cortex of the monkey. *J. Neurophysiol.,* 78:2226–2230

Murata A, Gallese V, Luppino G, Kaseda K, Sakata H. (2000). Selectivity for the shape, size and orientation of objects for grasping in neurons of monkey parietal area AIP. *J. Neurophysiol.,* 83:339–365

Morrey BF, Chao EY. (1976). Passive motion of the elbow joint. *J. Bone Joint Surg. Am.,* 58:501–508

Napier JR. (1956). The prehensile movements of the human hand. *J. Bone Joint Surg.,* 38B:902–913

Pause M, Kunesch E, Binkofski F, Freund HJ. (1989). Sensorimotor disturbances in patients with lesions in parietal cortex. *Brain,* 112:1599–1625

Pedreño-Molina JL, Molina-Vilaplana J, López-Coronado J, Gorce P. (2005). A modular neural network linking hyper RBF and AVITE models for reaching moving objects. *Robotica,* 23:625–633

Poggio T, Girosi F. (1989). *A theory of networks for approximation and learning.* AI Memo 1140, MIT

Pons JL, Ceres R, Pfeiffer F. (1999). Multifingered dextrous robotic hands design and control: a review. *Robotica,* 17(6):661–674

Porter R, Lemon RN. (1993). *Corticospinal function and voluntary movement.* Oxford University Press, New York

Pouget A, Snyder LH. (2000). Computational approaches to sensorimotor transformations. *Nature,* 3:1192–1198

Rizzolatti G, Camarda R, Fogassi L, Gentilucci M, Luppino G, Matelli M. (1988). Functional organization of inferior area 6 in the macaque monkey.II.area F5 and the control of distal movement. *Exp. Brain Res.,* 71:491–507

Rizzolatti G, Gentilucci M, Camarda R, Gallese V, Luppino G, Matelli M, Fogassi R. (1990). Neurons related to reaching-grasping arm movements in the rostral part of area 6 (area 6 beta). *Exp. Brain Res.,* 82:337–350

Sakata H, Taira M, Murata A, Mine S. (1995). Neural mechanisms of visual guidance of hand action in the parietal cortex of the monkey. *Cereb. Cortex,* 5:429–438

Santello M, Flanders M, Soechting JF. (2002). Patterns of hand motion during grasping and the influence of sensory guidance. *J. Neurosci.,* 22(4):1426–1435

Shadmehr R, Wise SP. (2005). *Computational neurobiology of reaching and pointing: a foundation for motor learning.* MIT, Cambridge MA

Sheng L, Yiqing W, Qingwei Ch, Weili H. (2006). A new geometrical method for the inverse kinematics of the hyper-redundant manipulators. *IEEE Int. Conf. Robot. Biomim.,* pp. 1356–1359

Snyder LH, Batista AP, Andersen RA. (1997). Coding of intention in the posterior parietal cortex. *Nature,* 386:167–170

Swett BA, Contreras-Vidal JL, Birn R, Braun AR. (2010). Neural substrates of graphomotor sequence learning: A combined fMRI and kinematic study. *J Neurophysiol.*, 103(6):3366–3377

Taha Z, Brown R, Wright D. (1997). Modeling and simulation of the hand grasping using neural networks. *Med. Eng. Physiol.*, 19(6):536–538

Tikhonov AN, Arsenin VY. (1977). *Solutions of ill-posed problems.* WH Winston (Ed). Washington DC

Todorov E, Jordan MI. (2002). Optimal feedback control as a theory of motor coordination. *Nat. Neurosci.*, 5(11):1226–1235

Todorov E. (2004). Optimality principles in sensorimotor control. *Nat. Neurosci.*, 7(9):907–915

Ulloa A, Bullock D. (2003). A neural network simulating human reach-grasp coordination by updating of vector positioning commands. *Neural Netw.*, 16:1141–1160

Ungerleider LG, Mishkin M. (1982) Two cortical visual systems. In DJ Ingle, MA Goodale and RJW Mansfield (Eds), *Analysis of visual behavior.* Cambridge, MA: MIT, pp. 549–586

Uno Y, Kawato M, Suzuki R. (1989). Formation and control of optimal trajectory in human multijoint arm movement. Minimum torque-change model. *Biol. Cybern.*, 61(2):89–101

Uno Y, Fukumura N, Suzuki R, Kawato M. (1995). A computational model for recognizing objects and planning hand shapes in grasping movements. *Neural Netw.*, 8(6):839–851

Wimmers RH, Savelsbergh GJ, Peek PJ, Hopkins B. (1998). Evidence for a phase transition in the early development of prehension. *Dev. Psychobiol.*, 32:235–248

Wise SP, Boussaoud D, Johnson PB, Caminiti R. (1997). Premotor and parietal cortex: corticocortical connectivity and combinatorial computations. *Annu. Rev. Neurosci.*, 20:25–42

Woelfl K, Pfeiffer F. (1995). Grasp strategies for a dextrous robotic hand. In V Graefe (Ed), *Intelligent robots and systems.* Amsterdam: Elsevier, pp. 259–277

Chapter 7
Schemata Learning

Ryunosuke Nishimoto and Jun Tani

Abstract This chapter describes a possible brain model that could account for neuronal mechanisms of schemata learning and discusses the results of robotic experiments implemented with the model. We consider a dynamic neural network model which is characterized by their multiple time-scales dynamics. The model assumes that the slow dynamic part corresponding to the premotor cortex inter-acts with the fast dynamics part corresponding to the inferior parietal lobe (IPL). Using this model, the robotics experiments on developmental tutoring of a set of goal-directed actions were conducted. The results showed that functional hierarchical structures emerge through stages of developments where behavior primitives are generated in the fast dynamics part in earlier stages, and their compositional sequences of achieving goals appear in the slow dynamics part in later stages. It was also observed that motor imagery is generated in earlier stages compared to actual behaviors. We discuss that schemata of goal-directed actions should be acquired with gradual development of the internal image and compositionality for the actions.

7.1 Introduction

How can humans acquire diverse skills for complex goal-directed actions in a flex-ible, fluent, robust, and context-dependent manner? We know that human infants develop such skills by having rich sensory-motor interaction experiences day by day. Then, questions are what are the underlying developmental principles of trans-forming such experiences to skills?

Our group has investigated possible neuronal mechanisms of learning goal-directed skilled actions by conducting synthetic neuro-robotics experiments and by analyzing their results using the dynamical systems framework (Schoner and Kelso 1988; Smith and Thelen 1994; Tani and Fukumura 1994; Beer 1995).

J. Tani (✉)
Laboratory for Behavior and Dynamic Cognition, RIKEN Brain Science Institute, 2-1 Hirosawa, Wako-shi, Saitama, 351-0198 Japan
e-mail: tani@brain.riken.jp

V. Cutsuridis et al. (eds.), *Perception-Action Cycle: Models, Architectures, and Hardware*, Springer Series in Cognitive and Neural Systems 1, DOI 10.1007/978-1-4419-1452-1_7, © Springer Science+Business Media, LLC 2011

Especially, the studies have focused on the possibility that the anticipatory learning paradigm (Jordan and Rumelhart 1992; Wolpert and Kawato 1998; Tani 1996; Pezzulo 2008) embedded in neuro-dynamics with rich sensory-motor interactions could result in acquiring generalized dynamic structures for performing a set of desired goal-directed actions (Tani et al. 2004, 2008b). The essential idea is that anticipatory learning of direct sensory feedbacks associated with each intended action would result in self-organization of internal images those are truly grounded to the actual experiences of the agents. And this idea is quite analogous to Piaget's theories on developmental psychology (Piaget 1954), which consider that any representations which children might have should have developed through sensory-motor level environmental interactions accompanied by goal-directed actions.

In general views, human skilled actions look too diverse and too complex to be constructed by single level mechanisms. They might require certain hierarchy. The motor schemata theory by Arbib (1981) postulates that a complex goal-directed action can be decomposed into sequence of reusable behavior primitives. On the other way around, the theory says that diverse actions can be generated by means of the higher level combining the reusable primitives stored in the lower level in a compositional way. If this type of hierarchical mechanism actually accounts for human skilled actions, essential questions might be how the levels can be organized and then how each level can be developed with interacting with other levels.

Recently, we proposed the so-called Multiple Time-scales RNN(MTRNN) in which functional hierarchy can be self-organized using the time-scale difference of local networks. The current text will describe how the schemata learning with hierarchy could proceed in the proposed model by showing our humanoid robotics experiments in the task of object manipulation which emphasizes interactive tutoring between human tutors and robots. The experimental results will clarify the structural relationship among developments of the sensory-motor primitives and their manipulations as well as developments of physical behaviors and motor imagery. Our analysis and discussions will show a possible neuropsychological mechanisms of how manipulatable representations with compositionality could naturally develop solely through sensory-motor experiences in the context of schemata learning.

7.2   Review of Prior Models and Neuroscience Evidences

There are several computational models that aim to achieve motor schemata learning in robots (Billard and Mataric 2001; Demiris and Hayes 2002; Schaal et al. 2003; Inamura et al. 2001; Inamura et al. 2004). Billard and Mataric (2001) use their invented connectionist-type network called DRAMA for imitation learning. Inamura et al. (2001) use a hidden markov model for representing behavior primitives and their sequential combinations. Demiris and Hayes (2002) use a set of gated-modular networks for representing behavior primitives whose architecture is similar to MOSAIC (Wolpert and Kawato 1998) and the mixture of RNN experts

(Tani and Nolfi 1999). Schaal et al. (2003) proposed to describe discrete and cyclic movement primitives in terms of two parameterized canonical equations of fixed point dynamics and limit cycling dynamics, respectively. Inamura et al. (2004) proposed the so-called mimesis space for representing behavior patterns in a certain metric space using artificially designed manifolds. The representation can be used both for generation and recognition. Tani et al. proposed the so-called recurrent neural network with parametric biases (RNNPB) in which a set of behavior primitive can be learned distributedly in a single RNN as attractors. Each primitive can be acquired by bifurcation mechanism with PB vector.

In the brain science perspective, we speculate that structures responsible for generating or mentally simulating skilled behaviors might be acquired in the inferior parietal lobe (IPL) through repeated sensory-motor experiences. Our ideas have been inspired by Ito (2005), who suggested that information from daily sensory-motor experiences is first consolidated in parietal cortex and then further consolidated into the cerebellum as internal models (Kawato et al. 1987) related to actions. Conventionally, parietal cortex has been viewed as a core site to associate and integrate the multi-modality of the sensory inputs. However, the neuropsychological studies investigating various apraxia cases, including ideomotor apraxia and ideational apraxia (Liepmann 1920; Heilman 1973), have suggested that IPL should be also an essential site to represent a class of behavior skills, especially related to object manipulations. We speculate that this region might function especially as a predictor for future sensory inputs for particular goals of actions based on neuroscience evidence that will be detailed in later sections. Furthermore, it has been speculated that IPL might function both for generating and recognizing the goal-directed behaviors (Fogassi et al. 2005) by having dense interactions with ventral premotor (PMv) cells which are known as mirror neurons (Rizzolatti et al. 1996).

Conventionally, parietal cortex has been regarded as a cortex region for associating multimodal sensory information including visual, auditory, tactile, and proprioceptive sensations (Colby et al. 1993). The sensory information integrated in parietal cortex is considered to be sent to PMv for organizing the corresponding motor programs, which are further sent to the primary motor area (M1) for generating more detailed programs (Sakata et al. 1995).

Fagg and Arbib (1998) introduced the so-called FARS model which attempts to explain how PMv and AIP (the anterior intra-parietal sulcus) in parietal cortex can generate object grasping behaviors. In this model, the visual information of an object is sent from the IT area in the visual ventral pathway to AIP. AIP extracts "affordance" information of the object which is a set of important visual properties of the object to be grasped. This affordance information is sent to canonical neurons in PMv where the corresponding motor program of grasping the object is generated by means of an inverse model.

Oztop and Arbib (2002) proposed a model of mirror neurons (Rizzolatti et al. 1996) found in PMv which recognizes the goal-directed behaviors of oneself and of others. They proposed that information about the positional relation between an object and the hand of an other person or animal is extracted and represented in AIP neurons. Then, this AIP neural representation is mapped to categorical activation of the mirror neurons in PMv. (They use a simple three-layered perceptron-type neural network to learn this mapping from the sensory feature inputs to the categorical outputs.)

One common idea in these two models by Arbib's group is that parietal cortex is considered to preprocess the sensory information before PMv in both cases of the behavior recognition and generation. It is assumed that parietal cortex may deal with sensory inputs merely as static patterns rather than temporally changing ones. On this account, we speculate that parietal cortex may function as forward models to anticipate coming sensory inputs.

Our interpretation originated from the literature of neuropsychological studies. It is well known that patients whose IPL are impaired often suffer from deficits in the usage of tools (Liepmann 1920; Geschwind and Kaplan 1962). The deficit is called ideomotor apraxia. Because their basic movements of limbs are not impaired, it is plausible to consider that IPL contains tacit knowledge about the skills to use the tools. More interestingly, some of ideomotor apraxia patients have deficits only in pantomiming but not in actually using the tools (Liepmann 1920; Geschwind and Kaplan 1962; McDonald et al. 1994; Ohshima et al. 1998). The pantomime requires a capability of generating mental imaginary of sensory inputs associated with the actual tool usages for their self-feedback. Because the chain of sensory imaginary can be generated by means of look-ahead prediction using the forward models (Tani 1996; Hesslow 2002), it is speculated that the deficit in pantomime might be originated from impairment of the sensory forward prediction mechanism assumed in IPL.

Eskandar and Assad (1999) investigated the role of the posterior parietal cortex (PPC) in the visual guidance of movements in monkeys trained to use a joystick to guide a spot to a target. In the electrophysiological experiments of the monkeys, they found cells in the lateral intraparietal area (LIP) which seem to encode a predictive representation of stimulus movement. Ehrsson et al. (2003) observed specific activity in PPC during control of fingertip forces for grasp stability in human fMRI scan experiments. Because this force coordination requires anticipation of the grip forces that match the requirements imposed by the self-generated load forces, PPC is assumed to implement such anticipatory mechanism.

Another clue came from recent studies on the involvement of the medial parietal regions of monkeys with goal-directed navigation by Sato et al. (2006). In their experiments, monkeys were trained to navigate to reach specific goal locations in a virtual office environment. The monkeys watched the computer-simulated egocentric view, which was projected into a front screen, and maneuvered the virtual workspace using a joystick controller. In the examination of the neural activities in the medial parietal region, large portions of neurons activate with specific movements (turning left or right) at specific positions. Moreover, some neurons respond in a goal-dependent manner, i.e., they respond depending not only on specific position and movement but also on specific goals. This result suggests that the medial parietal region stores route-based navigation knowledge or, in other words, the internal models of the workspace.

In summary, the evidence suggests that the parietal cortex implements different types of sensory anticipation functions at different local regions depending on the goal of behavior. In the following, our modeling focuses on sensory anticipation mechanisms assumed especially in IPL in the parietal cortex. IPL is known to have a dense connectivity with mirror neurons in the PMv. Our modeling assumes that the anticipation mechanisms in IPL might be different from the one considered in the conventional forward model (Kawato et al. 1987; Wolpert and Kawato 1998) hypothesized in the cerebellum. The conventional forward model predicts the next step sensory outcomes regarding the current action taken and the current sensory state given. In our model, IPL predicts the next step sensory state only with the current sensory state given. We may call this modified version of the forward model as the sensory forward model.

The sensory forward model, however, requires one more piece of information, that is goals for current actions. For example, let us consider a situation in which there is a coffee mug in front of us that is to be manipulated. Our anticipation of sensory sequences is based on the visual image of the mug and its relative position among the fingers and the arm. It is also based on the proprioception of each joint angle for the fingers and arms. The anticipation of sensory experiences might be different depending on our current goals of grasping the mug for drinking coffee or for throwing it toward somebody. Once the goal is set, the sensory sequence associated with this goal-directed behavior can be predicted. Therefore, in our formulation the sensory forward model is provided with the goal state as well as the current sensory state in its inputs, and then the next step sensory state is predicted in the outputs. It is important to note that the sensory forward model cannot play an equivalent role of an internal model or a world model that can tell sensory consequences for whatever motor command sequences. The function of the sensory forward model is much limited in a sense that it can predict sensory sequences only for a set of task goals which are frequently experienced. Therefore, it can be said that the sensory forward model can work only for skilled goal-directed behaviors rather than arbitrary motor behaviors.

It is also noted that the current sensory input could be provided by feeding back its own sensory prediction without having the actual sensory input. This enables the sensory imaginary loop required for the pantomime behaviors, as described previously.

An obvious question is then where the current goal information comes from. We assume that it comes from the mirror neurons in PMv. This assumption accords with the main arguments by the Rizzolatti group (Rizzolatti et al. 1996) that the mirror neurons do not encode exact movement profiles but encode the underlying goals of movements. Now, the sensory sequence is predicted in IPL with the goal information provided from PMv. Here, we show another assumption which could account for how actual motor behaviors can be generated from the predicted sensory sequences. The basic idea is that the predicted sensory sequences for given goals contain enough information about the corresponding motor sequences. We assume that the predicted proprioception is sent to primary sensory cortex (S1) and further sent to primary motor cortex (M1) for generation of actual motor command.

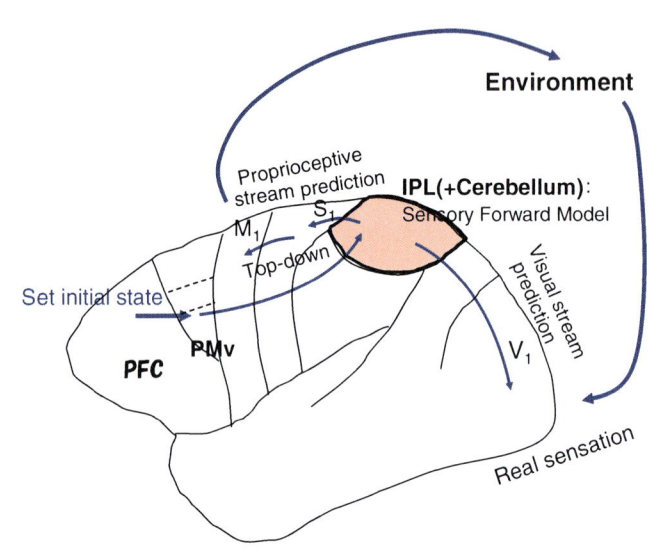

Fig. 7.1 The information flow assumed in brains to generate skilled goal-directed actions

The analogy in robot joint control is that the position encoder value at the next time step is predicted (IPL) and set as the desired position (S1). Then, the motor controller (M1) computes necessary motor torque command. This part of the assumption is inspired by observations by Fetz and his colleagues (Fetz et al. 1980; Soso and Fetz 1980) about S1 involvement in motor generation. They found that some monkey S1 cells are activated immediately before the actual movements of limbs, as if preparing for the movements. Figure 7.1 summarizes the behavior generation pathway, which is assumed in this chapter.

7.3 Proposed Model

This section describes how the ideas of the sensory forward model can be implemented in the so-called MTRNN (Yamashita and Tani 2008). Because of presumptions of general readers as well as limited space in the current special issue, the model is described intuitively with abstraction in details. For the details of the model refer to Yamashita and Tani (2008).

7.3.1 General

The model assumes that a humanoid robot with a simple vision system learns multiple goal-directed tasks of manipulating an object under tutor supervision. The goal for each task trajectory is provided by the experimenter to the robot by setting the

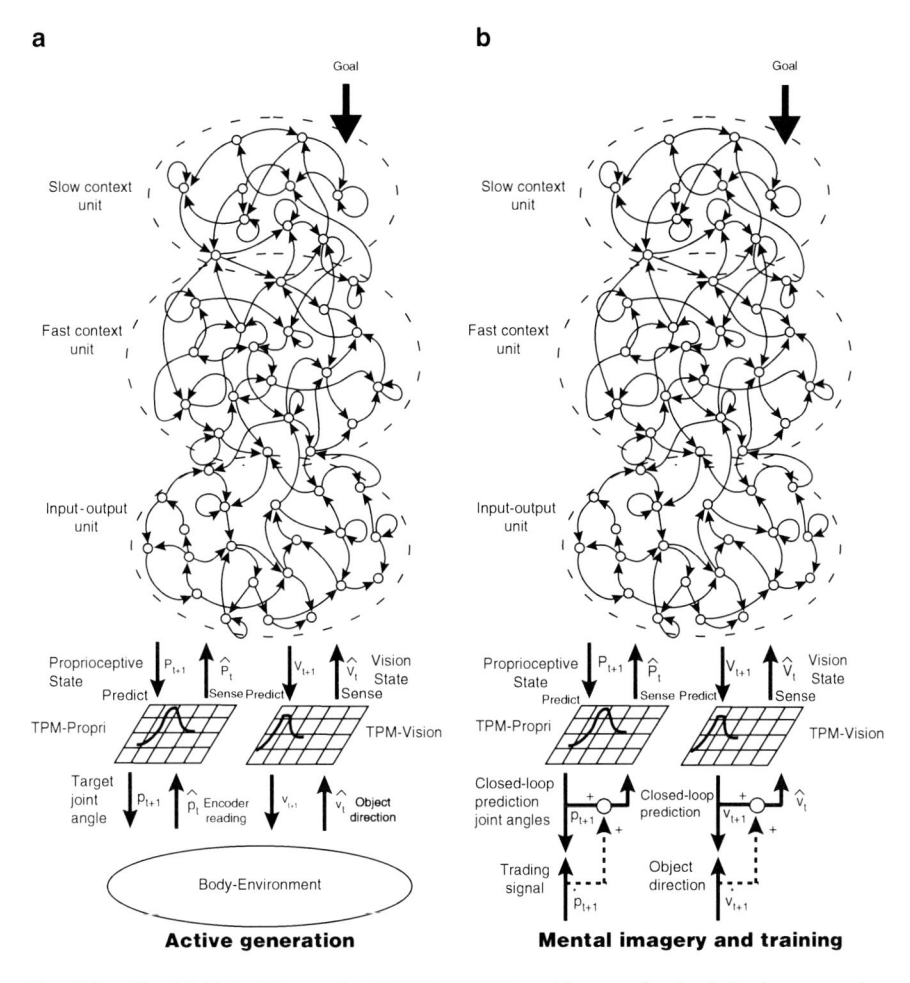

Fig. 7.2 The Multiple Time-scales RNN(MTRNN) architecture in the behavior generation mode (**a**) and in the motor imagery and the training mode (**b**)

initial state of some neurons in the network model used. Inputs to the system are the arm joints encoder readings $\hat{p}_t$ (eight dimensional vector of normalized) and two dimensional vector of the camera head angle $\hat{v}_t$ representing object position (Fig. 7.2). The camera head is programmed to target a red point marked on the frontal surface of the object. Those two modalities of inputs are sparsely encoded in the form of a population coding using the topology preserving map (TPM) where $\hat{P}_t$ proprioceptive state and $\hat{V}_t$ vision state are obtained. This topology preserving sparse encoding of visuo-proprioceptive (VP) trajectories, which resembles information processing in the primary sensory cortices such as VI and SI, reduced overlap between VP sequences and improved the learning capacity of the MTRNN.

Based on the current $\hat{p}_t$ and $\hat{v}_t$, the system generate predictions of proprioception p_{t+1} and the vision sense v_{t+1} for the next time step. This prediction of the proprioception p_{t+1} is sent to the robot in the form of target joint angles, and actual joint movements are made by a built-in PID controller. Changes in the environment, including changes in object position and changes in the actual position of joints, were sent back to the system as sensory feedback.

The main component of the system modeled by the MTRNN receives the current input of VP state and it outputs the prediction of its next step state. The goal for each task trajectory is given as the initial state in terms of the potential states of slow context units at the initial step. The generation of each task trajectory is made possible by the capacity of the RNN to preserve the intentionality toward the corresponding goal as the internal dynamics using the slow context units activities.

A conventional firing rate model, in which each unit's activity represents the average firing rate over a group of neurons, is used to model neurons in the MTRNN. In addition, every unit's membrane potential is assumed to be influenced not only by current synaptic inputs but also by their previous state. In the MTRNN, each neural unit activation is defined with continuous-time dynamics (Doya and Yoshizawa 1989) of which characteristic is described by the following differential equation, which uses a parameter τ referred to as the time constant:

$$\tau_i \frac{du_{i,t}}{dt} = -u_{i,t} + \sum_j w_{ij} a_{i,t}$$

$$a_{i,t} = \text{Sigmoid}(u_{i,t})$$

where $u_{i,t}$ is the membrane potential of each ith neuronal unit at time step t, $a_{j,t}$ is an activation of jth unit, and w_{ij} is synaptic weight from the jth unit to the ith unit. The current activation state of each unit is obtained as a sigmoidal output of its potential. The time constant τ mostly determines the time scale of the unit activation dynamics. When it is set with large values, the dynamics becomes slow and otherwise quick. Some modeling studies (Nolfi 2002; Nishimoto et al. 2008) have shown that τ affects strength of context-dependent memory effect in adaptive behavior.

The network that was used in the current model consisted of input–output and non-input–output units, the latter referred to as context units. Context units were divided into two groups based on the value of time constant τ. The first group consisted of fast context units with small time constant ($\tau = 5$) whose activity changed quickly, whereas the second group consisted of slow context unit with a large time constant ($\tau = 70$) whose activity, in contrast, changed much more slowly. Among the input–output units, units corresponding to proprioception and units corresponding to vision are not connected to each other directly. The slow context units and the fast context units are fully connected each other, and the input–output units and the fast context units do so as well while the slow context units and the input–output units are not directly connected.

The slow neural activation dynamics assumed in the model could be accounted by electrophysiological observation by Isomura and his colleagues (Isomura et al. 2003). They found that some groups of neurons in the PFC can have slow build-up of firing frequency in the order of some seconds. Sakai et al. (2006) and Okamoto et al. (2007) showed that the phenomena of the slow build-up of firing frequency can be reconstructed by simulations of assembly of spiking neurons with recurrent connectivity. On the other hand, neurons in posterior cortices mostly exhibit fast build-up in the order of split of second. This may correspond to fast neural activation dynamics in the model.

7.3.2 Training

In order to obtain a teaching signal, the experimenter guides both hands of the robot along the trajectory of the goal action. As the robot hands are guided along the trajectory, the sensed VP sequences are recorded, and they were used as teaching sequences. For each behavior task the object was located in three different positions (center position, 2 cm left of the center and 2 cm right of the center). The objective of learning was to find optimal values of connective weights minimizing the error between teaching sequences and model outputs. At the beginning of training, synaptic weights of the network were set randomly, resulting in the network generating random sequences. Synaptic weights were modified based on the error between teaching signals and generated sequences. After many repetitions of this process, the error between teaching sequences and model outputs eventually reached a minimum level.

This training process is conducted in an off-line manner in the sense that all teaching sequences gathered at each tutoring session are assumed to be stored in a short-term memory (this part is out of the scope), and they are used as teacher sequences for consolidation learning of the sensory-forward model assumed in IPL. The tutoring session with gathering new training sequences will be iterated in the course of development. At each training process, lookahead prediction of the VP sequence is generated by means of so-called closed-loop operations (Fig. 7.2b) in which the current prediction of the VP state are used as input for the next time step. Then, the error between the teacher sequences and the lookahead sequences of imagery are taken by which error-driven training of the network is conducted. The purpose of using this closed-loop operation in training is to enhance generation of stable dynamic structures of the network by minimizing the error integrated during long steps of lookahead prediction. Our preliminary trials indicated that conventional training scheme of using one-step prediction instead of lookahead one has difficulty in acquiring stable long-time correlations because the error generated at each step becomes too small.

Using the characteristic of initial sensitivity, the network dynamics is trained to generate multiple behavior sequences through adaptation of the initial states of slow context units. In the proposed model, a network is trained by means of supervised learning using teaching sequences obtained through tutoring by the experimenter. The conventional back-propagation through time (BPTT) algorithm (Rumelhart et al. 1986) is used for adaptation of both connective weights common to all sequences and the initial state of slow context units for each sequence (Nishimoto et al. 2008). (The initial states of fast context units are not adapted but set as neutral.)

7.3.3   Action Generation in Physical Environment and Motor Imagery

Through the training process, the network learns to predict the VP inputs for the next time step. The prediction of proprioceptive state provides the target joint angles to the robot controller which enables the robot to generate movements.

Moreover, using the prediction of VP feedback as input to the next time step (closed loop operation), the network can be able to autonomously generate VP trajectories without producing actual movements. This process of closed loop generation may correspond to motor imagery in terms of mental simulation of actions (Jeannerod 1994; Decety 1996; Tani 1996). It is noted that the motor imagery in this chapter is defined as image of inseparable coming flows of kinesthetic one and egocentric visual one in terms of VP trajectory.

7.4   Setup of Humanoid Robot Experiments

A small humanoid robot was used in the role of a physical body interacting with actual environment. A workbench was set up in front of the robot, and a cubic object (approximately $9 \times 9 \times 9$ cm) placed on the workbench served as the target object for manipulations. The robot task is to learn to generate three different task behaviors. The goal of each task behavior is to generate a different sequence of behavior primitives of manipulating the object. All task behaviors start from the home position and end with going back to the same position (Fig. 7.3).

In the task-1, with starting from the home position, both hands grasp the object, move the object up and down (UD) for four times, do it for left and right (LR) for four times, and go back to the home position (BH). In the task-2, the object is moved forward and backward (FB) for four times and then it is touched by left and right hands bilaterally (TchLR) for four times and finally BH. In the task-3, the robot repeats grasping and releasing the object by both hands (BG) for four times and then BH. A tutor teaches the robot with these three task behaviors in three tutoring sessions with changing the object position three times from the center position to the left and to the right for each task behavior. In the first session, the robot guidance is conducted by disabling active movements of the robot by setting the motor control gain to zero, because the networks are not yet effective with the randomly set initial

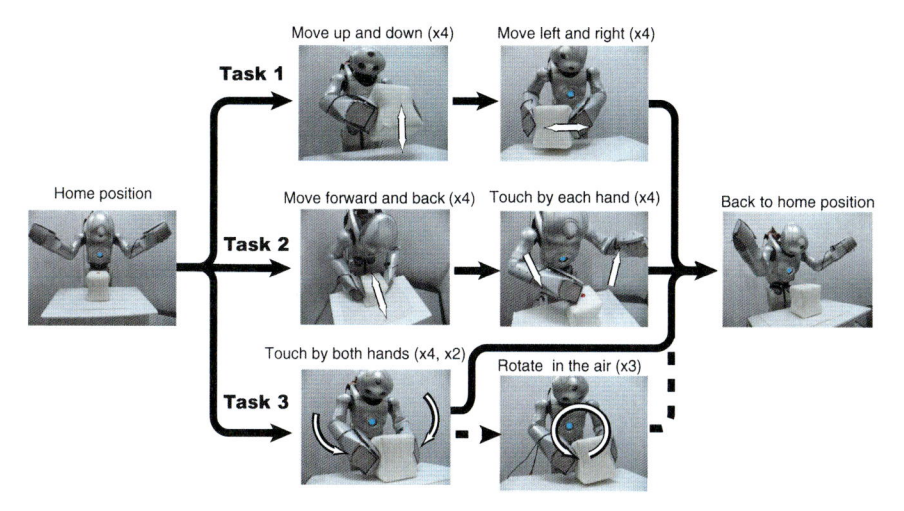

Fig. 7.3 Three task behaviors tutored to the robot. After the third session, the Task-3 is modified as illustrated by *dotted lines*

	Session 1	Session 2	Session 3	Session 4	Session 5
σ_{PropI}	300	1.5	0.75	0.75	0.75
σ_{PropF}	1.5	0.75	0.375	0.375	0.375
$\sigma_{\mathrm{VisionI}}$	150	0.75	0.375	0.375	0.375
$\sigma_{\mathrm{VisionF}}$	0.75	0.375	0.1875	0.1875	0.1875
CLr	0.6	0.8	0.9	0.6	0.9

Table 7.1 The parameter setting for each training session

synaptic weight values. In the second and third sessions, the tutoring is conducted interactively by enabling active movements of the robot with the control gain set to 20% of its normal operation value. The network is trained off-line using all tutoring sequence data obtained at each session. The network consists of 144 proprioceptive units, 36 vision units, 30 fast context units, and 20 slow context units.

After the basic tutoring of three sessions, the task-3 is modified with introducing a novel behavior primitive which is to rotate the object in the air (RO) by both hands. In the session 4 and 5 of the task-3, BG is repeated two times followed by three times repetitions of RO. This additional tutoring is conducted to examine the capability of the network to incrementally learn novel patterns. In the session 4, the training parameters are once relaxed to minimize the interference between the previously learned contents and the new one (see the 4th and 5th session in Table 7.1). It is noted that the interference could occur not only in cases of introducing novel primitives but also for novel sequential combinations of them, because this requires fine adjustments in both of the lower and higher levels to achieve end-to-end smooth connections between the primitives.

7.5 Experimental Results

7.5.1 Overall Task Performances in the End of Development

The developmental tutoring experiment was repeated twice with setting the initial synaptic weights of the networks as randomized. Figure 7.4 shows how the robot behaviors were generated in the test run after the five sessions of tutoring in one developmental case. Plots are shown for the VP trajectories (sequences of two representative arm joint angles denoted as "Prop #" and two camera head angles of normalized denoted as "Vision #") in the tutoring in the top row, the actually generated one in the second row, and the fast and slow context activations represented by the first four principal components denoted as "PC #" after their principal component analysis (PCA) in the third and the forth row, respectively, for all three tasks. It is observed that the actual VP trajectories are exactly reconstructed from the tutoring ones for all the tasks. Actually, the robot was successful in performing all tasks with the object position varied within the range of tutored after the five tutoring sessions. The profiles of the fast context activations and those of the slow

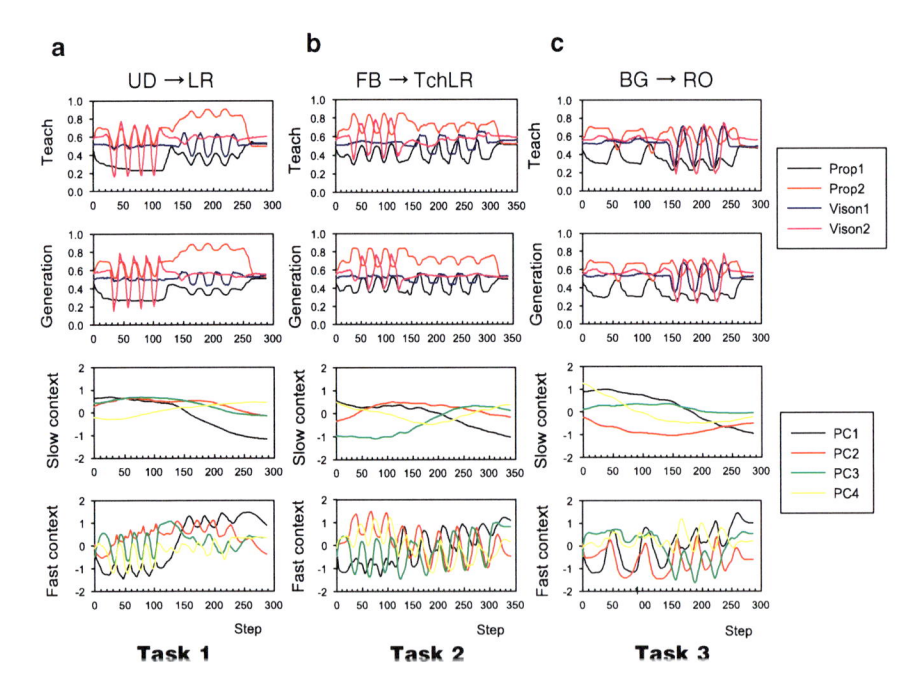

Fig. 7.4 Visuo-proprioceptive (VP) trajectories (two normalized joint angles denoted by Prop1 and Prop2 and camera head direction denoted by Vision1 and Vision2) in tutoring and in actual generation accompanied with fast and slow context profiles with principal component analysis (PCA) denoted by PC1, PC2, and PC3

ones can be contrasted. The fast ones mostly synchronize with the VP trajectories, while the slow one shows smoothly changing trajectory starting from different initial state of self-determined for each task. It is observed that the slow context profiles abruptly change when the cyclic pattern in the VP trajectories shift from one primitive to another. These observation suggests that each exact pattern of the primitives is embedded in the fast context activation dynamics, while each macro scenario of sequencing of the primitives is embedded in the slow context one. This result accords with the one in (Yamashita and Tani 2008).

7.5.2 Development Processes

Now, the development process is closely examined as the main focus of this chapter. Figure 7.5 shows one developmental case of the task-1 with the object located in the center from session 1 to session 3 before the novel task behavior is introduced in the task-3. Plots are shown for the VP trajectories of tutoring in the left, motor imagery in the middle, and actual robot generation in the right. The slow context units profiles in the motor imagery and the actual behavior are plotted for their first four principal components after the PCA. It is noted that the tutoring trajectories in the session 1 is quite distorted. The tutoring patterns of UD in the first half and LR in the second half are not regular cycles. This is a typical case when cyclic patterns are tutored to robots without using metronome-like devices. However, it can be seen that the cyclic patterns in the tutoring become much more regular as the session proceeds.

One interesting observation is that the motor imagery patterns develop faster than the actual ones over these three sessions. In the session 1, the cyclic pattern of UD is successfully generated (but not for LR) in the motor imagery, while neither UD nor LR are yet generated in the actual behavior generation. It is noted that the cyclic pattern of UD is more regular than the tutored one in the session 1. In the actual behavior generation, the robot hands touched the object but not accurate enough to grasp and hold it up, and after the failure the movements were frozen. One interesting observation was obtained by conducting an extra experiment using fake visual feedback. In this extra experiment, the tutor grasped the object and moved it up and down immediately after the robot touched the object. It turned out that the robot hands moved up and down correctly following the object movement of perceived. It can be understood that this arm movement was generated by means of the entrainment with the fake visual feedback. The same phenomena had been observed in the case using a modular network model of the mixture of RNN experts (Tani et al. 2008a).

In the session 2, both UD and LR cyclic patterns are generated in the correct sequence in motor imagery, while only UD pattern is generated which cannot be shifted to LR pattern in the actual behavior. This can be explained by an observation that the slow context profile in the motor imagery dynamically changes around 160 steps, while that of actual behavior does not show any significant changes around this transition period. It is considered that the dynamics of the slow context units

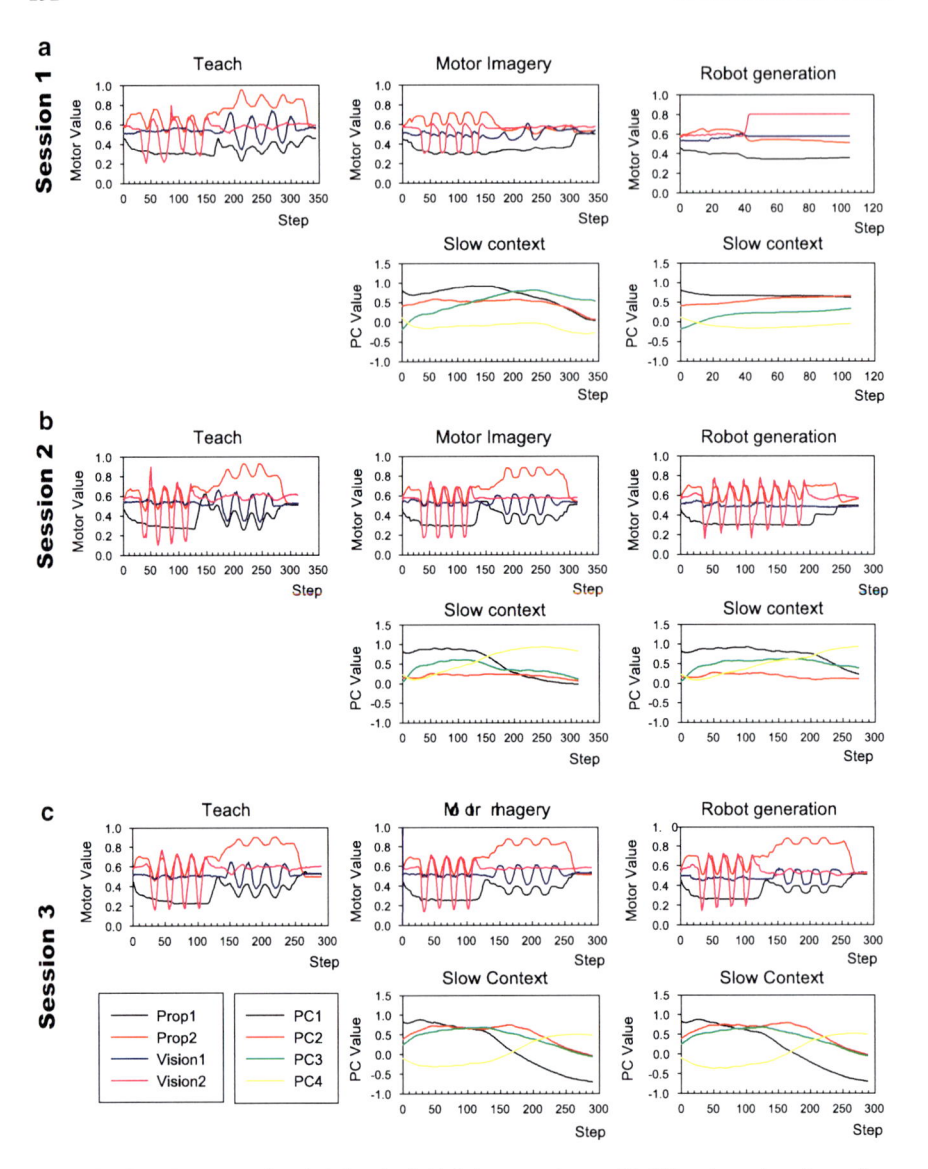

Fig. 7.5 Development of task-1 for the initial three sessions with VP trajectories of tutoring, motor imagery, and actual generation accompanied with slow context profiles by PCA

is not strong enough to generate the shifting in the actual behavior interacting with noisy environment. This consideration is supported by the fact that the robot could actually make the shift when the tutor assisted the robot to do so by guiding the arm trajectories with force in the transition period.

In the session 3, both UD and LR patterns are successfully generated both in the motor imagery and in the actual behavior generation. It was, however, observed in limited cases that counting of repetition times of primitives (as like UD four times) could go wrong within the range of plus or minus one time probably by perturbed by noise during physical execution of actions. An interesting observation here is that even when the counting goes wrong, smooth transition from one primitive to another is still preserved, e.g., moving object left and right always follows immediately after the object is once placed on the table. The transition never takes place by cutting through in the middle of on-going primitives. This observation implies first that counting in the higher level is more like implicit and analogical process rather than explicit and logical one and second that the lower level is successful in organizing fluidity in connecting primitives which could be expressed by Luria's metaphor of "kinetic melody" (Luria 1973).

7.5.3 Analyses

In this subsection, more detailed analyses are shown for examining the observed developmental processes. Figure 7.6 shows how the success rate in metal simulation and in actual behavior generation changes in average of all three task behaviors for the two developmental cases. Here, the success rate is defined as rate of how many primitive events can be successfully generated as in the trained order in mental simulation and actual behavior. For example, the number of primitive events in the task-1 is counted as 9 with 4 UDs, 4 LRs, and 1BH. The number is counted by looking at the robot behavior for actual behavior and by examining plots of VP trajectories generated compared with the trained one for mental simulation. In Fig. 7.6, it can be seen that the success rate of mental simulation is higher than the one of actual behavior at the least the first three sessions in both developmental cases. It is observed that the success rate becomes 1.0 as perfect after three sessions of tutoring for both mental simulation and actual behavior. Then the rate slightly decreases in

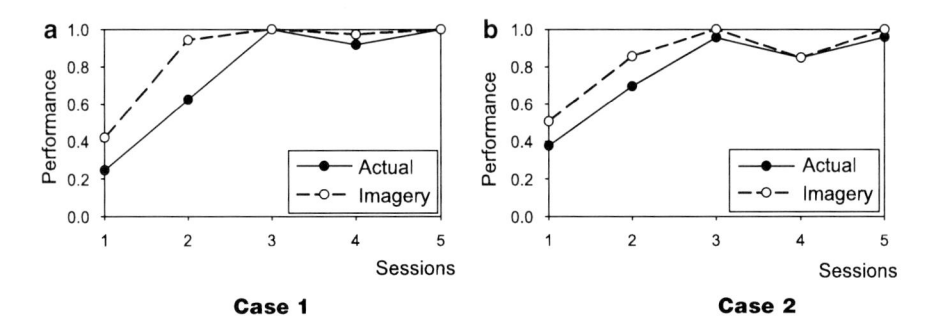

Fig. 7.6 The developments of success rate averaged over three task behaviors in motor imagery denoted by "Imagery" and in actual behavior denoted by "Actual" for two case runs

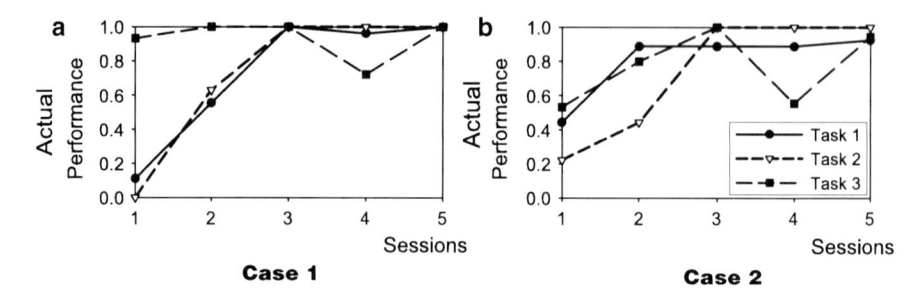

Fig. 7.7 The developments of success rate of each task behavior denoted as "Task1, task2 and task3" in actual behavior for two case runs

the session 4 when the novel task behavior is introduced in the task-3. It, however, goes back to nearly 1.0 in the session 5.

Figure 7.7 shows the success rate for each task behavior in actual behavior in developmental cases. It can be observed that the success rates of task-1 and task-2 stay near 1.0 after session 3 to the end while that of task-3 once decreases in session 4 when the novel behavior primitive RO is introduced and it reaches to 1.0 in the end. This result indicates that introduction of a novel behavior primitive in a task behavior does not affect the performances in other task behaviors unless they share the same behavior primitives. It was also observed that the behavior primitive of BG, which was followed by RO, was not distorted in the session 4 in both development cases. Only RO was immature. This means that once acquired, primitives can be used in generating different sequential combinations of the primitives. Such recombination capability for primitives was also shown in our prior study (Yamashita and Tani 2008).

Figure 7.8 illustrates how the encoding of the basic primitives by the fast context units develop during the first three sessions. The trajectory of the fast context units during each basic primitive pattern in mental simulation is plotted as a phase diagram using the first and the second principal components. Three colors in the plots denote three different object positions cases. It is observed that there are no regularities in the patterns shown for the session 1 except the BG case in which each trajectory shows a near cyclic pattern that is shifted with the position difference. In the session 2, such near cyclic patterns appear for all basic primitives. Finally, we see that the shapes of the patterns in the session 3 are mostly similar to the ones in the session 2. These results imply that basic primitives begin to be embedded in pseudo attractor of limit cycling by the fast context units with achieving the object position generalization in the second session. This development by fast context units seems to converge mostly in the second session.

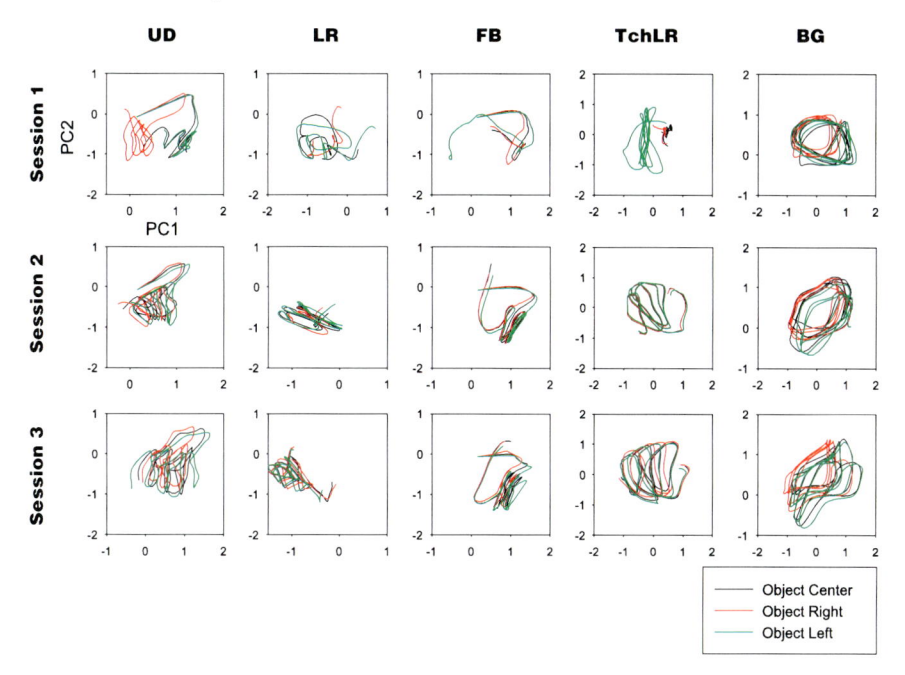

Fig. 7.8 The development of encoding of basic primitives by the fast context units with (PCA)

7.6 Discussion

7.6.1 Summary of the Robot Experiments

Now, the robotics experiment results are summarized with qualitative discussions. It was shown that the developmental learning processes of multiple goal-directed actions were successfully converged after several sessions of the teacher tutoring. The developmental process can be categorized in some stages. In the first stage which mostly corresponds to the session 1, no tasks are completed where most of the behavior primitives in actual generation are premature. In the second stage corresponding to the session 2, most of the behavior primitives can be actually generated although their sequencing is not yet completed. In the third stage corresponding to on and after the session 3, all tasks are successfully generated with correct sequencing of the primitives. From this observation, it can be said that there is an order in the formations of different levels of functionality. The level for behavior primitives is generated by the second stage, while the level for sequencing the primitives does by the third stage. It is natural that the primitive level as the lower level is organized earlier, and then the level for the sequencing as the higher level does later based on the prior formation of the lower level.

However, one interesting remark is that there is a time lag between the period of becoming able to generate motor imagery and actual behavior. The motor imagery is generated earlier than the actual behavior as it was seen that the motor imagery for all tasks are nearly completed by the session 2 as compared to the session 3 by the actual ones. This issue will be revisited with some psychological considerations later in this section. Another remark is that when a new task which composed of a prior-trained primitive and a novel one was introduced midway, the introduction affects the overall task performances only slightly. Although regeneration of the novel primitive is premature initially, the prior-trained primitive is well adopted in this new task and also performances of other tasks are intact.

7.6.2 Schemata Learning from Developmental Psychological Views

The above-mentioned qualitative observation in our robotics experiment for schemata learning could correspond to some observations and theories in developmental psychology. Among them, Piaget's constructivist accounts for infant developments might be most relevant. Piaget considered that if infants can have representations, they should self-organize through the dynamic interactions between subject and object. The dynamic interactions should involve with the one in sensory-motor level accompanied with goal-directed intentionality about the object. Then, operative and figurative aspects of intelligence should emerge as the results of self-organization through such dynamic interactions. There are two core concepts those compose the Piaget's theory. One is assimilation and the other is accommodation. Assimilation is a process that existing scheme of subject is exploited to establish structural coupling with object. On the other hand, accommodation is an adaptive process to modulate the scheme to establish another structural coupling with object.

If we look at our experiments, it is understood that scheme in Piaget's theory may correspond to a set of behavior primitives embedded in the fast context network. Depending on the top-down signal conveying the task goal information flowing from the upstream slow context network, different dynamic structures of the primitives are adopted which may explain dynamic mechanism of assimilation. These behavior primitives are actually the products of the neuronal self-organization with having rich sensory-motor interactions through iterative tutoring. This may account for accommodation. The case of introducing a new task behavior in the session 4 could be interpreted that both assimilation and accommodation occur because the pre-acquired primitive is used in the novel task while a novel behavior primitive is additionally self-organized. The fact that six different behavior primitives were compositionally manipulated to generate both actual behaviors and motor imagery for achieving multiple goals in the end of the developmental tutoring could be interpreted that certain operational representations are finally appeared through the long-term self-organization process. It is, however, argued that the operational representations appeared in this stage is not just compositional, as if composed of a set of discrete symbols, but "organically" composed (Tani et al. 2008b) by capturing fluid and contextual nature of human skilled behaviors in neuronal circuits of analog dynamical systems. This argument is supported by the current observations of various phenomena including implicit and analogical counting in repeating primitives and smooth transitions in the primitive sequences.

Our approach is also parallel to the ones by the so-called neo-Piagetian, especially who attempt to explain the infant development as time-development of complex systems (Smith and Thelen 1994). Smith and Thelen (1994) claim that infant development is better understood as the emergent product of many decentralized and local interactions that occur in real time where coherence among local parts is achieved. Our robotics experiments have been carefully designed such that local interactions can be enhanced in different levels. The MTRNN was designed such that neuronal dynamics can interact with row sensory-motor flow in the continuous space and time domain without introducing any a priori articulation mechanisms. Also, there is no algorithmic operations those act on segregated modules of the higher and the lower levels or independent modules of encoding primitives. All there exist are just a single network where different time-scale dynamics coexist, and their interactions result in self-organization of functional hierarchy. Furthermore, the tutoring procedure was designed such that the tutor and the robot can directly interact each other with force. It was observed that not only the robot trajectories develop but also the tutoring ones do across sessions to generate smooth and rhythmic patterns. The direct force level interactions enabled this sort of co-developments between the two sides.

The sensory forward model used in the current study should be distinguished from the conventional forward model (Kawato et al. 1990; Wolpert and Kawato 1998). The conventional forward model predicts the resultant future sensation for the current motor commands given. One notorious problem is that the forward model cannot predict all the outcomes of possible motor commands because of combinatorial complexity associated with their high dimensional space. This problem is related to the frame problem (McCarthy 1963) well known in artificial intelligence. It tells that an optimal action cannot be determined if infinite number of possible action consequences are examined at each step. Why does this happen? This is because the conventional forward model does not deal with goal-directedness. The conventional forward model attempts to predict consequences of arbitrary action sequences which may not be related to any goal achievements. On the other hand, the sensory forward model of our proposal attempts to predict coming sensory flow in the course of achieving each specified goal. Because the sensory forward model learns about only finite number of goal-directed paths through the actual tutoring experiences, it never faces with the combinatorial explosive problems. Indeed, Piaget's advocacy of goal-directedness is right in a sense that the burden of goal-directedness with enactments actually avoids unrealistic combinatorial computations in cognition.

Our experiments showed that motor imagery develops faster than actual behaviors. Does it correspond to any reality in human development and learning? Some contemporary developmental psychologists such as Karmiloff–Smith (1992) and Diamond (1991) claim that mental representation develops very earlier in life, or is even innate. It is said that infants of 2 months old already possess intentionality toward objects to deal with but just cannot reach properly to them because of immaturity in motor control skills. It might be plausible that motor imagery of reaching to objects develops easily if infants happen to reach to the objects of their interests by motor bubbling, and such episode is reinforced with joys. However, the actual motor actions on objects such as touching or grasping them are far more difficult because they involve with precise arm controls of making physical contacts to the objects, as had been shown in our robotics experiments. Because the generation of motor imagery, on the other hand, does not require such fineness, it could be achieved earlier. Flanagan et al. (2003) showed evidences that human subjects learn to predict sensory feedback faster than motor control in their psychophysics experiments on object manipulation under artificial force field. This finding might be related to the current results because the predictability of sensory feedback directly links to motor imagery.

Also it is known that generation of motor imagery has a positive role in consolidating memories (Feltz and Landers 1983; Vogt 1995; Jeannerod 1995) in schemata learning, as have been mentioned earlier. The robot-training scheme shown in the experiment is analogous to this evidence because in our scheme the network is trained to regenerate the teaching sequences in the closed-loop operation without receiving actual inputs as like rehearsing, and this explains why motor imagery develops earlier than the actual one.

The motor imagery by means of lookahead prediction can provide means for online monitoring of future perspective. If unexpected external changes happen during physical execution of goal-directed actions, the monitoring by the lookahead prediction can detect a future perspective gap as the error between the originally intended goal state and the currently predicted one. The detected error can be used to modify the original goal state to currently possible one by modulating the internal state that carries goal information. This online monitoring and the error-driven goal state modulation can be implemented by pairing the future lookahead prediction and the past regression as have been described elsewhere (Tani 2003; Ito et al. 2006). The motor imagery plays essential roles in accommodating cognitive behaviors in diverse ways including goal-directed motor planning, online monitoring of future perspective and resultant goal modulation, and enhancements of consolidation learning. Future studies should focus to integrate those different functions systematically in synthetic models.

7.7 Summary

The current text examined the problem of schemata learning by considering a brain-inspired model for generating skilled goal-directed actions. We showed a neuro-robotics experiment in which the processes of developmental learning for a set of goal-directed actions of a robot were examined. The robot was implemented with the MTRNN model which is characterized by co-existences of the slow dynamics and the fast dynamics in generating anticipatory behaviors. Through the iterative tutoring of the robot for multiple goal-directed actions, certain structural developmental processes emerged. It was observed that behavior primitives are self-organized in the fast dynamics part earlier and sequencing of them appears later in the slow dynamics part. It was also observed that motor imagery develops faster than the actual ones. The chapter discussed that the observed processes for the schemata learning by the robot are quite analogous to Piaget's ideas of the constructivism, which emphasis the roles of goal-directed sensory-motor interactions in acquiring operational representations in human development.

Acknowledgements The authors thank Sony Corporation for providing them with a humanoid robot as a research platform. The study has been partially supported by a Grant-in-Aid for Scientific Research on Priority Areas "Emergence of Adaptive Motor Function through Interaction between Body, Brain and Environment" from the Japanese Ministry of Education, Culture, Sports, Science and Technology.

References

Arbib, M.: Perceptual structures and distributed motor control. In: Handbook of physiology: the nervous system, II. motor control, pp. 1448–1480. MIT, Cambridge, MA (1981)

Beer, R.: A dynamical systems perspective on agent-environment interaction. Artificial Intelligence **72**(1), 173–215 (1995)

Billard, A., Mataric, M.: Learning human arm movements by imitation: evaluation of a biologically-inspired connectionist architecture. Robotics and Autonomous Systems **941**, 1–16 (2001)

Colby, C., Duhamel, J., Goldberg, M.: Ventral intraparietal area of the macaque: anatomic location and visual response properties. Journal of Neurophysiology **69**, 902–914 (1993)

Decety, J.: Do executed and imagined movements share the same central structures? Cognitive Brain Research **3**, 87–93 (1996)

Demiris, J., Hayes, G.: Imitation as a dual-route process featuring predictive and learning components: a biologically plausible computational model. In: Imitation in animals and artifacts, pp. 327–361. MIT, Cambridge, MA (2002)

Diamond, A.: Neuropsychological insights into the meaning of object concept development. In: The epigenesis of mind: essays on biology and cognition, pp. 67–110. Erlbaum, Hillsdale, NJ (1991)

Doya, K., Yoshizawa, S.: Memorizing oscillatory patterns in the analog neuron network. In: Proceedings of 1989 International Joint Conference on Neural Networks, pp. I:27–32. Washington, DC (1989)

Ehrsson, H., Fagergren, A., Johansson, R., Forssberg, H.: Evidence for the involvement of the posterior parietal cortex in coordination of fingertip forces for grasp stability in manipulation. Journal of Neurophysiology **90**, 2978–2986 (2003)

Eskandar, E., Assad, J.: Dissociation of visual, motor and predictive signals in parietal cortex during visual guidance. Nature Neuroscience **2**, 88–93 (1999)

Fagg, A.H., Arbib, M.A.: Modeling parietal-premotor interactions in primate control of grasping. Neural Networks **11**, 1277–1303 (1998)

Feltz, D.L., Landers, D.M.: The effects of mental practice on motor skill learning and performance: a meta-analysis. Journal of Sport Psychology **5**, 25–57 (1983)

Fetz, E., Finocchio, D., Baker, M., Soso, M.: Sensory and motor responses of precentral cortex cells during comparable passive and active joint movements. Journal of Neurophysiology **43**, 1070–1089 (1980)

Flanagan, J., Vetter, P., Johansson, R., Wolpert, D.: Prediction precedes control in motor learning. Current Biology **13**(2), 146–150 (2003)

Fogassi, L., Ferrari, P., Gesierich, B., Rozzi, S., Chersi, F., Rizzolatti, G.: Parietal lobe: from action organization to intention understanding. Science **308**, 662–667 (2005)

Geschwind, N., Kaplan, E.: Human cerebral disconnection syndromes. Neurology **12**, 675–685 (1962)

Heilman, K.: Ideational apraxia - a re-definition. Brain **96**, 861–864 (1973)

Hesslow, G.: Conscious thought as simulation of behaviour and perception. Trends in Cognitive Sciences **6**(6), 242–247 (2002)

Inamura, T., Nakamura, N., Ezaki, H., Toshima, I.: Imitation and primitive symbol acquisition of humanoids by the integrated mimesis loop. In: Proceedings of the IEEE International Conference on Robotics and Automation, pp. 4208–4213 (2001)

Inamura, T., Toshima, I., Tanie, H., Nakamura, Y.: Embodied symbol emergence based on mimesis theory. International Journal of Robotics Research **23**(44), 363–377 (2004)

Isomura, Y., Akazawa, T., Nambu, A., Takada, M.: Neural coding of "attention for action" and "response selection" in primate anterior cingulate cortex. The Journal of Neuroscience **23**, 8002–8012 (2003)

Ito, M.: Bases and implications of learning in the cerebellum – adaptive control and internal model mechanism. Progress in Brain Research **148**, 95–109 (2005)

Ito, M., Noda, K., Hoshino, Y., Tani, J.: Dynamic and interactive generation of object handling behaviors by a small humanoid robot using a dynamic neural network model. Neural Networks **19**, 323–337 (2006)

Jeannerod, M.: The representing brain: neural correlates of motor imitation and imaginary. Behavioral and Brain Science **17**, 187–245 (1994)

Jeannerod, M.: Mental imagery in the motor context. Neuropsychologia **33**(11), 1419–1432 (1995)

Jordan, M., Rumelhart, D.: Forward models: supervised learning with a distal teacher. Cognitive Science **16**, 307–354 (1992)

Karmiloff-Smith, A.: Beyond modularity. A developmental perspective on cognitive science. MIT, Cambridge, MA (1992)

Kawato, M., Furukawa, K., Suzuki, R.: A hierarchical neural network model for the control and learning of voluntary movement. Biological Cybernetics **57**, 169–185 (1987)

Kawato, M., Maeda, Y., Uno, Y., Suzuki, R.: Trajectory formation of arm movement by cascade neural network model based on minimum torque-change criterion. Biological Cybernetics **62**(4), 275–288 (1990)

Liepmann, H.: Apraxie. Ergebnisse der Gesamten Medizin **1**, 516–543 (1920)

Luria, A.: The working brain. Penguin Books, New York (1973)

McCarthy, J.: Situations, actions and causal laws. Stanford Artificial Intelligence Project, Memo2, (1963)

McDonald, S., Tate, R., Rigby, J.: Error types in ideomotor apraxia: a qualitative analysis. Brain and Cognition **25**(2), 250–270 (1994)

Nishimoto, R., Namikawa, J., Tani, J.: Learning multiple goal-directed actions through self-organization of a dynamic neural network model: a humanoid robot experiment. Adaptive Behavior **16**, 166–181 (2008)

Nolfi, S.: Evolving robots able to self-localize in the environment: The importance of viewing cognition as the result of processes occurring at different time scales. Connection Science **14**(3), 231–244 (2002)

Ohshima, F., Takeda, K., Bandou, M., Inoue, K.: A case of ideational apraxia -an impairment in the sequence of acts. Journal of Japanese Neuropsychology **14**, 42–48 (1998)

Okamoto, H., Isomura, Y., Takada, M., Fukai, T.: Temporal intagration by stochastic recurrent network dynamics with bimodal neurons. Journal of Neurophysiology **97**, 3859–3867 (2007)

Oztop, E., Arbib, M.A.: Schema design and implementation of the grasp-related mirror neuron system. Biological Cybernetics **87**, 116–140 (2002)

Pezzulo, G.: Coordinating with the future: the anticipatory nature of representation. Minds and Machines **18**, 179–225 (2008)

Piaget, J.: The construction of reality in the child. Basic Books, New York (1954)

Rizzolatti, G., Fadiga, L., Galless, V., Fogassi, L.: Premotor cortex and the recognition of motor actions. Cognitive Brain Research **3**, 131–141 (1996)

Rumelhart, D., Hinton, G., Williams, R.: Learning internal representations by error propagation. In: D. Rumelhart, J. McClelland (eds.) Parallel distributed processing, pp. 318–362. MIT, Cambridge, MA (1986)

Sakai, Y., Okamoto, H., Fukai, T.: Computational algorithms and neuronal network models underlying decision processes. Neural Networks **19**, 1091–1105 (2006)

Sakata, H., Taira, M., Murata, A., Mine, S.: Neural mechanisms of visual guidance of hand action in the parietal cortex of the monkey. Cerebral Cortex **5**, 429–438 (1995)

Sato, N., Sakata, H., Tanaka, Y., Taira, M.: Navigation-associated medial parietal neurons in monkeys. Proceedings of the National Academy of Sciences of USA **103**, 17,001–17,006 (2006)

Schaal, S., Ijspeert, A., Billard, A.: Computational approaches to motor learning by imitation. Philosophical Transaction of the Royal Society of London: Series B, Biological Sciences **358**(1431), 537–547 (2003)

Schoner, S., Kelso, S.: Dynamic pattern generation in behavioral and neural systems. Science **239**, 1513–1519 (1988)

Smith, L., Thelen, E.: A dynamic systems approach to the development of cognition and action. MIT, Cambridge, MA (1994)

Soso, M., Fetz, E.: Responses of identified cells in postcentral cortex of awake monkeys during comparable active and passive joint movements. Journal of Neurophysiology **43**, 1090–1110 (1980)

Tani, J.: Model-based learning for mobile robot navigation from the dynamical systems perspective. IEEE Transactions on Systems, Man, and Cybernetics B **26**(3), 421–436 (1996)

Tani, J.: Learning to generate articulated behavior through the bottom-up and the top-down interaction process. Neural Networks **16**, 11–23 (2003)

Tani, J., Fukumura, N.: Learning goal-directed sensory-based navigation of a mobile robot. Neural Networks **7**(3) (1994)

Tani, J., Ito, M., Sugita, Y.: Self-organization of distributedly represented multiple behavior schemata in a mirror system: reviews of robot experiments using RNNPB. Neural Networks **17**, 1273–1289 (2004)

Tani, J., Nishimoto, R., Namikawa, J., Ito, M.: Codevelopmental learning between human and humanoid robot using a dynamic neural network model. IEEE Transactions on Systems, Man, and Cybernetics **38**(1), 43–59 (2008)

Tani, J., Nishimoto, R., Paine, R.: Achieving "organic compositionality" through self-organization: reviews on brain-inspired robotics experiments. Neural Networks **21**, 584–603 (2008)

Tani, J., Nolfi, S.: Learning to perceive the world as articulated: an approach for hierarchical learning in sensory-motor systems. In: R. Pfeifer, B. Blumberg, J. Meyer, S. Wilson (eds.) From animals to animats 5. MIT, Cambridge, MA (1998). Later published in Neural Networks, vol 12, pp. 1131–1141, 1999

Vogt, S.: On relations between perceiving, imaging and performing in the learning of cyclical movement sequences. British Journal of Psychology **86**, 191–216 (1995)

Wolpert, D., Kawato, M.: Multiple paired forward and inverse models for motor control. Neural Networks **11**, 1317–1329 (1998)

Yamashita, Y., Tani, J.: Emergence of functional hierarchy in a multiple timescale neural network model: a humanoid robot experiment. PLoS Computational Biology **4**(11) (2008)

Chapter 8
The Perception-Conceptualisation-Knowledge Representation-Reasoning Representation-Action Cycle: The View from the Brain

John G. Taylor

Abstract We consider new and important aspects of brain processing in which it is shown how perception, attention, reward, working memory, long-term memory, spatial and object recognition, conceptualisation and action can be melded together in a coherent manner. The approach is based mainly on work done in the EU GNOSYS project to create a reasoning robot using brain guidance, starting with the learning of object representations and associated concepts (as long-term memory), with the inclusion of attention. Additional material on actions and internal simulation is taken from the EU MATHESIS project. The framework is thereby extended to the affordances of objects, so that effective action can be taken on the objects. The knowledge gained and the related rewards associated with the representations of the objects involved are used to guide reasoning, through the co-operation of internal models, to attain one or other of the objects. This approach is based on attention as a control system to be exploited to allow high level processing (in conscious thought) or lower level processing (in creative but unconscious thought); creativity is also considered as part of the abilities of the overall system.

8.1 Introduction

The list of faculties to be considered according to the title of this chapter is long, covering as it does most powers of the human brain. So it is natural to consider if it would be possible to shorten the list somewhat. After more detailed consideration, one realises that most of these components are needed for each other. Thus, perception is a crucial entry to the creation of object representations, leading to the formation of concepts, laid down in long-term memory and basic to reasoning. Attention is a mechanism allowing the development, through the basic process of

J.G. Taylor (✉)
Department of Mathematics, King's College, London, UK
e-mail: john.g.taylor@kcl.ac.uk

V. Cutsuridis et al. (eds.), *Perception-Action Cycle: Models, Architectures, and Hardware*, Springer Series in Cognitive and Neural Systems 1, DOI 10.1007/978-1-4419-1452-1_8, © Springer Science+Business Media, LLC 2011

attention filtering, of representations in the brain of single objects. Without such simplified representations, it would be very difficult to reason to solve important tasks, such as acquiring a highly rewarding object, such as a grape, by a monkey. Knowledge of objects is acquired by actions on them, such as grasping and eating the grape, so completing the list of faculties in the title of this chapter.

That there is need for inclusion of most of the brain's capacities in a reasoning system is to be expected. Reasoning is one of the highest faculties of the human mind and so would be expected to depend on the majority of the brain's components, even if only minimally. Thus we have to explore the manner in which the human brain is able to generate the amazing reasoning powers it does possess. There are other approaches to such high-level cognitive processes that do not depend on the architecture of the human brain. However, we consider it of value to attempt to use cues from the brain even if we may then discover more efficient ways of achieving some powers based on such an approach; the initial guidance from the brain involves a system that has evolved over a million or so years, so is to be expected to provide a reasonably optimised solution to many of the cognitive tasks under consideration.

It is thereby clear that the faculties used in reasoning have been assembled in an efficient manner in the human brain, as part of general evolutionary competition. We can extend from the human to some lower animal brains, especially to the primate brain. A range of primates are now well known to possess reasoning powers, as are various corvids, especially birds like Betty, the New Caledonian Crow. However, there is still not enough knowledge about the bird brain to be able to build suitably detailed neural architectures based on such knowledge, so we will limit ourselves here solely to primates and their brains.

Not only brains but bodies are also important to consider as part of cognitive processing. Without a body to control and use for a variety of responses, we do not expect much of a cognitive repertoire to be accessible. But it is well known that robotics presently has engineering difficulties over developing a truly all singing, all dancing, all games-playing robot. So in GNOSYS, there was simplification of the actions able to be taken to those enabling movement around an environment (so a wheeled robot) and possessing manipulative skills (grasping, putting down) by a gripper; for that was taken a simple two-fingered gripper.

It is arguable that knowledge and reasoning are the two activities at the highest cognitive level in the brain, so their careful discussion and analysis is of utmost importance. There is much progress in both of these areas, with considerable interaction between them. In this chapter, we will consider in what manner our understanding of the brain is adding to these two intertwined topics. The developments will be considered from both a broad experimental as well as a modelling/theoretical aspect, as well as considering the lower-level components needed to sustain such high-level processing.

Knowledge learning by neural networks is already a vast subject: there are 7.6M items to search on Google under that heading. There are at least 60 specific journals on the subject as well as at least 25 scientific organisations whose annual meetings are dedicated to studying the subject. It would thus be effectively impossible to try to cover the area in complete depth. However, the two approaches mentioned above – the experimental and the modelling/theoretical sides – in which new angles and advances are being recognised as important to move the subject forward and create ever better systems to learn and utilise knowledge. This plethora also holds for reasoning, in which, however, some specific advances have also been made.

In spite of the enormous amount of material available in the research community on these two topics and their neural "infrastructures", by restricting ourselves to brain-based approaches for both topics, we make them both more limited and more specific. Thus, knowledge is to be taken to consist of brain-based codes for the classification of objects and actions to be taken on them, and possible extensions to symbolic descriptors of such codes, as by means of language. However, this latter will not be of primary concern in this chapter (although it will be briefly described), since it requires at least a separate chapter (if not several books) to treat properly. Thus, knowledge is effectively non-linguistically defined, as it would be for most animals below humans and humans in early infancy, unless otherwise stated.

Such a restriction to working at a non-linguistic level might be thought of as a severe limitation on the generality of the resulting discussion. As seen from a cognitive science point of view, only a linguistic approach is appropriate for considering higher cognition. Yet, in the brain, the development and adult use of language rests on an underpinning of processing, which involves a large amount of non-linguistic processing. Direct speech input addresses most directly the speech centres, but even then, there are control components and non-linguistic codes that have to be accessed to give meaning and grounding to the language symbols – be they words, phrases or whole stories.

Seen from a different point of view, the most crucial architectural components of the brain to attain higher cognitive powers – those involving attention control systems, reward and valuation systems and coupled internal models – can be discussed for a range of processes that may or may not involve linguistic codes. This will be clear when we consider thinking as mental simulation, creativity as unattended mental simulation and reasoning as rewarded mental simulation.

We will also consider briefly how knowledge is acquired, especially through adaptive processes of neural network modular systems. Here, there is a large amount of data and modelling on the nature of codes for object and action representations, obtained by suitably causal learning laws going beyond the simple Hebbian law into spike-timing-dependent laws for communication between spiking neurons. Yet again, we will not pursue this very rich area of research in any detail but concentrate more on a global picture of how attention control is involved in learning and how the knowledge at a lower level can be combined into internal models that allow for higher cognitive processing. For without such possibilities, especially of fusing in attention control processes, and the brain-based architectures that may achieve such processing, we cannot attempt to explain higher cognition observable in animals as controlled by the brain. A similar approach will be taken to reasoning (although linguistic aspects will be discussed there more fully in association with logical reasoning).

So what is the basis we will use to attack the problems of higher order cognition in the brain? From the experimental side, there have been advances on the codes used in storing various forms of knowledge in the brain, both those involved with objects and those with actions. However, there is still considerable controversy over detailed neural aspects of such coding. This understanding is important in probing the manner in which attention is built as part of object codes. Such intertwining of simple object/action codes with attention has not been considered as an important part of the development of cognitive processing in the past, but in order to take account of the ability of attention to be focussed efficiently on any distributed object representation, the question as to how this facility arises is clearly crucial. Indeed the need for a system to be able to attend at will to a stimulus in its sensory field must be both fast and accurately employed in any attentive scanning of the environment for enemies or other stimuli warranting crucial fast response to enable survival.

At the same time, it is to be assumed that the attended object representations can be further manipulated in "interior planning". Such takes place in the system's brain, without any necessary action being made exteriorly. In this manner, the system can determine the effects of future actions without exposing itself to possible danger in its environment. This leads to the manner in which possible internal models (forward and inverse model controllers) can be trained so as to suitably use the object and action representation learnt as part of the knowledge basis of the brain.

The manner in which such internal models can be coupled and used for mental simulation must be a natural part of this discussion. Again the manner in which attention enters will be important, where it is to be used in a variety of ways: to prevent distracters from derailing any train of thought being followed, to enhance ongoing thinking and so to speed it up, to enhance the use of suitable rules of reasoning, and so on.

Simultaneously, there is the need to consider how attention can be switched off in following any train of thought. This could then allow creativity to enter through the use of lateral spreading of activity across the regions coding for object and action representations. Attention to relevant stimuli will be relaxed during this process of extended searching through semantic maps for suitable stimuli that may be of interest. Otherwise attention would constrain the search space; without it, the expansion of the space allows analogical reasoning to flourish. Such a process, involving reasoning by analogy, needs a careful discussion so as to develop a suitable architecture to achieve such a mixed level of processing (Table 8.1).

Finally, we need to consider reasoning, a process involving the creation of subgoals and their being followed. We discuss some simple non-linguistic examples of reasoning, especially the two-stick paradigm and how this could be simulated by means of a suitable neural architecture. Extension of this both to a general class of reasoning paradigms as well as to reasoning involving language will considered, although more briefly.

Table 8.1 Comparison between the GNOSYS cognitive architecture and that of other proposed such architectures

Name of cognitive architecture	Perception/ concept creation	Sensory attention	Goal creation	Motor attention	Reward learning	Internal models	Reasoning & "thinking"
GNOSYS	√	√	√	√	√	√	√
Global Workspace	X	X	√	X	√	√	X
Self-directed anticipated learning (SDAL)	X	X	√	X	X	√	X
Self-affecting self-aware (SASE)	X	X	√	X	X	√	X
Darwin Robots (Brain-based devices)	√	√	X	√	√	X	X
Humanoid Robot	X	X	√	X	X	X	X

A tick denotes the presence of the relevant cognitive component; a cross denotes its absence

8.2 The GNOSYS Model

8.2.1 The Basic GNOSYS Robot Platform and Environment

The robot used in GNOSYS was a Pioneer P3AT robot, with a Katana arm, a laser with resolution of 05° and a scan rate of at least 10 Hz. The stimuli in the environment consisted of sets of cylinders and sticks as well as a ball and a cube. Perceptual representations of the various objects was learnt by the software and then used in various active tasks, such as stacking the cylinders on top of each other, and of solving various reasoning tasks with the objects; these task solutions will be described later.

8.2.2 Information Flow and GNOSYS Sub-systems

The physically embodied robot was controlled by a remote software system that was running in a network of PCs. The software was the realisation of the GNOSYS cognitive architecture. The main modules that were present and the high level interactions are shown in Fig. 8.1.

The above modules can be partitioned into four major sub-systems:

- Perception (coarse or fine visual systems, attributes, threats)
- Memory (brain, place map, concepts)

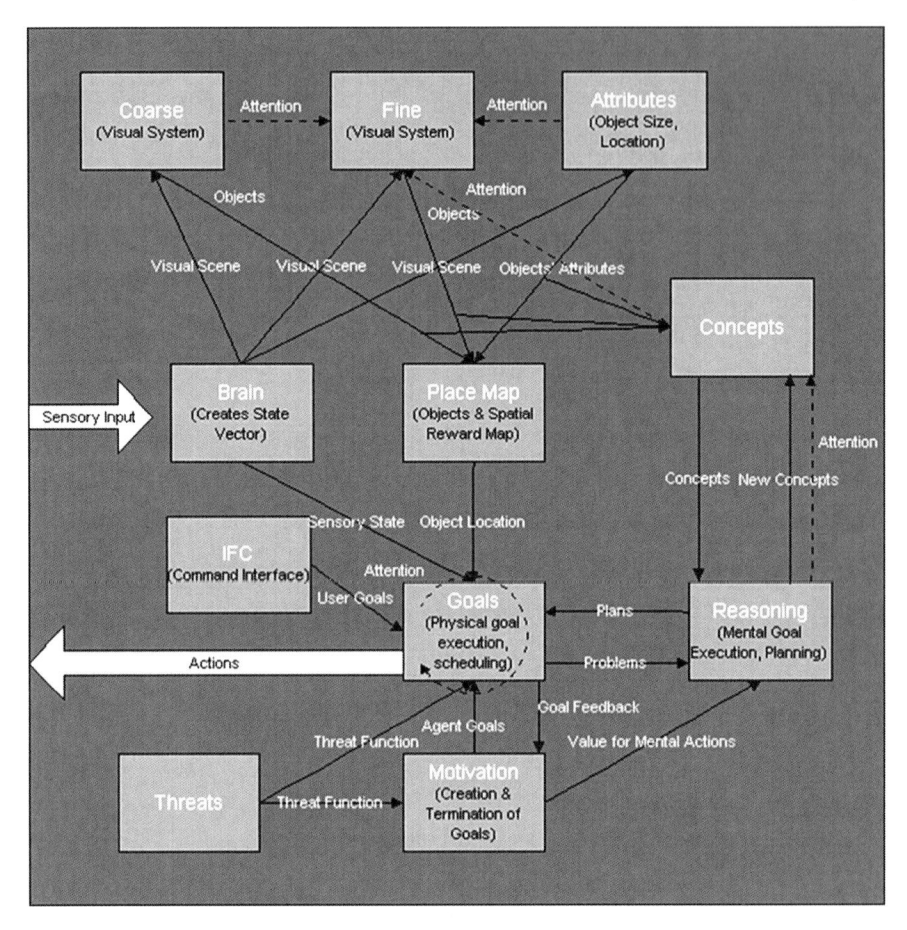

Fig. 8.1 Main components of the GNOSYS brain. The components shown in the figure are described more fully in the text

- Action execution (reasoning, goals, motivation)
- Interface (IFC)

The interface concerns the interaction of the agent with the users and it will not be discussed further here. Instead we will provide brief explanations of the modules of all other sub-systems.

8.2.2.1 Perception

The perception system analyses visual scenes and extracts coloured blobs (defined as a connected region of local intensity and colour different from its surroundings). The system pre-processes the image so as to remove visual noise; the subsequently discovered blobs are sent to the coarse and fine visual scene analysis systems. These provide additional information about the objects, based on the corresponding colour blob.

The coarse visual scene analysis provides object descriptors by using a sparse sampling approach. It sends its output (i.e., the blob descriptors) to the fine visual system for further processing and elimination of false descriptors (which are those interpreted by the software as arising from an object not present in the visual image). The descriptors for each camera are combined for the left and right cameras to produce a single object descriptor before they are exported.

The fine visual scene analysis provides object descriptors using a dense sampling approach, employing the input image for processing and the coarse visual system descriptors for directing attention to the possible objects present. Through attention, false descriptors (those incorrectly implying the presence of a given object) are eliminated. A reduced descriptor set per camera is thereby created. These are then combined across the left and right camera to produce the final fused descriptor set. The threat module uses laser input to track people or other moving agents in space, tracking a set of moving agents over time and providing a simple (linear) prediction of their paths. Its output is a threat map used to determine the threat level of given locations in space.

8.2.2.2 Memory

Brain

This keeps a global state vector for a number of time instances (typically five time-delayed instances of the state vector are kept). The length of each time slot is about 250 ms in real time, which is equivalent to the sampling rate of the laser sensor over a wireless connection. The state vector contains the information produced inside the agent (sensory signals, object descriptors after scene analysis, the set of goals being executed and their status, attention and other low-level hardware/software error events, user goal requests and other necessary information). Such a vector thereby gives a global view of any state in the current or previous time instances.

Place Map

This system keeps two maps internally. The first map is an occupancy map, where found objects are recorded and associated with a specific location. The second is a spatial reward map, enabling the transfer of rewards from specific objects to their spatial location, so as to return a set of probable locations to look for an object when the robot is physically searching for an object. This latter map is needed for when a recorded object is missing from its registered position in the occupancy map. The map also provides memory of rewards associated to places and not just to objects. The contents of the Place Map are updated every time there is visual processing.

Concepts

This stores representations of objects and goals at conceptual level. This information is more concerned with the characteristic properties and general knowledge of the objects. Its contents are used by the reasoning and goal execution processes. The contents are updated after (a) visual processing, (b) reasoning (which can discover and register new "virtual" or imagined objects during reasoning steps), and finally (c) after attempted motor actions on objects in the real world (during goal execution) to establish affordances. These latter affordances are features of objects of importance, allowing them to be manipulated by the system to solve tasks involving the objects (Gibson 1979; Heft 1989; Natsoulas 2004; Young 2006). Virtual objects correspond to either known object classes, but which are not physically present, (i.e., exist only in imagination) or to new concepts of an object (e.g., creating a new longer stick from previous shorter ones; the longer stick is assumed to be a new concept appearing for first time with its own parameters of length, etc).

8.2.2.3   Action Execution

Reasoning

This provides the mechanism, which discovers solution paths from an initial state to a desired state. It takes as input a goal (and its parameters), and the initial state, and then calculates a solution path (if this exists and it can be found). It provides a mental level execution of a goal without taking into account the dynamics of the real world (except that incorporated into the reasoning mechanism, such as the internal models employed in reasoning). It assumes simply a snapshot of the world in the moment of the problem; it then calculates a path using forward model predictions as to the consequences of any potential action (i.e., as in mental simulation). It returns a plan to the Goals module for execution in the real world. When a goal terminates successfully the Goals module provides feedback information in order to update the reward structures of the module.

Goals

This module provides an implementation of the Computational Model for Multiple Goals. These multiple goals are needed in a complex scene since some of them may have to be chosen as subgoals that are attained first, with only the overall goal achieved at the end of the reasoning. This thereby provides a real world execution of the actions of a Plan. The multiple goal manipulation system has been developed as an essential part of the reasoning system (Taylor and Hartley 2007; Mohan and Morasso 2007).

Every action is composed of a set of primitives and sub-plans. Primitive examples consist of requesting a new visual input and scene analysis, moving the robotic arm or the agent's body, etc. Sensory attention, termination conditions and unexpected events are handled at this level. When a plan is executed, typically the corresponding reasoning process is suspended in real time while it waits for the results of an action. When any feedback information is available, such as when an action terminates, fails, or cannot continue execution due to changed environmental conditions, the reasoning process is re-activated and is fed with the current state. A re-plan request is made in an effort to satisfy the original goal before reporting failure of the goal satisfaction. The Goals module takes input from the Motivation System regarding variables that control the lifetime of a goal. It also provides as output the actual implementation of an action plan. As a side effect of the latter, new information may be registered in the system, such as new concepts, location of new objects, new affordances to be established for known objects, detection of novel objects, etc.

Motivation

This module performs a dual function. First, it provides a mechanism that generates new goals, in the absence of any user goal (by activation of newly salient concepts), and second, it evaluates every goal's progress in order to determine its usefulness for the maximisation of the highest-level motivation variable determining the well-being of the agent. The system uses a hierarchical model of drives, which, through their collective dynamics, produce *internal* reward vectors for characterising the progress of a goal and for goal generation. For executing a goal, it outputs variables, which dynamically affect the lifetime of the goal. The system thereby eliminates goals that are not attainable or are not being executed successfully due to loss of any global goal competition.

8.3 The GNOSYS Model Processing Details

8.3.1 The GNOSYS Perception System

The brain basis of visual perception is by means of a hierarchical set of modules modelled on that of human visual perception. In the human brain, information to higher brain sites is carried along the two main visual streams flowing from the thalamus: the dorsal stream, coding for where objects are situated, and the ventral one for what the objects are that are detected at the various sites by the dorsal stream (Milner and Goodale 1995).

Beside the feed-forward flow of visual information in the brain, there is also a feedback of visual attention from higher sites in parietal lobes and the prefrontal cortex to the earlier sites in occipital cortex (Mehta et al. 2000). In this manner, attention control is exerted from the activity stored (for endogenous attention) or it rapidly attains (for exogenous attention) parietal and prefrontal areas. That there is

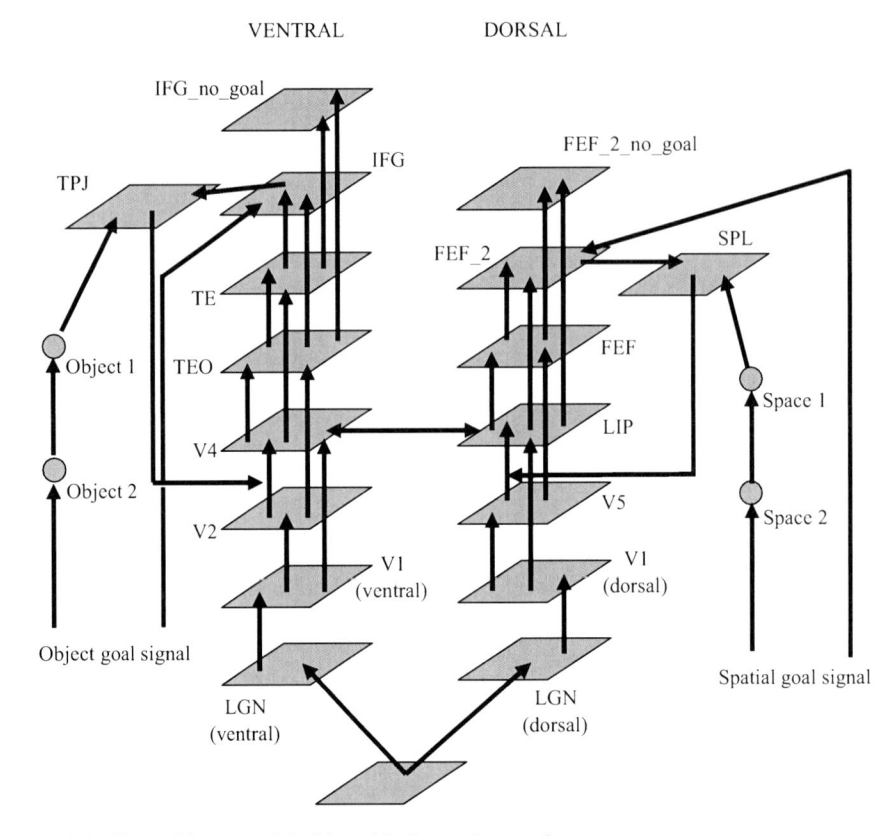

Fig. 8.2 The architecture of the hierarchical neural network

such a division of attention into the control networks in frontal and parietal regions and those regions undergoing attention feedback has also been decisively shown by numerous brain imaging experiments (Corbetta and Shulman 2002; Corbetta et al. 2005; Kanwisher and Wojciulik 2000).

The overall circuitry of the GNOSYS "fine" visual system is shown in Fig. 8.2. The modules have good basis in the brain as arises from many brain imaging and single cell studies, as well as by neuro-architectonic methods (based on different densities of the six cortical cell layers and other features) (Kandel et al. 2000). The GNOSYS software uses a simplified model of this hierarchical architecture, with far fewer neurons and associated trainable synapses.

A schematic of the architecture of visual modules, as used in the GNOSYS system, to generate the hierarchy of feature maps of object representations, is that shown in Fig. 8.2. Input enters at the bottom of the hierarchy and is sent to both the dorsal and ventral components of the geniculate nucleus (LGN), and then to the ventral and dorsal visual streams. It then enters temporal lobe processing in the ventral route (and then to the ventral frontal cortex) and parietal and frontal eye field processing in the dorsal route. Feedback attention signals are generated to these two routes by the TPJ (for the ventral route) and the TPJ (for the dorsal route). An added colour route parallel to the dorsal route, was included in the final processing. Note the reciprocal V4 ↔ TPJ connection between the two routes, used in moving rewards from the ventral route (so object-based) to the dorsal route. More details of the architecture and its functionality are given in the text.

The above neural network architecture was used in the visual perception/concept simulation in the GNOSYS brain. There is a hierarchy of modules simulating the known hierarchy of the ventral route of $V1 \rightarrow V2 \rightarrow V4 \rightarrow TEO \rightarrow TE \rightarrow PFC(IFG)$ in the human brain. The dorsal route is represented by $V1 \rightarrow V5 \rightarrow LIP \rightarrow FEF$, with a lateral connectivity from LIP to V4 to allow for linking the spatial position of an object with its identity (as known in the human brain). There are two sets of sigma–pi weights, one from TPJ in the ventral stream, which acts by multiplication on the inputs from V2 to V4, and the other from SPL, which acts similarly on the V5 to LIP inputs. These feedback signals thereby allow for the multiplicative control of attention.

8.3.2 Learning Attended Object Representations

As an infant develops its powers of attending, the period during which it actually does attend to its surroundings grows successively over the weeks and months from its birth. This growth of the time it can attend to the external world is no doubt supported by the growth of joint attention with its carer. This common attention guidance by the carer is expected to allow both the singling out of objects worthy of attention and the subsequent learning of their names. At the same time, the process of joint attention will lead to the infant learning both the object representation and its ability to attend to it.

It is unclear from present analyses of the joint attention process that attention is developed for sensory representations that have already been learnt at an unattended level or if these representations are learnt at the same time as the attention control feedback signals. However, the process of joint learning (by carer and infant) takes place from an early stage of infant development, even before the infant seems to be able to take much notice of its surroundings during its long periods of inattention (these being mainly sleep, especially in the infant's earliest stages). Thus it would be expected that the infant only learns the unattended sensory representations in its environment while it is attending to its surroundings. That it is through these joint attended/unattended sensory representations that are still the main manner of learning is unclear from the present data but will be assumed here. As noted in the recent paper of (Toda 2009), "These findings suggest that joint attention may be the basis for an infant's social and cognitive development".

Certainly, we all know from our adult experience that we learn about new objects while attending to them in our surroundings, both as sources of sensory stimuli as well as providing affordances for present or future manipulations. Thus we take as basic to the learning processes to expand a subject's knowledge base that such an expansion takes place in an attended manner.

We add that automatic responses to objects can be developed as an extension of the learnt processes under attention, so by over-learning (as is well known). Such automatic responses thereby allow sensory attention to be focussed on other areas of experience. An example of this is when gaining one's sea-legs on a ship, or on correcting such automatic processing itself, as in the case of coming back off the ship and having to attend to the non-movement of the surface of the quay.

Both of these processes can be modelled by a suitable neural architecture. The first process, which is that of developing automatic processing, can be modelled by a simple recurrent neural architecture, which can be implemented in the brain by means of the recurrent circuitry of

$$\text{Pre-frontal CX} \rightarrow \text{basal ganglia} \rightarrow \text{thalamus} \rightarrow \text{Pre-frontal CX}.$$

This can be defined by ever more complex structures involving increasingly complex neurons and neuro-chemicals, especially dopamine (Taylor and Taylor 2000a, b; Gurney et al. 2004). The architecture can thereby be used to model more complex processes involving knowledge representations, such as the process of rapid learning in pre-frontal cortex and the rapid learning of novel but rewarding stimuli (Taylor and Taylor 2007).

For GNOSYS, the development of attention was as follows. Input was created from visual input by suitable photo-detectors whose current was then passed into a spatially topographic array of models of neural cells sensitive to such current input and representing the lateral LGN of the thalamus (one cell layer up from the retina). This input then activated the most sensitive cells to that input, which subsequently sent activity sequentially up various routes in the hierarchy (dorsal, ventral, colour in the GNOSYS case) shown in Fig. 8.2. Attention feedback then occurred from the highest level activity (from FEF or IFG, in Fig. 8.2). There is a similar feedback process in neural models of attention (Mozer and Sitton 1998; Deco and Rolls 2005; Hamker and Zirnsak 2006), with similar amplificatory (of target activations) and inhibitory (of distracter activations) effects.

One of the novelties of our work was the employment of attention, which was crucial in object recognition and other processes involving the components of GNOSYS towards solving the reasoning tasks. The relevant control structure also allows the attention system to be extended to the more general CODAM model (Taylor 2000a, b, 2005, 2007) thereby allowing the introduction of a modicum of what is arguably awareness into the GNOSYS system.

In an earlier visual model of the ventral and dorsal streams, which did not include the recognition of colour and was smaller in size, we investigated the abilities of such a model with attention to help solve the problem of occlusion (Taylor et al. 2006, 2007a, b). This model was trained to recognise three simple shapes (square, triangle and circle) by overlapping two shapes (a square and a triangle) to differing degrees. We investigated how attention applied to either the ventral or dorsal stream could aid in recognising that a square and a triangle were present in such input stimuli. In the case of ventral stream processing, attention was directed to a specific object, which in our case was either the square or triangle. We found that for all the different levels of occlusion that were investigated, the firing rates of neurons in V4, TEO, TE and IFG, which had a preference towards the attended object, increased as against the case where no attention was present. At the object recognition level, in the IFG-no-goal site (in Fig. 8.4), the attended object was the most active. In general, a reduced activation was seen for those neurons, which preferentially respond to the non-attended object within the ventral stream from V4 upwards. This increased activity at V4 level for the attended object was transferred to the dorsal stream via the lateral connections between V4 excitatory layer and the LIP excitatory layer. Via the hierarchical dorsal model, this activity was then expressed as higher firing rates at the location of the attended object in all the FEF modules.

Attention could also be applied within the dorsal stream. In this case, attention was directed towards a small group of FEF2 nodes (3×3) (Fig. 8.2), which then, via the sigma–pi weights acting on the V5 to LIP inputs, led to increased activation at the attended location at the FEF2-no-goal site. Again using the lateral connections between the two streams at LIP and V4 level, this increased dorsal stream activity due to attention at a specific location was able to be transferred to the ventral route. Since V4 is not spatially invariant, this led to increased activation of those active nodes near the attended location. This resulted in correctly identifying in the IFG-no-goal region the object present at the attended location. So for both forms of attention, the model could correctly identify objects present within an occluded composite object (see Taylor et al. 2006, 2007a, b for a more complete investigation).

The various codes learnt by the synapses of the feed-forward connections of the modules of Fig. 8.2, on repeated stimulus presentations, were found to be close to those observed experimentally. Thus for V2, it was discovered that neurons were created, which were sensitive to particular angles between two input slits with their point of intersection at the centre of the receptive field of the cell. In V4, we discovered neurons sensitive to components of the boundary of input stimuli, as shown in Fig. 8.3 and in agreement with experimental results (Pasupathy and Connor 2001).

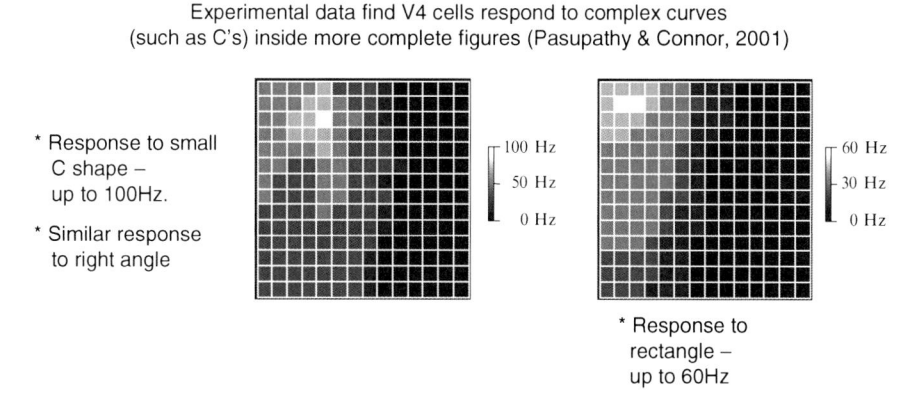

Experimental data find V4 cells respond to complex curves
(such as C's) inside more complete figures (Pasupathy & Connor, 2001)

* Response to small
C shape –
up to 100Hz.

* Similar response
to right angle

100 Hz
50 Hz
0 Hz

60 Hz
30 Hz
0 Hz

* Response to
rectangle –
up to 60Hz

Fig. 8.3 The sensitivity of V4 neurons to partial boundaries of shapes

The response of V4 neurons to the small C-shape in the left figure is up to 100 Hz for some of the neurons; that for a complete rectangle, shown in the right figure, is only up to 70 Hz (from Taylor et al. 2006, 2007a, b).

Finally, we used a fast adaptive method to train the attention feedback (only from TPJ to V4 for the ventral route and from SPL to LIP for the dorsal route). The method used was to insert sigma–pi feedback connections from a given active neuron in TPJ, for a given stimulus, to V4 onto an active neuron there so as to amplify the relevant input from V2 to that neuron in V4 to magnify the stimulus effect (so fitting data on the nature of attended responses of single cells in monkey V4, Reynolds et al. 1999). A similar method was used for introducing attention feedback from SPL to LIP with respect to its input from V5/MT.

In conclusion, we have presented a visual system composed of a hierarchy of neural modules similar to those observed in the brain (but much simpler). Further, we showed that after suitable training on sets of objects, the system has single cell activity similar to that observed in monkeys (specifically described for V2 and V4 neurons). Moreover, the resulting codes at the highest level of the hierarchy (TE and IFG) produced localised codes for the object used in training, which were roughly independent of the spatial position of the object for the ventral route and sensitive to the spatial position of the object for the dorsal route. Moreover, the attention feedback learnt as part of the high-level coding allowed for more efficient object recognition in a complex environment, as was shown by tests in which the coarse visual system was used to feed rough co-ordinates to the fine system (which possessed attention). This allowed attention to be directed to the appropriate position to obtain a more detailed set of stimulus activations, determining the presence and possible nature of the object. Thus, the coarse system acted as a saliency map to bring visual attention to bear, through the fine system.

8.3.3 Learning Expectation of Reward

An important component of a knowledge representation is that associated with the expected reward provided or coupled to the stimulus. This is now considered as being learnt by dopamine in the nucleus accumbens and orbito-frontal cortex (OFC). The basis of the learning process is considered to be the TD learning algorithm (Sutton 1988; Sutton and Barto 1998; Schultz et al. 1997; Schultz 1998, 2004), which can be formulated as an error-based learning process to enable prediction of future reward by an initial stimulus. It is the prediction of reward carried by such a stimulus that gives that stimulus intrinsic value. The overall predicted reward assignment to a stimulus is achieved by adding the value activity by which the stimulus is coded in the OFC to its ongoing activity representation in prefrontal CX. A possible architecture to achieve was given in Taylor et al. (2009).

The circuitry of Fig. 8.4 below is an extension of the standard TD architecture, using the spatial position map of rewards from that architecture. The architecture of Fig. 8.4 allows for conditioned place learning as demonstrated in the GNOSYS

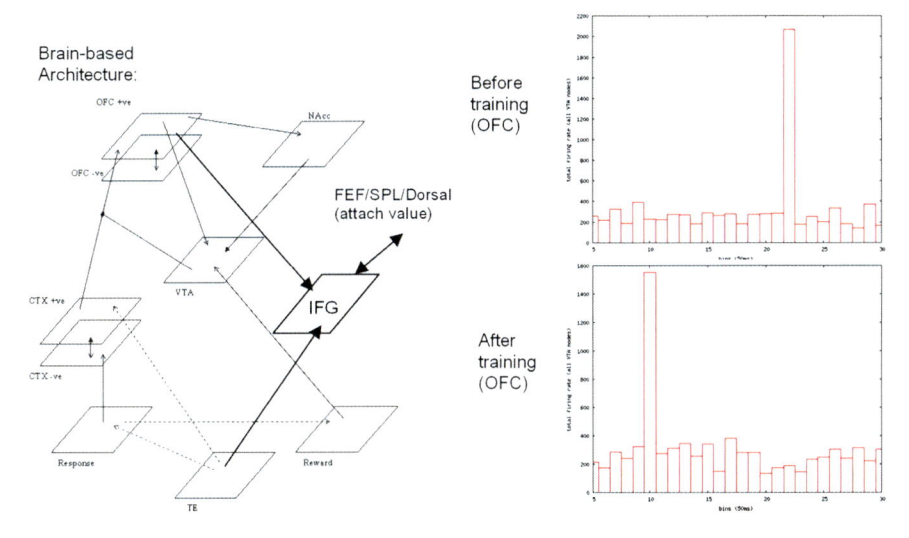

Fig. 8.4 The GNOSYS reward circuit

robot. The reward value for an object can be learnt onto the IFG nodes, via the weights indicated in Fig. 8.4; this value can then be transferred to a global map by locating the local positional information from the FEF dorsal modules in the global environment as indicated in the spatial reward map.

This circuit enables the transfer of the predicted value of an object to the spatial position at which the object is sited. A schematic of the simulation architecture is shown on the left-hand side of the figure. It is composed of a dopamine-based circuit (CTX, VTA, OFC, NAcc) carrying out TD learning of the reward to be attached to a given input stimulus through the CTX modules, having arrived from the object representation module denoted TE. This reward is then transferred by way of the input from OFC to the goal module denoted IFG so as to attach values to goals. On the right-hand side of the figure is shown the learning of the prediction of a reward at bin 22 (each bin is 50 ms long) by a cue at time ten; the top of the two figures shows the initial response pattern of the OFC nodes representing this cue, the bottom figure shows the reward prediction by the response of the cue stimulus representative in OFC when the cue appears. There is then no response to the later reward since it has been correctly predicted (as known to be part of standard TD-learning and noted by observation).

The reward model learns that certain positions within the global environment are more likely to have a specific object present that is also rewarded. The weights from OFC to the global value map are learnt using the value of the dopamine signal and provide for a direct link between the object's reward value and the object's location. The other learnt input weights come from object goal nodes: when the object is present, the object goal node is active and the weights are correspondingly learnt with a simple increment based on the dopamine signal from VTA. Both of these weight groups are normalised. The combination of learning rules means that the weights incorporate such information as frequency of the rewarded object at a specific location as well as how recent was that reward. Normalisation allows for the decay of locations, which may initially contain the rewarded object but, however, after more trials involve an absent object. The global value map can be used in a search mode for the robot, where it is ordered to find a specific object within the environment, so activating the specific object goal node. Since the weights from the goal node to the global value map give a history of the frequency and recency of the object being at certain locations, the map can be used to extract high value locations; an optimal search route can be planned taking in these highly valued positions. If the object is not present at the location, the weight from the goal node to the value map is reduced and weight normalisation follows.

The use of goal nodes as input to the global value map allows for a single global value map for this environment for all objects found within it, rather than multiple copies for each object class. Thus multiple goals can be catered for in this architecture. Nor any other system can transfer reward prediction from the code for a specific object, in a simplified model of OFC, to the spatial map assumed in dorsal part of the prefrontal cortex (such as in FEF).

8.3.4 The GNOSYS Concept System

In this chapter, we take the notion of "concepts" as representations that exist in real biological agents and not those in the statistical/pattern recognition literature, which refers more specifically to clustering. Clustering is one of the operations that is carried out by a real concept system but is not the only one. Some properties, which we ascribe to statistical concepts, do not necessarily hold in all contexts for real concepts. For example, the property of transitive inference does not always hold, such as in the case "all people think chairs are furniture items but most people believe that car seats are not furniture".

The GNOSYS concept system creates concepts out of perceptual representations in an incremental manner (Kasderidis 2007). The system uniformly represents object and goal concepts, calculating the similarity of a new stimulus to a known concept by using spreading activation in a semantic network. Internally, it uses two levels of representation:

- Prototypes (as result of a generalisation process through clustering)
- Exemplars (for boundary cases between concepts of different classes).

It supports the following operations:

- Addition of a new concept
- Incremental change of a concept
- Forgetting rarely used concepts
- Providing default values for missing attribute information
- Activation of concepts with missing perceptual information

- Support for multi-modal percepts
- Activation of multiple concepts due to a given percept, and concept selection by guidance of the reasoning system.

The concept system provides a representation in which objects, goals and relations can be represented. It uses as input the perceptual system (in more detail the descriptors coming out of the fine visual system and the attributes module described earlier) as well as the reasoning system. The latter returns new concepts that have been found/constructed through mental simulation during solution of a goal. Its output is used by the reasoning system primarily in order to use higher-level representation of objects, which are independent of perceptual noise. For example, object affordances are part of object concept representations, thus providing useful information in reasoning processes about possible initial actions to consider taking on an object. A secondary use of concepts in the GNOSYS system is the biassing of perceptual processes for recognition of objects at the perceptual level. In other words, the fine visual system not only takes into account the information coming from the Attributes and coarse visual system modules but also is biassed by the Concept system through the mechanism of attention. The latter is biassed in its turn by the reasoning system when we consider suitable concepts, during the solution of a goal object task, which might be useful in helping advancing our current (partial) solution.

8.4 The Development of Internal Models in the Brain

In order to be able to make predictions as to the effect of future actions, it is necessary for internal control models to be developed in the brain. These may be expected to occur through attention-based learning, with over-learning leading to automatic internal "thinking". The latter process, of becoming automatic through possible sequencing by recurrent internal models (with attention or consciousness), has been recently proposed as basic to creative thinking (Taylor and Hartley 2008); we will consider this later in the section.

We note that internal models in the brain have been heavily researched and developed in association with motor control (Desmurget and Grafton 2000). The existence of forward models (such as predictors of the effects of actions to be taken on the state of the system, or of actions needed to take the system from one state to another) has been detected in the brain in numerous situations and even corollary discharges of motor signals observed in higher order cortex (Sommer and Wurtz 2002; Diamond et al. 2000). This has been extended by the CODAM model to attention control (Taylor et al. 2007b,), where the model has led to the possibility of explaining consciousness as arising from the exploitation of the corollary discharge of the signal to move the focus of attention (Taylor 2009 and references therein).

There are also a number of approaches to understand action models in the brain, such as by means of recurrent networks able to model sequences. However, these models are not as specific as the internal model approach, since they do not specify how actions on state representations lead to new states, only in general how one state will lead to another in a given sequence. But it is the former – of imagining what would happen to X if one did action Y on it – that is needed in reasoning. In other words, the results of actions on given objects, and to what they lead, are crucial to internal reasoning. Stored sequences are clearly of importance to speed up that reasoning if a given state has, many times before, been embedded in a sequence of states or is nearly always the first state in a given sequence of states. But such stored sequences cannot replace the reasoning apparatus proposed here as composed of coupled internal models.

We will start by defining more specifically the internal models to be considered. They will be the simplest possible, but using them, say in parallel, can lead to very powerful control systems such as that of HMOSAIC (Wolpert and Kawato 1998). However, we are considering here not only direct action control in the HMO-SAIC manner but also the use of recurrence, with external action output decoupled from the motor response system, in order to achieve internal thinking and mental simulation (which we discuss in the following section), and later (in the next section), the process of reasoning. Such internal recurrence by coupled internal models is, we have suggested, the basis of mental simulation in higher species (Hartley and Taylor 2009) as well as reasoning (Taylor and Hartley 2008; Mohan and Morasso 2007).

To start with, then, let us define the internal models we will consider. The first is the forward model, defined as

$$X' = F(X, u, w) \tag{8.1}$$

where X is the sensory state of the system, and u is the action taken on the sensory state X to produce the sensory state X'. Equation (8.1) has been extended by inclusion of the parameter set w. It is proposed that through suitably adapting this set of parameters, it is possible to modify the forward model so that it gives a good representation of the action of the overall dynamical system of the higher animal whose body the brain inhabits. This uses the fact that neural networks can represent, to within any desired approximation, any given input–output function.

Such learning as is required can be achieved by use of the simple LMS algorithm

$$\Delta w = (\text{error}).w \tag{8.2}$$

where the error is defined as

$$|X(\text{actual}) - X'| \tag{8.3}$$

with $X(\text{actual})$ defined as the actual value of the sensory state after the action u has been taken in (8.1) and X' is as defined in (8.1).

This process of training looks biologically feasible. Both the actual final sensory state $X(\text{actual})$ and the predicted state X' can be sent to the same area of parietal lobe from early visual cortex (for $X(\text{actual})$) and from the site of the forward models (for X'). This latter site has been suggested a being in the parietal-frontal network involving premotor cortex (Miall 2003). Thus the two values of X(actual) and X' can be combined (with one being inhibitory) so as to lead to the error as defined in (8.3). Finally, the learning arising in (8.2) can be obtained as a Hebbian learning rule, with the neuron output being proportional to the error (8.3) and the input being proportional to the input weight w (as, say, arising from the output of the neuron for a linear response nerve cell). Overall, this leads to the learning rule of (8.2).

The inverse model IM is rather different. This is defined as:

$$u = \text{IM}(X(\text{initial}), X(\text{final}), v) \qquad (8.4)$$

where v are a further set of adaptive parameters so as to allow the output action u in (8.4) to take the sensory state of the system from X(initial) to X(final).

The inverse model would appear to have a similar structure to the forward model, although its training is more complicated. This is because the natural training set for the IM (8.4) only will have input states consisting of pairs of $\{X(\text{initial}), X(\text{final})\}$ and output state the corresponding action u needed to make the transition from $(X(\text{initial}) \rightarrow X(\text{final})$. In using this training set, it would be necessary to encode the necessary actions as well as the pair of initial and final sensory state. It is the former of these that will in general be difficult to encode, at least at the high level at which the IM is supposed to be functioning.

An alternative training scenario is by the development of a coupled FM/IM system. In that case, the error of (8.3) can be used directly to train both the parameters w and v, provided that the action u in the FM (8.1) is the same as that used in the IM (8.4).

In GNOSYS, we took from the EU MATHESIS project the architecture developed to model (successfully) some of the brain processes occurring in observational learning. This was based on brain activities observed in monkeys during their observations of the actions of an experimenter (Raos et al. 2004). The observation process employed for the macaque was for it to watch an actor or experimenter execute a learnt reach and grasp movement on a familiar object. The subject then had to match the observed grasp to one of the set of learnt grasps and associate that grasp with the object. The subject's task was then to reproduce that grasp when the object was later presented. It was assumed that grasps are learnt by learning the affordances of stimuli, in other words those features of the to-be-grasped object that allowed for its being accurately grasped. This could occur either by trial and error or by observation. We can construct a neural architecture to model these processes, as shown in Fig. 8.5.

In the figure, blue = observe only; red = execute only; purple = shared circuit. Activity passes from visual input via processing to activate an object module. In the execution case, this object then activates a grip, and the result is passed to motor planning where it can be integrated into a full movement. The inverse model controller (IMC) and forward model (FM) internal model control loop ensures the movement is executed correctly. In the observation case, additional circuitry is used to analyse the grip used and help to match against the internal list of grasps.

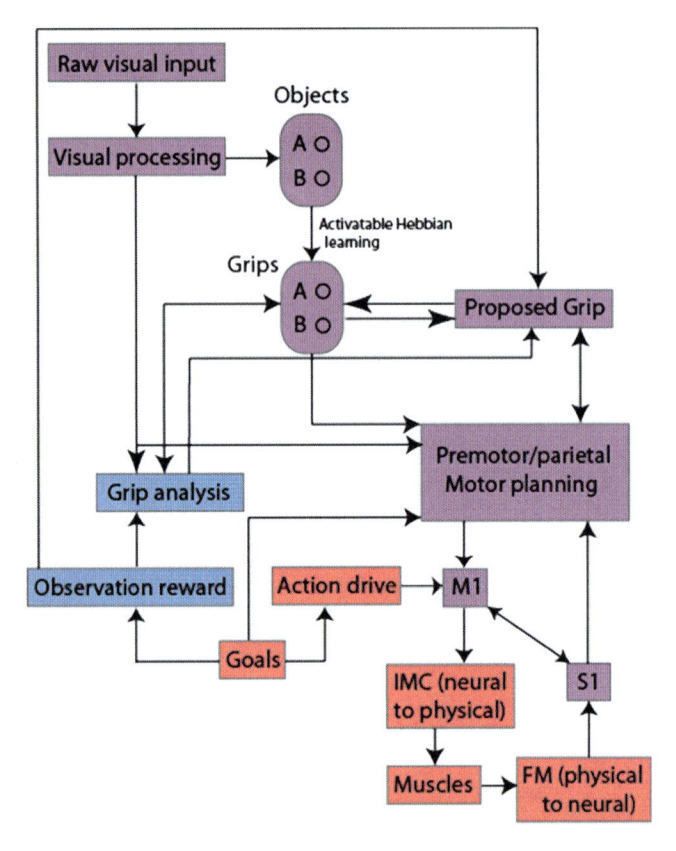

Fig. 8.5 The circuits involved in observation and execution of grasping an object

The architecture of Fig. 8.5 uses two internal motor models: an IMC, which generates a desired action to achieve movements from one state to a desired goal state, and a forward or predictor model (FM), which changes the estimated state of a system given its present state and the given action about to be taken.

We note that in terms of the architecture of Fig. 8.5,

1. Both reach and grasp movements are being modelled, and that the circuit for them has considerable overlap.
2. We have included an internal model FM/IMC pair for high-level motor planning.
3. We may determine the grasp parameters by a number of methods, such as by PCA or by time-dependent exponentials (Santelli et al. 2002). However, we used a very simple dedicated coding of requisite grasps, which only required differentiation between the few grasps assumed to be used by the macaques by the activations of the different dedicated neurons in the various modules of Fig. 8.5; this obviated the need to computing detailed parameters of the grasp trajectories.
4. We have included an action drive module in the architecture. This is expected to be present, turned on by various primary drives such as hunger, thirst, etc in a

macaque and its equivalent drive for energy and safety in a robot. It would also help the robot to distinguish the actions of self from those of others, so as to avoid acting in an externally driven manner (as occurs in humans who have lost certain parts of prefrontal cortex and repeat all actions they observe).

We then simulated simple grasps on objects (using dedicated neurons in M1 and S1 to represent particular actions and proprioceptive feedback). The results agreed well with those of Raos et al. (2004), as seen in Hartley et al. (2008).

Given suitably trained coupled FM/IM pairs, it is now possible to consider higher cognitive processes such as thinking and reasoning. We turn to each of these in the following two sections.

8.5 Thinking as Mental Simulation

Mental simulation is basic to thinking. It is presently of great interest in numerous areas: philosophy, psychology, business, defence/attack military planning, and so on. It involves the processing in the brain of internal images of the outside world and determining how they might be changed if certain actions were made on the contents of these images. In the case of observational learning, it allows for the acquiring of new skills (new strategies, new affordances and new ways of acting on objects) in a manner that leapfrogs trial and error learning. It is known to occur in chimpanzees (careful and well-known studies have been made of the learning by younger chimps of nut cracking by means of "hammer and anvil" set of stones by practiced elders). In infants, similar studies have been done and neural models built of how this could occur in developing infants (Hartley et al. 2008). There it is suggested that this uses the mental simulation loop to learn new strategies by observation, and ultimately new actions and affordances that would otherwise take a long time to learn by trial and error.

We show in Fig. 8.6 a simple neural architecture into which the mental simulation loop is embedded to allow actions on representations of object stimuli to be taken. The other modules have been introduced in other parts of this paper: the vision system including an object recognition module (for feature analysis of input stimulus activation), an object codes module (coding for separate objects by a form of self-clustering in the object feature space) and affordance extraction (by trial and error or observational learning); the executive control system as highest order control module (to guide the attention system to function in visual form for object and affordance coding or in motor form to achieve the processing of motor control), also activation of the goals module (with associated sub-goals as they are discovered by exploration, either actually or mentally through the mental simulation loop); the motor control system (involving action planning processes and action codes at a lower level; in GNOSYS, this was implemented as part of the robot control system outside the GNOSYS brain); and finally the sensorimotor system, not shown in the figure (involving proprioceptive feedback codes), coupled with the hand position vector in the sensorimotor module to bias the motor planning module.

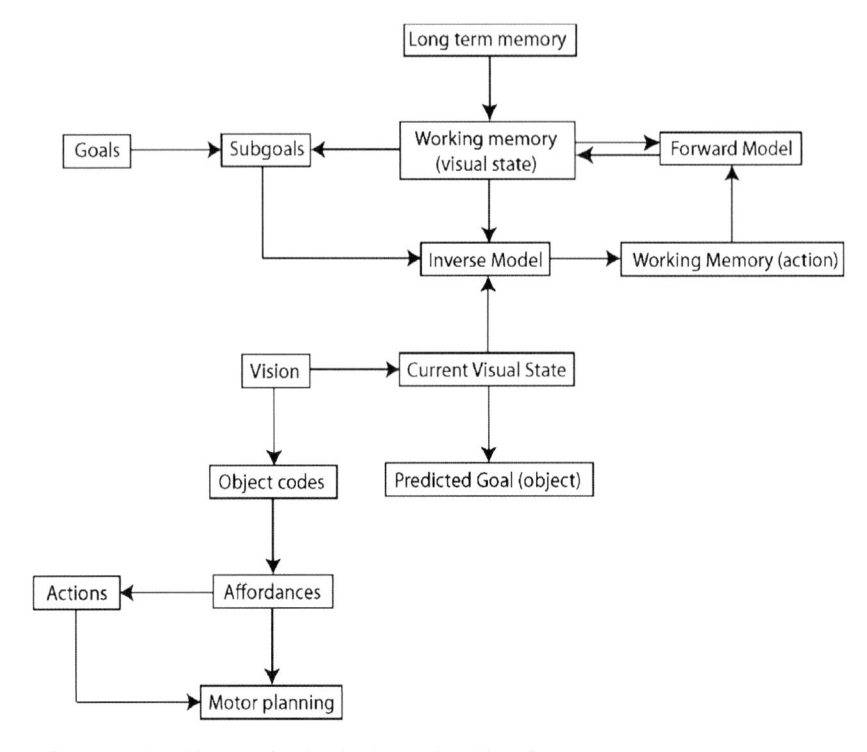

Fig. 8.6 Neural architecture for simple observational learning

The mental simulation loop itself consists of short-term memories (denoted working memory in Fig. 8.6 for visual states and motor acts) coupled to a forward model and an IMC. The inverse model produces the required action to lead from a given initial to a final state; the forward model is a predictor of the next state from a given state and an action on it, as discussed in the previous section.

Two forms of activity of the mental simulation loop are possible. The first, to be further discussed in this section, is that under attention control. As such, the two working memory buffers are used to having consciousness of the relevant states and actions. The second is subliminal, with no attention being applied so that the working memory buffers are inactive. This second form of mental simulation has been conjectured to be that of the creative thinking process, which is often interspersed with the conscious process of thinking in every-day life. We address this simulation process in the next section.

These various modules have been simulated in a very simple manner for infant development of observational learning (Hartley et al. 2008) and its successful duplication of experimental infant development data; we refer the interested reader to that publication.

8.6 Creativity as Unattended Mental Simulation

Let us first consider how we might, in principle, implement the several creative processes in the model of creativity of (Wallas 1926). This has been explored by numerous further researchers, such as (Wertheimer 1945), and in a brain-based neural model in (Vandervert 2003). The first of the processes in this model involves hard work in preparing the ground about the problem at hand. This hard work is expected to be under attention control with the subject conscious of their gathering apposite knowledge. The second process involves the incubation period. During this, attention is directed away from the original task, possibly to another task, or the subject just relaxes in total with no specific focus of attention. In the third process, attention is suddenly switched back on in the "eureka" moment. Finally, the subject has to get back to the hard work of verifying that the illuminating thought could solve the problem after all.

Thus we have the three stages as far as attention is concerned:

1. Attention is applied and the database of the subject's knowledge of particular relevance to the problem is expanded maximally.
2. Attention is relaxed (directed elsewhere) and the creative process occurs.
3. Attention is then switched back on by the illuminating thought appearing valuable, and the hard work in the verification has to be done (if it is needed).

Our model will assume that stage 1, the hard work of developing the relevant database, has been completed. It will allow the thinking process to generate a sequence of thoughts that is finally blocked, so that the solution to the given problem is not reached. Attention must then be switched. How that occurs is not relevant here, except that a switch is triggered by the failure of the model to reach a solution. The process of creativity then takes over to generate further (unconscious) thoughts, one of which finally leads to a mental state recognised as having value (say by being able to roughly extrapolate to a solution of the problem). Attention is then switched on by the reward value thereby given to this illuminating thought, so that the verification process can be started.

We consider the architecture of Fig. 8.7 as supporting the process of thinking at the two levels we have just described and in the introduction: at conscious and at unconscious levels. In order to switch between these levels, it is necessary to consider in more detail than hitherto the visual attention components in the architecture of Fig. 8.7, especially the visual attention IMC and the further attention connections included in the figure. It is through these, in concert with the other modules already present and some additional ones to be mentioned, that it will be possible to see how two levels of processing, conscious and unconscious, will be possible with the architecture.

We described briefly earlier the manner in which visual activity can be used as part of the motor control system. In addition, and as seen from the architecture of Fig. 8.7, it is also possible to see how the position of the working memory (visual state) module as sandwiched between the forward and inverse models allows there to be consciousness of the set of visual states in a mental simulation loop.

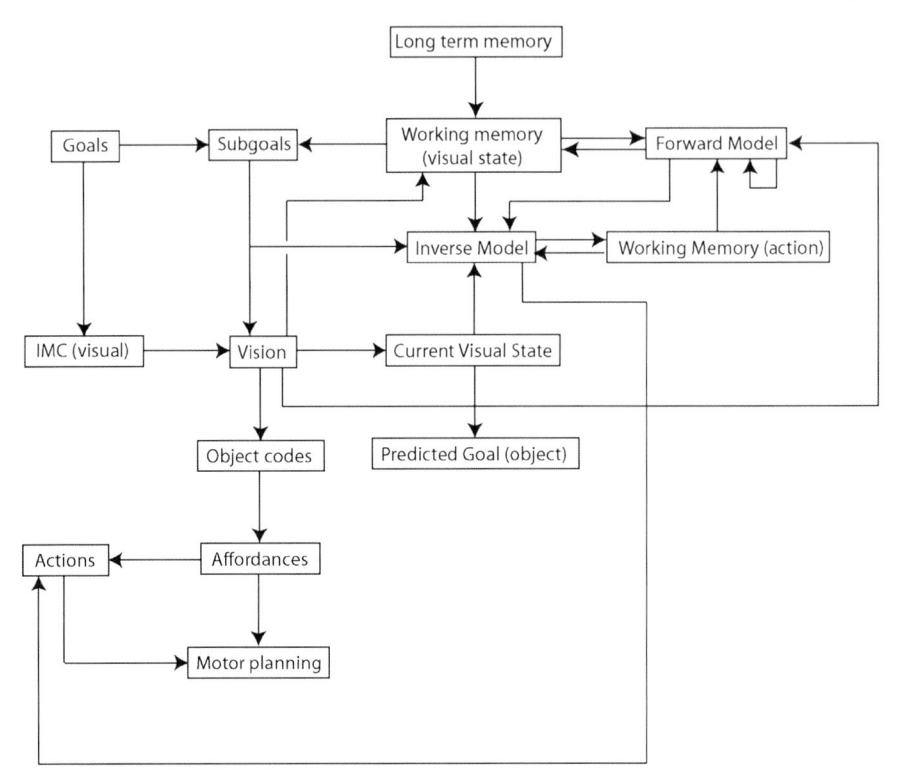

Fig. 8.7 Extension of the architecture of Fig. 8.6 to include attention in an explicit manner. The most important component added to the architecture of Fig. 8.6 is the visual attention signal generator, denoted IMC (visual). This causes attention amplification of specific attended target material in lower visual areas (in the module denoted 'Vision' in Fig. 8.7). This causes attended activity to be amplified sufficiently to attain the Working memory (visual state) in Fig. 8.7, and hence reach awareness. This attended, conscious route of using the mental simulation loop can be sidetracked, as shown in the architecture of Fig. 8.8

We need now to consider how the visual states in a mental simulation loop can be taken out of consciousness but yet be part of the mental simulation loop. This can be achieved by the insertion of a switching device to allow output from the forward model to avoid the working memory (visual state) module, so totally avoid conscious report. This switching device will be based on an error module, as in the CODAM model of attention (Taylor et al. 2007).

The mechanism to achieve mental simulation at a non-conscious level is by means of the connection lines in Fig. 8.7 described in the previous section, which avoid the working memory (visual state) module:

(a) The direct connection from the forward model to the inverse model. This enables the inverse model to produce the next action to achieve the sub-goal.

(b) The direct connection from the visual state module to the forward model. This will allow generation of the next state brought about by the new output of the inverse model and the visual state.
(c) Recurrent connection of the FM to itself if there is a sequence of virtual states to be traversed.

However, more is needed to be considered in the overall creative process. Let us turn to the example of giving unusual uses for an object: we take a cardboard box as an example. We can say "As a hat" as one such unusual use. That could arise from the flow of information in our brains:

Cardboard box (in picture or as words) → input processing → box nodes in object map → hat nodes in object map (by learnt lateral connections) → hat nodes in affordance map (by direct connections from the hat node in the object map and by lateral connections from the box representation of affordances to the hat representation there) → test of viability of putting on the box as a hat.

If the test of viability works, then the "putting on hat" action becomes attended to and there is a report, either by putting on the box as a hat or saying "As a hat". If the box is too large to fit stably on our head, then we put it on our head and keep our hands on it to steady it; if the box is too small, then we may desist from saying it could be used as a hat, or try it on as a little "pillar box" hat.

These various responses indicate that we try out subliminally what happens if we try to put the box on our head, using the simulation loop. If successful and the action is viable, then we attend to it and hence report it. If it is not, we move on to another subliminally analysed use.

To achieve the subliminal processing stage as well as the final report, there must be an attention switch, generated as part of the IMC (visual), so that when there is an attention control signal output, there is normal transmission from the forward and inverse models to their relevant working memory modules shown in Fig. 8.7. When there is no attention, then the mental simulation loop circuit functions without the relevant working memory modules. It thus functions in a subliminal or unconscious manner. There will need to be an extra module for assessing the relevance of states achieved during this unconscious activation of the loop; that will be fed by the forward model in parallel with the self-recurrence (or external running of the FM) and the signal to the inverse model. Given an error-based output from this assessment module, then its output would be used to bring attention to the final state and the sequence of intermediate states (assumedly not many) so as to attain the sub-goal more explicitly.

The reason for the presence of the switch itself is that of allowing the reasoning process to go "underground" when an apparently insuperable obstacle is met by the conscious reasoning system. This may be seen as part of the extended reasoning system discussed, for example, in Clarke (2004). However, such a switching process plays a crucial role in the truly creative cognitive process. When a blockage is met in "simpler" logical reasoning, then the attention control of processing has to lose its iron grip on what is allowed to follow what in the processing, with increased reasoning and recall efficiency by subliminal-level processing. This feature is well known, for example, in answering quiz questions and solving puzzles of a variety of sorts. So the switch into the subliminal mode may be achieved in the case of quizzes or creative processes such as painting or other artistic acts from the start of the search or creation process. In more general reasoning, the creative and subliminal component need only be used at points where logic gives out and more general "extended" and creative reasoning has to step in.

In the case of our example of unusual uses of the cardboard box, the attention switch is assumed to be turned off by the goal "unusual uses", since we know that going logically (and consciously) through a list of all possible uses of anything will not get us there, nor will any other logically based search approach. We have learnt that we need to speak "off the top of our head", in an unattended manner. So we can regard, in a simulation of this task, that we are not using attention at all after the switch has turned it elsewhere or reduced it to a very broad focus.

From this point of view, there may well be access by the internal models during this creative phase to a considerable range of neural modules for memory of both episodic and semantic form right across the cortex. The best approach to model this would thus be to have these connections develop as part of earlier learning processes, but such that they can function initially in an attentive phase and then be useable in a subliminal one. But the presence of the unattended learning of the required lateral connections may also be possible and need to be considered.

8.6.1 Simulation Results for Unusual Uses of a Cardboard Box

To simulate the paradigm involving imagining unusual uses of a cardboard box, we emphasise certain aspects of the model described above. In particular, we need to allow the use of lateral spreading with object and affordance codes and look at the more specific effects of attention. We can see the architecture of the model to be used here in Fig. 8.8 (as an extension of parts of Fig. 8.7 to handle the switch between attended and unattended processing).

Here we detail the function of the specific modules used in Fig. 8.8.

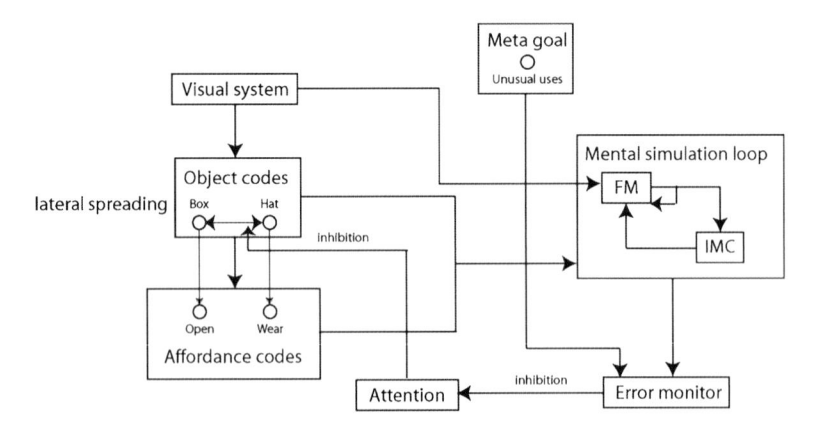

Fig. 8.8 Simplified brain-based architecture for creativity: solving the "Unusual Uses Task". The functions of the modules in Fig. 8.8 are specified in the text

8.6.1.1 Meta Goal

The overall goal of the simulation is to find unusual uses of the cardboard box. As part of this process, simple action/object goal pairs are created, so we need to code both the overall goal of imagining unusual uses and the immediate goals that are tested to see if they are unusual. We have not specified the immediate goals, since we are unclear if these are used in the creative, unattended lateralisation processing. If the processing is automatic from the affordance/action codes module to the action being taken, and then used by the mental simulation loop, then no such immediate goals module is needed; that is what has been used in our architecture and simulation. If needed it can be included between the affordance codes and the mental simulation loop without any expected change in the results we report below. The meta goal is coded as a single dedicated node (with the possibility of adding more nodes for expansion, either as a distributed representation or to include other meta goals).

8.6.1.2 Object Codes

Here objects are represented by single nodes. Lateral connections between the nodes allow similar objects to be activated by this spreading, in addition to visual stimulation. It is these lateral connections that allow the use of analogy. It is assumed that these lateral connections had already been learnt during the earlier "hard-working" attended phase of the (Wallas 1926) model mentioned earlier, with attention control in the lateral spreading as shown in Fig. 8.8 being learnt simultaneously.

8.6.1.3 Affordance Codes

The affordance module contains nodes representing specific affordant actions that can be used on objects (such as the action of opening a box). These are primed by the object code module using pre-selected connections.

8.6.1.4 Mental Simulation Loop

In the full model shown in Fig. 8.7, the mental simulation loop incorporates a forward model (FM), inverse model (IM) and buffer working memories. In unattended mental simulation, we suggest that these working memories are not active, such that activity passes straight between the FM and IM. The forward model generates an expected result of carrying out the action (these are pre-coded in this simplified model) while the inverse model determines the action necessary to achieve a suggested state. The function of the mental simulation loop in this simulation is to test subliminally the pairs of objects and affordances/actions generated by the lateral spreading to see if they are considered "unusual". If the use is considered unusual, then the attention is brought back to the system. We have not included these working memory buffers in Fig. 8.8, for simplicity.

8.6.1.5 Error Monitor

The error monitor is needed to determine whether a given object/action pair tested by the mental simulation loop has fulfiled the goal criterion of being unusual. If this criterion is met, it then activates the attention control module such that attention is restored to the goal of finding an unusual use for the box. In this simulation, the error monitor compares the selected action result (passed on from the mental simulation loop) against an internally maintained list of those considered novel.

8.6.1.6 Attention

Here we use a more specific property of the attention control system than that used so far. In particular, we now require the attention system to control lateral spreading in both object and affordance modules by inhibition of lateral connections. In our model, this occurs by output from the attention module stimulating the inhibitory connections present in the object code module. When attention is focussed, representations will be activated singly in each region, while after the removal of attention, activity can spread to similar representations (we assume that the organisation of the module is such that similar objects are laterally connected). How this attentional attenuation of lateral connection takes place at the neurobiological level is indicated to some extent by studies of visual attention (Fang et al. 2008; Friedman-Hill et al. 2003). We have not included the working memory buffers, present in Fig. 8.7, in Fig. 8.8, so as to keep the architecture as simple as possible, although they should be there; they play no direct role in our simple simulation.

We can see the flow of activations of the simulation areas in the following chart of Fig. 8.9. Activation can be split into two phases, where the first activates the goal of finding an unusual use and tries the action of opening the box, which is found not to be unusual. The second, after attention is relaxed, spreads activity such that the extra object (the hat) and its affordances become involved.

The flow paths in the upper diagram carry attention-controlled processing. That in the lower diagram have no attention focussed on them, so allowing more lateral spreading between concepts, as shown in the first line of that flow.

Recent brain imaging results (Kounios et al. 2008; Christoff et al. 2009; Bhattacharya et al. 2009) have also been directed to probing what areas of the brain are involved in creative or insightful solutions to reasoning problems and the timing of the activity involved. Thus (Kounios et al. 2008) have observed differences in levels of EEG frequencies (in the alpha, beta and gamma ranges) across cortical areas when insightful solutions were being obtained in solutions of anagrams; in particular, there was an increase in the level of gamma oscillations (in the range of 30–80 Hz) in the right hemisphere as compared to the left. In Christoff et al. (2009), considerable activity was observed by fMRI in the brain during creative phases. Similarly, in Bhattacharya et al. (2009), strong gamma activity in the right prefrontal cortex was reported as being observed up to 8 s before the creative solution to a problem.

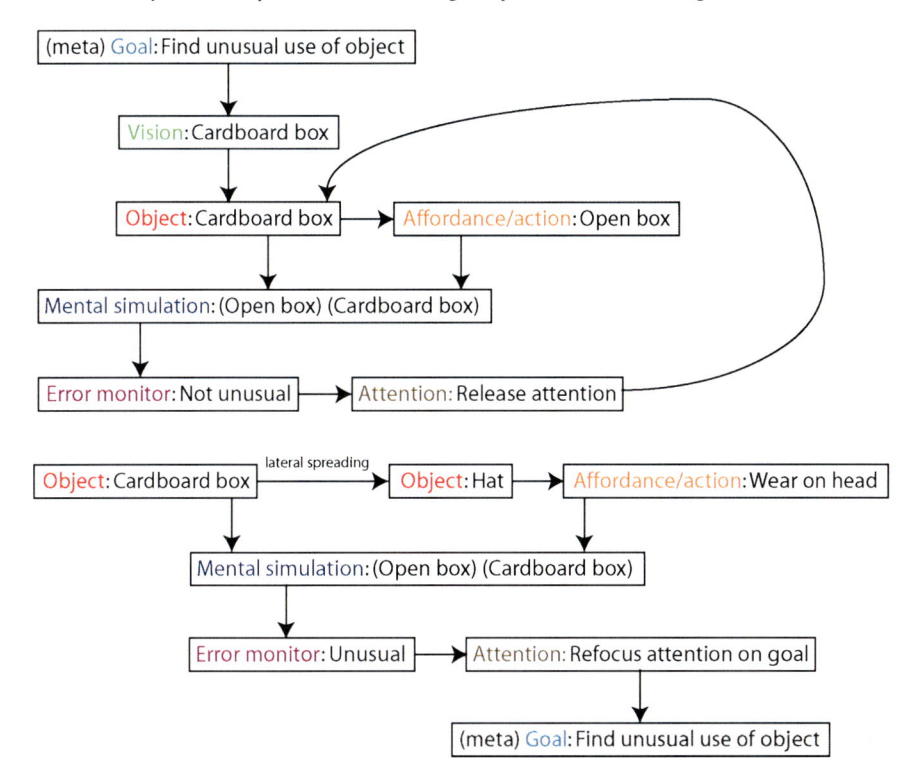

Fig. 8.9 The information flow during the creative act

These results support the overall architecture of Fig. 8.8 and the dynamical flow of the processes as shown in Fig. 8.9. Thus the right hemisphere gamma activity of (Kounios et al. 2008), observed as occurring about one third of a second before the subject's conscious experience of insight, is expected to occur in the information flow of Fig. 8.9 during the process of lateral spreading from the object representation for the cardboard box to one for hat. This use would have attention drawn to it when the error monitor showed the use as a hat was unusual. Consciousness of such a use would then arise accordingly, and the attention circuitry correspondingly re-activated. The observation of (Bhattacharya et al. 2009) of prefrontal activity up to 8 s prior to reaching a creative solution could then correspond to various alternative solutions being tried but failing, as shown by the error monitor activity. There may have been associated prefrontal goal activity to hold tentative solutions as goals for trying their usage out. Finally, the (Christoff et al. 2009) fMRI activity observed in a number of different brain area is more difficult to pin down due to the lack of accurate temporality of the observed activity but is also to be expected from the overall architecture of Fig. 8.7 when extended by Fig. 8.8 to the attentive and creative circuitry.

8.7 Reasoning as Rewarded Mental Simulation

8.7.1 Non-linguistic Reasoning

We present in Fig. 8.10 a simple modular architecture for reasoning in the two-stick paradigm, used on chimpanzees. There are two sorts of sticks: S1 (short) and S2 (long), which are present on a given trial.

A chimpanzee wants to reach the food, but cannot do so by stick S1 alone; it can only be reached by using S2. However, S2 can only be reached by use of S1 (since the chimpanzee is in a cage).

The relevant neural architecture is proposed to be as in Fig. 8.10. The modules are as follows:

(a) *Drives*: Basic drives that cause the system to attempt actions; in the present case: hunger (satisfied by pressing button on distant wall, so food delivered).
(b) *Goal list*: Goals are available to the system (independent of available actions within simulated world). Goals for us are represented by stimuli. For this paradigm, have three goals: button/S1/S2.
(c) *Vision*: This module provides the simulation with information about current state of the world. An IMC can then calculate movements to achieve selected goals, or it returns "NOGO" if not achievable.
(d) *Motor IMC*: The IMC (inverse model controller) allows simulation to determine if goals are achievable or not, given current state of world.

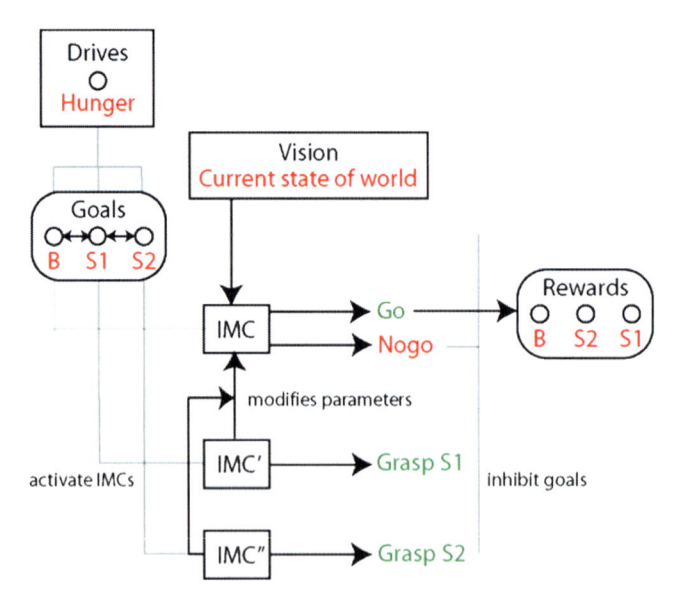

Fig. 8.10 The architecture for non-linguistic reasoning. See the text for details

(e) *Rewards*: System's reward values are modified by the results of mental simulation, and this modification allows correct actions to be executed to solve the paradigm.
(f) *Forward models*: We have not included these specifically in the architecture of Fig. 8.10. Each IMC will have an FM associated with it, which is easier to train than the IMC. This is because the FM has an immediately generated error as comparison of predicted next state (after an action) and the actual state as determined by sensors. This error can be used to train both the FM and IMC.

We present also in Fig. 8.10 a list of available actions (which may or may not be physically possible – for example, grasping the food reward initially can be attempted, but is not attainable) as goals within the goals module (so in this case, goal "B" represents the action "Push Button"). The model is composed of a drive module (a continually active node) representing continued hunger causing persistent activity emitted from it to activate the motor drive, and hence other module (until the hunger node is satisfied and turned off by obtaining the reward by pressing the button). There is a goals module coding for the three stimuli of button, stick S1 and stick S2 (to press the button, pick up the stick S1 and then the stick S2, respectively). There are three IMC modules, the first (denoted IMC) being for pushing the button by the gripper, the second is for grasping stick S1 (and denoted by IMC$'$, with consequent alteration of the length of the gripper in IMC to correspond to carrying stick S1), and the third (denoted IMC$''$) is for performing a similar action with stick S2 (and consequent change of parameters using in the gripper IMC). There is also a reward module in which there is modifiable steady activity corresponding to the current reward value of either the button B or the sticks S1 and S2 (all the observable objects in the environment).

At each stage of the simulation, the system is presented with a range of possible actions and must choose an action to perform. The IMC and FM (the latter not shown in Fig. 8.10) allow the system both to mentally simulate these actions (with no external actions) or alternatively to instantiate the actions (the former by inhibition of any output from the IMC, the latter when this input is allowed to activate the effectors by switching off the inhibition).

In Fig. 8.11 we show the flow of goal and reward activation within the system. After an initial attempt to press the button, a NOGO result is obtained. After this, but with the hunger drive still activating motor activity, S1 is attempted to be gripped, which, having been achieved, still fails to allow the button to be reached. After S2 is attempted, however, the button is now reachable. This causes S2 to be rewarded and moves the goal backward to achieving S2. This requires S1 to be gripped, so once that is obtained, S1 is rewarded and the simulation has correctly rewarded all of the potential goals and hold their relative activities with $r(S1) > r(S2) > r(button)$.

There are two key features critical to the operation of our model – the ability to rapidly transfer rewards between goals and the capacity to consider both spatial and spatially invariant goal representations. In any system in which in-

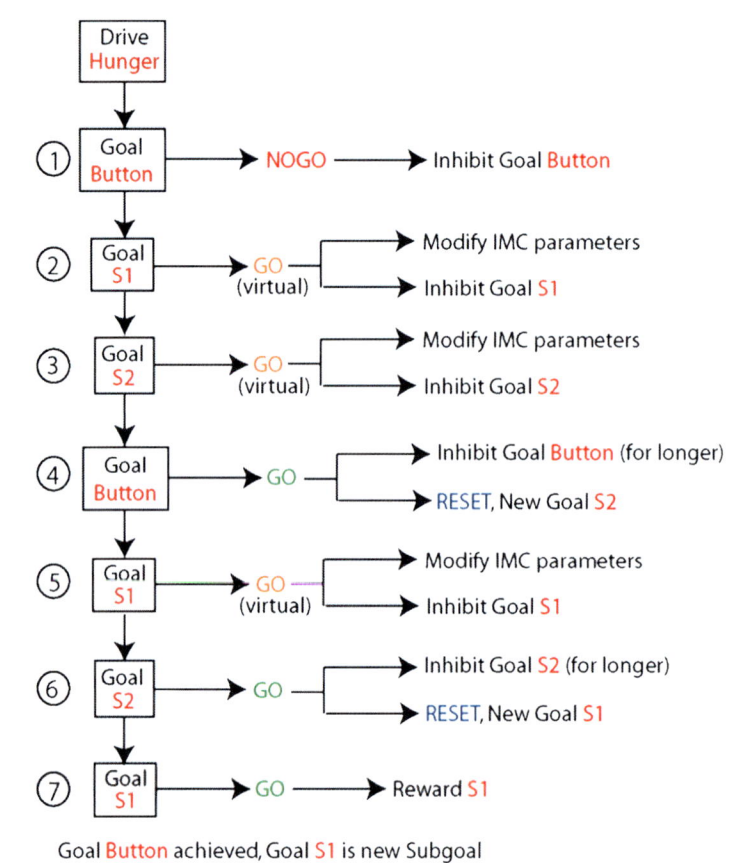

Fig. 8.11 Flow of information during reasoning in the two-stick paradigm. See text for details

termediate goals must be used to reach an ultimate goal, some mechanism must exist to make the reasoner want to perform the intermediate goals. We suggest that rapid movement of reward can do this, by attaching higher rewards to subgoals, which then gives reason to perform them. Without this movement of reward, some other mechanism must fulfil the same function – otherwise, the original maximally rewarded task (such as pushing the button, in our example paradigm) would be continually attempted, neglecting all sub-tasks. It is possible that the brain can do this in some other way. However, rewarding these intermediate actions provides a relatively simple mechanism. It may be possible to examine this hypothesis experimentally by observing neurotransmitters and neuromodulators known to be related to reward (such as dopamine) within the brain during reasoning (as in OFC or amygdala) or by seeing how agonists/antagonists to these neuro-chemicals affect reasoning.

8.7.2 Setting Up the Linguistic Machinery

Non-linguistic reasoning employs most of the important components of the brain. However, there is one missing, that of language. This is a very complex faculty, involving many of these same brain components but now with different coding. We can envisage parts of how this works through our ability to create neural networks of self-organising maps. Sets of these, learnt in conjunction with the associated visual stimulus representations, and with appropriate lateral connections learnt through the similarity of feature maps of similar stimulus inputs, leads to somewhat similar representations maps as expected to arise in the word semantic maps in the brain.

Maps for action representations require the object being acted on or its associated affordances to be defined as part of the representations. This requires suitable connections from the site for action coding to that for affordance coding. Such action codes are very likely in prefrontal cortex (especially in the premotor area) so as to allow for recurrent activity, leading to a succession of movements in effectors. Finally, verbs would then be similarly encoded, connected specifically to the action they represent.

Language coding requires mechanisms to extend beyond the initial one and two-word stumbling statements of the infant ("kick ball", "more cereal") to the much more complex phrase-inserted and grammatically correct sentences such as those in the papers in this book. Such grammar has been supposed by some to be impossible for a child to learn, leading to the idea of the "language instinct" encoded in the brain.

However, more recently, there has been an upsurge of a naturalistic approach to language that supposes that the language faculty is learnt gradually by children, given a general recurrent structure of the frontal lobes, along with crucial sub-cortical sites such as cerebellum and basal ganglia. Various models have been developed of how insertions may be constructed so as to change singular nouns into plural ones or change the tense of a verb, or even insert a given phrase grammatically correctly into a developing sentence. These processes specifically involve structures such as the cerebellum, hippocampus and basal ganglia. These latter systems are known to be involved in such activity, and their loss causes deficits in learning these various parts of grammar.

8.7.3 Linguistic Reasoning

There are several types of linguistic reasoning, such as logistic reasoning, spatial reasoning, induction (deducing laws from a finite set of events) and various others. Logistic reasoning is exemplified by syllogistic reasoning:

$$\text{"}p \text{ implies } q. p \text{ is true. Therefore } q \text{ is true."} \tag{8.5}$$

This can be solved in at least two different ways, by rote learning or by spatial maps.

Rote learning involves being able to manipulate the symbols p, q, "implies" and so on so as to be able to rewrite (8.5) as

$$\text{"All } p\text{'s are } q\text{'s. } Z \text{ is a } p, \text{ therefore } Z \text{ is a } q\text{."} \tag{8.6}$$

This can then be used by the identification of p, q and Z with specific cases. For the old stand-by: p = men, q = "mortal" and Z = Socrates, then there results the expected syllogistic argument "All men are mortal. Socrates is a man. Therefore Socrates is mortal". We need to consider how that, and similar syllogisms, could be implemented in a neural architecture.

The spatial approach would be as follows: rapidly to introduce a spatial activation neural disc (on a suitable spatial map) with the disc identified as the set of all men, and another such disc as the set of all mortals. The basic premise of the syllogism is that the space of all men is contained in that of all mortals. This can be coded by enclosing the first set inside the second (by suitable flexible movement of the discs). Then any particular man, whatever his name, would be a local region set inside the set of men, and hence inside the larger set of mortal animals. Thus spatial reasoning approach to logic, where possible, can be implemented in a spatially distributed set of neural modules.

The linguistic approach is not so obvious. Its main steps involve the neural architectures to:

(a) Identify Socrates as a man
(b) Transfer the "man" property to the property of "mortality"
(c) Identify Socrates thereby as possessing the property of mortality.

The first of these stages (a) is a problem of categorisation of a visual object and its categorical name (in Socrates case a "man"). We assume that there is a well-developed visual system and its associated semantic map to achieve that. Similarly, the third stage (c) should be achievable by given connectivity, with the ability to set up a lateral connection from Socrates to the adjective "mortal". It is the second stage (b) that is less obvious. It could be achieved by laying down the syllogism in long-term memory, having a permanent connection from nodes coding for "man" in the appropriate semantic map to the region coding for "mortal" in the region of the semantic map for adjectives. But such a possibility does not possess the flexibility that we observe available to us in arguing on the syllogism.

An alternative is to use prefrontal to posterior connectivity rapidly to lay down activations for "man" and "mortal" in that brain region, well connected to the relevant nodes in their appropriate posterior semantic maps. The syllogism would then arise by again rapidly setting up connections from the prefrontal nodes for "man" to those for "mortal". The realisation that "Socrates is a man" would then cause some activation of the "mortal" prefrontal region. This prefrontal activation would then be carried back to suitable semantic maps and hence ultimately lead to the activation of the "mortal" adjective. It is more likely that the second prefrontal route is used, as follows from numerous brain imaging results.

8.8 Overall Results of the System

So far, we have described the separate components of the GNOSYS/MATHESIS system and their response results for two reasons:

(a) To support their intrinsic value to the system (such as allowing the creation of an attention feedback-controlled visual system able to handle visual detection in noisy environments).
(b) Allowing relation to results from brain observations, thus showing how our system can reproduce these, and thus justify the use of the epithet "brain guidance".

However, all of these separate components have to be fused together into the overall GNOSYS software system or the "brain". In this section, we present the results obtained when the whole GNOSYS system was used to solve the various tasks it had been set.

The GNOSYS system has been tested using a number of scenarios in order to establish its capabilities in each cognitive faculty separately and in unison. From all these experiments, we describe next a set of tests that we find as most interesting in showing the reasoning capabilities and especially the tool construction strategy that it used.

8.8.1 Experiments

In simple environments furnished with the appropriate objects, we have tested the system with the "Trapping Groove" and "N-sticks" reasoning paradigms among other things. To use both paradigms together we asked the agent to retrieve a ball from a table that has a trapping groove in its middle (as shown in Fig. 8.12). The size of the table is such that the robot cannot accomplish this task directly when the ball is placed at its centre: it has to search for a tool of appropriate size. This can take place either by searching the spatial memory or by simply exploring the environment. At the same time the agent reasons as to whether it is possible to construct a tool of appropriate size by using components that have been found during mental or physical search. What solution will be selected depends on which solution's pre-requisites are fulfiled first. In Fig. 8.12 (bottom) we see a situation where the robot uses two small red sticks to construct a longer one and then uses the latter to retrieve the ball. If a longer blue stick blue is found first it is used instead (top).

Other complex commands were also given. An example is the request for the construction of a stack of cylinders at a specific point on the floor when the initial conditions of the environment were similar to those shown used in Fig. 8.13. This task required the ability to develop a high-level plan of forming a stack by the development of sub-plans such as finding and transporting appropriate cylinders to the neighbourhood of the target point before attempting the stacking action. To achieve this, the system had to either retrieve from (spatial) memory the location of

Examples

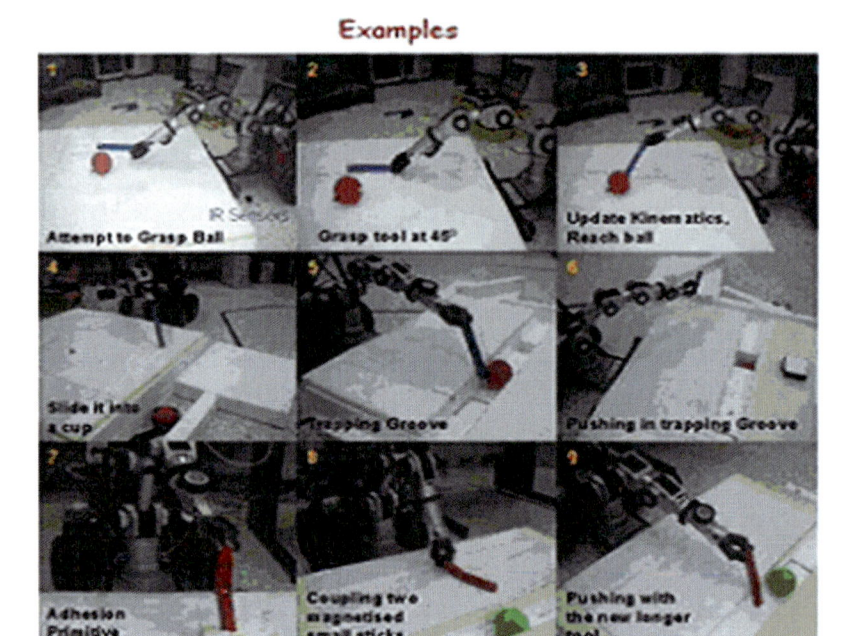

Fig. 8.12 Solution of the trapping groove paradigm and retrieval of a ball. The agent has first constructed a stick and then used it for reaching successfully the ball. The tool construction process is not hard-wired but rather is the result of reflections during "mental imagination". What is known are the forward/inverse model pairs for various objects. These have been acquired during training and familiarisation with the objects

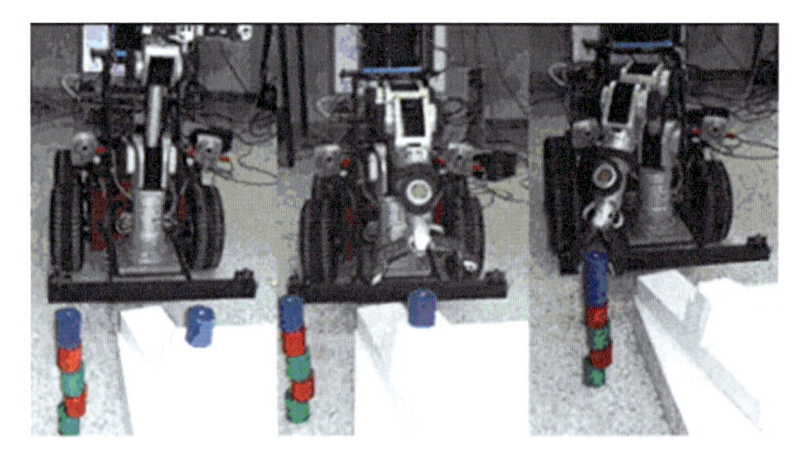

Fig. 8.13 Stack of six cylinders. Five cylinders were initially fetched in the neighbourhood of the target location of the stack from various initial locations on the floor. The last cylinder positioned to a nearby table. The robot collects the last cylinder and completes the stack

appropriate cylinders or to enter in an exploration phase, where it was surveying the space in order to locate the needed objects (the actual system used to achieve these results was that of UGDIST, which is very similar to that of KCL outlined earlier). Figure 8.13 (third panel) shows a typical stacking situation where the accuracy of the forward/inverse model for the arm can be seen in action.

8.8.2 Results

The aforementioned experiments and their variations were conducted as described above. In all cases, the system performed successfully in solving various reasoning problems. The GNOSYS system had a high success rate (>80%) in object recognition tasks from various viewpoints. This was achieved using the two-stage visual processing system mentioned earlier (first the coarse system and then the fine one), where in the second stage attention was applied to the visual input, in order to infer the class of occluded or badly illuminated objects, combined with the goal-biassing influence coming from the concept system. Activated concepts (either due to current reasoning processes or due to activation arising from previous stimuli) modulated the attention mechanism present in the fine visual-processing module. It is important to note that the visual system was based on two off-the-shelf web cameras operating at 160×120 pixel resolution. The only visual pre-processing that took place was cleaning up the camera image by using a colour segmentation algorithm. Object manipulation performance was very accurate (>90%) due to the highly accurate forward-inverse models (Morasso et al. 2005). Recognition of concepts was also highly accurate (>90%) as well as detecting novel objects (Taylor and Taylor 2007) in the environment (>70%).

8.8.3 Extensions Needed

There are numerous directions in which extensions to the system could be taken:

1. Addition of more detail in the motor system, using the known recurrence in the frontal system (frontal cortex, basal ganglia, cerebellum) to enable both attended and automatic actions to be learnt, with a smooth transition from the former to the latter.
2. Extension of the linguistic component to an enlarged semantic net, with learnt connections between the word codes and the action and object representations. This would thereby ground the words in the models of the world built in the GNOSYS brain. Note that we are only ever able to ground language that far, and not in the outside world, since the words can only be used to represent the outside world through the objects and actions, and more generally, the concepts that the words are trying to replace. The verbs, in particular, can only be represented

by the syntactic structures given by the representations of the action sequences carried by the frontal lobes and by the objects on which the actions are being taken.

3. A more complete version of the hippocampus, allowing both memory codes in the well-connected CA3 component (through learning lateral connections) and the resulting download (through learning) onto the neural modules nearby, representing neural structures in nearby cortical sites (temporal and related lobes).

4. More complex attention control circuitry, especially in the addition of a corollary discharge of the attention control signal (as in the CODAM model), so as to make attention control faster and more efficient.

5. The neural coding is developed by space–time-dependent learning, using the neuronal spiking transfer of information between modules. Such learning allows in particular the learning of the causal structure of inputs as well as in the use of causal flows of neural spikes in the attention control system of CODAM.

We discuss possible further work in Sect. 8.10.

8.9 Relation to Other Cognitive System Architectures

We have already noted the specific components of the GNOSYS Cognitive System, which, in general, distinguishes it from many other Cognitive System Architectures. Most specifically, it crucially depends on an attention control system both for its sensory input processing (presently specifically for vision) and for its motor action responses. The sensory attention system also allows for the learning of new concepts of objects in its environment, which can then be extended by abstraction into higher level concepts, connected by a web of lateral connections so as to enable reasoning and stimulus meaning to be extended almost in a symbolic manner (although no explicit symbols are employed in the GNOSYS system).

The system also crucially possesses a set of internal motor models (arising from fused work of the GNOSYS and MATHESIS projects, although we will still term the system the GNOSYS system for simplicity) enabling it to perform virtual (internal "mental") processes in which the results of the various steps of the internal models' activity do not cause any internal responses but modify internal state values so as most specifically to achieve desired goals. Such desire is itself set up by a set of predicted reward values encoded for each of the external stimuli the GNOSYS system has already encountered in its intercourse with its environment. Thus the GNOSYS brain possesses cognitive powers of perception and concept formation, learnt sensory and motor attention, goals creation in a given environment, reasoning and internal "mental simulation". Finally, we should add that we have implemented these powers in a robot embodiment so that they are observed in action in the robot itself.

We tabulate these powers and compare them with other leading proposals on cognitive systems in the table. We only consider in our comparison those architectures that are purely connectionist, since the symbolic or hybrid systems bring in symbolism, which is not clearly related to the sub-symbolic activities of the overall system, and limit the possible learning capabilities of the system as well as the use of brain guidance. This latter aspect is particularly important if we are considering those cognitive architectures that have a chance, at some later date, of the inclusion of a conscious component. This is, as we noted in the introduction, a crucial component of cognition but one mainly ignored so far in the AI/Machine intelligence approach to cognition. We also leave out the important and influential SOAR/EPIC/ACT-R/ICARUS class of models, since these use production rules and a symbolic component as well as not possessing suitably adaptive dynamics and active perception.

We have not considered more brain-based architectures such as Cog, Kismet or Cerebus, since these are still in developmental stages (as in Scassellati's Theory of Mind for Cog) or have a strong symbolic component (as in the case of Cerebus and Kismet).

8.10 Conclusions

In this chapter, we have attempted to address the tasks raised by the title to this paper: to create a neural architecture able to carry out the perception–conceptualisation–knowledge representation–reasoning–action cycle. It is possible to recognise such a cycle as indeed being carried out in the human brain: the flow of input into the brain leads to the generation of a set of percepts of objects in the environment being viewed, concepts in the knowledge representation system are then activated, and reasoning can be carried out as to the various action solutions to possible tasks that might be carried out to gain rewards from the objects. Actions to achieve such rewards are then determined and can be carried out.

We attacked this daunting problem by recognising that brain guidance in constructing an effective neural network architecture for this task would be appropriate. From that point of view, we then noted that the overall cycle involves brain processes at different levels:

(a) At the lowest level: perception, concept formation and action were involved, all under the control of attention.
(b) At a second level: the creation of knowledge representations, including internal models based on them, all under attention control. At the same time, reward prediction models are created for biassing the importance of perceived objects.
(c) At the highest level: the use of the lower and second level faculties to enable reasoning to take place for the solution of suitably rewarded tasks.

It can be seen from the different levels (a), (b) and (c) above that the two lower level faculties can be treated as being able to be learnt separately, without the need for consideration of reasoning proceeding. Such were the methods used to develop the GNOSYS architecture.

The final problem of higher cognitive processing is that there is a lack of experimental data on brain processes involved in non-linguistic reasoning. There is a growing literature on cortical areas involved in linguistic reasoning, including, as expected, numerous linguistic sites through the brain as well as those involved in executive functions. However, these do not necessarily move forward the architecture involved in the critical stages of reasoning, involved in setting up sub-goals. The manner of involvement of value map adjustment, as proposed in the model described earlier, needs careful discussion alongside augmented brain data to begin to validate it.

Finally, we consider points where further work needs to be done (emphasising and extending points already made in Sect. 8.3):

(a) Extension of the above architectures for thinking, reasoning and creativity explicitly to language. This has already been done implicitly, in terms of using words to code for given neurons in the simulations; details of how these word codes might arise through learning need to be explored.

(b) However, there is the more important question as to how language itself is learnt. Is that through more of the same sort of architectural components (attention using object and action codes, value maps, internal models)? Or are there different mechanisms at work to achieve a full linguistic repertoire? Given the rather different architectures of cerebellum and basal ganglia, it is very likely that the latter is true (although the former has been suggested as using internal models, and the recurrence of the frontal lobes using the basal ganglia has been used for forward model construction as sequence learners, as discussed earlier). Then what are the new principles involved, if any?

(c) In relation to the further analysis of the nature of linguistic processing of (b), are there similar coupled forward-inverse models in language that are developed especially by those who think linguistically rather than those who think visually? Or is inner speech driven, in all humans, by sequences of inner visual thoughts, but clothed in language after the event? This latter possibility does not seem likely but may occur in lower animals. How would that be tested?

(d) Further development of the internal model structures to enable thinking to be achieved at a non-attended level when the initial learning was at an attended level. This has been discussed in the chapter under the heading of creativity, but aspects such as associations between internal models and their extension to linguistic codes needs to be more fully explored as does the question of the details of the switches between the attended and unattended levels to allow for unconscious thoughts.

(e) Further development of the creativity switching architecture, especially the monitor system for assessing the value of a new concept activated by lateral connectivity in the manner suggested in the discussion in the chapter.

(f) Further exploration of the manner in which learning of lateral connections can be achieved so as to allow for creativity itself to be most effective. This is a basic question of creativity research: how best to study so that a deep problem can most efficiently be solved using the lateral connections thereby created? But how were these lateral connections created? If under attention, then taking attention away from any concept map might allow lateral spreading, and so reasoning by analogy. But how had the lateral spreading been achieved during the learning under attention in the first place? Was it during gaps in the attended learning process when the hard work was being done? Or was it during the attended learning process itself where suitably widespread reading brought these lateral connections about?

References

Bhattacharya J et al. (2009) Posterior beta and anterior gamma oscillations predict cognitive insight. Journal of Cognitive Neuroscience 21(7):1269–1279

Christoff K et al. (2009) Experience sampling during fMRI reveals default network and executive system contributions to mind wandering. Proceedings of the National Academy of Sciences of the United States of America 106(21):8719–8724

Clarke A (2004) Being There. Cambridge MA: MIT Press

Corbetta M & Shulman GL (2002) Control of goal-directed and stimulus-driven attention in the brain. Nature Reviews Neuroscience 3:201–215

Corbetta M, Tansy AP, Stanley CM, Astafiev SV, Snyder AZ, & Shulman GL (2005) A functional MRI study of preparatory signals for spatial location and objects. Neuropsychologia 43:2041–2056

Deco G & Rolls ET (2005) Attention, short-term memory, and action selection: a unifying theory. Progress in Neurobiology 76, 236–256

Desmurget M, Grafton S (2000) Forward modeling allows feedback control for fast reaching movements. Trends in Cognitive Sciences 6(11):423–431

Diamond MR, Ross J, & Morrone MC (2000) Extraretinal control of saccadic suppression. The Journal of Neuroscience 20(9):3449–3455

Fang F, Boyaci H, Kersten D & Murray SO (2008) Attention-dependent representations of a size illusion in V1. Current Biology 18(21):1707–1712

Friedman-Hill SR, Robertson LC, Desimone R & Ungerleider LG (2003) Posterior parietal cortex and the filtering of distracters. Proceedings of the National Academy of Sciences of the United States of America 1999(7):4263–4268

Gibson JJ (1979) The Ecological Approach to Visual Perception. Boston: Houghton-Mifflin

Gurney K, Prescott TJ, Wickens JR & Redgrave P (2004) Computational models of the basal ganglia: from robots to membranes. Trends in Neurosciences 27:453–459

Hamker FH & Zirnsak M (2006) V4 receptive field dynamics as predicted by a systems-level model of visual attention using feedback form the frontal eye field. Neural Networks 19(9):1371–1382

Hartley M, Fagard J, Esseily R & Taylor JG (2008) Observational versus trial and error effects in an infant learning paradigm – modelling and experimental data. Proceedings of the International Conference on Artificial Neural Networks 5164:277–289

Hartley M & Taylor JG (2009) A neural network model of creativity. Proceedings of the International Conference on Artificial Neural Networks (in press)

Heft H (1989) Affordances and the body: an intentional analysis of Gibson's ecological approach to visual perception. Journal for the Theory of Social Behaviour 19(1):1–30

Kandel ER, Schwartz JH & Jessell TM (2000) Principles of Neuroscience (4th edition) New York: McGraw Hill

Kanwisher N & Wojciulik E (2000) Visual attention: insights from brain imaging. Nature Reviews Neuroscience 1:91–100

Kasderidis S (2007) Developing concept representations In Proceedings of International Conference on Artificial Neural Network (ICANN 2007), Porto, 10–14 Sep. 2007, pp. 922–933

Kounios J, Fleck JI, Green DL, Payne L, Stevenson JL, Bowdend EM & Jung-Beeman M (2008) The origin of insight in resting state behaviour. Neuropsychologia 46(1):281–291

Mehta AD, Ulbert I, Schroeder CD (2000) Intermodal selective attention in monkeys. Cerebral Cortex 10:343–358

Miall RC (2003) Connecting mirror neurons and forward models. Neuroreport 14(17):2135–2137

Milner AD & Goodale MA (1995) The Visual Brain in Action Oxford: Oxford University Press

Mohan V & Morasso P (2007) Towards reasoning and coordinating action in the mental space. International Journal of Neural Systems 17(4):1–13

Morasso P, Bottaro A, Casadio M & Sanguineti V (2005) Preflexes and internal models in biomimetic robot systems. Cognitive Processing 6(1):25–36

Mozer MC & Sitton M (1998) Computational modelling of spatial attention In H. Pashler (Ed.) Attention (pp. 341–393). New York: Taylor & Francis

Natsoulas T (2004) "To see is to perceive what they afford" James J Gibson's concept of affordance. Mind and Behaviour 2(4):323–348

Pasupathy A & Connor CE (2001) Shape representation in area V4: position-specific tuning for boundary configuration. Journal of Neurophysiology 86:2505–2519

Raos V, Evangeliou MN & Savaki HE (2004) "Observation of action: grasping with the mind's hand". Neuroimage 23:191–201

Reynolds JH, Chelazzi I & Desimone R (1999) Competitive mechanisms subserve attention in macaque areas V2 and V4 Journal of Neuroscience 19:1736–1753

Santelli M, Flanders M & Soechting JF (2002) Patterns of hand motion during grasping and the influence of sensory guidance. Journal of Neuroscience 22(4):1426–1435

Schultz W (1998) Predictive reward signal of dopamine neurons Journal of Neurophysiology 80 (1):1–27

Schultz W (2004) Neural coding of basic reward terms of animal learning theory, game theory, microeconomics and behavioural ecology. Current Opinion in Neurobiology 14:139–147

Schultz W, Dayan P & Montague PR (1997) A neural substrate of prediction and reward. Science 275(5306): 1593–1599

Sommer MA & Wurtz RH (2002) A pathway in primate brain for internal monitoring of movements Science 296(5572):1480–1482

Sutton R (1988) Learning to predict by the methods of temporal differences Machine Learning 3(1):9–44

Sutton R & Barto A (1998) Reinforcement Learning Cambridge: MIT

Taylor JG (2000a) A control model for attention and consciousness. Society for Neuroscience Abstract 26:2231#839.3

Taylor JG (2000b) Attentional movement: the control basis for consciousness. Society for Neuroscience Abstracts 26:2231#839.3

Taylor JG (2005) Mind and consciousness: towards a final answer? Physics of Life Reviews 2(1):1–45

Taylor JG (2007) CODAM: a model of attention leading to the creation of consciousness. Scholarpedia 2(11):1598

Taylor JG (2009) A neural model of the loss of self in schizophrenia. Schizophrenia Bulletin (in press)

Taylor JG & Hartley MR (2007) Through reasoning to cognitive machines. IEEE Computational Intelligence Magazine 2(3):12–24

Taylor JG & Hartley M (2008) Exploring cognitive machines – neural models of reasoning, illustrated through the 2-sticks paradigm. Neurocomputing 71:2411–2419

Taylor JG & Taylor NR (2000a) Analysis of recurrent cortico-basal ganglia-thalamic loops for working memory. Biological Cybernetics 82:415–432

Taylor NR & Taylor JG (2000b) Hard-wired models of working memory and temporal sequence storage and generation. Neural Networks 13:201–224

Taylor NR & Taylor JG (2007) A novel novelty detector, in: J. Marques de Sa, L. A. Alexandre, W. Duch and D. Mandic (eds), Proceedings of the International Conference on Artificial Neural Networks, Lecture Notes in Computer Science #4669, Springer, Berlin, pp 973–983

Taylor NR, Panchev C, Kasperidis S, Hartley M & Taylor JG (2006) "Occlusion, attention and object representations", ICANN'06

Taylor NR, Panchev C, Hartley M, Kasperidis S, Taylor JG (2007a) Occlusion, attention and object representations. Integrated Computer-Aided Engineering 14(4):283–306

Taylor JG, Freeman W, Cleeremans A (eds) (2007b) Brain and consciousness. Neural Networks 20(9):929–1060

Taylor JG, Hartley M, Taylor NR, Panchev C & Kasperidis S (2009) A hierarchical attention-based neural network architecture, based on human brain guidance, for perception, conceptualisation, Action and Reasoning. Image and Vision Computing 27(11):1641–1657

Toda S (2009) Joint Attention between Mother and Infant in Play Situations Paper Presented at the Annual Meeting of the XVth Biennial International Conference on Infant Studies

Vandervert L (2003) How working memory and cognitive modelling functions of the cerebellum contribute to discoveries in mathematics. New Ideas in Psychology 21:159–175

Wallas G (1926) The Art of Thought. New York: Harper

Wertheimer M (1945) Productive Thinking. New York: Harper

Wolpert DM & Kawato M (1998) Multiple paired forward and inverse models for motor control. Neural Networks 11:1317–29

Young G (2006) Are different affordances subserved by different neural pathways? Brain and Cognition 62:134–142

Chapter 9
Consciousness, Decision-Making and Neural Computation

Edmund T. Rolls

Abstract Computational processes that are closely related to conscious processing and reasoning are described. Evidence is reviewed that there are two routes to action, the explicit, conscious, one involving reasoning, and an implicit, unconscious route for well-learned actions to obtain goals. Then a higher order syntactic thought (HOST) computational theory of consciousness is described. It is argued that the adaptive value of higher order syntactic thoughts is to solve the credit assignment problem that arises if a multi-step syntactic plan needs to be corrected. It is then suggested that it feels like something to be an organism that can think about its own linguistic and semantically based thoughts. It is suggested that qualia, raw sensory and emotional feels, arise secondarily to having evolved such a HOST processing system, and that sensory and emotional processing feels like something because it would be unparsimonious for it to enter the planning, HOST system and *not* feel like something. Neurally plausible models of decision-making are described, which are based on noise-driven and therefore probabilistic integrate-and-fire attractor neural networks, and it is proposed that networks of this type are involved when decisions are made between the explicit and implicit routes to action. This raises interesting issues about free will. It has been argued that the confidence one has in one's decisions provides an objective measure of awareness, but it is shown that two coupled attractor networks can account for decisions based on confidence estimates from previous decisions. In analyses of the implementation of consciousness, it is shown that the threshold for access to the consciousness system is higher than that for producing behavioural responses. The adaptive value of this may be that the systems in the brain that implement the type of information processing involved in conscious thoughts are not interrupted by small signals that could be noise in sensory pathways. Then oscillations are argued to not be a necessary part of the implementation of consciousness in the brain.

E.T. Rolls (✉)
Oxford Centre for Computational Neuroscience, Oxford, UK
e-mail: Edmund.Rolls@oxcns.org

V. Cutsuridis et al. (eds.), *Perception-Action Cycle: Models, Architectures,*
and Hardware, Springer Series in Cognitive and Neural Systems 1,
DOI 10.1007/978-1-4419-1452-1_9, © Springer Science+Business Media, LLC 2011

9.1 Introduction

In the perception–reason–action cycle, the reasoning step may produce an error. For example, in a reasoned plan with several steps to the plan, if there is an error, how do we know which step has the error? There is a credit assignment problem here, for the occurrence of an error does not tell us which step had a fault. I argue that in this situation, thoughts about the steps of the plan, that is thoughts about thoughts, namely "higher order thoughts," can help us to detect which was the weak step in the plan, with perhaps weak premises, so that we can correct the plan and try again. I argue that having thoughts about our previous thoughts, reflecting on them, may be a computational process that, when it occurs, is accompanied by feelings of consciousness, that it feels like something, and this general approach to the phenomenal aspects of consciousness is shared by a number of philosophers and scientists (Rosenthal 1990, 1993, 2004, 2005; Weiskrantz 1997). I then go on to argue that when this "higher order syntactic thought" (HOST) brain processor is actively processing sensory states or emotions (about which we may sometimes need to reason), then conscious feelings about these sensory states or emotions, called qualia, are present. I argue that this reasoning (i.e. rational) explicit (conscious) system is propositional and uses syntactic binding of symbols.

I contrast this rational or reasoning, explicit, system with an implicit (unconscious) system that uses gene-specified goals [e.g. food reward when hungry, water when thirsty, social interaction (Rolls 2005b, 2011)] for actions and can perform arbitrary actions to obtain the genotypically defined goals. The explicit system, because of the reasoning, can perform multiple-step planning which might lead to goals that are in the interest of the phenotype but not of the genotype (e.g. not having children in order to devote oneself to the arts, philosophy, or science). I argue that when decisions are made between the implicit and explicit computational systems (the genotype vs. the phenotype), then noise produced by neuronal spiking can influence the decision-making in an attractor network. Decisions made that are based on our subjective confidence in our earlier decisions are also influenced by noise in the brain (Rolls and Deco 2010). The computations involved in the implicit vs. the explicit system, and the effects of noise in decision-making, raise the issues of free will and of determinism. Finally, I consider some related computational issues, such as the role of oscillations and stimulus-dependent synchrony in consciousness, and why the threshold for consciousness is set to be at a higher level than the threshold for sensory processing and some implicit behavioural responses.

What is it about neural processing that makes it feel like something when some types of information processing are taking place? It is clearly not a general property of processing in neural networks, for there is much processing, for example, that concerned with the control of our blood pressure and heart rate, of which we are not aware. Is it then that awareness arises when a certain type of information processing is being performed? If so, what type of information processing? And how do emotional feelings, and sensory events, come to feel like anything? These feels are called qualia. These are great mysteries that have puzzled philosophers for centuries, and many approaches have been described (Dennett 1991, 2005; Chalmers 1996;

Block 2005; Rosenthal 2005; Davies 2008). They are at the heart of the problem of consciousness, for why it should feel like something at all is the great mystery. Other aspects of consciousness, such as the fact that often when we "pay attention" to events in the world, we can process those events in some better way, that is process or access as opposed to phenomenal aspects of consciousness, may be easier to analyze (Allport 1988; Block 1995; Chalmers 1996).

The puzzle of qualia, that is of the phenomenal aspect of consciousness, seems to be rather different from normal investigations in science, in that there is no agreement on criteria by which to assess whether we have made progress. So, although the aim of what follows in this paper is to address the issue of consciousness, especially of qualia, what is written cannot be regarded as being establishable by the normal methods of scientific enquiry. Accordingly, I emphasize that the view on consciousness that I describe is only preliminary, and theories of consciousness are likely to develop considerably. Partly for these reasons, this theory of consciousness should not be taken to have practical implications.

9.2 A Higher Order Syntactic Thought Theory of Consciousness

9.2.1 Multiple Routes to Action

A starting point is that much perception and action can be performed relatively automatically without apparent conscious intervention. An example sometimes given is driving a car. Another example is the identification of a visual stimulus that can occur without conscious awareness as described in Sect. 9.6. Another example is much of the sensory processing and actions that involve the dorsal stream of visual processing to the parietal cortex, such as posting a letter through a box at the correct orientation even when one may not be aware of what the object is (Milner and Goodale 1995; Goodale 2004; Milner 2008). Another example is blindsight, in which humans with damage to the visual cortex may be able to point to objects even when they are not aware of seeing an object (Weiskrantz 1997, 1998). Similar evidence applies to emotions, some of the processing for which can occur without conscious awareness (De Gelder et al. 1999; Phelps and LeDoux 2005; LeDoux 2008; Rolls 2005b, 2008a, b). Further, there is evidence that split-brain patients may not be aware of actions being performed by the "non-dominant" hemisphere (Gazzaniga and LeDoux 1978; Gazzaniga 1988, 1995; Cooney and Gazzaniga 2003). Further evidence consistent with multiple including non-conscious routes to action is that patients with focal brain damage, for example to the prefrontal cortex, may perform actions, yet comment verbally that they should not be performing those actions (Rolls et al. 1994a; Rolls 1999a, 2005b; Hornak et al. 2003, 2004). The actions, which appear to be performed implicitly, with surprise expressed later by the explicit system, include making behavioural responses to a no-longer rewarded visual stimulus in a visual discrimination reversal

(Rolls et al. 1994a; Hornak et al. 2004). In both these types of patient, confabulation may occur, in that a verbal account of why the action was performed may be given, and this may not be related at all to the environmental event that actually triggered the action (Gazzaniga and LeDoux 1978; Gazzaniga 1988; Rolls et al. 1994a; Gazzaniga 1995; Rolls 2005b; LeDoux 2008).

This evidence (see further Sect. 9.3.1) suggests that there are multiple routes to action, only some of which involve conscious processing (Rolls 2005a, b). It is possible that sometimes in normal humans when actions are initiated as a result of processing in a specialized brain region such as those involved in some types of implicit behaviour, the language system may subsequently elaborate a coherent account of why that action was performed (i.e. confabulate). This would be consistent with a general view of brain evolution in which as areas of the cortex evolve, they are laid on top of existing circuitry connecting inputs to outputs, and in which each level in this hierarchy of separate input–output pathways may control behaviour according to the specialized function it can perform (see schematic in Fig. 9.1). (It is of interest that mathematicians may get a hunch that something

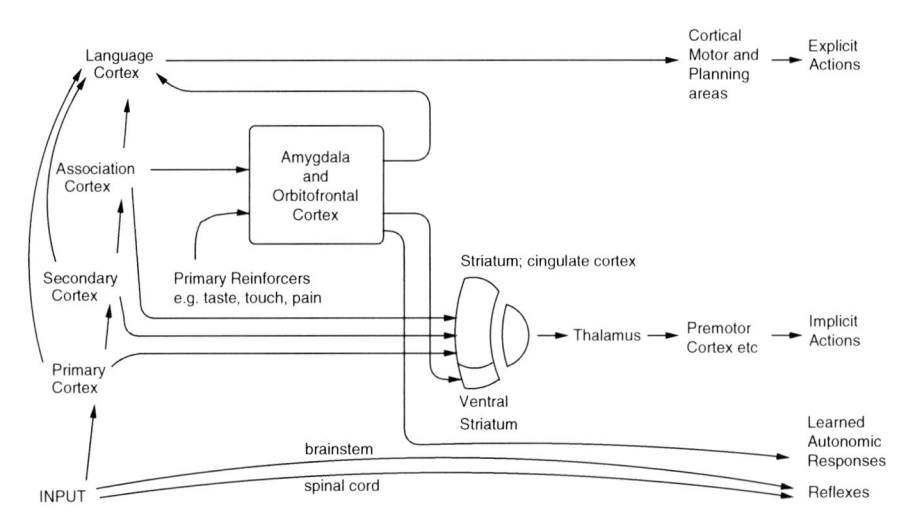

Fig. 9.1 Dual routes to the initiation of action in response to rewarding and punishing stimuli. The inputs from different sensory systems to brain structures such as the orbitofrontal cortex and amygdala allow these brain structures to evaluate the reward- or punishment-related value of incoming stimuli or of remembered stimuli. The different sensory inputs enable evaluations within the orbitofrontal cortex and amygdala based mainly on the primary (unlearned) reinforcement value for taste, touch and olfactory stimuli and on the secondary (learned) reinforcement value for visual and auditory stimuli. In the case of vision, the "association cortex" which outputs representations of objects to the amygdala and orbitofrontal cortex is the inferior temporal visual cortex. One route for the outputs from these evaluative brain structures is via projections directly to structures such as the basal ganglia (including the striatum and ventral striatum) to enable implicit, direct behavioural responses based on the reward or punishment-related evaluation of the stimuli to be made. The second route is via the language systems of the brain, which allow explicit decisions involving multi-step syntactic planning to be implemented

is correct, yet not be able to verbalize why. They may then resort to formal, more serial and language-like theorems to prove the case, and these seem to require conscious processing. This is an indication of a close association between linguistic processing and consciousness. The linguistic processing need not involve an inner articulatory loop.)

We may next examine some of the advantages and behavioural functions that language, present as the most recently added layer to the above system, would confer. One major advantage would be the ability to plan actions through many potential stages and to evaluate the consequences of those actions without having to perform the actions. For this, the ability to form propositional statements and to perform syntactic operations on the semantic representations of states in the world would be important. Also important in this system would be the ability to have second-order thoughts about the type of thought that I have just described (e.g. I think that he thinks that...), as this would allow much better modelling and prediction of others' behaviour, and therefore of planning, particularly planning when it involves others.[1] This capability for HOSTs would also enable reflection on past events, which would also be useful in planning. In contrast, non-linguistic behaviour would be driven by learned reinforcement associations, learned rules, etc., but not by flexible planning for many steps ahead involving a model of the world including others' behaviour. [For an earlier view which is close to this part of the argument see Humphrey (1980).] The examples of behaviour from non-humans that may reflect planning may reflect much more limited and inflexible planning. For example, the dance of the honey-bee to signal to other bees the location of food may be said to reflect planning, but the symbol manipulation is not arbitrary. There are likely to be interesting examples of non-human primate behaviour, perhaps in the great apes, that reflect the evolution of an arbitrary symbol-manipulation system that could be useful for flexible planning, cf. Cheney and Seyfarth (1990). It is important to state that the language ability referred to here is not necessarily human verbal language (though this would be an example). What it is suggested is important to planning is the syntactic manipulation of symbols, and it is this syntactic manipulation of symbols which is the sense in which language is defined and used here.

I understand *reasoning*, and *rationality*, to involve syntactic manipulations of symbols in the way just described. Reasoning thus typically may involve multiple steps of "if. then" conditional statements, all executed as a one-off or one-time

---

[1] Second order thoughts are thoughts about thoughts. Higher order thoughts refer to second order, third order etc. thoughts about thoughts... (A thought may be defined briefly as an intentional mental state, that is a mental state that is about something. Thoughts include beliefs, and are usually described as being propositional (Rosenthal DM (2005) Consciousness and Mind. Oxford: Oxford University Press). An example of a thought is "It is raining". A more detailed definition is as follows. A thought may be defined as an occurrent mental state (or event) that is intentional - that is a mental state that is about something - and also propositional, so that it is evaluable as true or false. Thoughts include occurrent beliefs or judgements. An example of a thought would be an occurrent belief that the earth moves around the sun/ that Maurice's boat goes faster with two sails/ that it never rains in southern California.)

process (see below), and is very different from associatively learned conditional rules typically learned over many trials, such as "if yellow, a left choice is associated with reward".

9.2.2 A Computational Hypothesis of Consciousness

It is next suggested that this arbitrary symbol-manipulation using important aspects of language processing and used for planning but not in initiating all types of behaviour is close to what consciousness is about. In particular, consciousness may *be* the state which arises in a system that can think about (or reflect on) its own (or other peoples') thoughts, that is in a system capable of second or higher order thoughts (HOTs) (Rosenthal 1986, 1990, 1993, 2004, 2005; Dennett 1991; Rolls 1995, 1997a, b, 1999b, 2004b, 2005b, 2007c, 2008a; Carruthers 1996; Gennaro 2004). On this account, a mental state is non-introspectively (i.e. non-reflectively) conscious if one has a roughly simultaneous thought that one is in that mental state. Following from this, introspective consciousness (or reflexive consciousness, or self consciousness) is the attentive, deliberately focused consciousness of one's mental states. It is noted that not all of the HOSTs need themselves be conscious (many mental states are not). However, according to the analysis, having a higher-order thought about a lower order thought is necessary for the lower order thought to be conscious. A slightly weaker position than Rosenthal's (and mine) on this is that a conscious state corresponds to a first order thought that has the *capacity* to cause a second order thought or judgement about it (Carruthers 1996). [Another position which is close in some respects to that of Carruthers and the present position is that of Chalmers (1996), that awareness is something that has *direct availability for behavioural control*, which amounts effectively for him in humans to saying that consciousness is what we can report (verbally) about.] This analysis is consistent with the points made above that the brain systems that are required for consciousness and language are similar. In particular, a system that can have second or HOSTs about its own operation, including its planning and linguistic operation, must itself be a language processor, in that it must be able to bind correctly to the symbols and syntax in the first order system. According to this explanation, the feeling of anything is the state that is present when processing is being performed by this particular neural system that is capable of second or HOSTs.

It might be objected that this captures some of the process aspects of consciousness, what is being performed in the relevant information processing system, but does not capture the phenomenal aspect of consciousness. I agree that there is an element of "mystery" that is invoked at this step of the argument, when I say that it feels like something for a machine with HOSTs to be thinking about its own first or lower order thoughts. But the return point (discussed further below) is the following: *if a human with second order thoughts is thinking about its own first order thoughts, surely it is very difficult for us to conceive that this would NOT feel like something?* (Perhaps the HOSTs in thinking about the first order thoughts would need to have in doing this some sense of continuity or self, so that the first order thoughts would be related to the same system that had thought of something else a few minutes ago. But even this continuity aspect may not be a requirement for consciousness. Humans with anterograde amnesia cannot remember what they felt a few minutes ago; yet their current state does feel like something.)

As a point of clarification, I note that according to this theory, a language processing system (let alone a working memory, LeDoux 2008) is not *sufficient* for consciousness. What defines a conscious system according to this analysis is the ability to have HOSTs, and a first order language processor (that might be perfectly competent at language) would not be conscious, in that it could not think about its own or others' thoughts. One can perfectly well conceive of a system that obeyed the rules of language (which is the aim of some connectionist modelling) and implemented a first-order linguistic system that would not be conscious. [Possible examples of language processing that might be performed non-consciously include computer programs implementing aspects of language, or ritualized human conversations, e.g. about the weather. These might require syntax and correctly grounded semantics and yet be performed non-consciously. A more complex example, illustrating that syntax could be used, might be "If A does X, then B will probably do Y, and then C would be able to do Z." A first order language system could process this statement. Moreover, the first order language system could apply the rule usefully in the world, provided that the symbols in the language system (A, B, X, Y, etc.) are grounded (have meaning) in the world.]

A second clarification is that the plan would have to be a unique string of steps, in much the same way as a sentence can be a unique and one-off (or one-time) string of words. The point here is that it is helpful to be able to think about particular one-off plans and to correct them, and that this type of operation is very different from the slow learning of fixed rules by trial and error or the application of fixed rules by a supervisory part of a computer program.

9.2.3 Adaptive Value of Processing in the System That Is Related to Consciousness

It is suggested that part of the evolutionary *adaptive significance* of this type of HOST is that it enables correction of errors made in first order linguistic or in non-linguistic processing. Indeed, the ability to reflect on previous events is extremely important for learning from them, including setting up new long-term semantic structures. It is shown elsewhere that the hippocampus may be a system for such "declarative" recall of recent memories (Rolls 2008b). Its close relation to "conscious" processing in humans [Squire and Zola (1996) have classified it as a declarative memory system] may be simply that it enables the recall of recent memories, which can then be reflected upon in conscious, higher order, processing (Rolls and Kesner 2006; Rolls 2008b). Another part of the adaptive value of a HOST system may be that by thinking about its own thoughts in a given situation, it may be able to better understand the thoughts of another individual in a similar situation and therefore predict that individual's behaviour better (cf. Humphrey 1980, 1986; Barlow 1997).

In line with the argument on the adaptive value of HOSTs and thus consciousness given above, which they are useful for correcting lower order thoughts, I now suggest that correction using HOSTs of lower order thoughts would have adaptive value primarily if the lower order thoughts are sufficiently complex to benefit from correction in this way. The nature of the complexity is specific: that it should involve syntactic manipulation of symbols, probably with several steps in the chain, and that the chain of steps should be a one-off (or in American, "one-time", meaning used once) set of steps, as in a sentence or in a particular plan used just once, rather than a set of well-learned rules. The first or lower order thoughts might involve a linked chain of "if"... "then" statements that would be involved in planning, an example of which has been given above. It is partly because complex lower order thoughts such as these which involve syntax and language would benefit from correction by HOSTs that I suggest that there is a close link between this reflective consciousness and language. The *computational hypothesis* is that by thinking about lower order thoughts, the HOSTs can discover what may be weak links in the chain of reasoning at the lower order level, and having detected the weak link, might alter the plan, to see if this gives better success. In our example above, if it transpired that C could not do Z, how might the plan have failed? Instead of having to go through endless random changes to the plan to see if by trial and error some combination does happen to produce results, what I am suggesting is that by thinking about the previous plan, one might, for example using knowledge of the situation and the probabilities that operate in it, guess that the step where the plan failed was that B did not in fact do Y. So by thinking about the plan (the first or lower order thought), one might correct the original plan, in such a way that the weak link in that chain, that "B will probably do Y", is circumvented.

I draw a parallel with neural networks: there is a *"credit assignment"* problem in such multi-step syntactic plans, in that if the whole plan fails, how does the system assign credit or blame to particular steps of the plan? [In multilayer neural networks, the credit assignment problem is that if errors are being specified at the output layer, the problem arises about how to propagate back the error to earlier, hidden, layers of the network to assign credit or blame to individual synaptic connections; see Rumelhart et al. (1986), Rolls and Deco (2002) and Rolls (2008b).] The suggestion is that this is the function of HOSTs and is why systems with HOSTs evolved. The suggestion I then make is that if a system were doing this type of processing (thinking about its own thoughts), it would then be very plausible that it should feel like something to be doing this. I even suggest to the reader that it is not plausible to suggest that it would not feel like anything to a system if it were doing this.

9.2.4 Symbol Grounding

A further point in the argument should be emphasized for clarity. The system that is having syntactic thoughts about its own syntactic thoughts (higher order syntactic thoughts or HOSTs) would have to have its symbols grounded in the real world for it to feel like something to be having HOSTs. The intention of this clarification is to exclude systems such as a computer running a program when there is in addition some sort of control or even overseeing program checking the operation of the first program. We would want to say that in such a situation it would feel like something to be running the higher level control program only if the first order program was symbolically performing operations on the world and receiving input about the results of those operations and if the higher order system understood what the first order system was trying to do in the world. The issue of symbol grounding is considered further by Rolls (2005b) . The symbols (or symbolic representations) are symbols in the sense that they can take part in syntactic processing. The symbolic representations are grounded in the world in that they refer to events in the world. The symbolic representations must have a great deal of information about what is referred to in the world, including the quality and intensity of sensory events, emotional states, etc. The need for this is that the reasoning in the symbolic system must be about stimuli, events and states, and remembered stimuli, events and states, and for the reasoning to be correct, all the information that can affect the reasoning must be represented in the symbolic system, including, for example, just how light or strong the touch was, etc. Indeed, it is pointed out in *Emotion Explained* (Rolls 2005b) that it is no accident that the shape of the multi-dimensional phenomenal (sensory, etc.) space does map so clearly onto the space defined by neuronal activity in sensory systems, for if this were not the case, reasoning about the state of affairs in the world would not map onto the world and would not be useful. Good examples of this close correspondence are found in the taste system, in which subjective space maps simply onto the multi-dimensional space represented by neuronal firing in primate cortical taste areas. In particular, if a three-dimensional space reflecting the distances between the representations of different tastes provided by macaque neurons in the cortical taste areas is constructed, then the distances between the subjective ratings by humans of different tastes are very similar (Yaxley et al. 1990; Smith-Swintosky et al. 1991; Kadohisa et al. 2005). Similarly, the changes in human subjective ratings of the pleasantness of the taste, smell and sight of food parallel very closely the responses of neurons in the macaque orbitofrontal cortex (see *Emotion Explained*).

The representations in the first order linguistic processor that the HOSTs process include beliefs (e.g. "Food is available", or at least representations of this), and the HOST system would then have available to it the concept of a thought [so that it could represent "I believe (or there is a belief) that food is available"]. However, as argued by Rolls (1999b, 2005b), representations of sensory processes and emotional states must be processed by the first order linguistic system, and HOSTs may be about these representations of sensory processes and emotional states capable of taking part in the syntactic operations of the first order linguistic processor. Such sensory and emotional information may reach the first order linguistic system from many parts of the brain, including those such as the orbitofrontal cortex and amygdala implicated in emotional states (see Fig. 9.1 and *Emotion Explained*, Fig. 10.3). When the sensory information is about the identity of the taste, the inputs to the first order linguistic system must come from the primary taste cortex, in that the identity of taste, independently of its pleasantness (in that the representation is independent of hunger), must come from the primary taste cortex. In contrast, when the information that reaches the first order linguistic system is about the pleasantness of taste, it must come from the secondary taste cortex, in that there the representation of taste depends on hunger (Rolls and Grabenhorst 2008).

9.2.5 Qualia

This analysis does not yet give an account for sensory qualia ("raw sensory feels", for example why "red" feels red), for emotional qualia (e.g. why a rewarding touch produces an emotional feeling of pleasure), or for motivational qualia (e.g. why food deprivation makes us *feel* hungry). The view I suggest on such qualia is as follows. Information processing in and from our sensory systems (e.g. the sight of the colour red) may be relevant to planning actions using language and the conscious processing thereby implied. Given that these inputs must be represented in the system that plans, we may ask whether it is more likely that we would be conscious of them or that we would not. I suggest that it would be a very special-purpose system that would allow such sensory inputs, and emotional and motivational states, to be part of (linguistically based) planning and yet remain unconscious. It seems to be much more parsimonious to hold that we would be conscious of such sensory, emotional and motivational qualia because they would be used (or are available to be used) in this type of (linguistically based) HOST processing, and this is what I propose.

The explanation for perceptual, emotional and motivational subjective feelings or qualia that this discussion has led towards is thus that they should be felt as conscious because they enter into a specialized linguistic symbol-manipulation system that is part of a HOST system that is capable of reflecting on and correcting its lower order thoughts involved, for example, in the flexible planning of actions. It would require a very special machine to enable this higher-order linguistically based thought processing, which is conscious by its nature, to occur without the sensory, emotional and motivational states (which must be taken into account by the HOST system) becoming felt qualia. The qualia are thus accounted for by the evolution of the linguistic system that can reflect on and correct its own lower order processes and thus has adaptive value.

This account implies that it may be especially animals with a higher order belief and thought system and with linguistic symbol manipulation that have qualia. It may be that much non-human animal behaviour, provided that it does not require flexible linguistic planning and correction by reflection, could take place according to reinforcement-guidance [using e.g. stimulus-reinforcement association learning in the amygdala and orbitofrontal cortex (Rolls 2004a, 2005b, 2008b)] and rule-following [implemented e.g. using habit or stimulus-response learning in the basal ganglia (Rolls 2005b)]. Such behaviours might appear very similar to human behaviour performed in similar circumstances, but would not imply qualia. It would be primarily by virtue of a system for reflecting on flexible, linguistic, planning behaviour that humans (and animals with demonstrable syntactic manipulation of symbols, and the ability to think about these linguistic processes) would be different from other animals and would have evolved qualia.

It is of interest to comment on how the evolution of a system for flexible planning might affect emotions. Consider grief which may occur when a reward is terminated and no immediate action is possible [see Rolls (1990, 2005b)]. It may be adaptive by leading to a cessation of the formerly rewarded behaviour and thus facilitating the possible identification of other positive reinforcers in the environment. In humans, grief may be particularly potent because it becomes represented in a system which can plan ahead and understand the enduring implications of the loss. (Thinking about or verbally discussing emotional states may also in these circumstances help, because this can lead towards the identification of new or alternative reinforcers and of the realization that, for example, negative consequences may not be as bad as feared.)

9.2.6 Pathways

In order for processing in a part of our brain to be able to reach consciousness, appropriate pathways must be present. Certain constraints arise here. For example, in the sensory pathways, the nature of the representation may change as it passes through a hierarchy of processing levels, and in order to be conscious of the information in the form in which it is represented in early processing stages, the early processing stages must have access to the part of the brain necessary for consciousness (see Fig. 9.1). An example is provided by processing in the taste system. In the primate primary taste cortex, neurons respond to taste independently of hunger, yet in the secondary taste cortex, food-related taste neurons (e.g. responding to sweet taste) only respond to food if hunger is present and gradually stop responding to that taste during feeding to satiety (Rolls 2005b, 2006). Now the quality of the tastant (sweet, salt, etc.) and its intensity are not affected by hunger, but the pleasantness of its taste is decreased to zero (neutral) (or even becomes unpleasant) after we have eaten it to satiety (Rolls 2005b). The implication of this is that for quality and intensity information about taste, we must be conscious of what is represented in the primary taste cortex (or perhaps in another area connected to it which bypasses the secondary taste cortex) and not of what is represented in the secondary taste cortex. In contrast, for the pleasantness of a taste, consciousness of this could not reflect what is represented in the primary taste cortex, but instead what is represented in the secondary taste cortex (or in an area beyond it).

The same argument arises for reward in general and therefore for emotion, which in primates is not represented early on in processing in the sensory pathways (nor in or before the inferior temporal cortex for vision), but in the areas to which these object analysis systems project, such as the orbitofrontal cortex, where the reward value of visual stimuli is reflected in the responses of neurons to visual stimuli (Rolls 2005b, 2006). It is also of interest that reward signals (e.g. the taste of food when we are hungry) are associated with subjective feelings of pleasure (Rolls 2005b, 2006). I suggest that this correspondence arises because pleasure is the subjective state that represents in the conscious system a signal that is positively reinforcing (rewarding), and that inconsistent behaviour would result if the representations did not correspond to a signal for positive reinforcement in both the conscious and the non-conscious processing systems.

Do these arguments mean that the conscious sensation of, for example, taste quality (i.e. identity and intensity) is represented or occurs in the primary taste cortex and of the pleasantness of taste in the secondary taste cortex, and that activity in these areas is sufficient for conscious sensations (qualia) to occur? I do not suggest this at all. Instead the arguments I have put forward above suggest that we are only conscious of representations when engage a system capable of HOSTs. The implication then is that pathways must connect from each of the brain areas in which information is represented about which we can be conscious to the system that has the HOSTs, which as I have argued above requires a brain system capable of HOSTs. Thus, in the example given, there must be connections to the language areas from the primary taste cortex, which need not be direct, but which must bypass the secondary taste cortex, in which the information is represented differently (Rolls 2005b). There must also be pathways from the secondary taste cortex, not necessarily direct, to the language areas so that we can have HOSTs about the pleasantness of the representation in the secondary taste cortex. There would also need to be pathways from the hippocampus, implicated in the recall of declarative memories, back to the language areas of the cerebral cortex (at least via the cortical areas which receive backprojections from the amygdala, orbitofrontal cortex and hippocampus, see Fig. 9.1, which would in turn need connections to the language areas).

9.2.7 Consciousness and Causality

One question that has been discussed is whether there is a causal role for consciousness [e.g. Armstrong and Malcolm (1984)]. The position to which the above arguments lead is that indeed conscious processing does have a causal role in the elicitation of behaviour, but only under the set of circumstances when HOSTs play a role in correcting or influencing lower order thoughts. The sense in which the consciousness is causal is then it is suggested that the HOST is causally involved in correcting the lower order thought, and that it is a property of the HOST system that it feels like something when it is operating. As we have seen, some behavioural responses can be elicited when there is not this type of reflective control of lower order processing nor indeed any contribution of language (see further Rolls (2003, 2005a) for relations between implicit and explicit processing). There are many brain processing routes to output regions, and only one of these involves conscious, verbally represented processing which can later be recalled (see Fig. 9.1).

I suggest that these concepts may help us to understand what is happening in experiments of the type described by Libet and many others (Libet 2002), in which consciousness appears to follow with a measurable latency the time when a decision was taken. This is what I predict, if the decision is being made by an implicit perhaps reward/emotion or habit-related process, for then the conscious processor confabulates an account of or commentary on the decision, so that inevitably the conscious account follows the decision. On the other hand, I predict that if the rational (multi-step, reasoning) route is involved in taking the decision, as it might be during planning, or a multi-step task such as mental arithmetic, then the conscious report of when the decision was taken, and behavioural or other objective evidence on when the decision was taken, would correspond much more. Under those circumstances, the brain processing taking the decision would be closely related to consciousness, and it would not be a case of just confabulating or reporting on a decision taken by an implicit processor. It would be of interest to test this hypothesis in a version of Libet's task (Libet 2002) in which reasoning was required. The concept that the rational, conscious, processor is only in some tasks involved in taking decisions is extended further in the section on dual routes to action below.

9.2.8 Consciousness, a Computational System for Higher Order Syntactic Manipulation of Symbols, and a Commentary or Reporting Functionality

I now consider some clarifications of the present proposal, and how it deals with some issues that arise when considering theories of the phenomenal aspects of consciousness.

First, the present proposal has as its foundation the type of computation that is being performed and suggests that it is a property of a HOST system used for correcting multi-step plans with its representations grounded in the world that it would feel like something for a system to be doing this type of processing. To do this type of processing, the system would have to be able to recall previous multi-step plans and would require syntax to keep the symbols in each step of the plan separate. In a sense, the system would have to be able to recall and take into consideration its earlier multi-step plans, and in this sense *report* to itself, on those earlier plans. Some approaches to consciousness take the ability to report on or make a *commentary* on events as being an important marker for consciousness (Weiskrantz 1997), and the computational approach I propose suggests why there should be a close relation between consciousness and the ability to report or provide a commentary, for the ability to report is involved in using HOSTs to correct a multi-step plan.

Second, the implication of the present approach is that the type of linguistic processing or reporting need not be verbal, using natural language, for what is required to correct the plan is the ability to manipulate symbols syntactically, and this could be implemented in a much simpler type of mentalese or syntactic system (Fodor 1994; Jackendoff 2002; Rolls 2004b) than verbal language or natural language which implies a universal grammar.

Third, this approach to consciousness suggests that the information must be being processed in a system capable of implementing HOSTs for the information to be conscious and in this sense is more specific than global workspace hypotheses (Baars 1988; Dehaene and Naccache 2001; Dehaene et al. 2006). Indeed, the present approach suggests that a workspace could be sufficiently global to enable even the complex processing involved in driving a car to be performed, and yet the processing might be performed unconsciously, unless HOST (supervisory, monitory, correcting) processing was involved.

Fourth, the present approach suggests that it just is a property of HOST computational processing with the representations grounded in the world that it feels like something. There is to some extent an element of mystery about why it feels like something, why it is phenomenal, but the explanatory gap does not seem so large when one holds that the system is recalling, reporting on, reflecting on and reorganizing information about itself in the world in order to prepare new or revised plans. In terms of the physicalist debate (see for a review Davies 2008), an important aspect of my proposal is that it is a *necessary* property of this type of (HOST) computational processing that it feels like something (the philosophical description is that this is an absolute metaphysical necessity), and given this view, then it is up to one to decide whether this view is consistent with one's particular view of physicalism or not (Rolls 2008a). Similarly, the possibility of a zombie is inconsistent with the present hypothesis, which proposes that it is by virtue of performing processing in a specialized system that can perform higher order syntactic processing with the representations grounded in the world that phenomenal consciousness is necessarily present.

An implication of these points is that my theory of consciousness is a computational theory. It argues that it is a property of a certain type of computational processing that it feels like something. In this sense, although the theory spans many levels from the neuronal to the computational, it is unlikely that any particular neuronal phenomena such as oscillations are necessary for consciousness, unless such computational processes happen to rely on some particular neuronal properties not involved in other neural computations but necessary for higher order syntactic computations. It is these computations and the system that implements them that this computational theory argues are necessary for consciousness.

These are my initial thoughts on why we have consciousness and are conscious of sensory, emotional and motivational qualia, as well as qualia associated with first-order linguistic thoughts. However, as stated above, one does not feel that there are straightforward criteria in this philosophical field of enquiry for knowing whether the suggested theory is correct; so it is likely that theories of consciousness will continue to undergo rapid development, and current theories should not be taken to have practical implications.

9.3 Selection Between Conscious vs. Unconscious Decision-Making and Free Will

9.3.1 Dual Routes to Action

According to the present formulation, there are two types of route to action performed in relation to reward or punishment in humans (see also Rolls 2003, 2005b). Examples of such actions include emotional and motivational behaviour.

The first route is via the brain systems that have been present in non-human primates such as monkeys, and to some extent in other mammals, for millions of years. These systems include the amygdala and, particularly well-developed in primates, the orbitofrontal cortex. These systems control behaviour in relation to previous associations of stimuli with reinforcement. The computation which controls the action thus involves assessment of the reinforcement-related value of a stimulus. This assessment may be based on a number of different factors. One is the previous reinforcement history, which involves stimulus-reinforcement association learning using the amygdala, and its rapid updating especially in primates using the orbitofrontal cortex. This stimulus-reinforcement association learning may involve quite specific information about a stimulus, for example of the energy associated with each type of food, by the process of conditioned appetite and satiety (Booth 1985). A second is the current motivational state, for example whether hunger is present, whether other needs are satisfied, etc. A third factor which affects the computed reward value of the stimulus is whether that reward has been received recently. If it has been received recently but in small quantity, this may increase the reward value of the stimulus. This is known as incentive motivation or the "salted peanut" phenomenon. The adaptive value of such a process is that this positive feedback of reward value in the early stages of working for a particular reward tends to lock the organism onto behaviour being performed for that reward. This means that animals that are, for example, almost equally hungry and thirsty will show hysteresis in their choice of action rather than continually switching from eating to drinking and back with each mouthful of water or food. This introduction of hysteresis into the reward evaluation system makes action selection a much more efficient process in a natural environment, for constantly switching between different types of behaviour would be very costly if all the different rewards were not available in the same place at the same time. (For example, walking half a mile between a site where water was available and a site where food was available after every mouthful would be very inefficient.) The amygdala is one structure that may be involved in this increase in the reward value of stimuli early on in a series of presentations, in that lesions of the amygdala (in rats) abolish the expression of this reward incrementing process which is normally evident in the increasing rate of working for a food reward early on in a meal (Rolls 2005b). A fourth factor is the computed absolute value of the reward or punishment expected or being obtained from a stimulus, e.g. the sweetness of the stimulus (set by evolution so that sweet stimuli will tend to be rewarding, because they are generally associated with energy sources), or the pleasantness of touch (set by evolution to be pleasant according to the extent to which it brings animals of the opposite sex together, and depending on the investment in time that the partner is willing to put into making the touch pleasurable, a sign which indicates the commitment and value for the partner of the relationship). After the reward value of the stimulus has been assessed in these ways, behaviour is then initiated based on approach towards or withdrawal from the stimulus. A critical aspect of the behaviour produced by this type of system is that it is aimed directly towards obtaining a sensed or expected reward by virtue of connections to brain systems such as the basal ganglia and cingulate cortex (Rolls 2009), which are concerned with the initiation of actions (see Fig. 9.1). The expectation may of course involve behaviour to obtain stimuli associated with reward, which might even be present in a chain.

Now part of the way in which the behaviour is controlled with this first route is according to the reward value of the outcome. At the same time, the animal may only work for the reward if the cost is not too high. Indeed, in the field of behavioural ecology, animals are often thought of as performing optimally on some cost-benefit curve [see e.g. Krebs and Kacelnik (1991)]. This does not at all mean that the animal thinks about the rewards and performs a cost-benefit analysis using a lot of thoughts about the costs, other rewards available and their costs, etc. Instead, it should be taken to mean that in evolution, the system has evolved in such a way that the way in which the reward varies with the different energy densities or amounts of food and the delay before it is received can be used as part of the input to a mechanism, which has also been built to track the costs of obtaining the food (e.g. energy loss in obtaining it, risk of predation, etc.), and to then select given many such types of reward and the associated cost, the current behaviour that provides the most "net reward". Part of the value of having the computation expressed in this reward-minus-cost form is that there is then a suitable "currency", or net reward value, to enable the animal to select the behaviour with currently the most net reward gain (or minimal aversive outcome).

The second route in humans involves a computation with many "if... then" statements to implement a plan to obtain a reward. In this case, the reward may actually be *deferred* as part of the plan, which might involve working first to obtain one reward, and only then to work for a second more highly valued reward, if this was thought to be overall an optimal strategy in terms of resource usage (e.g. time). In this case, syntax is required, because the many symbols (e.g. names of people) that are part of the plan must be correctly linked or bound. Such linking might be of the form: "if A does this, then B is likely to do this, and this will cause C to do this...". The requirement of syntax for this type of planning implies that an output to language systems in the brain is required for this type of planning (see Fig. 9.1). Thus the explicit language system in humans may allow working for deferred rewards by enabling use of a one-off, individual, plan appropriate for each situation. Another building block for such planning operations in the brain may be the type of short-term memory in which the prefrontal cortex is involved. This short term memory may be, for example, in non-human primates of where in space a response has just been made. A development of this type of short term response memory system in humans to enable multiple short term memories to be held in place correctly, preferably with the temporal order of the different items in the short term memory coded correctly, may be another building block for the multiple step "if.... then" type of computation in order to form a multiple step plan. Such short term memories are implemented in the (dorsolateral and inferior convexity) prefrontal cortex of non-human primates and humans (Goldman-Rakic 1996; Petrides 1996; Rolls 2008b) and may be part of the reason why prefrontal cortex damage impairs planning (Shallice and Burgess 1996).

Of these two routes (see Fig. 9.1), it is the second which I have suggested above is related to consciousness. The hypothesis is that consciousness is the state which arises by virtue of having the ability to think about one's own thoughts, which has the adaptive value of enabling one to correct long multi-step syntactic plans. This latter system is thus the one in which explicit, declarative, processing occurs. Processing in this system is frequently associated with reason and rationality, in that many of the consequences of possible actions can be taken into account. The actual computation of how rewarding a particular stimulus or situation is or will be probably still depends on activity in the orbitofrontal and amygdala, as the reward value of stimuli is computed and represented in these regions, and in that it is found that verbalized expressions of the reward (or punishment) value of stimuli are dampened by damage to these systems. (For example, damage to the orbitofrontal cortex renders painful input still identifiable as pain, but without the strong affective, "unpleasant", reaction to it.) This language system which enables long-term planning may be contrasted with the first system in which behaviour is directed at obtaining the stimulus (including the remembered stimulus) which is currently most rewarding, as computed by brain structures that include the orbitofrontal cortex and amygdala. There are outputs from this system, perhaps those directed at the basal ganglia, which do not pass through the language system, and behaviour produced in this way is described as implicit, and verbal declarations cannot be made directly about the reasons for the choice made. When verbal declarations are made about decisions made in this first system, those verbal declarations may be confabulations, reasonable explanations or fabrications, of reasons why the choice was made. These reasonable explanations would be generated to be consistent with the sense of continuity and self that is a characteristic of reasoning in the language system.

The question then arises of how decisions are made in animals such as humans that have both the implicit, direct reward-based, and the explicit, rational, planning systems (see Fig. 9.1) (Rolls 2008b). One particular situation in which the first, implicit, system may be especially important is when rapid reactions to stimuli with reward or punishment value must be made, for then the direct connections from structures such as the orbitofrontal cortex to the basal ganglia may allow rapid actions (Rolls 2005b). Another is when there may be too many factors to be taken into account easily by the explicit, rational, planning, system, when the implicit system may be used to guide action. In contrast, when the implicit system continually makes errors, it would then be beneficial for the organism to switch from automatic, direct, action based on obtaining what the orbitofrontal cortex system decodes as being the most positively reinforcing choice currently available to the explicit conscious control system which can evaluate with its long-term planning algorithms what action should be performed next. Indeed, it would be adaptive for the explicit system to regularly be assessing performance by the more automatic system and to switch itself in to control behaviour quite frequently, as otherwise the adaptive value of having the explicit system would be less than optimal.

There may also be a flow of influence from the explicit, verbal system to the implicit system, in that the explicit system may decide on a plan of action or strategy, and exert an influence on the implicit system which will alter the reinforcement evaluations made by and the signals produced by the implicit system (Rolls 2005b).

It may be expected that there is often a conflict between these systems, in that the first, implicit, system is able to guide behaviour particularly to obtain the greatest immediate reinforcement, whereas the explicit system can potentially enable immediate rewards to be deferred and longer-term, multi-step, plans to be formed. This type of conflict will occur in animals with a syntactic planning ability, that is in humans and any other animals that have the ability to process a series of "if... then" stages of planning. This is a property of the human language system, and the extent to which it is a property of non-human primates is not yet fully clear. In any case, such conflict may be an important aspect of the operation of at least the human mind, because it is so essential for humans to correctly decide, at every moment, whether to invest in a relationship or a group that may offer long-term benefits or whether to directly pursue immediate benefits (Rolls 2005b, 2008b).

The thrust of the argument (Rolls 2005b, 2008b) thus is that much complex animal including human behaviour can take place using the implicit, non-conscious, route to action. We should be very careful not to postulate intentional states (i.e. states with intentions, beliefs and desires) unless the evidence for them is strong, and it seems to me that a flexible, one-off, linguistic processing system that can handle propositions is needed for intentional states. What the explicit, linguistic, system does allow is exactly this flexible, one-off, multi-step planning ahead type of computation, which allows us to defer immediate rewards based on such a plan.

This discussion of dual routes to action has been with respect to the behaviour produced. There is of course in addition a third output of brain regions, such as the orbitofrontal cortex and amygdala involved in emotion, that is directed to producing autonomic and endocrine responses (see Fig. 9.1). Although it has been argued by Rolls (2005b) that the autonomic system is not normally in a circuit through which behavioural responses are produced (i.e. against the James–Lange and related somatic theories), there may be some influence from effects produced through the endocrine system (and possibly the autonomic system, through which some endocrine responses are controlled) on behaviour or on the dual systems just discussed that control behaviour.

9.3.2 The Selfish Gene vs. the Selfish Phene

I have provided evidence in Sect. 9.3.1 that there are two main routes to decision-making and action. The first route selects actions by gene-defined goals for action and is closely associated with emotion. The second route involves multi-step planning and reasoning which requires syntactic processing to keep the symbols involved at each step separate from the symbols in different steps. (This second route is used by humans and perhaps by closely related animals.) Now the "interests" of the first and second routes to decision-making and action are different. As argued very convincingly by Richard Dawkins in The Selfish Gene (Dawkins 1989), and by others (Hamilton 1964, 1996; Ridley 1993), many behaviours occur in the interests of the survival of the genes, not of the individual (nor of the group), and much behaviour can be understood in this way. I have extended this approach by arguing that an important role for some genes in evolution is to define the goals for actions that will lead to better survival of those genes; that emotions are the states associated with these gene-defined goals; and that the defining of goals for actions rather that actions themselves is an efficient way for genes to operate, as it leaves flexibility of choice of action open until the animal is alive (Rolls 2005b). This provides great simplification of the genotype as action details do not need to be specified, just rewarding and punishing stimuli, and also flexibility of action in the face of changing environments faced by the genes. Thus the interests that are implied when the first route to action is chosen are those of the "selfish genes" and not those of the individual.

However, the second route to action allows, by reasoning, decisions to be taken that might not be in the interests of the genes, might be longer term decisions and might be in the interests of the individual. An example might be a choice not to have children, but instead to devote oneself to science, medicine, music or literature. The reasoning, rational, system presumably evolved because taking longer-term decisions involving planning rather than choosing a gene-defined goal might be advantageous at least sometimes for genes. But an unforeseen consequence of the evolution of the rational system might be that the decisions would, sometimes, not be to the advantage of any genes in the organism. After all, evolution by natural selection operates utilizing genetic variation like a Blind Watchmaker (Dawkins 1986). In this sense, the interests when the second route to decision-making is used are at least sometimes those of the "selfish phenotype". Indeed, we might euphonically say that the interests are those of the "selfish phene" (where the etymology is Gk phaino, "appear", referring to appearance, hence the thing that one observes, the individual). Hence the decision-making referred to in Sect. 9.3.1 is between a first system where the goals are gene-defined and a second rational system in which the decisions may be made in the interests of the genes, or in the interests of the phenotype and not in the interests of the genes. Thus we may speak of the choice as sometimes being between the "Selfish Genes" and the "Selfish Phenes".

Now what keeps the decision-making between the "Selfish Genes" and the "Selfish Phenes" more or less under control and in balance? If the second, rational, system chose too often for the interests of the "Selfish Phene", the genes in that phenotype would not survive over generations. Having these two systems in the same individual will only be stable if their potency is approximately equal, so that sometimes decisions are made with the first route and sometimes with the second route. If the two types of decision-making, then, compete with approximately equal potency, and sometimes one is chosen, and sometimes the other, then this is exactly the scenario in which stochastic processes in the decision-making mechanism are likely to play an important role in the decision that is taken. The same decision, even with the same evidence, may not be taken each time a decision is made, because of noise in the system.

The system itself may have some properties that help to keep the system operating well. One is that if the second, rational, system tends to dominate the decision-making too much, the first, gene-based emotional system might fight back over generations of selection and enhance the magnitude of the reward value specified by the genes, so that emotions might actually become stronger as a consequence of them having to compete in the interests of the selfish genes with the rational decision-making process.

Another property of the system may be that sometimes the rational system cannot gain all the evidence that would be needed to make a rational choice. Under these circumstances, the rational system might fail to make a clear decision, and under these circumstances, basing a decision on the gene-specified emotions is an alternative. Indeed, Damasio (1994) argued that under circumstances such as this, emotions might take an important role in decision-making. In this respect, I agree with him, basing my reasons on the arguments above. He called the emotional feelings gut feelings, and, in contrast to me, hypothesized that actual feedback from the gut was involved. His argument seemed to be that if the decision was too complicated for the rational system, then send outputs to the viscera, and whatever is sensed by what they send back could be used in the decision-making and would account for the conscious feelings of the emotional states. My reading of the evidence is that the feedback from the periphery is not necessary for the emotional decision-making, or for the feelings, nor would it be computationally efficient to put the viscera in the loop given that the information starts from the brain, but that is a matter considered elsewhere (Rolls 2005b).

Another property of the system is that the interests of the second, rational, system, although involving a different form of computation, should not be too far from those of the gene-defined emotional system, for the arrangement to be stable in evolution by natural selection. One way that this could be facilitated would be if the gene-based goals felt pleasant or unpleasant in the rational system and in this way contributed to the operation of the second, rational, system. This is something that I propose is the case.

9.3.3 Decision-Making Between the Implicit and Explicit Systems

Decision-making as implemented in neural networks in the brain is now becoming understood and is described in Sect. 9.4. As shown there, two attractor states, each one corresponding to a decision, compete in an attractor single network with the evidence for each of the decisions acting as biases to each of the attractor states. The non-linear dynamics, and the way in which noise due to the random spiking of neurons makes the decision-making probabilistic, makes this a biologically plausible model of decision-making consistent with much neurophysiological and fMRI data (Wang 2002; Deco and Rolls 2006; Deco et al. 2009; Rolls and Deco 2010).

I propose (Rolls 2005b, 2008b) that this model applies to taking decisions between the implicit (unconscious) and explicit (conscious) systems in emotional decision-making, where the two different systems could provide the biasing inputs λ_1 and λ_2 to the model. An implication is that noise will influence with probabilistic outcomes which system takes a decision.

When decisions are taken, sometimes confabulation may occur, in that a verbal account of why the action was performed may be given, and this may not be related at all to the environmental event that actually triggered the action (Gazzaniga and LeDoux 1978; Gazzaniga 1988, 1995; Rolls 2005b; LeDoux 2008). It is accordingly possible that sometimes in normal humans when actions are initiated as a result of processing in a specialized brain region such as those involved in some types of rewarded behaviour, the language system may subsequently elaborate a coherent account of why that action was performed (i.e. confabulate). This would be consistent with a general view of brain evolution in which, as areas of the cortex evolve, they are laid on top of existing circuitry connecting inputs to outputs, and in which each level in this hierarchy of separate input–output pathways may control behaviour according to the specialized function it can perform.

9.3.4 Free Will

These thoughts raise the issue of free will in decision-making.

First, we can note that in so far as the brain operates with some degree of randomness due to the statistical fluctuations produced by the random spiking times of neurons, brain function is to some extent non-deterministic, as defined in terms of these statistical fluctuations. That is, the behaviour of the system, and of the individual, can vary from trial to trial based on these statistical fluctuations, in ways that are described in this book. [Philosophers may wish to argue about different senses of the term deterministic, but is it being used here in a precise, scientific and quantitative way, which has been clearly defined (Rolls and Deco 2010).]

Second, do we have free will when both the implicit and the explicit systems have made the choice? Free will would in Rolls' view (2005b) involve the use of language to check many moves ahead on a number of possible series of actions and their outcomes and then with this information to make a choice from the likely outcomes of different possible series of actions. (If in contrast choices were made only on the basis of the reinforcement value of immediately available stimuli, without the arbitrary syntactic symbol manipulation made possible by language, then the choice strategy would be much more limited, and we might not want to use the term free will, as all the consequences of those actions would not have been computed.) It is suggested that when this type of reflective, conscious, information processing is occurring and leading to action, the system performing this processing and producing the action would have to believe that it could cause the action, for otherwise inconsistencies would arise, and the system might no longer try to initiate action. This belief held by the system may partly underlie the feeling of free will. At other times, when other brain modules are initiating actions (in the implicit systems), the conscious processor (the explicit system) may confabulate and believe that it caused the action, or at least give an account (possibly wrong) of why the action was initiated. The fact that the conscious processor may have the belief even in these circumstances that it initiated the action may arise as a property of it being inconsistent for a system that can take overall control using conscious verbal processing to believe that it was overridden by another system. This may be the reason why confabulation occurs.

The interesting view we are led to is thus that when probabilistic choices influenced by stochastic dynamics are made between the implicit and explicit systems, we may not be aware of which system made the choice. Further, when the stochastic noise has made us choose with the implicit system, we may confabulate and say that we made the choice of our own free will and provide a guess at why the decision was taken. In this scenario, the stochastic dynamics of the brain plays a role even in how we understand free will.

9.4   Decision-Making and "Subjective Confidence"

Animals including humans can not only take decisions, but they can then make further decisions based on their estimates of their confidence or certainty in the decision just taken. It has been argued that the ability to make confidence estimates "objectively measures awareness" (Koch and Preuschoff 2007; Persaud et al. 2007). This process is sometimes called "subjective confidence", referring to the fact that one can report on the confidence one has in one's decisions. But does estimating the confidence in a decision really provide a measure of consciousness or require it? The process of confidence estimation has been described in animals including monkeys and rodents, who may, for example, terminate a trial if they estimate that a wrong decision may have been made so that they can get on to the next trial (Hampton 2001; Hampton et al. 2004; Kepecs et al. 2008). Does this really imply subjective (i.e. conscious) awareness (Heyes 2008)?

We have now developed an understanding of how probabilistic decision-making may be implemented in the brain by a single attractor network, and how adding a second attractor network allows the system to take a decision based on the confidence in the decision that emerges in the neuronal firing during the decision-making in the first network (Insabato et al. 2010). We describe these developments next. We note that there is no reason to believe that a system with two attractor networks is consciously aware, yet this system can account for confidence estimation and decisions made based on this, that is, what is described as "subjective confidence".

9.4.1 Neural Networks for Decision-Making That Reflect "Subjective Confidence" in Their Firing Rates

In spite of the success of phenomenological models for accounting for decision-making performance (Smith and Ratcliff 2004), a crucial problem that they present is the lack of a link between the model variables and parameters and the biological substrate. Recently, a series of biologically plausible models, motivated and constrained by neurophysiological data, have been formulated to establish an explicit link between behaviour and neuronal activity (Wang 2002; Deco and Rolls 2006; Wong and Wang 2006; Rolls and Deco 2010; Rolls et al. 2010a, b). The way in which these integrate-and-fire neuronal network models operate is as follows.

An attractor network of the type illustrated in Fig. 9.2a is set up to have two possible high firing rate attractor states, one for each of the two decisions. The evidence for each decision (1 vs. 2) biases each of the two attractors via the external inputs λ_1 and λ_2. The attractors are supported by strengthened synaptic connections in the recurrent collateral synapses between the (e.g. cortical pyramidal) neurons activated when λ_1 is applied or when λ_2 is applied. (This is an associative or Hebbian process set up during a learning stage by a process like long-term potentiation.) Inhibitory interneurons (not shown in Fig. 9.2a) receive inputs from the pyramidal neurons and make negative feedback connections onto the pyramidal cells to control their activity. When inputs λ_1 and λ_2 are applied, there is positive feedback via the recurrent collateral connections and competition implemented through the inhibitory interneurons so that there can be only one winner. The network starts in a low spontaneous state of firing. When λ_1 and λ_2 are applied, there is competition between the two attractors, each of which is pushed towards a high firing rate state, and eventually, depending on the relative strength of the two inputs and the noise in the network caused by the random firing times of the neurons, one of the attractors will win the competition, and it will reach a high firing rate state, with the firing of the neurons in the other attractor inhibited to a low firing rate. The process is illustrated in Fig. 9.3. The result is a binary decision, with one group of neurons due to the positive feedback firing at a high firing rate, and the neurons corresponding to the other decision firing with very low rates. Because it is a non-linear positive feedback system, the final firing rates are in what is effectively a binary decision state, of high firing rate or low firing rate, and do not reflect the exact relative values of the two inputs λ_1 and λ_2 once the decision is reached. The noise in the network due to the random spiking of the neurons is important to the operation of the network, because it enables the network to jump out of a stable spontaneous rate of firing to a high firing rate and to do so probabilistically, depending on whether on a particular trial there is relatively more random firing in the neurons of one attractor than the other attractor. This can be understood in terms of energy landscapes, where each attractor (the spontaneous state and the two high firing rate attractors) is a low energy basin, and the spiking noise helps the system to jump over an energy barrier into another energy minimum, as illustrated in Fig. 9.2c. If λ_1 and λ_2 are equal, then the decision that is taken is random and probabilistic, with the noise in each attractor determining which decision is taken on a particular trial. If one of the inputs is larger than

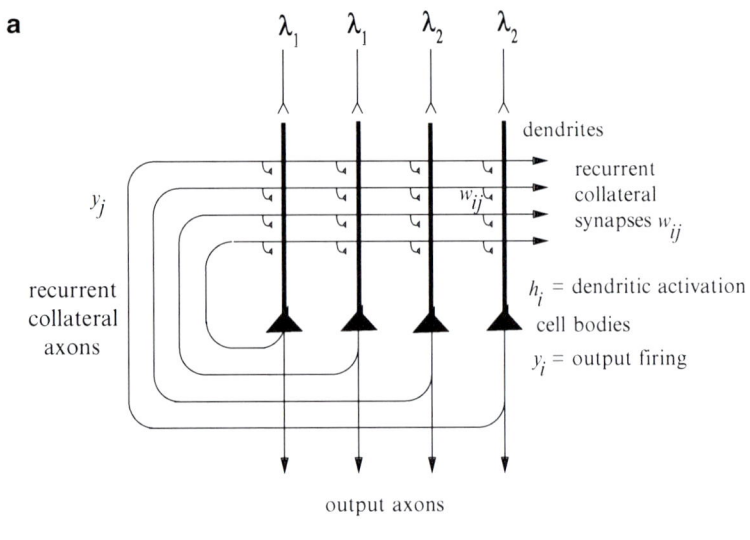

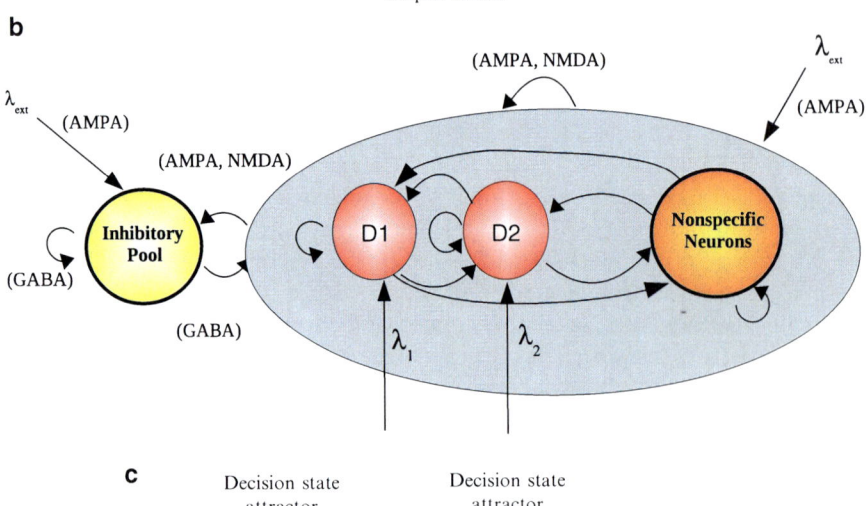

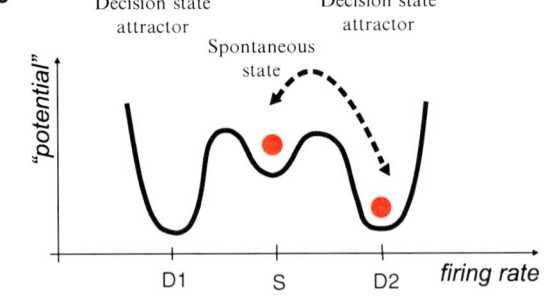

Fig. 9.2 (a) Attractor or autoassociation single network architecture for decision-making. The evidence for decision 1 is applied via the λ_1 inputs and for decision 2 via the λ_2 inputs. The synaptic weights $w_i j$ have been associatively modified during training in the presence of λ_1 and at a different time of λ_2.

the other, then the decision is biased towards it but is still probabilistic. Because this is an attractor network, it has short term memory properties implemented by the recurrent collaterals, which tend to promote a state once it is started, and these help it to maintain the firing once it has reached the decision state, enabling a suitable action to be implemented even if this takes some time.

In the multi-stable regime investigated by Deco and Rolls (2006), the spontaneous firing state is stable even when the decision cues are being applied, and noise-related fluctuations are essential for decision-making (Deco et al. 2009; Rolls and Deco 2010).

The process of decision-making in this system, and how the certainty or confidence of the decision is represented by the firing rates of the neurons in the network, is illustrated in Fig. 9.3. Figure 9.3a, e shows the mean firing rates of the two neuronal populations D1 and D2 for two trial types, easy trials ($\Delta I = 160$ Hz) and difficult trials ($\Delta I = 0$) (where ΔI is the difference in spikes/s summed across all synapses to each neuron between the two inputs, λ_1 to population D1 and λ_2 to population D2). The results are shown for correct trials, that is, trials on which the D1 population won the competition and fired with a rate for > 10 spikes/s for the last 1,000 ms of the simulation runs. Figure 9.3b shows the mean firing rates of the four populations of neurons on a difficult trial, and Fig. 9.3c shows the rastergrams for the same trial. Figure 9.3d shows the firing rates on another difficult trial ($\Delta I = 0$) to illustrate the variability shown from trial to trial, with on this trial prolonged competition between the D1 and D2 attractors until the D1 attractor finally won after approximately 1,100 ms. Figure 9.3f shows firing rate plots for the four neuronal populations on an example of a single easy trial ($\Delta I = 160$), Fig. 9.3g shows the synaptic currents in the four neuronal populations on the same trial, and Fig. 9.3h shows rastergrams for the same trial.

Three important points are made by the results shown in Fig. 9.3. First, the network falls into its decision attractor faster on easy trials than on difficult trials. Reaction times are thus shorter on easy than on difficult trials. Second, the mean

---

Fig. 9.2 (continued) When λ_1 and λ_2 are applied, each attractor competes through the inhibitory interneurons (not shown), until one wins the competition, and the network falls into one of the high firing rate attractors that represents the decision. The noise in the network caused by the random spiking of the neurons means that on some trials, for given inputs, the neurons in the decision 1 (D1) attractor are more likely to win, and on other trials the neurons in the decision 2 (D2) attractor are more likely to win. This makes the decision-making probabilistic, for, as shown in (**c**), the noise influences when the system will jump out of the spontaneous firing stable (low energy) state S, and whether it jumps into the high firing state for decision 1 (D1) or decision 2 (D2). (**b**) The architecture of the integrate-and-fire network used to model decision-making (see text). (**c**) A multi-stable "effective energy landscape" for decision-making with stable states shown as low "potential" basins. Even when the inputs are being applied to the network, the spontaneous firing rate state is stable, and noise provokes transitions into the high firing rate decision attractor state D1 or D2 (see Rolls and Deco 2010)

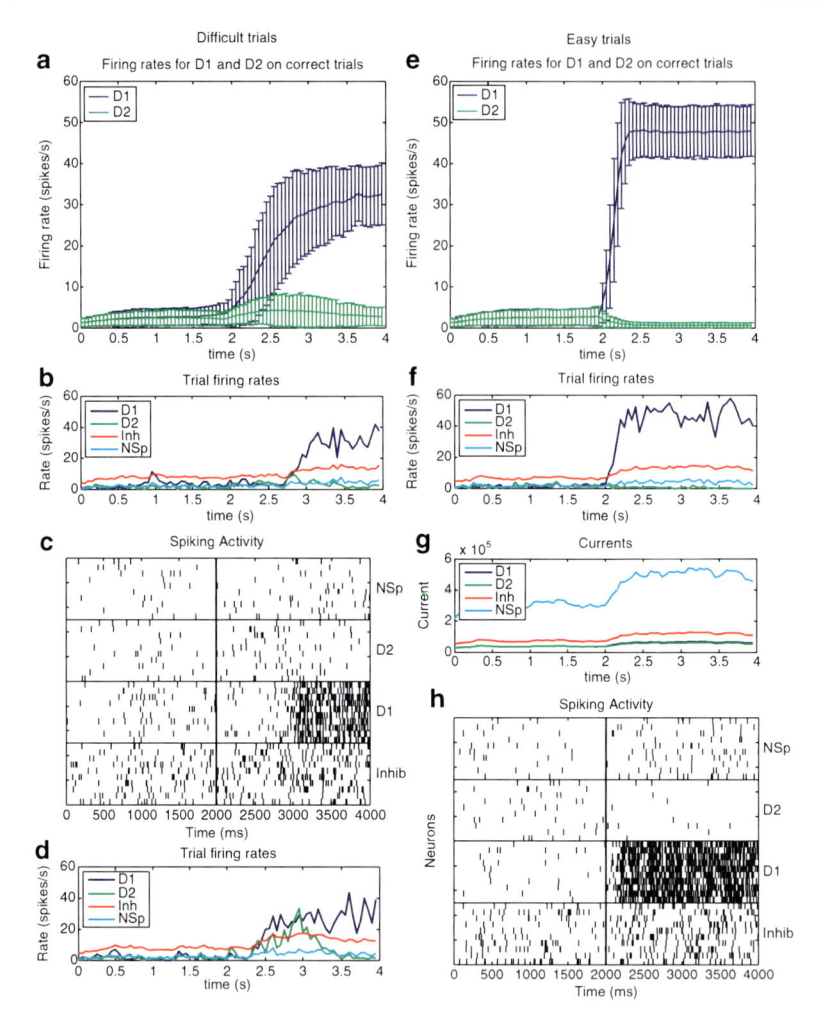

Fig. 9.3 (**a** and **e**) Firing rates (mean ± sd) for difficult ($\Delta I = 0$) and easy ($\Delta I = 160$) trials. The period 0–2 s is the spontaneous firing, and the decision cues were turned on at time = 2 s. The mean was calculated over 1,000 trials. D1: firing rate of the D1 population of neurons on correct trials on which the D1 population won. D2: firing rate of the D2 population of neurons on the correct trials on which the D1 population won. A correct trial was one in which in which the mean rate of the D1 attractor averaged >10 spikes/s for the last 1,000 ms of the simulation runs. (Given the attractor nature of the network and the parameters used, the network reached one of the attractors on >90% of the 1,000 trials, and this criterion clearly separated these trials, as indicated by the mean rates and standard deviations for the last s of the simulation as shown.) (**b**) The mean firing rates of the four populations of neurons on a difficult trial. Inh is the inhibitory population that uses GABA as a transmitter. NSp is the non-specific population of neurons (see Fig. 9.2). (**c**) Rastergrams for the trial shown in (**b**) 10 neurons from each of the four pools of neurons are shown. (**d**) The firing rates on another difficult trial ($\Delta I = 0$) showing prolonged competition between the D1 and D2 attractors until the D1 attractor finally wins after approximately 1,100 ms. (**f**) Firing rate plots for the 4 neuronal populations on a single easy trial ($\Delta I = 160$). (**g**) The synaptic currents in the four neuronal populations on the trial shown in (**f**). (**h**) Rastergrams for the easy trial shown in (**f** and **g**)

firing rate of the winning attractor after the network has settled into the correct decision attractor is higher on easy trials (with large ΔI, and when certainty and confidence are high) than on difficult trials. This is because the exact firing rate in the attractor is a result not only of the internal recurrent collateral effect, but also of the external input to the neurons, which in Fig. 9.3 is 32 Hz to each neuron (summed across all synapses) of D1 and D2, but in Fig. 9.3a is increased by 80 Hz to D1 and decreased by 80 Hz to D2 (i.e. the total external input to the network is the same, but $\Delta I = 0$ for Fig. 9.3a and $.\Delta I = 160$ for Fig. 9.3b). Third, the variability of the firing rate is high, with the standard deviations of the mean firing rate calculated in 50 ms epochs indicated in order to quantify the variability. The large standard deviations on difficult trials for the first second after the decision cues are applied at $t = 2s$ reflects the fact that on some trials the network has entered an attractor state after 1,000 ms, but on other trials it has not yet reached the attractor, although it does so later. This trial-by-trial variability is indicated by the firing rates on individual trials and the rastergrams in the lower part of Fig. 9.3.

The effects evident in Fig. 9.3 are quantified and elucidated over a range of values for ΔI elsewhere (Rolls et al. 2010a, b). They show that a continuous-valued representation of decision certainty or decision confidence is encoded in the firing rates of the neurons in a decision-making attractor.

9.4.2 A Model for Decisions About Confidence Estimates

We have seen that a continuous-valued representation of decision certainty or decision confidence is encoded in the firing rates of the neurons in a decision-making attractor. What happens if instead of having to report or assess the continuous-valued representation of confidence in a decision one has taken, one needs to take a decision based on one's confidence estimate that one has just made a correct or incorrect decision? One might, for example, wait for a reward if one thinks one's decision was correct, or alternatively stop waiting on that trial and start another trial or action. We suggest that in this case, one needs a second decision-making network that takes decisions based on one's decision confidence (Insabato et al. 2010).

The architecture has a decision-making network, and a separate confidence decision network that receives inputs from the decision-making network, as shown in Fig. 9.4. The decision-making network has two main pools or populations of neurons, D1 which become active for decision 1, and D2 which become active for decision 2. Pool D1 receives sensory information about stimulus 1 (e.g. odor A), and Pool D2 receives sensory information about stimulus 2 (e.g. odor B). Each of these pools has strong recurrent collateral connections between its own neurons, so that each operates as an attractor population. There are inhibitory neurons with global connectivity to implement the competition between the attractor subpopulations. When stimulus 1 is applied, pool D1 will usually win the competition and end up with high firing indication that decision 1 has been reached. When stimulus 2 is applied, pool D2 will usually win the competition and end up

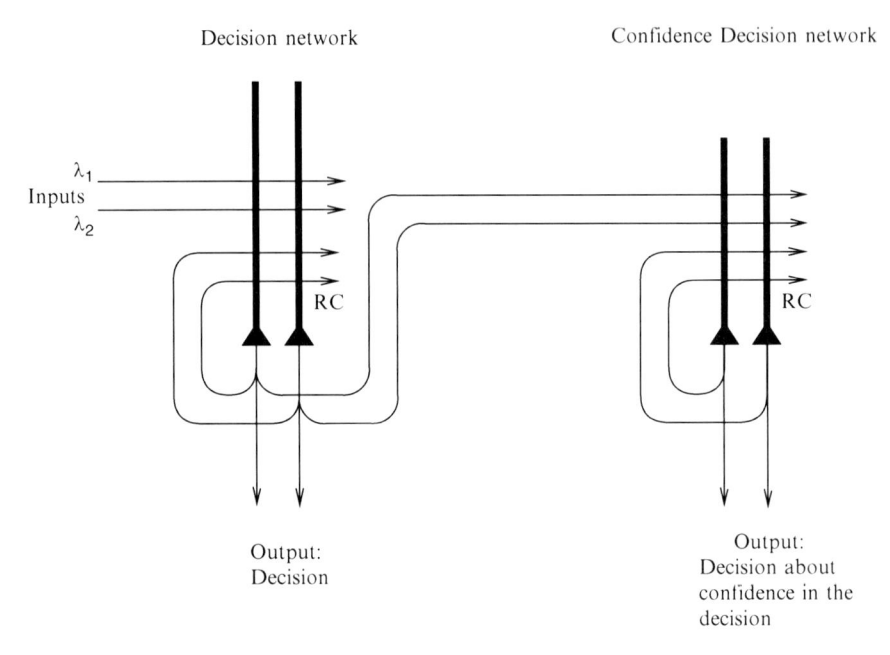

Fig. 9.4 Decisions about confidence estimates. The first network is a decision-making network, and its outputs are sent to a second network that makes decisions based on the firing rates from the first network, which reflect the decision confidence. In the first network, high firing of population D_1 represents decision 1, and high firing of population D_2 represents decision 2. The second network is a confidence decision network and receives inputs from the first network. The confidence network has two selective pools of neurons, one of which, C, responds to represent confidence in the decision, and the other of which responds when there is little or a lack of confidence in the decision (LC). In each network of the integrate-and-fire simulations, the excitatory pool is divided into three subpopulations: a non-specific one and two stimulus selective populations. The selective pools are endowed with strong recurrent connections (w_+), while the connections between the two selective pools are weak (w_-). All other connections are set to the default value 1. All neurons in the network receive an input (λ_{ext}) emulating the spontaneous firing of neurons in other cortical areas. Pools D_1 and D_2 receive also a stimulus related input (respectively λ_1 and λ_2 which is the information that will drive the decision) (After Insabato et al. 2010)

with high firing indication that decision 2 has been reached. When a mixture is applied, the decision-making network will probabilistically choose stimulus 1 or 2 influenced by the proportion of stimulus 1 and 2 in the mixture. The decision-making is probabilistic because the neurons in the network have approximately Poisson spike time firings which are a source of noise, which, in a finite size system, cause coherent statistical fluctuations, as described in more detail elsewhere (Deco et al. 2009; Rolls and Deco 2010). The proportion of correct decisions increases as the proportion of stimulus 1 and 2 in the mixture is altered from 50% (corresponding to $\Delta I = 0$) to 100% of one and 0% of the other (corresponding to a large ΔI). The firing rates of the neurons in the two selective populations as a function of the proportion of stimulus 1 and 2 in the mixture are similar to those illustrated in Fig. 9.3. The neurons that win on correct trials have higher firing rates as the difference in the proportion of A (stimulus 1) and B (stimulus 2) in the mixture, which alters ΔI, becomes larger in magnitude. (ΔI is defined as $(A - B)/((A + B)/2)$. ΔI is thus large and positive if only A is present, is large and negative if only B is present, and is 0 if A and B are present in equal proportions.) The reason that the firing rates of the winning pool become higher as ΔI becomes larger in magnitude is that the external inputs from the stimuli 1 or 2 then support the winning attractor and add to the firing rates being produced by the recurrent collateral connections in the winning attractor. On the other hand, the firing rates of the winning pool become lower on error trials as ΔI increases, because then the external sensory inputs are inconsistent with the decision that has been taken and do not support and increase the firing rate of the winning pool.

The decision network sends outputs from its decision-making selective pools D1 and D2 to the confidence network. The confidence network has two selective pools of neurons, one of which (C) responds to represent confidence in the decision, and the other of which responds when there is little or a lack of confidence in the decision (LC). If the output firing of D1 and D2 is high because the decision just taken has sensory inputs consonant with the decision, then the confidence network acting as a second level network takes the decision, probabilistically as before, to have confidence in the decision, and the C population wins the competition. If the output firing of D1 and D2 is low because the decision just taken has sensory inputs that are not consonant with the decision, then the confidence network takes the decision, probabilistically as before, to have a lack of confidence in the decision, and the LC population wins the competition. The confidence network thus acts as a decision-making network to make confident decisions if the firing rates from the first, decision-making, network are high and to make lack of confidence decisions if the firing rates from the first, decision-making network are low.

In this situation, we find in integrate-and-fire simulations (Insabato et al. 2010) that on correct trials with high ΔI (easy decisions), C has a high firing rate, whereas it has a lower rate for $\Delta I = 0$, that is difficult decisions. Conversely, on error trials when the firing rates in the first level, decision-making, networks are lower, and the confidence neurons C have firing rates that decrease as the magnitude of ΔI increases. The LC attractor neurons do not receive inputs from the first level, decision-making, network, and thus through the inhibitory neurons has the opposite type of firing to the confidence pool C. That is, the firing rates of the LC are in general high on error trials (when the firing rates of the first-level neurons are low) and increase as ΔI increases (Insabato et al. 2010).

This new theoretical approach to confidence-related decision-making accounts for neuronal responses in the rat orbitofrontal cortex related to confidence in decisions (Kepecs et al. 2008). The rats had to perform a binary categorization task with a mixture of two pure odourants (A, caproic acid; B, 1-hexanol) by entering one of two ports to indicate that the mixture was more like odour A or odour B. Correct choices were rewarded after a delay of 0.3–2 s. Varying the relative concentration of the odourants allowed the difficulty of the trial to be altered. Neural activity related to decision confidence should occur just after the decision is taken and before the trial outcome. Kepecs et al., therefore, analyzed recordings of neuronal activity during a delay period after a decision had been taken before a reward was given. (The single neuron recordings were made in the rat orbitofrontal cortex, though exactly what area in primates and humans corresponds to the area in which recordings were made is not yet clear.) The neurons were divided into two groups based on whether they fired faster on correct or on error trials. Kepecs et al. found that the group of neurons with an increased firing rate on error trials had higher firing rates with easier stimuli. The same neurons fired at a substantially lower rate on correct trials, and on these trials the firing rates were lower when the decision was made easier. This produced opposing V-shaped curves. The authors argued that this pattern of activity encoded decision confidence.

In the experiment of Kepecs et al. (2008), C corresponds to a decision to stay and wait for a reward, i.e. what they call the positive outcome population, though it really represents confidence or a prediction that the decision just taken will have a positive outcome. LC corresponds to a decision to abort a trial and not wait for a possible reward, i.e. what they call the negative outcome population, though it really represents lack of confidence that the decision just taken will have a positive outcome.

Kepecs et al. (2008) then performed a second experiment to investigate if rats were able to make use of this information encoded by orbitofrontal cortex neurons. The delay period was prolonged up to 8 s in order to allow the rat to reinitiate the trial. The subject could decide to leave the choice port and repeat the trial or could wait for the reward. It was found that when the likelihood of a reward was low, due to the decision difficulty and the choice just made, the rat returned to the odour port. The probability that the rat would restart a trial as a function of stimulus difficulty and the choice just made were consistent with the responses of the neurons with activity described as encoding decision confidence: in our terminology, in taking decisions based on one's confidence in an earlier decision. The model makes predictions about the firing rates of these neuronal populations when the second, confidence decision, network itself makes an error due to the noise in the system (Insabato et al. 2010).

These results indicate that confidence estimates, previously suggested to "objectively measure awareness" (Koch and Preuschoff 2007; Persaud et al. 2007), can be computed with relatively simple operations, involving the firing rates of the neurons when a decision-making network falls into an attractor. Moreover, adding a second attractor decision-making network even enables decisions to be made about the confidence in the first decision. There would seem to be no awareness in either of the networks. The implication is that we need to be very careful in ascribing awareness to processes that at first seem complex and closely tied to awareness. The implication in turn is that activity in a special and different processing system, argued above to be that capable of thoughts about thoughts (higher order syntactic thoughts, HOSTs), is what is occurring when we are aware, that is when we have phenomenal consciousness.

9.5 Oscillations and Stimulus-Dependent Neuronal Synchrony: Their Role in Information Processing in the Ventral Visual System and in Consciousness

We now turn in Sects. 5–7 to some other computational issues related to the implementation of consciousness in the brain. The first is on whether oscillations are important in consciousness.

It has been suggested that syntax in real neuronal networks is implemented by temporal binding (see Malsburg 1990), which would be evident as, for example, stimulus-dependent synchrony (Singer 1999). According to this hypothesis, the binding between features common to an object could be implemented by neurons coding for that object firing in synchrony, whereas if the features belong to different objects, the neurons would not fire in synchrony. Crick and Koch (1990) postulated that oscillations and synchronization are necessary bases of consciousness. It is difficult to see what useful purpose oscillations per se could perform for neural information processing, apart from perhaps resetting a population of neurons to low activity so that they can restart some attractor process (see e.g. Rolls and Treves 1998), acting as a reference phase to allow neurons to provide some additional information by virtue of the time that they fire with respect to the reference waveform (Huxter et al. 2003) or increasing spike numbers (Smerieri et al. 2010). Neither putative function seems to be closely related to consciousness. However, stimulus-dependent synchrony, by implementing binding, a function that has been related to attention (Treisman 1996) might perhaps be related to consciousness. Let us consider the evidence on whether stimulus-dependent synchrony between neurons provides significant information related to object recognition and top-down attention in the ventral visual system.

This has been investigated by developing information theoretic methods for measuring the information present in stimulus-dependent synchrony (Panzeri et al. 1999; Rolls et al. 2003; Franco et al. 2004) and applying them to the analysis of neuronal activity in the macaque inferior temporal visual cortex during object recognition and attention. This brain region represents both features, such as parts of objects and faces, and whole objects in which the features must be bound in the correct spatial relationship for the neurons to respond (Rolls and Deco 2002; Rolls 2007b, 2008b). It has been shown that simultaneously recorded single neurons do sometimes show stimulus-dependent synchrony, but that the information available is less than 5% of that available from the spike counts (Rolls et al. 2003, 2004; Franco et al. 2004; Aggelopoulos et al. 2005; Rolls 2008b).

The neurophysiological studies performed have included situations in which feature binding is likely to be needed, that is when the monkey had to choose to touch one of two simultaneously presented objects, with the stimuli presented in a complex natural background in a top-down attentional task (Aggelopoulos et al. 2005). We found that between 99 and 94% of the information was present in the firing rates of inferior temporal cortex neurons, and less than 5% in any stimulus-dependent synchrony that was present, as illustrated in Fig. 9.3. The implication of these results is that any stimulus-dependent synchrony that is present is not quantitatively important as measured by information theoretic analyses under natural scene conditions.

The point of the experimental design used was to test whether when the visual system is operating normally, in natural scenes and even searching for a particular object, stimulus-dependent synchrony is quantitatively important for encoding information about objects, and it was found not to be in the inferior temporal visual cortex. It will be of interest to apply the same quantitative information theoretic methods to earlier cortical visual areas, but the clear implication of the findings is that even when features must be bound together in the correct relative spatial positions to form object representations, and these must be segmented from the background, then stimulus-dependent synchrony is not quantitatively important in information encoding (Aggelopoulos et al. 2005; Rolls 2008b). Further, it was shown that there was little redundancy (less than 6%) between the information provided by the spike counts of the simultaneously recorded neurons, making spike counts an efficient population code with a high encoding capacity (Rolls 2008b).

The findings (Aggelopoulos et al. 2005) are consistent with the hypothesis that feature binding is implemented by neurons that respond to features in the correct relative spatial locations (Elliffe et al. 2002; Rolls and Deco 2002; Rolls 2008b) and not by temporal synchrony and attention (Malsburg 1990; Singer 1999). In any case, the computational point is that even if stimulus-dependent synchrony was useful for grouping, it would not without much extra machinery be useful for binding the relative spatial positions of features within an object or for that matter of the positions of objects in a scene which appears to be encoded in a different way by using receptive fields that become asymmetric in crowded scenes (Aggelopoulos and Rolls 2005). The computational problem is that synchronization does not by itself define the spatial relations between the features being bound, so is not as a binding mechanism adequate for shape recognition. For example, temporal binding might enable features 1, 2 and 3, which might define one stimulus to be bound together and kept separate from, for example, another stimulus consisting of features 2, 3 and 4, but would require a further temporal binding (leading in the end potentially to a combinatorial explosion) to indicate the relative spatial positions of the 1, 2 and 3 in the 123 stimulus, so that it can be discriminated from, for example, 312 (Rolls 2008b). However, the required computation for binding can be performed by the use of neurons that respond to combinations of features with a particular spatial arrangement (Elliffe et al. 2002; Rolls and Deco 2002; Rolls and Stringer 2006; Rolls 2008b).

Another type of evidence that stimulus-dependent neuronal synchrony is not likely to be crucial for information encoding, at least in the ventral visual system, is that the code about which visual stimulus has been shown can be read off from the end of the visual system in short times of 20–50 ms, and cortical neurons need fire for only this long during the identification of objects (Tovee et al. 1993; Rolls and Tovee 1994; Rolls et al. 1994b, 2006; Tovee and Rolls 1995; Rolls 2008b). These are rather short time windows for the expression of multiple separate populations of synchronized neurons.

If large populations of neurons become synchronized, oscillations are likely to be evident in cortical recordings. In fact, oscillations are not an obvious property of neuronal firing in the primate temporal cortical visual areas involved in the representation of faces and objects when the system is operating normally in the awake behaving macaque (Tovee and Rolls 1992). The fact that oscillations and neuronal synchronization are especially evident in anaesthetized cats does *not* impress as strong evidence that oscillations and synchronization are critical features of consciousness, for most people would hold that anaesthetized cats are not conscious. The fact that oscillations and synchronization are much more difficult to demonstrate in the temporal cortical visual areas of awake behaving monkeys (Aggelopoulos et al. 2005) might just mean that during evolution to primates the cortex has become better able to avoid parasitic oscillations, as a result of developing better feedforward and feedback inhibitory circuits (see Rolls and Deco 2002; Rolls and Treves 1998).

However, in addition there is an interesting computational argument against the utility of oscillations. The computational argument is related to the speed of information processing in cortical circuits with recurrent collateral connections. It has been shown that if attractor networks have integrate-and-fire neurons and spontaneous activity, then memory recall into a basin of attraction can occur in approximately 1.5 time constants of the synapses, i.e. in times in the order of 15 ms (Treves 1993; Simmen et al. 1996; Battaglia and Treves 1998; Rolls and Treves 1998; Panzeri et al. 2001). One factor in this rapid dynamics of autoassociative networks with brain-like integrate-and-fire membrane and synaptic properties is that with some spontaneous activity, some of the neurons in the network are close to threshold already before the recall cue is applied, and hence some of the neurons are very quickly pushed by the recall cue into firing, so that information starts to be exchanged very rapidly (within 1–2 ms of brain time) through the modified synapses by the neurons in the network. The progressive exchange of information starting early on within what would otherwise be thought of as an iteration period (of perhaps 20 ms, corresponding to a neuronal firing rate of 50 spikes/s) is the mechanism accounting for rapid recall in an autoassociative neuronal network made biologically realistic in this way. However, this process relies on spontaneous random firings of different neurons, so that some will always be close to threshold when the retrieval cue is applied. If many of the neurons were firing synchronously in an oscillatory pattern, then there might be no neurons close to threshold and ready to be activated by the retrieval cue, so that the network might act much more like a discrete time network with fixed timesteps, which typically takes 8–15 iterations to settle, equivalent to perhaps 100 ms of brain time, and much too slow for cortical processing within any one area (Rolls and Treves 1998; Panzeri et al. 2001; Rolls 2008b). The implication is that oscillations would tend to be detrimental to cortical computation by slowing down any process using attractor dynamics. Attractor dynamics are likely to be implemented not only by the recurrent collateral connections between pyramidal neurons in a given cortical area but also by the reciprocal feedforward and feedback connections between adjacent layers in cortical processing hierarchies (Rolls 2008b). However, if oscillations increase spike numbers in a process like stochastic resonance, this could speed information processing (Smerieri et al. 2010).

Another computational argument is that it is possible to account for many aspects of attention, including the non-linear interactions between top-down and bottom-up inputs, in integrate-and-fire neuronal networks that do not oscillate or show stimulus-dependent synchrony (Deco and Rolls 2005a, b, c; Rolls 2008b).

The implication of these findings is that stimulus-dependent neuronal synchro-nization and oscillatory activity are unlikely to be quantitatively important in cortical processing, at least in the ventral visual stream. To the extent that we can be conscious of activity that has been processed in the ventral visual stream (made evident, for example, by reports of the appearance of objects), stimulus-dependent synchrony and oscillations are unlikely to be important in the neural mechanisms of consciousness.

9.6   A Neural Threshold for Consciousness: The Neurophysiology of Backward Masking

Damage to the primary (striate) visual cortex can result in blindsight, in which patients report that they do not see stimuli consciously, yet when making forced choices can discriminate some properties of the stimuli such as motion, position, some aspects of form, and even face expression (Azzopardi and Cowey 1997; Weiskrantz 1997, 1998; De Gelder et al. 1999). In normal human subjects, back-ward masking of visual stimuli, in which another visual stimulus closely follows the short presentation of a test stimulus, reduces the visual perception of the test visual stimulus, and this paradigm has been widely used in psychophysics (Humphreys and Bruce 1991). In this section, I consider how much information is present in neuronal firing in the part of the visual system that represents faces and objects, the inferior temporal visual cortex (Rolls and Deco 2002; Rolls 2008b), when human subjects can discriminate face identity in forced choice testing but cannot consciously per-ceive the person's face, and how much information is present when they become conscious of perceiving the stimulus. From this evidence, I argue that even *within* a particular processing stream the processing may not be conscious yet can lead to behaviour, and that with higher and longer neuronal firing, events in that system be-come conscious. From this evidence, I argue that the threshold for consciousness is normally higher than for some behavioural response. I then suggest a computational hypothesis for why this might be adaptive. A fuller account of this issue is available elsewhere (Rolls 2007a).

9.6.1   The Neurophysiology and Psychophysics of Backward Masking

The responses of single neurons in the macaque inferior temporal visual cortex have been investigated during backward visual masking (Rolls and Tovee 1994; Rolls et al. 1994b). Recordings were made from neurons that were selective for faces, using distributed encoding (Rolls and Deco 2002; Rolls 2007b, 2008b), during pre-sentation of a test stimulus, a face, that lasted for 16 ms. The test stimulus was followed on different trials by a mask with stimulus onset asynchrony (S.O.A.)

values of 20, 40, 60, 100 or 1,000 ms. (The S.O.A. is the time between the onset of the test stimulus and the onset of the mask.) The duration of the pattern masking stimulus (letters of the alphabet) was 300 ms, and the neuron did not respond to the masking stimulus.

One important conclusion from these results is that the effect of a backward masking stimulus on cortical visual information processing is to limit the duration of neuronal responses by interrupting neuronal firing. This persistence of cortical neuronal firing when a masking stimulus is not present is probably related to cortical recurrent collateral connections which could implement an autoassociative network with attractor and short-term memory properties (see Rolls and Treves 1998; Rolls and Deco 2002; Rolls 2008b), because such continuing post-stimulus neuronal firing is not observed in the lateral geniculate nucleus (K. Martin, personal communication).

Information theoretic analyses (Rolls et al. 1999) showed that as the S.O.A. is reduced towards 20 ms the information does reduce rapidly, but that nevertheless at an S.O.A. of 20 ms there is still considerable information about which stimulus was shown. Rolls et al. (1994b) performed human psychophysical experiments with the same set of stimuli and with the same apparatus used for the neurophysiological experiments so that the neuronal responses could be closely related to the identification that was possible of which face was shown. Five different faces were used as stimuli. All the faces were well known to each of the eight observers who participated in the experiment. In the forced choice paradigm, the observers specified whether the face was normal or rearranged (i.e. with the features jumbled) and identified whose face they thought had been presented. Even if the observers were unsure of their judgement they were instructed to respond with their best guess. The data were corrected for guessing. Forced choice discrimination of face identity was better than chance at an S.O.A. of 20 ms. However, at this S.O.A., the subjects were not conscious of seeing the face, or of the identity of the face, and felt that their guessing about which face had been shown was not correct. The subjects did know that something had changed on the screen (and this was not just brightness, as this was constant throughout a trial). Sometimes the subjects had some conscious feeling that a part of a face (such as a mouth) had been shown. However, the subjects were not conscious of seeing a whole face or of seeing the face of a particular person. At an S.O.A. of 40 ms, the subjects' forced choice performance of face identification was close to 100%, and at this S.O.A., the subjects became much more consciously aware of the identity of which face had been shown (Rolls et al. 1994b).

Comparing the human performance purely with the changes in firing rate under the same stimulus conditions suggested that when it is just possible to identify which face has been seen, neurons in a given cortical area may be responding for only approximately 30 ms (Rolls and Tovee 1994; Rolls et al. 1994b; Rolls 2007a). The implication is that 30 ms is enough time for a neuron to perform sufficient computation to enable its output to be used for identification. When the S.O.A. was increased to 40 ms, the inferior temporal cortex neurons responded for approximately 50 ms and encoded approximately 0.14 bits of information. At this S.O.A., not only was face identification 97% correct, but the subjects were much more likely to be able to report consciously seeing a face and/or whose face had been shown. One further way in which the conscious perception of the faces was measured quantitatively was by asking subjects to rate the clarity of the faces. This was a subjective assessment and therefore reflected conscious processing and was made using magnitude estimation. It was found that the subjective clarity of the stimuli was low at 20 ms S.O.A., was higher at 40 ms S.O.A. and was almost complete by 60 ms S.O.A (Rolls et al. 1994b; Rolls 2003, 2005a, 2007a).

It is suggested that the threshold for conscious visual perception may be set to be higher than the level at which small but significant sensory information is present so that the systems in the brain that implement the type of information processing involved in conscious thoughts are not interrupted by small signals that could be noise in sensory pathways. It is suggested that the processing related to consciousness involves a HOST system used to correct first order syntactic thoughts, and that these processes are inherently serial because of the way that the binding problems associated with the syntactic binding of symbols are treated by the brain. The argument is that it would be inefficient and would not be adaptive to interrupt this serial processing if the signal was very small and might be related to noise. Interruption of the serial processing would mean that the processing would need to start again, as when a train of thought is interrupted. The small signals that do not interrupt conscious processing but are present in sensory systems may nevertheless be useful for some implicit (non-conscious) functions, such as orienting the eyes towards the source of the input, and may be reflected in the better than chance recognition performance at short S.O.A.s even without conscious awareness.

9.6.2   The Relation to Blindsight

These quantitative analyses of neuronal activity in an area of the ventral visual system involved in face and object identification which show that significant neuronal processing can occur that is sufficient to support forced choice but implicit (unconscious) discrimination in the absence of conscious awareness of the identity of the face is of interest in relation to studies of blindsight (Azzopardi and Cowey 1997; Weiskrantz 1997, 1998; De Gelder et al. 1999). It has been argued that the results in blindsight are not due just to reduced visual processing, because some aspects of visual processing are less impaired than others (Azzopardi and Cowey 1997; Weiskrantz 1997,1998, 2001). It is though suggested that some of the visual capacities that do remain in blindsight reflect processing via visual pathways that are alternatives to the V1 processing stream (Weiskrantz 1997, 1998, 2001). If some of those pathways are normally involved in implicit processing, this may help to give an account of why some implicit (unconscious) performance is possible in blindsight patients. Further, it has been suggested that ventral visual stream processing is especially involved in consciousness, because it is information about objects and faces that needs to enter a system to select and plan actions (Milner and Goodale 1995; Rolls 2008b), and the planning of actions that involves the operation and correction of flexible one-time multiple-step plans may be closely related to conscious processing (Rolls 1999b, 2005b, 2008b). In contrast, dorsal stream visual processing may be more closely related to executing an action on an object once the action has been selected, and the details of this action execution can take place implicitly (unconsciously) (Milner and Goodale 1995; Rolls 2008b), perhaps because they do not require multiple step syntactic planning (Rolls 1999b, 2005b, 2008b).

One of the implications of blindsight thus seems to be that some visual pathways are more involved in implicit processing and other pathways in explicit processing. In contrast, the results described here suggest that short and information-poor signals in a sensory system involved in conscious processing do not reach consciousness and do not interrupt ongoing or engage conscious processing. This evidence described here thus provides interesting and direct evidence that there may be a threshold for activity in a sensory stream that must be exceeded in order to lead to consciousness, even when that activity is sufficient for some types of visual processing such as visual identification of faces at well above chance in an implicit mode. The latter implicit mode processing can be revealed by forced choice tests and by direct measurements of neuronal responses. Complementary evidence at the purely psychophysical level using backward masking has been obtained by Marcel (1983a, b) and discussed by Weiskrantz (1998, 2001). Possible reasons for this relatively high threshold for consciousness are considered above.

9.7 The Speed of Visual Processing Within a Cortical Visual Area Shows That Top-Down Interactions with Bottom-Up Processes Are Not Essential for Conscious Visual Perception

The results of the information analysis of backward masking (Rolls et al. 1999) emphasize that very considerable information about which stimulus was shown is available in a short epoch of, for example, 50 ms of neuronal firing. This confirms and is consistent with many further findings on the speed of processing of inferior temporal cortex neurons (Tovee et al. 1993; Tovee and Rolls 1995; Rolls et al. 2006; Rolls 2008b) and facilitates the rapid read-out of information from the inferior temporal visual cortex. One direct implication of the 30 ms firing with the 20 ms S.O.A. is that this is sufficient time both for a cortical area to perform its computation and for the information to be read out from a cortical area, given that psychophysical performance is 50% correct at this S.O.A. Another implication is that the recognition of visual stimuli can be performed using feedforward processing in the multi-stage hierarchically organized ventral visual system comprising at least V1–V2–V4–Inferior Temporal Visual Cortex, in that the typical shortest neuronal response latencies in macaque V1 are approximately 40 ms, and increase by approximately 15–17 ms per stage to produce a value of approximately 90 ms in the inferior temporal visual cortex (Dinse and Kruger 1994; Nowak and Bullier 1997;

Rolls and Deco 2002; Rolls 2008b). Given these timings, it would not be possible in the 20 ms S.O.A. condition for inferior temporal cortex neuronal responses to feed back to influence V1 neuronal responses to the test stimulus before the mask stimulus produced its effects on the V1 neurons. (In an example, in the 20 ms S.O.A. condition with 30 ms of firing, the V1 neurons would stop responding to the stimulus at $40 + 30 = 70$ ms, but would not be influenced by backprojected information from the inferior temporal cortex until $90 + (3 \text{ stages} \times 15 \text{ ms per stage}) = 135$ ms. In another example for conscious processing, in the 40 ms S.O.A. condition with 50 ms of firing, the V1 neurons would stop responding to the stimulus at $40 + 50 = 90$ ms, but would not be influenced by backprojected information from the inferior temporal cortex until $90 + (3 \text{ stages} \times 15 \text{ ms per stage}) = 135$ ms.) This shows that not only recognition, but also conscious awareness, of visual stimuli is possible without top-down backprojection effects from the inferior temporal visual cortex to early cortical processing areas that could interact with the processing in the early cortical areas.

The same information theoretic analyses (Tovee et al. 1993; Tovee and Rolls 1995; Rolls et al. 1999, 2006; Rolls 2008b) show that from the earliest spikes of the anterior inferior temporal cortex neurons described here after they have started to respond (at approximately 80 ms after stimulus onset), the neuronal response is specific to the stimulus, and it is only in more posterior parts of the inferior temporal visual cortex that neurons may have an earlier short period of firing (of perhaps 20 ms) which is not selective for a particular stimulus. The neurons described by Sugase et al. (1999) thus behaved like more posterior inferior temporal cortex neurons and not like typical anterior inferior temporal cortex neurons. This evidence thus suggests that in the anterior inferior temporal cortex, recurrent processing may help to sharpen up representations to minimize early non-specific firing (cf Lamme and Roelfsema 2000).

9.8 Comparisons with Other Approaches to Consciousness

The theory described here suggests that it feels like something to be an organism or machine that can think about its own (linguistic and semantically based) thoughts. It is suggested that qualia, raw sensory and emotional subjective feelings arise secondary to having evolved such a HOST system, and that sensory and emotional processing feels like something because it would be unparsimonious for it to enter the planning, HOST system and *not* feel like something. The adaptive value of having sensory and emotional feelings, or qualia, is thus suggested to be that such inputs are important to the long-term planning, explicit, processing system. Raw sensory feels, and subjective states associated with emotional and motivational states, may not necessarily arise first in evolution. Some issues that arise in relation to this theory are discussed by Rolls (2000, 2004b, 2005b); reasons why the ventral visual system is more closely related to explicit than implicit processing (because reasoning about objects may be important) are considered by Rolls (2003) and by

Rolls and Deco (2002); and reasons why explicit, conscious, processing may have a higher threshold in sensory processing than implicit processing are considered by Rolls (2003, 2005a, b).

I now compare this approach to consciousness with those that place emphasis on working memory (LeDoux 2008). LeDoux (1996), in line with Johnson-Laird (1988) and Baars (1988), emphasizes the role of working memory in consciousness, where he views working memory as a limited-capacity serial processor that creates and manipulates symbolic representations (p. 280). He thus holds that much emotional processing is unconscious, and that when it becomes conscious, it is because emotional information is entered into a working memory system. However, LeDoux (1996) concedes that consciousness, especially its phenomenal or subjective nature, is not completely explained by the computational processes that underlie working memory (p. 281).

LeDoux (2008) notes that the term "working memory" can refer to a number of different processes. In attentional systems, a short term memory is needed to hold on-line the subject of the attention, for example the position in space at which an object must be identified. There is much evidence that this short term memory is implemented in the prefrontal cortex by an attractor network implemented by associatively modifiable recurrent collateral connections between cortical pyramidal cells, which keep the population active during the attentional task. This short term memory then biases posterior perceptual and memory networks in the temporal and parietal lobes in a biased competition process (Miller and Cohen 2001; Rolls and Deco 2002; Deco and Rolls 2005a, b; Rolls 2008b). The operation of this type of short term memory acting using biased competition to implement top-down attention does not appear to be central to consciousness, for as LeDoux (2008) agrees, prefrontal cortex lesions that have major effects on attention do not impair subjective feelings of consciousness. The same evidence suggests that attention itself is not a fundamental computational process that is necessary for consciousness, as the neural networks that implement short term memory and operate to produce biased competition with non-linear effects do not appear to be closely related to consciousness (Deco and Rolls 2005a; Rolls 2008b), though of course if attention is directed towards particular perceptual events, this will increase the gain of the perceptual processing (Deco and Rolls 2005a, b; Rolls 2008b), making the attended phenomena stronger.

Another process ascribed to working memory is that items can be manipulated in working memory, for example, placed into a different order. This process implies at the computational level some type of syntactic processing, for each item (or symbol) could occur in any position relative to the others, and each item might occur more than once. To keep the items separate yet manipulable into any relation to each other, just having each item represented by the firing of a different set of neurons is insufficient, for this provides no information about the order or more generally the relations between the items being manipulated (Rolls and Deco 2002; Rolls 2008b). In this sense, some form of syntax, that is a way to relate to each other the firing of the different populations of neurons each representing an item, is required. If we go this far (and LeDoux (1996) p. 280 does appear to), then we see that this aspect of working memory is very close to the concept I propose of syntactic thought in my HOST theory. My particular approach though makes it clear what the function is to be performed (syntactic operations), whereas the term working memory can be used to refer to many different types of processing and is in this sense less well defined computationally. My approach of course argues that it is thoughts about the first order thoughts that may be very closely linked to consciousness. In our simple case, the HOST might be "Do I have the items now in the correct reversed order? Should the X come before or after the Y?" To perform this syntactic manipulation, I argue that there is a special syntactic processor, perhaps in cortex near Broca's area, that performs the manipulations on the items, and that the dorsolateral prefrontal cortex itself provides the short-term store that holds the items on which the syntactic processor operates (Rolls 2008b). In this scenario, dorsolateral prefrontal cortex damage would affect the number of items that could be manipulated, but not consciousness or the ability to manipulate the items syntactically and to monitor and comment on the result to check that it is correct.

A property often attributed to consciousness is that it is *unitary*. LeDoux (2008) might relate this to the limitations of a working memory system. The current theory would account for this by the limited syntactic capability of neuronal networks in the brain, which render it difficult to implement more than a few syntactic bindings of symbols simultaneously (McLeod et al. 1998; Rolls 2008b). This limitation makes it difficult to run several "streams of consciousness" simultaneously. In addition, given that a linguistic system can control behavioural output, several parallel streams might produce maladaptive behaviour (apparent as e.g. indecision) and might be selected against. The close relation between, and the limited capacity of, both the stream of consciousness, and auditory-verbal short term memory, may arise because both require implementation of the capacity for syntax in neural networks. My suggestion is that it is the difficulty the brain has in implementing the syntax required for manipulating items in working memory, and therefore for multiple step planning, and for then correcting these plans, that provides a close link between working memory concepts and my theory of higher order syntactic processing. The theory I describe makes it clear what the underlying computational problem is (how syntactic operations are performed in the system, and how they are corrected), and argues that when there are thoughts about the system, i.e. HOSTs, and the system is reflecting on its first order thoughts (cf. Weiskrantz 1997), then it is a property of the system that it feels conscious. As I argued above, first order linguistic thoughts, which presumably involve working memory (which must be clearly defined for the purposes of this discussion), need not necessarily be conscious.

The theory is also different from some other theories of consciousness (Carruthers 1996; Gennaro 2004; Rosenthal 2004, 2005) in that it provides an account of the evolutionary, adaptive, value of a HOST system in helping to solve a credit assignment problem that arises in a multi-step syntactic plan, links this type of processing to consciousness and therefore emphasizes a role for syntactic processing in consciousness. The type of syntactic processing need not be at the natural language level (which implies a universal grammar), but could be at the level of mentalese or simpler, as it involves primarily the syntactic manipulation of symbols (Fodor 1994; Rolls 2004b, 2005b).

The current theory holds that it is HOSTs that are closely associated with consciousness, and this may differ from Rosenthal's HOTs theory (Rosenthal 1986, 1990, 1993, 2004, 2005), in the emphasis in the current theory on language. Language in the current theory is defined by syntactic manipulation of symbols and does not necessarily imply verbal or "natural" language. The reason that strong emphasis is placed on language is that it is as a result of having a multi-step flexible "on the fly" reasoning procedure that errors which cannot be easily corrected by reward or punishment received at the end of the reasoning need "thoughts about thoughts", that is some type of supervisory and monitoring process, to detect where errors in the reasoning have occurred. This suggestion on the adaptive value in evolution of such a higher order linguistic thought process for multi-step planning ahead, and correcting such plans, may also be different from earlier work. Put another way, this point is that *credit assignment* when reward or punishment are received is straightforward in a one layer network (in which the reinforcement can be used directly to correct nodes in error, or responses), but is very difficult in a multi-step linguistic process executed once "on the fly". Very complex mappings in a multilayer network can be learned if hundreds of learning trials are provided. But once these complex mappings are learned, their success or failure in a new situation on a given trial cannot be evaluated and corrected by the network. Indeed, the complex mappings achieved by such networks (e.g. backpropagation nets) mean that after training they operate according to fixed rules and are often quite impenetrable and inflexible (Rolls and Deco 2002). In contrast, to correct a multi-step, single occasion, linguistically based plan or procedure, recall of the steps just made in the reasoning or planning, and perhaps related episodic material, needs to occur, so that the link in the chain which is most likely to be in error can be identified. This may be part of the reason why there is a close relation between declarative memory systems, which can explicitly recall memories and consciousness.

Some computer programs may have supervisory processes. Should these count as higher order linguistic thought processes? My current response to this is that they should not to the extent that they operate with fixed rules to correct the operation of a system which does not itself involve linguistic thoughts about symbols grounded semantically in the external world. If on the other hand it were possible to implement on a computer such a higher order linguistic thought supervisory correction process to correct first order one-off linguistic thoughts with symbols grounded in the real world, then this process would prima facie be conscious. If it were possible in a thought experiment to reproduce the neural connectivity and operation of a human brain on a computer, then prima facie it would also have the attributes of consciousness. It might continue to have those attributes for as long as power was applied to the system.

Another possible difference from other theories of consciousness is that raw sensory feels are suggested to arise as a consequence of having a system that can think about its own thoughts. Raw sensory feels, and subjective states associated with emotional and motivational states, may not necessarily arise first in evolution.

Finally, I provide a short specification of what might have to be implemented in a neural network to implement conscious processing. First, a linguistic system, not necessarily verbal, but implementing syntax between symbols grounded in the environment would be needed (e.g. a mentalese language system). Then a HOST system also implementing syntax and able to think about the representations in the first order language system, and able to correct the reasoning in the first order linguistic system in a flexible manner, would be needed. So my view is that consciousness can be implemented in neural networks of the artificial and biological type, but that the neural networks would have to implement the type of higher order linguistic processing described in this paper.

Acknowledgements The author acknowledges helpful discussions with Martin Davies, Marian Dawkins and David Rosenthal.

References

Aggelopoulos NC, Rolls ET (2005) Natural scene perception: inferior temporal cortex neurons encode the positions of different objects in the scene. European Journal of Neuroscience 22:2903–2916.

Aggelopoulos NC, Franco L, Rolls ET (2005) Object perception in natural scenes: encoding by inferior temporal cortex simultaneously recorded neurons. Journal of neurophysiology 93:1342–1357.

Allport A (1988) What concept of consciousness? In: Consciousness in Contemporary Science (Marcel AJ, Bisiach E, eds), pp 159–182. Oxford: Oxford University Press.

Armstrong DM, Malcolm N (1984) Consciousness and Causality. Oxford: Blackwell.

Azzopardi P, Cowey A (1997) Is blindsight like normal, near-threshold vision? Proceedings of the National Academy of Sciences USA 94:14190–14194.

Baars BJ (1988) A Cognitive Theory of Consciousness. New York: Cambridge University Press.

Barlow HB (1997) Single neurons, communal goals, and consciousness. In: Cognition, Computation, and Consciousness (Ito M, Miyashita Y, Rolls ET, eds), pp 121–136. Oxford: Oxford University Press.

Battaglia FP, Treves A (1998) Stable and rapid recurrent processing in realistic auto-associative memories. Neural Computation 10:431–450.

Block N (1995) On a confusion about a function of consciousness. Behavioral and Brain Sciences 18:227–247.

Block N (2005) Two neural correlates of consciousness. Trends in Cognitive Sciences 9:46–52.

Booth DA (1985) Food-conditioned eating preferences and aversions with interoceptive elements: learned appetites and satieties. Annals of the New York Academy of Sciences 443:22–37.

Carruthers P (1996) Language, Thought and Consciousness. Cambridge: Cambridge University Press.

Chalmers DJ (1996) The Conscious Mind. Oxford: Oxford University Press.

Cheney DL, Seyfarth RM (1990) How Monkeys See the World. Chicago: University of Chicago Press.

Cooney JW, Gazzaniga MS (2003) Neurological disorders and the structure of human consciousness. Trends in Cognitive Sciences 7:161–165.

Crick FHC, Koch C (1990) Towards a neurobiological theory of consciousness. Seminars in the Neurosciences 2:263–275.

Damasio AR (1994) Descartes' Error. New York: Putnam.

Davies MK (2008) Consciousness and explanation. In: Frontiers of Consciousness (Weiskrantz L, Davies MK, eds), pp 1–54. Oxford: Oxford University Press.

Dawkins R (1986) The Blind Watchmaker. Harlow: Longman.

Dawkins R (1989) The Selfish Gene, 2nd Edition. Oxford: Oxford University Press.

De Gelder B, Vroomen J, Pourtois G, Weiskrantz L (1999) Non-conscious recognition of affect in the absence of striate cortex. Neuroreport 10:3759–3763.

Deco G, Rolls ET (2005a) Neurodynamics of biased competition and co-operation for attention: a model with spiking neurons. Journal of Neurophysiology 94:295–313.

Deco G, Rolls ET (2005b) Attention, short-term memory, and action selection: a unifying theory. Progress in Neurobiology 76:236–256.

Deco G, Rolls ET (2005c) Synaptic and spiking dynamics underlying reward reversal in orbitofrontal cortex. Cerebral Cortex 15:15–30.

Deco G, Rolls ET (2006) Decision-making and Weber's Law: a neurophysiological model. European Journal of Neuroscience 24:901–916.

Deco G, Rolls ET, Romo R (2009) Stochastic dynamics as a principle of brain function. Progress in Neurobiology 88:1–16.

Dehaene S, Naccache L (2001) Towards a cognitive neuroscience of consciousness: basic evidence and a workspace framework. Cognition 79:1–37.

Dehaene S, Changeux JP, Naccache L, Sackur J, Sergent C (2006) Conscious, preconscious, and subliminal processing: a testable taxonomy. Trends in Cognitive Sciences 10:204–211.

Dennett DC (1991) Consciousness Explained. London: Penguin.

Dennett DC (2005) Sweet Dreams: Philosophical Obstacles to a Science of Consciousness. Cambridge, MA: MIT.

Dinse HR, Kruger K (1994) The timing of processing along the visual pathway in the cat. Neuroreport 5:893–897.

Elliffe MCM, Rolls ET, Stringer SM (2002) Invariant recognition of feature combinations in the visual system. Biological Cybernetics 86:59–71.

Fodor JA (1994) The Elm and the Expert: mentalese and its semantics. Cambridge, MA: MIT.

Franco L, Rolls ET, Aggelopoulos NC, Treves A (2004) The use of decoding to analyze the contribution to the information of the correlations between the firing of simultaneously recorded neurons. Experimental Brain Research 155:370–384.

Gazzaniga MS (1988) Brain modularity: towards a philosophy of conscious experience. In: Consciousness in Contemporary Science (Marcel AJ, Bisiach E, eds), pp 218–238. Oxford: Oxford University Press.

Gazzaniga MS (1995) Consciousness and the cerebral hemispheres. In: The Cognitive Neurosciences (Gazzaniga MS, ed), pp 1392–1400. Cambridge, Mass.: MIT.

Gazzaniga MS, LeDoux J (1978) The Integrated Mind. New York: Plenum.

Gennaro RJ, ed (2004) Higher Order Theories of Consciousness. Amsterdam: John Benjamins.

Goldman-Rakic PS (1996) The prefrontal landscape: implications of functional architecture for understanding human mentation and the central executive. Philosophical Transactions of the Royal Society B 351:1445–1453.

Goodale MA (2004) Perceiving the world and grasping it: dissociations between conscious and unconscious visual processing. In: The Cognitive Neurosciences III (Gazzaniga MS, ed), pp 1159–1172. Cambridge, MA: MIT.

Hamilton W (1964) The genetical evolution of social behaviour. Journal of Theortical Biology 7:1–52.

Hamilton WD (1996) Narrow Roads of Gene Land. New York: W. H. Freeman.

Hampton RR (2001) Rhesus monkeys know when they remember. Proceedings of the National Academy of Sciences of the United States of America 98:5359–5362.

Hampton RR, Zivin A, Murray EA (2004) Rhesus monkeys (Macaca mulatta) discriminate between knowing and not knowing and collect information as needed before acting. Animal Cognition 7:239–246.

Heyes CM (2008) Beast machines? Questions of animal consciousness. In: Frontiers of Consciousness (Weiskrantz L, Davies M, eds), pp 259–274. Oxford: Oxford University Press.

Hornak J, Bramham J, Rolls ET, Morris RG, O'Doherty J, Bullock PR, Polkey CE (2003) Changes in emotion after circumscribed surgical lesions of the orbitofrontal and cingulate cortices. Brain 126:1691–1712.

Hornak J, O'Doherty J, Bramham J, Rolls ET, Morris RG, Bullock PR, Polkey CE (2004) Reward-related reversal learning after surgical excisions in orbitofrontal and dorsolateral prefrontal cortex in humans. Journal of Cognitive Neuroscience 16:463–478.

Humphrey NK (1980) Nature's psychologists. In: Consciousness and the Physical World (Josephson BD, Ramachandran VS, eds), pp 57–80. Oxford: Pergamon.

Humphrey NK (1986) The Inner Eye. London: Faber.

Humphreys GW, Bruce V (1991) Visual Cognition. Hove, East Sussex: Erlbaum.

Huxter J, Burgess N, O'Keefe J (2003) Independent rate and temporal coding in hippocampal pyramidal cells. Nature 425:828–832.

Insabato A, Pannunzi M, Rolls ET, Deco G (2010) Confidence-related decision-making. Journal of Neurophysiology 104:539–547.

Jackendoff R (2002) Foundations of Language. Oxford: Oxford University Press.

Johnson-Laird PN (1988) The Computer and the Mind: An Introduction to Cognitive Science. Cambridge, MA: Harvard University Press.

Kadohisa M, Rolls ET, Verhagen JV (2005) Neuronal representations of stimuli in the mouth: the primate insular taste cortex, orbitofrontal cortex, and amygdala. Chemical Senses 30:401–419.

Kepecs A, Uchida N, Zariwala HA, Mainen ZF (2008) Neural correlates, computation and behavioural impact of decision confidence. Nature 455:227–231.

Koch C, Preuschoff K (2007) Betting the house on consciousness. Nature Neuroscience 10:140–141.

Krebs JR, Kacelnik A (1991) Decision Making. In: Behavioural Ecology, 3rd Edition (Krebs JR, Davies NB, eds), pp 105–136. Oxford: Blackwell.

Lamme VAF, Roelfsema PR (2000) The distinct modes of vision offered by feedforward and recurrent processing. Trends in Neuroscience 23:571–579.

LeDoux J (2008) Emotional coloration of consciousness: how feelings come about. In: Frontiers of Consciousness (Weiskrantz L, Davies M, eds), pp 69–130. Oxford: Oxford University Press.

LeDoux JE (1996) The Emotional Brain. New York: Simon and Schuster.

Libet B (2002) The timing of mental events: Libet's experimental findings and their implications. Consciousness and Cognition 11:291–299; discussion 304–333.

Malsburg Cvd (1990) A neural architecture for the representation of scenes. In: Brain Organization and Memory: Cells, Systems and Circuits (McGaugh JL, Weinberger NM, Lynch G, eds), pp 356–372. New York: Oxford University Press.

Marcel AJ (1983a) Conscious and unconscious perception: an approach to the relations between phenomenal experience and perceptual processes. Cognitive Psychology 15:238–300.

Marcel AJ (1983b) Conscious and unconscious perception: experiments on visual masking and word recognition. Cognitive Psychology 15:197–237.

McLeod P, Plunkett K, Rolls ET (1998) Introduction to Connectionist Modelling of Cognitive Processes. Oxford: Oxford University Press.

Miller EK, Cohen JD (2001) An integrative theory of prefrontal cortex function. Annual review of neuroscience 24:167–202.

Milner AD (2008) Conscious and unconscious visual processing in the human brain. In: Frontiers of Consciousness (Weiskrantz L, Davies M, eds), pp 169–214. Oxford: Oxford University Press.

Milner AD, Goodale MA (1995) The Visual Brain in Action. Oxford: Oxford University Press.

Nowak LG, Bullier J (1997) The timing of information transfer in the visual system. In: Extrastriate visual cortex in primates (Rockland KS, Kaas JH, Peters A, eds), pp 205–241. New York: Plenum.

Panzeri S, Schultz SR, Treves A, Rolls ET (1999) Correlations and the encoding of information in the nervous system. Proceedings of the Royal Society of London B 266:1001–1012.

Panzeri S, Rolls ET, Battaglia F, Lavis R (2001) Speed of feedforward and recurrent processing in multilayer networks of integrate-and-fire neurons. Network: Computation in Neural Systems 12:423–440.

Persaud N, McLeod P, Cowey A (2007) Post-decision wagering objectively measures awareness. Nature Neuroscience 10:257–261.

Petrides M (1996) Specialized systems for the processing of mnemonic information within the primate frontal cortex. Philosophical Transactions of the Royal Society B 351:1455–1462.

Phelps EA, LeDoux JE (2005) Contributions of the amygdala to emotion processing: from animal models to human behavior. Neuron 48:175–187.

Ridley M (1993) The Red Queen: Sex and the Evolution of Human Nature. London: Penguin.

Rolls ET (1990) A theory of emotion, and its application to understanding the neural basis of emotion. Cognition and Emotion 4:161–190.

Rolls ET (1995) A theory of emotion and consciousness, and its application to understanding the neural basis of emotion. In: The Cognitive Neurosciences (Gazzaniga MS, ed), pp 1091–1106. Cambridge, Mass.: MIT.

Rolls ET (1997a) Brain mechanisms of vision, memory, and consciousness. In: Cognition, Computation, and Consciousness (Ito M, Miyashita Y, Rolls ET, eds), pp 81–120. Oxford: Oxford University Press.

Rolls ET (1997b) Consciousness in neural networks? Neural Networks 10:1227–1240.

Rolls ET (1999a) The functions of the orbitofrontal cortex. Neurocase 5:301–312.

Rolls ET (1999b) The Brain and Emotion. Oxford: Oxford University Press.

Rolls ET (2000) Précis of The Brain and Emotion. Behavioral and Brain Sciences 23:177–233.

Rolls ET (2003) Consciousness absent and present: a neurophysiological exploration. Progress in Brain Research 144:95–106.

Rolls ET (2004a) The functions of the orbitofrontal cortex. Brain and cognition 55:11–29.

Rolls ET (2004b) A higher order syntactic thought (HOST) theory of consciousness. In: Higher-Order Theories of Consciousness: An Anthology (Gennaro RJ, ed), pp 137–172. Amsterdam: John Benjamins.

Rolls ET (2005a) Consciousness absent or present: a neurophysiological exploration of masking. In: The First Half Second: The Microgenesis and Temporal Dynamics of Unconscious and Conscious Visual Processes (Ogmen H, Breitmeyer BG, eds), pp 89–108, chapter 106. Cambridge, MA: MIT.

Rolls ET (2005b) Emotion Explained. Oxford: Oxford University Press.

Rolls ET (2006) Brain mechanisms underlying flavour and appetite. Philosophical Transactions of the Royal Society London B 361:1123–1136.

Rolls ET (2007a) A computational neuroscience approach to consciousness. Neural Networks 20:962–982.

Rolls ET (2007b) The representation of information about faces in the temporal and frontal lobes. Neuropsychologia 45:125–143.

Rolls ET (2007c) The affective neuroscience of consciousness: higher order linguistic thoughts, dual routes to emotion and action, and consciousness. In: Cambridge Handbook of Consciousness (Zelazo P, Moscovitch M, Thompson E, eds), pp 831–859. Cambridge: Cambridge University Press.

Rolls ET (2008a)Emotion, higher order syntactic thoughts, and consciousness. In: Frontiers of Consciousness (Weiskrantz L, Davies MK, eds), pp 131–167. Oxford: Oxford University Press.

Rolls ET (2008b) Memory, Attention, and Decision-Making: A Unifying Computational Neuroscience Approach. Oxford: Oxford University Press.

Rolls ET (2009) The anterior and midcingulate cortices and reward. In: Cingulate Neurobiology and Disease (Vogt BA, ed), pp 191–206. Oxford: Oxford University Press.

Rolls ET (2011) Neuroculture. Oxford: Oxford University Press.

Rolls ET, Tovee MJ (1994) Processing speed in the cerebral cortex and the neurophysiology of visual masking. Proceedings of the Royal Society of London B 257:9–15.

Rolls ET, Treves A (1998) Neural Networks and Brain Function. Oxford: Oxford University Press.

Rolls ET, Deco G (2002) Computational Neuroscience of Vision. Oxford: Oxford University Press.

Rolls ET, Stringer SM (2006) Invariant visual object recognition: a model, with lighting invariance. Journal of Physiology Paris 100:43–62.

Rolls ET, Kesner RP (2006) A computational theory of hippocampal function, and empirical tests of the theory. Progress in Neurobiology 79:1–48.

Rolls ET, Grabenhorst F (2008) The orbitofrontal cortex and beyond: from affect to decision-making. Progress in Neurobiology 86:216–244.

Rolls ET, Deco G (2010) The Noisy Brain: Stochastic Dynamics as a Principle of Brain Function. Oxford: Oxford University Press.

Rolls ET, Tovee MJ, Panzeri S (1999) The neurophysiology of backward visual masking: information analysis. Journal of Cognitive Neuroscience 11:335–346.

Rolls ET, Grabenhorst F, Deco G (2010a) Choice, difficulty, and confidence in the brain. NeuroImage 53:694–706.

Rolls ET, Grabenhorst F, Deco G (2010) Decision-making, errors, and confidence in the brain. Journal of Neurophysiology 104:2359–2374.

Rolls ET, Hornak J, Wade D, McGrath J (1994a) Emotion-related learning in patients with social and emotional changes associated with frontal lobe damage. Journal of Neurology, Neurosurgery and Psychiatry 57:1518–1524.

Rolls ET, Franco L, Aggelopoulos NC, Reece S (2003) An information theoretic approach to the contributions of the firing rates and correlations between the firing of neurons. Journal of Neurophysiology 89:2810–2822.

Rolls ET, Aggelopoulos NC, Franco L, Treves A (2004) Information encoding in the inferior temporal cortex: contributions of the firing rates and correlations between the firing of neurons. Biological Cybernetics 90:19–32.

Rolls ET, Franco L, Aggelopoulos NC, Perez JM (2006) Information in the first spike, the order of spikes, and the number of spikes provided by neurons in the inferior temporal visual cortex. Vision Research 46:4193–4205.

Rolls ET, Tovee MJ, Purcell DG, Stewart AL, Azzopardi P (1994b) The responses of neurons in the temporal cortex of primates, and face identification and detection. Experimental Brain Research 101:473–484.

Rosenthal DM (1986) Two concepts of consciousness. Philosophical Studies 49:329–359.

Rosenthal DM (1990) A Theory of Consciousness. Bielefeld, Germany: Zentrum für Interdisziplinaire Forschung.

Rosenthal DM (1993) Thinking that one thinks. In: Consciousness (Davies M, Humphreys GW, eds), pp 197–223. Oxford: Blackwell.

Rosenthal DM (2004) Varieties of Higher-Order Theory. In: Higher Order Theories of Consciousness (Gennaro RJ, ed). Amsterdam: John Benjamins.

Rosenthal DM (2005) Consciousness and Mind. Oxford: Oxford University Press.

Rumelhart DE, Hinton GE, Williams RJ (1986) Learning internal representations by error propagation. In: Parallel Distributed Processing: Explorations in the Microstructure of Cognition (Rumelhart DE, McClelland JL, Group TPR, eds). Cambridge, Mass.: MIT.

Shallice T, Burgess P (1996) The domain of supervisory processes and temporal organization of behaviour. Philosophical Transactions of the Royal Society B 351:1405–1411.

Simmen MW, Treves A, Rolls ET (1996) Pattern retrieval in threshold-linear associative nets. Network (Bristol, England) 7:109–122.

Singer W (1999) Neuronal synchrony: A versatile code for the definition of relations? Neuron 24:49–65.

Smerieri A, Rolls ET, Feng J (2010) Decision time, slow inhibition, and theta rhythm. Journal of Neuroscience 30:14173–14181.

Smith-Swintosky VL, Plata-Salaman CR, Scott TR (1991) Gustatory neural encoding in the monkey cortex: stimulus quality. Journal of Neurophysiology 66:1156–1165.

Smith PL, Ratcliff R (2004) Psychology and neurobiology of simple decisions. Trends in Neurosciences 27:161–168.

Squire LR, Zola SM (1996) Structure and function of declarative and nondeclarative memory systems. Proceedings of the National Academy of Sciences USA 93:13515–13522.

Sugase Y, Yamane S, Ueno S, Kawano K (1999) Global and fine information coded by single neurons in the temporal visual cortex. Nature 400:869–873.

Tovee MJ, Rolls ET (1992) The functional nature of neuronal oscillations. Trends in Neurosciences 15:387.

Tovee MJ, Rolls ET (1995) Information encoding in short firing rate epochs by single neurons in
the primate temporal visual cortex. Visual Cognition 2:35–58.

Tovee MJ, Rolls ET, Treves A, Bellis RP (1993) Information encoding and the responses of single
neurons in the primate temporal visual cortex. Journal of Neurophysiology 70:640–654.

Treisman A (1996) The binding problem. Current Opinion in Neurobiology 6:171–178.

Treves A (1993) Mean-field analysis of neuronal spike dynamics. Network (Bristol, England)
4:259–284.

Wang XJ (2002) Probabilistic decision making by slow reverberation in cortical circuits. Neuron
36:955–968.

Weiskrantz L (1997) Consciousness Lost and Found. Oxford: Oxford University Press.

Weiskrantz L (1998) Blindsight. A Case Study and Implications, 2nd Edition. Oxford: Oxford
University Press.

Weiskrantz L (2001) Blindsight – putting beta (β) on the back burner. In: Out of Mind: Varieties of
Unconscious Processes (De Gelder B, De Haan E, Heywood C, eds), pp 20–31. Oxford: Oxford
University Press.

Wong KF, Wang XJ (2006) A recurrent network mechanism of time integration in perceptual decisions. Journal of Neuroscience 26:1314–1328.

Yaxley S, Rolls ET, Sienkiewicz ZJ (1990) Gustatory responses of single neurons in the insula of
the macaque monkey. Journal of Neurophysiology 63:689–700.

Chapter 10
A Review of Models of Consciousness

John G. Taylor

Abstract We review the main models of consciousness and develop various criteria to assess them. One of the most important criteria is that of the existence of the inner self of Western phenomenology (following Hussar, Sartre, Merleau-Ponty and colleagues). Only one model survives that criterion, that of CODAM (Taylor, 2007; ibid, 2010a; ibid, 2010b). Another test is that of agreeing with relevant experimental data (to be considered in detail later), especially a few data sets of specific relevance to consciousness (Taylor and Fragopanagos, 2007). A number pass this test over a limited range of data (not necessarily the most relevant data for consciousness). Very few models pass the third test of explicitly involving attention (regarded as the gateway to consciousness). A fourth and final test is that of helping explain defects of experience in various diseases of the mind, especially schizophrenia. Only one model survives that test (again CODAM, Taylor, 2007, 2010a). Our conclusion is that although many models emphasise various components of consciousness creation, very few yet include all relevant components.

10.1 Introduction

To begin with let us define consciousness: it is the experience of the external world, such as the smelling of the rose, the tasting of the wine and the seeing of the beautiful sunset. It is a process taking place over time and crucially involves activity in the brain of the human who is having the experience. The experience itself takes place as centered on the person having the experience and has a degree of privacy consistent with the complexity of brain activity (so difficult to be probed by another person). But yet the real nature of consciousness is unclear: is it solely arising

J.G. Taylor (✉)
Department of Mathematics, King's College, London, UK
e-mail: john.g.taylor@kcl.ac.uk

V. Cutsuridis et al. (eds.), *Perception-Action Cycle: Models, Architectures, and Hardware*, Springer Series in Cognitive and Neural Systems 1,
DOI 10.1007/978-1-4419-1452-1_10, © Springer Science+Business Media, LLC 2011

from the subtle activity of millions to billions of neurons in our brains or is there some non-material aspect involved? In this paper, we look solely at brain-based models of consciousness; any immaterial components will thus have to be ignored, although they are in any case difficult to define, being ineffable in form due to their immateriality.

There is presently much controversy over how consciousness is to be explained. Even amongst those who accept that it is only through brain activity that consciousness arises, there is a strong divergence of opinion as to exactly how the brain achieves this subtle and remarkable feat. In this paper, we wish to compare and contrast the most important of these models with each other. More crucially we also present a set of natural criteria which allow us to test how well each model passes a simple set of criteria.

The criteria we consider for each model to be considered are to fit:

- All relevant data, that is data especially associated with the presence or absence of conscious experience in a controlled fashion (Taylor and Fragopanagos 2007)
- Data arising from the experiences of those with mental ill-health, and especially those with schizophrenia, this being a disease which most strongly affects the minds of the sufferers (Sass and Parnas 2003; Cermolacce et al. 2007)
- The present experimental data on attention (Corbetta et al. 2008; Gregoriou et al. 2009; Bressler et al. 2008)
- The existence and nature of the "inner self" of Western phenomenology (Zahavi 2005)

The first and second criteria are those that would be applied by any hard-nosed scientist to any model of a part of the world – try to fit the extant and relevant data wherever that is, either for the normal or diseased brain, respectively. Both of these criteria must be applied successively and with increasing rigour, so that either the model fails and a new model is to be put in its place which should fit the recalcitrant data or it succeeds to fit the data and be tested another day on further data. All of the models clearly have a long way to go to fit all possible and relevant data; the symbol "P" in table of the test results below indicates only a partial fit to available data. We describe in detail later what is to be considered as "relevant" data, since it is not possible to fit all the possible data from brain science by a model in one fell swoop.

The third criterion can be seen as also coming under the headings of tests 1 and 2. However, it is of a different character, involving as it does the specific mental faculty of attention. The inclusion of such a processing component is well-founded and requires insertion of a certain level of detail into the model itself. Some of the models of the list discussed in Sect. 10.2 do not possess enough detail to allow such inclusion, and so the relevant model is denoted as "incomplete" (I in the table).

The fourth criterion is the most subtle of all, but yet is the most important to help "bridge the gap" between outer and inner experience (Levine 1983), to help solve the "hard problem" (Chalmers 1995), to give the sense of "what it is like to be" (Nagel 1974) and also to provide the "immunity to error through misidentification of the first person pronoun (Shoemaker 1968). It is that component associated with the creation of the inner self, corresponding to the sense of "I". Without that experience there is no consciousness, as will be argued [and has been analysed most persuasively and carefully over the last century through the Western phenomenology, as eloquently explained in Zahavi (2005)]. We consider the criterion that such an inner self be part of the dynamics of the model as crucial for consciousness creation. Only one model again survives that test, that of CODAM (Taylor 2007, 2010a, b).

In the next section, we give a brief review of the various models which we will explore with their different varieties being considered where possible. In the following, we develop a set of criteria we apply to these various theories so as to be able to begin to explore their pros and cons in an appropriate scientific form. In Sect. 10.4, these criteria are applied to the eight classes of models presented in Sect. 10.2, with the results summarised in a table. Finally, we draw conclusions in Sect. 10.5.

10.2 The Models of Consciousness

In this section, we present brief reviews of the variety of neural network models of the creation of consciousness of interest to us. At the same time, we consider some of the intrinsic difficulties that are possessed by certain of the models (such as logical problems that arise in the interpretation of a given model). Only in Sect. 10.4 will the four main criteria of testing be applied to the models reviewed here.

We note here that some of the models are not formulated so as to see how, even in principle, they would necessarily bridge the gap between the first and third person description. In other words, the model under discussion may be a good descriptor of brain processing – the third person description – when consciousness is occurring in the subject, but there is no clear indication of how the activity being described would give rise to a first person description. Thus such a gap has to be assumed to be bridged as a further part of building the relevant model.

10.2.1 The Higher Order Thought Model

This class of models has been of great interest to those studying the philosophy of mind, causing (and still causing) considerable controversy (as discussed in more detail below). It is also being employed by various working neuroscientists (see e.g. the chapter by Rolls in this book). Its main thesis is that a mental state becomes conscious if another mental state is focussed on (or is about) the first thought (Rosenthal 1986, 1993). Thus Rosenthal puts it: "The core of the theory, then, is that a mental state is a conscious state when and only when it is accompanied by a suitable 'higher order' thought" (Rosenthal 1993). The contents of that higher order state are that the subject is in that first mental state in order for there to be consciousness of it. It is usual to consider the first state as one at a lower level and the second, consciousness-producing state as at a higher (in this case second) level.

Hence the name "higher order theory" or HOT. Such an approach is very simple and has attracted much discussion, as noted, some of which will be discussed below. A less restrictive variant of the above HOT model is that of Carruthers (1996) who proposes that a state is conscious if there is a disposition to have a state consisting of a thought that one is in that state. This can be regarded as an extension of the idea of 1st and 2nd intentions of the scholastics.

One of the main problems with the HOT approach is that it seems difficult, in order to describe consciousness by HOT in a complete manner, to avoid the need for an infinite sequence of levels of thoughts (Rowlands 2001; Zahavi 2005) or the possibility of circularity, in which one is trying to define consciousness by using it to define itself (Rowlands 2001). The former of these difficulties arises from the problem of the specification of the second order state as knowing that one is in that lower order mental state. How can that be unless the second order state itself is conscious (especially of being in that first order state)? Thus consciousness is already assumed as given in the definition of the second order state, so leading to circularity. On the other hand, if the second order state is not a conscious one then how its relation to the first order state confers consciousness onto the latter is unclear.

Another crucial problem is of getting into an infinite regress by the HOT definition. Such an infinite regress arises in HOT by the attempt to unravel the circularity just identified. Thus if the second order state is conscious (of being in that first order mental state), then a third order state is needed, by the HOT principle quoted from Rosenthal above, to be focussed on this second order state to make it conscious. But then a fourth order state will be needed to ... and so on. This leads to the infinite regress mentioned and to the associated loss of any useful approach to consciousness. Various arguments from a dispositional viewpoint have been attempted to avoid this (and the circularity) problem (Rowlands 2001).

However, the reason for the continued existence of the HOT approach to consciousness in spite of these difficulties is that it has somewhat of a correct "feel" to it, and so should be present in some form in any final theory of consciousness. Such a possibility will be discussed at the end of the penultimate section. Moreover, those proposing the various HOT approaches have come back with quite sensible answers or modifications to handle these criticisms.

10.2.2 The Working Memory Model

This is one of the neuro-physiologically simplest of the models of consciousness presently on offer. It is based on the idea that if neural activity in some region of the brain converges to a stable attractor level of activity, even for a limited time, then there is consciousness of such activity. Thus if the activity of a trained Hopfield network converges to an attractor A, then consciousness will arise of the previous input stimulus which caused the attractor A to be set up in the network in the first place.

There is considerable support for the notion that working memory (WM) buffers form the basis for the conscious experience of content (Taylor 1999b). These modules are expanded under the GW heading (Sect. 10.2.3). Each buffer is in general expected to have its WM character (of holding activity over several seconds in the brain) supported by some form of recurrence. This might be internal or involve parietal–frontal recurrent loops.

An important extension of the WM model of consciousness to include remembered actions on external objects, termed a "probabilitistic automaton", has been pursued especially strongly in simulation by Aleksander and colleagues (Aleksander and Morton 2007). This model allows crossing of the boundaries between various classical memories subdivisions. The model develops internal memories of remembered sequences of visual images and the actions of moving from one image to the next. Each of the internal visual states is coded in recurrent nets, which are able to encode stable states as well as transitions between them. Such activities are regarded by the authors as conscious.

WM and its extension to other forms of memory encoding (episodic, semantic, etc.) are well known to occur in the brain and so are to be accepted as an important component of any model of consciousness creation.

10.2.3 The Global Workspace Model

A further attractive model is that of Baars in the global workspace (GW) (Baars 1998, 2002). That model proposes that there exists a GW (such as a set of coupled WMs) which if suitably accessed by stimulus activity will lead to consciousness of the resulting activity. The original version of this, introduced in the late 1980s by Baars, has been modified into a more sophisticated version (Baars 2002, p. 47): "Consciousness facilitates widespread access between otherwise independent brain functions". GW theory thus supposes that there is a set of fleeting memories of which only one can be dominant at any one time, and such that the dominant one can be distributed across many areas of the brain. The GW is thus a pivotal region with global access to other brain sites; such a style of architecture is used in various computer models in terms of a "global blackboard". In Baars (2002), numerous lines of experimental support were presented for the GW model. It appears on the surface to be an updated model of multi-WM form, but in the work of Franklin on the software system IDA (Baars and Franklin 2007) has been extended considerably beyond that simple description.

It is to be noted that the GW model is a broad-brush approach to consciousness and does not provide what is more specifically a detailed neural architecture for the creation of conscious experience. Various attempts have been made since to develop such an architecture, as well as, develop a mathematical analysis backing up the GW approach from a dynamical systems viewpoint (Shanahan 2005; Wallace 2005).

The GW approach is an appealing and basic framework that must be seriously considered as part of the overall attack on the details of the consciousness-creating brain processes.

10.2.4   The Complexity Model

It is interesting to note that the notion of complexity is related to the GW approach above. As Edelman and Tononi (Edelman and Tononi 2000) state: "When we become aware of something ... it is as if suddenly many different parts of our brain were privy to some specialised subsystem .... The wide distribution of information is guaranteed." This distribution process is close to that being suggested by Baars in his GW model. However, Edelman and his colleagues decided that it was more important to emphasise the "specialised subsystem" – it was termed the "dynamic core", involving information flow with high complexity, and so to follow the search for regions of complexity in the brain. It would be these, it could be conjectured, which would be the sites of the creation of consciousness. This is closely related to the idea of reafference or recurrence, to be discussed in Sect. 10.2.5, in which neuronal group selection plays an important role (through lateral interactions between neuronal groups in a given module so as to select the optimally active one). As part of this approach, there is the "key claim ... that conscious qualia are these high-dimensional discriminations" (Seth et al. 2006, p 10799).

In this latter paper, the authors considered several mathematical measures of the complexity of neural activity in a neural network: Thus neural complexity expresses the extent to which a system is both dynamically segregated and integrated and defined more specifically as the sum of the average mutual information across all bipartitions of the system. A further measure of complexity is that of "information integration", defined as the "effective information" across the informational weakest link of the system, so is the minimum information bipartition. However, a brain with a high value of the parameter but displaying no activity at all would still be conscious. It is also difficult to calculate since the number of bipartitions increases approximately as N^N (where N is the number of neurons in the network). Furthermore, there is a simple Hopfield network that has an equivalent or greater value of the parameter for any network, so implying that such a Hopfield network would be conscious.

A somewhat independent development of the notions of complexity has been developed more recently in Tononi (2008). In this approach, a new measure of "information integration" was introduced, in terms of the relative entropy of the state in comparison with other possible states of the same system. This measure has allowed (Tononi 2008) to relate to possible brain sites and manners of processing, although proved difficult to allow for the value of this measure for specific regions of the brain.

An alternative parameter can be defined as the causal density (cd) of the network, obtained as proportional to the number of Granger-causal interaction in the net (determined by a multi-factor time series regression analysis). A difficulty with calculating cd is that it requires a dynamical analysis of the time series of activities of the neurons of the network in order to calculate, so is increasingly difficult to calculate as the number of elements of the network increases (due to the increasing numbers of variables required for the Granger causality evaluation). The overall conclusion of the authors (Seth et al. 2006) is that it appears difficulty to calculate any of these complexity parameters for large systems (as the human brain).

An alternate approach, but still preserving some notion of complexity, is to use the notion of "traffic" introduced in Taylor et al. (2000) in a structural model analysis of brain fMRI data in a memory retrieval task. After the regions of interest (ROIs) were determined, the structural model was evaluated and the number of significantly weighted lines emanating from a given ROI was calculated; those ROIs with relatively large amounts of traffic were considered as possessing greatest complexity. Such an approach needs to be done for a variety of paradigms and then the average traffic per ROI calculated: those ROIs with greatest traffic would be considered putatively as involved in the creation of consciousness. Such an evaluation leaves to one side any causal notion, but a Granger causality approach could also be taken as part of the structural modelling if needed. The traffic approach avoids high complexity by reducing the network elements to ROIs, so as aggregates of neurons.

These various approaches to brain complexity are all of relevance in attempting to get a dynamical complexity handle on consciousness.

10.2.5 The Recurrent Model

There is a considerable variety of such models and of neuroscientists who are adherent to them. One in particular has had considerable publicity under the terminology of the "reafference principle" (Edelman 1992). Another is termed ART: a powerful concept of a local neuronal architecture for cortex introduced in the 1970s (Grossberg 1976). Others have also had considerable impact on the attempt to found consciousness in recurrence (Lamme 2006; Pollen 2003).

There is an enormous amount of recurrence in the brain: any module well connected to another brain area will have reciprocal connections with it. This is particularly noticeable in the hierarchy of visual systems, where the hierarchy is seen to run both ways: up from V1 to V2 to V3, etc., as well as back down the set of visual cortices. Such connectivity has been used in developing numerous effective vision processing models. However, the detailed fashion as to how consciousness could thereby be created is unclear, since no associated specific neural model of consciousness, based on recurrence, has been proposed so as to generate conscious experience itself and test it on the criteria developed in Sect. 10.4.

It is correct to say that neural recurrence in connectivity between any pair of brain modules is an important part of brain processing (recurrence in connectivity seems to be the rule rather than the isolated instance between any two brain areas), so it will thereby be expected to be important to be included in the creation of consciousness as part of an overall brain architecture.

10.2.6 The Neural Field Model

Neural fields come in a variety of shapes and sizes, with different levels and types of physiological basis. Thus a "neural field" (Amari 1977; Taylor 1999b) can be defined as a continuum of neurons in a given module; this is a relatively good approximation to neural distributions in cortex provided one is not working at a level approaching the single neuron scale. It has been especially employed by Freeman (2007) to help describe the synchronisation and phase transformations he has observed by use of surface electrodes in humans as well as lower animals. Such phase analysis in Fourier space, with its implication of interesting switches of phase, has been developed into a fascinating story by Freeman, who distinguishes between "the neural point processes dictated by the neuron doctrine versus continuously variable neural fields generated by neural masses in cortex" (Freeman 2007, p 1021). It is through dynamical processes in the latter continuous neural fields that Freeman sees the emergence of consciousness.

An alternative definition of neural field is that of an actual field in 3D space around a set of neurons (Laberge and Kasevich 2007). In particular the authors claim (p 1004) "The neural basis of consciousness is theorized here to be the elevated activity of the apical dendrite within a thalamocortical circuit." This is an interesting approach to consciousness by allowing it to expand into the surroundings of an active neuron rather than have it created from purely the interactions between spiking neurons in some complex manner. It thereby allows for a continuum approach to consciousness so fitting some aspects of our own experience.

10.2.7 The Relational Mind

This was introduced by the author in the 1970s, based on the notion that conscious experience is strongly determined by the closeness which a given percept has to past experiences in long-term memory (Taylor 1999a). The relations a given input stimulus has to present percepts held either in the earlier processing regions in cortex and/or in long-term memory was proposed, in this approach, as the basis of conscious experience. Such elicitation of past experiences and their influence on present consciousness is undoubtedly an important component of processing leading to consciousness in the brain. Yet it has to be admitted that this approach is limited to only explaining parts of experience (those biased by the past) and does not help in getting to terms with the deeper aspects of the conscious experience itself.

10.2.8 The Attention-Based CODAM Model

Attention is, in the CODAM approach, to be regarded as a control system. As such an attention signal, to move the focus of attention to another place or stimulus, is generated by a suitable control signal generator module (such as in parietal lobe). Thereby the attention focus is moved to the new place/stimulus, as coded by the activity of various lower level visual modules.

Fig. 10.1 The CODAM
model

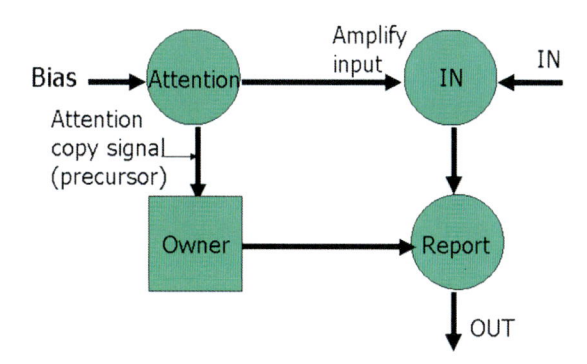

The CODAM model involves the use of a corollary discharge (a copy) of this signal causing the movement of the attention focus so as to provide greater efficiency in the process of moving the focus of attention. The existence of this attention copy signal is conjectural but provides attention control with the best means of ensuring speed and accuracy in attention movement. It thereby saves time (and related overall neuron energy) in processing, as well as working on the attended state of the environment (so reducing complexity by a large amount in complex environments). The resulting architecture is shown in Fig. 10.1.

Here a goal module, which provides the bias to where the attention focus must move, is indicated by the bias input on the left. This causes the "Attention" module to generate the movement of the focus of attention to the relevant biased position and so to amplify the input stimulus representing the attended stimulus. The IN module denotes the visual cortex for visual attention as before.

The Attention copy signal is shown proceeding from the "Attention" module to a module denoted "Owner" in Fig. 10.1. This acts as a buffer to hold the content of this signal for a short time. The signal will thereby be available for a number of things: to speed up the access of the attended stimulus activity into the WM module and also to correct any errors that might be made (such as allowing distracters to creep into the working memory module, so into the content of consciousness). The copy signal can also be used to increase the posterior-going attention signal if there is a possible problem with a distracter. This extension thereby allows corrections to be made to attention movement, in comparison with the rather inflexible ballistic control, in the absence of the attention copy signal.

It is the attention copy signal which, it is claimed in CODAM, provides the "owner" content of the relevant neural activity. It generates the experience of the "inner self", that of ipseity. As such the CODAM extension of Fig. 10.1 allows for the inclusion of the necessary complexity of consciousness in attention to begin, for example, to tackle the experiences of schizophrenics.

We add that the modules in Fig. 10.1 are partly observable in specified parts of the brain, as follows from brain imaging. Thus the "IN" module involves lower order visual cortices (for visual attention); the "Attention" module (generating the movement of the focus of attention) is observed in parietal lobes, as are various WM "Report" sites. The "Bias" input arises from the prefrontal cortex, where goals are being held. Only the site of the "Owner" module is uncertain, but is very likely in parietal as well. These sites all have associated object or spatial codes (at a high level, other than the "IN" component, which is coded in a hierarchical level of feature codes of observed objects).

The various modules are explained in the text by the figure. For a more detailed discussion, the interested reader is directed to Taylor (2007, 2010a, b).

We note that the "Owner" module codes for the inner self or ipseity of the subject. This includes information arising from the Attention signal generator as well as other high-level components of the brain's information processing. Since these high-level codes have no content of external objects per se, then the inner self is itself only experiencing such high levels codes and so no content as such. This agrees with the "nothingness" of Sartre, for example.

10.2.9   Further Models of Consciousness

There are other suggested models of consciousness that are difficult to analyse due to lack of clarity or of being rejected by their creators. Thus, the 40 Hz model of Crick and Koch (1990) has been criticised not only by other researchers, but also avowed as incorrect by themselves (Gold 1999; Crick 2004).

The later approach to consciousness of Crick and Koch (2003) presented no detailed neural model but only a general survey of overall brain processing and a set of ten selected features. The most important aspect in their latest "theory" appears to be that of requiring some neural activity "for a feature has to cross a threshold for consciousness of that feature". But the approach is too unclear to be usefully inserted in the WM or GW models since those have additional criteria associated with attention amplification and length of time of neural activity.

Further models are considered in Seth (2007), to which the reader is referred.

10.3   Criteria for the Review

Given such a plethora of models we have to begin to explore how well these models fit the expectations we should have of them. To do that we develop a set of criteria which are appropriate to use and which codify a range of these expectations. These criteria should apply to any model, so are beyond any intrinsic criticism of a model as developed in the brief reviews given in Sect. 10.2. The criteria are themselves chosen to emphasise what are to the author of this paper the most crucial aspects of experience and experimental data that a neural model of consciousness should satisfy. They build on standard scientific criteria built up over the last three centuries but are especially adapted to consciousness in terms of the paradigms that are especially selected. To keep the discussion sensible in size we have to exclude analysis of tests for animal consciousness, although that is an important problem.

In particular we consider the following criteria:

10.3.1 Fits to Experimental Data

Any scientific theory worth its salt must fit relevant data. It may be a difficult task to determine if such fitting is taking place, as instanced by the recent construction of the Large Hadron Collider (costing billions of dollars and involving thousands of scientists) to test the existence of the Higg's boson, supposedly generating mass for our observable universe.

In the case of consciousness, there is a large (and increasing) amount of data relevant to such testing. Indeed one can say that almost any human activity provides such data. However, in general such data is too complex, so the tasks set and the measurements arising have been slimmed down to provide simpler and more useful testing domains. At the same time there has occurred the development of brain imaging machines and of single cell/multi-unit recording techniques requiring such simpler test paradigms.

Such data provides measurements of both "outer" brain activities at different scales according to the measuring apparatus as well as "inner" experience. Most of such inner activity involves the report of a subject as to what they saw or heard of a set of input stimuli as part of an experimental paradigm set up by the relevant experimental team. But more recently the reported experience has been of broader nature, such as of day-dreaming or similar resting processes (Kounios et al. 2009).

It is essential that any scientific model of consciousness must be able to fit or at least explain (if only qualitatively) such experimental data as noted just above. Thus data, involving bi-stable visual experiences, such as arise in the Necker cube, or processing tasks like the attentional blink (Shapiro et al. 1994), need to be given if possible a quantitative explanation. Without that it is unclear that the proposed model of consciousness is able to be acceptable as a model properly based on brain action.

The attentional blink (Shapiro et al. 1994) was so called because stimuli were being observed by a subject at such a fast rate that when one "target" stimulus had been detected in the fast stream of distracter stimuli, then a further one – the "second target" – proved most difficult to detect. The time lapse between the first and second such targets for greatest such difficulty was observed to be about 270 ms, being the time that attention "blinks shut", being fully occupied with the first target.

The Necker cube is composed of the edges of a cube so drawn that when it is viewed the subject will observe on corner as being towards the viewer. After a second or so there is an automatic switch so that the opposite corner of the skeletal cube now projects forward according to the viewer. Whilst this is an important paradigm (with claims that certain features of such competitive processing imply the separation of attention and consciousness, Pastukhov and Braun 2005), it is not dwelt on further here since it could arise from low-level competitive maps between low-level visual codes for the two cube structures and so only arise exogenously into consciousness after the hard work was done. Phenomena such as the attentional blink are more important as arising from an attempted breakdown of attention control by the second target during the blink. This approach is parallel to the high-energy physics experimental methodology in breaking the articles down into their basic constituents, for attention control as to how the control can be broken down, under the harshest conditions, to find out how it is put together.

Further relevant data, beyond the attentional blink, are described and simulated more carefully in Taylor and Fragopanagos (2007). These include object substitution, in which an object, presented and then suitably masked (to be outside of awareness by the subject) is found to still be able to create a brain signal at a reasonably high level in the processing hierarchy. In particular it was observed that there is still the so-called N2 EEG-measured signal for the masked object (the N2 is about 180–280 milliseconds after the presentation of the visual stimulus).

Another paradigm relevant to attention and consciousness is that of change-blindness. In a quantitative study of Fernandez-Duque and Thornton (2000; see also Landman et al. 2003) stimuli of visual objects formed an equidistant set round a circle, and after a brief mask one of the objects was replaced by another. The subject had to report if this substitution had occurred or not. A cue was presented as to where the object substitution might occur either before, during or after the appearance of the mask. Data were collected on the accuracy levels of the subjects under these three different cueing conditions.

Further paradigms were also considered qualitatively, especially on blindsight (Kentridge et al. 1999), where an otherwise blind subject GY was able to employ attention cueing in his blind-field to improve his ability to detect a cue in that same field.

The bottom line for the models is that all such data should be able to be explained by a given over-arching brain-based model. That can be regarded as an extension of the Turing test for consciousness. The mechanical system purporting to be conscious is not now hidden behind a screen. The system is now being tested much more thoroughly by probing the inner workings of its "brain" at the same time as listening to its report as to what it is "experiencing". It must pass the test of all such results, with "brain" activity as well as responses being similar to those of a human. However, at the present state of available models, only a limited range of experimental data is being tested for as a first stage (although models based on an architecture arising from a broader range of data than another such model will clearly have an advantage in explaining more general data at a further stage of model testing). The results of such testing will be considered in more detail in Sect. 10.4, when the various appropriate tests are considered for the various models already discussed in the previous section.

10.3.2 The Presence of Attention

It is now pretty well universally accepted in neuroscience that attention is crucial for consciousness [see, however, the opposing view taken in Koch and Tsuchiya (2007); in any case, the claimed places where attention is not needed for consciousness tend to be abnormal, so can be neglected in any attempt to consider normal brain processes and in the distortions experienced in schizophrenia]. Thus various phenomena: inattentional blindness, the attentional blink (as discussed in the previous section), neglect (where a stroke causes a patient to be unable to attend to one or other side of their visual field or body and be unaware of what happens there), extinction (when a limited form of neglect takes place, with a patient unable to attend to and be aware of a stimulus, say in their left visual field if a competing stimulus is present also in their right visual field), etc. indicate that without attention to a stimulus there is no conscious experience of the stimulus itself. That is not to say that there may not be after-effects of the input of a subliminal stimulus, i.e. one is not able to gain awareness. Thus a stimulus exposed to a subject for only a few milliseconds, with no awareness of what was presented to the subject, may still lead to priming effects which can bias later processing (Bar and Biederman 1998).

We note that in Koch and Tsuchiya (2007) they define attention as top-down (endogenous); in this paper it is defined to include the exogenous or bottom-up form as well. There is considerable overlap between the circuits involved in both, as well as use by the TPJ-led circuit to allow for directing attention exogenously to the same endogenous circuitry (Corbetta et al. 2008).

It is generally accepted, as stated earlier, that consciousness requires attention (using attention, as we have noted, to denote either the exogenous or endogenous form). However, there are still phenomena claimed in Koch and Tsuchiya (2007) as being examples of "consciousness without attention". Thus natural scenes can be perceived in the "near absence" of attention. Various experiments reported on this use training to achieve reporting of these peripheral scenes during a dual task being performed at the centre of attention (Reddy et al. 2004). However, the operative phrase is "near absence": if there is no control of the amount of attention being allowed to stray from the centre (and with special training to achieve such splitting of the focus of attention such splitting may well be possible), the experiment is not a counter-example to the claim that consciousness needs attention. Even a small amount of attention applied to a salient target (such as a face) could well amplify the target enough for it to become conscious.

A further experiment was quoted in Tsuchiya and Koch (2010) to justify the authors' claim of consciousness without attention. This involved the process of reversal of ambiguous figures with or without attention being paid to the ambiguous figure, using a central demanding task to manipulate peripheral attention (Pastukhov and Braun 2005). The interesting result is inconclusive due to the possible division of the focus of attention, so that there could still be such division even though the central task was being performed at a 100% level of accuracy. That only results from three (practised) subjects were used in this case is a hint that they might have developed control powers to achieve such an attention division.

It is to be noted that in Sect. 10.5 on "awareness without attention" in Braun (2008) it is claimed that:

(1) The gist of natural scenes can be extracted before attention could be directed to them (such as in the first 30 ms after stimulus onset). However, whilst that first short glance may contain enough information for a gist to be extracted it need

not be the case that the gist comes into awareness 30 ms post-stimulus onset. There is no clear data on such timing but it very likely happens some several hundred milliseconds later.

(2) There is a relative independence of the need for attention to a salient object for it to enter into a subject's awareness. However, the remaining level of attention needed for the salient stimulus to become conscious is not clear and is very likely non-zero. But such attention movement is difficult to assess, being exogenous and in any case could function to bring about momentary attention focus to bear on the salient stimulus.

(3) Studies claiming to show discriminability of various features (luminance, chromatic contrast and some other features) at low attention levels also suffer the criticism of possible divided attention, as well as the further drawing of exogenous attention to stimulus causing the exogenous circuitry to be activated (Corbetta et al. 2008).

We thus conclude that, in spite of the above valiant but as yet incomplete attempts to show the contrary, attention acts as the gateway to consciousness. Without an attention control system as part of the model then only very simple experiences would be accessible to consciousness. In particular only in environments with a single stimulus would a subject be able to separate the sensory inputs to make any sense of them as single stimuli. Since such a criterion of single stimuli would be far too limiting for the claims made for all of the various models we considered in the previous section, then the possession of attention would seem critical for any model. If the model does not posses this faculty at a quantitative level (not purely qualitatively), then it can only be considered as incomplete.

10.3.3 As Providing an Explanation of Mental Diseases

This set of tests can be more properly considered as part of the tests under the heading of Sect. 10.3.1, since they involve experimental data both from the reports of sufferers of the various diseases and of the relevant damaged regions of their brain. In particular, we consider the experiences of schizophrenics to be the example bar none for a model of consciousness to be expected to explain. For schizophrenia has the greatest set of distortions to the self amongst mental diseases. One might consider that dementia possesses such a property, although the main defects there appear more in the light of memory loss and the associated emotional disturbances this causes. However in schizophrenia the sufferer talks about "losing my self, my I" and being "increasingly distanced from the world" (Sass and Parnas 2003; Cermolacce et al. 2007). It is such distortions which would appear initially to be directed at the inner self as an important but vulnerable component of the sufferers' experience. Any model of consciousness should be able to attempt to explain that loss of "I" in the schizophrenic. We will turn in Sect. 10.3.4 to a more complete discussion of the nature of "I", of the inner self of Western phenomenology.

10.3.4 Existence of an Inner Self

This is the most controversial of the set of criteria to be applied to models of consciousness. It relates closely to the remarks just made about the schizophrenic experience in Sect. 10.3.3, which is being increasingly and effectively interpreted as critically involving the distortion of the subject's inner self, and thereby being able to be analysed by any model of consciousness in which such a component of consciousness is created (Taylor 2010b). However, aspects to be explained in any model of the creation of consciousness must especially include the experiences of the normal subject (Taylor 2010a).

The "inner self" or pre-reflective self, as known by the Western phenomenologists (Husserl, Sartre, Merleau-Ponty and many others, as surveyed in Zahavi 2005), is the crucial component still missing from the neural network models discussed so far for consciousness and also from most Western analytic philosophers of Mind's discussion of the topic. The latter group regards the presence of some form of pre-reflective self or "ipseity" as crucial". Ipseity denotes the inner self, sometimes termed the "pre-reflective self" of Western phenomenology. In this paper, we identify ipseity with the inner self, and that with the pre-reflective self, although there may be subtle differences between them according to different definitions in Western phenomenology (Zahavi 2005); these differences will be ignored here. Thus we expect a satisfactory model of consciousness to include a description of how it leads to the experience of the inner self or ipseity.

We use the definition of ipseity (inner self) of p 428 (Sass and Parnas 2003): "…the experiential sense of being a vital and self-coinciding subject of experience or first person perspective on the world". Ipseity arises from a part of the self which is devoid of the components of the reflective self, which itself is composed of those characteristics of the self that can be obtain by reflecting, such as whether or not one has a beard or is impatient, and so forth. The pre-reflective or inner self or ipseity appears instead as content-free, and its existence provides a centre of gravity in which the ownership of ones' experiences is gathered. We will use only the phrase "inner self" to denote it from now on in this paper. We add finally that the inner self is more to be considered as only a part of the total experience of consciousness, so one cannot properly answer the question "Is the inner self conscious or not"? It contributes crucially to the conscious experience but cannot be that experience on its own.

Besides the claimed existence of the inner self is the proposed temporal flow of conscious experience. It is claimed to consist of three components, termed "pretention", the "primal moment" and "protention" (Sokolowski 2000). The first of these, pretention, is a part of preparatory brain activity just before the experience of consciousness. The second, the primal moment, is the short period of the experience of consciousness itself. The third component of the temporal flow of consciousness, protention, involves a decaying activity record of the primal moment in the brain.

It is clear that without an inner self there is no-one to experience the consciousness of any content on a WM site. Thus the inner self is the crucial component of the answer to the question raised by Nagel: "what is it like to be?" The immunity to error of Shoemaker (1968) (I can have no error about what "I" think) will be achieved in a situation in which the inner self has surety of experiencing exactly what it has expected. Finally, the "explanatory gap" (Levine 1983: the gap between first and third person accounts of experience) and Chalmer's hard problem (Chalmers 1995: how to get consciousness out of suitably designed matter structures) are solved by understanding in what manner the inner self interacts with the outer world through the WM sites holding the content of experience. Thus the inner self, according to these ideas, must be a crucial component of any self-respecting model of consciousness.

In order for a model of consciousness thus to be able to solve the problems which have been plaguing it in the West (as noted above as raised by Nagel 1974; Shoemaker 1968, Levine 1983 and Chalmers 1995), it needs:

(a) To have also a neural model of the inner self
(b) To have a model of the manner in which this inner self interacts with the attended WM containing the content of stimuli entering consciousness
(c) To be able to define the temporal flow of activity so it agrees with the three successive components: pretention – primal moment – protention, as specified above (Sokolowski 2000)

We add that none of the models but CODAM in Seth (2007) can be claimed to have an inner self in the sense use so far in this paper. There are claims of self-monitoring in some of the models, but that is not the same as possession of such an inner self. The claim is made in Seth (2007) that "High-level conceptual models can provide insights into the processes implemented by the neural mechanisms underlying consciousness, without necessarily specifying the mechanisms themselves. Several such models propose variations of the notion that consciousness arises from brain-based simulation of organism–environment interactions. These models illuminate in particular two fundamental aspects of phenomenology: the attribution of conscious experience to an experiencing 'self', and the first-person perspective that structures each conscious scene." However, such knowledge does not provide an experiencing self at the level of the inner or pre-reflective self being discussed and developed in CODAM. The self described by Seth is that of the reflective self, corresponding to knowledge of what one looks like in the mirror and what one is doing currently. But the pre-reflective self is a monitor only of the corollary discharge of the attention control signal, being a pre-amplifier of contentful input about to enter a WM site. This is a concept foreign to the models discussed by Seth (2007); these models fail all the criteria of Sect. 10.3.

10.4   The Test Results

We present the test results for the application of each of the four criteria of the last section in tabular form in Table 10.1. We have, as the first column, whether or not the model/approach fits relevant data. As data we single out a set of experimental paradigms which are of relevance to the creation of consciousness: blindsight,

Table 10.1 Results of the tests

Test criterion	Fits relevant data	Contains attention	Explains mental disease	Exists inner self
Model name				
1. HOT	X	X	X	X
2. WM	I	I	X	X
3. GW	I	I	X	X
4. Complexity	I	X	X	X
5. Recurrence	I	I	X	X
6. Neural field	I	I	X	X
7. Relational mind	I	I	X	X
8. CODAM	$\checkmark$	$\checkmark$	P	$\checkmark$

P for partial, denoting some relevant paradigms explained in depth; *I* denotes "incomplete" as mainly a descriptive account; *X* denotes absent or NO; $\checkmark$ denotes present or YES

inattentional blindness, object substitution, attentional blink and several more (Taylor and Fragopanagos 2007). It is numerical fits to these paradigms (where numerical data is present) which are to be calculated by any of the models under discussion (as done in the paper just referred to). If the relevant model does not have a specific neural architecture to enable this to be done, and these phenomena have not even been discussed in the literature, then the model fails on this task. In the second column are entries indicating the degree to which the model is based on attention control. The third column explores how far the model explains distortions in schizophrenic and other mental disease patients. Finally, the fourth column states if the inner self can be glimpsed in any manner through the model.

We now discuss each of the entries in more detail.

Let us go through the list of eight models whose results are presented in Table 10.1 to justify and expand on the results given there.

10.4.1   Higher Order Thought

This model has, in the hands of Rolls (this book and 2007) been especially extended to the linguistic domain by means of what Rolls has termed HOST (higher order syntactic thought). He proposes specifically (Rolls 2007, p 962) that it "feels like something to be an organism that can think about its own linguistic and semantically-based thoughts". The neural basis of this approach is possible, but still would seem to suffer from the problems of circularity and regress, criticisms already raised in our review of the HOT approach in Sect. 10.2.1. A strongly linguistic basis for HOT does not seem to avoid these earlier problems.

At the same time there is little to be seen in either the past literature on HOT or in the HOST update to put any success under the first column of Table 10.1 and similarly there is no success for the remaining columns. There is no clear explanation of distorted experiences in mental diseases. There is also a complete absence of any ability of the model to create an inner self. Altogether the model does poorly on these four tests. However, we will see shortly that it fits on neatly to CODAM.

10.4.2 Working Memory

Reasonably clear entries arise from the review in Sect. 10.2.2: there are various models over the past of WM applied to the phonological loop and various memory phenomena such as recency and primacy effects. Yet these paradigms are not those proposed as most relevant to consciousness (Taylor and Fragopanagos 2007), so explaining the "I" entry. Nor have there been attempts to apply these ideas to mental diseases. Nor is there any hint of how the inner self could arise from purely continued brain activity over whatever length of time. This is also true of the stochastic automata model of Aleksander and Morton (2007) that has extended the WM approach.

10.4.3 Global Workspace

Similar features apply to the GW model as to the WM model, so explaining the entries in the table for this approach. There is a broader range of relevant experimental data which GW explains as compared to a single WM, although this is still discussed only in a qualitative manner (Baars 2002). Thus the GW approach does not succeed in getting through the more down-to-brain tests of the previous section.

10.4.4 Complexity

Some data has been explained (on complexity) by the proponents of the approach, but no data relevant to the key experiments discussed in Taylor and Fragopanagos (2007). Moreover, there is no hint presently of attention, distorted mental disease experience or the inner self being explained through complexity.

10.4.5 Recurrence

The various approaches to recurrence mentioned in Sect. 10.2.5 have led to specific models of pattern processing in vision and improved understanding of difficulties for such models assuming only simple feed-forward visual processing. Attention can be included if desired and many models of attention, using recurrence at various stages, inside neural models, have been developed and used in vision tasks. However, this inclusion is incomplete in terms of explaining relevant phenomena such as the attentional blink and the related phenomena mentioned at the beginning of this section and associated with the entry in Table 10.1 (Taylor and Fragopanagos 2007). Mental disease and the nature of the inner self have so far been outside these model approaches emphasising recurrence to approach consciousness.

10.4.6 Neural Field Theory

The neural field can be recognised as a framework inside which to develop a broad range of neural architectures to explain various phenomena in brain processing (visual coding, illusions of colour and other visual processing, etc.). The use of the continuum has been especially emphasised by Freeman (2007) and important progress made thereby in understanding the more general dynamical systems approach to neural processing. However, the nature of consciousness and its generation through such processing is still not completely clear, so the approach is still in early days of development. That is why only incomplete entries are given for fitting data and attention inclusion. No progress has occurred in mental diseases and the inner self. Similar assessments apply for the neuronal model of the brain's electric field generated by neuronal activity, as reported in Laberge and Kasevich (2007).

10.4.7 Relational Mind

This approach has the same level of progress as the neural field approach (in spite of their qualitative dissimilarity), so has similar entries in the table.

10.4.8 CODAM

CODAM has presently the best set of marks in the table of all entries. It can be used to fit the important experimental data of the puzzling attention/conscious paradigms (Taylor and Fragopanagos 2007) and is automatically embedded in the latest attention architectures. CODAM automatically contains the inner self, as discussed in various recent publications on it and especially in Taylor (2010b).

Experiences of subjects suffering a variety of mental diseases have also been considered, although only in a qualitative manner as yet (Taylor 2006), except for schizophrenia. In that case, a more complete discussion has been developed recently of how distortions of the attention control system of CODAM can help explain the main symptoms and stages of schizophrenia (Taylor 2010a).

10.4.9 Possible Model Fusion

With the set of models of consciousness laid out before us how might we best proceed? One useful approach would be to see how we might fuse together the insights of as many of the models as possible. If we take the most successful model, that of CODAM, we can see how that can be done with the HOT, WM, GW, Recurrence, Relational and Complexity models.

Thus for HOT we have already noted that the corollary discharge (attention copy) signal of CODAM is to be regarded as a second-order thought/state in the HOT version. However, its presence does not cause the infinite regress (or circularity) as a danger noted for the original version of the HOT since the corollary discharge is content free and is not a "thought" in the standard way: the consciousness of the corollary discharge is fleeting and of a different character (being contentless) as compared to the first order state (composed of the content of experience: the red of the rose, etc., through the relevant working memory activation along with correlated feature map activations).

The WM approach is already built into CODAM (as the sensory buffer). The GW has been noted earlier as being complimentary to CODAM, with CODAM entering as the mechanism to provide the inner self and phenomenological experience to the set of activations entering the GW, each through their own relevant WM. Recurrence is also built into CODAM. Complexity along the lines of the traffic mechanism or the original complexity factors can be used to check for the presence of what has been proposed in Taylor (2010b) as the "inner self network". This net is expected to have high complexity as part of brain activity during consciousness due to the need for extended activity alerting distant parts of the network as to what was happening elsewhere. The Relational Mind model can also be added on to CODAM by addition of a long-term memory system to the CODAM architecture of Fig. 10.1.

10.5   Conclusions

In this paper, we have tried to compare and contrast the various main neural network models of the creation of consciousness in the human brain that are presently at the centre of considerable discussion. Having briefly reviewed some of these models in Sect. 10.2, we considered how they could be assessed. In Sect. 10.3, we introduced a set of four criteria: fitting relevant data, containing attention, explaining mental diseases and finally being able to give a hint on the nature of the creation of the inner self.

On applying these four criteria we find only one model (that of CODAM) able to presently be relatively successful on all four of these counts. That does not mean that none of the other models could be as effective as CODAM, but only those they presently have not necessarily reached the same stage of development. An example of this is the HOT model, which has little specific neural architecture proposed to back it up and allow more quantitative criteria, such as in the table, be applied. However, some of the models would need considerable conceptual development to be able to reach the same level of applicability to the problems of consciousness (such as those of schizophrenia) as CODAM. That is possible in terms of the fusion methodology proposed at the end of the previous section. Other approaches are also certainly possible.

It may be argued that the criteria in the table are not relevant to assessing such models, especially the last one of the creation of the inner self. Even the existence of an inner self may be denied by some workers in the field of consciousness. The first three criteria of the table are part of the ongoing approach of science: to explain the relevant part of the world in a quantitative manner. It is only the last, the existence of an inner self as created by the dynamics of a given model, which does not fit that solely quantitative approach. Yet the inner self is relevant to be taken account of scientifically because it involves report. Report has been a mainstay of most paradigms in experimental psychology. Such report on the inner self is about a subtle component of experience. But it cannot be excluded from the class of inner reports, in spite of its more qualitative rather than quantitative aspect. It might ultimately be possible to bring the inner self into the quantitative universe by quantifying the level of the interaction between the inner self and the external world. If it is reduced in schizophrenia why not ask sufferers to quantify by how much their inner selves had been reduced? That may be very hard to assess but should be tried. In any case discussion of the inner self is still valid from a scientific point of view, since it would appear to have strong evidence for its existence from reports of its distortion in schizophrenia as well as from our own experience throughout each of our lives. To claim that we live without an inner self is as much as stating "I am a zombie". Indeed without an inner self we would indeed be in such as state. For then, as already mentioned earlier, there would be no experiencer, no-one for whom "it is like to be". Such is the fate of all of the models of Seth (2007) except for the briefly mentioned CODAM model.

From the model viewpoint, consciousness research still has a multitude of models to contend with. Those who have proposed their own pet model strongly hold on to it, in spite of possible evidence to the contrary or internal logical difficulties. The number of citations to a given model also depends on how strongly the progenitor shouts out from the roof-tops (for that read conferences) about the greatness of their model. Yet real progress will be made by both theoretical confrontations between models and also their confrontation with experimental data, as presented, for example, in Table 10.1. Only through such confrontations can we hope to make real progress in the important subject of understanding consciousness.

References

Aleksander I and Morton H (2007) Phenomenology and digital neural architectures. Neural Networks 20:932–937

Amari S-I (1977) Dynamics of pattern formation in lateral-inhibition type neural fields. Biological Cybernetics 27:77–87

Baars B (1998) A cognitive theory of consciousness. Cambridge: Cambridge University Press

Baars B (2002) The conscious access hypothesis: origins and recent evidence. Trends in Cognitive Sciences 6(1):47–52

Bar M and Biederman I (1998) Sublimal visual priming. Psychological Science 9(6):464–469

Braun J (2008) Attention and awareness. In T Bayne, A Cleeremans and P Wilken. Oxford: Companion to Consciousness Oxford

Baars BJ and Franklin S (2007) An architectural model of conscious and unconscious model of global brain functions: Global workspace theory and IDA Neural Networks 20(9):955–961

Bressler SL, Tang W, Sylvester CM, Shulman GL and Corbetta M (2008) Top-Down control of human visual cortex by frontal and parietal cortex in anticipatory visual spatial attention. Journal of Neuroscience 28(40):10056–10061

Cermolacce M, Naudin J and Parnas J (2007) The "minimal self" in psychopathology: Re-examining the self-disorders in the schizophrenia spectrum. Consciousness and Cognition 16:703–714

Carruthers P (1996) Natural theories of consciousness European Journal of Philosophy 6:203–222

Chalmers D (1995) The conscious mind: in search of a fundamental theory. Oxford: Oxford University Press

Corbetta M, Patel G and Shulman G (2008) The Reorienting System of the Human Brain: From Environment to Theory of Mind. Neuron 58:306–324

Crick FHC and Koch C (1990) Towards a neurobiological theory of consciousness. Seminars in the Neurosciences 2:263–275

Crick FHC (2004) The astonishing hypothesis. Cambridge: Cambridge University Press

Crick FHC and Koch C (2003) A framework for consciousness. Nature neuroscience 6:119–126

Edelman GM and Tononi G (2000) A universe of consciousness. New York: Basic Books

Edelman GM (1992) Bright air brilliant fire. New York: Basic Books

Fernandez-Duque D, Thornton JM (2000) Change detection without awareness. Visual Cognition 7:323–344

Freeman WJ (2007) Indirect biological measures of consciousness from field studies of brains as dynamical systems. Neural Networks 20:1021–1031

Gold I (1999) Does 40Hz oscillation play a role in visual consciousness? Consciousness & Cognition 8:186–195

Gregoriou GG, Gotts SJ, Zhou H and Desimone R (2009) High-Frequency, Long-Range Coupling Between Prefrontal and Visual Cortex During Attention. Science 324:1207–1210

Grossberg, S (1976) Adaptive pattern classification and universal recoding, I: Parallel development and coding of neural feature detectors & II: Feedback, expectation, olfaction, and illusions. Biological Cybernetics 23:121–134 & 187–202

Kentridge RW, Heywood CA and Weiskrantz L (1999) Attention without awareness in blindsight. Proceedings of Biological Sciences 266:1805–1811

Koch C and Tsuchiya N (2007) Attention and consciousness: two distinct brain processes Trends in Cognitive Sciences 11(1):16–22

Kounios et al (2009) The origin of insight in resting state behaviour. Neuropsychologia 46(1):281–291

Laberge D and Kasevich R (2007) The apical dendritic theory of consciousness. Neural Networks 20:1004–1020

Lamme VAF (2006) Towards a true neural stance on consciousness Trends in Cognitive Sciences 10(11):494–501

Landman R, Spekreijse H and Lamme VAF (2003) Large capacity storage of integrated objects before change blindness. Vision Research 43:149–164

Levine J (1983) Materialism and qualia: the explanatory gap. Pacific Philosophical Quarterly 64:354–361

Nagel T (1974) What is it like to be a bat? Philosophical Review 83:434–450

Pastukhov A and Braun J (2005) Perceptual reversals need no prompting from attention. Journal of Vision 5:1–13

Pollen DA (2003) Explicit neural representations, recursive neural networks and conscious visual perception. Cerebral Cortex 13(8):807–814.

Reddy L, Wilken P and Koch P (2004) Face-gender discrimination is possible in the near-absence of attention. Journal of Vision 4:106–117

Rolls ET (2007) A computational neuroscience approach to consciousness. Neural Networks 20:962–982

Rosenthal D (1986) Two concepts of consciousness. Philosophical Studies 49:329–359

Rosenthal D (1993) Thinking that one thinks. In G Humphreys and M Davies (eds) Consciousness. Oxford: Blackwell

Rowlands M (2001) Consciousness and higher-order thoughts. Mind & Language 16(3):290–310

Sass LA and Parnas J (2003) Schizophrenia, consciousness and the self. Schizophr Bull 29(3):427–444

Seth A (2007) Models of consciousness. Scholarpedia 2(1):1328

Seth A, Izhikevich E, Reeke GN and Edelman GM (2006) Theories and measures of consciousness: An extended framework. Proceedings of the National Academy of Science USA 103(28):10799–10804

Shanahan MP (2005) Global access, embodiment, and the conscious subject. Journal of Consciousness Studies 12(12):46–66

Shapiro KL, Raymond JE and Ansell KM (1994) Attention to visual pattern information produces the attentional blink in rapid serial visual presentation. Journal of Experimental Psychology: Human Perception & Performance 20:357–371

Shoemaker S (1968) Self Reference & Self-Awareness, Journal of Philosophy 65:555–567

Sokolowski R (2000) Introduction to phenomenology. Cambridge: Cambridge University Press

Taylor JG, Horwitz B, Shah NJ, Fellenz WA, Mueller-Gaertner H-W and Krause JB (2000) Decomposing memory: functional assignments and brain traffic in paired word associate learning. Neural Networks 13(8–9):923–940

Taylor JG and Fragopanagos N (2007) Resolving some confusions over attention and consciousness. Neural Networks 20:993–1003

Taylor JG (2007) The CODAM model: Through attention to consciousness. Scholarpedia 2(11):1598

Taylor JG (2006) The mind: a user's manual. Chichester: Wiley

Taylor JG (1999a) The race for consciousness Cambridge, MA: MIT

Taylor JG (1999b) Neural bubble dynamics in two dimensions: foundations. Biological Cybernetics 80:5167–5174

Taylor JG (2010a) A neural model of the loss of self in Schizophrenia Bulletin (in press)

Taylor JG (2010b) The "I" s eye view of itself. Journal of Consciousness Studies 17(1/2), Jan–Feb, 2010

Tononi G (2008) Consciousness as integrated information: a provisional manifesto. Biological Bulletin 215:216–242

Tsuchiya N and Koch C (2010) Attention and consciousness/opposite effects Scholarpedia 5(1):4173

Wallace R (2005) Consciousness: A mathematical treatment of the global neuronal workspace model. Berlin & New York: Springer

Zahavi D (2005) Subjectivity and selfhood: investigating the first-person perspective Cambridge, MA: MIT

Part II
Cognitive Architectures

Vassilis Cutsuridis, Amir Hussain, and John G. Taylor

In this part, leading computational scientists present architectures, algorithms and systems empowered with cognitive capabilities of the various components of the perception–action cycle, namely perception, attention, cognitive control, decision making, conflict resolution and monitoring, knowledge representation and reasoning, learning and memory, planning and action, and machine consciousness. These systems are *minimally* guided by knowledge of the human and animal brain. Instead, they make use of knowledge from the areas of cognitive science, computer vision, cognitive robotics, information theory, machine learning, computer agents and artificial intelligence.

In the chapter entitled "Vision, attention control and goals creation system", Rapantzikos, Avrithis and Kollias present computation solutions to the four functions of the attentional process: (1) the *bottom-up* process, which is responsible for the *saliency* of the input stimuli, (2) the *top-down* process that bias attention towards known areas or regions of pre-defined characteristics, (3) the *attentional selection* that fuses information derived from the two previous processes and enables focus and (4) the *dynamic evolution* of the attentional selection process.

In the chapter entitled "Semantics extraction from multimedia data: an ontology-based machine learning approach", Petridis and Perantonis present a machine learning method for extracting complex semantics stemming from multimedia sources. The method is based on transforming the inference problem into a graph expansion problem, expressing graph expansion operators as a combination of elementary ones and optimally seeking elementary graph operators. The latter issue is then reduced to learn a set of soft classifiers, based on features each one corresponding to a unique graph path. The advantages of the method are demonstrated on an athletics web-pages corpus, comprising images and text.

In the chapter entitled "Cognitive algorithms and systems of episodic memory, semantic memory and their learnings", Zhang reviews cognitive systems that mimic human explicit memory and its impairments in anterograde, retrograde and developmental amnesias, and semantic learning deficit.

In the chapter entitled "Motivational processes within the perception-action cycle", Sun and Wilson present the CLARION cognitive architecture. The CLARION is an integrative cognitive architecture, consisting of a number of distinct subsystems with specific functionalities: action control, general knowledge maintenance, motivation and drives, and action control and modification based on the action's success or failure.

In the chapter entitled "Error monitoring, conflict resolution and decision making", Lima addresses the problem of plan representation, analysis and execution in multi-robot systems using a well-known formal model of computation: Petri nets. He reviews some of the Petri net-based approaches to robot task modelling described in the literature, the formal models they introduced and some of the results obtained. He then introduces his proposed multi-robot task model, the corresponding plan representation by Petri nets, how to analyse plan performance in the presence of uncertainties and some examples of application to robot soccer. He concludes with a summary on what has been accomplished so far under this line of research, the success stories, limitations found, future challenges and some suggestions on how to address their solution.

In the chapter entitled "Developmental learning of cooperative robot skills: a hierarchical multi-agent architecture", Karigiannis, Rekatsinas and Tzafestas present a new framework of developmental skill learning process by introducing a hierarchical multi-agent architecture. The model is then tested and evaluated in two numerical experiments, one related to dexterous manipulation, and the other to cooperative mobile robots.

In the chapter entitled "Actions and imagined actions in cognitive robotics", Mohan, Morasso, Metta and Kasderidis describe the various internal models for real and mental action generation developed in the GNOSYS Cognitive architecture and demonstrate how their coupled interactions can endow the GNOSYS robot with a preliminary ability to virtually manipulate neural activity in its mental space to initiate flexible goal-directed behaviour in its physical space. The performance of these models is then tested against various experimental conditions and environments.

In the chapter entitled "Cognitive algorithms and systems: reasoning and knowledge representation", Garcez and Lamb describe computational models with integrated reasoning capabilities, where the neural networks offer the machinery for cognitive reasoning and learning, while symbolic logic offers explanations to the neural models facilitating the necessary interaction with the world and other systems.

In the chapter entitled "Information theory of decisions and actions", Tishby and Polani address the question of in what sense the "flow of information" in the perception–action cycle can be described by Shannon's measures of information introduced in his mathematical theory of communication. They describe that decision and action sequences turn out to be directly analogous to codes in communication, and their complexity – the minimal number of (binary) decisions required for reaching a goal – directly bounded by information measures, as in communication. They consider the future expected reward in the course of a behaviour sequence towards a goal (value-to-go) by estimating the cumulated information processing cost or bandwidth required to specify the future decision and action sequence (information-to-go). They conclude by obtaining new algorithms for calculating the optimal trade-off between the value-to-go and the required information-to-go, unifying the ideas behind the Bellman and the Blahut-Arimoto iterations.

In the final chapter entitled "Artificial consciousness", Chella and Manzotti provide answers to what is consciousness, whether it is a physical phenomenon and how can it be replicated by an artificial system designed and implemented by humans.

Chapter 11
Vision, Attention Control, and Goals Creation System

Konstantinos Rapantzikos, Yannis Avrithis, and Stefanos Kolias

Abstract Biological visual attention has been long studied by experts in the field of cognitive psychology. The Holy Grail of this study is the exact modeling of the interaction between the visual sensory and the process of perception. It seems that there is an informal agreement on the four important functions of the attention process: (a) the *bottom-up* process, which is responsible for the *saliency* of the input stimuli; (b) the *top-down* process that bias attention toward known areas or regions of predefined characteristics; (c) *the attentional selection* that fuses information derived from the two previous processes and enables focus; and (d) the *dynamic evolution* of the attentional selection process. In the following, we will outline established computational solutions for each of the four functions.

11.1 Overview

Most of our impressions and memories are based on vision. Nevertheless vision mechanisms and functionalities are still not apparent. How do we perceive shape, color, or motion and how do we automatically focus on the most informative parts of the visual input? It has been long established that primates, including human, use focused attention and fast saccades to analyze visual stimuli based on the current situation or the desired goal. Neuroscientists have proven that neural information related to shape, motion, and color is transmitted through, at least, three parallel and interconnected channels to the brain rather than a single one. Hence a second question arises related to how these channels are "linked" in order to provide useful information to the brain.

The Human Visual System (HVS) creates a perceptual representation of the world that is quite different than the two dimensional depiction of the retina.

K. Rapantzikos (✉)
Image, Video and Multimedia Systems Laboratory, Computer Science Division,
School of Electrical and Computer Engineering, National Technical University of Athens,
Iroon Polytexneiou 9, 15780 Zografou, Greece
e-mail: rap@image.ece.ntua.gr

V. Cutsuridis et al. (eds.), *Perception-Action Cycle: Models, Architectures,*
and Hardware, Springer Series in Cognitive and Neural Systems 1,
DOI 10.1007/978-1-4419-1452-1_11, © Springer Science+Business Media, LLC 2011

This perceptual representation enables us to, e.g., identify same objects under quite different conditions (illumination, noise, perspective distortion, etc.), especially when we are searching for a target object (*goal*). These abilities led the experts to distinguish between *bottom-up* (stimuli-based) and *top-down* (goal-oriented) processes in the HVS. Bottom-up attentional selection is a fast, and often compulsory, stimulus-driven mechanism. Top-down attentional selection initiates from the higher cognitive levels in the brain that influence the attentional system to bias the selection in favor of a particular (or a combination of) feature(s). Only information about the region that is preattentively extracted can be used to change the preferences of the attentional system.

11.2   Computational Models of Visual Attention

From the computer vision point of view, all aspects related to the visual attention mechanism are quite important. Creating constraints about what and where to "look" is a great benefit for, e.g., object detection and recognition tasks, visual surveillance, and learning of unknown environments (robotic vision). Computational modeling of visual attention is mainly related to *selective visual attention* that includes *bottom-up* and *top-down* mechanisms, *attentional selection*, and *dynamic evolution* (Sternberg 2006).

11.2.1   Bottom-Up Visual Attention

The *Feature Integration Theory* (*FIT*), introduced in 1980 (Treisman and Gelade 1980), is considered as the seminal work for computational visual attention based on a *master map*. The theory inspired many researchers and evolved toward current research findings. The main idea is that "different visual features are registered automatically and in parallel across the visual field, while objects are identified separately and only thereafter at a later stage, which requires focused attention" (Treisman and Gelade 1980).

   According to this theory, the visual features (e.g., intensity, color, orientation, size, and shape) are linked after proper attentional processing. This observation is in accordance with the two-stage attentional process of James (James 1890/1981), namely the preattentive and attentive ones. Information from the *feature maps* – topographical maps that highlight saliency according to the respective feature – is collected in the master map. This map specifies *where* (in the image) the entities are situated, but not *what* they are. Scanning serially through this map directs the focus of attention toward selected scene entities and provides data useful for higher perception tasks. Information about the entities is gathered into so-called *object files*. An object file is a midlevel visual representation that "sticks" to a moving object over time on the basis of spatiotemporal properties and stores (and updates) information about that object's properties.

Based on the FIT model, Koch and Ullman introduced a complete computational architecture of visual attention (Koch and Ullman 1985). The idea is that several features are computed in parallel, and their conspicuities are collected in a saliency map (the equivalent of a master map). A *Winner-Take-All* (WTA) network determines the most salient location in this map, which is routed to a central representation, where more complex processing might take place. One of the most widely exploited computational models is the one introduced by Itti et al. (1998), which is based on the model of Koch and Ullman and hypothesize that various visual features feed into a unique saliency map (Koch and Ullman 1985) that encodes the importance of each minor visual unit. Among the wealth of methods based on saliency maps, the differences are mainly due to the selection of feature maps to be included in the model [contrast (May and Zhang 2003), color (Itti et al. 1998), orientation (Itti et al. 1998), edges (Park et al. 2002), size, skin maps (Rapantzikos and Tsapatsoulis 2005), texture, motion (Rapantzikos and Tsapatsoulis 2005; Milanese et al. 1995; Abrams and Christ 2003), etc.], the linear, or nonlinear operations [multiscale center-surround (Itti et al. 1998), entropy (Kadir and Brady 2001), etc.] applied to these maps and the final fusion to produce the saliency map (min/max/sum fusion, learning, etc.). Recent methods use more advanced features like wavelet responses learned through Independent Component Analysis (ICA) (Bruce and Tsotsos 2009; Torralba et al. 2006).

Computing saliency involves processing the individual feature maps to detect distinguished regions and fusing them to produce the final saliency distribution of the input. A well known biologically motivated operator is the center-surround one that enhances areas that pop out from the surroundings (Itti et al. 1998). This operator presumes a multiscale representation of the input with the center lying on a fine level and the surround on a coarser one. Kadir et al. (2001) use an entropy measure to compute saliency and derive salient regions that represent well the underlying scene. Their detector is based on the entropy measure in a circle around each pixel (spatial saliency) and in a range of scales (scale saliency). Improvements of this detector have been proposed focusing mainly on making it robust under affine transformations of the input or computationally lighter (Shao et al. 2007).

Recently, information theory has been used to model the importance, saliency, or surprise [a notion introduced by Itti and Baldi (2009)]. As the authors claim "Life is full of surprises, ranging from a great Christmas gift or a new magic trick, to wardrobe malfunctions,..." (Itti and Baldi 2009), and therefore Itti and Baldi proposed a mathematical model of surprise based on Bayesian theory. Based on knowledge or experience, an observer is watching a scene having a priori thoughts about it. Hence, surprise is defined as a great change between what is expected and what is observed. The expectations of the observer could either be modeled using a priori information or by the temporary scene context. Toward this direction, Bruce and Tsotsos (2006, 2009) compute saliency based on information maximization. They define saliency by quantifying Shannon's self information of a local image patch. In order to produce a reasonable estimate of the probability distribution – necessary to compute the information measure – the authors perform ICA to a large sample of patches from natural images. If a patch can be adequately predicted by this distribution, then it is not considered salient. Overall, entropy-based approaches put emphasis on the structural complexity of the scene, while information maximization methods enhance regions that differ from their context. Both approaches are biologically plausible as shown by experiments in the field (Treisman and Gelade 1980; Leventhal 1991; Sillito and Jones 1996).

A biologically plausible solution of attentional selection has been proposed by Tsotsos et al. (1995) that is also based on the spatial competition of features for saliency. The *selective tuning model* exploits a hierarchy of WTA networks that represent the input at increasing scales. The higher levels (coarse) guide the lower ones (finer) toward selecting salient regions of increasing detail, which in turn provide feedback to their "ancestors." Practically, each level is a saliency map that communicates with the neighboring levels to define the saliency value of each neuron. The main advantage of this model is the fact that there is no need for a single – and often oversimplifying – centralized saliency map, since saliency is computed through *distributed competition.*

11.2.2   Top-Down Visual Attention

Bias toward specific features or areas of interest can be incorporated through a top-down channel independently of the bottom-up architecture. Top-down influence can be modeled either by biasing specific areas of the saliency map or by adapting the weights of the intermediate conspicuity maps. Correspondingly in the selective tuning model, the top-down weights of the hierarchy can be used to bias attention toward a known target. Such an approach has certain similarities with the Guided-Search Model (GSM) proposed by Wolfe et al. (Wolfe 1994, 2007; Wolfe et al. 1989), where the weights of the features or the neurons are changed according to existing knowledge. Unlike FIT this model does not follow the idea of separate maps for each feature type, it defines only one map for each feature dimension, and within each map different feature types are represented. Comparable to the saliency map of location in FIT, there is an activation map in which the feature maps are fused. But in contrast to at least the early versions of FIT, in GSM the attentive part profits from the results of the preattentive one. The fusion of the feature maps is done by summing them. Additionally to this bottom-up behavior, the model includes the influence of a priori knowledge about the target by maintaining a top-down map that selects the feature type which distinguishes the target best from its distracters. Frintrop et al. have recently proposed similar attention models and applied them to real applications (Frintrop et al. 2005; Frintrop and Cremers 2007). Lee et al. (2005) use a neural network-based model with higher level cues guiding the attention (shapes, faces, etc.).

11.2.3   Attentional Selection: Attention as a Controller

The simpler approach to select the region to attend is the use of a fixed shape (e.g., a circle of fixed radius) located at the maximum of the saliency distribution.

Moving the focus-of-attention (FOA) is an important part of the models. In order to avoid revisiting the same area, most of them use an *inhibition-of-return* approach that does not permit successive focus of the same region. This is the approach, e.g., of the standard model of Itti and Koch, where FOA is the only active one at each iteration (the rest are ignored). This is the approach adopted by most computational models so far.

Nevertheless, experimental evidence exists that supports the direct selection of objects to attend, rather than a fixed shape region around them (Duncan 1984) (Duncan's Integrated Competition Hypothesis). This is related to the notion of proto-objects introduced by Walther and Koch (2006), where the FOA is adapted to the objects we expect to find in the scene. The shape of the FOA is formed by thresholding the conspicuity map of the Itti's model that contributes more to the saliency map. The resulting FOA mask is then used as a top-down bias to the attentional selection process. Sun and Fisher (2003) use a different notion of attentional selection based on Gestalt groupings. When a potential object is attended, the model extracts its groupings (collections of pixels, features, patches, etc.) and compares them against the next FOA. If the groupings of the attended areas remain similar, then the model considers the underlying FOA as belonging to the same object. On the contrary, if the groupings are dissimilar, then another object is attended.

Recently, Cutsuridis proposed a cognitive model of saliency overt attention, and natural picture scanning that unravels the neurocomputational mechanisms of how human gaze control operates during active real-world scene viewing. It is based on both the resonance of top-down and bottom-up processes (Cutsuridis 2009). The resonance is accomplished by a value module (dopamine) that ensures the similarity of the top-down and bottom-up salient representation by increasing/decreasing the SNR of neural representations. The model is heavily supported by the neuroscientific evidence and addresses important questions in active vision (overt attention). It also provides the insights on the neural mechanisms of inhibition of return and focus of attention.

11.2.4   CODAM: COrollary Discharge of Attention Movement

Narrowing down the role of attention to controlling the evolution of the FOA after creating a saliency map may work well for few applications, but it is far from the real. Attention is the result of a complex interplay between numbers of brain modules across parts of the brain, as well as complex temporal dynamics involved in this interplay, which are not necessarily modeled in the previous systems. As said, attention acts as a filter to concentrate processing resources on a particular salient stimulus and remove distracters. The attended stimulus, held for a period on the attended state estimator buffer, can then be processed by further higher level mechanisms as might be involved, for example, in thinking, reasoning, comparison, imagining, and so on.

CODAM proposed by Taylor et al. (2000, 2003, 2007) is a neural network model for the brain-based creation of awareness or consciousness that uses the attention copy signal as the crucial component. The attention movement signal generator (inverse model controller or IMC), biased by a set of predefined goals, sends a new attention signal to lower level cortical stimulus activity; this can be summarized as a two-stage model, in which the higher level control system generators (goals and IMC) send attention signals to lower level cortical representations (Taylor et al. 2007). The Control model is composed of the following modules: (a) input modules (early visual hierarchy in vision); (b) inverse model controller (IMC), as the generator of a feedback attention control signal to amplify the attended stimulus activity and reduce that of distracters (acting as a saliency map, running a competitive process to attend to the most salient stimulus input); and (c) goals module, to allow for endogenous bias to the IMC for goals created from other sources in a top-down manner.

11.3 Applications

Applications are numerous, since saliency fits well fit with many of the existing approaches in computer vision. We present a concise report on applications of both spatial and spatiotemporal saliency models.

11.3.1 Scene/Object Recognition

Rutishauer et al. (2004) investigate empirically to what extent pure bottom-up attention can extract useful information about objects and how this information can be utilized to enable unsupervised learning of objects from unlabeled images. They show that recognition performance for objects in highly cluttered scenes can be improved dramatically and that other problems, such as learning multiple objects from single images, are only possible using attention. This evidence is further confirmed in Walther et al. (2004, 2005). Specifically, the authors combine Itti's model with a recognition model based on Lowe's SIFT features (Lowe 1999). The final process includes a top-down inhibition-of-return method that prohibits areas of nonproto objects (Walther and Koch 2006) to become salient. The notion of *proto objects* is described by Rensink as the volatile units of visual information that can be bound into a coherent and stable object when accessed by focused attention (Rensink 2000).

Torralba (Koch and Ullman 1985; Treisman and Gelade 1980) integrates saliency with context information (task driven focus-of-attention) and introduces a simple framework for determining regions-of-interest within a scene. They use context learned by training to guide attention (Torralba 2003; Torralba et al. 2006). Various features are extracted from a large collection of images, which are used to train a classifier. The top-down process consists of the classifier that guides the attention toward scene parts that most probably contain the target object. The main concept is that the location of certain natural objects is constrained by context, since, e.g., a car is expected to be found on the road rather than on the sky. A top-down process that biases bottom-up feature weights has been also incorporated to further enhance target objects.

Navalpakkam and Itti (2005, 2006) propose a modification of the basic Itti's model inspired by the Guided Search of Wolfe. They carry out a range of experiments both for visual search (fast search in pop-out tasks, slow search in conjunction tasks, slow search when the target is similar to the distracters, etc.) and visual recognition. With little computational cost incurred through multiplicative top-down weights on bottom-up saliency maps, their model combines both stimulus-driven and goal-driven attention to optimize speed of guidance to likely target locations, while simultaneously being sensitive to unexpected stimulus changes.

In all methods discussed so far, the attention model operates as a front-end to an existing recognition one. Nevertheless, Rothenstein and Tsotsos (2008) claim, based on experimental evidence, that attention and recognition should be interdependent in a bidirectional feedback process which results both in the detection and recognition of the object. An early example of such a model is the one proposed by Rybak et al. (1998), where the set of edges extracted at each fixation provides potential targets for the next gaze fixation. Recognition may then be achieved hierarchically and even from a part of the image (from a fraction of the scan-path belonging to this part) when the image is partly perturbed or the object is occluded. Obviously, the stability of recognition increases with the number of fixations.

11.3.2 Novelty Detection and Video Summarization

Attention selection and saliency computation methods are often implicitly used in common computer vision tasks like novelty detection or visual summarization of an image sequence. Generally, unusual activities like the rarity of an event, the sudden appearance/disappearance, the abrupt behavioral change of an object, etc. are defined as salient in various applications. We should differentiate between two definitions of unusual activities: (a) activities dissimilar to regular ones and (b) rare activities with low similarity among other usual ones.

Most attempts that are built around the first definition tackle the problem by predefining a particular set of activities as being usual, model it in some way, and then detect whether an observed activity is anomalous (Boiman and Irani 2005; Stauffer and Grimson 2000). Researchers working on rare event detection assume that unusual events are sparse, difficult to describe, hard to predict, and can be subtle, but given a large number of observations it is easier to verify if they are indeed unusual. Measures on a prototype-segment cooccurrence matrix reveal unusual formations that correspond to rare events. Using a similar notion of unusual activity, but under a quite different framework, Adam et al. automatically analyze the video stream from multiple cameras and detect unusual events by measuring the likelihood of the observation with respect to the probability distribution of the observations stored in the buffer of each camera (Adam et al. 2008). Intuitively similar are the methods proposed in Hamid et al. (2005), Itti and Baldi (2005), and Zhong et al. (2004).

Going one step further toward human perception of saliency, Ma et al. (2002) propose a framework for detecting the salient parts of a video based on user attention models. They use motion, face, and camera attention along with audio attention models (audio saliency and speech/music) as cues to capture salient information and identify the audio and video segments to compose the summary. Rapantzikos et al. (Evangelopoulos et al. 2008, 2009) build further on visual, audio, and textual attention models for visual summarization. The authors form a multimodal saliency curve integrating the aural, visual, and textual streams of videos based on efficient audio, image, and language processing and employ it as a metric for video event detection and abstraction. The proposed video summarization algorithm is based on the fusion of the three streams and the detection of salient video segments. The algorithm is generic and independent of the video semantics, syntax, structure, or genre. Subjective evaluation (user attention curves as ground-truth) showed that informative and pleasing video skims can be obtained using such multimodal saliency indicator functions.

11.3.3 Robotic Vision

If vision is important for humans, the same holds for robots. Robotic systems of different categories are equipped with visual sensory to interact with the environment and achieve predefined goals. Naturally, such robotic systems benefit from the incorporation of visual attention mechanisms, since time and efficiency are important in the field.

Attention mechanisms prove beneficial for significant robotic applications like *Simultaneous Localization and Mapping*, where an unknown environment should be mapped and the location of the robot should be efficiently detected. Visual attention is used to limit the amount of visual input and focus on the most informative parts of the scene. VOCUS (Visual Object detection with a Computational attention System) (Frintrop et al. 2007; Siagian and Itti 2007) is a successful example that includes a bottom-up and top-down process, while incorporating further sensory input (laser, etc.) (Frintrop et al. 2005). The experimental evaluation of the system shows the robot's behavioral improvement when using the attention mechanism.

The CODAM model has been recently used as part of an autonomous cognitive system called GNOSYS (Taylor et al. 2009). The basic platform is a Pioneer P3AT robot with a Katana arm and the visual stimuli consist of sets of cylinders and sticks as well as a ball and a cube. The robot had to accomplish several simple and complex tasks related to the objects (e.g., find and grasp and stack the cylinders). Many useful conclusions were drawn that highlight the crucial role of perception, as modeled by a group of modules including CODAM, in performing tasks. More details are given in Taylor et al. (2009).

11.4 Volumetric Saliency by Feature Competition

Apparently computational models, which are either based on strict or relaxed correspondences to well-founded neurophysiological counterparts of the visual attention system, have been proposed and applied to several computer vision fields. Saliency in a natural image sequence can be computed either in a frame-by-frame basis using either of the above models or by incorporating the temporal dimension to the model and let it interact tightly with the spatial one. Computational modeling of spatiotemporal saliency has recently become an interesting topic in the literature, since most video analysis algorithms may benefit from the detection of salient spatiotemporal events (e.g., events characterized by strong variations of the data in both the spatial and temporal dimensions).

Most models are inspired by biological mechanisms of motion-based perceptual grouping and compute saliency by extending operators previously proposed for static imagery. Specifically, we extend the spatial center-surround operator of Itti et al. in a straightforward manner by using volumetric neighborhoods in a spatiotemporal Gaussian scale-space (Rapantzikos et al. 2007). In such a framework, a video sequence is treated as a volume, which is created by stacking temporally consequent video frames. Hence, a moving object in such a volume is perceived as occupying a spatiotemporal area. Evaluation is performed for a surveillance application using a public available dataset. Large-scale volume representation of a video sequence, with the temporal dimension being long, has not been used often in the literature. Indicatively, Ristivojević et al. have used the volumetric representation for 3D segmentation, where the notion of "object tunnel" is used to describe the volume carved out by a moving object in this volume (Ristivojević and Konrad 2006). Okamoto et al. used a similar volumetric framework for video clustering, where video shots are selected based on their spatiotemporal texture homogeneity (Okamoto et al. 2002). Quite recently, Mahadevan and Vasconcelos (2009) proposed a biologically plausible model for spatiotemporal saliency. Under their formulation, the saliency of a location is equated to the power of a predefined set of features to discriminate between the visual stimuli in a center and a surround window, centered at that location. The features are spatiotemporal video patches and are modeled as dynamic textures to achieve a principled joint characterization of the spatial and temporal components of saliency. Their model is evaluated on a background subtraction task and compares well with the competition. Nevertheless, as the authors claim, computationally efficiency is an issue.

Most existing models do not count in efficiently the competition among different features, which according to experimental evidence has its biological counterpart in the HVS (Kandel et al. 1995) (interaction/competition among the different visual pathways related to motion/depth (M pathway) and gestalt/depth/color (P pathway), respectively). The early work of Milanese et al. (1995), one of the first computational visual saliency models, was based on such a competition that was implemented in the model as a series of continuous inter- and intrapixel differences. Toward this direction we implement competition by a constrained minimization approach that involves interscale, intrafeature, and interfeature constraints

(Rapantzikos et al. 2009). A centralized saliency map along with an inherent feature competition scheme is to provide a computational solution to the problem of region-of-interest (ROI) detection/selection in video sequences. As stated before, a video shot is represented as a solid in the three-dimensional Euclidean space, with time being the third dimension extending from the beginning to the end of the shot. Hence, the equivalent of a saliency map is a volume where each voxel has a certain value of saliency. This saliency volume is computed by defining cliques at the voxel level and uses an optimization/competition procedure with constraints coming both from inter-, intrafeature, and interscale level. Our method is based on optimization, though in this case competition is performed across features, scales, and voxels in our volumetric representation. The competition is implemented through a constrained minimization method with the constraints being inspired by the Gestalt laws.

Gestalt theory refers to a unified configuration of objects/events that has specific attributes, which are greater than the simple sum of its individual parts that share common properties (Koffka 1935; Wertheimer 1923). For example, we automatically perceive a walking event as a complete human activity rather than a set of individual parts like moving legs, swinging arms, etc. Each subaction is clearly an individual unit, but the greater meaning depends on the arrangement of the sub-actions into a specific configuration (walking action). Specifically, the constraints are related to the figure/ground (identify objects as distinct from their background), proximity (group elements that exhibit spatial or temporal proximity), closure (visually close gaps in a form), similarity (group similar elements depending on form, color, size, or brightness), and the common fate (elements with similar movements are perceived as a whole) Gestalt laws. The output of our model is a spatiotemporal distribution with values related to the saliency of the individual visual units.

11.5 Problem Formulation

Saliency computation in video is a problem of assigning a measure of interest to each spatiotemporal visual unit. We propose a volumetric representation of the visual input where features interact and compete to attain a saliency measure. Figure 11.1 depicts a schematic diagram of the method. The input is a sequence of frames represented in our model as a volume in space-time. This volume is decomposed into a set of conspicuity features, each decomposed into multiple scales.

A three-way interaction scheme is implemented, which allows voxel competition in the following ways: (a) intrafeature (proximity), between voxels of the same feature and same scale; (b) interscale (scale), between voxels of the same feature but different scale; and (c) interfeature (similarity), between voxels of different features. The stable solution of the energy minimization leads to the final saliency volume.

Let V be a volume representing a set of consequent input frames, defined on a set of points Q with $q = (x, y, t)$ being an individual space-time point. Points $q \in Q$ form a grid in the discrete Euclidean 3D space defined by their Cartesian

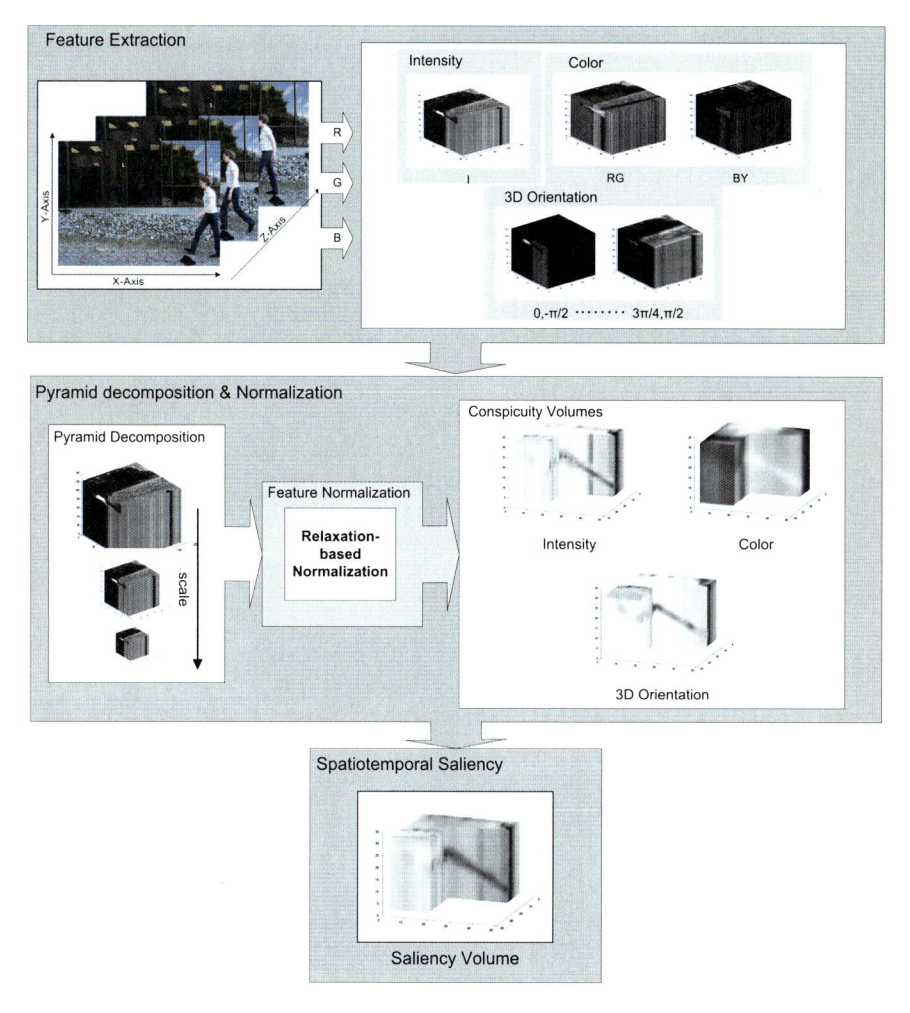

Fig. 11.1 Schematic diagram of the proposed method

coordinates. Under this representation point q becomes the equivalent to a voxel in this volume. Let $V(q)$ be the value of volume V at point q.

V is decomposed into a set of conspicuity volumes C_i with $i = 1, \ldots, M$ corresponding to three different features, namely intensity, color, and motion. Intensity and color features are based on color opponent theory, and spatiotemporal orientation (motion) is computed using 3D steerable filters. Each conspicuity volume is further decomposed into multiple scales ℓ and a set $\mathbf{C} = C_{i\ell}$ is created with $i = 1, \ldots, M$ and $\ell = 1, \ldots, L$ representing a Gaussian volume pyramid.

The final saliency distribution is obtained by minimizing an energy function E composed of a data term E_D and a smoothness term E_S:

$$E(\mathbf{C}) = \lambda_D \cdot E_D(\mathbf{C}) + \lambda_S \cdot E_S(\mathbf{C}). \tag{11.1}$$

The data term models the interaction between the observation and the current solution, while the smoothness term is composed of the three following constraints each related to a different saliency dimension.

E_1 models intrafeature coherency, i.e., defines the interaction among neighboring voxels of the same feature at the same scale and enhances voxels that are incoherent with their neighborhood:

$$E_1(\mathbf{C}) = \sum_i \sum_f \sum_q \left(C_{i,\ell}(q) - \frac{1}{|N_q|} \sum_{r \in N_q} C_{i,\ell}(r) \right)^2 . \qquad (11.2)$$

E_1 produces small spatiotemporal blobs of similar valued voxels.

E_2 models interfeature coherency, i.e., it enables interaction among different features so that voxels being conspicuous across all feature volumes are grouped together and form coherent regions. It involves competition between a voxel in one feature volume and the corresponding voxels in all other feature volumes:

$$E_2(\mathbf{C}) = \sum_i \sum_\ell \sum_q \left(C_{i,\ell}(q) - \frac{1}{M-1} \sum_{j \neq i} C_{j,\ell}(q) \right)^2 . \qquad (11.3)$$

E_3 models interscale coherency among ever coarser resolutions of the input, i.e., aims to enhance voxels that are conspicuous across different pyramid scales. If a voxel retains a high value along all scales, then it should become more salient.

$$E_3(\mathbf{C}) = \sum_i \sum_\ell \sum_q \left(C_{i,\ell}(q) - \frac{1}{L-1} \sum_{n \neq \ell} C_{i,n}(q) \right)^2 . \qquad (11.4)$$

To minimize (11.1), we adopt a steepest gradient descent algorithm. Detailed information is given in Rapantzikos et al. (2009). Visual examples of saliency volumes are given throughout this chapter (Figs. 11.2 and 11.6b).

11.6 Saliency-Based Video Classification

We choose video classification as a target application to obtain objective, numerical evaluation of the proposed model. The experiment involves multiclass classification of several video clips where the classification error is used as a metric for comparing a number of approaches either using saliency or not, thus providing evidence that the proposed model provides a tool for enhancing classification performance. The motivation is that if classification based on features from salient regions is improved

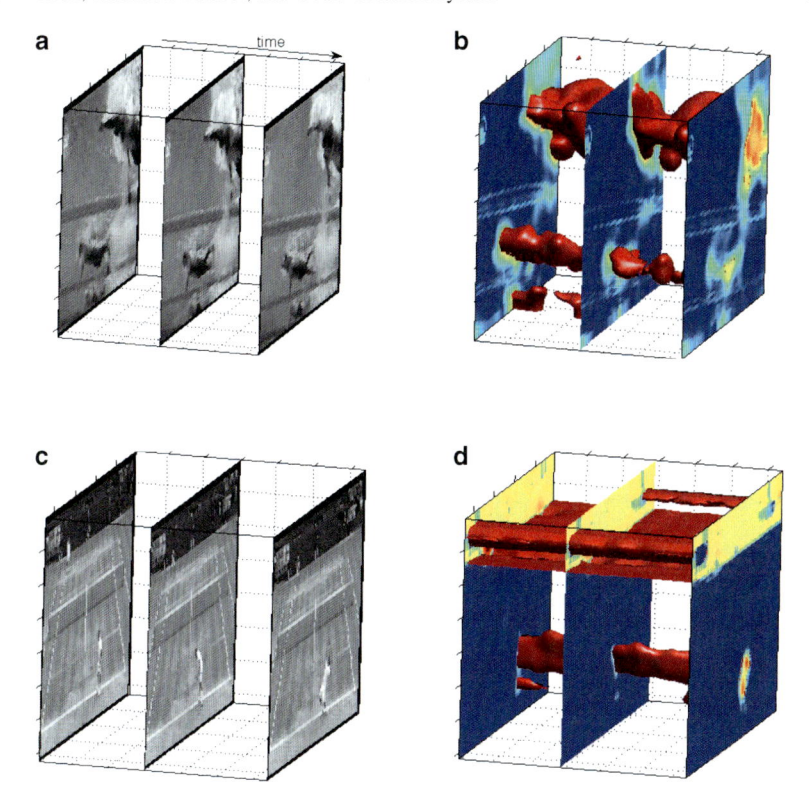

Fig. 11.2 (**a**) (**c**) Examples of slices from the original volume and the corresponding slices from the computed saliency volume. (**b**) (**d**) Corresponding isosurfaces, where the red color corresponds to high values. The salient regions become evident (better viewed in color)

when compared to classification without saliency, then there is strong evidence that the selected regions represent well the input sequence. In other words, we assume that if we could select regions in an image or video sequence that best describe its content, a classifier could be trained on such regions, and learn to differentiate efficiently between different classes. This would also decrease the dependency on feature selection/formulation.

We evaluate the performance of the spatiotemporal saliency method by setting up a multiclass video classification experiment and observing the classification error's increase/decrease when compared against other techniques. Input data consists of several sports clips, which are collected and manually annotated by the authors (Rapantzikos et al. 2009). Obtaining a meaningful spatiotemporal segmentation of a video sequence is not a simple and straightforward task. Nevertheless, if this segmentation is saliency driven, namely if regions of low (or high) saliency should be treated similarly, segmentation becomes easier. The core idea is to incrementally discard regions of similar saliency starting from high values and watch the impact on the classification performance. This procedure may seem contradictory, since the goal of attention approaches is to focus on high- rather than low-saliency areas. In this paper, we exploit the dual problem of attending low saliency regions. These regions are quite representative since they are consistent through the shot and are therefore important for recognizing the scene (playfield, slowly changing events, etc.). In order to support this approach, we have to place a soft requirement: regions related to background of the scene should cover a larger area than regions belonging to the foreground. Under this requirement, low salient regions are related to the background or generally to regions that do not contribute much to the instantaneous interpretation of the observed scene.

The feature extraction stage calculates histograms of the primary features used for computing saliency, namely color, orientation, and motion. To keep the feature space low, we calculate the histograms by quantizing them in a small number of bins and form the final feature vector. We use Support Vector Machines (SVMs) for classifying the data. We train the classifiers using a radial basis function (RBF) kernel after appropriately selecting a model using fivefold cross-validation estimation of the multiclass generalization performance. After obtaining the parameter that yields the lowest testing error, we perform a refined search in a shorter range and obtain the final parameters.

To sum up in a few words, the input video sequence is segmented into one or more regions after discarding a percentage of high saliency voxels, and histograms of precalculated features are extracted for each of them. Feature vectors feed an SVM classifier, and the outputs of all methods are compared.

11.7 Evaluation of Classification Performance

In order to test the robustness and efficiency of the proposed model, we compare it against a method based on a simple heuristic, two methods that share a common notion of saliency, against our early spatiotemporal visual attention model and a fifth one, which is based on PCA and has proven its efficiency in background subtraction approaches (Oliver et al. 2000).

Our early visual attention model shared the same notion of spatiotemporal saliency, but without the feature competition module. This model has proven its efficiency in enhancing performance of a video classification system (Rapantzikos and Avrithis 2005). The two other saliency-based methods are the state-of-the art static saliency-based approach of Itti et al. (1998) and an extension using a motion map (Rapantzikos and Tsapatsoulis 2005). Both methods produce a saliency measure per pixel. The static saliency-based approach processes the videos in a per frame basis. After producing a saliency map for each frame, we generate a saliency volume by stacking them together. We filter this volume with a 3D median filter to improve temporal coherency. The motion map of the extended approach is derived using a motion estimation technique, which is based on robust statistics (Black and Anandan 1996). The same procedure for producing a saliency volume is followed for the PCA-based technique. For the sake of completeness, we also provide results

a   **b**

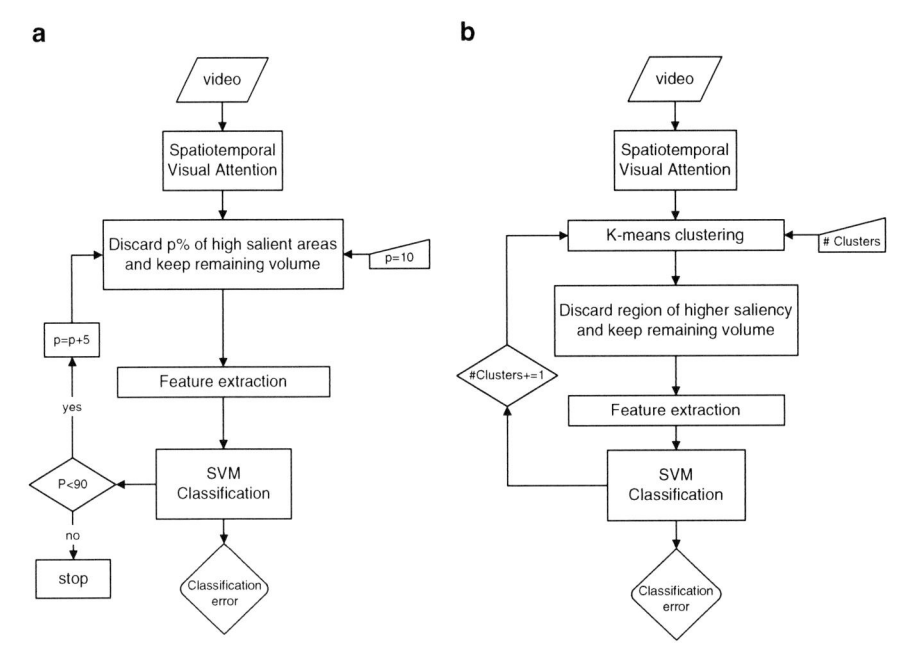

Fig. 11.3 Saliency-based classification: (**a**) based on foreground/background detection; (**b**) based on >1 salient regions

of a method that operates in a heuristic way and is based on the fact that people pay often more attention to the region near the center of the view. At each iteration, the initial video volume is incrementally reduced by $p\%$ and a classification error is produced. The reduction is done spatially in a uniform way, which means that we reduce the extent of $x-y$ axes from the edges to the center and leave the temporal dimension intact.

We prove the benefit obtained using saliency by two different experiments. Each of them is carried out on the same dataset and exploits each methods' results in a different way. The first approach is illustrated in Fig. 11.3a and is composed of three steps: (1) discard voxels of high saliency until a volume of $p\%$ of the whole volume remains; (2) extract histograms of precalculated features; and (3) feed the features to a classifier and obtain an error for each p value. The saliency volume is segmented into two regions, namely a high- and a low-salient one using automatic threshold-ing driven by the percentage of the high-saliency pixels to retain. Practically, the volume is iteratively thresholded using a small threshold step until the desired per-centage of discarded pixels is approximately obtained. At each step, a salient and a nonsalient region are produced. The feature vector generated from features bound to the less salient region is always of the same size and is formed by encoding the color histograms using 32 bins per color channel (i.e., 96 elements per region) and the motion/2D-orientation features using 16 bins.

Intuitively, there exist a number of regions that represent best the underlying scene. For example, in case of sport clips, one region may be representative of the playfield, another one may include the players, the advertisements, the audience, etc. Each of these regions corresponds to a single scene property, but not all of them are required in order to provide a complete scene description. If we follow the reasoning of the previous experiment, we expect that if the appropriate regions are selected, the classification error would be further reduced. Hence the second experiment segments the saliency volume into a varying number of regions (# clusters) as shown in Fig. 11.3b. The same incremental procedure is applied with the saliency volume being segmented into more than two regions at each iteration. After segmenting the input, the resulting regions are ordered in terms of saliency and the most salient one is discarded. This scenario has an intrinsic difficulty, since, if the number of regions is not constant for each video clip, the size of the feature vector will not be constant. Thus, direct comparison between vectors of different clips would not be straightforward. To overcome this problem, we segment the saliency volume into a predetermined number of regions using unsupervised clustering. In this framework, we use a clustering technique that allows for nonhard thresholding and labeling. K-means is used to partition the saliency volume into regions of different saliency. Voxels are clustered in terms of their saliency value and a predefined number of clusters are extracted. Afterwards, we order the clusters in increasing order of saliency, discard the last one, and label the rest using 3D connectivity. The optimal number of clusters, in terms of classification error minimization, is found using ROC curve analysis. At this scenario, 8 bins per color channel (i.e., 24 elements per region) and 4 bins for motion/2D-orientation are used.

Figure 11.4 shows the classification error along with standard error intervals when varying the size of the discarded region. The plot shows the results of evaluated methods. Each point on the graphs should be interpreted as follows: we discard, e.g., 10% of the high salient pixels (x-axis) and obtain a classification error (y-axis) using fivefold cross validation (standard error interval for each point). In case of the heuristic method, the ratio represents the portion of the discarded regions starting from the borders. Classification driven by spatiotemporal saliency provides improved statistics over the other methods, since for all measured values of sensitivity it achieves lower false positive rates. Although differences in magnitude are not tremendous, two facts become evident: first, salient-based approaches seem to provide more consistent and overall better results than the nonsalient one; second, the proposed model compares well against the other three-commonly established-approaches.

The second experiment illustrates the effect on classification performance when using features bound to more than one salient region. This experiment corresponds to the flow diagram shown in Fig. 11.3b. Figure 11.5 shows the obtained classification error vs. the number of segmented regions. The proposed and the PCA-based techniques perform overall better than the rest with the first having lower fluctuations and achieving the lowest average error.

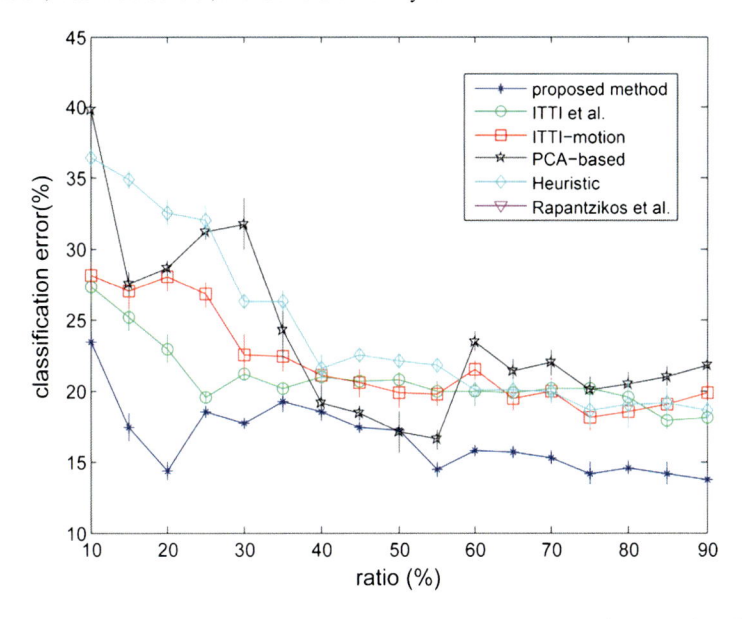

Fig. 11.4 Experiment I – Classification error along with standard error intervals for all tested methods when varying the size of the discarded region (ratio represents the percent of discarded high saliency voxels)

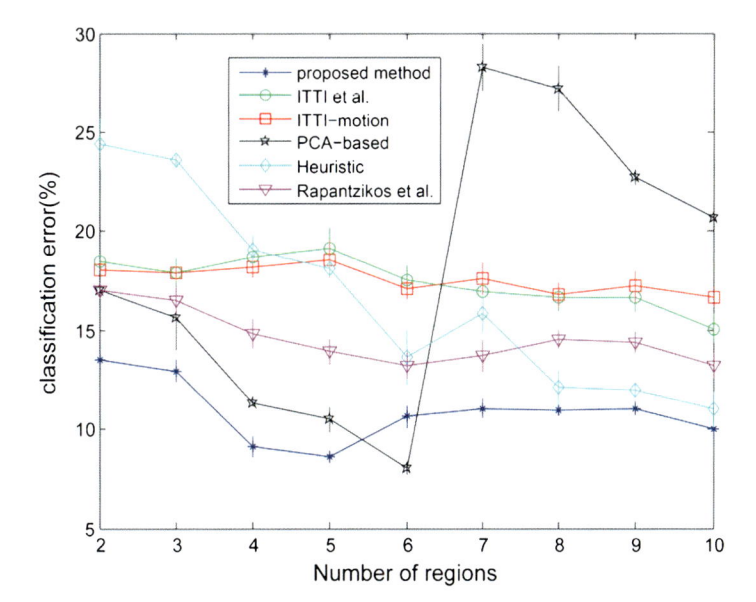

Fig. 11.5 Classification error along with standard error intervals when varying the number of regions (error vs. number of regions used to segment the volumes)

11.8   Action Recognition

Although not always explicitly said, an important computer vision field, namely interest point detection, is also based on a saliency-related representation of the input. For example, the famous point detector of Harris et al. detects points as local maxima of a distribution that measures the image intensity change in multiple directions (Harris and Stephens 1988). Similarly, the Hessian-affine detectors involving computation of second derivatives give strong (salient) response on blobs and ridges (Lindeberg 1998). The detector of Kadir and Brady is explicitly based on the notion of saliency and aims at detecting representative, discriminative, and therefore salient regions in the image (Kadir and Brady 2001). A review and a well-grounded comparison of spatial detectors is given by Mikolajczyk et al. (2005). Recently, points of interest combined with bag-of-words approaches have been also used for object/event detection and recognition. Such methods represent the visual input by a set of small visual regions (words) extracted around salient points like the ones referred before. Csurka et al. use visual words around regions extracted with the Harris affine detector (Mikolajczyk and Schmid 2002) to represent and detect visual objects (Csurka et al. 2004). Their approach demonstrates robustness to background clutter and good classification results. Wang et al. proposed a similar recognition framework and consider the problem of describing the action being performed by human figures in still images (Wang et al. 2006). Their approach exploits edge points around the shape of the figures to match pairs of images depicting people in similar body poses. Bosch et al. also propose a bag-of-words based method to detect multiple object categories in images (Bosch et al. 2006). Their method learns categories and their distributions in unlabeled training images using probabilistic Latent Semantic Analysis (pLSA) and then uses their distribution in test images as a feature vector in a supervised k-NN scheme.

Spatiotemporal event representation and action recognition have recently attracted the interest of researchers in the field. One of the few interest point detectors of a spatiotemporal nature is an extension of the Harris corner detection to 3D, which has been studied quite extensively by Laptev and Lindeberg (2003) and further developed and used in Laptev et al. (2007). A spatio-temporal corner is defined as an image region containing a spatial corner whose velocity vector is changing direction. They proposed also a set of image descriptors for representing local space-time image structures as well as a method for matching and recognizing events and activities based on local space-time interest points. Dollár et al. proposed a framework, which is based on a bag-of-words approach, where a visual word is meant as a cuboid (Dollár et al. 2005). They developed an extension of a periodic point detector to the spatio-temporal case and test the performance on action recognition applications. They show how the use of cuboid prototypes, extracted from a spatiotemporal video representation, gives rise to an efficient and robust event descriptor by providing statistical results on diverse datasets. Niebles et al. use the same periodic detector and propose an unsupervised learning method for human action categories (Niebles et al. 2006). They represent a video sequence by a collection of spatiotemporal words based on the extracted interest points and learn the probability distribution of the words using pLSA.

11.9 Spatiotemporal Point Detection

Salient points are extracted as the local maxima of the spatiotemporal saliency distribution obtained after minimization of (11.1). Such points are located at regions that exhibit high compactness (proximity), remain intact across scales (scale), and pop-out from their surroundings due to feature conspicuity (similarity). Hence we expect that the points will not be only located around spatiotemporal corners, but also around smoother space-time areas with distinguishing characteristics that are often important for action recognition. Figure 11.6 shows an example of such points.

We evaluate the proposed model by setting up experiments in the action recognition domain using two action datasets, namely the KTH dataset[1] and the Hollywood Human Actions (HOHA) one.[2] Both are public and available online. We provide a qualitative evaluation of the proposed detector, short descriptions of the datasets, and the corresponding recognition frameworks and devote the rest of the section to quantitative analysis. For comparison purposes we use two state-of-the-art detectors, namely the periodic one proposed by Dollár et al. (2005) and the space-time point detector of Laptev and Lindeberg (2003), which are publicly available. In the following, we will denote the first one by "periodic" and the second one by "stHarris" (Fig. 11.7).

a **b**

Fig. 11.6 (**a**) Indicative slices of a handwaving sequence and an ISO surface with the detected points overlaid for the (**b**) proposed

---

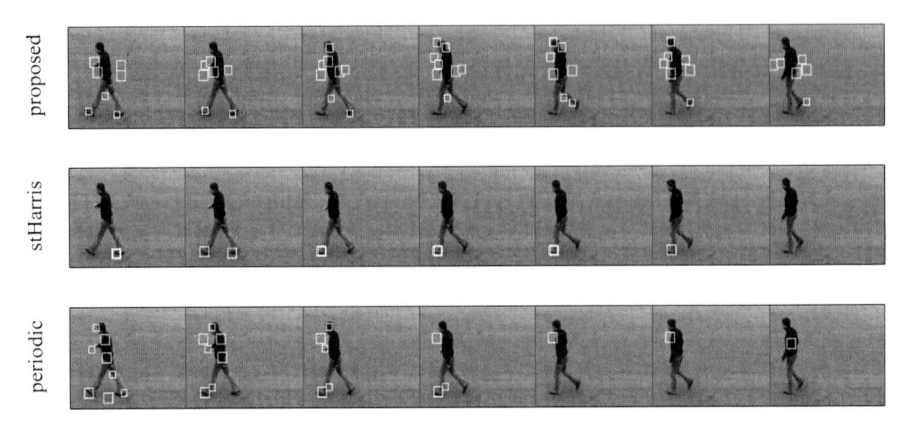

Fig. 11.7 Cuboids detected on a walking sequence from KTH dataset

Overall, saliency is obtained as the solution of an energy minimization problem that is initiated by a set of volumetric feature conspicuities derived from intensity, color, and motion. The energy is constrained by terms related to spatiotemporal proximity, scale, and similarity, and feature points are detected at the extrema of the saliency response. Background noise is automatically suppressed due to the global optimization framework, and therefore the detected points are dense enough to represent well the underlying actions. We demonstrate these properties in action recognition using two diverse datasets. The results reveal behavioral details of the proposed method and provide a rigorous analysis of the advantages and disadvantages of all methods involved in the comparisons. Our detector performs quite well in all experiments and either outperforms the state-of-the-art techniques it is compared to or performs among the top of them depending on the adopted recognition framework. More details can be found in Rapantzikos et al. (2009).

11.10   Discussion

Saliency-based image and video processing contribute in several aspects to solve common computer vision problems. The proposed volumetric saliency-based model for saliency computation exploits the spatiotemporal structure of a video sequence and produces a per voxel saliency measure based on a feature competition approach. This measure provides evidence about important and nonimportant regions in the sequence. Saliency is obtained as the solution of an energy minimization problem that is initiated by a set of volumetric feature conspicuities derived from intensity, color, and motion. The energy is constrained by terms related to spatiotemporal proximity, scale, and similarity. Background noise is automatically suppressed due to the global optimization framework, and therefore the salient regions or points represent well the underlying sequence/actions. The experimental results reveal behavioral details of the proposed method and provide a rigorous analysis of the advantages and disadvantages of all methods involved in the comparisons. In the future, motivated by recent works, we will focus on computational efficiency issues and more applications in the field. Furthermore, it would be interesting to explore top-down approaches in order to guide our saliency computation framework toward desired goals.

References

Abrams, R.A., Christ, S.E., "Motion onset captures attention", Psychological Science, vol. 14, pp. 427–432, 2003.

Adam, A., Rivlin, E., Shimshoni, I., Reinitz, D., "Robust Real-Time Unusual Event Detection using Multiple Fixed-Location Monitors", IEEE Transactions on Pattern Analysis and Machine Intelligence, vol. 30, no. 3, pp. 555–560, Mar 2008.

Black, M.J., Anandan, P., "The Robust Estimation of Multiple Motions: Parametric and Piecewise-Smooth Flow Fields", CVIU, vol. 63, no. 1, pp. 75–104, 1996.

Boiman, O., Irani, M., "Detecting Irregularities in Images and in Video", IEEE International Conference on Computer Vision (ICCV), Beijing, 2005.

Bosch, A., Zisserman, A., Munoz, X., "Scene Classification via pLSA", ECCV06, pp. 517–530, 2006.

Bruce, N.D.B., Tsotsos, J.K., Saliency, attention, and visual search: An information theoretic approach. Journal of Vision, vol. 9, no. 3, pp. 1–24, 2009.

Bruce, N., Tsotsos, J., "Saliency based on information maximization", Advances in Neural Information Processing Systems, vol. 18, pp. 155–162, 2006.

Csurka, G., Bray, C., Dance, C., Fan, L., "Visual categorization with bags of key-points", pp. 1–22, Workshop on Statistical Learning in Computer Vision, ECCV, 2004.

Cutsuridis, V., "A Cognitive Model of Saliency, Attention, and Picture Scanning", Cognitive Computation, vol. 1, no. 4, pp. 292–299, Sep. 2009.

Dollár, P., Rabaud, V., Cottrell, G., Belongie, S., "Behavior Recognition via Sparse Spatio-Temporal Features", VS-PETS, pp. 65–72, Oct 2005.

Duncan, J., "Selective attention and the organization of visual information", Journal of Experimental Psychology: General, vol. 113, no. 4, pp. 501–517, 1984.

Evangelopoulos, G., Rapantzikos, K., Potamianos, A., Maragos, P., Zlatintsi, A., Avrithis, Y., "Movie Summarization Based On Audio-Visual Saliency Detection", Proceedings International Conference on Image Processing (ICIP), San Diego, California, 2008.

Evangelopoulos, G., Zlatintsi, A., Skoumas, G., Rapantzikos, K., Potamianos, A., Maragos, P., Avrithis, Y., "Video event detection and summarization using audio, visual and text saliency", IEEE International Conference on Acoustics, Speech and Signal Processing (ICASSP), pp. 3553–3556, 2009.

Frintrop, S., Cremers, A., "Top-down attention supports visual loop closing", In. Proceedings Of European Conference On Mobile Robotics (ECMR'05), 2007.

Frintrop, S., Backer, G., Rome, E., "Goal directed search with a top-down modulated computational attention system", LCNS, vol. 3663, no. 117, 2005.

Frintrop, S., Rome, E., Nuchter, A., Surmann, H., "A bimodal laser-based attention system", Computer Vision and Image Understanding, vol. 100, no. 1–2, pp. 124–151, 2005.

Frintrop, S., Klodt, M., Rome, E., "A real-time visual attention system using integral images", In Proceedings Of the 5th International Conference on Computer Vision systems, ICVS, 2007.

Hamid, R., Johnson, A., Batta, S., Bobick, A., Isbell, C., Coleman, G., "Detection and explanation of anomalous activities: representing activities as bags of event n-grams", CVPR'05, vol. 1, pp. 1031–1038, Jun 2005.

Harris, C., Stephens, M., "A combined corner and edge detector", Alvey Vision Conference, pp. 147–152, 1988.

Itti, L., Baldi, P., "A Principled Approach to Detecting Surprising Events in Video", CVPR'05, 2005, vol. 1, pp. 631–637, 2005.

Itti, L., Baldi, P., "Bayesian surprise attracts human attention", Vision Research, vol. 49, no. 10, pp. 1295–1306, 2009.

Itti, L., Koch, C., Niebur, E., "A model of saliency-based visual attention for rapid scene analysis", IEEE Transactions on Pattern Analysis and Machine Intelligence, vol. 20, no. 11, pp. 1254–1259, 1998.

James, W., "The principles of psychology", Cambridge, MAL Harvard UP, 1890/1981.

Kadir, T., Brady, M., Saliency, scale and image description, International Journal of Computer Vision, vol. 45, no. 2, pp. 83–105, 2001.

Kandel, E.R., Schwartz, J.H., Jessell, T.M., "Essentials of Neural Science and Behavior", Appleton & Lange, Stamford, Connecticut, 1995.

Koch, C., Ullman, S., "Shifts in selective visual attention: towards the underlying neural circuitry", Human Neurobiology, vol. 4, no. 4, pp. 219–227, 1985.

Koffka, K., Principles of Gestalt Psychology, Harcourt, New York, 1935.

Laptev, I., Lindeberg, T., "Space-Time Interest Points", in Proceedings of the ICCV'03, Nice, France, pp. 432–443, 2003.

Laptev, I., Caputo, B., Schuldt, C., Lindeberg, T., "Local Velocity-Adapted Motion Events for Spatio-Temporal Recognition", Computer Vision and Image Understanding, vol. 108, pp. 207–229, 2007.

Lee, K., Buxton, H., Feng, J., "Cue-guided search: A computational model of selective attention", IEEE Transactions On Neural Networks, vol. 16, no. 4, pp. 910–924, 2005.

Leventhal, A., "The neural basis of visual function: vision and visual dysfunction", Nature Neuroscience, vol. 4, 1991.

Lindeberg, T., "Feature detection with automatic scale selection", International Journal of Computer Vision, vol. 30, no. 2, pp. 79–116, 1998.

Lowe, D., "Object recognition from local scale-invariant features", In Proceedings of ICCV, pp. 1150–1157, 1999.

Mahadevan, V., Vasconcelos, N., "Spatiotemporal Saliency in Dynamic Scenes", IEEE Transactions on Pattern Analysis and Machine Intelligence, 2009.

Ma, Y.F., Lu, L. Zhang, H.J., Li, M., "A user attention model for video summarization", ACM Multimedia Conference, pp. 533–542, 2002.

May, Y., Zhang, H., "Contrast-based image attention analysis by using fuzzy growing", In Proceedings ACM International Conference on Multimedia, pp. 374–381, 2003.

Mikolajczyk, K., Schmid, C., "An affine invariant interest point detector", ECCV, pp. 128–142, 2002.

Mikolajczyk, K., Tuytelaars, T., Schmid, C., Zisserman, A., Matas, J., Schaffalitzky, F., Kadir, T., Van Gool, L., "A comparison of affine region detectors", International Journal of Computer Vision, vol. 65, no. 1/2, pp. 43–72, 2005.

Milanese, R., Gil, S., Pun, T., "Attentive mechanisms for dynamic and static scene analysis", Optical Engineering, vol. 34 no. 8, pp. 2428–2434, 1995.

Navalpakkam, V., Itti, L., "An integrated model of top-down and bottom-up attention for optimal object detection", Computer Vision and Pattern Recognition (CVPR), pp. 1–7, 2006.

Navalpakkam, V., Itti, L., "Modeling the influence of task on attention", Vision Research, vol. 45, no. 2, pp. 205–231, 2005.

Niebles, J.C., Wang, H., Fei-Fei, L., "Unsupervised Learning of Human Action Categories Using Spatial-Temporal Words", British Machine Vision Conference (BMVC), Edinburgh, 2006.

Okamoto, H., Yasugi, Y., Babaguchi, N., Kitahashi, T., "Video clustering using spatiotemporal image with fixed length", ICME'02, pp. 2002–2008, 2002.

Oliver, N.M., Rosario, B., Pentland, A.P., "A Bayesian computer vision system for modeling human interactions", IEEE Transactions on Pattern Analysis and Machine Intelligence, vol. 22, no. 8, Aug 2000.

Park, S., Shin, J., Lee, M., "Biologically inspired saliency map model for bottom-up visual attention", Lectrure Notes in Computer Science, pp. 418–426, 2002.

Rapantzikos, K., Avrithis, Y., "An enhanced spatiotemporal visual attention model for sports video analysis", International Workshop on Content-based Multimedia indexing (CBMI'05), Riga, Latvia, Jun 2005.

Rapantzikos, K., Tsapatsoulis, N., "Enhancing the robustness of skin-based face detection schemes through a visual attention architecture", Proceedings of the IEEE International Conference on Image Processing (ICIP), Genova, Italy, vol. 2, pp. 1298–1301, 2005.

Rapantzikos, K., Tsapatsoulis, N., Avrithis, Y., Kollias, S., "A Bottom-Up Spatiotemporal Visual Attention Model for Video Analysis", IET Image Processing, vol. 1, no. 2, pp. 237–248, 2007.

Rapantzikos, K., Avrithis, Y., Kollias, S., "Dense saliency-based spatiotemporal feature points for action recognition", Conference on Computer Vision and Pattern Recognition (CVPR), 2009.

Rapantzikos, K., Tsapatsoulis, N., Avrithis, Y., Kollias, S., "Spatiotemporal saliency for video classification", Signal Processing: Image Communication, vol. 24, no. 7, pp. 557–571, 2009.

Rensink, R.A., "Seeing, sensing, and scrutinizing", Vision Research, vol. 40, no. 10–12, pp. 1469–1487, 2000.

Ristivojević, M., Konrad, J., "Space-time image sequence analysis: object tunnels and occlusion volumes", IEEE Transactions Of Image Processings, vol. 15, pp. 364–376, Feb. 2006.

Rothenstein, A., Tsotsos, J., "Attention links sensing to recognition", Image and Vision Computing, vol. 26, no. 1, pp. 114–126, 2008.

Rutishauer, U. Walther, D., Koch, C., Perona, P., "Is bottom-up attention useful for object recognition?", Computer Vision and Pattern Recognition (CVPR), vol. 2, 2004.

Rybak, I., Gusakova, V., Golovan, A., Podladchikova, L., Shevtsova, N., "A model of attention-guided visual perception and recognition", Vision Research, vol. 38, no. 15, pp. 2387–2400, 1998.

Shao, L., Kadir, T., Brady, M., "Geometric and photometric invariant distinctive regions detection", Information Sciences 177, vol. 4, pp. 1088–1122, 2007.

Siagian, C., Itti, L., "Biologically inspired robotics vision monte-carlo localization in the outdoor environment, In Proceedings of the IEEE/RSJ International Conference on Intelligent Robots and Systems, IROS, 2007.

Sillito, A., Jones, H., "Context-dependent interactions and visual processing in V1", Journal of Physiology-Paris, vol. 90, no. 3–4, pp. 205–209, 1996.

Stauffer, C., Grimson, E., "Learning Patterns of Activity Using Real-Time Tracking", IEEE Transactions on Pattern Analysis and Machine Intelligence, vol. 22, no. 8, pp. 747–757, Aug 2000.

Sternberg, R., "Cognitive Psychology," Wadsworth Publishing, 2006.

Sun, Y., Fisher, R., "Object-based visual attention for computer vision", Artificial Intelligence, vol. 146, no. 1, pp. 77–123, 2003.

Taylor, J.G., "Attentional movement: the control basis for consciousness", Society for Neuroscience Abstracts, vol. 26, no. 2231, 2000.

Taylor, J.G., "CODAM: A neural network model of consciousness", Neural Networks, vol. 20, no. 9, pp. 983–992, Nov 2007.

Taylor, J.G., "On the neurodoynamics of the creation of consciousness", Cognitive Neurodynamics, vol. 1, no. 2, Jun 2007.

Taylor, J.G., "Paying attention to consciousness", Progress in Neurobiology, vol. 71, pp. 305–335, 2003.

Taylor, J.G., Hartley, M., Taylor, N., Panchev, C., Kasderidis, S., "A hierarchical attention-based neural network architecture, based on human brain guidance, for perception, conceptualisation, action and reasoning", Image and Vision Computing, vol. 27, no. 11, pp. 1641–1657, 2009.

Torralba, A., "Contextual priming for object detection", International Journal of Computer Vision, vol. 53, no. 2, pp. 169–191, 2003.

Torralba, A., Oliva, A., Castelahno, M., Henderson, J., "Contextual guidance of eye movements and attention in real-world scenes: the role of global features in object search", Psychological Review, vol. 113, no. 4, pp. 766–786, 2006.

Treisman, A.M., Gelade, G., "A feature integration theory of attention", Cognitive Psychology, vol. 12, no. 1, pp. 97–136, 1980.

Tsotsos, J.K., Culhane, S.M., Wai, W.Y.K., Lai, Y., Davis, N., Nuflo, F., "Modelling visual attention via selective tuning", Artiffcial Intelligence, vol. 78, pp. 507–545, 1995.

Walther, D., Koch, C., "Modelling attention to salient proto-objects", Neural Networks, vol. 19, no. 9, pp. 1395–1407, 2006.

Walther, D., Rutishauer, U., Koch, C., Perona, P., "On the uselfuness of attention for object recognition", In Workshop of Attention for Object Recognition at ECCV, pp. 96–103, 2004.

Walther, D., Rutishauer, U., Koch, C., Perona, P., "Selective visual attention enables learning and recognition of multiple objects in cluttered scenes, Computer Vision and Image Understanding (CVIU), vol. 100, no. 1–2, pp. 41–63, 2005.

Wang, Y., Jiang, H., Drew, M.S., Li, Z., Mori, G., "Unsupervised Discovery of Action Classes". In Proceedings of CVPR'06, vol. 2, pp. 17–22, 2006.

Wertheimer, M., "Laws of Organization in Perceptual Forms", First published as "Untersuchungen zur Lehre von der Gestalt II, in Psycologische Forschung, vol. 4, pp. 301–350, 1923.

Wolfe, J.M., "Guided search 2.0: A revised model of visual search", Psychonomic Bulletin & Review 1, vol. 2, pp. 202–238, 1994.

Wolfe, J.M., "Guided search 4.0: current progress with a model of visual search", Integrated Models of Cognitive Systems, pp. 99–119, 2007.

Wolfe, J.M., Cave, K.R., Franzel, S.L., "Guided search: an alternative to the feature integration model for visual search", Journal of Experimental Psychology: Human Perception and Performance, vol. 15, no. 3, pp. 419–433, 1989.

Zhong, H., Shi, J., Visontai, M., "Detecting Unusual Activity in Video", CVPR'04, Washington, DC, vol. 2, pp. 819–826, Jun 2004.

Chapter 12
Semantics Extraction From Multimedia Data: An Ontology-Based Machine Learning Approach

Sergios Petridis and Stavros J. Perantonis

Abstract It is often the case that related pieces of information lie in adjacent but different types of data sources. Besides extracting such information from each particular type of source, an important issue raised is how to put together all the pieces of information extracted by each source, or, more generally, what is the optimal way to collectively extract information, considering all media sources together. This chapter presents a machine learning method for extracting complex semantics stemming from multimedia sources. The method is based on transforming the inference problem into a graph expansion problem, expressing graph expansion operators as a combination of elementary ones and optimally seeking elementary graph operators. The latter issue is then reduced to learn a set of soft classifiers, based on features each one corresponding to a unique graph path. The advantages of the method are demonstrated on an athletics web-pages corpus, comprising images and text.

12.1 Introduction

It is often the case that related pieces of information lie in adjacent but different types of data sources, such as text, images, audio and video streams. Automatically extracting such information involves being able to analyse each particular type of source, which is undeniably a challenging task and as such the focus of devoted research. However, an additional issue raised in this case is how to put together all the pieces of information extracted by each source, or, more generally, what is the optimal way to collectively extract information, considering all media sources together.

S. Petridis (✉)
Institute of Informatics and Telecommunications, NCSR "Demokritos", Patriarchou
Grigoriou and Neapoleos St. GR-15310, Aghia Paraskevi, Attiki, Greece
e-mail: petridis@iit.demokritos.gr

V. Cutsuridis et al. (eds.), *Perception-Action Cycle: Models, Architectures,*
and Hardware, Springer Series in Cognitive and Neural Systems 1,
DOI 10.1007/978-1-4419-1452-1_12, © Springer Science+Business Media, LLC 2011

Being of wide applicability, this issue has been tackled from several different angles and under different constraints. This accounts for a variety of names used to identify it in the literature, such as the combinations of any of the terms below:

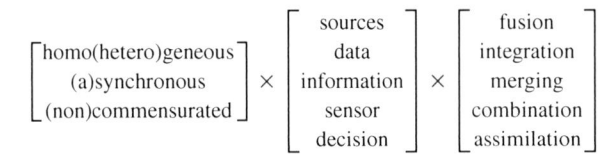

$$
\begin{bmatrix} \text{homo(hetero)geneous} \\ \text{(a)synchronous} \\ \text{(non)commensurated} \end{bmatrix} \times \begin{bmatrix} \text{sources} \\ \text{data} \\ \text{information} \\ \text{sensor} \\ \text{decision} \end{bmatrix} \times \begin{bmatrix} \text{fusion} \\ \text{integration} \\ \text{merging} \\ \text{combination} \\ \text{assimilation} \end{bmatrix}
$$

In this chapter, we use the term *fusion* and deal with the most general setting, where data sources are not necessary homogeneous, neither do they need to be of a particular type. As will be discussed later, this prompts for a methodology that deals with fusion at high level, i.e. where a substantial analysis of each data source has been preceded before bringing the results together.

Our approach uses effectively the machine learning/soft classification paradigm to drive approximate inference in ontologies. It allows to extract semantics from multimedia data, though the core algorithm is general and may be applied to any kind of data, once these are in the form of ontological assertions. It has three significant features. First, it deals inherently with uncertainty, allowing to represent both uncertain data and rules. Second, it is completely learnable, allowing for its adaptation to any domain and any ontology. Third, its application has polynomial complexity which makes it suitable for processing large scales of data.

To the best of the authors' knowledge, this is the first approach in the literature that uses machine learning classification algorithms to infer semantics for data extracted from multimedia content. The significant benefits of using this method are that:

- It allows the model to be learned using a training set of input–output assertions set, instead of manually building axioms and rules.
- It takes into account the uncertainty of assertions during inference.
- It benefits from existing mature soft classification literature to infer complex structured ontological knowledge.
- It balances complexity vs. accuracy by means of the size of subgraphs considered during evaluation and the classification models used, as will be explained later.
- It guarantees polynomial complexity.

Note that, independently of the approach to jointly handle multimedia data, a very important part of the semantics extraction from multimedia content task is delegated to media-specific analysis algorithms. The effectiveness of these methods are, needless to say, crucial for the overall performance of the fusion approach. In this chapter, we will not, however, go into details of any such algorithm, but rather concentrate on the fusion of their outcomes.

This chapter has the following structure. Section 12.2 describes the general setting of the fusion methodology, puts forward the advantages of semantic fusion and explains the steps required for expressing extracted multimedia information in the semantic level. Section 12.3 analyses our methodology for extracting high level semantic knowledge from multimedia data within the machine learning framework. In Sect. 12.4, we describe an evaluation procedure and analyze the performance of our methodology. Finally, Sect. 12.5 gives a brief review of related work and Sect. 12.6 summarizes our concluding remarks.

12.2 Fusing at the Semantic Level

This section describes the semantic setting under which we consider multimedia data fusion. In Sect. 12.2.1, the differences of low-, mid- and high-level fusion are highlighted. Section 12.2.2 analyses the concepts of redundancy and complementarity when fusing at the semantic level, and Sect. 12.2.3 describes a method of representing location-related information between medium segments by means of logical relations. Finally, Sect. 12.2.4 proposes a decomposition-fusion approach to efficiently deal with practical implications of a multimedia analysis system.

12.2.1 Low-, Mid- and High-Level Fusion

Fusion approaches that may be used to jointly extract information from multimedia data are generally categorized in three types: low-level fusion, mid-level fusion and high-level fusion. These differ at the level of representation in which fusion takes place. The reader may refer to Fig. 12.1 for an overview of their differences.

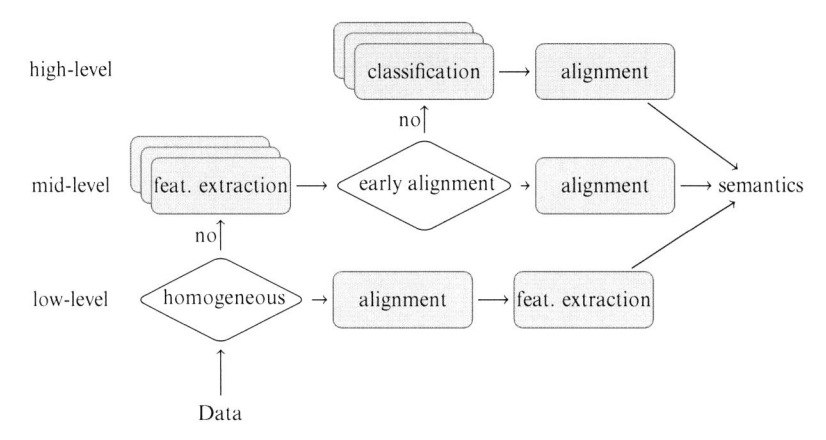

Fig. 12.1 Levels of fusion. Notice that the difference lies in which order does the alignment step takes place

12.2.1.1 Low-Level Fusion

When sources are *homogeneous*, i.e. of the same type, the fusion approach may take advantage of very specific techniques, developed for the specific data structure and data structural elements. A characteristic case is *image fusion* (Goshtasby and Nikolov 2007), where the correspondence between images can be done on a pixel basis either in the spatial or in the spectral domain. This enables the use of medium-specific algorithms (e.g. the *discrete wavelet transform* in images), to be applied on the fused data.

Note that the low-level fusion demands data to be prior aligned. Sometimes, this task has a difficulty of its own and appropriate techniques should be devised. As a particular example, in image fusion, the process of spatially aligning images to correctly fuse them is referred to as *image registration* (Lucas and Kanade 1981).

12.2.1.2 Mid-Level Fusion

Another case arises when the considered data, although not homogeneous, are alignable, i.e. one can establish a correspondence between their structural elements. A typical case of alignable data are audiovisual documents, where video and audio are aligned on their time dimension. These are typically stored using a combination of compressions techniques and a container format (e.g. MP4). Decomposing an audiovisual document into its audio and video parts is a rather straightforward procedure, if one knowns the particular container format that is used.

A possible approach here is to homogenize the elements by extracting features from all different media (Aarabi and Dasarathy 2004). Since feature-vectors are medium-neutral, they can be concatenated, thus producing a unified feature vector for the joint multimedia element. Subsequently, one may apply algorithms specific to the common multimedia document structure.

Note that modality feature extraction rates may differ, due to different sampling rates. In this case, aligning the data may require some effort, such as element-wise interpolating the features of one modality to the others modality frame rate. On the other hand, this approach has the advantage of allowing the application of medium-specific techniques to extract useful features from each medium, while retaining the structural alignment of the media.

This approach is commonly referred to as *mid-level* fusion or *feature-level* fusion. Note, however, that as an alternative to feature extraction, one may as well use kernel-based techniques to combine data from different sources, as long as these have been aligned.

12.2.1.3 High-Level Fusion

Another way to proceed is to defer the fusion between corresponding elements to a common, medium neutral, symbolic level. This implies that a complete analysis is done in each medium, aiming to achieve information at a symbolic level of representation before fusion. This method is referred to as *high-level* fusion, though the terms *symbolic-level* fusion and *decision-level* fusion are sometimes used. A characteristic of this approach is that it does not impose feature vector generation from each medium, but allows for particular methods to be applied independently until the symbolic level is reached.

An important advantage of high-level fusion is that it can handle the case of weaker alignment of documents. In particular, it is possible that alignment of documents can be done in coarse documents elements, within which the generated features by each medium can change significantly. As an example, in audiovisual speech recognition, synchronization may be achieved in the phoneme or syllable level, rather than in static sub-phonemic audio elements from which the features are traditionally extracted. A number of approaches relying on this weak alignment, such as *Coupled Hidden Markov Models* and *Factorial Hidden Markov Models*, have been successfully applied (Nefian et al. 2002).

The methodology presented in Sect. 12.3 falls into the high-level fusion type, since it relies on symbolic results obtained by separately analyzing each medium. In particular, fusion takes place at the semantic level, since it is assumed that the extracted results are described as a set of description logics assertions conforming to an assumed ontology.

12.2.2 Redundancy and Complementarity of Multimedia Information

There are two good reasons for jointly extracting information from multimedia sources at the semantic level: *complementarity* and *redundancy*. We will briefly review here these two concepts.

12.2.2.1 Complementarity

Complementarity of information refers to the case where different pieces of related information are expressed in different media. Hence, the integral information, together with further implied information it induces, may be gained only by their joint consideration.

Here is an example. Assume that a system can detect and analyze, though not recognize, persons in photos and person names in text. Then by jointly analyzing a photo showing a person face together with the caption "John Lennon", the system may additionally infer that John Lennon was wearing glasses, something that cannot be deduced neither solely by the text nor solely by the photo.

To fuse concepts across modalities, an indirect matching of instances extracted across modalities takes place. Namely, each medium-specific algorithm aims at extracting instances of medium-specific concepts. These are typically

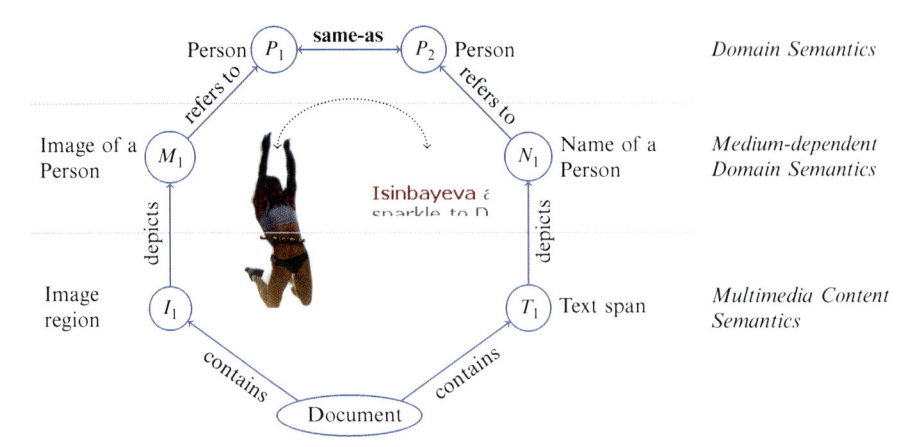

Fig. 12.2 Fusing multimedia concepts. The image and text samples correspond to a web-page containing a captioned image. The concepts and relation names used in this figure are indicative. Note that data of different modalities are indirectly linked at the fusion level via the same-as relation

results of classifications among a set of possible labels. For example, an image processing algorithm may detect a photograph of a *person's face*, whereas the text processing algorithm may detect a written form of a *person's name*. The existence of medium-specific instances imply the existence of referring medium-independent instances. For example, both *person's face* and *person's name* imply the existence of a *person*, although not necessarily the same one. In these terms, finding correct pairs of medium-independent instances (i.e. persons) is the goal of the fusion algorithm. The relation used in ontology to denote instance matching is termed *same-as*. Figure 12.2 depicts the instance matching procedure.

12.2.2.2 Redundancy

Redundancy of information refers to the case where information regarding the same object is expressed in different media sources, and hence information pieces may be either confirming or contradicting to each other in various degrees. In this case, jointly extracting the information may lead to increasing the accuracy of the extracted information.

Here is another example. Assume that a system recognize an object in a photo, but gives similar probabilities that is a candy or a bow tie. Adjacent to the photo, there is a caption displaying the text "a red bow", which the system may relate it to either a weapon, a music instrument, a ship's bow, etc. By combining the information from the text and image sources, the system may infer with great confidence what could not be inferred by considering each media in isolation: that the referenced object is a red bow tie.

12.2.3 Physical and Logical Document Structure

An inherent characteristic of multimedia data is that they can be localized by means of spatiotemporal locations within a specific context, such as a document. A specifically crafted ontology may provide adequate means for specifying such a *physical* structure of a multimedia document. In particular, consider the *decompose* relation to denote a segmentation of image into image regions, an audio stream to audio segment, etc. Clearly, each localisation is medium-specific and hence particular concepts have been used to allow describing the location of a segment in function of the medium used. Note also that information within a given segment may as well be represented in different modality, which infact accounts for the usage of the *embed* relation in the ontology. As an example, an image segment may contain textual information, which once detected and extracted by means of image optical character recognition (OCR) techniques is further subject to text analysis.

On the other hand, one should also stress that any document can be decomposed into a *logical* structure of elements which represent how distinct components of the document are logically related to each other. Such logical relations may be denoted via a distinct relation *contains*. For instance, a web-page can be decomposed into a set of text items and captioned images items, where each captioned image is further decomposed into an image and a caption. What is worth noticing about the logical structure, as opposed to the physical one, is that it can be defined in a medium-independent way. This holds because the logical structure between components, although induced by the physical structure, does no longer require medium-specific location information.

For instance, most of the times, the caption is a piece of text found *spatially* nearby an image. This physical relation can be exploited to derive a logical relation between these two items, namely that the first has the role of a caption and the second the role of an image in a captioned image entity. Once this relation has been discovered, we can consider to omit from further consideration how the text and image items are related spatially, and hence retain only a cross-media uniform logical representation.

Based on the above, before applying the fusion methodology described in Sect. 12.3, the physical relation between data items is taken into account to discover higher level logical relations between them (see also the graph at Fig. 12.3). Namely, the following steps are applied:

1. Analysis results corresponding either to the entire document under examination or to its particular parts of it are taken into account. For example, when dealing with web-pages, the results of text analysis of textual content, the results of image analysis of images and the results of decomposing the web-page in coherent segments are considered.
2. Common logical relation between components of the image are derived based on physical relation of segments. To give a simplified example, when text analysis results refer to the fact that there is a person name between the characters say 1,000 and 1,010 of the html page, and the html analysis results refer to the fact

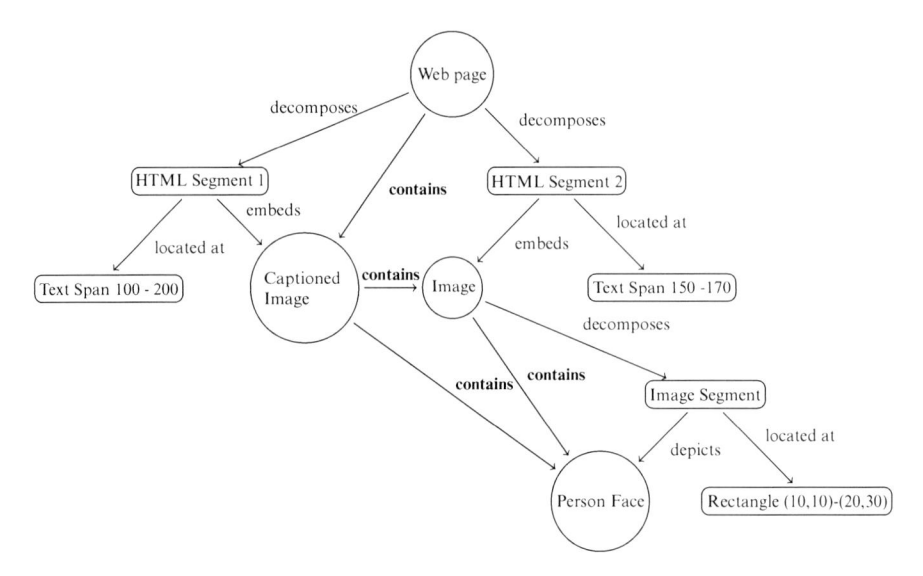

Fig. 12.3 Abstracting structural relation. This figure depicts an example of deriving a logical relation between data based on their physical relation. Note that, once the *contains* relations are inferred using location information, only the concepts denoted with circles remain important for consideration

that there is a caption between characters 900 and 1,020, then the fact that the given person name is contained in the given caption is derived.

3. Once logical relations are extracted, all other medium-specific location information regarding the analysis is discarded.

12.2.4 Practical Considerations

In several occasions, such as in OCR-embedded text in images or web-pages, multimedia data are not given as separate sources but as an integral multimedia document. In this section, we present a general engineering approach to deal with practical issue when dealing particularly with such multimedia documents. This approach, referred to as the *decomposition-fusion* approach, has been successfully applied in the BOEMIE system (Petridis and Tsapatsoulis 2006) to semantically annotate audiovisual streams and web-pages.

The decomposition-fusion approach consist of three steps, namely:

Decompose   In a first step, the multimedia document is decomposed in its constituent parts using an appropriate algorithm, depending on the nature of the document. Decomposing a multimedia document refers to both finding its single medium elements and their relative structural position.

Analyze   Then, each constituent part of the multimedia document is given for analysis to appropriate single medium processing techniques.

Fuse   Finally, the results produced by individually processing the single-medium elements of the document, together with the information regarding the decomposition of the multimedia document, are jointly given as input into the "fusion" module that aims at providing the final outcome, i.e. the extracted semantics for the whole multimedia document.

Importantly, the interaction between the decomposition, the analysis and the fusion step is accomplished through the usage of the same semantic model, i.e. as ontological assertions conforming to a specially crafted ontology that accounts for both the semantics needed to describe the domain of application and those to describe the multimedia content structure.

Note that an important feature in this approach is that it decouples the media-specific analysis algorithms from the multimedia decomposition and fusion approaches. This has the significant advantage of allowing the overall semantics extraction process to be open to new achievements in single-medium data analysis, while also enabling to focus on enhancing the algorithms that perform the fusion step.

To give an example of the procedure, the initially given multimedia document (e.g. a web-page) is processed by a *decomposition* module that separates it into sub-documents corresponding to different media formats (i.e. video, audio, image and text).

Each one of these sub-documents is then given for analysis to the corresponding analysis module. The analysis modules, based on the semantic model, analyze the sub-documents so as to identify and classify elements within, such as two-dimensional regions in images, word sequences in text, temporal segments in audio and spatio-temporal segments in video. The set of allowable classification labels corresponds to a subset of the ontology concepts which are directly identifiable in the media. The analysis also identifies relations between the extracted elements, such as *adjacent to* in image, *before* in audio or *subject of* in text. The output of the analysis modules is a set of assertions containing the list of the extracted elements and element relations together with sufficient properties to describe the related instances, the position of the elements in the sub-document, the extracted features used to conduct the analysis, as well as the confidence estimation of the classification. Then, a *result analysis* module analyses the results to re-route potentially identified encapsulated media (such as OCR extracted text) to the cross-modality coordinator for further analysis. Finally, once all single media sources have been processed, fusion takes place to extract further semantics based on the complementarity and redundancy of the sources.

On-line/Off-line modes A common use of semantics extraction systems is to enable answering complex queries over the multimedia content. To give an example, an end-user of such a system may wish to search for particular multimedia content that satisfy some criteria. Since multimedia analysis may be a time-consuming task,

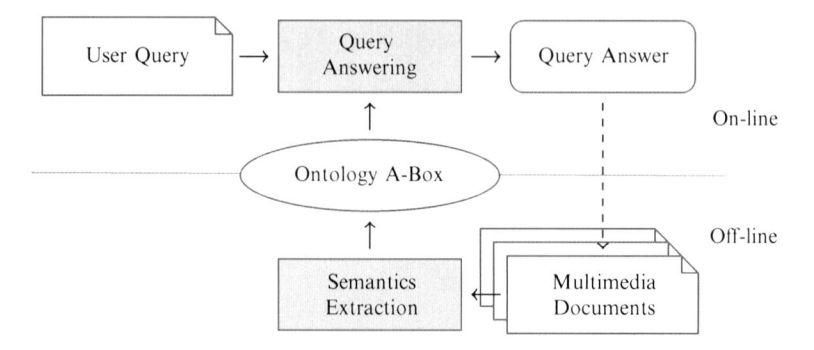

Fig. 12.4 Querying semantics extraction results. A principal goal, that has to be taken into account by the fusion approach, is to allow answering semantically rich user queries

it is not recommended to perform it once the user has placed a particular query, Rather, it is better to have previously analyzed the content and match the query against the saved analysis results. Figure 12.4 depicts this approach, distinguishing the steps that take place in an off-line mode, from those taking place in an on-line mode.

12.3 Methodology

In this section, we analyze our approach to extract semantics from fused multimedia data. Section 12.3.1 explains our motivation for using the machine learning framework for inferring structure knowledge. In Sect. 12.3.2, we formally define the semantics extraction problem as a system identification problem, and in Sect. 12.3.3, we show how it may be translated as a graph expansion problem. In Sect. 12.3.4, we study how to decompose the problem in elementary graph expansion operators, and, finally, in Sect. 12.3.5, we propose metrics to learn optimal ones.

12.3.1 Motivation

The approach presented is based on knowledge expressed in terms of ontological assertions. However, it avoids using exact reasoning. Indeed, clarity and unambiguity in ontologies have allowed reasoning to be embedded in ontologies in the form of the description logics formalism. Exact reasoning may, however, hinder using ontologies for large scale of data, due to the high computational times it demands: problems to be solved with exact reasoning have typically EXPTIME or even NEXPTIME complexity. This becomes a critical issue when extracting semantics from multimedia content, which may require reasoning over thousands of instances. To overcome this issue, the approach followed here is based on approximate inference.

On the other hand, approximate inference sacrifices soundness or completeness for a significant speed-up of reasoning (Rudolph et al. 2008). In addition, it fits more with data, such as results of a multimedia extraction algorithm, which are inherently uncertain. Indeed, it does not make perfect sense to perform exact reasoning with input that are not exact: (a) valuable information regarding the uncertainty of data is lost during exact reasoning and (b) the fact that reasoning results are exact does not make them correct, since the input to reasoning may not be correct itself.

Another critical factor for successfully applying ontologies in real-world problems is the complexity of building ontologies. Research on this issue has made several advances, such as constructing subsumption hierarchies from text corpora (Zavitsanos et al. 2007), though, research for inducing axioms and/or safe rules based on the supervised machine learning paradigm has not yet received significant interest. In what follows, we will show how to build ontologies based on a reference set of assertions, where a subset of them, the *input*, would be some given explicit knowledge and another subset, the *output*, would be the knowledge induced by the axioms/rules to be learned.

Our approach has a significant difference in respect to the traditional approach for inference within ontologies. Namely, ontologies are commonly viewed as systems that contain implicit information, whereas our approach is based on considering them as input–output systems. In particular, we consider DL axioms, DL safe rules and abduction rules as particular ways of performing a single task, given a set of assertions A, and a model T, generate a set of assertions A^+, which is a superset of the original set of assertions A. Considering ontologies as a particular class of input–output systems may not be very convenient in cases where there is a need to keep knowledge only in its implicit form. However, it may be convenient as a way to integrate methods that share this input–output view, such as machine learnable classifiers. The issue then reduces to investigating how such methods may fit into the inference within the ontology framework.

12.3.2 Problem Formulation

In this section we formulate the approximate inference in ontologies problem as an input–output problem.

Let an A-box A be a set of assertions which are valid according to a DL vocabulary of terms (concepts and roles) and let $\mathscr{A}$ be the set of all such A-boxes. Let also a family of functions:

$$\mathscr{T} = \{t : \mathscr{A} \rightarrow \mathscr{A} : \forall A \in \mathscr{A}, t(A) \in S(A)\}, \tag{12.1}$$

i.e. the family of "operators" that, given an A-box, they generate an A-box that belongs to $S(A)$, where $S(A)$ is the set of all A-boxes that are supersets of $A \in \mathscr{A}$, i.e. they contain *at least* the same assertions as A. Let also a family of approximation functions,

$$\mathscr{F} = \{f : \mathscr{A} \rightarrow \mathscr{A}\}. \tag{12.2}$$

Given a reference set of N A-boxes $\{A_i\}_0^N$, $A_i \in \mathscr{A}$, a function $t \in \mathscr{T}$ and a distance metric between A-boxes $d : \mathscr{A} \times \mathscr{A} \rightarrow \mathfrak{R}$, find the optimal function $\hat{f} \in \mathscr{F}$, such that the expected distance between A-boxes is minimal:

$$\hat{f} = \arg\min_{f \in \mathscr{F}} E_i \left[d \left(f \left(A^i \right), t \left(A^i \right) \right) \right] \tag{12.3}$$

12.3.2.1 Reference Functions

Note that family of functions $\mathscr{T}$ may be constrained to particular procedures that are valid in the DL formalism:

- A first family is one that allows making explicit the implicit information deduced through the DL axioms to the existing assertions.
- A second family is the one that corresponds to the application of a set of DL-safe rules.
- A third family is the one that corresponds to the application of abduction rules in ontologies, which results in the addition of new concept instances.

In all cases, any assertion box $t(A)$ is assumed to be valid in respect to the given ontology. Note also that many DL axioms and/or rules correspond to operators t, where $t(A)$ may have infinite elements for some A.

12.3.2.2 Approximation Functions

On the other hand, $\hat{\mathscr{F}}$ is a family of functions that approximate functions in $\mathscr{T}$, but need not be the same as $\mathscr{T}$. Any set of operators that, given a set of assertions, produce a superset of these assertions, according to the given vocabulary of concepts and roles, may be used. The family of functions that is described later in this section is of quite different nature than the ones used in logic-based systems.

Approximating $\mathscr{T}$ by $\mathscr{F}$ may lead to sacrificing soundness or completeness of the results: although $t(A)$ is always sound and complete, $\hat{f}(A)$ may not be. However, a careful selection of $\mathscr{F}$ may have other benefits, such as:

- $f(A)$ may be much faster to evaluate than $t(A)$.
- Functions in $\mathscr{F}$ may handle uncertainty regarding the input assertions and postpone the thresholding step to until after the inference.
- $\hat{f}(A)$ may have nice properties, such that the optimal function $\hat{f}$ can be found through *learning*, i.e. through the reference set of A-boxes.

12.3.2.3 Distance Between A-Boxes

Note that defining the problem requires providing a distance metric that measures the distance between two set of assertions. In what follows, we discuss such measures based on the isomorphism between DL A-boxes and directed graphs. As a preliminary comment, one should take care that the distance should reflect the degree of semantic equivalence between the two sets of assertions.

12.3.3 Using Directed Graphs

12.3.3.1 Set of DL Assertions as Directed Graphs

In this section, we establish a one-to-one correspondence between set of DL assertions A, commonly referred to as an Assertion box (A-box) and a directed graph $G = (V, A)$ with vertices V and directed edges A belonging to the family of directed graphs $\mathcal{G}$, where vertices take values in $\mathcal{V}$ and edges take values in $\mathcal{E}$. We hence show that sets of description logics assertions are isomorphic to directed graphs. This will allow us to deal with the problem formulated in Sect. 12.3.2 by means of functions where both their input and outputs are graphs.

Let a pair of mappings M and M^{-1} from A-boxes to graphs such that

$$\forall A \in \mathcal{A} : \quad A = M^{-1}[M[A]] \tag{12.4}$$

The mapping discussed here involves only the assertions, leaving out the terminological part of the ontology. The mapping is as follows:

- Every individual is mapped to exactly one vertex in the graph. The value of the vertex is (a) the unique identifier of the individual and (b) information regarding the concept that the individual is member of. In the simplest case, this is the name of the concept. In this case, $\mathcal{V}$ is the set of all concepts of the ontology.
- Every statement that a *relation* holds between *two individuals* is mapped to exactly one edge in the graph, namely the edge that connects the respective vertices. The edge is directed, reflecting the fact that the relation between instances is not necessarily symmetric. The value of the edge is information regarding the type of relation holding between individuals. In the simplest case, this is the unique name of the relation, and thus $\mathcal{E}$ is the set of all relation types in the ontology.

Figure 12.5 shows an example of some DL assertions expressed as OWL statements and the respective graph. There are overall two instances (`001, Person` and `002, PersonName`) and one relation that holds between the instances (`hasPersonName`).

The inverse mapping It is quite straightforward to define the inverse mapping: each vertex maps to an individual of identifier and concept membership as provided through the vertex values, while edges are mapped to relation between two individuals, as indicated by the edge value.

Notice that the mapping described allows to preserve the instances unique identifiers. Application of the inverse mapping to the mapped A-box results then to exactly the same A-box. Were the information regarding the unique identifiers of

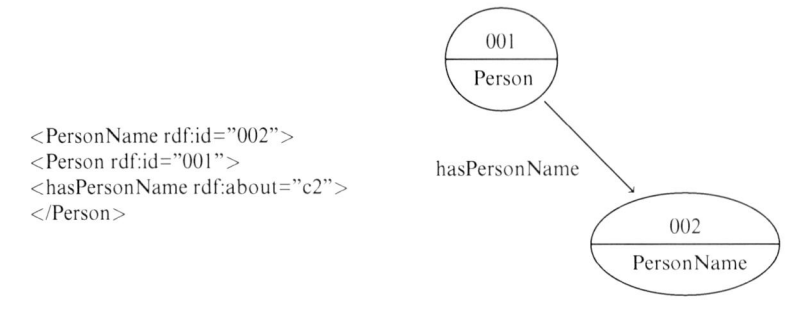

```
<PersonName rdf:id="002">
<Person rdf:id="001">
<hasPersonName rdf:about="c2">
</Person>
```

Fig. 12.5 A set of DL assertions as a graph. PersonName, Person and hasPersonName are terms included in the T-BOX of the ontology

the individuals omitted, the inverse mapping would generate *semantically* equivalent assertions as the original ones, although with different unique identifiers for the individuals.

Problem reformulation The problem discussed in Sect. 12.3.2 involves sets of description logic assertions. Since sets of description logics assertions and graphs are isomorphic, a way to proceed, to solve it, is to first formulate it using graphs, find optimal graph operators and then transform back the optimal graphs into set of assertions. The first and last step correspond to applying the mapping and inverse mapping, respectively. The core problem, then, reduces to the following:

Let $\mathscr{G}$ be a family of directed graphs. Let $S(G)$ be the set of graphs that are supersets of G, i.e. graphs which have *at least* the same vertices and edges with G. Let also the family of graph functions

$$\mathscr{T} = \{t : \mathscr{G} \to \mathscr{G} : \forall G \in \mathscr{G}, t(G) \in S(G)\}, \qquad (12.5)$$

i.e. specific "operators" that generate supersets of graphs. Let also a family of approximation functions,

$$\mathscr{F} = \{f : \mathscr{G} \to \mathscr{G}\}. \qquad (12.6)$$

Given a reference set of N graphs $\{G\}_{i=0}^N$ and a distance metric $d : \mathscr{G} \times \mathscr{G} \to \mathfrak{R}$ between directed graphs, find the optimal operator $\hat{f} \in \mathscr{F}$, such that the expected distance between graphs is minimal:

$$\hat{f} = \arg\min_{f \in \mathscr{F}} E_i \left[d \left(f(G^i), t(G^i) \right) \right] \qquad (12.7)$$

Note that we have used the same symbols as in Sect. 12.3.2 to denote functions and family of functions, since they are practically equivalent. Moreover, all observations regarding the family of functions $\mathscr{T}$ and $\mathscr{F}$ apply to the graph formulation. Note also that the issue of defining a distance metric between set of assertions translates into defining a distance metric between directed graphs.

Since all operators f considered produce graphs that are supersets of the original graph, it makes sense to call them *graph expansion* operators. In what follows, we describe a way to find optimal such operators of a particular family.

12.3.4 Optimal Graph Expansion Operators

12.3.4.1 Elementary Operators

Before examining how to find optimal graph expansion operators, let us first consider how the graph expansion functions may look like. Following the graph-based problem formulation, the graph functions considered result in graphs that have more edges and possibly more vertices than the ones given as input. It is easy to see that any such function can be decomposed as a recursive application of elementary graph expansions operators with two variants:

- An operator that adds an edge between two existing vertices that are not connected

$$h_1 = (e_{\text{new}} \leftarrow \text{addEdge}(v_i, v_j, e)) \tag{12.8}$$

  where v_i, v_j are two vertices of the graph and e is the value of the new edge.
- An operator that adds a vertex to the graph, along with an edge that connects it to an existing vertex

$$h_2 = (v_{\text{new}} \leftarrow \text{addVertex}(v)), (e_{\text{new}} \leftarrow \text{addEdge}(v_i, v_{\text{new}}, e)) \tag{12.9}$$

where v is the value of the new vertex, v_{new} is the newly inserted vertex, v_i is an existing vertex of the graph and e is the value of the edge to be inserted. In other words, denoting the family of elementary graph expansion operators as $\mathcal{H}$, any graph expansion operator may be expressed as

$$\forall f \in \mathcal{F}, \exists h_t \in H, t = 1 \dots T : f = h_1 \circ h_2 \circ \cdots h_T, \tag{12.10}$$

where T is the number of edges or vertices + edges added. In what follows, we exploit the decomposition of graph expansion operators in elementary ones to find optimal ones with respect to our problem.

Note that, by translating these operators back to the initial problem formulation, these variants correspond to adding a relation between two instances of the a-box or adding a new instance to the assertions box and connecting it to an existing one.

12.3.4.2 Greedy Search for Optimal Operators

Expressing a graph expansion operator as a composition of elementary ones also suggests a way to find near-optimal graph expansion operators. In this work, we apply a greed approach which consists of the following steps:

- Step 0 (*initialization*): $t = 0, G^t \leftarrow G$.
- Step 1 (*optimization*): Find an optimal operator $\hat{h}^t$ in respect to G^t.
- Step 2 (*expansion*): Set $G^{t+1} \leftarrow \hat{h}^t(G^t)$
- Step 3 (*recursion*): If $t < T$ set $t \leftarrow t + 1$ and go to Step 1; else stop

where T is the number of steps, which may be either specified beforehand or optimally decided using a threshold. Using this procedure, the optimal function found is

$$\hat{f}^{\text{greedy}} = \hat{h}_1 \circ \hat{h}_2 \circ \cdots \hat{h}_T. \qquad (12.11)$$

Using the greedy approach has the benefit of simplicity and speed, though $\hat{f}^{\text{greedy}}$ may *not* be as good as the optimal one $\hat{f}$. Alternative methods, as for instance those of the forward–backward family, could also be considered to improve the optimality of the final operator. The core approach, though, stays the same, taking into account the decomposition of the optimal operator as a sequence of elementary operators and trying to iteratively find optimal values of each one of them.

12.3.4.3 Optimal Elementary Operators

We now describe how to find the optimal elementary operator h that adds a single edge or a vertex and an edge to the graph. Our approach consists of three steps:

- Define a scoring function that evaluates how good the result of an elementary operator is. This step is where *learning* takes place. In our case, learning is conducted once using the training material, whereas the scoring function may be used repeatedly every time we search for the optimal elementary operator
- Exhaustively search at the domain of operators for an optimal elementary operators
- Keep the elementary operator that scores most.

The scoring function differentiates in respect to the operator variant.

12.3.4.4 Optimal Edge Addition

For the first elementary operator h_1 that adds an edge between two existing vertices, the search space is $V \times V \times \mathcal{E}$. Optimal choice of the pair of vertices v_i and v_j as well as of the value of the new edge can be done by defining a *scoring* function S_1

$$S_1 : V \times V \times \mathcal{E} \to \mathfrak{R} \qquad (12.12)$$

Once this function is defined, the optimal operator can be chosen as

$$\hat{h}_1 = h_1(\hat{v}_i, \hat{v}_j, \hat{e}) = \arg\max_{v_i, v_j, e} S_1 \left[h_1(v_i, v_j, e) \right] \qquad (12.13)$$

12.3.4.5 Optimal Vertex Addition

Similarly, for the second elementary operator h_2 that adds a vertex and connects it to an existing vertex with an edge, the search space is $V \times \mathcal{V} \times \mathcal{E}$. Optimal choice of the pair of vertex v_i, the value of the new vertex v as well as of the value of the new edge can be done by defining a *scoring* function S_2

$$S_2 : V \times \mathcal{V} \times \mathcal{E} \to \Re \qquad (12.14)$$

Once this function is defined, the optimal operator can be chosen as

$$\hat{h}_2 = h_2(\hat{v}_i, \hat{v}, \hat{e}) = \operatorname*{argmax}_{v_i, v, e} S_2 \left[h_2(v_i, v, e) \right] \qquad (12.15)$$

12.3.4.6 Complexity Issues

Since an exhaustive search is conducted, the complexity of performing a single elementary graph operation is bounded by $O(|V|^2 \cdot |\mathcal{E}|)$ for the first operator and $O(|V| \cdot |\mathcal{V}| \cdot |\mathcal{E}|)$ for the second one, where

- $|V|$ is the number of vertices of the graph.
- $|\mathcal{V}|$ is the size of vocabulary for vertices, i.e. the number of possible concepts.
- $|\mathcal{E}|$ is the size of vocabulary for edges, i.e. the number of possible relations.

Note that complexity is polynomial in respect to the graph quantities. This compares favourably, with the EXPTIME complexity which is commonly associated with exact reasoning approaches. However, the above complexity does not take into account the complexity of calculating the scoring function. To restrain the complexity in polynomial number steps, one should take care that calculating the scoring function is also bounded in polynomial number of steps. The issue of defining a scoring function is discussed in the next section.

12.3.5 Scoring Functions for Graph Expansion Operators

So far, we have re-formulated the problem as an input–output system identification problem, defined a mapping between set of assertions and directed graphs and stated the problem as finding an optimal graph expansion operator. We have then proposed to find such optimal operators by decomposing them in elementary ones and greedily find optimal elementary operators by exhaustively evaluating a scoring function for each one of them. Therefore, we have described the way approximate inference in ontologies may be reduced to defining scoring functions for elementary graph expansion operators. We now describe a way to define scoring functions.

The definition of a scoring function is crucial for the success of the approach. Since a scoring function is evaluated in each step, its properties affect significantly the properties of the overall approach. Good properties of a scoring function are:

Learnability   It should be possible to adjust the scoring function parameters given a reference set of input and output a-boxes.

Low complexity   Evaluation of the function should be bounded by polyno-
                          mial number of steps in respect to the graph quantities.
Uncertainty handling   Scoring should take into account uncertainty values that
                          are possibly associated with the graph vertices and edges.

12.3.5.1 Graph Local Representations

The choice of a scoring function should be such that the resulting graph maximally
resembles the one that would have been obtained through exact reasoning. By con-
straining the range of the scoring function to $[0..1]$, the score for an operator that
adds an edge and/or vertex that would have been added by exact reasoning should
be maximum, i.e. 1.0, whereas the score for an addition that would *not* have been
should be minimum, i.e. 0.0. In other terms, the ideal scoring function for a proposed
expanded graph should be inverse monotone to the *distance* between the proposed
graph and the optimal graph. Note that learning and applying the ideal scoring function may be possible but prohibitively expensive in terms of complexity. Here, we
propose a family of scoring functions that approximates the ideal ones by taking
into account vertices and edges that are *close*, in terms of graph distance, to the
considered edge/vertex addition.

Namely, to insert an edge between two specific vertices v_1 and v_2, one could
take into account information from the whole graph. However, it makes sense to
take *more* into account vertices and edges that are close to v_1 and v_2, e.g. directly
connected, and *less* into account vertices and edges that are connected via a long
path. More formally, let $\mathcal{X}$ be a convenient local representation of the graph and
let the family of approximate local mappings of the graph *at the neighborhood of* a
vertex as:

$$\mathcal{U} = \{u : \mathcal{G} \times V \times N \rightarrow \mathcal{X}\} \tag{12.16}$$

where

- $G \in \mathcal{G}$ is the input directed graph under consideration.
- $v \in V$ is a particular vertex of the graph.
- $n \in N$ is an integer that specifies the maximum path length for any other vertex of
  the graph, so that it is taken into account at u. For example, $n = 1$ is equivalent to
  say that only direct edges and neighbors are consider at the local representation.
- $X \in \mathcal{X}$ is the local representation of the graph.

Depending on our choices, $\mathcal{X}$ may be simple or sophisticated. Two trivial cases are
the following:

1. X is the subgraph of G that contains v, its n-close neighbors and the relevant
   edges.
2. X is a "bag of words" consisting of the names of concepts and names of relations
   of vertices and edges that are around v.

The importance in defining the graph local representation is that the complexity of
the learning and evaluation of the scoring function is controlled by the degree n of

locality considered in combination with the structure of the representation space $\mathcal{N}$. By letting n take small values, e.g. $n < 3$, and simplifying the structure, e.g. bag of words, the scoring function complexity is significantly reduced

12.3.5.2 Representing Graph Paths as Features

The particular graph local representation we have devised and used for our experiments consists of converting the local neighborhood of a graph around a vertex or edge into a vector of features, each one reflecting the path from the center vertex or edge to a vertex or edge within the neighborhood. Each path is considered distinct if the values of the vertices and edges that form it, in their particular order, are unique. Each distinct path corresponds to a distinct feature. The value of the feature, in a particular neighborhood, corresponds to the number of cases the respective type of path connects the central vertex to other vertices or edges.

Namely, let v and e be a vertex and an edge that correspond to a concept instance and a relation, respectively. Let $c[v]$ be the value of the vertex, i.e. the concept of the instance, and $r[e]$ be the value of the edge, i.e. the type of relation. Then, the path

$$v_1 \xrightarrow{e_1} v_2 \xleftarrow{e_2} \cdots v_n \tag{12.17}$$

corresponds to the feature

$$c[v_1], r[e_1], c[v_2], \bar{r}[e_2] \cdots , c[v_n], \tag{12.18}$$

where $\bar{r}$ denotes the inverse relation. For convenience, since the concepts and relations are commonly given as unique alphanumerical strings, the unique name of the feature is made as a concatenation of the names they take part, in their specific order. In particular, the relation names are suffixed distinctively so as to denote whether the direct or inverse relation is part of the path.

12.3.5.3 Example

Consider the concepts `PoleVaulter`, `Pole` and `Pillar` and the relations `isAboveRight` and `isRightOf`. Let the vertices v_1, v_2 and v_3 have value `PoleVaulter`, `Pole` and `Pillar`, respectively. and the edges e_1 from v_1 to v_2 and e_2 from v_3 to v_1 value have values `isAboveRight` and `isRightOf`. Also assume that v_1 is not connected to any other vertices in the graph. Then, the local neighborhood of vertex v_1 is represented as follows:

- The value of the center vertex, `PoleVaulter`.
- The feature `isAboveRightS_Pole` has value 1.
- The feature `isRightOfO_Pillar` has value 1.
- All other features have zero value

Note that relation names have been suffixed with "S" and "0" to denote the direction of relation. Note also that the `PoleVaulter`, i.e. the center value of the neighborhood, is not part of the names of the features. Rather, this value is to be *predicted* based on the values of all other features. Assuming that the number of all other unique paths in the graph is M, and the features above correspond to the third and fifth, then the above information is compactly represented as:

$$[0, 0, 1, 0, 1, \cdots 0] \rightarrow \texttt{PoleVaulter} \tag{12.19}$$

12.3.5.4 Representing Uncertainty

The uncertainty values of vertices and edges are used to derive an uncertainty value for each path and therefore for each feature. Namely, by letting $w(v)$ and $w(e)$ be the uncertain values of vertices and edges, the uncertainty value of (12.17) is evaluated by multiplying the uncertainties of elements in the path:

$$w(v_1) \cdot w(e_1) \cdot c(v_2) \cdot r(e_2) \cdots w(v_n).$$

Example

To continue the example above, if $w(v_1) = 0.9, w(v_2) = 0.9, w(v_3) = 0.1, w(e_1) = 0.5$ and $w(e_2) = 0.9$, then the uncertainty of the first feature is $0.5 \cdot 0.9 = 0.45$, and the uncertainty of the second feature is $0.9 \cdot 0.1 = 0.09$. In result, the overall information is represented as:

$$[0, 0, 0.45, 0, 0.09, \cdots 0] \rightarrow \{\texttt{PoleVaulter}, 0.9\} \tag{12.20}$$

12.3.5.5 Complexity Issues

The graph path feature vector representation has the following complexity characteristics:

- The number of features depends on the size of the vocabulary and the size of the graph neighborhood considered. Namely, the number of features is bounded by $(|\mathcal{V}|^n \cdot |\mathcal{E}|^n)$ where

  - $|\mathcal{V}|$ is the size of vocabulary for vertices, i.e. the number of possible concepts,
  - $|\mathcal{E}|$ is the size of vocabulary for edges, i.e. the number of possible relations and
  - n is the longest path considered.

Although this bound is not polynomial in respect to n, the actual existing distinct paths in graphs of real-world problem will tend to be of a much smaller number. Moreover, as experiments have shown us, very good performances may be attained for $n < 4$.

- The complexity of evaluating each feature is bounded by $O(n \cdot \bar{E})$, where E is the maximum degree of any vertex in the graph. Again, the typical situation in a real world graph is that most vertices are connected to few others wile few vertices are connected to many. Therefore, the average complexity for evaluating each feature should typically be much less than its bound.

Based on the above, the overall worst complexity for evaluating the feature vector is $O(n \cdot |\mathcal{V}|^n \cdot |\mathcal{E}|^n \cdot \bar{E})$. The average complexity though is expected to be much smaller. In any case, though, for a fixed n, the complexity remains polynomial.

What is more n serves as a complexity/accuracy control parameter. For large n value, one obtains highly accurate feature vector representation. For smaller n values, the representation is less accurate, though much quicker to learn and apply. Typically, a value of $n = 3$ has been adequate for all problems we have evaluated the algorithm.

12.3.5.6   Soft Classifiers as scoring functions

Having specified how feature vectors associated with a vertex are generated, we now return back to the question of defining the scoring function. Remember that the scoring function should evaluate the optimality of adding an edge and/or a vertex to the graph into a particular point of the graph. In what follows, we concentrate on the elementary operator regarding a vertex, though the same arguments hold for edges.

Namely, in a somehow different way, the same problem may be stated as finding the probability that the vertex v_{new} is inserted to the graph with any of value $c \in \mathcal{V}$. Based on the graph path feature vector representation, the problem specializes further as follows:

> Given the graph path feature vector representation around a point of the graph, find the optimal value of vertex.

This is a typical case of soft classification problem in machine learning literature, where

- The input is the feature vector representation around the vertex to be inserted.
- The output is (a) the optimal value of the vertex to be inserted and (b) a confidence score for attributing the particular value.

The following remarks apply:

- The confidence score is not necessarily a probability value of the concept class in respect to the others. However, methods that can adapt scores found by classifiers into probability measures exist and have been applied in the literature.
- The confidence scores of classifiers are more accurately extracted as probability values of generative models for the optimal value found by classification. In this way, the score of the winning class for one class may still be low, even if has a high score in respect to all other competing values.

Soft classifiers may thus be used as scoring functions for elementary graph expansion operators. This has the significant advantage of delegating the adaptation of scoring functions for particular cases of ontologies to standard machine learning algorithms that use the feature vector representation. K-nearest neighbors, support vector machines and artificial neural networks are some of the approaches one may choose to adopt. Each one of these methods has a particular mathematical model to implement the scoring function and a specific way to learn the scoring function from training material, consisting of a set of input–correct output samples of form similar to (12.20).

Since the scoring function approximates part of the "reasoning", the choice of the classifier may affect the accuracy of the overall approach. Typically, classifiers with complex mathematical models will be more accurate, though they will demand a large number of training samples as inputs. Therefore, accuracy will be obtained at the cost of higher training time and larger sizes of ground-truth data to be provided.

12.4 Evaluation

In this section, we present the evaluation results of the method presented in this chapter. In Sect. 12.4.1.1, the data used for evaluation are described, while in Sect. 12.4.1.2 the evaluation procedure is explained. In Sect. 12.4.2, we present the results obtained for image, text and fused data, also showing the sensitivity of the method to the sub-graph size and classifier used, respectively.

12.4.1 Experimental Setting

12.4.1.1 Data

The proposed method has been evaluated on a corpus of approximatively 600 web-pages containing text and image from the BOEMIE athletics experiment (Petridis and Tsapatsoulis 2006). The assertion boxes resulting from per-media analysis tools have been used as the *input*. The output (ground truth) has been formed using the BOEMIE inference engine, adapted with hand-written DL-axioms, DL-safe rules and abduction rules (Peraldi et al. 2007). The fact that output may be somehow different than the ones produced manually is not significant in respect to our goal, which is to evaluate the way of the proposed approach to learn how to learn to infere based on a given reference set.

We have conducted separate experiments to test the suitability of the method for assertions sets related to image, text as well as the fused web-pages content. Depending on the medium, the set of values vertices and edges to be predicted, as well as the feature vector space differ. Note that the classes to be predicted are concepts and relations that exist only in the output assertions sets and not in the

Table 12.1 The class set used for evaluating the assertion boxes extracted from image and text. The class sets corresponding to the assertion boxes extracted from web-pages are the union of the respective one for text and image

	#	Class set
Image Vertices	7	HighJump, HighJumper, JavelinThrow, JavelinThrower, Person, PoleVault, PoleVaulter
Image Edges	16	Contains, hasPart, hasParticipant, isAbove, isAboveLeft, isAboveRight, isAtBothSides, isBehind, isBelow, isBelowLeft, isBelowRight, isLeft, isNear, isOverlapping, isRight
Text Vertices	17	DiscusThrowCompetition, HammerThrowCompetition, HighJumpCompetition, Hurdling110mCompetition, JavelinThrowCompetition, LongJumpCompetition, MarathonCompetition, Person, PoleVaultCompetition, Running100mCompetition, SportsCompetition, SportsEvent, SportsRound, SportsTrial, Walking10kmCompetition, Walking20kmCompetition, Walking50kmCompetition
Text Edges	25	Contains, hasAge, hasGender, hasNationality, hasPart, hasParticipant, hasPerformance, hasPersonName, hasRanking, hasSportsEventName, hasSportsName, hasSportsRoundName, hasStartDate, personToPerformance, personToRanking, sportsCompetitionToSportsRoundName, sportsCompetitionhasPartRound, sportsEventToSportsName, sportsRoundToPerformance, sportsRoundToRanking, sportsRoundhasPartSportsTrial, takesplaceInCity, takesplaceInCountry, takesplaceInSportsPOI

respective input ones. These correspond to the high-level concepts, the relations between high-level concepts and the relations between high-level concepts and mid-level concepts. The set of these concepts is given in Table 12.1.

The features generated from the input data depend both on the medium and on the size of the subgraphs considered. Table 12.2 shows an indicative set of the features used for some cases. The paths are formed using concepts and relation that are part of both the input and the output assertions. Note that the size of the feature set grows as the longest paths in the graph grow. For instance, the feature set to predict image vertices grows from 17 to 118 features. However, the number of features may be much less than the combinations of values, as is the case for the text vertices related features, where the feature set grows from 30 to 105. This happens because the path considered are not all the possible combinations, but only those that occur in the reference material.

12.4.1.2 Methodology

The evaluation aimed at measuring the generalization accuracy of the method. To this end, the model was trained with a part of the reference material and tested with another part of the reference material. For image-related assertions, a tenfold cross-validation test has been conducted. For text and web-page-related assertions, which number is significantly larger, a small part of the reference material (1–5%) has been used for training, whereas the other part has been used for testing.

Table 12.2 Indicative features used for evaluating the assertion boxes extracted from image and text

	Path size	#	Indicative feature set
Image Vertices	1	18	hasPartS, hasParticipantO, hasParticipantS, isAboveLeftS, isAboveRightS, isAboveS, isAtBothSidesS, isBehindS, isBelowLeftS, isBelowRightS, isBelowS, isLeftS
	2	112	hasPartS_HorizontalBar, hasParticipantS_PoleVaulter, isAboveS_Hammer, isAtBothSidesS, isBelowLeftS_Discus, isBelowRightS_HorizontalBar, isBelowS_Javelin, isLeftS_PersonBody, isNearS_PersonFace, isOverlappingS_Pillar
Text Vertices	1	30	hasAgeS, hasGenderS, hasNationalityS, hasPartOhasPartS, hasParticipantO, hasParticipantS, hasPerformanceS, hasPersonNameS, hasRankingS, hasSportsNameS, hasStartDateS, personToRankingS, takesplaceInCityS
	2	105	hasGenderS_Female, hasPartS_DiscusThrowCompetition, hasParticipantO_SportsTrial hasSportsEventNameS_SportsEventName, hasSportsNameS_Running100mName, hasSportsNameS_Walking50kmName, hasSportsRoundNameS_SportsRoundName, personToPerformanceS_Performance, sportsCompetitionhasPartRoundS, takesplaceInCountryS_Country, takesplaceInSportsPOIS_Stadium

To measure the success of the method, one has to compare the ground truth assertion A-boxes (reference A-boxes) against the ones predicted by the method. To that end, we apply the following steps:

- Find the optimal correspondence between each assertion of the reference A-box with each assertions of the predicted A-box.
- Measure the *recall* performance, by counting the number of vertices and edges of the *reference* A-box that have been matched by value with vertices and edges of the *predicted* A-box, in respect to the total number of vertices of the reference A-box.
- Measure the *precision* performance, by counting the number of vertices and edges of the *predicted* A-box that have been matched by value with vertices and edges of the *reference* A-box, in respect to the total number of vertices of the predicted A-box.

Once the alignment has been established, the steps for counting the recall and precision are straightforward. On the other hand, the optimal alignment uses a more elaborate procedure.

Optimal graph alignment Namely, an optimal alignment could be possible by exhaustively searching all possible alignments and select the one that offers maximum performance in respect to the evaluated measures. Nevertheless, this approach is of prohibited complexity. Instead, we propose here a greedy approach to iteratively align each assertion of the reference a-box with an assertion of the predicted a-box.

The steps are the following:

Step 0   Represent the two assertions sets as two directed graphs, using the approach that has been described in this chapter.
Step 1   Represent each edge of both graphs with the feature vector extracted using the approach described in this chapter, i.e. through the paths that connect the edge to its graph neighbourhood. The neighbourhood size may be controlled to balance the accuracy vs. complexity of the alignment. Indicatively, a path length of 3 should be adequate for most cases.
Step 2   For each edge of the first graph, find the closest edge of the second graph. Distance between edges is measured as follows:

- If the identifiers of adjacent vertices are identical, the distance is 0, else
- The distance between edges is measure as the L_2 distance of their respective feature vectors.

Step 3   Match the pair of edges that have the smaller distance and remove them from the set of edges to be examined.
Step 4   Go back to Step 2 to find the closest edges among those that are not yet matched, until all edges of either of the graph are matched.
Step 5   Match the vertices of the two graphs based on matching of the edges.

12.4.2 Evaluation Results

This set of experiments focus on evaluating the feature extraction approach and measure its sensitivity to the graph path length used and the machine learning classification algorithm used. Namely, Table 12.3 summarizes the performance achieved for the image modality. The results correspond to hiding one vertex of the graph to be predicted and try to recover it using the proposed algorithm. The first column specifies whether the task was to predict the value of a vertex or of an edge. The second column specifies the maximum graph path considered. The rest of the

Table 12.3 Evaluation results for the image and text-related data

Case study	Path	Naive Bayes	LogitBoost	K-NN	SVM
Image vertices	1	63.75	64.54	58.57	61.35
	2	**99.80**	99.80	93.82	99.80
Image edges	1	60.75	66.68	66.61	66.64
	2	70.04	**94.18**	89.80	92.62
Text vertices	1	92.74	93.78	**93.78**	93.32
	2	94.76	**96.63**	95.01	95.76
Text edges	1	95.48	96.67	95.12	**96.67**
Html vertices	1	82.35	82.86	80.95	72.51
	2	89.11	**91.25**	90.08	89.88
Html edges	1	96.37	97.03	97.36	**97.69**

columns give the accuracy obtained using the Naive Bayes, Logit Boost, K Nearest Neighbors and SVM classifier, respectively. Note that for the text and html edges, the graph path length examined was up to 1, since the values were satisfactory for that value.

Sensitivity to the class value Table 12.4 show that performance varies with the particular class to predict. Note that these correspond to using the Naive Bayes and thus are not the optimal one. Rather they are given to illustrate the sensitivity of methods to each class. Similar conclusion may be drawn by looking to the confusion matrix depicted in Table 12.5. Note, for instance, that though the

Table 12.4 Analysis of the Naive Bayes results for classifying edges in image assertions

Relation	Precision	Recall	f-measure	roc-area
hasPart	0.996	0.635	0.776	0.962
hasParticipant	1	1	1	1
isAbove	0.463	0.503	0.482	0.88
isAboveLeft	0.173	0.31	0.222	0.916
isAboveRight	0.353	0.526	0.423	0.918
isAtBothSides	0.091	0.2	0.125	0.84
isBehind	0.41	0.762	0.533	0.916
isBelow	0.601	0.688	0.642	0.889
isBelowLeft	0.135	0.28	0.182	0.936
isBelowRight	0.207	0.545	0.3	0.903
isLeft	0.218	0.267	0.24	0.825
isNear	0.178	0.31	0.226	0.929
isOverlapping	0.451	0.36	0.4	0.811
isRight	0.294	0.2	0.238	0.831

Table 12.5 Confusion matrix for the results of the Naive-Bayes classifier for edge classification task using graph path length 1

a	b	c	d	e	f	g	h	i	j	k	l	m	n	o	¡ – classified as
500	0	0	0	0	0	0	0	0	0	0	0	0	0	0	a = contains
0	546	0	16	6	21	9	104	59	13	21	15	35	6	9	b = hasPart
0	0	186	0	0	0	0	0	0	0	0	0	0	0	0	c = hasParticipant
0	0	0	81	11	6	9	18	7	2	2	3	6	16	0	d = isAbove
0	0	0	8	9	0	1	0	1	3	0	5	1	1	0	e = isAboveLeft
0	0	0	4	2	30	3	7	2	0	0	1	0	3	5	f = isAboveRight
0	1	0	7	5	5	6	1	0	2	0	1	1	1	0	g = isAtBothSides
0	0	0	2	0	3	1	112	4	0	11	2	1	9	2	h = isBehind
0	0	0	21	3	1	6	13	170	8	2	4	7	11	1	i = isBelow
0	1	0	0	1	0	1	3	4	7	0	1	0	7	0	j = isBelowLeft
0	0	0	1	0	1	2	0	0	1	12	0	0	4	1	k = isBelowRight
0	0	0	8	5	0	1	2	6	4	1	12	1	5	0	l = isLeft
0	0	0	6	0	0	0	0	7	0	2	2	13	12	0	m = isNear
0	0	0	16	8	10	13	5	23	12	4	9	8	64	6	n = isOverlapping
0	0	0	5	1	7	14	8	0	0	2	0	0	3	10	o = isRight

relation `hasParticipant` is always predicted with maxim accuracy, the relation `isLeft` is predicted with a not satisfactory level of accuracy. The study of this table could lead to several optimizations, including the addition of supplemental material regarding these relation, to enhance the training process.

Sensitivity to the path length By looking at Table 12.3, one can note that the method accuracy depends on the graph path length. Clearly, for graph path length equal to 1, i.e. considering only direct neighbors, the accuracy is not satisfactory. However, by considering a graph path length equal 2, a very good performance is obtained. This is particularly noticeable for predicting image edges. Since the results have been satisfactory for graph length equal to 2, there has been no need to experiment with longest graph paths. Nevertheless, the conclusion that graph length up to 2 is adequate applies only to the particular case we have studied.

Sensitivity to the classification method By looking at Table 12.3, one sees that results depend on the soft classification method used. This is a typical situation in the machine learning, since the performance of the underlying classification model and learning algorithm may depend on the domain of application. Moreover, one may choose, depending on the complexity requirements, to use a "cheap" classifier, such as the Naive Bayes, or a more complicated one.

12.5 Related Work

Extending strict reasoning approaches so that they are amenable to learnability and uncertainty handling has been the focus of much recent research.

Approaches such as fuzzy description logics (Straccia 2001) or (Lukasiewicz 2007) are a direct attempt to extend description logics so as to allow for uncertainty handling. However, the complexity of such approaches is prohibitive for the large scale of multimedia data we are dealing here.

On the other hand, there has been important work on the emerging field of statistical relational learning, such as probabilistic relational models (Friedman et al. 1999), relational dependency networks (Neville and Jensen 2007), markov logic (Richardson and Domingos 2006) and logical hidden markov models (Kersting et al. 2006). Although allowing both learning and uncertainty handling, these approaches differ to ours in two ways. First, they all considering specific network-like models which connections weights among nodes reflect the degree of association between the corresponding concepts. In contrast, the approach presented here is blind in this respect, since modeling is deleguated to a pluggable soft classifier (such as LogitBoost), and hence not necessarily a network-like structure. Second, our problem formulation allows to expand the A-box in an arbitrary way, given that representative training material has been provided. This allows, for example, adding individuals not existing at the input, such as those inferable by abductive reasoning (Peraldi et al. 2007).

12.6 Conclusions

In this chapter, we have presented a complete methodology that allows for extracting semantic knowledge from multimedia data sources. We have first presented that steps allow to express multimedia information by means of semantic assertions.

We have then described a machine learning method that allows to extract complex semantics, through performing inexact inference in ontologies. The method is based on transforming the inference problem into a graph expansion problem, expressing graph expansion operators as a combination of elementary ones and optimally seeking elementary graph operators. The latter issue is reduced to learn a set of soft classifiers, based on features each one corresponding to a unique graph path.

Our method has been evaluated in a web-page corpus comprising images and test. A specific evaluation procedure has been devised to evaluate the corpus. The evaluation results show that predicting the concepts and relations that structure together multimedia data is possible by inexact inference. Since the method is completely learnable, its adaptability to domains other than the one presented here is an issue of further investigation by the authors.

Acknowledgements This study is partly supported by the research projects "BOEMIE, Bootstrapping Ontology Evolution with Multimedia Information Extraction". FP6-027538/STREP, 2006–2009, http://www.boemie.org and "CASAM, Computer-Aided Semantic Annotation of Multimedia". ICT-217061/STREP, 2008, http://www.casam-project.eu/

References

P. Aarabi and B.V. Dasarathy. Robust speech processing using multi-sensor multi-source information fusion – an overview of the state of the art. *Information Fusion*, 5(2):77–80, 2004.

A. Goshtasby and S. Nikolov. Image fusion: Advances in the state of the art. *Information Fusion*, 8(2):114–118, 2007.

N. Friedman, L. Getoor, D. Koller, and A. Pfeffer. Learning probabilistic relational models. In *International Joint Conference on Artificial Intelligence*, volume 16, pages 1300–1309. Citeseer, 1999.

K. Kersting, L. De Raedt, and T. Raiko. Logical hidden markov models. *Journal of Artificial Intelligence Research*, 25(1):425–456, 2006.

B.D. Lucas and T. Kanade. An iterative image registration technique with an application to stereo vision. In *International joint conference on artificial intelligence*, volume 3, pages 674–679. Citeseer, 1981.

T. Lukasiewicz. Probabilistic description logic programs. *International Journal of Approximate Reasoning*, 45(2):288–307, 2007.

A.V. Nefian, L. Liang, X. Pi, X. Liu, and K. Murphy. Dynamic Bayesian Networks for Audio-Visual Speech Recognition. *EURASIP Journal on Applied Signal Processing*, 2002(11): 1274–1288, 2002.

J. Neville and D. Jensen. Relational dependency networks. *The Journal of Machine Learning Research*, 8:692, 2007.

S.E. Peraldi, A. Kaya, S. Melzer, R. Moller, and M. Wessel. Multimedia interpretation as abduction. In *Proc. DL-2007: International Workshop on Description Logics*, 2007.

S. Petridis and N. Tsapatsoulis. Semantics Extraction from Multimedia Content: The BOEMIE Architecture. In *Proceeding of the first international conference on Semantics and digital Media Technology (SAMT 2006)*, pages 6–8, 2006.

M. Richardson and P. Domingos. Markov logic networks. *Machine Learning*, 62(1):107–136, 2006.

S. Rudolph, T. Tserendorj, and P. Hitzler. What Is Approximate Reasoning? In *Proceedings of the 2nd International Conference on Web Reasoning and Rule Systems*, pages 150–164. Springer, 2008.

U. Straccia. Reasoning within Fuzzy Description Logics. *Journal of Artificial Intelligence Research*, 14:137–166, 2001.

E. Zavitsanos, G. Paliouras, G.A. Vouros, and S. Petridis. Discovering subsumption hierarchies of ontology concepts from text corpora. In *Proceedings of the IEEE/WIC/ACM International Conference on Web Intelligence*, pages 402–408. IEEE Computer Society Washington, DC, USA, 2007.

Chapter 13
Cognitive Algorithms and Systems of Episodic Memory, Semantic Memory, and Their Learnings

Qi Zhang

Abstract Explicit (declarative) memory, the memory that can be "declared" in words or languages, is made up of two dissociated parts: episodic memory and semantic memory. This dissociation has its neuroanatomical basis–episodic memory is mostly associated with the hippocampus and semantic memory with the neocortex. The two memories, on the other hand, are closely related. Lesions in the hippocampus often result in various impairments of explicit memory, e.g., anterograde, retrograde and developmental amnesias, and semantic learning deficit. These impairments provide opportunities for us to understand how the two memories may be acquired, stored, and organized. This chapter reviews several cognitive systems that are centered to mimic explicit memory, and other systems that are neuroanatomically based and are implemented to simulate those memory impairments mentioned above. This review includes the structures of the computational systems, their learning rules, and their simulations of memory acquisition and impairments.

13.1 Introduction

Memory is a single term referring to a multitude of human capacities. Although a universally accepted categorization scheme does not exist, human memory can be divided into short-term memory (working memory) and long-term memory. Long-term memory can be fractionated into explicit (declarative) memory and implicit (nondeclarative) memory (Graf and Schacter 1985). Implicit memory encompasses priming, perceptual learning, and procedural skills, etc. (Squire 2004), and explicit memory can be further divided into semantic memory and episodic memory (Tulving 1983).

Episodic memory is defined as memory for events; one must retrieve the time and place of occurrence in order to retrieve the event, as in answering the question,

Q. Zhang (✉)
Sensor System, 8406 Blackwolf Drive, Madison, WI, USA
e-mail: qizhang_sensor@yahoo.com

V. Cutsuridis et al. (eds.), *Perception-Action Cycle: Models, Architectures, and Hardware*, Springer Series in Cognitive and Neural Systems 1,
DOI 10.1007/978-1-4419-1452-1_13, © Springer Science+Business Media, LLC 2011

"What did you do this morning?" The retrieval query specifies the time, but in order to recall the events, the person must retrieve the place where the events occurred. Semantic memory refers to relatively permanent knowledge of the world or factual knowledge. Our knowledge that bird has wings, that fire burns, and that Lance Armstrong is a cycling legend, constitutes our factual knowledge or semantic memory.

One idea about the relationship between episodic and semantic memory is that repeatedly experienced events may become represented in a decontextualized form in semantic memory. For example, one should come up with a conceptual knowledge that all birds have wings, after seeing ducks, chickens, robins, crows, and hummingbirds, etc. One should be able to answer the question "Do all birds have wings?" without having to retrieve any specific episode (a particular time and place) in which one encountered a bird.

The relationship between the acquisition of semantic memory and episodic memory is twofold. One is that episodic memory may be a prerequisite for semantic memory. Empirical studies often reveal that amnesic patients, who lost their capacity to retain episodic memory, become almost impossible to acquire new semantic knowledge (e.g., Squire and Zola 1998), but their previously acquired semantic memory may still be preserved (e.g., Cohen and Squire 1980). The other aspect is that semantic memory (factual knowledge) is most likely abstracted and generalized from stored past experiences (i.e., episodic memories) through a cognitive process, named memory consolidation.

Besides the difference in properties, episodic memory and semantic memory are also associated with different cortical regions in the brain. It is generally agreed that semantic memory is associated with general neocortex, while episodic memory mainly with the medial temporal lobe (MTL), which includes the hippocampus and its surrounding cortices (e.g., Eichenbaum 2004). Lesions isolated within the MTL, especially the hippocampus, always lead to various amnesias. For example, a patient who lost his entire hippocampal function would not remember if he had had breakfast or recognize a person he had spoken to minutes ago. However, the same patient was still able to live his daily life and make intelligent conversations with semantic knowledge acquired before his onset of amnesia (e.g., Scoville and Milner 1957).

Given the importance of the hippocampus to semantic learning, how episodic memory is stored and activated becomes one of the central issues of memory formation. The hippocampus is characterized with sequential learning and spatial navigation capacities (e.g., Levy 1989, 1996; Granger et al. 1996; Wallenstein et al. 1998; McNaughton and Morris 1987), which allow us to retrieve a specific episode with particular sequence of time and coordination of space. Hippocampal cells that are activated when studied subjects (human and other mammals) perform given tasks are reactivated when the subjects are asleep, especially in rapid eye movement (REM) stage of sleep, which is considered the sign of dreaming. In some studies (e.g., Pavlides and Winson 1989; Schwartz 2003; Zhang 2009a), the hippocampal reactivation during dream sleep is considered the reactivation of episodic memory and is part of the process of memory consolidation. In dream sleep, episodic memory is likely activated in the form of segments instead of whole episodes (e.g., Fosse et al. 2003), and the segments are likely activated randomly, based on the conclusion that dreaming is the result of random impulse (Hobson and McCarley 1977; Foulkes 1985; Wolf 1994).

Furthermore, the information pathways between the hippocampus and neocortex may provide some important clues for how long-term memory is organized and how semantic knowledge is acquired. The hippocampus has a major input pathway from the neocortex (i.e., the entire spectrum of sensory modalities and multimodal association areas) to the perirhinal and parahippocampal cortices, to the entorhinal cortex, and to the dentate gyrus. And, the hippocampus also has a major output pathway from the subiculum to the entorhinal cortex, and back to the neocortex (e.g., Aggleton and Brown 1999; Gluck et al. 2003).

The properties of episodic memory and semantic memory, and their relations, are briefly introduced above. A cognitive system in mimicking human memory and learning, with regards to explicit memory (both episodic memory and semantic memory), is expected to reflect all these aspects. In the following sections, several computational systems are reviewed, and a new approach is then described in detail.

13.2 Computational Systems of Episodic Memory, Semantic Memory, and Their Learnings

Sun (2004) puts forward four essential criteria for the architecture of cognitive system: ecological realism, bio-evolutionary realism, cognitive realism, and eclecticism of methodologies and techniques. The central point of the realisms is that, in architecturing such a system, we should "aim to capture the essential characteristics of human behavior and cognitive processes, as we understand them from psychology, philosophy, and neuroscience." He further specifies the essential characteristics as bottom-up learning, modularity (specialized and separate cognitive faculties), dichotomy of implicit and explicit memories/processes, and synergistic interaction of the two memories/processes. In other words, these characteristics are centered on how memories are acquired, stored, utilized and how memories may interact. The author would like to extend these essential criteria and characteristics to any computational system of either cognitive or connectionist modeling that is aimed to mimic human cognition.

13.2.1 Cognitive Systems of Learning and Memory

Many cognitive architectures have been proposed, including Collins and Quillian's Model (Collins and Quillian 1969), ACT-R (Anderson 1983), SOAR (Newell 1990), EPIC (Meyer and Kieras 1997), PRODIGY (Minton 1990), DEM (Drescher 1991), COGNET (Zachary et al. 1996), and CLARION (Sun et al. 2001), etc. Each of the architectures is essentially a system of learning and memory, regardless what cognitive tasks it may perform. In the following, we will review some of the architectures and compare them against the essential criteria and characteristics given above.

It is noted that, in all these architectures/systems, explicit/declarative memory only refers to semantic (symbolic) memory, and episodic memory has rarely been considered (Sun 2004). The lack of episodic memory is clearly one of the shortcomings existing in these cognitive systems, given the fact that episodic memory is the prerequisite of semantic knowledge as described previously. It is also noted that not all of the cognitive systems are symbolic architectures, and some of them are symbolic-connectionist hybrid systems, e.g., CLARION.

13.2.1.1 Collins and Quillian's Hierarchical Network Model

Collins and Quillian's Hierarchical Network Model has been one of the most influential models in the symbolic approach of memory research. This model is a structure of symbolic knowledge (knowledge tree) that is thought to reflect how people represent and retrieve semantic information, and allow for inferential reasoning; thus, a computational system may be able to comprehend human language.

The hierarchical network of semantic memory structure is a network of concepts, as shown in Fig. 13.1. The concepts are connected together by labeled relations. The meaning of a concept in this system is therefore represented by the total configuration of relations it has to other concepts. The relations have two kinds: one is category membership labeled by "*is a*," and the other is property relation labeled by "*is*," "*has*," and "*can*." The organization of the network is hierarchical in both membership and property relations. Category members have a direct "*is a*" link to their immediate superordinates (e.g., *robin* is directly linked to *bird*), and properties are only stored at the highest concept level to which they apply (e.g., "*can eat*" is stored with *animal*, rather than with *bird*).

When receiving a statement (e.g., *A robin can fly*), the structure can verify if the statement is true or false, by first entering the network at the node corresponding to

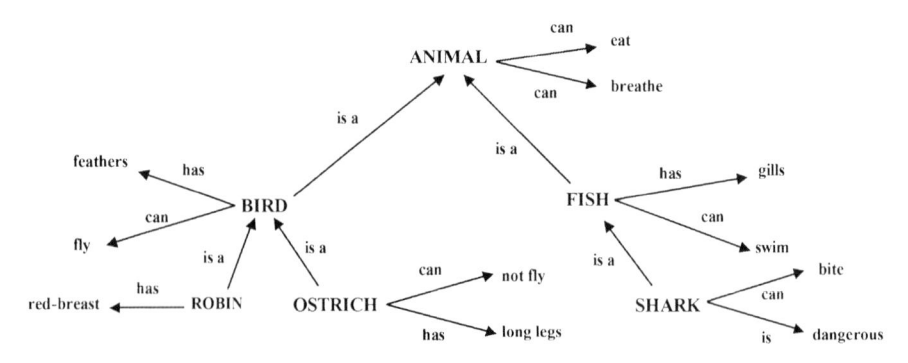

Fig. 13.1 Average verification time for true sentences as a function of number of levels in the hierarchy separating the subject and predicate terms. (After Collins and Quillian 1969)

the subject term and then searching for the predicate term. The search process first examines the relations that are directly lined to the subject term. If the predicate is found, the search stops and the subject responds *"true."* Otherwise, the search process moves up the hierarchy to next level and examines those relations.

This structure of semantic knowledge is one of the earliest attempts and has outlined the foundation for subsequent developments of symbolic memory structure in many other models. Since this model is not a fully developed cognitive architecture, it is not compared with the four criteria and characteristics. However, there are two issues that need to be mentioned. One is that there is no learning mechanism implemented, and as a result the system can only respond and cannot learn. The same issue exists in a few other cognitive architectures, such as PRODIGY (Minton 1990). The other is that in the semantic network, the relations are hand-coded. The same issue exists in many current semantic networks in which conceptual hierarchies require a priori determination through hand-coding, and slots need to be determined also through hand-coding.

13.2.1.2 ACT-R

ACT-R (Adaptive Control of Thought–Rational, Anderson 1983; Anderson et al. 1998) is arguably the most successful cognitive architecture in existence. The model has been applied in capturing a variety of human data in many different task domains, including, e.g., simulations of primacy and recency effects of working memory (Anderson et al. 1998; Anderson and Matessa 1997), and modeling of language acquisition and understanding (Anderson et al. 2004; Budiu and Anderson 2004), etc. The core of the ACT-R modeling is its learning and processing algorithm of declarative memory. Based on this central algorithm, several versions of ACT-R system have been developed for various applications with added modules that are intended to match human anatomical cortical areas for executive reasoning, visual perception, and motor control, etc.

Figure 13.2a is the architecture of ACT-R 6.0 of the latest version in which the cognitive center is the combination of the "Declarative module" for declarative memory and "Productions" module for procedural memory. In this modeling, declarative memory consists of symbolic facts such as "Washington, DC is the capital of USA" or "$3 + 4 = 7$", and procedural memory is made of production rules, which is intended to mimic human implicit memory about how we do things such as driving car or writing words. The procedural module serves as a switchboard function and connects every module through a designated buffer.

According to ACT-R, declarative knowledge is represented in terms of chunks of different types, and each type has an associated set of pointers encoding its contents; procedural knowledge is represented by production rules that are "procedurals" by which the pointers of an associated chunk (declarative knowledge) can be reassembled. Therefore, learning involves both declarative and procedural knowledge (i.e., both chunks and rules). Figure 13.2b is a graphical display of a chunk that encodes "$3 + 4 = 7$" with pointers to *three* (B_i), *four* (W_j), and *seven* (S_{ji}). The production

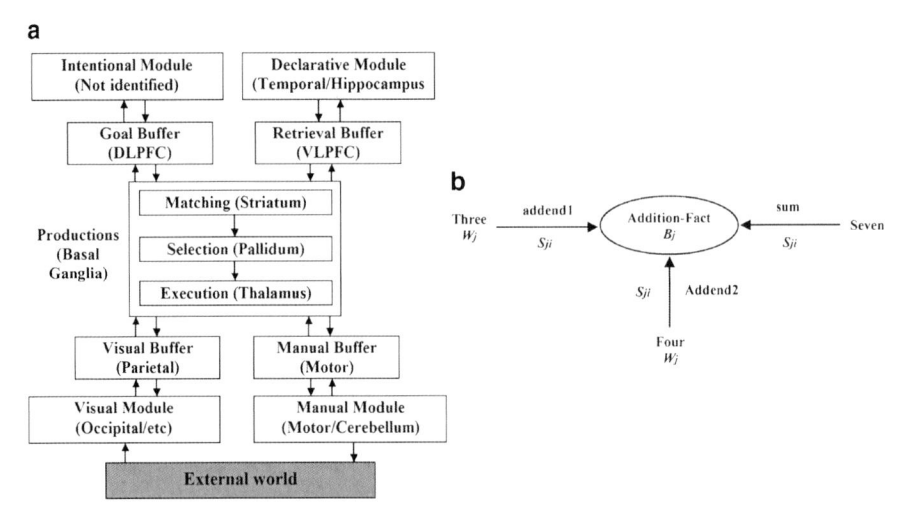

Fig. 13.2 (**a**) The structure of ACT–R version 5.0. Semantic knowledge (chunks of pointers or symbols) is stored in the declarative module; the associations among the pointers are stored in the procedural (productions) module. (**b**) A chunk encoding the fact that $3 + 4 = 7$. After Anderson et al. (2004)

rule of this addition is as follows: IF the goal is to process a string containing digits d1 and d2, and d3 is the sum of d1 and d2, THEN set a subgoal to write out d3. Each production consists of a condition (IF) that consists of a specification of the current goal, and an action (THEN). When this learning occurs, a chunk containing the three pointers is stored in the declarative module, and the production rule is stored in the procedural module. The retrieval of a declarative knowledge is driven by production rule. This rule is excited by input from external world and then applied to the declarative module through its buffer to activate and retrieve a specific chunk. The retrieved elements (pointers) of the chunk are "reassembled" in the manual module based on the production rule.

It can be seen that ACT-R modeling does not match the four criteria and characteristics very well. Two important issues are elaborated in here. One is about the problematic division between declarative and procedural knowledge. For example, the association among pointers (e.g., "d3 is the sum of d1 and d2") is not only about how the pointers may be manipulated but also how we understand the world. Without the association, the pointers are meaningless symbols. In other words, the combination of declarative and procedural knowledge in ACT-R refers to semantic memory of declarative memory of general understanding, and the procedural knowledge in ACT-R is not the procedural memory (like how to drive or how to ride bicycle) of general understanding. This issue leads to the second issue that ACT-R does not address bottom-up learning, because the production rules are only abstracted at top level. For example, the production rule acquired from the semantic input, "London is the capital of UK," is a manipulation rule at top-level.

13.2.1.3   CLARION

CLARION stands for Connectionist Learning with Adaptive Rule Induction ON-line (Sun 1999; Sun et al. 2001). CLARION is a hybrid system with a combination of localist and distributed representation. It has a dual-representational structure and consists of two levels: the top level captures explicit processes/knowledge and the bottom level implicit processes/knowledge (see Fig. 13.3a, b). In CLARION version 1, the action and non-action-centered explicit (or implicit) representations are combined in one block (Sun et al. 2001). Different from existing models of mostly high-level skill learning that use a top-down approach (i.e., turning declarative knowledge into procedural knowledge through practice as reviewed in ACT-R modeling), CLARION uses a bottom-up approach toward low-level skill learning, where procedural knowledge develops first and declarative knowledge develops later.

In the bottom level, the learning algorithm is called Q-learning-backpropagation algorithm, which is a supervised and/or reinforcement learning algorithm adopted from Q-learning algorithm (Watkins 1989). Such learning is to acquire Q values. Each Q value is an evaluation of the "quality" of an action in a given state. During learning, Q values are gradually tuned to enable reactive sequential behavior to emerge in the bottom level. The calculation of Q values for the current input with respect to all the possible actions is done in a connectionist fashion through parallel spreading activation. In the system, a four-layered connectionist network is used in which the first three layers form a backpropagation network for computing Q values and the fourth layer (with only one node) performs stochastic decision making. The output of the third layer indicates the Q value of each action (represented by an

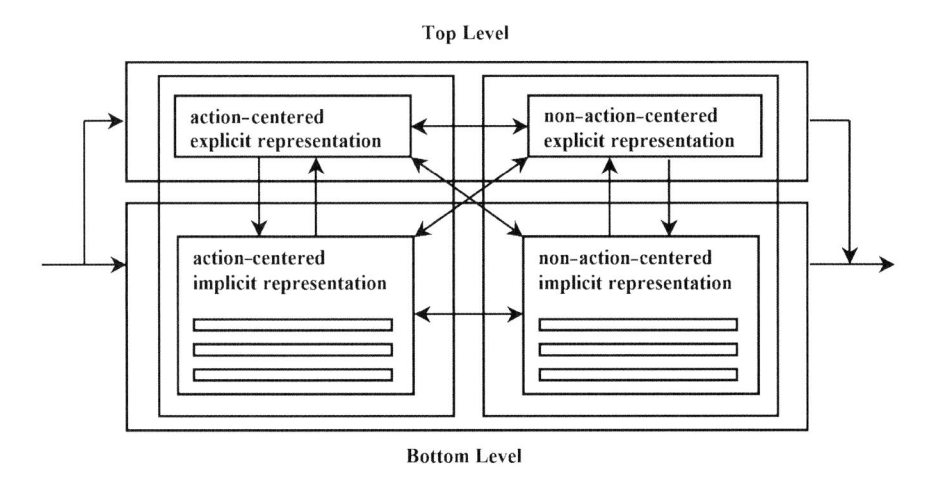

Fig. 13.3 The CLARION architecture. After Sun 2004

individual node), and the node in the fourth layer determines probabilistically the action to be performed based on the Boltzmann distribution, given as follows

$$p(a|x) = \frac{e^{Q(x,a)/\alpha}}{\sum_i e^{Q(x,a_i)/\alpha}}.$$

Here, α controls the degree of randomness of the decision-making process (Watkins 1989), and x is a state of the network.

In the top level, declarative knowledge is in a simple prepositional rule form that captures a bottom-up learning process by using information generated in the bottom level. The correlation between top-level rule and bottom-level output is a set of preset correspondences based on a localist connectionist model with which a set of rules is translated into the network. Assume that an input state x is made up of a number of dimensions (e.g., $x1, x2, \ldots, xn$). Each dimension can have a number of possible values (e.g., $v1, v2, \ldots, vm$). Rules are in the following form: *current-state* 3 *action*, where the left-hand side is a conjunction of individual elements each of which refers to a dimension xi of the (sensory) input state x, specifying a value or a value range (i.e., xi [vi, vi] or xi [$vi1$, $vi2$]), and the right-hand side is an action recommendation a. The top-level learning algorithm is as follows: If an action decided by the bottom level is successful then the agent constructs a rule (with its action corresponding to that selected by the bottom level and with its condition specifying the current sensory state), and adds the rule to the top-level rule network.

The fundamental difference between CLARION and ACT-R is the plausible bottom-up learning algorithm in which a connectionist network is expected to learn implicit knowledge, which becomes the bases of declarative knowledge. This model is a result of the effort to resolve the fundamental and long-standing problem of symbol grounding (Harnad 1990; Searle 1980; Smolenshy 1997) by connecting symbols to their meanings that may be acquired by neural network, or by associating rule-based knowledge to similarity-based knowledge (Sun 1995). However, in practice, this architecture has not simulated as many human behaviors and cognitive processes as ACT-R has, and the effectiveness of symbol grounding remains to be demonstrated in terms of robustness and flexibility in using acquired knowledge. It is noted that the original version of CLARION has a subsystem of episodic (or instance) memory to store recent experiences in the form of "input, output, result" (i.e., stimulus, response, and consequence), but this subsystem is removed in later versions.

13.2.2 Connectionist Systems of Episodic Memory, Semantic Memory and Their Learnings

Some connectionist systems are briefly reviewed for three reasons. One is that, as we have seen, cognitive architectures generally ignore episodic memory, even if episodic memory is the prerequisite of semantic memory/knowledge. The lack of episodic memory is clearly against the essential criteria of cognitive system given by Sun (2004). On the other hand, episodic memory is broadly considered and implemented in connectionist memory systems. The second reason is about symbol grounding. Many believe that, in order for a cognitive system to be robust and flexible, the system has to learn the meanings of symbols. Symbolic-connectionist hybrid system is the best candidate, and neural network is expected to capture the meanings (e.g., Harnad 1990; Sun 1995). Thus, one may want to know the progress in capturing meanings through connectionist memory systems. Finally, these systems are able to simulate cognitive process and behaviors of memory consolidation and amnesias, which are also simulated by a cognitive system developed by the author (to be introduced later).

The three connectionist systems (McClelland et al. 1995; Murre 1996; Meeter and Murre 2005; O'Reilly et al. 1998; Squire and Alvarez 1995) to be reviewed are considered neuroanatomically based systems as noted by Meeter and Murre (2005) because of the structural similarity to human brain. They all have the same view that the hippocampus and neocortex play distinct, but complementary, roles in long-term memory (i.e., episodic vs. semantic memory, and fast learning vs. slow learning). They all are able to simulate memory consolidation and retrograded amnesia, and the simulated results are examined by cued recall. Cued recall is one of the standard tests in human memory study. In the test, the subject is first presented with an information pair (i.e., picture-word pair), and then is asked to recall the word when promoted by a cue (i.e., the picture).

The system, presented by Alvarez and Squire (1994), consists of two "cortical" areas that are reciprocally interconnected with the MTL area, and the proposed MTL (representing the hippocampus and its surrounding areas) is a temporary connection that binds two separated cortical areas of the proposed neocortex (see Fig. 13.4). Each of the cortical area is made up of two groups of four simplified neurons, whereas the MTL consists of four neurons. During the training phase of episodic learning, the two cortical areas store externally presented patterns, and the MTL stores the "indexes" of the patterns stored in the neocortical areas. The "index"

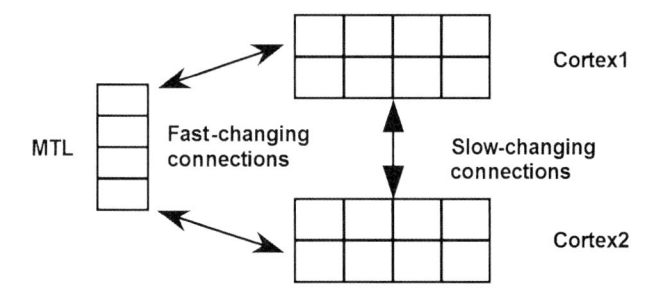

Fig. 13.4 The structure of the binding model (Alvarez and Squire 1994). Cortex1 and Cortex2 represent two neocortical areas, and each consists of eight neurons. The MTL is made up of four neurons. Each neuron in all three areas is reciprocally connected to each neuron in the other areas. There is no connection within areas, only a form of winner-take-all inhibition

is proposed to point to relevant neocortical cells and activate them (Teyler and DiScenna 1986) into specific pattern that has been learned in episodic learning. Under the guidance of the indexes, cortical–cortical connections associated with the stored patterns are slowly strengthened during memory consolidation. Simulations show that the system is at chance to perform cued recall when the MTL unit is lesioned soon after the training phase. With sufficient memory consolidation, however, the same damage no longer affects the recall because the connections among the stored information have been established and the binding function of the MTL is no longer necessary. During such tests, the connections between the MTL and cortical areas can be lesioned (inactivated) or normal (kept active). The former is called "lesioned" state that corresponds to amnesic state, and the latter is called "normal" state. These tests yield two curves of activation error vs. consolidation time. The "normal" curve matches the direction and shape of cued recall performed by healthy people, while the "lesioned" curve matches that of cued recall performed by patients with retrograde amnesia, namely Ribot gradient (1881), or temporal gradient. It is explained that, in amnesic state, remote memory is better retained than recent memory because of longer and more sufficient consolidation.

The TraccLink system, initially proposed by Murre (1996) and further developed by Meeter and Murre (2005), is based on a similar concept as the binding system, i.e., episodic memory is initially stored in the neocortical basis, and consolidation binds the traces of the stored information. The TraceLink system has three subsystems: a trace system (a layer of 200 nodes, i.e., highly simplified neurons) representing neocortical areas, a link system (a layer of 42 nodes) representing MTL, and a modulatory system representing basal forebrain, etc. Similar as the binding model, it is assumed that the formation of associations between neuron groups within the trace system is a slow process compared with the formation between the trace and link system.

This system undergoes four stages for long-term learning. In stage 1, an external pattern activates a set of trace nodes. In stage 2, the activated trace nodes activate a set of link nodes. Stage 3 is considered the initial consolidation process in which the link system is given a burst of random activation to initiate a random search for the nearest representation in the trace system. After a representation is found, the representation remains active until the next burst of random activation in the link system. Consolidation occurs through the formation and strengthening of connections within the trace system at a fixed base rate. In the final stage of consolidation, stage 4, trace–trace connections have become very strong, and retrieval of the initial memory becomes independent of the link system.

Normal learning and temporally graded retrograde amnesia are simulated and tested by meanings of cued recall, and the results are similar as those of the binding model/system reviewed earlier. The simulations of retrograde amnesia is implemented by entirely disabling the link system after initial learning, thus the trace nodes activated by the initial learning in one group cannot form strong associations to the trace nodes in the other group Anterograde amnesia is the opposite of retrograde amnesia. Patients with pure anterograde amnesia show strong deficit in recalling events experienced after their amnesic onset. TraceLink model simulates such impairment with two kinds of causes. One is a lesioned link system that is similar as in retrograde amnesic simulation. The other is a lesioned modulatory system. As a result, the link system loses its fast learning function and no longer assists the association between groups of trace nodes.

McClelland and colleagues (e.g., McClelland et al. 1995; O'Reilly et al. 1998) present a different model from the previous two, in which episodic memory is considered to be initially stored in hippocampus. Memory consolidation is considered a "training process," in which the hippocampus slowly teaches the hippocampal representations into the "neocortex." In the simulations of memory consolidation and retrograde amnesia, McClelland et al. (1995) only implemented a network system for semantic memory (the proposed neocortex), but not a hippocampal system. The hippocampal functions of rapid learning and information interleaving are assumed through data feeding to the input layer of the semantic system. In the simulations, a three-layer network system (generic three-layered feed-forward network, McCloskey and Cohen 1989), consisting of 16 input units, 16 hidden units, and 16 output units, is used. The system is first fully trained on a set of 20 random input–output pairs. These pairs are considered as previously acquired experiences. Then, the system continues to be exposed to these pairs throughout subsequent learning of 15 more input–output pairs. Thus, one additional training pair can be "interleaved" into previously learned associations during the new learning. After introduction of the new pair, training continues as before, which is assumed to be consolidation process. And, other new learnings continue in the same fashion for a total of 15 pairs. After all of the new pairs have been learned, the system is examined by being presented with a newly learned input to its input layer at given time intervals of consolidation. The output of the system is compared with the learned output that is assigned to the tested input.

13.3 A Multileveled Network System of Episodic Memory, Semantic Memory, and Their Learnings

Human explicit memory (declarative memory) consists of two dissociated components: episodic and semantic memory. Episodic memory is the memory for events that are featured with temporal sequence and spatial coordination of occurrence, whereas semantic memory is about factual and conceptual knowledge that is independent of a specific past experience. Semantic memory resides in the general neocortex and is independent of the MTL, but newly acquired episodic memory is dependent of the MTL. The development of semantic memory relies on the retention of episodic memory, and semantic memory is most likely acquired from episodic memory in a cognitive process named memory consolidation. Lesioned MTL may lead to various amnesias and result in the acquisition deficit of new semantic memory.

As reviewed earlier, in most cognitive systems, episodic memory is not considered. The lack of the episodic memory indicates the lack of neurobiological realism and cognitive realism in capturing the "essential characteristics of human behavior and cognitive processes." If semantic knowledge has to be consolidated from episodic memory, the consolidation process must selectively consolidate certain information from the episodic memory and ignore others, thus results in the robustness and flexibility of acquired knowledge. Such robustness and flexibility of human knowledge has yet to be demonstrated by computational systems.

On the other hand, the reviewed connectionist systems have fairly captured the relation between episodic memory and semantic memory. However, they fall short in demonstrating the temporal/spatial features in simulated episodic memory, and especially in demonstrating the factual/conceptual properties in the simulated semantic memory. During training phase, the systems are often presented with a series of patterns, but only one pattern, rather than an "episode" of trained patterns (like serial recall), may be recalled at a time. Although the systems are implemented with a semantic subsystem, the recalled materials are those trained patterns, which are arbitrary patterns and are clearly not factual/conceptual knowledge, despite much hope has been given to neural network in capturing meanings of knowledge (Harnad 1990; Sun 1995).

A cognitive system of learning and memory is introduced next (see Fig. 13.6). At structure level, the system is like a typical cognitive architecture in many aspects. It consists of a symbol subsystem and representation subsystem, which are equivalent to the declarative memory and implicit memory in the reviewed cognitive architectures. It employs a bottom-up learning mechanism, which has similar purpose as the subsymbolic learning in ACT-R or the similarity learning in CLARION. It will be seen that the system almost agrees with the four criteria and characteristics put forward by Sun (2004). However, there is a fundamental difference. It is not a rule-based system; rather it is a meaning-based system. The bottom-up learning is centered on abstracting and generalizing meanings (common features) from episodic memory. Such a learning mechanism is intended to find a practical solution to resolve the open question of symbol grounding problem. The effectiveness of symbol grounding will be demonstrated.

The presented system also has some similarity to the reviewed connectionist systems. It has cognitive areas that are equivalent to the hippocampus and neocortex and is able to repeat what have been achieved by the reviewed connectionist systems. In addition, it can also simulate developmental amnesia and direct semantic learning. The same cognitive system and some of these simulations have been reported previously (Zhang 2005, 2009a, b). The combination of the symbol and representation subsystems makes up the center of the presented system, each of which is a multilevel cognitive construct from base level to subsystem level. This system is introduced in a bottomup fashion i.e., from the base level to the top level of the overall system.

Fig. 13.5 The schematics of (**a**) single memory and (**b**) memory triangle, after Zhang 2005

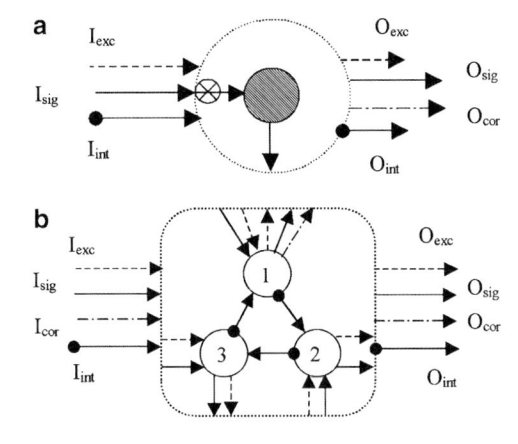

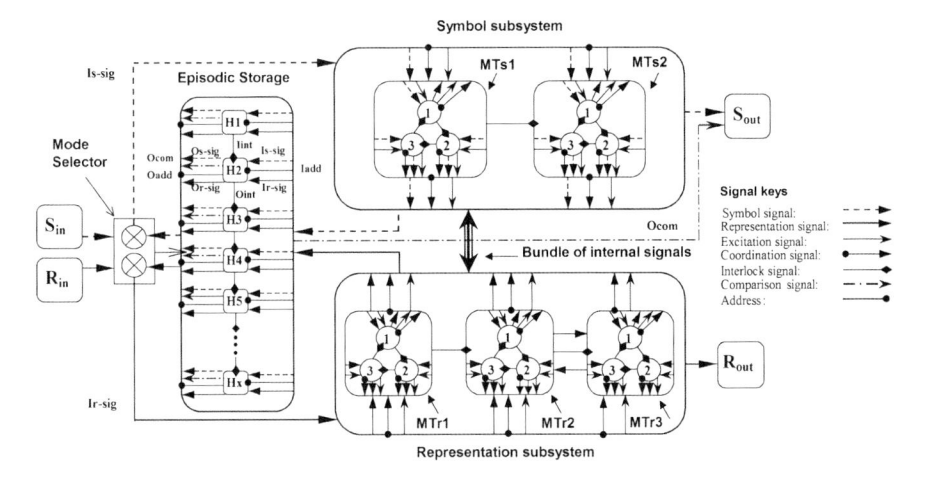

Fig. 13.6 The cognitive structure of a computational system that consists of a semantic system (the combination of the two subsystems) and an Episodic Storage. Each of the subsystem has three layers arranged in a hierarchal order from single memory (each of the small circles marked with 1, 2, and 3) whose function states are given in Table 13.1, to memory triangle (e.g., *MTs*1 and *MTr*1), and to subsystem. The symbol subsystem learns only symbols, and the representation subsystem learns only common features. Conceptual knowledge is acquired when a learned symbol is associated with a learned common feature. The Episodic Storage consists of a number of memory cells whose function states are given in Table 13.2. All of the cells are interlocked so that the Episodic Storage is able to store and retrieve a sequence of presented events. The Storage can also activate its stored information randomly. During episodic learning, a signal input (*Is-sig*) and a representation input (*Ir-sig*) are presented at *Sin* and *Rin*, respectively. Since the semantic system is a slow learner, it forwards them to the Storage for immediate episodic learning. During sequential recall, the Storage, triggered by the Mode Selector, activates a series of stored events along the interlocked direction. During recognition test, a representation input is presented at *Rin* and is forwarded to the Storage for comparison. A "*yes*" or "*no*" signal of *Ocom* is fired as the result of comparison and is projected to the *Sout*. During memory consolidation, the stored information is randomly and repeatedly activated from the Storage and becomes the source of semantic learning of the semantic system. After Zhang 2009b

13.3.1 Single Memory: To Locally Store Information

A single memory (SM) is the basic cognitive unit of the presented system in this study as shown in Fig. 13.5a, which can store and process either one symbol or one numerical value. A SM has three types of input: signal input (*Isig*), excitation input (*Iexc*), and interlock input (*Iint*). It has four types of output: signal output (*Osig = Isig*), excitation output (*Oexc*), interlock output (*Oint*), and coordination output (*Ocor*). The signal input is associated with external information and is the subject to learn. Its activation becomes the signal output that is used as a feedback to external world. All other signals are internal signals and are used to organize a dynamic knowledge structure, and to activate the stored external information as well. A SM may learn an incoming signal input only when it also receives a positive interlock input and an excitation input; it may activate its signal output only when it receives its designated excitation input. The function of the coordination output is to associate a signal input learned in one subsystem with a signal input learned in the other subsystem. The learning rules and activation rules are given in Table 13.1.

A SM is a storage unit to store an *Isig-Iexc* pair, as well as, a comparator that fires its stored information accordingly after comparing an arriving signal with what has been stored. Table 13.1 indicates that a SM learns in two consecutive steps. In step 1,

Table 13.1 Three important states of the single memory in the semantic system, after Zhang 2009a

State		Input	Output
Learning	Step 1: firing Ocor signal	Isig = "Io" Iexc = null Iint = "yes"	Osig = null Oexc = null Oint = "no" Ocor = "yes"
	Step 2: storing "Io" and "Iexco" permanently	Isig = "Io" Iexc ="Iexco" Iint = "yes"	Osig = null Oexc = null Oint = "yes" Ocor = "yes"
Firing stored "Io" upon receiving "Iexco" after the single memory has learned		Isig = any Iexc = "Iexco" Iint = "yes"	Osig = "Io" Oexc = "Iexco" or null[a] Oint = "yes" Ocor = "no"
Firing stored "Iexco" upon receiving "Io" after the single memory has learned		Isig = "Io" Iexc = any Iint = "yes"	Osig = "Io" or null[b] Oexc = "Iexco" Oint = "yes" Ocor = "no"

[a] Depending on Isig
[b] Depending on Iexc

if condition allows, the SM fires an *Ocor* and waits. In step 2, after the acting SM receives a unique *Iexc* it has been waiting for, it stores the *Iexc* together with the *Isig*. The generation of the unique *Iexc* is a result of the *Ocor*, and this correlation is best explained at system level later.

Memory formation in a biological system is thought to associate with the changes in synaptic efficiency that permits strengthening of associations between neurons, and the synaptic efficiency is related to two phases, short-lived and long-lasting, of synaptic modifications. The long-lasting modification may mostly (although not always) be induced by a series of tetanic stimulations over a long period of time in laboratory condition and is considered an attractive candidate for the molecular analog of long-term memory (see Lynch 2004). In order to cooperate with the long-lasting modification, a delay, T, is added to the learning mechanism of the SM at step 2. At step 2, therefore, a SM can fire a positive interlock signal only after it has been stimulated by the same signal input, *Isig* (= "*Io*"), for a given number of times over a period of time.

13.3.2 Memory Triangle: To Learn Meanings or Common Features

The cognitive capacity of one SM is very limited, and the capacity can be extended when a number of SMs are organized into a group. Three SMs are organized into such a group, named memory triangle (MT), in which three single memories form a loop via interlock signals (see Fig. 13.5b). The function of a MT is to learn a data point (*Isig*) for three times, in case the data point is the only common feature in a number of external representations.

According to Immanuel Kant and John Locke, a concept is a common feature or characteristic, and concepts are abstracts in that they omit the differences of the things in their extension, treating them as if they were identical. In the concept "Bird has wings" "wing" is the common feature of all birds, whereas specific characteristics such as color, size, and sound possessed by a specific bird, can be omitted. The MT is designed to capture a common data point (common feature) and generalize it. In a MT, only one SM is activated to capture the "common data point" at a time. After a MT has stored the "common data point" for three times into each of its SMs in the order from 1 to 3, the common feature is considered learned and generalized because of the existence of the loop. In here the order of learning is regulated by the *Iint,* and the extension of the common feature is realized by the loop.

A loop formed by three SMs, instead of four or five, may be best explained in terms of the principle of minimum potential energy. This principle is one of the fundamental principles we understand about nature. This principle says that a system always intends to configure itself into a formation that has minimum potential energy. The act of the principle is everywhere: the shape of star is always sphere, river runs to ocean, and one oxygen atom bonds to two hydrogen atoms instead of one or three. A loop consisting of three, four or more SMs can perform the same function of common feature extension, but a MT is the smallest loop that requires the least energy to maintain, thus becomes the first choice.

13.3.3 Organizing Memory Triangles: To Learn a Knowledge Structure

"Knowledge is an integrated phenomenon; every piece of knowledge depends on every other one"; what an intelligent system "has to do is to slowly accumulate information, and each new piece of information has to be lovingly handled in relation to the pieces already in there" (Schank 1995). Similarly, an external representation may come with only one common feature, but often it comes with more features that may be interlaced and correlated. A number of memory triangles may be organized into a subsystem that can learn more common features that are logically interrelated.

Either one of the two subsystems in Fig. 13.6, the symbol subsystem or representation subsystem, is formed to learn several interrelated common features. In the subsystems, information is locally stored. Where a given external input may be stored is the key for an overall knowledge structure and is regulated by interlock signals. The rule of interlocking is simple, same as how it works on three SMs within a MT: only when a MT has acquired a common feature, this MT unlocks the next MT. Under this mechanism, *MTr1* must first learn, then *MTr2*, and finally *MTr3*, if the to-be-learned common features are interrelated in a logically hierarchal structure. It learns in a similar way as people do: we have to know the meaning of "zero" before knowing the meaning of "one"; we have to understand what "one" is before knowing what "many" is.

The outline of either the symbol or representation subsystem is called "interface," which is the top cognitive layer of its subsystem and connects all its MTs. An interface is the information gateway of a subsystem, which delivers external stimuli to its MTs, exchanges information between the two subsystems, and projects signal output to the external world and other subsystems. When an interface receives an external signal, it disassembles the signal into a sequence of data points, and distributes the sequenced data points, one by one, to its MTs and SMs. It also collects and organizes activated information and forwards them to the opposite subsystem or external world.

13.3.4 Conceptual Learning: To Ground Symbols to Their Meanings

A concept is an abstract idea or a common feature, and a word is a symbol for concept. A cognitive system should learn a common feature together with its symbol to complete a knowledge acquisition. It is well known that symbolic approach of cognitive modeling has the advantage in learning symbols, while connection modeling is effective in learning patterns. How to ground symbol to meaning is still an open question. Connectionist modeling can also be developed to learn pure symbols (i.e., the network of symbolic knowledge tree presented by McClelland et al. 1995); however, there has been no substantial progress to ground symbols to their meanings (instead of patterns) in a network. Researchers have made great efforts to answer the question of symbol grounding (e.g., see Sun and Alexandre 1997). The system presented here can be seen as one of the efforts.

The system illustrated in Fig. 13.6 has two subsystems of symbol and representation, one is dedicated to learn symbols and the other is to learn common features. These two subsystems communicate with each other via the "bundle of internal signals." When the system learns, it abstracts common feature from external representation and stores it in the representation subsystem and does the equivalent to symbol in the opposite subsystem. When learning occurs, the symbol stored in one subsystem is paired up with the common feature stored in the other subsystem. The pairing is realized by excitation input ($Iexc$). Excitation input is one of the four inputs of a SM and is generated by the "bundle of internal signals." The generation only occurs when this bundle receives one $Ocor$ from either subsystem. When the acting SM that has fired the $Ocor$ in either subsystem receives a newly generated $Iexc$, it stores this $Iexc$ together with an $Isig$ (see Table 13.1). Every $Iexc$ is unique and is acting like a dynamic "address." So even if every SM in a subsystem is queried by a $Iexc$, only the SM containing same "address" can be excited to fire its $Osig$ (see also Table 13.1).

This combination of subsystems is considered the semantic system of the overall system shown in Fig. 13.6. This combination is inspired by the finding of split-brain (Myers and Sperry 1953) that indicates each brain half appears "to have its own, largely separate, cognitive domain," and to "have its own learning processes and its own separate chain of memories" as described by Sperry (1982). Sperry further noted that our left hemisphere is capable of comprehending printed and spoken word, and our right hemisphere is word-deaf and word-blind but capable of comprehending spatial and imagistic information.

13.3.5 Episodic Storage: To Store Episodic Memory

The episodic storage, in Fig. 13.6, is a storage site for external information. Due to the modification delay implemented in the SM, the semantic system is unable to learn any external information rapidly, and it always redirects the information to the episodic storage for immediate and direct storage. The direct storage function is same as the notion of "hippocampal system" proposed by McClelland et al. (1995).

The Episodic Storage consists of a number of memory cells (MC) that are enclosed by an interface that delivers inputs in parallel to all MCs and collects

Table 13.2 Three function states of the memory cell in the Episodic Storage, after Zhang 2009b

State	Input	Output
Learning: To store "Iso" and "Iro"	Is-sig = "Iso" Ir-sig = "Iro" Iint = "yes" Iadd = null	Os-sig = null Or-sig = null Oint = "yes" Oreco = null Oadd = "address-o"
Firing: To fire stored "Iso" and "Iro"	Is-sig = null Ir-sig = null Iint = null Iadd = "address-o"	Os-sig = "Iso" Or-sig = "Iro" Other signals = null

Comparing: To compare incoming signal with stored "Iso" and "Iro"	When Is-sig = "Iso";
	   other signals = null
	
	 When Ir-sig = "Iro";
	   other signals = null
	
	 When Is-sig ≠
	   "Iso"; or Ir-sig
	   ≠ "Iro"; other
	   signals = null
		Oreco = "yes";
		 Or-sig = "Iro";
		   other signals = null
		
		 Oreco = "yes";
		 Os-sig = "Iso";
		   other signals = null
		
		 Oreco = "no";
		   other signals = null

outputs from them. Each MC has four inputs (symbol input, *Is-sig*, representation input, *Ir-sig,* interlock input, *Iint*, and address input, *Iadd*) and five outputs (symbol output, *Os-sig*, representation output, *Or-sig*, interlock output, *Oint*, comparison output, *Ocom,* and address output *Oadd*). The function states of a MC are given in Table 13.2. When a MC learns, it stores a pair of *Is-sig* and *Ir-sig*, and sends its "address," *Oadd,* to the Storage interface. Since all MCs are interlocked by interlock signals in one direction, a sequence of external events can be both stored and activated in the original order of arrival. When a MC receives an external signal that matches any one of the two stored signals, it fires an *Ocom* of "yes," otherwise, "no." The interface can activate the MCs to fire stored signals along the interlocked sequence by sending all MCs a sequence of specific addresses, or activate them to fire randomly regardless of existing sequence. This function is to mimic the sequential learning function of the hippocampus that has been concluded in many studies. The episodic storage receives both symbol and representation inputs from the semantic system as shown in Fig. 13.6, which coincides with the fact that the hippocampus mainly receives inputs from the neocortex.

This storage has its designated information pathways to and from the semantic system of the paired subsystems, which coincide with the major pathways concluded in the studies by Aggleton and Brown (1999) and Gluck et al. (2003).

Other components of the system in Fig. 13.6 are explained as follows. *Sin/Rin* are external input interfaces and *Sout/Rout* are external output interfaces of the system.

The Mode Selector is a switch to select input source for the semantic system. The input source can be external information or internal information coming from the Storage during "dream sleep." The Selector can also send a simple triggering signal to the Storage's interface to stimulate sequential or random firing from there.

13.4 Simulating Episodic Memory, Semantic Memory, and Their Learnings

The presented multileveled memory system is employed to simulate serial recall, memory consolidation, dreaming, retrograde amnesia, developmental amnesia, and direct semantic learning. In the simulations, the episodic memory is demonstrated to be episodic-like, e.g., it may recall an "episode" of past experiences. The semantic memory is demonstrated to be conceptual, e.g., the acquired semantic knowledge can be utilized to process unfamiliar external information.

In the simulations, results are examined in terms of serial recall, recognition, object naming (may also be seen as cued recall), and object drawing. These are all standard tests in psychological studies for learning and memory. It is noted that object naming is similar to, but beyond cued recall. When the system is presented with an experienced input, the result is equivalent to cued recall, but when it is presented with an unfamiliar input, the result is an object naming. Cued recall is almost the only testing method used in the three reviewed computational systems.

13.4.1 Episodic Learning, Serial Recall, and Recognition

Episodic memory is the explicit memory for events. One must retrieve the time and place of occurrence in order to retrieve the event. The sequential learning and spatial navigation capacities of the hippocampus (e.g., Levy 1989, 1996; Granger et al. 1996; Wallenstein et al. 1998; McNaughton and Morris 1987) play an important part in episodic memory, and allow one to retrieve a specific episode with particular sequence in time and coordination in space.

The first step in all following simulations is episodic learning, i.e., let the system learn a sequence of external "events." In episodic learning, the Mode selector is set for the system to receive externally presented "events" given in either Table 13.3a or b. Each "event" in the table is a symbol–representation pair, and the representation input contains the meaning assigned to the symbol. All input pairs in either Table 13.1a or b contain three concepts, "zero," "one," and "tally," which are represented by the common features carried by the representation inputs.

It is noted that the common features for "one" and "tally" are peak/peaks in Table 13.3a and are pit/pits in Table 13.3b. In episodic learning phase, we only let the system learn from either Table 13.3a or b. After the system has gone through

Table 13.3 (a) Input pairs for episodic learning (b) Input pairs for direct semantic learning

(a)

Sequence	Symbol	Representation
1st	**III**	
2nd	**I**	
3rd	**I**	
4th	**Z**	
5th	**IIII**	
6th	**Z**	
7th	**Z**	
8th	**I**	
9th	**II**	
10th	**Z**	

(b)

Sequence	Symbol	Representation
1st	**Z**	
2nd	**Z**	
3rd	**Z**	
4th	**I**	
5th	**I**	
6th	**I**	
7th	**II**	
8th	**III**	

semantic learning phase, the system is expected to be able to tally either peaks or pits. The purpose is to show the system's flexibility in learning different common features. For simplicity, however, in most of the simulations to follow, the episodic learning is the sequenced pairs in Table 13.3a, and only once the pairs in Table 13.3b.

In the episodic learning, each input pair is presented to the system based on the sequence indicated in the table. The symbol input is at the *Sin* and the representation is at the *Rin*. This learning is a one-time experience, and the system is expected to remember the sequenced event thereafter.

Table 13.4 Simulations of sequential recall and recognition after episodic learning

Task	Input	Output
Sequential recall	Triggered by the Mode Selector	III I I Z IIII Z I II Z
Recognition		yes
		yes
		no

The input pairs are first transported to the semantic system for learning. However, in most of the cases, the semantic system is not able to learn external information due to the "modification delay" and the complexity of external information. As a result, the input pairs are sent to the Episodic Storage, one after another, for immediate storage into different MCs along the interlocked direction.

The system may be "asked" to recall the sequenced events it has just experienced. During the sequential or serial recall, the Mode Selector sends a trigger signal to the Episodic Storage. In turn, the Storage's interface sends a sequence of *Iadd* to all MCs to activate appropriate memory input pairs. Since the *Iadd* is associated with the interlocked chain of MCs, a past experience is recalled in the same sequence as what has been experienced in episodic learning. The first simulation in Table 13.4 is such a sequential recall.

The system may also recognize the input if a newly presented item is an experienced one. In this process, externally presented information is forwarded to every MC in the Episodic Storage for comparison. When a match is found, a "yes" output is fired from the specific MC. The last three simulations in Table 13.4 are recognition tests. In these three tests, the first two representation inputs are included in Table 13.3a and have been "memorized" in the Storage. The system "recognizes" them and shows "yes" to confirm. The last presented input is not included in Table 13.3a, thus the Storage has no memory of it and fires a "no" for the unrecognized input.

13.4.2 Dreaming, Learning, and Memory Consolidation

Dreaming refers to the subjective conscious experience we have during sleep. Numerous studies have concluded that dream deprivation always causes poor mastering of knowledge (explicit memory) and skill (implicit memory) that have been learned in the previous day. Findings of the correlation between dream sleep and waking learning have suggested that dream sleep may play an important role in learning and memory consolidation (e.g., Bloch et al. 1979; Fishbein 1970; Greenberg and Pearlman 1974; Pearlman 1971).

The relation between dream sleep and memory consolidation is also proposed in the studies of neuronal recording, which reveal the replaying of recent waking patterns of neuronal activity within the hippocampus during sleep, especially dream sleep (e.g., Pavlides and Winson 1989; Wilson and McNaughton 1994; Staba et al. 2002; Poe et al. 2000; Louie and Wilson 2001). Importantly, this replay, or hippocampal firing, is synchronized with activities in the neocortex, rather than an isolated activity. Such synchronization is attributed to be the evidence of memory consolidation from the hippocampus into the neocortex (Battaglia et al. 2004).

It is often concluded that dreams are more or less random thoughts and are caused by random signals (Hobson and McCarley 1977; Foulkes 1985; Wolf 1994). In reviewing the correlation between daily experiences and dream contents, it is found that daily experiences are often replayed in the form of segments, rather than entire episodes during REM (rapid eye movement) sleep (Fosse et al. 2003).

In short, dreaming may be a learning and memory consolidation process in which the segments of daily experience are randomly activated from the hippocampus, and the neocortex is synchronized to incorporate with the randomly arriving information. Such a process is simulated with the system as shown in Fig 13.6.

In the simulation of dreaming, the Mode Selector is set for the Episodic Storage to randomly activate its MCs to fire stored information pairs. A stream of activated events flows to the semantic system for further process and learning, and this stream can be recorded as "dream report" (Zhang 2009a). At the subsystem level, every randomly arriving representation "event" or symbol "event" is disassembled into a sequence of smallest information pieces, and these pieces are, one after another, delivered to every SMs. Each SM may react to, ignore, or learn from the arriving external signal according to the rules given in Table 13.1. The four levels of hierarchal structure from SM to semantic system regulate whether a SM has a potential to learn. The regulations decide whether an external input fits an existing memory structure, and where the external input should be stored. All of the regulations are simply reflected at the interlock signal inputs among SMs and among MTs. When learning occurs, a symbol is stored in the symbol subsystem and its associated common feature(s) stored in the representation subsystem spontaneously. In the meantime, the stored symbol is paired up with the stored common features by a unique excitation signal that is automatically assigned by the semantic system.

The system has to experience thousands of random activations before it is able to fully consolidate those memorized episodic events into the semantic system. After full consolidation, the system can be set to "waking mode" to process other external information. The last three simulations in Table 13.5 summarize how the system may respond to external information after the consolidation with a disabled Episodic Storage. It is noted that the three external representation inputs in these three tests are not the ones that have been presented during episodic learning. However, the system is able to count how many peaks exist in the given external inputs. It is able to do so because the semantic system has acquired the common features or conceptual knowledge and is able to flexibly use the knowledge to process either familiar or unfamiliar information.

Table 13.5 Simulations of impaired sequential recall, and intact recognition and semantic learning

Condition	Input	Output	Comment
Tested after episodic learning, but before consolidation	Triggered by Mode Selector	I Z	Impaired experience recall
		yes	Intact recognition
		yes	Intact recognition
		no	Intact recognition
Tested after consolidation and with disabled Storage		Z	Intact tally
		II	Intact tally
		IIIII	Intact tally

13.4.3 *Retrograde Amnesia and Anterograde Amnesia*

Patients with severe bilateral lesions in the hippocampus are often unable to remember events from moment to moment and show a mild loss of old memories extending back in time for years (e.g., Anon 1996; Scoville and Milner 1957; Squire and Zola 1998). The former is named anterograde amnesia and the latter is named retrograde amnesia. Most patients with retrograde amnesia show a temporal gradient (Ribot gradient) in memory retrieval, i.e., episodic memory acquired long before the lesion is better recalled than that of newer memory, which is also named temporal graded retrograde amnesia.

In the three computational systems reviewed earlier, both anterograde and retrograde amnesias are explained in terms of loss of the hippocampal function to bind (Alvarez and Squire 1994), to link (Murre 1996) episodic memory that are stored in the neocortex, or to rapidly store episodic memory (McClelland et al. 1995). Memory consolidation is exclusively proposed to be the key reason to cause the "temporal gradient" that is observed in most patients with retrograde amnesia. Old episodic memory has more chances than newer memory to be consolidated into the neocortex and to become independent of the hippocampus, thus to be better retrieved in the absence of functional hippocampus. Retrograde amnesia is almost accompanied by anterograde amnesia in cases of the bilateral lesions. So the cause of retrograde amnesia also applies to anterograde amnesia, because without a functional hippocampus, episodic memory after the onset of the bilateral lesions cannot be established and retrieved.

The key simulation of retrograde amnesia is to exhibit the temporal gradient. Those three computational systems have demonstrated this property under the same mechanism that the temporal gradient reflects the progress of memory consolidation. A similar property is also simulated using the system presented in this chapter, as shown in Fig 13.7. This temporal gradient curve is a relationship between the number of random activations and number of episodic events that have

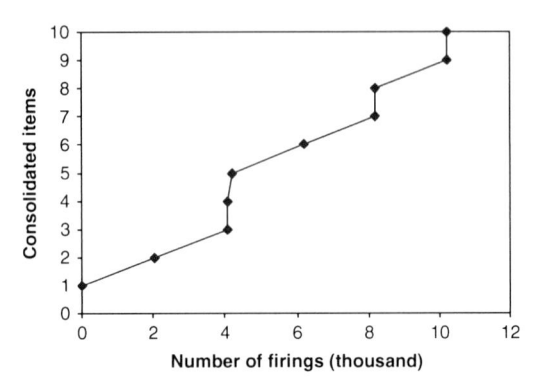

Fig. 13.7 A simulation of memory consolidation showing the relationship between consolidated items and the number of random firings. Here, the ratio of T/t (modification delay/random firing interval) is set at 2000. After Zhang 2009b

been consolidated into the semantic system. Since the random firings are activated at a fixed time interval, this curve also represents a relationship between consolidation rate and time. The curve is obtained from scores of cued recall tests, the same method used in the simulations in other studies. In the tests, the cues are those representation inputs listed in Table 13.3, and the targeted recall materials are those symbols that are paired with the representations. During such a test, the system is set to go through a given number of random activations, then the Episodic Storage is disabled, finally it is presented with a cue and its symbol output at *Sout* is examined.

However, if the Episodic Storage is disabled before episodic learning, the system is unable to recall any of the experienced events, which is a similar mechanism of anterograde amnesia simulated by Meeter and Murre (2005).

13.4.4 Developmental Amnesia

Developmental amnesia is an atypical form of memory deficit that has been discovered to occur in children with hippocampal atrophy. A clear dissociation has been revealed between relatively preserved semantic memory and badly impaired episodic memory. Such patients always suffer bilateral damage to the hippocampal formation at very early ages with sparing of surrounding cortical areas. The badly impaired episodic memory is mostly shown in delayed sequential recall and spatial recall. The patients may score anywhere from a few percent up to 25%, compared with control groups, in both delayed storytelling and delayed reproduction of geometric designs. However, they have compatible IQs as those of control groups and their recognition ability appears to be normal or close to normal (Vargha-Khadem et al. 1997).

It seems that such early loss of episodic memory may impede cognitive development and result in severe mental retardation (Baddeley et al. 2001), since many believe that semantic memory is mainly acquired from episodic memory through memory consolidation. Several explanations have been suggested. One is that the recollective process of episodic memory is not necessary either for recognition or for acquisition of semantic knowledge (Baddeley et al. 2001; Vargha-Khadem et al. 1997). However, this explanation does not really offer a mechanism for why such patients may still presumably learn semantic knowledge from memory consolidation and recognize presented items but perform poorly when recalling a sequence of events or spatial-related information. Another explanation (Squire and Zola 1998) is that since none of the patients have entirely lost their "recall memory," the residual "recall memory" may be enough to explain the near normal semantic memory performance, although no detailed mechanism is offered either.

While the mechanism of developmental amnesia still remains unclear, the system shown in Fig. 13.6 is the only one that is able to simulate an impaired sequential recall vs. an intact capacities of semantic learning and recognition (Zhang 2009b). These simulations are based on the proposal that limited hippocampal atrophy (27–56% compared with healthy subjects, see, Isaacs et al. 2003) may only impair its sequential and spatial learning capacities, but spare its storage function. It is considered that, in order to memorize sequential or spatial information, a system needs to memorize not only the elements in the information but also the associations among the elements. When a lesioned hippocampus is no longer able to store the associations, it is problematic to recall the sequential or spatial information entirely, even if every element of the information has been stored. The system may still be able to recognize a past event and learn semantic knowledge from those disconnected events but is unable to recall them in their originally presented order.

Two mechanisms are implemented to cause the impaired sequential learning or impaired associations of information in the simulations. One is that the interface of the Episodic Storage is unable to register or encode most of the associations, and the other is that most of the associations have been encoded wrongly, e.g., incorrect addresses are provided during episodic learning. These two mechanisms imply that the lost associations are fixed at the moment when episodic learning occurs and are not randomly selected during recall. In other words, the patients may show the same pattern of memorized elements vs. lost elements in retelling of the same story and redrawing of the same picture in repeated tests. However, in all reported empirical studies, the repeatability is not reported.

The system implemented with either lesioned situation mentioned above is used to simulate developmental amnesia and some of the results are shown in Table 13.5. The first simulation is a sequential recall, which is apparently an incomplete recall compared with the same recall in Table 13.4. This partial recall is comparable to the data given in the initial study (Vargha-Khadem et al. 1997) in which recalled materials are in the range of 20–25% of controls in either delayed storytelling or redrawing of presented geometric design. The pattern of the simulated performance is similar to the redrawing of a geometric design performed by three patients in the initial study. In the study, the geometric design is a single structure consisting of many interlaced triangles, rectangles, and lines. The patients are only able to redraw a small portion of the whole design. Interestingly, the redrawn portions are mostly detached triangles and rectangles and the associations among the patterns are lost. The same feature of detached elements is also shown in the simulations of impaired sequential recalls.

The damaged sequential learning mechanism does not necessarily impair the recognition ability of the Storage, because recognition process utilizes the comparison function that is a different mechanism from sequential recall. Information can be recognized as long as it has been stored even with a wrongly registered address. The three simulations show that experienced events can always be recognized (the two recognition tests that generated "yes" output in Table 13.5), while nonexperienced events cannot (the one recognition test that generated "no" output).

Random activation of the proposed hippocampus has also been implemented for the memory consolidation simulations in the three reviewed systems. When semantic knowledge can be learned from randomly activated episodic memory in memory consolidation process, the related semantic learning should be less affected by an impaired sequential learning function. The last three simulations in Table 13.5 show that the semantic knowledge acquired from randomly activated information is not only equivalent to, but also beyond what has been learned in episodic learning. In these simulations, the semantic knowledge has been utilized to process "unfamiliar" external inputs even if they are seemingly more complex. This demonstration of semantic knowledge property is considered the basic requirement for the simulation of developmental amnesia, since normal IQ is the key characteristic of the patients with years of developmental amnesia.

13.4.5 Dense Amnesia and Direct Semantic Learning

Densely amnesic patients not only show a total loss of episodic learning capacity but also become almost impossible to acquire new semantic knowledge (Squire and Zola 1998). When such patients are tested for semantic learning, e.g., new words, over a relatively short period of time (e.g., days or weeks) and with infrequent encounters of learning materials, the results are always negative (Gabrieli et al. 1988; Postle and Corkin 1998), although the learning conditions are adequate for healthy subjects. The understanding is that such patients are not able to hold new episodic memory, which can be re-accessed for numerous times over a period of time in memory consolidation process to acquire semantic knowledge.

However, the same patients may very slowly acquire semantic knowledge, if they have repeatedly encountered the same information over years of time. The most significant case of the slow semantic learning reported (O'Kane et al. 2004) is about the famous patient, H.M. He is the most studied amnesic patient, and his case has a special position in the understanding of human memory system because of his well-known and well-localized MTL lesion that has left him with no hippocampal function (Scoville and Milner 1957). In tests, H.M. is able to tell the last names of more than one-third of people who became famous after his amnesic onset, when whose first names are provided as cues. He is able to describe John Glenn as "the first rocketeer" and Lee Harvey Oswald as a man who "assassinated the president." This new knowledge is demonstrated to be flexible and semantic (O'Kane et al. 2004) because H.M. is able to retrieve the same knowledge promoted by different cues. Slow semantic learning, over a long period of time (e.g., 13 years), has also been observed in other densely amnesic patients (Butters et al. 1993; Kitchener et al. 1998; Tulving et al. 1991).

Given the fact of H.M.'s well-known hippocampal lesion, the semantic knowledge he is able to demonstrate is unlikely acquired through the mechanisms identical to the ones that healthy adults use to acquire semantic knowledge. It is suggested that H.M.'s mechanism for semantic learning appears to be via slow learning, whereby following extended and repeated encounters of the same information (O'Kane et al. 2004). Other similar studies (Butters et al. 1993; Kitchener et al. 1998; Tulving et al. 1991) have also come to the same conclusion that the demonstrated semantic knowledge may have been acquired directly and gradually by the neocortex in years of extensive exposures to information. On the other hand, no computational system has been reported previously to simulate the proposed direct and gradual learning mechanism, and more importantly, to demonstrate the learned material is semantic knowledge.

The semantic system shown in Fig 13.6 can be employed to simulate the direct and gradual process of semantic learning. In the simulations, the Episodic Storage is removed to incorporate with an entirely nonfunctional hippocampus. The learning materials are the external input pairs listed in Table 13.3b. The learning procedure is to present the input pairs, one after another, along the given sequence, to the semantic system for a great number of repetitions. After a given number of repetitions, the learning progress is examined in terms of object naming and object drawing.

The first four simulations in Table 13.6 are the test results after 300 direct learning repetitions, which show that the semantic system is able to use its knowledge

Table 13.6 Naming and drawing after given numbers of repetitions

R^*	External stimulus	Output
300	Z	
		Z
	I	
500	I	
	Z	
		I
	II	
550	II	
		II
		I
	III	

R^*: number of repetitions

about "zero" to process external information but is unable to understand "one" or "many" When the repetition further progresses, it is able to understand "one" but not "two" after 500 repetitions (the second group of four simulations), and then "two" but not "three" after 550 repetitions (the third groups of four simulations). The representation inputs in the tests are similar in concept to, but different in details from the ones that have been repeatedly presented to the system. The system is able to perceive the meanings from them by flexibly using its acquired knowledge and giving correct answers. One may have noticed that the meanings for tallying given in Table 13.3b are pit/pits, instead of the ones in Table 13.3a of peak/peaks. This new kind of meaning is used for the purpose of demonstrating the relative flexibility of the semantic system in learning different concepts.

13.4.6 Robustness and Flexibility

Human knowledge is meaning based and is robust and flexible. Similar robustness and flexibility has been sought in various computational systems, e.g., CLARION can be considered as one of the approaches. It is believed that only when a system is able to acquire meanings from external information, it may exhibit strong robustness and flexibility (Harnad 1990; Sun 1995). This presented system is architectured to acquire (abstract and generalize) meanings from external information. As a result, it has exhibited strong flexibility in using its acquired knowledge in many aspects.

The flexibility is summarized. First, the system can perform variety of cognitive tasks that are often employed in human memory study, e.g., serial recall, cued recall, object naming, object drawing, and recognition. Second, it can tally any given number of objects and "draw" any number of objects, although it has only learned a maximum of "three" (III) as shown in Table 13.3a or b. This flexibility demonstrates that the system has truly acquired and generalized the related meanings from given examples. Third, it can tally unfamiliar object that is different from any learned example. Finally, it has certain fuzzy capacity to deal with irregular input. These flexibilities match the criteria outlined by Sun (1995), including *generalization* from examples, *similarity-based* cognition, handling *inexact matches*, and handling *fuzzy* information.

13.5 Future Challenges

This multileveled network system succeeds in mimicking many properties of episodic memory and semantic memory, and their relationships. It interprets and simulates more phenomena about human episodic and semantic memories and their learnings, than many other reported systems. It suggests a mechanism for the cause of developmental amnesia, and predicts a pattern of forgetting vs. remembering in repeated recall tests. In a recent communication with one of the principle researchers who reported development amnesia, it is said that the prediction is most likely true based on some existing data although the patients have not been tested purposely for repeated recalls. On the other hand, this system has been tested with a number of testing methods, such as, serial recall, cued recall, recognition, object naming and object drawing, which are commonly used in the study of human learning and memory. As a comparison, the three reviewed systems may only be tested by one method, e.g., cued recall. When testing method changes, those systems would cease to function as reviewed previously. Furthermore, by grounding symbols to their meanings, this multileveled system is able to flexibly use its conceptual knowledge to process unfamiliar information, compared with the reviewed computational systems that only process and recall arbitrary patterns.

Although this multileveled system/algorithm has shown a number of promising cognitive capacities, its basic cognitive unit, SM, is not neuron-like. Thoughts have been given for how to make the SM compatible to biological neurons. It is likely that one SM can be formed with a number of artificial neurons of different kinds. Such a possibility is obvious because SM is a generic cognitive unit in the cognitive system, regardless what external information it may associate with, a symbol or a meaning, also because biological neurons are believed to be generic cognitive units in the brain.

The multileveled system is able to abstract and generalize a few numerical concepts from given examples, and to tally either peaks or pits from externally presented representations thereafter. A similar system has been trained to learn Arabic numerals that are used as alternatives of the symbols for tallying (Zhang 2005). How to extend the system to learn nonnumerical conceptual knowledge will be one of future efforts, which may involve a number of cognitive aspects. This multileveled system is designed to process sequenced information, like a person who may only make sense of the surrounding world by continuously touching. Mechanisms are needed for the system to abstract and learn common features from parallel information (e.g., vision). Fortunately, substantial progresses have been made in visual perceptions that may help to overcome this issue. Furthermore, the spatial learning capacity of the hippocampus may also shed light on this effort. Another aspect is about the boundaries of concepts. Many concepts are true only within given boundaries. We may learn a concept from positive examples, and we may also learn its limitation from negative examples. This multileveled system is only able to learn conceptual knowledge from positive examples. Thus, further development is also needed in the respect of concept boundaries.

References

Anderson, J.R. (1983). *The architecture of cognition*. Cambridge, MA: Harvard University Press.

Anderson, J.R., Bothell, D., Lebiere, C., and Matessa, M. (1998). An integrated theory of list memory. *Journal of Memory and Language* 38: 341–380.

Anderson, J.R., and Matessa, M.P. (1997). A production system theory of serial memory. *Psychological Review* 104: 728–748.

Anderson, J.R., Bothell, D., Byrne, M.D., Douglass, S., Lebiere, C., and Qin, Y. (2004). An integrated theory of the mind. *Psychological Review* 111: 1036–1060.

Alvarez, P., and Squire, L. (1994). Memory consolidation and the medial temporal lobe: a simple network model. *Proceedings of the National Academic Science USA* 91: 7041–7045.

Aggleton, J.P., and Brown, M.W. (1999). Episodic memory, amnesia and the hippocampal–anterior thalamic axis. *Behavioral Brain Science* 22: 425–489

Anon (1996). Previous cases: hippocampal amnesia. *Neurocase* 2: 259–298.

Baddeley, A.D., Vargha-Khadem, F., and Mishkin, M. (2001). Preserved recognition in a case of developmental amnesia: implications for the acquisition of semantic memory. *Journal of Cognitive Neuroscience* 13: 357–369

Battaglia, F.P., Sutherland, G.R., and McNaughton, B.L. (2004). Hippocampal sharp wave bursts coincide with neocortical "up-state" transitions. *Learning & Memory* 11: 697–704.

Bloch, V., Hennevinm, E., and Leconte, P. (1979). Relationship between paradoxical sleep and memory processes. In: Brazier M.A. (ed), *Brain mechanisms in memory and learning: from the single neuron to man* (pp. 329–343), Raven: New York.

Budiu, R., and Anderson, J.R. (2004). Interpretation-based processing: a unified theory of semantic sentence processing. *Cognitive Science* 28: 1–44.

Butters, M.A., Glisky, E.L., and Schacter, D.L. (1993). Transfer of new learning in memory-impaired patients. *Journal of Clinical Experimental Neuropsychol* 15: 219–230.

Drescher, G. (1991). *Made-up minds*. Cambridge, MA: MIT.

Fishbein, W. (1970). Interference with conversion of memory from short-term to long-term storage by partial sleep deprivation. *Communications in Behavioral Biology* 5: 171–175.

Gluck, M.A., Meeter, M., and Myers, C.E. (2003). Computational models of the hippocampal region: linking incremental learning and episodic memory. *TRENDS in Cognitive Neuroscience* 7: 269–276.

Greenberg, R., and Pearlman, C. (1974). Cutting the REM nerve: An approach to the adaptive role of REM sleep. *Perspectives in Biology & Medicine* 17: 513–521.

Cohen, N.J., and Squire, L.R. (1980). Preserved learning and retention of pattern analyzing skill in amnesia: Dissociation of knowing how and knowing that. *Science* 210: 207–209.

Collins, A.M., and Quillian, M.R. (1969). Retrieval time from semantic memory. *Journal of Learning and Verbal Behavior* 8: 240–247.

Eichenbaum, H. (2004). Hippocampus: cognitive processes and neural representations that underlie declarative memory. *Neuron* 44: 109–120.

Fosse, M.J., Fosse, R., Hobson, J.A., and Stickgold, R.J. (2003). Dreaming and episodic memory: a functional dissociation? *Journal of Cognitive Neuroscience* 15: 1–9.

Foulkes, D. (1985). *Dreaming: a cognitive-psychological analysis*. Erlbaum: Hillsdale.

Gabrieli, J.D.E., Cohen, N.J., and Corkin, S. (1988). The impaired learning of semantic knowledge following bilateral medial temporal-lobe resection. *Brain Cognitive* 7: 151–177.

Graf, P., and Schacter, D.L. (1985). Implicit and explicit memory for new associations in normal and amnesic subjects. *Journal of Experimental Psychology: Learning, Memory & Cognition* 11: 501–518.

Granger, R., Wiebe, S.P., Taketani, M., and Lynch, G. (1996). Distinct memory circuits composing the hippocampal region. *Hippocampus* 6: 567–578.

Harnad, S. (1990). The symbol grounding problem. *Physica D*, 42: 335–346.

Hobson, J.A., and McCarley, R.W. (1977). The brain as a dream-state generator: An activation-synthesis hypothesis of the dream process. *American Journal of Psychiatry* 134: 1335–1348.

Kitchener, E.G., Hodges, J.R., and McCarthy, R. (1998). Acquisition of post-morbid vocabulary and semantic facts in the absence of episodic memory. *Brain* 121: 1313–1327.

Isaacs, E.B., Vargha-Khadem, F., Watkins, K.E., Lucas, A., Mishkin, M., and Gadian, D.G. (2003). Developmental amnesia and its relationship to degree of hippocampal atrophy. *PNAS* 100: 13060–13063.

Levy, W.B. (1989). A computational approach to hippocampal function. In: Hawkins R.D., Bower G.H., (eds.), *Computational models of learning in simple neural systems* (pp. 243–305). Orlando, FL: Academic.

Levy, W.B. (1996). A sequence predicting CA3 is a flexible associator that learns and uses context to solve hippocampal-like tasks. *Hippocampus* 6: 579–590.

Louie, K., and Wilson, M.A. (2001). Temporally structured replay of awake hippocampal ensemble activity during rapid eye movement sleep. *Neuron* 29: 145–156.

Lynch, M.A. (2004). Long-Term Potentiation and Memory. *Physiol Reviews* 84: 87–136.

Meyer, D. and Kieras, D. (1997). A computational theory of executive cognitive processes and human multiple-task performance: part 1, basic mechanisms. *Psychological Reviews* 104: 3–65.

McClelland, J.L., McNaughton, B.L., and O'Reilly, R.C. (1995). Why there are complementary learning systems in the hippocampus and neocortex: insights from the successes and failures of connectionist models of learning and memory. *Psychol Reviews* 102: 419–457.

McCloskey, M., and Cohen, N.J. (1989). Catastrophic interference in connectionist networks: The sequential learning problem. In Bower G.H. (ed.), *The psychology of learning and motivation* (Vol. 24, pp. 109–165). New York: Academic.

McNaughton, B.L., and Morris, R.G.M. (1987). Hippocampal synaptic enhancement and information storage. *Trends Neuroscience* 10: 408–415.

Meeter, M., and Murre, J.M.J. (2005). Tracelink: A model of consolidation and amnesia. *Cognitive Neuropsychology* 22: 559–587.

Minton, S. (1990). Quantitative results concerning the utility of explanation-based learning. *Artificial Intelligence* 42: 363–391.

Murre, J.M. (1996). TraceLink: a model of amnesia and consolidation of memory. *Hippocampus* 6: 675–684.

Myers, R.E., and Sperry, R.W. (1953). Interocular transfer of a visual forma discrimination habit in cats after section of the optic chaism and corpus callosum. *Anatomy Record* 115: 351–352.

Newell, A. (1990). *Unified theories of cognition*. Cambridge, MA: Harvard University Press.

O'Kane, G., Kensinger, E.A., and Corkin, S. (2004). Evidence for semantic learning in profound amnesia: an investigation with patient H.M. *Hippocampus* 14: 417–425.

O'Reilly, R.C., Norman, K., and McClelland, J.L. (1998). A hippocampal model of recognition memory. In: Jordan M.I., Kearns M.J., Solla S.A., (eds.), *Neural information processing systems*, (Vol. 10, pp. 73–79). Cambridge, MA: MIT.

Pavlides, C., and Winson, J. (1989). Influences of hippocampal place cell firing in the awake state on the activity of these cells during subsequent sleep episodes. *Journal of Neuroscience* 9: 2907–2918.

Pearlman, C. (1971). Latent learning impaired by REM sleep deprivation. *Psychonomic Science* 25: 135–136.

Poe, G.R., Nitz, D.A., McNaughton, B.L., and Barnes, C.A. (2000). Experience dependent phase-reversal of hippocampal neuron firing during REM sleep. *Brain Research* 855: 176–180.

Postle, B.R., and Corkin, S. (1998). Impaired word-stem completion priming but intact perceptual identification priming with novel words: evidence from the amnesic patient H.M. *Neuropsychologia* 36: 421–440.

Ribot, T. (1881). *Les maladies de la memoire*. Paris: Germer Baillare.

Schank, R. (1995). Information is suprises. In: Brockman J. (ed.), *The third culture*. New York: Simon and Schuster.

Schwartz, S. (2003). Are life episodes replayed during dreaming? *Trends in Cognitive Sciences* 7: 325–327.

Scoville, W.B., and Milner, B. (1957). Loss of recent memory after bilateral hippocampal lesions. *Journal of Neurology, Neurosurgery and Psychiatry* 20: 11–21.

Searle, J. (1980). Minds, brains, and programs. *Behavioral and Brain Sciences* 3: 417–424.

Smolenshy, P. (1997), Connectionist modeling, In Haugeland, J. (ed.), *Mind design II*. London: MIT.

Sperry, R. (1982). Some effects of disconnecting the cerebral hemispheres. *Science* 217: 1223–1226.

Squire, L.R. (2004). Memory systems of the brain: A brief history and current perspective. *Neurobiology of Learning and Memory* 82: 171–177.

Squire, L.R., and Alvarez, P. (1995). Retrograde amnesia and memory consolidation: a neurobiological perspective. *Current Opinion in Neurobiology* 5: 169–177.

Squire, L.R., and Zola, S.M. (1998). Episodic memory, semantic memory, and amnesia. *Hippocampus* 8: 205–211.

Staba, R. J., Wilson, C. L., Fried, I., and Engel, J. J. (2002). Single neuron burst firing in the human hippocampus during sleep. *Hippocampus* 12: 724–734.

Sun, R. (2004). Desiderata for cognitive architectures. *Philosophical Psychology* 17: 343–374.

Sun, R., and Alexandre, F. (eds.) (1997). *Connectionist: symbolic interpretation; from unified to hybrid approaches.* Lawrence Erlbaum Associates.

Sun, R., Merrill, E., and Peterson, T. (2001). From implicit skills to explicit knowledge: a bottom–up model of skill learning. *Cognitive Science* 25: 203–244.

Sun, R. (1995). Robust reasoning: integrating rule-based and similarity-based reasoning. *Artificial Intelligence* 75(2): 241–296.

Sun, R. (1999). Accounting for the computational basis of consciousness: a connectionist approach. *Consciousness and Cognition* 8: 529–565.

Tulving, E. (1983). *Elements of episodic memory.* Cambridge: Oxford University Press.

Tulving, E., Hayman, C.A.G., and Macdonald, C.A. (1991). Long-lasting perceptual priming and semantic learning in amnesia: a case experiment. *Journal of Experimental Psychology: Learning, Memory & Cognition* 17: 595–617.

Teyler, T.J., and DiScenna, P. (1986). The hippocampal memory indexing theory, *Behavioral Neuroscience* 100: 147–154.

Vargha-Khadem, F., Gadian, D.G., Watkins, K.E., Connelly, A., Van Paesschenm, W., and Mishkin, M. (1997). Differential effects of early hippocampal pathology on episodic and semantic memory. *Science* 277: 376–380.

Wallenstein, G.V., Eichenbaum, H.B., and Hasselmo, M.E. (1998). The hippocampus as an associator of discontiguous events. *Trends Neuroscience* 21: 317–323.

Watkins, C. (1989). *Learning with delayed rewards.* Ph.D. Thesis, Cambridge University, Cambridge, UK.

Wilson, M.A., and McNaughton, B.L. (1994). Reactivation of hippocampal ensemble memories during sleep. *Science* 265: 676–679.

Wolf, F.A. (1994). *The dreaming Universe.* New York: Simon & Schuster.

Zachary, W., Lementec, J., and Ryder, J. (1996). Interface agents in complex systems. In Nituen, C. and Park, E. (eds.), *Human interaction with complex systems: conceptual principles and design practice.* Needham, MA: Kluwer.

Zhang, Q. (2005). An artificial intelligent counter. *Cognitive Systems Research* 6: 320–332.

Zhang, Q. (2009a). A computational account of dreaming: learning and memory consolidation. *Cognitive Systems Research* 10: 91–101.

Zhang, Q. (2009b). A consequence of failed sequential learning: A computational account of sdevelopmental amnesia. *Cognitive computation* 1: 244–256.

Chapter 14
Motivational Processes Within the Perception–Action Cycle

Ron Sun and Nick Wilson

Abstract The present chapter discusses psychologically well-justified models of motivational processes within the human perception–action cycle (in particular, the CLARION cognitive architecture). First, some background relevant to studying and modeling motivational processes, structures, and representations is discussed. Then, the CLARION cognitive architecture is described. Some simulation results of human motivation and personality from CLARION are then briefly reviewed, and their implications and the future directions outlined.

14.1 Overview

Within the perception–action cycle (e.g., Fuster 2002; Sun et al. 2001), motivational processes, representations, and structures are needed for a complete model of cognitive agents (in a psychological sense) or for a sophisticated artificial intelligence system. Direct mapping from perception to action is often insufficient for either of the two purposes above. Direct mapping from perception to action will not be sufficient, for example, for accounting for the following phenomena:

(1) Internal needs and desires (which control an organism's responses to the environment to a significant extent; Maslow 1943; Weiner 1992; Sun 2009)
(2) Action variability across individuals and/or across times (i.e., different people may choose different actions given the same situation due to motivational differences, or the same person may choose different actions at different times given the same situation due to motivational differences; Read and Miller 2002; Sun 2009)
(3) Perceptual variability (people may even see different things given the same situation, because of the differences in motivations; Maner et al. 2005)

R. Sun (✉)
Cognitive Science Department, Rensselaer Polytechnic Institute, 110 Eighth Street, Carnegie 302A, Troy, NY 12180, USA
e-mail: rsun@rpi.edu

V. Cutsuridis et al. (eds.), *Perception-Action Cycle: Models, Architectures, and Hardware*, Springer Series in Cognitive and Neural Systems 1, DOI 10.1007/978-1-4419-1452-1_14, © Springer Science+Business Media, LLC 2011

Maslow (1943) argued that "the situation or the field in which the organism reacts must be taken into account but the field alone can rarely serve as an exclusive explanation for behavior.... Field theory cannot be a substitute for motivation theory." As Maslow (1943) put it, "man is a perpetually wanting animal."

A complete model of human psychology (or human-level intelligence) clearly needs not only decision making, reasoning, problem solving, and so on, in between perception and action, but also motivational variables impinging on decision making, reasoning, problem solving, and so on, to help to determine the very processes of decision making, reasoning, problem solving, and so on (at least to a significant extent), and thus to help to determine the final action outcomes.

In the present chapter, we will discuss a psychologically well-justified model of motivational variables within the human perception–action cycles, namely, the CLARION cognitive architecture. CLARION is a comprehensive framework of a variety of psychological processes that has been implemented computationally. It has been described in detail and justified psychologically in Sun (2002, 2003) (see also Sun et al. 2001, 2005; Sun 2009). It is particularly worth noting that CLARION is an integrative cognitive architecture, consisting of a number of distinct subsystems (with a dual-representational structure in each subsystem: implicit versus explicit representations). Its subsystems include the action-centered subsystem (the ACS), the non-action-centered subsystem (the NACS), the motivational subsystem (the MS), and the meta-cognitive subsystem (the MCS). The role of the ACS is to control actions (implicitly or explicitly), regardless of whether the actions are for external physical movements or for internal mental operations. The role of the NACS is to maintain general knowledge (either implicit or explicit). The role of the MS is to provide underlying motivations (implicit and explicit) for perception, action, and cognition, in terms of providing impetus and feedback. The role of the MCS is to monitor, direct, and modify the operations of the ACS dynamically (as well as the operations of the other subsystems).

CLARION, with the aforementioned motivational subsystem, is capable of displaying a broad and variable range of behavior, that is, there is clearly sufficient flexibility in CLARION to capture a variety of motivations, which are important for the perception–action cycle in an organism. It is also capable of capturing a variety of human personalities – in fact, it is our contention that personality lies fundamentally in the MS (but involves also other subsystems; more on this point later, and see also Sun 2009; Sun and Wilson 2009).

Therefore, based on CLARION, our general research agenda (as sketched in the present chapter) addresses the following issues that are of a great deal of interest to a number of different research communities (within artificial intelligence, as well as within psychology):

(1) The motivation–action interaction (given any situation); that is, how motivation may help to determine actions (given any particular situation)
(2) The relation between motivation and personality, and the claim that personality lies fundamentally in the MS (but involves also other subsystems)
(3) The personality–action interaction; that is, how personality may help to explain or predict actions given any situation (i.e., the usefulness of hypothesizing the personality–motivation connection within a general psychological framework)

There are also many other related issues and questions: for example, how motivation affects decision making and reasoning, how motivation affects perception, what the fundamental motivational elements that determine personality are, and so on. We will not address all of these questions above in this chapter; beyond the present chapter, the interested reader is referred to Caprara and Cervone (2000), Sun (2003, 2009), and Sun and Wilson (2009) regarding these additional issues.

In the remainder of this chapter, first, some background relevant to studying and modeling motivational processes, structures, and representations will be discussed. Then, the CLARION cognitive architecture will be described. Some simulation results from CLARION will then be briefly reviewed, and their implications and the future directions outlined. This chapter will be sketchy by necessity, as it summarizes a large amount of work, especially the authors' own work from the past decade and half.

14.2 Background: Data and Models Relevant to Motivational Representations, Processes, and Structures

14.2.1 Previous Work on Motivation

Work on human motivations has had a long history. Some particularly relevant work will be highlighted here, in relation to our own theory of human motivation as embodied in the CLARION cognitive architecture (Sun 2003, 2009).

First of all, Murray (1938) proposed a pertinent set of basic needs (or primary drives in our terminology, as in CLARION). Murray's proposal (1938) included the need for conservance (covered by the drive for conservation in CLARION, as will be detailed later), the need for order (also covered by the drive for conservation in CLARION), the need for retention (derived from the drive for conservation in CLARION), the need for acquisition (derived from the need for conservation in CLARION), the need for inviolacy (attributable to the drive for recognition and achievement, as well as the drive for dominance and power, according to CLARION), and so on. Some other needs identified by Murray, such as contrarience, aggression, abasement, rejection, succorance, exposition, construction, and play, are not fundamental needs (or primary drives) in our view – they are likely the results of more fundamental (i.e., primary) drives or their combinations. For example, according to CLARION, the need for play may be attributed sometimes to the drive of curiosity and sometimes to the physiological drive of avoiding boredom or avoiding repulsive or unpleasant stimuli (e.g., when overwork leads to work-related stimuli becoming unpleasant). For another example, Murray's contrarience need, if exists, may be attributed to the drive for recognition and achievement and/or the drive for dominance and power, according to CLARION. Murray's proposal also included some low-level (physiological, or viscerogenic in Murray's term) needs (which may be attributed to some combinations of the low-level primary drives in CLARION).

Reiss (2004) proposed another set of basic needs (primary drives), which was highly similar to Murray's, but with some differences. For example, as proposed by Reiss (2004), there are the need for saving, the need for order (both included in the drive for conservation in CLARION), the need for family (derived, according to CLARION, from the drive for affiliation and belongingness, as well as the drive for nurturance and the drive for honor), vengeance (which includes the desire to get even, derived from the drive for fairness and the drive for honor in CLARION; which also includes the desire to compete and win, derived from the drive for recognition and achievement, the drive for honor, and so on, in CLARION), the need for "idealism" (derived from other drives in CLARION, such as affiliation and belongingness, honor, fairness, nurturance, etc.), the need for status (derived from the drive for dominance and power and the drive for recognition and achievement in CLARION), the need for acceptance (derived from the drive for affiliation and belongingness, the drive for honor, and the drive for recognition and achievement in CLARION), as well as the need for eating, the need for tranquility, the need for physical exercises, the need for romance, and so on.

As alluded to above, in Sun (2003, 2009), a detailed theory of human motivation as embodied in CLARION was presented. CLARION incorporates multiple, interacting subsystems: the ACS, the NACS, the MS, and the MCS. It is also distinguished by its focus on the separation and the interaction of implicit and explicit knowledge (in these different subsystems). Within the MS, there are implicit drives and explicit goals (with the goals being primarily generated based on drives). While some drives, denoting essential needs and desires, are primary and built-in, some other drives may be acquired and secondary. The primary drives include: affiliation and belongingness, dominance and power, recognition and achievement, autonomy, deference, similance, fairness, honor, nurturance, conservation, curiosity, as well as some low-level primary drives (Sun 2009). With these architectural mechanisms, especially the motivational mechanisms, CLARION is able to capture, account for, and explain many psychological data and phenomena related to human motivation. [See further discussions later, and also see Sun (2003, 2009)].

Relatedly, Schwartz's (1994) ten universal values, although addressing a different aspect of human behavior (i.e., human "values"), bear some resemblance to the essential needs (i.e., primary drives) identified earlier (see Sun 2009). Moreover, each of these values can be derived from some primary drive (i.e., some essential need) or some combination of these primary drives (Sun 2009).

McDougall (1936) proposed a framework that was concerned with "instincts." Instincts, in our framework, refer to (more or less) evolutionarily hard-wired (i.e., innate) behavior patterns/routines that can be relatively easily triggered by pertinent stimuli in pertinent situations. On the other hand, basic needs (or primary drives as we termed them in CLARION; Sun 2009) are essential driving forces of behaviors. Instincts are different from basic needs, because one does not have to follow instincts when there is no pertinent stimulus, and even when pertinent stimuli are present, one may be able to refrain from following them (at least more easily than from basic needs or primary drives). In other words, they are pre-set routines – while they are relatively easily triggered, they are not inevitable. For example, McDougall listed the following instincts: imitation, emulation or rivalry, pugnacity/anger/resentment, sympathy, hunting, fear, appropriation/acquisitiveness, constructiveness, play, curiosity, sociability and shyness, secretiveness, cleanliness, modesty and shame, love, jealousy, parental love, and so on (James 1890). As evident from the list above, many of these instincts result from what we call primary drives or basic needs (such as "curiosity" and "parental love"), or are derived, by some means, from primary drives or basic needs (such as "play" and "constructiveness"). Some other instincts are not because they do not represent basic needs (e.g., "hunting" or "jealousy"). See more discussions of primary drives within CLARION later. Note that there have been various efforts at verifying those drives through experiments and data analysis (see, e.g., Reiss 2004; Sun 2009).

There have also been some less psychologically oriented models of motivation. Among such models, Doerner's model and Sloman's model stand out. In Sloman's motivational model (see, e.g., Wright and Sloman 1997), goals ("motives") are generated from a suite of modules ("generactivators"), each of which expresses a single "concern" (such as caring for dependents or removing damaged dependents). Each of these modules may search through a database of beliefs; if it finds a match, a declarative representation of a goal (a "motive") is generated. On that basis, the resource management system takes goal representations and generates intentions for action. Although the model bears some resemblance to CLARION (see more discussions of CLARION later; see also Sun 2009), the model has not been used to capture or explain psychological data in any detail. In addition, computationally speaking, searching through databases is costly and may not be cognitively realistic.

Doerner (2003; see also Bach 2009) described the PSI theory, which included internal deficits, displeasure signals (due to deficits), negative reinforcement (from displeasure signals), urges, goals, action learning through random exploration (based on reinforcement), and so on. At an abstract level, the model is similar to CLARION to some extent (see Sun 2002, 2003), but it appears less psychologically grounded or justified. In addition, its computational mechanisms appear less well developed algorithmically.

14.2.2 Previous Work on Personality

Models of personality, in our view, may be based on models of motivation. In the psychology literature, there have been a few computational models of personality on the basis of motivation. For example, Shoda and Mischel (1998) presented a constraint satisfaction model of human personality (based on the cognitive social learning theory of personality). In the model, a set of input units (representing situational features) is connected to a recurrently connected set of mediating "cognitive affective units" (representing motivations), which are then connected to behavior nodes. However, although their model was supposed to represent human personality, they did not attempt to capture the known structures/types of human personality (e.g., the Big Five; Digman 1990). In their model, different personalities are represented by different, randomly connected patterns of mediating units, and therefore they did not specify the structure of the underlying motivational (sub)system (e.g., Sun 2003, 2009). (In contrast, note that the research described in the present chapter is based on demonstrating how the known structures/types of human personality can be the result of the complex interactions among underlying motivational and cognitive representations and processes; see Sun and Wilson 2009 for details.)

Read and Miller (2002) developed goal-based personality models and mapped goal structures onto the Big Five personality structures (Digman 1990). Read and Miller (2002) used a simple multilayer feed-forward backpropagation neural network. However, there are some significant shortcomings in their models. First, a better model should be based on a more solid foundation of a well-developed cognitive architecture and thus based on a more comprehensive and better grounded view of the architecture of the mind. Furthermore, a better model needs to capture personality types based on both drive and goal representations, in which drives are more fundamental and serve as the basis for setting goals and choosing actions (Sun 2009), to avoid arbitrary goal setting (see the discussions of motivation earlier). Finally, a better model should be based on a well-developed motivational (sub)system that synthesizes past work on motivational processes, representations, and structures (Murray 1938; Sun 2003, 2009; Reiss 2004).

Poznanski and Thagard (2005) developed a neural network model of personality (specifically for virtual characters). The model is a feed-forward neural network where personality is represented by a set of nodes each of which corresponds to an extreme end of one of the Big Five dimensions (Digman 1990). Specific personalities are created by assigning baseline activations to each node. In the model, personality nodes send activation to behavior nodes, which also receive inputs from emotion nodes, situation nodes, mood nodes, and relationship nodes. Thus, activation of a given behavior is jointly determined by the inputs from the personality nodes, the situation nodes, the relationship nodes, and the emotion nodes. Besides incorporating the Big Five personality structure model (which is useful in its own right), this work did not provide new insight into human personality per se.

Doerner (2003) described personality within the PSI theory, which, as described before, included internal deficits, displeasure signals, negative reinforcement (from displeasure signals), urges, goals, action learning through random exploration, and so on. The variation of parameter settings in their model generated different "personality" patterns, which differed in the way of coping with specific situations. To some extent, they showed that the behavior patterns produced by the model were consistent with empirical data from human subjects (although at a very abstract level). However, the problems with the work included the lack of detailed psychological justifications of the model details and the lack of connections to much of the existing personality literature.

Sun and Wilson (2009) proposed a more thoroughly developed model of personality that remedied some of the aforementioned shortcomings. The model is based on CLARION and maps CLARION motivational representations, along with other representations and parameters, onto the personality structures (i.e., the Big Five; Digman 1990). It successfully captured some human data and demonstrated some of the Big Five personality dimensions. The CLARION model will be discussed in detail later.

14.2.3 Previous Work on Cognitive Architectures

It has been suggested (Newell 1990) that cognitive theories (including computational models) should be developed that satisfy multiple criteria to avoid theoretical myopia. There have been steady developments of generic cognitive models, that is, cognitive architectures, for the past three decades since Newell's suggestion. Early cognitive architectures often took the form of production systems and were (more or less) concerned with psychological phenomena. However, other forms of cognitive architectures have also been developed over the years – they may be in the form of a connectionist model, a constraint satisfaction network, a loosely organized library of models, a hybrid system of different models, and so on (some of them may be more concerned with applications to building artificial systems than explaining psychological phenomena).

Soar, a symbolic production architecture, has been developed over the past 30 years, mostly for the purpose of building application systems (Rosenbloom et al. 1993; Newell 1990). In Soar, based on the framework of a state space and operators for searching the state space, decisions are made by different productions proposing different operators, when there is a goal on a goal stack. When a sequence of productions leads to achieving a goal, chunking occurs, which creates a single production that summarizes the process (using so-called explanation-based learning). A large amount of initial (a priori) knowledge about states and operators is required for Soar to work. In addition, it lacks sophisticated motivational structures and processes.

Another series of similar architectures were also proposed: in particular, ACT* and ACT-R (e.g., Anderson and Lebiere 1998). ACT* is made up of declarative knowledge (captured through a semantic network) and procedural knowledge (captured in a production system). Procedural knowledge (in productions) is acquired through "proceduralization" of declarative knowledge, modified through use by generalization and discrimination (i.e., specialization), and has strengths associated with them (which are used for firing). ACT-R is a descendant of ACT*, in which procedural learning is limited to production formation through mimicking and production firing is based on log odds of success. There are additions to ACT-R, including visual and motor modules, but there have never been any sufficiently complex motivational structures.

CLARION is an integrative cognitive architecture (Sun 2002, 2003). The CLARION cognitive architecture, as mentioned earlier, consists of multiple, interacting subsystems: the ACS, the NACS, the MS, and the MCS. It is also distinguished from other existing cognitive architectures by its focus on the separation and the interaction of implicit and explicit knowledge (in these different subsystems). More importantly, in relation to motivational issues, compared with other cognitive architectures, CLARION is unique in that it contains built-in motivational constructs and built-in meta-cognitive constructs. These features are not commonly found in other existing cognitive architectures. Nevertheless, these features are crucial to the enterprise of cognitive architectures, as they capture important elements in the interaction between an agent and its physical and social world (Sun 2002, 2009). With these mechanisms, especially the motivational and meta-cognitive mechanisms,

CLARION has something unique to contribute to cognitive modeling – it attempts to capture the motivational (and meta-cognitive) aspects of cognition and to explain their functioning in concrete computational terms.

14.2.4 Essential Desiderata

Judging from the literature, as discussed in Sun (2003, 2009), in order to survive, an agent in its activities must meet the following criteria (among others), which rely on motivational processes:

- Sustainability: An agent must attend to its essential needs, such as hunger and thirst, and also know to avoid physical dangers, and so on (Toates 1986).
- Purposefulness: The action of an agent must be chosen in accordance with some criteria, instead of completely randomly (Hull 1943; Anderson 1993), and those criteria are related to enhancing sustainability of an agent (Toates 1986).
- Focus: An agent must be able to focus its activities with respect to specific purposes. That is, its actions need to be somehow consistent, persistent, and contiguous, with respect to its purposes (Toates 1986). However, an agent needs to be able to give up some of its activities, temporally or permanently, when necessary (e.g., when a much more urgent need arises; more later; Simon 1967; Sloman 1987).
- Adaptivity: An agent must be able to adapt its behavior (i.e., to learn) for the sake of improving its purposefulness, sustainability, and focus.

We contend that, to meet these criteria, specific motivational representations need to be formed that can address issues related to purpose and focus. Motivational representations and their dynamics are an essential part of human (or animal) behavior. It is ever-present in such behavior.

14.3 The CLARION Cognitive Architecture: The Role of Motivational Variables

14.3.1 Overview of CLARION

CLARION is a generic and comprehensive model of psychological processes (of a wide variety), specified and implemented computationally (Sun 2002, 2003; see also Sun et al. 2001, 2005 for psychological justifications). CLARION is an integrative "cognitive architecture," consisting of a number of distinct subsystems. The role of the ACS is to control actions (regardless of whether the actions are for external physical movements or for internal mental operations). The role of the NACS is to maintain general knowledge. The role of the MS is to provide underlying motivations for perception, action, and cognition. The role of the MCS is to monitor, direct, and modify the operations of the other subsystems dynamically.

Each of these interacting subsystems consists of two "levels" of representation (i.e., a dual-representational structure) as theoretically posited in Sun (2002) and Sun et al. (2005). Generally speaking, in each subsystem, the top level encodes explicit knowledge (using symbolic/localist representations) and the bottom level encodes implicit knowledge (using distributed representations; see Sun 1994). The two levels interact, for example, by cooperating in action decision making, through a combination of the action recommendations from the two levels, as well as by cooperating in learning through a "bottom-up" and a "top-down" learning process (as will be discussed below). See Fig. 14.1 for a sketch.

Note that an important characteristic of this cognitive architecture is its focus on the cognition–motivation–environment interaction. Essential motivations of an agent (its biological needs in particular) arise naturally, prior to cognition (but interact with cognition). Such motivations are the foundation of an agent's action and cognition. In a way, cognition is evolved to serve the essential needs of an agent. Cognition, in the process of helping to satisfy needs and following motivational forces, has to take into account environments – their regularities and structures. Thus, cognition bridges the needs and motivations of an agent and its environments (be it physical or social), thereby linking all three in a complex nexus (Sun 2003, 2009).

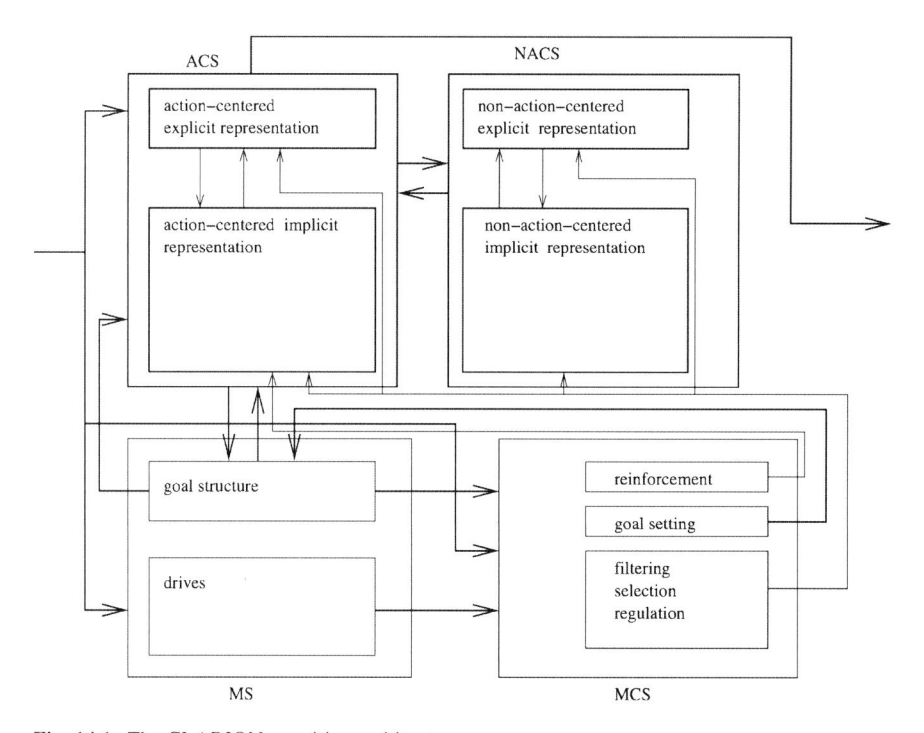

Fig. 14.1 The CLARION cognitive architecture

14.3.2   The Action-Centered Subsystem

The action-centered subsystem (the ACS) in CLARION is meant to capture the step-by-step action decision making of an agent in its interaction with the world.

In the ACS, the process for action decision making is essentially the following: Observing the current state of the world, the two levels of processes within the ACS (implicit or explicit) make their separate decisions in accordance with their own knowledge, and their outcomes are somehow "integrated." Thus, a final selection of an action is made and the action is then performed. The action changes the world in some way. Comparing the changed state of the world with the previous state somehow, the agent learns (e.g., in accordance with Q-learning of Watkins 1989; more later). The cycle then repeats itself.

Thus, the overall algorithm for action decision making during the interaction of an agent with the world is as follows:

1. Observe the current state x (including the current goal).
2. Compute in the bottom level the "value" of each of the possible actions (a_i's) associated with the state x : $Q(x, a_1)$, $Q(x, a_2)$, ..., $Q(x, a_n)$. Stochastically choose one action according to these values.
3. Find out all the possible actions ($b_1, b_2, \ldots, b_m$) at the top level, based on the current state x (which goes up from the bottom level) and the existing explicit rules in place at the top level. Stochastically choose one action.
4. Choose an appropriate action, by stochastically selecting the outcome of either the top level or the bottom level.
5. Perform the action and observe the next state y and (possibly) a reinforcement signal r.
6. Update knowledge at the bottom level in accordance with an appropriate learning algorithm (e.g., Q-learning; to be detailed later), based on the feedback information.
7. Update the top level using an appropriate learning algorithm (for extracting, refining, and deleting explicit rules; to be detailed later).
8. Go back to Step 1.

In this subsystem, the bottom level is implemented with backpropagation neural networks involving distributed representations (Sun 1994), and the top level is implemented using symbolic/localist representations (Sun 1994).

The input (x) to the bottom level consists of three sets of information: (1) sensory input, (2) working memory items, and (3) the current goal. The input state x is represented as a set of dimension–value pairs: $(d_1, v_1)(d_2, v_2)\ldots(d_n, v_n)$. The output of the bottom level is the action choice. Note that the current goal and the current sensory information are both important in deciding on an action by the ACS.

In each neural network encoding implicit knowledge at the bottom level, actions are selected based on their Q values. A Q value is an evaluation of the "quality" of an action in a given state: $Q(x, a)$ indicates how desirable action a is in state x. At each step, given state x, the Q values of all the actions [i.e., $Q(x, a)$ for all a's] are computed. Then the Q values are used to decide stochastically on an action to be performed, through a Boltzmann distribution of Q values:

$$p(a|x) = e^{Q(x,\,a)/\tau} \Big/ \sum_i e^{Q(x,\,a_i)/\tau}$$

where τ controls the degree of randomness (temperature) of the action decision-making process, and i ranges over all possible actions (this method is also known as Luce's choice axiom; Watkins 1989.)

For learning implicit knowledge at the bottom level (i.e., the Q values), the Q-learning algorithm (Watkins 1989), a reinforcement learning algorithm, is used. With the algorithm, $Q(x, a)$ approximates the maximum (discounted) total reinforcement that can be received from the current state x on. Q values are gradually tuned, on-line, through successive updating, in accordance with the Q-learning algorithm, which enables reactive sequential behavior to emerge through trial-and-error interaction with the world (see Watkins 1989; Sun et al. 2001 for further details).

To learn explicit rules at the top level with a "bottom-up" learning process (Sun et al. 2001), the *Rule-Extraction-Refinement* algorithm (RER) constructs explicit rules using information from the bottom level. The basic idea of bottom-up learning of action-centered knowledge is as follows: If an action chosen (by the bottom level) is successful (i.e., it satisfies a certain criterion), then an explicit rule is extracted at the top level. Then, in subsequent interactions with the world, the rule is refined by considering the outcome of applying the rule: If the outcome is successful, the condition of the rule may be generalized to make it more universally applicable; if the outcome is not successful, then the condition of the rule should be made more specific.

To provide a rational basis for making these above decisions, numerical criteria (measuring whether a result is successful or not) are used in deciding whether or not to apply these operations. The details of the numerical criteria can be found in Sun et al. (2001) and Sun (2003). Essentially, at each step, a statistical measure, "information gain," is computed. The aforementioned rule learning operations (extraction, generalization, and specialization) are decided and performed based on the information gain measure (see Sun et al. 2001 for further details).

On the other hand, in the opposite direction, the dual representation in the ACS also enables "top-down" learning. With explicit knowledge (in the form of explicit rules) in place at the top level, the bottom level learns under the guidance of the explicit rules. That is, initially, the agent relies mostly on the explicit rules at the top level for its action decision making. But gradually, when more and more implicit knowledge is acquired by the bottom level through "observing" actions directed by the top-level explicit rules (based on the same reinforcement learning mechanism at the bottom level as described before), the agent becomes more and more reliant on the bottom level (given that the inter-level stochastic selection mechanism is adaptable). Hence, top-down learning takes place.

For the stochastic selection of the outcomes of the two levels, at each step, with probability P_{BL}, the outcome of the bottom level is used. Likewise, with probability

P_{RER}, if there is at least one RER rule indicating a proper action in the current state, the outcome from that rule set (determined through competition based on rule utility) is used; otherwise, the outcome of the bottom level is used (which is always available). There exists some psychological evidence for such intermittent use of rules; see, for example, Sun et al. (2001).[1]

Note that the inter-level selection probabilities may be variable, determined through a process known as "probability matching," that is, the probability of selecting a component is determined based on the relative success ratio of that component.

14.3.3 *The Non-Action-Centered Subsystem*

The non-action-centered subsystem (the NACS) is for representing general knowledge that is not action centered, and for the purpose of making inferences about the world. It stores such knowledge in a dual-representational form (the same as in the ACS), that is, in the form of explicit "associative rules" (at the top level), as well as in the form of implicit "associative memory" (at the bottom level). (Note that its operation is under the control of the ACS.)

First, at the bottom level of the NACS, "associative memory" networks encode non-action-centered implicit knowledge. (Note that associations are formed by mapping an input to an output; the backpropagation learning algorithm, for example, may be used to establish such associations between pairs of input and output). On the other hand, at the top level of the NACS, a general knowledge store encodes explicit non-action-centered knowledge (cf. Sun 1994). As in the ACS, "chunks" (concepts) at the top level are specified by and linked to dimensional values represented at the bottom level. The chunk node connects to its constituting features (i.e., dimension–value pairs) represented as separate nodes at the bottom level (i.e., the distributed representation in the bottom level). Additionally, in the top level, links between chunks encode explicit associations between pairs of chunk nodes, which are known as associative rules. Such explicit associative rules may be formed (i.e., learned) in a variety of ways in the top level of the NACS (see Sun 2003).

As shown by Sun (1994), in the NACS, different sequences of mixed similarity-based and rule-based reasoning capture essential patterns of human everyday (mundane, commonsense) reasoning.

As in the ACS, top-down or bottom-up learning may take place in the NACS, either to extract explicit knowledge in the top level from the implicit knowledge in the bottom level or to assimilate the explicit knowledge of the top level into the implicit knowledge in the bottom level.

---

[1] Note that other components may be included in the stochastic selection in a like manner.

14.3.4 The Motivational Subsystem

The motivational subsystem (the MS) of CLARION is concerned with why an agent does what it does. Simply saying that an agent chooses actions within the ACS to maximize rewards or reinforcement leaves open the question of what determines rewards or reinforcement. The relevance of the MS to the main part of the cognitive architecture, the ACS, lies primarily in the fact that it provides the context in which the goal and the reinforcement of the ACS are determined. It thereby influences the working of the ACS (and by extension, the working of the NACS).

A dual motivational representation scheme is as follows (see Sun 2009). The explicit goals (such as "finding food") of an agent (which is essential to the working of the ACS, as explained before) may be generated based on internal drive states (for example, "being hungry") of the agent. This explicit representation of goals derives from, and hinges upon, the (implicit) drive states. See Fig. 14.2 for a sketch of the MS (for detailed theoretical justifications, see Sun 2009).

Primary Drives. Specifically, we refer to as "primary drives" those drives that are essential to an agent and are most likely built-in (hard wired) to a significant extent to begin with.[2] The primary drives in the MS of CLARION may be roughly explained as follows (see Sun 2009 for further details):

- Food: The drive to consume nourishment.
- Water: The drive to consume fluid.

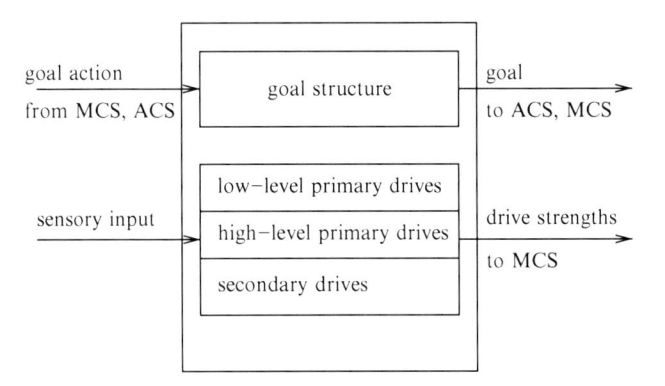

Fig. 14.2 Structure of the motivational subsystems

---

[2] Note that a generalized notion of "drive" is adopted here, different from the stricter interpretation of drives (e.g., as physiological deficits that require to be reduced by corresponding behaviors; Hull 1943). In our sense, drives denote internally felt needs of all kinds that likely may lead to corresponding behaviors, regardless of whether the needs are physiological or not, whether the needs may be reduced by the corresponding behaviors or not, or whether the needs are for end states or for processes. Therefore, it is a generalized notion that transcends controversies surrounding the stricter notions of drive.

- Sleep: The drive to rest or sleep.
- Reproduction: The drive to mate.
- Avoiding danger: The drive to avoid situations that have the potential to be or already are physically harmful.
- Avoiding unpleasant stimuli: The drive to avoid situations that are uncomfortable or negative in nature.
- Affiliation and belongingness: The drive to associate with other individuals and to be part of social groups.
- Dominance and power: The drive to have power over other individuals or groups.
- Recognition and achievement: The drive to excel and be viewed as competent at something.
- Autonomy: The drive to resist control or influence by others.
- Deference: The drive to willingly follow and serve a person of a higher status of some kind.
- Similance: The drive to identify with others, to imitate others, and to go along with their actions.
- Fairness: The drive to ensure that one treats others fairly and is treated fairly by others.
- Honor: The drive to follow social code of behavior and to avoid blames.
- Nurturance: The drive to care for, or to attend to the needs of, others who are in need.
- Conservation: The drive to conserve, to preserve, to organize, or to structure (e.g., one's environment).
- Curiosity: The drive to explore, to discover, and to gain new knowledge.

[Note that this set of primary drives has been explored and justified in detail in a series of prior writings (e.g., Sun 2003, 2009), based on existing work in social psychology as well as early work of ethology. The similarity and differences of this set with other existing sets in the literature were discussed earlier in this chapter.]

Besides primary drives, there are also derived (or secondary) drives. See Sun (2009) and Sun and Wilson (2009) for further details.

Details of Drive Processing. The computational processes of drive processing involve the following components within the MS (named "processors" below):

(1) The drive preprocessor: For each drive d, there is a "preprocessor" that picks out and evaluates relevant input information for determining a drive-specific stimulus level (an evaluation), which is to be used in calculating drive activation (drive strength):

$$state\ x \rightarrow stimulus_d\ (for\ all\ d's)\,.$$

This mapping represents the built-in detector for relevant information in relation to drive d. This mapping is to be performed for each state-action step.[3]

---

[3] This mapping may include generalizations from some familiar scenarios to other scenarios (accomplished, e.g., by using neural networks or similarity matching; Sun 2003).

(2) The drive initialization processor: It carries out the following two mappings:

$$(a)\, personality\ type \rightarrow deficit_d\ (for\ all\ d\text{'}s),$$

that is, the mapping from the personality type to the initial deficit of a drive d (which, however, may later change);

$$(b)\, personality\ type \rightarrow baseline_d\ (for\ all\ d\text{'}s),$$

that is, the mapping from the personality type to the baseline drive strength of a drive d. These two mappings are relatively fixed, that is, they are often performed only once for the entire duration of a simulation. (More details and justifications of these mappings may be found in Sun and Wilson 2009).

(3) The drive core processor: It generates drive strengths based on:

$$ds_d = gain_u \times gain_s \times gain_d \times stimulus_d \times deficit_d + baseline_d,$$

where ds_d is the strength of drive d, $gain_d$ is the individual gain parameter for drive d, $gain_u$ is the universal gain affecting all drives, $gain_s$ is the gain affecting all the drives of one type (the approach or the avoidance type; see Sun and Wilson 2009), $stimulus_d$ is a value representing the relevance of the current situation to drive d (which measures how pertinent the current situation is to drive d), $deficit_d$ indicates the perceived deficit in relation to drive d (which represents an agent's intrinsic sensitivity and inclination toward activating a particular drive d), and $baseline_d$ is the baseline value of drive d.[4] This mapping is to be performed at each state-action step.[5]

(4) The deficit change processor: This processor determines how "$deficit_d$" changes over time, for each state-action step, in relation to the goals adopted, the states encountered, the actions performed, and so on (for the previous steps and the current step):

$$goal\, g,\, state\, x,\, action\, a \rightarrow change\ of\ deficit_d\ (for\ all\ d\text{'}s)\ [6]$$

These "processors" above are implemented as neural networks. The justifications for mappings (1) and (3) may be found in a variety of literatures, ranging from ethological research and modeling (e.g., Tyrell 1993; Toates 1986; McClelland 1951;

---

[4] Note that drive strengths actually could be a function of the above equation; in the simplest case, an identity function may be assumed, as shown above.

[5] Note that although $gain_d$, $gain_u$, $gain_s$, $deficit_d$, and $baseline_d$ may be considered a core part of the MS, they are treated as input here because they may be tunable (by processes outside the core processor). In this regard, the "stimulus" may also be tunable (i.e., learnable). We deal with tuning/learning in Sun and Wilson (2009).

[6] Note that, for a simplified simulation, one may, for example, use simple "decay" (at each step) of the deficits of the drives that are being addressed by the current goal (which helps to decide the current action).

McFarland 1989) to cognitive modeling (e.g., Sun 2003, 2009; Doerner 2003). In particular, the multiplicative combination of $stimulus_d$ and $deficit_d$ has been argued extensively (see, e.g., Sun 2003, 2009; Tyrell 1993 for detailed justifications based on a set of desiderata). The details and justifications of mapping (2) were discussed in relation to personality modeling in Sun and Wilson (2009).

14.3.5 The Meta-Cognitive Subsystem

The existence of a large variety of drive states and explicit goals resulting from them leads to the need for meta-cognitive control and regulation. Meta-cognition refers to one's knowledge (implicit or explicit) concerning one's own cognitive processes and their outcomes. Meta-cognition also includes the active monitoring and consequent regulation and orchestration of these processes through goal setting (Mazzoni and Nelson 1998). In CLARION, the meta-cognitive subsystem (the MCS) is closely tied to the MS. The MCS monitors, controls, and regulates cognitive processes for the sake of cognitive performance (Simon 1967; Sloman 1987). Control and regulation may be in the forms of setting goals (which are then used by the ACS) on the basis of drive states, interrupting and changing on-going processes in the ACS and the NACS, setting essential parameters of the ACS and the NACS, and so on. Control and regulation are also carried out through setting reinforcement functions for the ACS (on the basis of drive states and so on).

Structurally, this subsystem may be subdivided into a number of functional modules, including:

- The goal setting module
- The reinforcement function module
- The processing mode selection module
- The input filtering/selection module
- The output filtering/selection module
- The parameter setting module (for setting learning rates, temperatures, etc.)

Let us look into some details. First of all, the goal setting module of the MCS determines the goal strength for some or all of the goals, based on information from the MS (e.g., the drive strengths, and the current goal) as well as the current state:

$$drives, goal, state \rightarrow goal$$

This mapping may be implemented through a backpropagation neural network. See Tolman (1932) for the general idea of goal setting on the basis of implicit motives (i.e., drives). In the simplest case, the following calculation may be performed by this module:

$$gs_g = \sum_d relevance_{d \rightarrow g} \times ds_d,$$

where gs_g is the activation of goal g, $relevance_{d \to g}$ is a measure of how relevant drive d is to goal g (which represents the support that drive d provides to goal g), and ds_d is the strength of drive d as reported by the MS. Once calculated, the goal activations are turned into a Boltzmann distribution and the new goal is chosen stochastically from that distribution.

Separately, the reinforcement function module produces an evaluation of the current state in terms of the current goal, in the context of the currently active drives: whether it satisfies the goal and/or the drives or not, or how much it satisfies the goal and/or the drives. The current state (environmental and internal sensory information), the drives, and the current goal are all input to the module. That is, this module maps sensory information, along with the drives and the goal, to an evaluation that is used as reinforcement to the agent (in particular, used in the ACS for reinforcement learning; see, for example, Sun et al. 2001 regarding reinforcement functions in human skill learning in general).

As yet another example, the processing mode selection module determines how much each level of the ACS will be used for action decision making (i.e., how implicit or explicit the ACS processing will be; see the discussion of the ACS earlier; see Sun 2003 for further details).

14.3.6 Model of Personality Within CLARION

Our model of personality was developed essentially on a foundation of motivational representations – specifically, in the context of CLARION, on a foundation of drive strength distributions (and then further on goal setting and action selection). That is, fundamentally, the CLARION personality model is based on drive "deficits" (because, within CLARION, given the current state and hence given the drive-specific stimulus "$stimulus_d$" as described earlier, the drive strengths are primarily determined by drive deficits, along with drive gain parameters: $gain_d$, $gain_u$, and $gain_s$). On that basis, goal setting and action selection take place. Using (mostly) the MS to account for personality is based on some principled justifications – in particular, of how various personality types are mapped to various distributions of deficits (and vice verse). We may justify this approach from a variety of perspectives: philosophy, the literature on generic computational cognitive architectures, the literature on personality, and the existing computational models of personality; see Sun and Wilson (2009) for details of justifications.

Beyond motivation, personality may also involve a variety of other psychological mechanisms and processes, although these mechanisms and processes may be secondary in terms of their importance to personality compared with motivation. Therefore, personality types, besides being mapped onto motivational structures, may also be mapped onto other mechanisms and processes (although to a lesser degree). For instance, within the CLARION framework, personality may involve a variety of parameters within the ACS, the NACS, the MS, and the MCS (although the MS is the most important part as argued earlier).

This model of personality (based on CLARION) captures more detailed aspects of psychological processes than previous work (e.g., Read and Miller 2002; Shoda and Mischel 1998; and so on). We go beyond overly abstract (and somewhat ungrounded) notions of goals, plans, resources, and beliefs, and so on. It is one thing to argue that personality traits consist of configurations of goals, plans, resources, and beliefs (Read and Miller 2002) or cognitive affective units (Shoda and Mischel 1998), it is another to map personality traits to lower-level, more specific, more detailed, and grounded cognitive processes and mechanisms found in generic computational cognitive architectures (such as CLARION). We would argue that it is more useful to ground personality traits in computational cognitive architectures, so that they are explained in a more unified way, along with many other psychological phenomena, data, and constructs, based on the primitives of cognition as envisaged within a generic model of cognition. Such explanations may be more detailed, more specific, and therefore deeper, in addition to the fact that such explanations lead to unified accounts of a wide variety of psychological phenomena, data, and constructs within a generic cognitive architecture.

A general outline of our personality model is as follows (Sun and Wilson 2009): Within the framework of CLARION, the MS, the MCS, the ACS, as well as the NACS, interact with each other. Within the MS, a set of drives exists and may be activated under various circumstances. Individual differences may be accounted for (for the most part) by the differences in relative drive strengths in different situations by different individuals. These drives lead to different setting of goals as well as cognitive parameters by the MCS. Individual differences in terms of relative drive strengths are consequently reflected in the resulting goals and major cognitive parameters. On the basis of the goals set and the cognitive parameters chosen, an agent makes action decisions (within the ACS). Thus, the actions of the agents (chosen within the ACS) reflect their fundamental individual differences (as well as situational factors) as a result. Therefore, on this view, personality is the result of very complicated interactions among a large set of entities, elements, and components. Computational modeling and simulations enable us to see how exactly these elements, entities, and components interact with each other in ways that are precise and detailed, that is, in exact mechanistic and process-based ways (Sun 2009). [See Sun and Wilson (2009) for further details of the personality model, including all relevant parameter values].

14.4 Results, Successes, Limitations, and Future Challenges

14.4.1 Some Simulation Results

Many psychological simulations using CLARION have been undertaken, are currently under way, or have been planned. Let us look briefly into some of those involving motivational processes.

First of all, the afore-discussed motivational representations have been used in developing a comprehensive personality simulation within CLARION. Various tests show that the CLARION personality model (as described earlier) is capable of demonstrating stable personality traits but at the same time showing sufficient variability of behaviors in response to different situations (Sun and Wilson 2009). It maps onto, and computationally demonstrates, the well-known Big Five personality structures, among other things.

For example, a set of 15 scenarios was used to assess the above point (cf. Read and Miller 2002). Figures 14.3–14.5 show the results of the simulation with the CLARION personality model using three pairs of personality types, where each pair consists of two personality types at the opposite ends of one of the personality dimensions. One pair consists of the shy and the sociable, at the two ends of the extroversion dimension. Another pair consists of the anxious and the confident, at the two ends of the neuroticism dimension. The third pair consists of the lazy and the conscientious, at the two ends of the conscientiousness dimension. As shown by Figs. 14.3–14.5, the two personality types in each pair behave quietly differently across this set of 15 scenarios.

This CLARION personality model has also been used to simulate some relevant human data (from social-personality psychology). For example, Moskowitz et al. (1994) examined the influence of social roles on interpersonal behavior in a work environment. Social role reflects the status of the person with whom an individual is interacting. It was hypothesized that the social role/status (as opposed to gender) would have an effect on behavior. Participants were expected to behave more submissively, e.g., when interacting with a boss versus a coworker or a subordinate. Participants were expected to be more dominant, e.g., when with a subordinate or a coworker than with a boss. Event contingent recording was used to gather data, and the data analysis confirmed all the expected effects. The CLARION simulation setup was identical to the personality model: it used the same behaviors from the

Top Behaviors (behavior indices are shown)

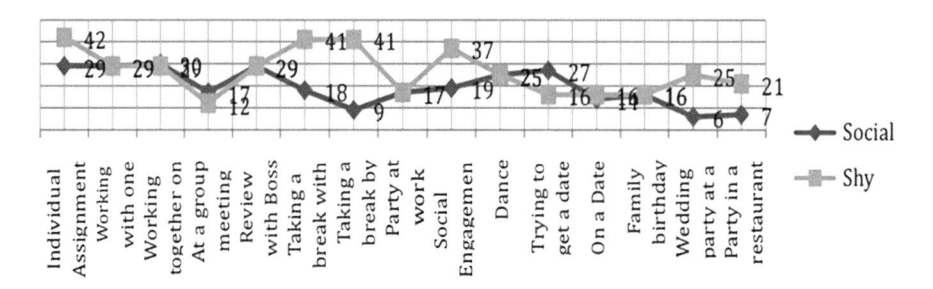

Fig. 14.3 The most frequent behaviors of the sociable and the shy personality across 15 scenarios. The *Y*-axis shows the behavior indices (as detailed in Fig. 14.6)

Top Behaviors

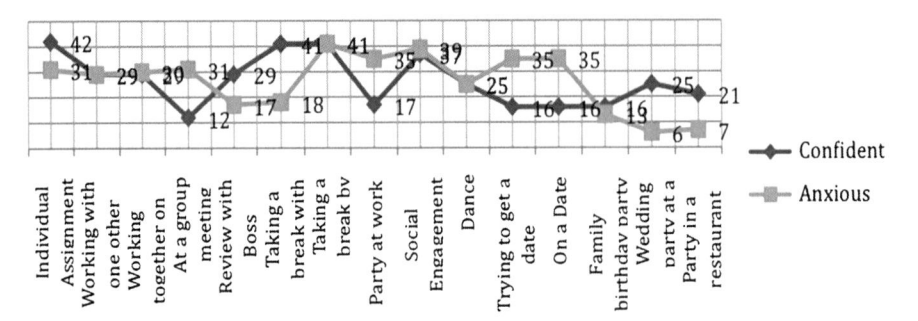

Fig. 14.4 The most frequent behaviors of the confident and the anxious personality across 15 scenarios. The Y-axis shows the behavior indices (as detailed in Fig. 14.6)

Top Behaviors

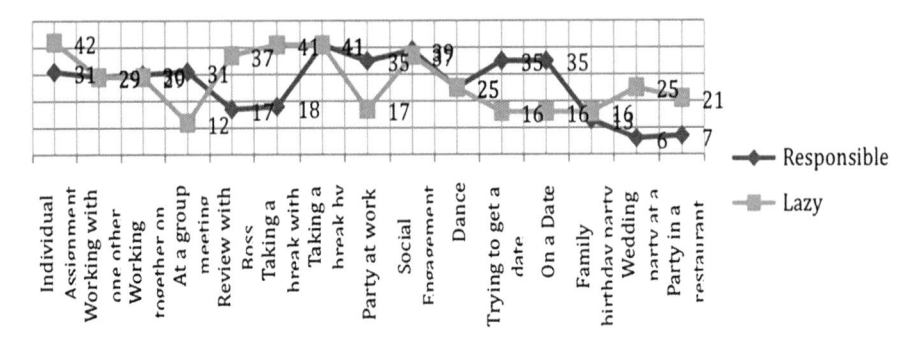

Fig. 14.5 The most frequent behaviors of the responsible and the lazy personality across 15 scenarios. The Y-axis shows the behavior indices (as detailed in Fig. 14.6)

previous simulations and used two of the scenarios from the earlier simulations ("Urgent project" and "Work with 1 other"). The simulation result captured perfectly all the major effects exhibited within the human data: for example, simulated participants behaved more submissively when interacting with a boss versus a coworker or a subordinate, and they behaved more dominantly with a subordinates or a coworker than with a boss (cf. Quek and Moskowitz 2007). Various other simulations of human data have also been carried out (see Sun and Wilson, 2009).

In a separate strand, Lambert et al. (2003) showed that in socially stressful situations, social stereotyping was more pronounced. To demonstrate this point, they examined the task of the recognition of tool versus gun, with priming by black or white faces. The results show that, in socially stressful situations, when paired with a black face, tools are much more likely to be mistaken as guns. This phenomenon has been captured, explained, and simulated using the motivational representations

Eat/drink E/D (1)	Stay at periphery SP (12)	Help others with work HOW (23)	Ensure work distributed fairly EDF (34)
Drink alcohol DA (2)	Self-disclose SD (13)	Order others what to do OO (24)	Wear something distinctive WSD (35)
Relax R (3)	Ask others about self AO (14)	Dance D (25)	Steal S (36)
Play practical joke PPJ (4)	Talk politics TP (15)	Ask other to dance AOD (26)	Kiss up KU (37)
Tease/make fun of T/M (5)	Gossip/Talk About Others G/T (16)	Ask for date AD (27)	Be cheap BC (38)
Try new dance steps TND (6)	Talk about work (job-related) TAW (17)	Kiss K (28)	Mediate M (39)
Intro self to others ISO (7)	Tell jokes TJ (18)	Do job DJ (29)	Give in GI (40)
Surf web SW (8)	Compliment others CO (19)	Extra effort job EEJ (30)	Procrastinate P (41)
Explore environment EE (9)	Ignore others IO (20)	Find new way to do job FNJ (31)	Pretend to work PW (42)
Leave L (10)	Insult others InO (21)	Improve skills IS (32)	Stay with comfortable others SCO (43)
Be silent BS (11)	Clean up CU (22)	Confront other about slacking COS (33)	

Fig. 14.6   A list of behavior indices

within CLARION (along with the other mechanisms of CLARION). When certain (avoidance-oriented) drive strengths become extremely high, the processing within the ACS of CLARION becomes very implicit, which is controlled and determined by the MCS of CLARION on the basis of the drive strength levels within the MS. The implicit processing within the ACS is more likely to be subject to stereotyping effects. The simulation using CLARION captured the corresponding human data well (Wilson et al. 2009), including, in particular, the stereotyping effect under stress, and provided a detailed, mechanistic, and process-based explanation for the data.

Likewise, skilled performance may deteriorate when individuals are under pressure. For example, in terms of mathematical skills, Beilock et al. (2004) showed that performance worsened when pressure was high. To demonstrate this point, they used a modular arithmetic problem set of the form $A = B \pmod{C}$ and tested participants either under pressure (with monetary incentives, peer pressure, and social evaluation) or not. The result showed clear differences between with and without pressure. This task has been simulated using the MS, the MCS, and the ACS within CLARION, which provide detailed, mechanistic, and process-based explanations. When certain (avoidance-oriented) drive strengths are very high, the processing within the ACS becomes very implicit (controlled by the MCS on the basis of the drive strength levels). Overly implicit processing leads to worsened performance (see, e.g., Sun et al. 2005). The simulation using CLARION captured the corresponding human data (Wilson et al. 2009).

The same phenomenon has been captured by CLARION simulations in low-level skill domains (involving mostly sensorimotor skills). Beilock and Carr (2001) showed that golf putting performance was worse when participants were under pressure (due to video taping and other setups that induced pressure). This phenomenon has been simulated within CLARION using the same mechanisms and processes as the simulations of the corresponding phenomena in high-level skill domains (as discussed above). The simulation successfully captured the human data (Wilson et al. 2009).

14.4.2 Implications, Limitations, and Future Work

This work addresses the essential motivational processes, structures, and representations necessary for a comprehensive cognitive architecture. The need for implicit drive representations, as well as explicit goal representations, has been hypothesized. Drive representations consist of primary drives (both low-level and high-level primary drives), as well as derived (secondary) drives. On the basis of drives, explicit goals may be generated on the fly during an agent's interaction with various situations.

The afore-discussed motivational representations and their resulting dynamics help to make a computational cognitive architecture more complete and functioning in a more psychologically realistic way. We believe that this work constitutes a requisite step forward in making computational cognitive architectures more realistic models of the human mind taking into consideration all of its complexity and intricacy, especially in terms of its motivational dynamics.

Detailed, mechanistic, and process-based modeling of what is underlying human motivation and personality, as has been done in the present work based on CLARION, may have the potential to eventually provide a coherent account of a wide range of phenomena in motivation and personality (as well as many other psychological aspects; see, e.g., Sun 2002; Sun et al. 2005; Sun and Zhang 2004), and meanwhile provides a useful tool for hypothesis generation and testing, as well as for theory building and integration, in the research on motivation, personality, and beyond (Sun 2009). This work thus advances understanding of human motivation and personality (as well as the science of human behavior in general because of the generality of CLARION).

Limitations of the work thus far include limited simulations of human data, limited validation of the model, and limited applications (e.g., to robotics). More detailed validation has yet to be carried out that can validate more detailed aspects of the model. This requires a great deal of detailed technical work in social-personality psychology. The application of the model to building cognitive robots and other intelligent systems also needs to be addressed in the future.

Significant future challenges in furthering this line of work include applying this framework to the building of intelligent systems that can display intelligent behavior with more flexibility and versatility. Another significant challenge is to further validate, through empirical work (especially psychological empirical work), this framework and its implications for understanding human motivation and personality. Many more experiments and tests will be needed and shall be pursued in the future.

Acknowledgments This work has been supported in part by ONR grant N00014-08-1-0068 (to Ron Sun), as well as by ARI contract W74V8H-05-K-0002 (to Ron Sun and Robert Mathews). Thanks to all the reviewers who provided comments on an earlier version.

References

Anderson, J. R. (1993). *Rules of the Mind.* Lawrence Erlbaum Associates, Hillsdale, NJ.

Anderson, J. and C. Lebiere (1998). *The Atomic Components of Thought.* Lawrence Erlbaum Associates, Mahwah, NJ.

Bach, J. (2009). *Principles of Synthetic Intelligence.* Oxford University Press, New York.

Beilock, S., C. Kulp, L. Holt, and T. Carr (2004). More on the fragility of performance: Choking under pressure in mathematical problem solving. *Journal of Experimental Psychology,* 133, 584–600.

Beilock, S. and T. Carr (2001). On the fragility of skilled performance: What governs choking under pressure? *Journal of Experimental Psychology,* 130, 701–725.

Caprara, G. V. and D. Cervone (2000). Personality: *Determinants, Dynamics, and Potentials.* Cambridge University Press, New York.

Digman, J. M. (1990). Personality structure: Emergence of the five-factor model. *Annual Review of Psychology,* 41, 417–440.

Doerner, D. (2003). The mathematics of emotions. In: Frank Detje, D. D. and Schaub, H. (eds.), *Proceedings of the Fifth International Conference on Cognitive Modeling,* Bamberg, Germany, pp. 75–79.

Fuster, J. M. (2002). Physiology of the executive functions: The perception-action cycle. In: Stuss, D. T. and Knight, R. T. (eds.), *Principles of Frontal Lobe Function.* Oxford University Press, Oxford, pp. 96–108.

Hull, C. (1943). *Principles of Behavior: An Introduction to Behavior Theory.* D. Appleton-Century Company, New York.

James, W. (1890). *The Principles of Psychology.* Dover, New York.

Lambert, A., B. Payne, L. Jacoby, L. Shaffer, A. Chasteen, and S. Khan (2003). Stereotypes and dominant responses: On the "social facilitation" of prejudice in anticipated public contexts. *Journal of Personality and Social Psychology,* 84, 277–295.

Maner, J. K., D. T. Kenrick, S. L. Neuberg, D. V. Becker, T. Robertson, B. Hofer, A. Delton, J. Butner, and M. Schaller (2005). Functional projection: How fundamental social motives can bias interpersonal perception. *Journal of Personality and Social Psychology,* 88, 63–78.

Maslow, A. (1943). A theory of human motivation. *Psychological Review,* 50, 370–396.

Mazzoni, G. and Nelson, T. (eds.), (1998). *Metacognition and Cognitive Neuropsychology.* Erlbaum, Mahwah, NJ.

McClelland, D. (1951). *Personality.* Dryden, New York.

McDougall, W. (1936). *An Introduction to Social Psychology.* Methuen & Co., London.

McFarland, D. (1989). *Problems of Animal Behaviour.* Longman Publishing, Singapore.

Moskowitz, D. S., E. J. Suh, and J. Desaulniers (1994). Situational influences on gender differences in agency and communion. *Journal of Personality and Social Psychology,* 66, 753–761.

Murray, H. (1938). *Explorations in Personality.* Oxford University Press, New York.

Newell, A. (1990). *Unified Theories of Cognition.* Harvard University Press, Cambridge, MA.

Poznanski, M. and P. Thagard (2005). Changing personalities: towards realistic virtual characters. *Journal of Experimental and Theoretical Artificial Intelligence,* 17, 221–241.

Quek, M. and D. S. Moskowitz (2007). Testing neural network models of personality. *Journal of Research in Personality,* 41, 700–706.

Read, S. J. and L. C. Miller (2002). Virtual Personalities: A Neural Network Model of Personality. *Personality and Social Psychology Review,* 6, 357–369.

Reiss, S. (2004). Multifaceted nature of intrinsic motivation: The theory of 16 basic desires. *Review of General Psychology,* 8(3), 179–193.

Rosenbloom, P., J. Laird, and A. Newell (1993). *The SOAR Papers: Research on Integrated Intelligence.* MIT, Cambridge, MA.

Schwartz, S. (1994). Are there universal aspects of human values? *Journal of Social Issues,* 50, 19–45.

Shoda, Y. and W. Mischel (1998). Personality as a stable cognitive–affective activation network: Characteristic patterns of behavior variation emerge from a stable personality structure. In Read, S. J. and Miller, L. C. (eds.), *Connectionist models of social reasoning and social behavior.* Lawrence Erlbaum Associates, Inc., Mahwah, NJ, pp. 175–208.

Simon, H. A. (1967). Motivational and emotional controls of cognition. *Psychological Review,* 74, 29–39.

Sloman, A. (1987). Motives, mechanisms and emotions. *Emotion and Cognition,* 1, 217–234.

Sun, R. (1994). *Integrating Rules and Connectionism for Robust Commonsense Reasoning.* New York. John Wiley and Sons.

Sun, R. (2002). *Duality of the Mind: A Bottom-up Approach Toward Cognition.* Lawrence Erlbaum Associates, Mahwah, NJ.

Sun, R. (2003). *A Tutorial on CLARION 5.0.* Technical Report, Cognitive Science Department, Rensselaer Polytechnic Institute. http://www.cogsci.rpi.edu/rsun/sun.tutorial.pdf.

Sun, R. (2009). Motivational representations within a computational cognitive architecture. *Cognitive Computation,* 1(1), 91–103.

Sun, R., E. Merrill, and T. Peterson (2001). From implicit skills to explicit knowledge: A bottom-up model of skill learning. *Cognitive Science,* 25, 203–244.

Sun, R., Slusarz, P. and C. Terry (2005). The interaction of the explicit and the implicit in skill learning: A dual-process approach. *Psychological Review,* 112, 159–192.

Sun, R. and N. Wilson (2009). A computational personality model within a comprehensive cognitive architecture. Technical Report, *Cognitive Science Department,* Rensselaer Polytechnic Institute.

Sun, R. and X. Zhang (2004). Top-down versus bottom-up learning in cognitive skill acquisition. *Cognitive Systems Research,* 5(1), 63–89.

Toates, F. (1986). *Motivational Systems.* Cambridge University Press, Cambridge, UK.

Tolman, E.C. (1932). *Purposive Behavior in Animals and Men.* Century, New York.

Tyrell, T. (1993). *Computational Mechanisms for Action Selection.* Ph.D Thesis, Oxford University, Oxford, UK.

Watkins, C. (1989). *Learning with Delayed Rewards.* Ph.D Thesis, Cambridge University, Cambridge, UK.

Weiner, B. (1992). *Human Motivation: Metaphors, Theories, and Research.* Sage, Newbury Park, CA.

Wilson, N., R. Sun, and R. Mathews (2009). A motivationally-based simulation of performance degradation under pressure. Neural Network, 22, 502–508.

Wright, I. P., and A. Sloman (1997). MINDER1: An Implementation of a Proto-emotional Agent Architecture. Technical Report CSRP-97-1, University of Birmingham, School of Computer Science. (Available from ftp://ftp.cs.bham.ac.uk/pub/tech-reports/1997/CSRP-97-01.ps.gz).

Chapter 15
Cognitive Algorithms and Systems of Error Monitoring, Conflict Resolution and Decision Making

Pedro U. Lima

15.1 Overview

There are currently several approaches to decision making in complex systems, particularly in robotics. In most cases, the decision-making process resembles the well-known control or sense–think–act loop: the process output or state is sensed, its deviation (error) from the desired value is continuously monitored and, based on some appropriate algorithm, a control action is picked from the available action set to be applied to the process, so that the loop is closed and the decision-making process moves to its next iteration.

The strict-sense control approach typically handles continuous state-space systems, driven by time (either discrete/sampled or continuous), where the controller algorithm implements a closed-form function that maps errors in continuous state-space to control signals from a continuous control-space. On the other hand, the wider-sense approach to decision making often handles discrete state-space, event-driven systems, where the monitored error is used to resolve a conflict among actions from a discrete set, typically following an optimization algorithm that attempts to maximize a payoff function over some finite or infinite horizon of steps. Examples of the former are the control of chemical processes or of mechanical devices, such as robot joints, while the latter concerns, e.g. scheduling in manufacturing systems or conflict resolution while executing conditional plans in robotics.

In this chapter, we will address the problem of plan representation, analysis and execution in multi-robot systems using a well-known formal model of computation: Petri nets. Therefore, we will focus on the discrete state-space, event-driven model, so as to view a multi-robot plan as a discrete event system (DES) (Cassandras and Lafortune 2007). Most of the existing robot task models are not based on formal approaches but tailored to the task at hand, usually leading to task plans with few primitive actions, simply because increasing their number and the plan complexity may lead to unexpected results, not predicted by any analysis studies. Applying DES

P.U. Lima (✉)
Institute for Systems and Robotics, Instituto Superior Técnico, Av. Rovisco Pais,
1 1049-001 Lisboa, Portugal
e-mail: pal@isr.ist.utl.pt

V. Cutsuridis et al. (eds.), *Perception-Action Cycle: Models, Architectures, and Hardware*, Springer Series in Cognitive and Neural Systems 1, DOI 10.1007/978-1-4419-1452-1_15, © Springer Science+Business Media, LLC 2011

concepts and theory to model (multi-)robot tasks provides a systematic approach to modelling, analysis and design, scaling up to realistic applications, and enabling analysis of formal properties, as well as design from specifications.

In particular, representing multi-robot plans by Petri nets enables tackling a considerable number of issues:

- A plan is seen as control policy that maps states onto actions
- Control policies are, in general, non-deterministic, as there may exist more than one possible action for a given state
- A plan represented as a Petri net can be executed by simply following Petri net firing rules, where (sequential) decision-making algorithms are used for conflict resolution whenever more than one action are available for a given state
- Using a *qualitative untimed Petri net view*, plans can be analysed regarding their formal properties, e.g. using algorithms that address Petri net analysis problems (such as conservation, blocking, liveness, invariants)
- Using a *quantitative stochastic timed Petri net view*, plans can be analysed regarding their performance under uncertainty, e.g. using closed form algorithms and/or Monte Carlo simulations that address Petri net stochastic performance (such as plan success probability, plan robustness)

We will review existing Petri net-based approaches to robot plan representation, both from a qualitative untimed view and from a quantitative stochastic view. Details on the two Petri net views, analysis problems and existing techniques for their solution will be provided. We will then introduce a multi-robot task model composed of primitive actions, behaviours, predicates to represent the system state and events to trigger state transitions. These concepts will be defined and mapped onto Petri nets. Next, we will explain how the qualitative and quantitative views help analysing different problems regarding plan formal properties and performance in the presence of uncertainty in the action effects. Then, we will state when does a stochastic Petri net representation of a plan map onto a Markov chain, enabling the use of Markov chain theory to analyse plans performance in closed form. Adding controllable actions to the multi-robot task model finally leads to a controlled Markov chain or Markov Decision Process (MDP), for which different solutions of the planning problems (seen as the determination of a control policy) exist in the literature. Among those, we will describe how to use reinforcement learning to resolve control policy conflicts between alternative controllable actions for the same state.

The chapter is organized as follows: in the next section, we will review some of the Petri net-based approaches to robot task modelling described in the literature, the formal models they introduced and some of the results obtained. Section 15.3 introduces our proposed multi-robot task model, the corresponding plan representation by Petri nets, how to analyse plan performance in the presence of uncertainties and some examples of application to robot soccer. The last section will summarize what has been accomplished so far under this line of research, the success stories, limitations found, future challenges and some suggestions on how to address their solution.

Throughout the chapter, most provided details concern theoretical aspects, while results are presented only for the purpose of illustrating the main concepts, as some of them are still preliminary. Petri net analysis techniques, Markov chain and sequential decision making under uncertainty (including reinforcement learning) topics will assume basic knowledge about them by the reader and introduce only definitions required for the sake of understanding their relation to the robot task model also introduced here.

15.2 Algorithm/System Justification

The motivation for using Petri nets as our formal model of computation could well apply to other formal models, in particular, to finite state automata. However, there are several practical and theoretical arguments for preferring Petri nets (Cassandras and Lafortune 2007):

- Petri net languages (languages marked by Petri nets) are a superset of regular languages (languages marked by finite state automata), mainly due to Petri net memory and concurrency distinctive features. Therefore, the set of modelled roles and behaviours is potentially richer when using Petri nets
- Petri nets enable distributed state modelling, i.e. one can start with simple models (e.g. a primitive action and its pre-conditions) and build more complex ones (e.g. a behaviour Petri net out of several primitive action Petri nets)
- Several tools for Petri net formal analysis are available. Formal verification is useful for programming, but stochastic performance evaluation can also be achieved supported on widely available bodies of theory, methods and tools for stochastic Petri nets: a stochastic model of primitive actions, their times between failures and success probabilities can be built readily from the non-stochastic model and analysed beforehand, even using closed-form Markov Chain analysis methods, under some conditions

Petri nets have been widely used in the literature to model DESs, namely manufacturing systems (Viswanadham and Narahari 1992). Petri net-based models of robot tasks have been reported in recent years, starting with the pioneer work of Wang and co-authors (Wang et al. 1993), where Petri nets are used to implement the coordination level of Saridis' 3-level hierarchy for intelligent machines (Saridis 1979), coordinating the execution of a robot plan. The coordination level is split into a dispatcher and several coordinators, corresponding to robot sub-systems. Coordination structures are introduced to link the dispatcher and the coordinators, and several conditions to ensure the preservation of qualitative properties, such as liveness and boundedness, of the coordination structure components are studied and proven. Lima and co-workers have been developing work on Petri net-based robot task model and plan representation since 1998, covering issues such as using the different views of Petri nets for plan representation, qualitative analysis, quantitative analysis and execution (Lima et al. 1998) or defining a language-based model of robotic tasks (Milutinovic and Lima 2002). The group has recently introduced the concept of closed-loop behaviour analysis for the robot controller situated in its environment. To accomplish that, Petri net models of the robot controller and of the robot world are defined. Methods and results of qualitative and quantitative performance analysis of the resulting multi-robot closed loop system (Costelha and Lima 2007) and their extension to multi-robot systems, which basically adds communication actions to the original model (Costelha and Lima 2008), are presented. Teamwork modelling for multi-robot systems, including synchronization and commitments (Cohen and Levesque 1991), within the different framework of Petri net plans (PNPs) (Ziparo et al. 2008) was also introduced.

Most of the work in Petri net models for robot tasks concerns the qualitative view, mainly addressing the usage of PNPs for behaviour programming and verification, including cooperative multi-robot teams (Ziparo and Iocchi 2006), with application to a real quadruped robot team, or the compilation of plans for multiple robots, generated by planning methods, into Petri nets for analysis, execution and monitoring (King et al. 2003). In this work, supervisory control techniques are applied to a Petri net controller to handle conflicts that arise due to the presence of shared resources. Novel supervisory control techniques are also introduced and applied to simulated and real sensor networks mixing static and mobile sensors in Giordano et al. (2006). Another formal framework for robotic cooperation based on an extension to Petri nets, known as workflow nets, is introduced by Kotb et al. (2007) to establish a protocol among mobile agents based on the task coverage they maintain. Petri nets are used to ensure the soundness of the framework and to quantify task performance and determine goal state reachability.

The quantitative, stochastic timed view is taken by Kim et al. (2005). The authors provide a formal navigation behaviour selection framework for a service robot and show its application to museum tour guide robot. Their approach handles modelling, analysis and performance evaluation using generalized stochastic Petri nets. Earlier work concerning a deterministic timed view addresses specification, validation and code generation in real-time robot-control applications (Montano et al. 2000).

For Petri net analysis, one can use available tools such as PIPE (Akharware 2000) or TimeNET (Zimmermann and Freiheit 1998) to study the task properties, both through simulation and closed-form equivalent Markov Chain analysis.

15.3 The Algorithm/System and How It Deviates from Its Predecessors

A (multi-)robot task model can be specified by defining the task resources (the robots and their devices), the task plan components (*roles*, *behaviours* and *primitive actions*) and the plan coordination components (*predicates* and *events* that represent the environment state and dynamics). The model must provide support for decision making, taking into account the main subsystems at the individual and collective level, as well as the requisites for cooperation, concerning different aspects, such as cooperative perception and teamwork, and enable different decision-making strategies (e.g. predicate-based, event-based, mixed).

In this section, we introduce such a multi-robot task model and define its main concepts and components. Then we explain how those resources and components are organized as building blocks of a *functional architecture*. Next we describe the information flow among the building blocks within the architecture and finally define the map between the above components, resources and building blocks, and a Petri net-based representation and execution of individual and cooperative plans.

15.3.1 Robot Task Model Components

We have split the robot task model components into three categories (*plan components*, *plan coordination components* and *resources*), and define them as follows.

Robot tasks are carried out through *plans*. Ultimately, plans are sequences of *primitive actions*. However, to reduce the complexity of sequential decision making when selecting the next primitive action, primitive actions can be grouped in *behaviours*, and only a subset of available behaviours is active when a given *role* is selected. Therefore, one can look at plans at higher abstraction levels and see them as a sequence of *roles* or as a sequence of *behaviours*. *Primitive actions*, *behaviours* and *roles* are the *plan components* and they are depicted in Fig. 15.1.

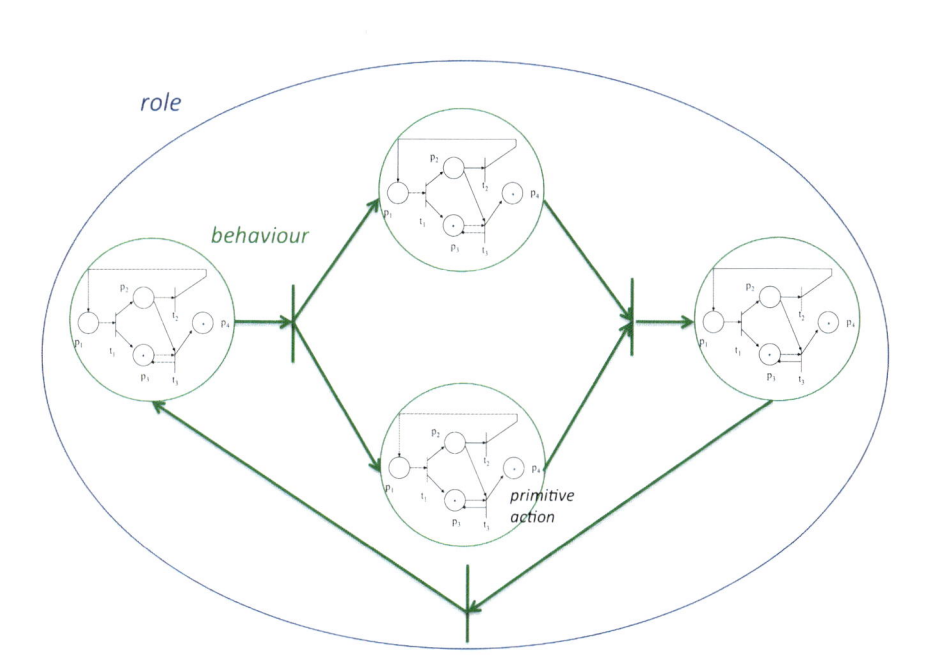

Fig. 15.1 Robotic task model plan components (*roles, behaviours and primitive actions*) and their interconnections. More roles, behaviours and primitive actions may exist and coincide or not with some of those in the diagram

Plans are not pre-defined sequences of primitive actions, but rather control policies. A control policy maps a state onto a primitive action or, more generally, onto a set of alternative primitive actions. The decision-making algorithm selects, at each plan execution step, one of the alternative primitive actions for a given state, according to the control policy established at planning time. The determined control policy can be *deterministic*, in which case the same primitive action is selected whenever the robot reaches a given state, or *stochastic*, if the selection performed by the decision-making algorithm is based on a probability mass function over the set of alternative actions. This is typically the case when reinforcement learning algorithms are used to improve the primitive action selection over time, as a result of an adaptation to the characteristics of the environment in which the robot(s) evolve.

World states are described in our model by the truth value of *predicates*. Transitions between world states result from the occurrence of *events*. Predicates and events are designated as *plan coordination components*.

Additionally, the model includes hardware interface components that encapsulate the hardware devices, or *resources*, with which the functional architecture interfaces at the individual robot level: actuators, transducers and sensors.

15.3.1.1 Plan Coordination Components

- *Predicates* are defined as Boolean relations over the domain of world objects. The set of predicates for a given plan is $D = \{d_1, \ldots, d_{n_d}\}$. If o_i is a world object, $W = \{o_1, \ldots, o_{n_o}\}$, any $d_i \in D$ is defined as $d_i = \rho(o_k, \ldots, o_l) \in \{\text{True,False}\}$, $o_k, \ldots, o_l \in W$. Examples in the robot soccer domain are $\text{see}(x)$ where x can be a ball, pole or field_line, or $\text{near}(r, x)$, where r is any of the team robots, and x can be any world object.
- An *Event* is, in general, an instantaneous occurrence which denotes a state change (e.g. of a variable, of a robot). In our multi-robot task model, we clearly define *event* as a change of (logical conditions over) a predicate value from True to False or False to True. The set of events for a given plan is $E = \{e_1, \ldots, e_{n_e}\}$. All events from E are associated with some (logical condition over a) predicate. An event $e(t)$ occurs at time t if its associated (logical condition over a) predicate $\lambda(t, d \in D)$ changes its logical value instantaneously, e.g. $\lambda(t^-, d) = \text{True}, \lambda(t^+, d) = \text{False}$. Technically, one might distinguish between *internal events*, corresponding to the above definition, and *external events*, triggered by instantaneous occurrences from sources external to a robot. Nevertheless, the latter do in fact meet the *internal event* definition, as one may associate the state of the associated external source with a predicate. Thus, even though in practice some events may be triggered by external sources, we do not distinguish internal and external events from a formal standpoint. Examples of internal events in robot soccer are: event lost_object occurs when the proposition has(object) changes its value from True to False, and vice versa for event got_object; event found_object occurs when the proposition

see(object) changes from or `False` to `True`. Examples of external events in robot soccer are signals sent by the referee box telling the robots to stop, execute a goal, corner kick or a throw-in.

15.3.1.2 Plan Components

- A *primitive action* is the atomic element of a plan, which cannot be further decomposed. The set of primitive actions for a given plan is $\Pi = \{\pi_1, \ldots, \pi_{n_p}\}$. A primitive action typically consists of some calculations (e.g. determination of the desired posture) plus a call to a navigation algorithm or the direct activation of an actuator. Desirably, it is designed as a control (or sense–think–act) loop, i.e. a generalized view of the closed-loop control system concept. This means that our model assumes primitives that move the robot towards its goal while avoiding obstacles, rather than having one primitive that moves towards the goal and another that avoids obstacles. Navigation algorithms are implemented as functions organized in a library. A navigation function implements a guidance algorithm which, based on the current and target robot postures (position plus orientation) and current self-localization estimate, computes the required wheel speeds to move the robot from the current to the target position avoiding obstacles on the way.
- *Behaviours* are defined as "macros" of primitive actions grouped together using some coordination mechanism. Coordination mechanisms use plan coordination components to determine when to switch among primitive actions in the course of executing the behaviour. For instance, a behaviour may be represented by a state machine whose states represent primitive actions, and transitions between states have associated events, but it could also be represented by a fuzzy decision-making algorithm based on fuzzy rules, used to select sequences of primitive actions to be executed. The set of behaviours for a given plan is given by $B = \{b_1, \ldots, b_{n_b}\}$. In general, one can mathematically define each behaviour b_i as a graph whose vertices have associated primitive actions $\pi_i \in \Pi$, and edges are labelled by functions of plan coordination components, $f : E \cup D \rightarrow \{\text{True,False}\}$. However, we will defer a more concrete definition to subsection 3.4, where we provide an instantiation of a coordination mechanism based on Petri nets.
- *Roles* can also be seen as "macros" of behaviours. They are defined as sets of behaviours $\{b|b \in B\}$ which are subsets of the set B of all available behaviours, grouped together using some appropriate coordination mechanism (see discussion of coordination mechanisms in the previous item). When a role is selected, a new set of behaviours becomes enabled for selection by the behaviour coordination mechanism. In practice, a role constrains the possible options for a robot selection of behaviours, effectively constraining the overall behaviour displayed by the robot. The set of roles for a given plan is $R = \{r_1, \ldots, r_{n_r}\}$. Role $r_i = \{b|b \in B\}_i$, as explained before. Note that, even though $\bigcup_{r_i \in R} r_i = B$, roles do not form a partition over the set of available behaviours, i.e. $r_i \cap r_j \neq \emptyset$

in general, since there are behaviours that may be shared by more than one role (similar to the correspondence between primitive actions and behaviours in the previous item).

15.3.1.3 Resources

- *Devices* to handle the low level interface with physical-world devices, both actuators and transducers (e.g. motors, sonars, cameras).
- *Sensors* to obtain information from the transducer devices (e.g. odometry, obstacle location, ball position).

Note that, in our model, one *transducer* may correspond to several *sensors*: a typical example is the vision camera transducer and the sensors one can develop by processing one or several acquired images from it, e.g. to recognize specific objects, to find straight lines and to discriminate colours. This distinction is made on purpose to decouple the physical device from the virtual sensors one can build from it.

15.3.2 Functional Architecture

The functional architecture underlying our multi-robot task model is divided into three major building blocks, each of them having several modules:

- ATLAS (i.e. the subsystem that supports the whole system) is responsible for the operations directly related with the robot's interaction with the environment: sensing and acting. Thus, ATLAS includes the following plan components: *devices*, *sensors* and *primitive actions*. It also includes one extra functional module, designated as Information Fusion, where information from several sensors (of the same robot or from different robots in the team) is fused.
- WISDOM (i.e. a very relevant requirement for intelligence to be displayed) acts as a central point of information storage in the format of raw, processed and symbolic data, as well as of data registration mechanisms, so that information updates are propagated through the web of relations established among stored items, e.g. predicates updates resulting from changes in the values of attributes of world objects. WISDOM is split into:
  - World Info: where general purpose high-level data are stored. This includes symbolical and numerical data, organized as objects $o \in W$ and their attributes, e.g. the robot and its current location. The information can either be generated locally, by the local ATLAS Information Fusion module, or remotely, by another robot's ATLAS Information Fusion module. It is typically relevant for predicate evaluation, though it can also be used for navigation purposes by ATLAS modules. World Info does not presently include any advanced data retrieval and maintenance methods, but those can be

freely implemented later. However, it does include historical data from stored objects, relevant for robot dynamic planning and control methods.

- Predicate and Event Manager: manages the dynamic instantiation of predicates by their Boolean values, based on the current data values, stored in the World Info. Predicates are registered on their related world objects attribute values, so that the Boolean value of the predicate relation is automatically updated when attribute value changes trigger predicate changes. Similarly, events are registered on their associated (logical conditions over) predicates, so as to be triggered when the corresponding Boolean values switch.

- CORTEX (from CoORdinator, TEam organizer, eXecutor) implements the hierarchical coordination mechanism that controls the selection of roles, behaviours and primitive actions (moving top down through the hierarchy). It naturally includes three modules:

  - Team Organizer: responsible for selecting roles.
  - Behaviour Coordinator: responsible for behaviour selection. Behaviours are selected from the set of behaviours available for the currently selected role.
  - Behaviour Executor: responsible for behaviour execution, i.e. for the switching between primitive actions which are part of the currently selected behaviour.

The details of information flow inter- and intra-modules are provided in the next subsection.

Figure 15.2 shows the functional architecture, expressed as a UML diagram of main components and other modules, including their connections.

15.3.3 Information Flow

A typical flow of information in our multi-robot task model proceeds as follows: raw information about the surrounding environment of the robots is acquired by transducer *devices*, processed by *sensors* and possibly fused among several sensors of the same robot and/or sensors from different robots in the Information Fusion either as numeric or symbolic data (examples in soccer robots are the estimated ball position, the estimated robot posture or proposition `see(ball)`). The Predicate and Event Manager module is in charge of updating a predicate value, upon notification by the World Info of changes on data relevant for the computation of the logical value of the predicate. This information, as well as changes in the predicate Boolean values (i.e. *events*), can be accessed by the CORTEX modules, so as to take decisions on selection among *roles* (Team Organizer), *behaviours* (Behaviour Coordinator) and *primitive actions* (Behaviour Executor). *Primitive actions* compute set points to be sent to the actuator *devices*, and the loop is closed through the environment.

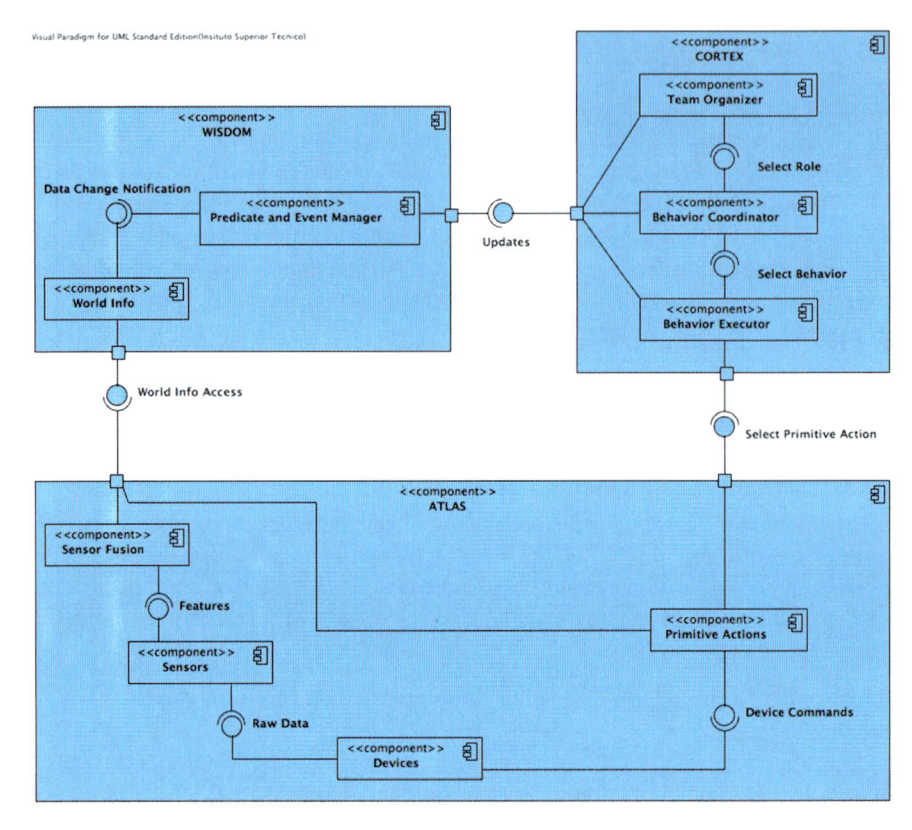

Fig. 15.2 Robotic task model functional architecture

Cooperation among team robots can occur in different forms. In our model, we distinguish three forms of displayed cooperation:

- *Team Organization* reflects the current distribution of roles per team-mates, e.g. a robot rescue team that composed of three aerial surveillance robots and two fire-extinguisher land robots displays an organization different from a team composed of one communication-relay aerial robot, two aerial surveillance robots, one fire-extinguisher land robot and one scout land robot (that communicates with the other land robots via the aerial relay robot).
- *Relational Behaviours* are the behaviours involving more than one team-mate (could be the entire team) in a *temporary* relation, so as to accomplish a momentary task that requires teamwork, e.g. a pass in robot soccer.
- *Individual Behaviours* are the behaviours involving one single robot (thus, no cooperation).

Typically, cooperative plan execution requires the exchange of synchronization and commitment signals among robots (Ziparo et al. 2008). Such signals lead to changes of the Boolean values of specific World Info propositions related to communication (e.g. `sent(msg-ack)`, `pass-made(robot1)`) in all the robots involved in the teamwork. The teamwork predicates are then be used by the Team Organizer and Behaviour Coordinator to take decisions on *role* and *behaviour* selection, respectively, therefore taking into account the current state of their team-mates.

15.3.4 Petri Net Model of Task Plans

In this subsection, we will formalize the map from the functional architecture building blocks onto Petri nets. Petri nets not only model the hierarchical selection mechanisms of CORTEX but also include components from WISDOM and ATLAS, so as to interact with the world. Different Petri net views (untimed vs stochastic timed) will be used, depending on the model objectives.

Petri nets (Girault and Valk 2003) represent a class of DES. These are systems with a discrete set of states, whose dynamics is driven by the occurrence of events which cause transitions between states (Cassandras and Lafortune 2007). The events occur asynchronously and typically DES do not model the event generation mechanism, i.e. they assume an event may happen in a given state, but do not care to model the dynamics of the process causing the event.

We will first focus on the *untimed qualitative Petri net plan representation*. Untimed Petri nets are defined by a 5-tuple (P, T, A, M_0, W), where:

- P is a set of places
- T is a set of transitions
- A is a set of arcs, connecting transitions to places and places to transitions
- $M_0 : P \rightarrow N$ is the initial marking, where markings denote the state of the Petri net, by assigning to each place $p \in P, n \in N$ tokens
- $W : A \rightarrow N^+$, known as the set of arc weights, assigns to each arc $a \in A$ some $n \in N^+$, denoting how many tokens are consumed from a place by a transition, or alternatively, how many tokens are produced by a transition and put into each place

Figure 15.3 shows an untimed Petri net and three consecutive markings, due to firing a transition sequence.

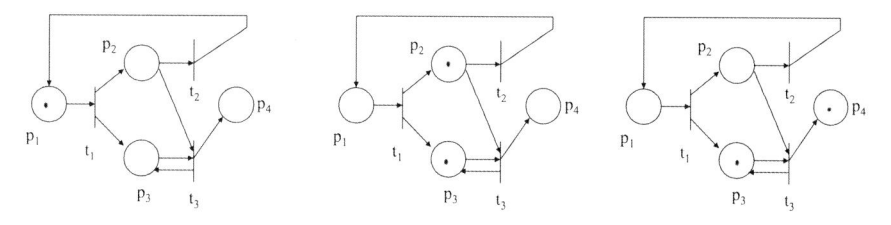

Fig. 15.3 Typical untimed Petri net and three consecutive markings (from *left* to *right*) due to the firing of transitions t_1 and t_3 (between the second and third markings, t_2 and t_3 were in conflict and a choice has been made to fire one of them)

The untimed Petri nets used to model robot tasks have all arc weights equalling 1; therefore, the weights are not represented in the Petri net graphs for the sake of simplified notation.

In our untimed model, *behaviours* are modelled by Petri nets as follows:

- Each place in the Petri net is labelled by an associated primitive action or by a predicate, i.e. $l_p : P \rightarrow \Pi \cup D$, where l_p is the (in general non-injective) place labelling function
- Each transition in the Petri net is labelled by an event, i.e. $l_t : T \rightarrow E \cup \{\varepsilon\}$, where l_t is the (in general non-injective) transition labelling function, and ε is the ever-occurring event

A token in a place means that the primitive action associated with that place is currently active (i.e. it is running), or that the predicate labelling that place is True. Transitions are *enabled* when all their input places have at least one token each, meaning that the pre-conditions for the next step are satisfied. A transition is *fired* if it is enabled and its labelling event (if any) occurs. A transition labelled by ε always fires when enabled. The firing of a transition provokes the switching from the current marking to another (possibly the same) marking, as the result of the token flow dictated by the arc weights (see definition above).

The above definitions state that switching from primitive action π_i to primitive action π_j through transition $t \in T$ occurs, for any $p_k, p_l \in P$, when

- p_k is an input place of t, with $l_p(p_k) = \pi_i$, and p_l is an output place of t, with $l_p(p_l) = \pi_j$
- All other input places of t have 1 token, meaning that all predicates representing *pre-conditions* of π_j are true (if concurrent primitive actions are allowed, some of the other input places of t might be primitive actions, but the same explanation applies, in this case for the switching from a set of primitive actions to another set of primitive actions)
- $l_t(t) = \varepsilon$ or event e, $l_t(t) = e \in E$, occurs

Cooperative plan execution can be modelled by Petri nets representing *behaviours*. In this case, the Petri nets represent *relational behaviours* and different Petri nets composing the relational behaviour run, for execution purposes, concurrently in the involved robots. For modelling purposes, the different Petri nets are linked by places representing communication predicates, which enable exchange of synchronization and commitment messages (Ziparo et al. 2008) among the team-mates. Those messages influence the flow of control in one robot as a consequence of actions taken by one or more of its teammates.

One simplified example of a pass between two soccer robots, modelled by a Petri net, is depicted in Fig. 15.4. No commitment mechanisms are represented, for the sake of simplifying reading the Petri net.

An interesting feature of untimed PNP representations is that both *predicate-based* production rules in the form **If** primitive action is running and pre-conditions hold **Then** switch to new primitive action and *event-driven* control mechanisms can be modelled.

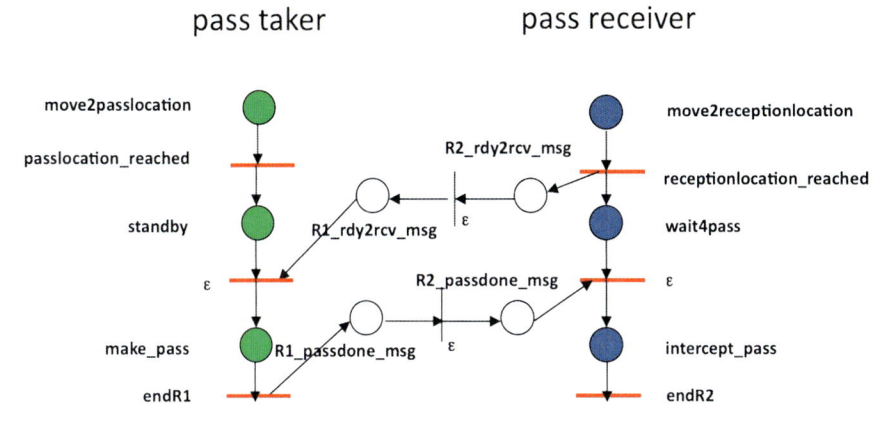

Fig. 15.4 Pass between two soccer robots modelled as a Petri net

Another relevant aspect is the hierarchical nature of Petri nets. *Behaviour* Petri nets implement the switching mechanism between *primitive actions*. Similarly, Petri nets can represent *roles*, if places are now labelled (besides predicates) by *behaviours*, instead of *primitive actions*:

- Each place in the Petri net is labelled by an associated behaviour or by a predicate, i.e. $l_p : P \rightarrow B \cup D$, where l_p is the (in general non-injective) place labelling function
- Each transition in the Petri net is labelled by an event, i.e. $l_t : T \rightarrow E \cup \{\varepsilon\}$, where l_t is the (in general non-injective) transition labelling function, and εgs the ever-occurring event

One can build a hierarchical Petri net multi-robot controller to model the full COR-TEX hierarchy in the previous subsection. *Predicates* and *events* are generated in WISDOM by appropriate algorithms, from the information stored in the World Info module. In turn, this information results from raw data acquired by transducer *devices* after processed by the Sensors and Information Fusion modules. On the other side, *primitive actions* interface the actuator *devices* to turn their decisions into actual actions over the world. Therefore, Petri nets interact with the world through predicates, events and primitive actions that label their places and transitions.

This is illustrated in Fig. 15.5 for a soccer robot team. In this simplified Petri net example, the Team Organizer starts by selecting the `Supporter` role, when predicate `ShouldSupport` becomes `True`. The Petri net for the role is then selected, and it starts with behaviour `BaseSupport` active. The Petri net switches to behaviour `Stop` if the referee box signals to stop the game. Then, it waits for a message from the referee box signalling a foul, and it switches to behaviour `FoulDefend` until the foul is taken (as signalled by the occurrence of an event). Once this happens, the Petri net returns to the initial `BaseSupport` behaviour. It may also get there from the `Stop` behaviour if the referee box signals a game (re)start. Behaviour `BaseSupport` Petri net model is depicted on the right of the

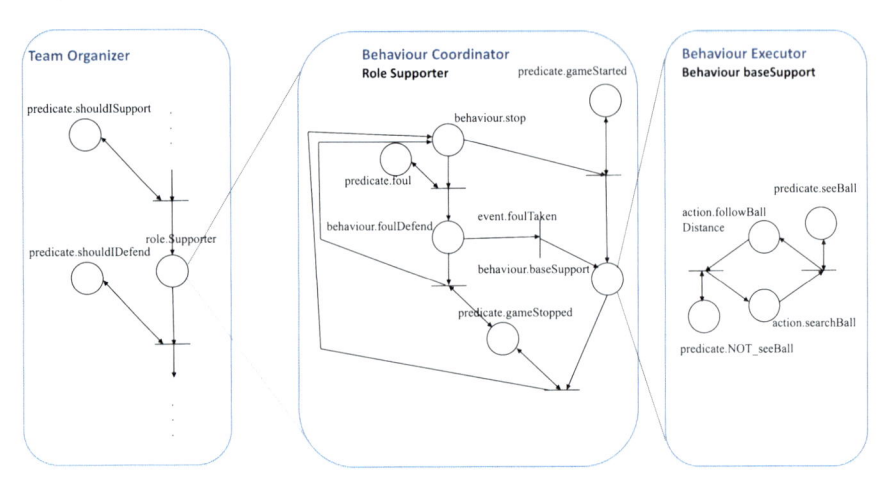

Fig. 15.5 Illustrating the Petri net formal model of the functional architecture of our multi-robot task model with a simplified hierarchical Petri net for a multi-robot soccer team

figure: the `Supporter` either follows the ball at a given distance or searches for it, depending on whether it sees or does not see the ball, respectively. Associated primitive actions and predicates should be clear from this description and form the figure.

Whenever analysis is a concern, the world dynamics must also be captured by a Petri net model. The world Petri net models the token game between places labelled by predicates, i.e. the conditions that change a predicate truth value. The closed loop between the Petri net controller and the world Petri net models the interaction between the multi-robot team and its surrounding environment. Events labelling world model transitions are not controllable by a robot individually, but rather the result of the world dynamics or of other robot's actions. An event labelling a controller transition, however, is controllable by the robot, corresponding to activating the selection of a primitive action labelling an output place of that transition.

Using both predicates and events, the decisions on *role*, *behaviour* and *primitive action* selection can be either event-based and/or predicate-based (logic-based) and/or on a mix of the two, as mentioned before.

The mapping between the multi-robot task model and untimed Petri nets enables *qualitative analysis* of a multi-robot plan by applying Petri net analysis techniques (Girault and Valk 2003) to the closed loop between the controller and the world nets. Such techniques determine Petri net properties that correspond to multi-robot plan properties. Some examples are as follows:

- *Liveness* is the property of a Petri net with initial marking M_0 for which there always exists some sample path such that any transition can eventually fire from any marking reached from M_0. Markings represent *states* of the system modelled by the Petri net. If a Petri net representing a robot plan in closed loop with its world model is live, it guarantees that one can start from an arbitrary state of the closed loop between the robot controller and its world model, and end up in any other arbitrary state reachable from the initial state.

- *k-Boundedness* is the property of a Petri net whose reachable markings from the initial marking never get more than k tokens deposited in any of its places. If a Petri net is *safe* (same as 1-*bounded*), one guarantees that the multi-robot system it models does not violate the rule that predicate places should only have 1 or 0 tokens.

The *generalized stochastic Petri net model* of a multi-robot plan provides a *stochastic timed view* over the corresponding multi-robot task model, by associating stochastic times with transitions and probability mass functions to Petri net *conflicts* (a set of transitions enabled by the current marking such that, if one fires, all the others will be disabled).

A *generalized stochastic Petri net* (GSPN) is a 7-tuple ($P, T = T_0 \cup T_D$, A, W, M_0, F, S) where (P, T, A, W, M_0) is a marked Petri net, and

- $F : R[M_0] \times T_D \to \Re$ is a function that associates with each *timed* transition $t \in T_D$ in each marking $R[M_0]$ reachable from M_0 a random variable. Each $t \in T_0$ has zero firing time for all reachable markings.
- S is a set (possibly empty) of elements called *random switches*, which associate probability distributions with subsets of conflicting immediate transitions.

For *exponential timed* GSPNs, $F : R[M_0] \times T_D \to \Re$ is a function that associates with each transition $t_j \in T_D$ in each reachable marking M an *exponential* random variable with rate $\lambda_j(M)$. The transitions in T_D are known as *exponential transitions* and refer to $\lambda_j(x)$ as the *firing rate* of t_j in M.

When there is *conflict* in marking M_i, if T_i is the set of enabled transitions in M_i, the probability of firing $t_j \in T_i$ is:

1. If T_i is composed by exponential transitions only, $\dfrac{\lambda_j(x_i)}{\sum\limits_{t_k \in T_i} \lambda_k(x_i)}$
2. If T_i includes one single immediate transition, this is the one that will fire
3. If T_i includes two or more immediate transition, a probability mass function will be specified over them by an element of S. The subset of immediate transitions plus the switching distribution is called a *random switch*

The state automaton, which state space is the set of all reachable markings of a Petri net, and which transitions are labelled by the events that cause the Petri net to switch between states (markings), is denoted as the *marking graph*. For GSPN, events are replaced by transition probabilities in the marking graph transition labelling.

One significant theorem for GSPNs states that the marking graph of an exponential timed Petri net is a continuous time Markov Chain (CTMC) (Viswanadham and Narahari 1992). This powerful result enables using Markov Chain theory to study the stochastic performance of multi-robot plans represented by exponential timed GSPN.

Under the stochastic view, places are still labelled by *predicates* and *primitive actions*, and immediate transitions are still labelled by *events*, as before. However, events may now occur with a given probability, if they are part of a *random switch*.

Alternatively, the time between two successive occurrences of the same event may be exponentially distributed. In either case (see items 1 and 3 about the probability of firing conflicting transitions above), one ends up with a probability mass function over the conflicting events, which models the probability of their occurrence in a given marking (state of the multi-robot system). The difference is that, for stochastic timed transitions, one describes the uncertainty using physical information about the process, namely the inter-event occurrence times. This means that the different primitive action sequences resulting from the actual plan execution have associated probabilities, and plan performance can be stochastically quantified. Examples of performance criteria are as follows:

- Plan *probability of success*, i.e. the probability that according to the multi-robot system model (including the GSPN-based world model), the plan ends in a goal state (corresponding to the sum of probabilities of primitive action sequences leading to goal states)
- Plan *robustness*, i.e. whether the plan probability of success remains above a given threshold when the success of some of its composing primitive actions drops by certain amounts

Whenever there are transition conflicts, two possibilities exist (assuming that, as stated in our model, one of the input places of all the conflicting transitions is the same for all of them and corresponds to the currently selected primitive action π_{curr}):

1. If the set of predicates labelling all the conflicting transitions is the same for all those transitions, the conflict corresponds to alternative choices for the selection of the next primitive action, and the events labelling them are *controllable events* and correspond to decisions to start some new primitive action (the one labelling the output place of the fired transition), e.g. t_5 and t_7 in Fig. 15.6 represent the alternative selection of primitive actions Kick2Goal or Dribble2Goal after primitive action CatchBall.
2. Otherwise, the conflict corresponds to different probabilities of outcome for π_{curr}, e.g. t_3 and t_2 in the same figure, for action Move2Ball. In this case, the events labelling the conflicting transitions are *uncontrollable* and occur due to the world dynamics.

Case 2 above models the uncertainty about the primitive action effect on the world. The conflicting transitions are, in this case, often stochastic timed. As an example, in Fig. 15.4, the communication transitions labelled with ε could be turned into stochastic transitions for the purpose of performance evaluation, so as to take into account the time taken to transmit synchronization messages between the two robots, which is stochastic due to packet collisions.

Case 1 corresponds to introducing decisions about the selection of primitive actions in our multi-robot task model. In this case, for exponential timed GSPN, the equivalent CTMC is in fact a *controlled Markov Chain* or a *Markov Decision Process* (MDP). MDPs are very well studied in the literature. Their solution corresponds to minimizing the expected (discounted or undiscounted) payoff an agent (in this case a robot team) will receive over a (finite or infinite) time horizon by appropriate choice of a sequence of actions. MDPs can be solved by dynamic programming

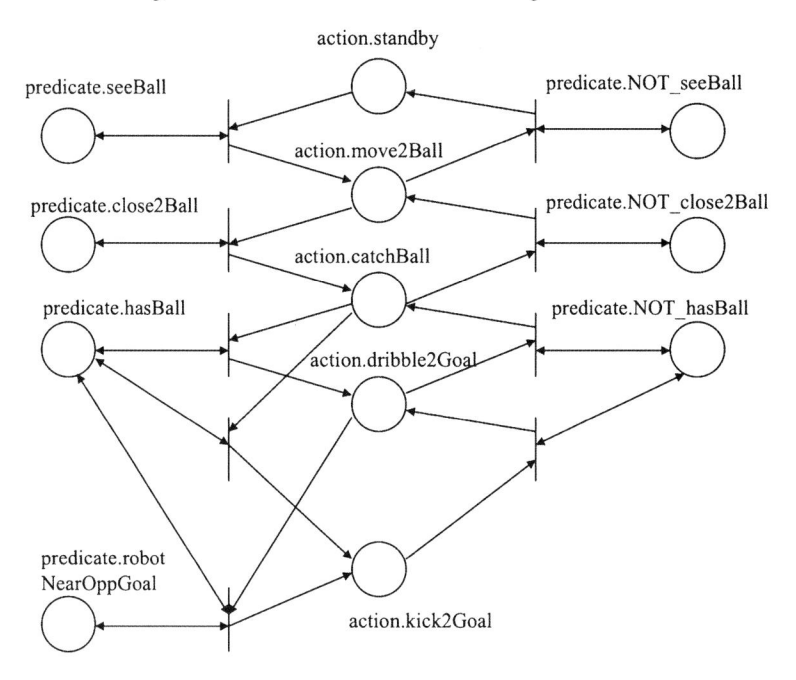

Fig. 15.6 Petri net model of ScoreGoal behaviour in robot soccer. [reprinted from (Costelha and Lima 2007)]

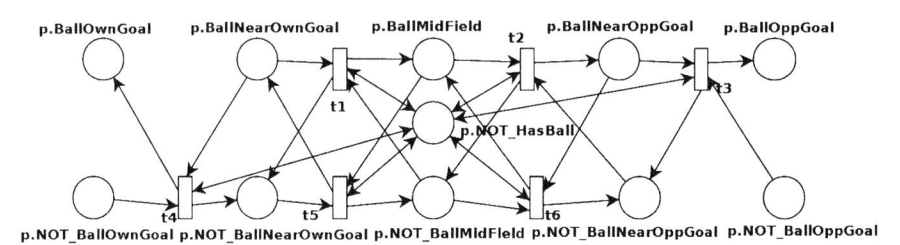

Fig. 15.7 Petri net modelling the ball position in the robot soccer field (one of the modules composing the world model)

methods, in which case the system model must be known or estimated (e.g. using Monte Carlo methods) beforehand, or using model-free reinforcement learning methods (Sutton and Barto 1998), such as the Q-learning algorithm (Watkins and Dayan 1992).

On-going work in this line of research (Costelha and Lima 2007) (Costelha and Lima 2008) has demonstrated the ability of the model to evaluate the performance of a realistic simulation of a single robot system. In Fig. 15.7, the ball position module of the Petri net modelling the world is represented. The full world model Petri net is composed with the robot behaviour Petri net to obtain the closed loop system model. Figure 15.8 depicts the results obtained by the analysis of

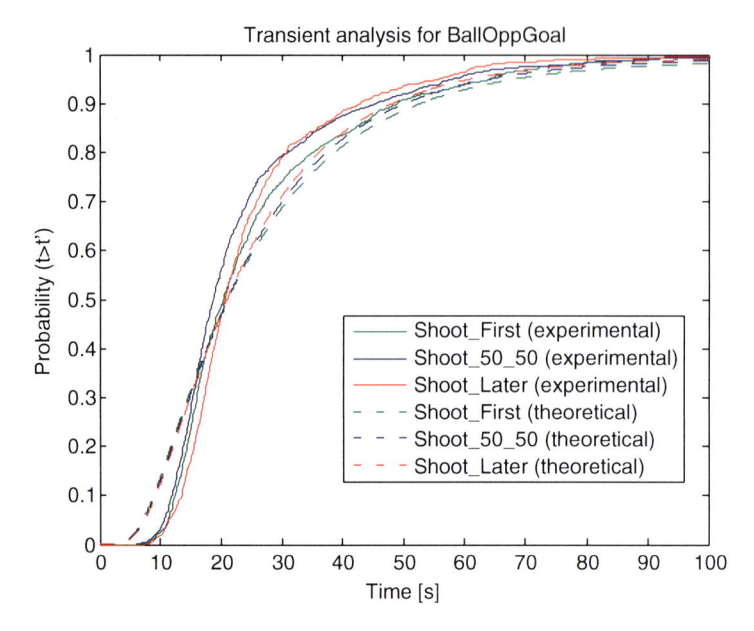

Fig. 15.8 Plots of the probability of robot soccer behaviour `score_goal_shoot_first` scoring a goal before time t' units, as obtained by solving the equivalent CTMC of the closed loop Generalized Stochastic Petri nets (GSPN) model (*dashed lines*) and running a realistic simulation of the behaviour (*solid lines*)

an equivalent Markov Chain model and by a realistic simulation of a behaviour (`score_goal_shoot_first`) in robot soccer, concerning the probability of scoring a goal before t' time units.

The simulation was performed in the Webots realistic simulator of a RoboCup Middle-Size League (MSL) scenario, where a robot moves to the ball, catches it and kicks to the goal as soon as it grabs the ball repeatedly (by assigning probability 1 to transition t_7 in Fig. 15.6, so that t_5 will never fire, meaning the robot switches from action `CatchBall` to action `Kick2Goal` without going through action `Dribble2Goal`), under full state (ball, robot location) observability, but including the possibility of uncertain action effects, like shooting in the wrong direction.

The behaviour GSPN was designed by the task designer, while the parameters, places and transitions of the GSPN world model were estimated from several runs of the simulator. The equivalent CTMC of the closed loop system was analysed by the TimeNet tool (Zimmermann and Freiheit 1998).

The two plots show a very good adjustment between the evolution of the probability of scoring a goal after some time units have elapsed, obtained from several simulator runs and from solving the closed loop equivalent CTMC.

Supported by this confidence on the model, three transient performance tests over the `score_goal_shoot_first` behaviour model shown in Fig. 15.6 were carried out using again with TimeNet. The tests are distinguished by considering different weights assigned to transitions t_5 and t_7 (the single random switch present in this model that corresponds to alternative choices for the selection of a primitive action):

- *Shoot First*: by assigning zero probability of firing to transition t_5 and probability of firing 1 to t_7, t_5 will never fire, and the robot directly switches from action `CatchBall` to action `Kick2Goal` without going through action `Dribble2Goal`, thus kicking to the goal as soon as it grabs the ball
- *Shoot 50–50*: by assigning equal probabilities (0.5) of firing to transitions t_5 and t_7, the robot always chooses one of the primitive actions `Dribble2Goal` and `Kick2Goal` with probability 0.5, as soon as it grabs the ball while running action `CatchBall`
- *Shoot Later*: by assigning probability of firing 1 to transition t_5 and probability of firing 0 to t_7, t_7 never fires, and the robot always runs action `Kick2Goal` after running action `Dribble2Goal` successfully. As such, the robot will only kick the ball when it has possession of the ball and is near the opponent goal

For each simulation run, the robot was positioned near its goal and the ball was positioned in the field centre. Each run consisted of analysing the closed loop PNP, running from the initial marking until a deadlock occurred (goal scored), computing the number of expected tokens in places `BallOwnGoal` and `BallOppGoal` (see Fig. 15.7) over time. This measure corresponds to the probability of having a goal scored in our team goal or in the opponent goal, respectively, yielding the results depicted in Fig. 15.9. The plots confirm that the ball always ends up inside one of the goals, given that the sum of the probabilities of scoring in either goal when the system moves to steady state tends to 1. They also provide evidence of the intuitive result that kicking as soon as the robot grabs the ball leads to a lower scoring probability in the long term, since the robot kicks from any position on the field, leading to more failures. Furthermore, shooting nearby the opponent goal after dribbling is the most reliable solution in this study.

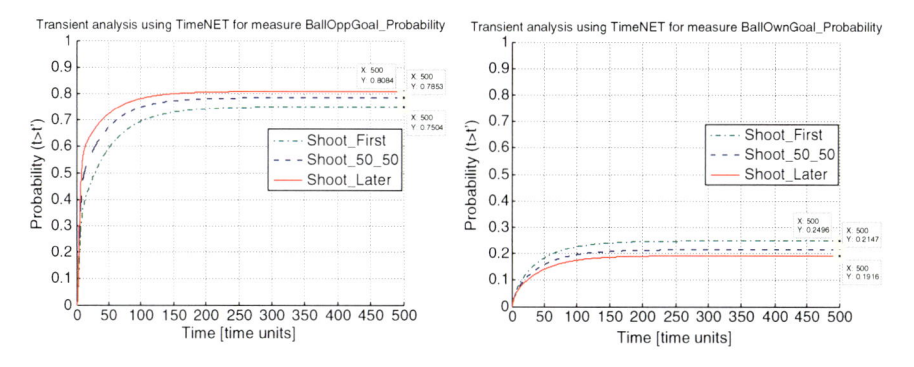

Fig. 15.9 Plots of the probability of scoring a goal in the opponent goal (*left*) or in our own goal (*right*) for the three alternative options in selecting a primitive action within behaviour `score_goal`

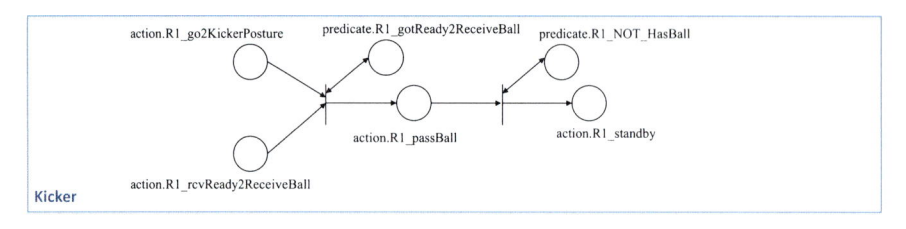

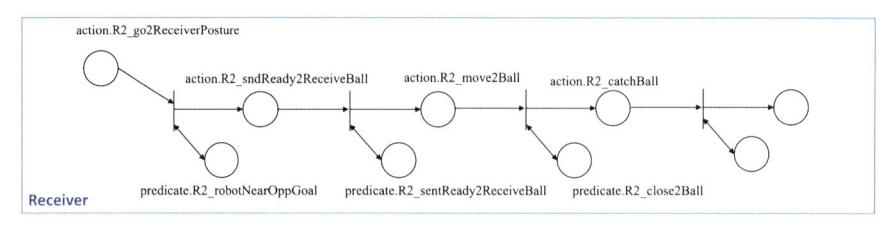

Fig. 15.10 Petri nets for kicker (R1) and receiver (R2) robots involved in the relational pass behaviour

Although not clearly visible in Fig. 15.9, the transient analysis also shows that this ordering of the three alternatives according to their performance is not the same in the initial steps of the simulation, where shooting first is slightly better. Furthermore, the performance may change for different game scenarios and opponent teams. This is where learning or situation-dependent adaptive algorithms can be used to adjust the transition probabilities so as to improve performance over time.

Figure 15.10 shows a simplified Petri net for a relational `pass` behaviour between soccer robots R1 (the kicker) and R2 (the receiver). Places labelled

- p.R2_Sent_Ready2ReceiveBall
- a.R2_SendReady2ReceiveBall
- a.R1_RecvReady2ReceiveBall
- p.R1_Got_Ready2ReceiveBall

are part of a more complex model of *communication actions*, not detailed here. Communication action models are introduced to take into account the required communications among robots from a team, for cooperation purposes.

Pass performance concerning the probability of success of the relational behaviour was determined by computing the ratio of tokens that reach place `R2_StandBy` with respect to the number of times the behaviour is executed. Note that robot R2 only reaches action `StandBy` if it was able to successfully receive the ball.

The setup used in the simulations of the relational behaviour consisted on placing both robots in the midfield area, with robot R1 holding the ball. Results are shown in Table 15.1 and were obtained by considering some of the world model transitions as stochastic (e.g. losing the ball or getting close to it) and other deterministic (such as those from the ball position model in Fig. 15.7), and by changing the success rates of the behaviour actions, as well as the success and failure rates of the communication

Table 15.1 Pass relational behaviour success probability as a function of action success and communication success and failure rates (expressed in number of occurrences per time unit)

Experiment	Action success	Comm. success	Comm. failure	Behaviour success probability
1	1	1	1	0.32
2	1	1	10	0.06
3	1	10	1	0.62
4	1	10	10	0.34
5	1	100	0.1	0.69
6	1	10,000	0.0001	0.69
7	10	10,000	0.0001	0.96

actions (a failure meaning that the message sent is not received). Increasing the communication success increases the behaviour success probability as expected, but only up to a certain value, as experiments 5 and 6 show. The behaviour success probability can only increase further (up to almost probability 1) by increasing the non-communication action transitions success rate, as in experiment 7.

Videos of real and simulated soccer robots running individual and cooperative behaviours designed as Petri nets are available at `http://socrob.isr.ist.utl.pt/videos/behaviors/behaviors.html`.

15.4   Successes, Limitations and Future Challenges

Multi-robot plan representation by Petri nets introduces a systematic way of modelling multi-robot tasks and, consequently, of analysing them. Furthermore, the analysis can be made either at a qualitative or at a quantitative level, by changing the view of the model.

Qualitative analysis is obtained by applying analysis techniques to the untimed Petri net view of the robot task model, and it enables formal verification of some of the logical properties of the plan used to accomplish the task, such as state reachability, return to the initial state, absence of deadlocks and/or livelocks, or bounded use of resources.

The timed stochastic view enables the quantitative analysis of performance, using GSPN models. Interesting outcomes arise from representing multi-robot plans and modelling the multi-robot system by GSPN:

- If timed transitions are associated with exponential random variables, an equivalent CTMC model is obtained, to which powerful Markov Chain results available in the literature can be applied for analysis purposes
- If, furthermore, events labelling transitions of the GSPN include controllable events, representing the decision about selecting some primitive action for

execution, an equivalent MDP is obtained, to which dynamic programming and reinforcement learning techniques can be applied to obtain optimal solutions with respect to maximizing infinite or finite horizon payoffs

- The advantage of starting with a GSPN rather than a straight MDP model is that the model can be built naturally from simpler modular Petri nets representing primitive action models and connecting them to form behaviours and to model the world. Moreover, complexity is decreased because the model naturally constrains the different possible alternatives (e.g. in the traditional rat-in-a-maze toy problem, for every state all four actions up–down–left–right are possible in typical MDP models, not taking into account possible features of the world, such as walls and other obstacles)
- Even if timed transitions are associated with other distributions rather than the exponential, the model can be analysed by Monte Carlo methods, though not in closed form anymore

There exist several assumptions underlying the proposed model. Perhaps the most relevant is that complexity, both concerning analysis and model dimension, can be handled. While it is obvious that the dimension of a Petri net model grows quickly, even for a relatively simple scenario, the whole point is that the advantage of formal methods comes into play exactly when plans become too large to be analysed intuitively. "Design-by-intuition" is perhaps the major drawback of current ad hoc approaches to multi-robot practical planning (beyond the vast body of important work regarding exclusively logic-based plan generation and verification). This is especially true for quantitative performance evaluation. Moreover, plan representation by Petri nets reduces the complexity comparing to starting from plain MDP models, because such models ignore details of the actual task structure. Stochastic performance analysis methods can be applied in reasonable time using the right tools. Markov Chain closed form analysis is especially relevant, because it does not require several simulation runs of the model.

Another strong assumption is that the world can be reasonably modelled by a Petri net. Although we have recently developed methods to identify the structure and parameters of world GSPN models from the realistic simulation of the multi-robot system, world model limitations are clear and must be improved, as the realism of the analysis results depends on the model realism. This is actually one of our current and future challenges: improving the quality of automatic identification of Petri net world models, and increase the speed of the process, so as to make it applicable to real robots as well.

Another future challenge is to obtain an equivalent partially observable MDP (POMDP) model from a Petri net model. This will require using Probabilistic Petri nets (Albanese et al. 2008) where fractions (instead of an integer number) of tokens flow across transitions, such that a probability density function over the reachable markings is defined, representing the *belief* of the multi-robot system about its state, resulting from noisy observations. This means also abandoning the current binary-logic predicate model, as one will no longer have a crisp measure of the predicate truth value.

Our goal is to provide, in the near future, analysis tools for multi-robot plans composed of many primitive actions, predicates and events that can automatically determine relevant logical properties of the task and evaluate its stochastic performance. In the long term, one further step beyond will be to achieve automatic plan generation from qualitative and quantitative specifications for a task, given the available set of primitive action, predicates and events.

Acknowledgements   The simulations whose results are presented in Sect. 15.3 were carried out by the PhD student Mr. Hugo Costelha. While some parts of his PhD thesis work have been published before and are cited throughout the chapter, most of the results in that section were still unpublished at the time of writing this text.

References

Akharware N (2000) Pipe2: Platform Independent Petri Net Editor, MSc Thesis, Imperial College of Science, Technology and Medicine, University of London, London, UK

Albanese M, Chellappa R, Moscato V, Picariello A, Subrahmanian VS, Turaga P, Udrea O (2008) A Constrained Probabilistic Petri Net Framework for Human Activity Detection in Video, IEEE Transactions On Multimedia, 10(6)

Cassandras C, Lafortune S (2007) Introduction to Discrete Event Systems, Springer

Cohen PR, Levesque HJ (1991) Teamwork, Special Issue on Cognitive Science and Artificial Intelligence, 25(4):486–512

Costelha H, Lima PU (2007) Modelling, Analysis and Execution of Robotic Tasks using Petri Nets, Proceedings of IEEE International Conference on Intelligent Robots and Systems, San Diego, CA, USA

Costelha H, Lima PU (2008) Modelling, Analysis and Execution of Multi-Robot Tasks using Petri Nets, Proceedings of 7th International Joint Conference on Autonomous Agents and Multi-Agent Systems, Estoril, Portugal

Giordano V, Ballal P, Lewis F, Turchiano B, Zhang JB (2006) Supervisory Control of Mobile Sensor Networks: Math Formulation, Simulation, and Implementation, IEEE Transactions on Systems, Man and Cybernetics — Part B:Cybernetics, 36(4)

Girault C, Valk R (2003) Petri Nets for Systems Engineering: A Guide to Modeling, Verification, and Applications, Springer

Kim G, Chung W, Park S-K, Kim M (2005) Experimental Research of Navigation Behaviour Selection Using Generalized Stochastic Petri Nets (GSPN) for a Tour-Guide Robot, Proceedings of IEEE International Conference on Intelligent Robots and Systems, Edmonton, Alberta, Canada

King J, Pretty RK, Gosine RG (2003) Coordinated Execution of Tasks in a Multiagent Environment, IEEE Transactions on Systems, Man and Cybernetics — Part A: Systems and Humans, 33(5)

Kotb YT, Beauchemin SS, Barron JL (2007) Petri Net-Based Cooperation in Multi-Agent Systems, 4th Canadian Conference on Computer and Robot Vision

Lima PU, Grácio H, Veiga V, Karlsson A (1998) Petri Nets for Modelling and Coordination of Robotic Tasks, Proceedings of 1998 IEEE International Conference on Systems, Man and Cybernetics, San Diego, USA

Milutinovic D, Lima PU (2002) Petri Net Models of Robotic Tasks, Proceedings of IEEE International Conference on Robotics and Automation, Washington DC, USA

Montano L, García FJ, Villarroel JL (2000) Using the Time Petri Net Formalism for Specification, Validation, and Code Generation in Robot-Control Applications, International Journal of Robotics Research, 19(1):59–76

Saridis GN (1979) Toward Realization of Intelligent Control, Proceedings IEEE, 27

Sutton R, Barto A (1998) Reinforcement Learning, The MIT

Viswanadham N, Narahari Y (1992) Performance Modeling of Automated Manufacturing Systems, Prentice Hall

Wang F-Y, Kyriakopoulos K, Tsolkas A, Saridis GN (1993) A Petri-Net Coordination Model for an Intelligent Mobile Robot, IEEE Transactions on Robotics and Automation, 9(3):257–271

Watkins CJCH, Dayan P (1992) Q-learning, Machine Learning, 8, 279–292

Zimmermann A, Freiheit J (1998) TimeNETMS-an Integrated Modeling and Performance Evaluation Tool for Manufacturing Systems, Proceedings of IEEE International Conference on Systems, Man and Cybernetics, San Diego, CA, USA

Ziparo V, Iocchi L (2006) Petri Net Plans, Proceedings of the 4th International Workshop on Modelling of Objects, Components, and Agents (MOCA06), Turku, Finland

Ziparo V, Ziparo A, Iocchi L, Nardi D, Palamara PF, Costelha H (2008) Petri Net Plans, A Formal Model for Representation and Execution of Multi-Robot Plans. Proc. of 7th Int. Conf. on Autonomous Agents and Multiagent Systems (AAMAS 2008), Padgham, Parkes, Müller and Parsons (eds.), May 12–16., 2008, Estoril, Portugal

Chapter 16
Developmental Learning of Cooperative Robot Skills: A Hierarchical Multi-Agent Architecture

John N. Karigiannis, Theodoros Rekatsinas, and Costas S. Tzafestas

Abstract Research activities targeting new methodologies, architectures and in general frameworks that will improve the design of intelligent robots attract significant attention from the research community. Self-organization problems, intrinsic behaviors as well as effective learning, and skill transfer processes in the context of robotic systems have been significantly investigated by researchers. This chapter presents a new framework of developmental skill learning process by introducing a hierarchical multi-agent architecture. More specifically, the methodology proposed is based on using reinforcement learning (RL) techniques in a fuzzified state-space, leading to a collaborative control scheme among the agents engaged in a continuous space, which enables the multi-agent system to learn, over a period of time, how to perform sequences of continuous actions in a cooperative manner without any prior task model. By organizing the agents in a nested architecture, as proposed in this work, a type of problem-specific recursive knowledge acquisition process is obtained. The agents may correspond in fact to independent degrees of freedom of the system and manage to gain experience over the task that they collaboratively perform by continuously exploring and exploiting their state-to-action mapping space. Two numerical experiments are presented, one related to dexterous manipulation and one simulated experiment concerning cooperative mobile robots. Two distinct problem settings are considered. The first one concerns the case of redundant and dextrous robot manipulation tasks, in the framework of which the problem of autonomously developing control skills is considered. Initially, a simulated redundant, four degrees-of-freedom (DoF) planar kinematic chain is considered, trying to develop the skill of accurately reaching a specified target position. In the same problem setting, a simulated three-finger manipulation example is subsequently presented, where each finger is comprised of 4 DoF performing a quasi-static grasp. For the second problem setting, the same theoretical framework is adapted in the case of two mobile robots performing a collaborative box-pushing task. This task involves

C.S. Tzafestas (✉)
National Technical University of Athens, School of Electrical and Computer Engineering,
Division of Signals, Control and Robotics, Zographou Campus, Athens 15773, Greece
e-mail: ktzaf@softlab.ntua.gr

V. Cutsuridis et al. (eds.), *Perception-Action Cycle: Models, Architectures,*
and Hardware, Springer Series in Cognitive and Neural Systems 1,
DOI 10.1007/978-1-4419-1452-1_16, © Springer Science+Business Media, LLC 2011

two moving robots actively cooperating to jointly push an object on a plane to a specified goal location. In this case, the actuated wheels of the mobile robots are considered as the independent agents that have to build up cooperative skills over time, for the robot to demonstrate intelligent behavior. Our goal in this experimental study is to evaluate both the proposed hierarchical multi-agent architecture and the methodological control framework. Such a hierarchical multi-agent approach is envisioned to be highly scalable for the control of robotic systems that are kinematically more complex, comprising multiple DoF and redundancies in open or closed kinematic chains, particularly dexterous robot manipulators and complex biologically inspired robot locomotion systems.

16.1 Introduction

Developmental Robotics is a scientific field situated in the intersection between robotics and developmental sciences (i.e., cognitive phycology, neuroscience). The goal of developmental robotics can been defined as: using robots to instantiate and investigate models originating from developmental science, but can also be seen as an attempt that seeks to design better robotic systems by applying insights gained from studies on ontogenetic development. Furthermore, developmental robotics motivates the usage of robots as a novel research tool to study and model the development of cognition and action. Ontogenetic development has many facets. For instance, it can been defined as a self-organizing, incremental process, but it can also be seen as comprising self-exploratory activities, and in many occasions cooperative activities. Thus, to understand better all these different facets of developmental learning, several research groups have been addressing their work to cognitive multi-agent robotic system. A complete survey can be found in (Lungarella et al. 2003).

Understanding human cooperative behavior has been a major concern in multi-agent robotic systems and has been addressed by work done on mobile robots (Cao et al. 1995), robotic hands, and multiple manipulators (Khatib et al. 1996; Nakamura 1990; Yoshikawa and Zheng 1993). In (Donald et al. 1997), manipulation protocols have been developed for a team of mobile robots that collaborate to push large boxes. In (Rus 1997), an algorithmic structure coordinates the reorientation of objects in a plane by independent robot-agents. In (Ahmadabadi and Nakano 2001), a study is presented where distributed cooperation strategies are required by a group of behavior-based mobile robots for handling an object. The common approach in all these works relies on the assumption that the motion of the object under pushing/manipulation is quasi-static, and that all the agents involved have predefined behavior models that they combine using certain architecture (like subsumption architecture Brooks 1986).

Cooperative behavior is one of the aspects that have been studied within a multi-agent manipulation framework. Human behavior also demonstrates evolutionary characteristics and self-organizing abilities. These unique attributes of human behavior have been extensively studied in the process of designing intelligent robots that need to operate/collaborate autonomously and adapt to their environment. In this context, the application and use of bio-inspired techniques, such as reinforcement learning (RL), evolutionary computation, and fuzzy systems, constitute an emergent research topic. More specifically, RL (Sutton and Barto 1998; Bertsekas and Tsitsiklis 1996; Dayan and Abbott 2001) is an active area of machine learning research that is also receiving attention from the fields of decision theory and control engineering. Various RL methods (Kok and Vlassis 2004; Zamora et al. 1997; Takahashi et al. 2001) have been used on multi-agent architectures that target the control of mobile robots operating within a fully or partially observable environment. Moreover, in (Doya 1996; Kondo and Ito 2004; Iida et al. 2004; Shibata et al. 2001) we have seen cases where single agent architectures use RL methods in a continuous three-dimensional space, implemented by neural networks. In (Liu et al. 2007), a three-layered architecture is introduced (namely, motion patterns, behavior models, planning component), which uses RL to control a robotic fish. In general, RL constitutes an approach used extensively for building a policy based on data acquired through exploration.

A different approach used to acquire robot manipulation skills is through Learning from Demonstration (LfD) (Argall et al. 2008), also referred to as Learning by Imitation (Schaal 1999). By LfD, instead of learning by exploration, a policy is learned from examples, or demonstrations, provided by a teacher. These examples are defined as sequences of state-action pairs that are recorded during the teacher's demonstration of the desired robot behavior. We note that a policy derived under LfD is necessarily defined only in those states encountered, and for those actions taken, during the example executions. A hierarchy of developmental stages and of respective skills that have to be acquired at each stage is presented in (Lopes and Santos-Victor 2007). The initial steps of skill acquisition in this hierarchy allow the robot to establish sensory motor coordination. The next step in the hierarchy allows building skills by interacting with the environment, while the last step introduces skill acquisition by imitation of teacher's demonstrating examples.

Within this general research framework, the work that is presented in this chapter addresses the problem of evolutionary learning on multi-agent architectures and skill acquisition, not from demonstration but through agents' exploration, without having build-in behavior models. The long-term objective of this research work is to contribute to evolutionary behaviors established within multi-agent systems. The short-term goal is to evaluate RL-based developmental mechanisms along with appropriate control architectures used in two distinct domains: (a) the domain of dexterous robot manipulation (a domain in which, to the best of our knowledge, multi-agent learning architectures have not been extensively studied yet), and (b) the domain of collaborative autonomous mobile robots.

In particular, in this work we propose a methodology that introduces a hierarchical multi-agent architecture, together with a corresponding skill-acquisition algorithm, where nested agents learn to explore their space and reach their common goal by going through a set of collaborative action-selection steps. Thus, an attempt is made to incorporate evolutionary processes, and more specifically RL methods, within a nested multi-agent architecture, and to evaluate the overall behavior of the multi-agent system both in the domain of dexterous manipulation control and in the domain of collaborative autonomous mobile robots. Initially, an attempt is made to perform a simulated quasi-static three-finger grasp using the proposed multi-agent framework. The joints (or links) of every kinematic chain, simulating a finger of the manipulator, are considered as distinct agents that use RL methods to explore their task-space and establish certain skills. Within the multi-agent environment that is formulated in the proposed system, every agent in the group is selecting an action independently of the rest by observing and performing an estimate of what the rest will do. The resulting cumulative action of the system is a joint effort (joint actions) of all the nested entities comprising the system. Although autonomous, the agents are closely coupled with each other due to the physical connectivity, making the precise cooperation and coordination among them extremely important to achieve stability of such a system. In the second part of the work presented, we analyze the application of the same theoretical framework to the case of collaborative mobile robots. The task here involves the wheels of the robots and the cooperation skills that have to evolve over time to build a collaborative behavior, which will enable to jointly drive the robots in a way that will result in pushing an object to a specified target location. A key issue to point out in this case is the absence of task model for the described activity, as well as the fact that the overall robot behavior is an outcome of the actions selected by the individual agents-wheels that operate autonomously.

The chapter is organized as follows. The following section describes in general terms the basic aspects and assumptions of the proposed multi-agent control framework. Section 16.3 then describes the multi-agent architecture, focusing on the case of dexterous manipulation learning tasks and the implementation of the proposed approach, both from the side of state space continuity and from the perspective of action selection. Section 16.4 focuses on the adaptation of this framework in the case of collaborative mobile robots, covering all aspects related to agent mapping and learning. Section 16.5 then presents the overall control architecture and Sect. 16.6 presents the experimental setups and the results obtained in the three experimental cases considered: (a) a simulated single kinematic chain, (b) a 3-finger manipulator with 4 DoF per finger, and (c) the case of two cooperating mobile robots. Finally, the chapter concludes, in Sect. 16.7, with a general discussion over the results and the related future research plans.

16.2 Hierarchical Multi-Agent Control Framework

The proposed control framework fits in the context of a continuous research effort aiming to explore architectures that would enable a complex robotic system to autonomously develop and progressively acquire control skills in a modular, scalable and robust manner, without the need for tedious task modelling and restrictive pre-programming. The methodology presented in this chapter can be also seen as an attempt to bridge the gap between high-level and low-level control, by means of a hybrid architecture that integrates both artificial intelligence (AI) learning techniques and classic control methods. In the proposed hierarchical architecture, the higher layer consists of a nested team of agents, formulating a system that incorporates RL components aiming to enable the agents to establish certain policies over time; the lower end consists of a classic local feedback controller, responsible to drive the corresponding actuators. The basic features and requirements of the proposed control framework are described in the sequel.

16.2.1 Mapping Agents to Degrees of Freedom

Each robot joint (actuated DoF) is to be assigned an agent having as a function to govern local control at that joint level. The challenge here is to build global dexterity through progressive acquisition of local skills at each local agent level. It should be noted here, however, that although every joint is to be assigned an agent, the reverse definition is different. Every agent could represent more than a single degree of freedom. So we could have an agent that is actually comprised of two or more degrees of freedom.

16.2.2 Hierarchical Architecture

Each agent functions locally by observing the rest of the agents and by making an estimate of the actions that these agents could potentially perform in a future horizon. However, a measure of global task performance is supposed to exist, provided by a higher-level agent and distributed to lower agents in a nested architecture. This hierarchical process guides, in that manner, the RL procedures through the computation of a reward function. This reward function must be computed on a continuous scale (instead of waiting for a discrete event of type success or failure to occur), leading to a continuous adaptive dynamic behavior for the system.

16.2.3 Continuous Problem Setting

The learning process must also be designed to function in a continuous state-space, for the system to establish efficient manipulation skills. In the methodological approach presented in this chapter, a fuzzification step is applied to the robot sensor readings, forming the system state. Learning is then accomplished in a discrete state-action mapping sense; a defuzzification step can then be used to perform action selection in a continuous domain. Let us consider a kinematic chain that is comprised of n dofs (agents $i = 1, \ldots, n$), nested in the manner presented in Fig. 16.1.

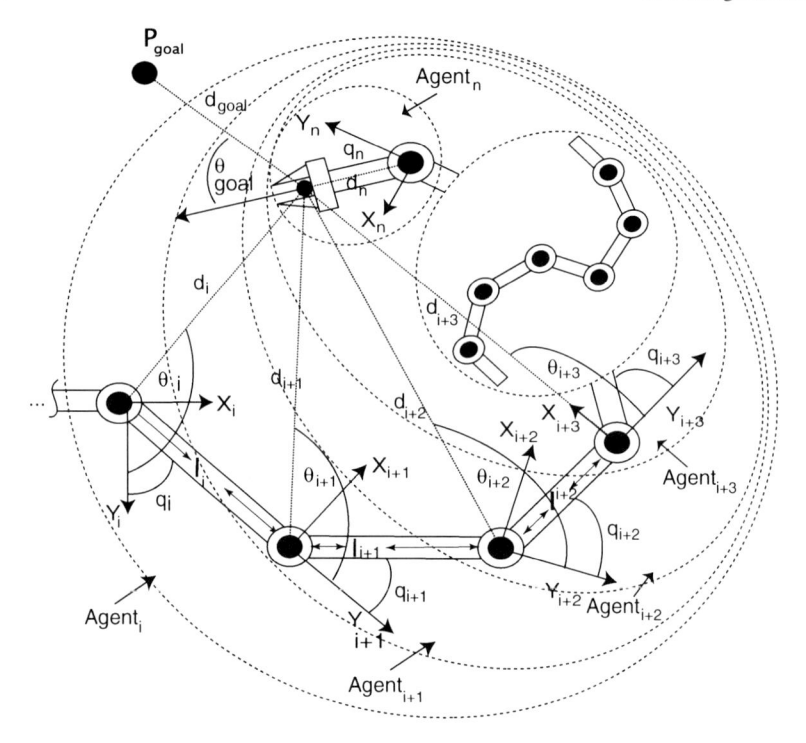

Fig. 16.1 $n-dof$ robot manipulator (open kinematic chain), possessing kinematic redundancies with respect to the considered manipulation task (in this case, positioning the end-effector at a given goal position)

We define the state of agent a_i as: $S_i = \langle q_i, \theta_i, d_i, \vec{g}_i \rangle$, where q_i is the current relative position of the ith joint (i.e., the angular displacement of the ith link with respect to the previous link in the kinematic chain), θ_i is the current angular position of the robot's end-effector with respect to the ith agent, d_i is the current Euclidian distance of the robot's end-effector from the ith agent, $\vec{g}_i$ is the current vector that describes the position of the goal at the task space with respect to the end-effector. We will be referring to agent a_i as root agent, and all the other agents are considered to be located below in the hierarchy. More details about the multi-agent architecture will be discussed in the following section.

The flow of operations is depicted in Fig. 16.2, which demonstrates the basic algorithmic structure at an agent-level, as proposed in this framework. In order to achieve continuity over the state space and the action space, both spaces will be fuzzified. This process will be described in a subsequent section. Every *agent$_i$* obtains information and identifies certain variables defining its state, and then forwards that information in a nested manner to the agent(s) of the next layer in the hierarchy, to facilitate them in their own process of identifying their parameters. This whole process is composed of two subprocesses that evolve in a recursive manner. The first one is evolving in a top-down manner, while the second one is evolving in a bottom-up approach. The top-down subprocess starts from the root agent of the hierarchy

```
Fuzzification of the State Space
Fuzzification of the Action Space

State Evaluation
      Agent = Agent(i), Tries to fully evaluate its state.
      If success
            Traverse back the Hierarchy of agents and provide them
            with all the available information in order for them to
            successfully define their state

            Go to Action Selection
      Else
            Pass Information gathered to Agent(i+1) and  i = i+1

            Go to State Evaluation

      Loop until all agents have evaluated their states.

Action Selection
      Agent = Agent(i),
      If NO experience exists in Agent
            Stochastically select Action = a(i)
            Stochastically estimate all other Agents actions
      Else if experience exist
            If Agent wants to explore
                  Select Action = not necessary the best action
                  Stochastically estimate all other Agents actions
            If Agent do not want to explore
                  Select Action = action that will generate the greater reward
                  Stochastically estimate all other Agents actions
            End
      End
      Agent = Agent(i+1)
      Loop until all agents have selected Action

Joint Action Execution
Reward Assigned to all Agents
Go Back to State Evaluation
```

Fig. 16.2 Basic algorithmic structure of the proposed multi-agent control framework

and travels to the lower ones. During this initial phase, the agent identifies its joint variables and physical configuration (i.e., $agent_i \langle q_i, l_i \rangle$, where q_i is the agent's joint angular displacement and l_i its length). It can be seen that q_i defines partially the state of $agent_i$. In order to fully define its state, $agent_i$ requires additional information about variables $\langle \theta_i, d_i, \vec{q}_i \rangle$. These variables for $agent_i$ cannot be computed at this phase since their computation requires additional information provided by the other agents that comprise the agent community; and this information is not available at the moment. So, since $agent_i$ cannot fully solve the problem of determining its state at the moment, it forwards the partially computed solution to the following $agent_{i+1}$ residing at the next layer in the hierarchy. Similarly, $agent_{i+1}$ calculates the variables that can be computed and forwards the partial solution to the agents below in the hierarchy. The subprocess iterates until an $agent_n$ is reached that succeeds in computing all those variables that allow the agent to define uniquely its state.

In the next phase, the recursive process continuous by traversing back from $agent_{i+1} \rightarrow agent_i$, providing the agents at the higher layer in the hierarchy with the information that they were missing. One agent after the other, starting from agent a_n (i.e., the end-effector) and moving up the hierarchical structure to the root agent a_i, gets enabled to compute the missing variables and define its state. When this phase concludes, every agent in the system has fully solved its state definition problem, simultaneously resulting in a multi-agent system with a fully defined state. This iterative/recursive process is repeated to define the fuzzified state of every agent in the multi-agent environment. We note here that throughout this top-down $\rightarrow$ bottom-up process, the system autonomously self-defines its state. It allows the agents that comprise it to communicate in a structured hierarchical way their partial knowledge of their state variables, achieving two goals at the same time. First, every agent in the system fully solves its own individual state definition problem, and second, through this iterative process a solution evolves for the whole multi-agent system's state definition problem.

Having completed the process of state evaluation, the system proceeds to the next phase, which comprises the joint action selection. Every agent in the system acts initially without having any prior knowledge (i.e., there is no previously defined state – joint action mapping), thus acting in a sense stochastically. The $agent_i$ decides to perform a random action a_i and at the same time computes an estimate of what the other agents in the system might choose as their potential actions. That is, each agent, independently of the rest, selects an action, while at the same time estimating (predicting) how the other agents are likely to act; each agent thus learns joint actions. This process is again recursive downward from the root agent $agent_i \rightarrow agent_{i+1}$ (where $agent_{i+1}$ is an agent at the lower level than $agent_i$). So, $agent_i$ selects action a_i and estimates that the rest of the agents will select $a'_{i+1}, a'_{i+2}, \ldots a'_n$, respectively. Subsequently, in the same manner, $agent_{i+1}$ selects an action a_{i+1} while performing an estimate of what the other agents might select (i.e., $a''_{i+2} \ldots a''_n$). After completion of this recursive process, a specific joint action is then formulated, which the multi-agent system then executes. The system subsequently provides reward or punishment to the agents for the joint effort that they have demonstrated.

16.3   Agent Architecture: The Case of Robot Kinematic Chains

The multi-agent control architecture proposed in this work possesses a nested, hierarchical structure, where uniformity and modularity, regarding the form and the representation of all agents, are key design principles, thus creating those conditions necessary to expand and further scale up the system in different application areas. Figure 16.3 presents a schematic representation of the nested form of the proposed hierarchical multi-agent architecture. More specifically, every agent "sees" only those agents that are one level below (in the case of Fig. 16.3, this means $agent_{i+1}$ only).

In the case of $agent_{i+2}$, the observed agents are only $agent_{i+3}$ and $agent_{i+4}$. This approach provides to the multi-agent system the ability of a recursive

Fig. 16.3 Multiagent
hierarchical framework
for dexterous manipulation

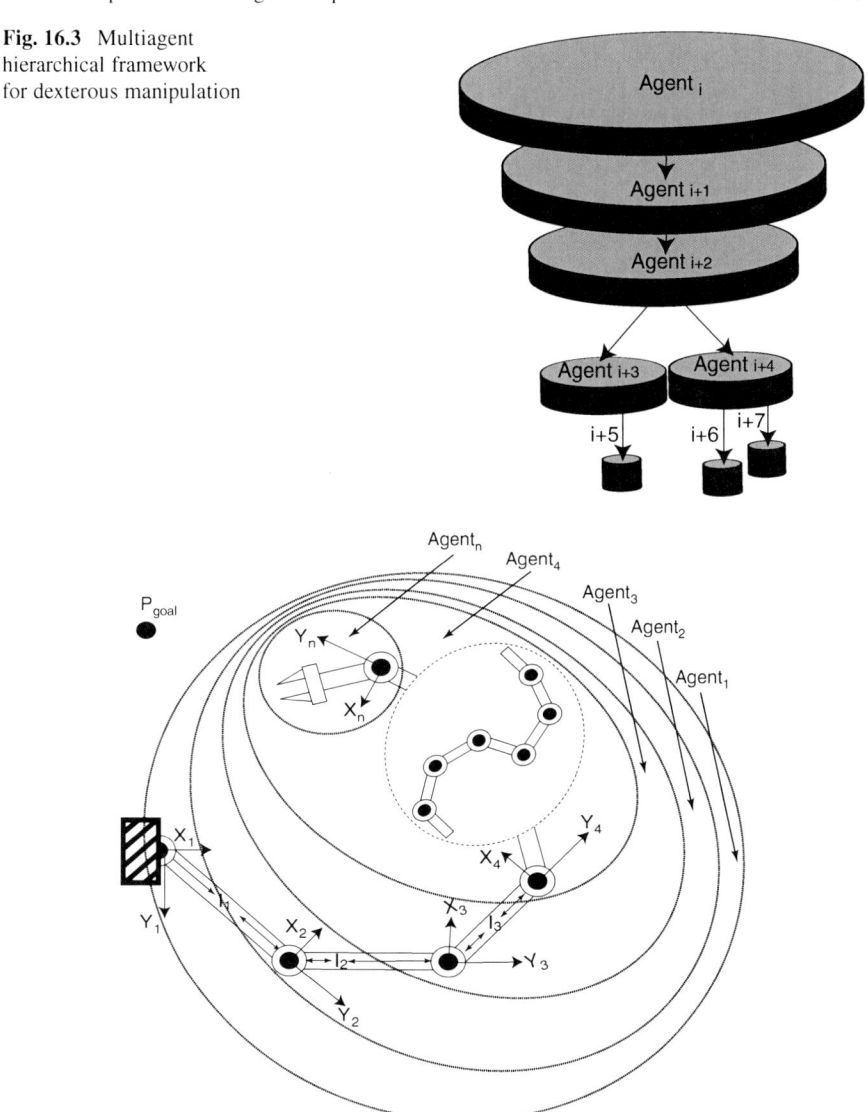

Fig. 16.4   Agents assigned to degrees of freedom (joints) of an open kinematic chain

solution-search approach downward from $agent_i \rightarrow agent_{i+1}$. If an agent cannot
solve a problem (or cannot contribute to its solution), then it passes the knowledge gathered to the agent(s) below. When an agent is reached where a possible
contribution to the problem is feasible, then the agent acts (contributes) and tra-
verses backward the chain of agents, distributing (propagating) back the knowledge
obtained. Figure 16.4 illustrates an example of how agents can be assigned to
the individual degrees of freedom (joints) of an n-dof open kinematic chain, thus

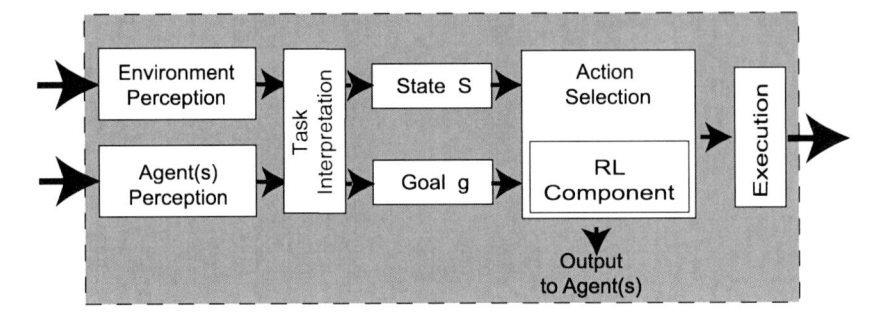

Fig. 16.5 Basic building blocks and interconnections of an agent

forming a nested (hierarchical) multi-agent system. Figure 16.5 then depicts the interconnections and communication signals exchanged between the different building blocks comprising an agent. These basic building blocks of an agent, with their input/output and requirements properties, are described in the sequel.

16.3.1 Basic Internal Functions of an Agent

The basic functional blocks within an agent are the following:

- **Environment Perception Block**

  - *Input*: Agents sensors input.
  - *Output*: The environment state interpreted from the specific sensors of the agent.
  - *Requirement(s)*: The signals coming from the specific sensors are mapped/translated to a certain state. That state gives specific information regarding the status of the agent, based on the input that is coming not only from the sensors of the agents (environment perception block) but also potentially from the agents perception block which is gathering information from the agent(s) in higher location(s) in the nested architecture.

- **Agent(s) Perception Block**

  - *Input*: The state information of the parent agent(s) that the specific agent has.
  - *Output*: The new global state that is formulated.

- **Task Interpretation Block**

  - *Input*: Environment & Parent Agent(s) Perception States
  - *Output*: Current State and Goal.
  - *Requirement(s)*: The architecture that is adopted for the control of this multi-agent system is behavior-based. There is no central task model within the agent, something that would result in a situation where moving from one task to another would be very difficult. There exists a local task model for every

agent that is created incrementally based on the feedback received after every iteration. Our architecture is using RL to optimize this dynamically created local task model.

- **Action Selection Block**

  - *Input*: Current State and Goal.
  - *Output*: Action to be Executed (triplet with the state, the action selected, and the goal to be achieved).
  - *Requirement(s)*: The architecture adopted for the control of this multi-agent system is behavior-based, which means that there is no global action-planning module. What exists is a local action plan encoded in the state-action mapping space. This local action plan of every agent is optimized after each iteration, using RL after each action is executed.

- **Execution Block**

  - *Input*: Action Selected.
  - *Output*: Control-Level Output.

16.3.2 *Continuous Reinforcement Learning: Kinematic Chain*

In the following subsections, the learning mechanism that has been used for the case of kinematic chain is analyzed. The corresponding learning approach used for the case of mobile robots is described in subsequent section.

16.3.2.1 Q: Learning Method

Learning in general is experience dependent and is often characterized by a relative permanent change of behavior resulting from exercise and practise. In this section, we describe the learning process that has been adopted in our system, and how through this process the multi-agent system builds up knowledge. RL methods have been applied in significant number of cases (Takahashi et al. 2001; Kondo and Ito 2004; Shibata and Okabe 1994), mostly on mobile robots. In our case, RL method is used in a quite different domain, namely skill learning and behavior-based multi-agent control of robotic (dexterous) manipulation. Back in 1992 (Matsui et al. 1992), a multi-agent architecture for controlling a multi-fingered hand was presented, but without incorporating any RL methods building skill learning on an agent base. In (Shibata and Ito 2000, 2003), some cases are presented where RL methods are used in dexterous manipulations, but on a single-agent system architecture. The research work presented in this chapter constitutes an attempt to develop a hierarchical, nested multi-agent architecture that aims to enable a robot control system, through RL methods, to acquire by itself skills and knowledge on how to perform agile manipulation.

In a more formal way, let us assume a collection of n (homogeneous) agents, each agent $i \in n$ having a finite set of individual actions A_i available to it. Agents operate repeatedly within the framework of the environment posed, where each individual agent independently selects an individual action to perform. In (Kaelbling et al. 1996), RL is defined as the problem faced by an agent that must learn a behavior through trial-and-error interactions with a dynamic environment. In terms of mathematical description, RL has been formalized as a Markov Decision Process (MDP). An MDP has four components: states, actions, transitions, and reward distributions. More precisely, an MDP is a 4-tuple (S, A, T, r), where S denotes a finite set of states, A denotes the action space, T is a probabilistic transition function $T : S \times A \times S \rightarrow [0, 1]$ that denotes the probability of transition from a state s to a new state s' when a certain action a is applied, and $r : S \times A \rightarrow \Re$ is a reward function that denotes the reward for applying a certain action a to a certain state s. At this stage, a formal definition for the state of our system has to be provided. Given the agent architecture formulated before, the state of every individual agent and the state of the entire multi-agent system are both expressed as $\langle q_i, \theta_i, d_i, \vec{g}_i \rangle$ (where i is an index referring to an individual agent $- i$). For a 4 dof manipulator, the corresponding state definition of each individual agent, and subsequently the state definition of the entire system, is $S_t = \{\langle q_1, \theta_1, d_1, \vec{g}_1 \rangle, \langle q_2, \theta_2, d_2, \vec{g}_2 \rangle, \langle q_3, \theta_3, d_3, \vec{g}_3 \rangle, \langle q_4, \theta_4, d_4, \vec{g}_4 \rangle\}$, at a specific time instance t. All agents wish to select actions that maximize the (expected) reward. Each agent contributes its own action component to the joint action that is eventually applied to the environment and determines the transition. The goal is to find a policy that maximizes the sum of discounted reward (Bertsekas and Tsitsiklis 1996).

Before proceeding further, we adopt some standard terminology from game theory, to facilitate the discussion below (Myerson 1991). A randomized policy for an agent i is a distribution $\pi \in \Delta(A_i)$ (where $\Delta(A_i)$ is a set of distributions over the action set A_i of the agent). Intuitively, $\pi(a^i)$ denotes the probability of agent i selecting the individual action a^i. A policy π is deterministic if $\pi(a^i) = 1$ for some $a^i \in A_i$. A collection of policies for each agent i is called policy profile, $\Pi = \{\pi_i : i \in n\}$, where n is the collection of agents. The expected value of acting according to a fixed profile can easily be determined. If each $\pi \in \Pi$ is deterministic, we can think of Π as a joint action. A reduced profile for agent i is a policy profile for all agents but i (denoted Π_{-i}). Given a profile Π_{-i}, a policy π_i is a best response for agent i if the expected value of the policy profile $\Pi_{-i} \cup \{\pi_i\}$ is maximal for agent i; that is, agent i could not do better using any other policy π_i'. In the following section, we refer to the requirement of continuous state-space, introducing an issue that has to be resolved, namely the infinite number of states (Sutton and Barto 1998), which is discussed hereafter.

16.3.2.2 State-Space Fuzzification for Continuous Problem Sets

In a continuous state-space, the number of parameters to be learned by the agent grows exponentially as the number of states increases. In order to achieve the desired continuity in the state-space without building huge lookup tables storing all the parameters of the agent, each of the parameters defining the state of each agent and the state of the system (joint angles, angular displacements, Euclidean distance, and all other signals required) are fuzzified using specific membership functions. Each joint (agent) continuous angular position, ranging from 0 to 2π, is divided into eight discrete states (but with assigned 'weight' for each state, from 0 to 1). For this reason, standard triangular (equidistant) fuzzy membership functions are used. The action-selection space is also fuzzified in the same manner. In the sequel, the action selection and reward computation functions are described.

16.3.2.3   Action Selection and Reward Function

Action selection is significantly difficult if there are multiple optimal joint actions. If the joint actions are chosen randomly, or in some way reflecting personal biases, then there is a risk of selecting a suboptimal or uncoordinated joint action. This general problem of equilibrium selection (or joint action selection) can be addressed in several ways. One way is the communication among the agents (Shoham and Tennenholtz 1992); another is to introduce conventions or rules that restrict behaviors, and so to ensure coordination. What we are proposing results in a coordination among the agents action through a repeated performance of the specific task by the same agents. In our action selection mode, each agent i keeps a count of the number of times a specific action has been performed in the past by the same agent (as well as by its collaborative agents). That concept, although simple, is some times quite effective and is known as fictitious play (Fundenberg and Kreps 1992; Brown 1951). More precisely, each agent a_i keeps a count $C^i(\alpha_k^j)$, for every agent a_j that is visible by a_i, indicating the number of times agent a_j has selected action $\alpha_k^j \in A_j$ in the past. When a task is assigned to our multi-agent system, each agent a_i treats the relative frequencies of the moves of all other agents a_j as indicative of their current policy. That is, agent a_i assumes that agent a_j performs action $\alpha_k^j \in A_j$ with probability:

$$P^i(\alpha_k^j) = \frac{C^i(\alpha_k^j)}{\sum_{b^j \in A_j} C^i(b^j)}$$

We note that most models (in game theory) assume that each agent can observe the actions executed by its counterparts with certainty. What we actually use is something more general that allows each agent to obtain an observation that is related stochastically to the actual joint action selected. Action selection is more difficult when agents are not aware of the rewards associated with various joint actions; hence, the expected reward associated with individual and joint actions has to be estimated based on previous experience.

Q-learning algorithm developed by Watkins (Watkins 1989) is the most frequently used RL algorithm. An agent estimates the utility for doing each of its actions, chooses an action based on a selection function of the expected values, observes the reward, and then updates the Q-value or the estimate of the utility of that action. In the case of a stateless setting, we have an agent updating its estimate $Q(a)$ as follows: $Q(a) \leftarrow Q(a) + \lambda(r - Q(a))$ where action a was performed resulting in reward r. Here λ is the learning rate ($0 \leq \lambda \leq 1$) governing to what extent the new sample replaces the current estimate. The next issue concerns the action selection function, which is particularly important, since effective learning requires sufficient exploration. The action selection mechanism that we employ in this work is a variant of ε-greedy, which is called ε-decreasing, where the probability of an exploration action decreases as trials progress. This action selection mechanism starts with an exploration probability: $\varepsilon \cdot (T(t) - 1)/(T_{\max} - 1)$. In order to estimate the probability of choosing an action, we employed Boltzmann distribution as shown below, and on each trial, with probability: $1 - \varepsilon \cdot (T(t) - 1)/(T_{\max} - 1)$, each agent a_i chooses the action α^i with the greatest estimated $\pi(\alpha^i)$:

$$\pi(\alpha^i) = \frac{e^{\frac{EV(\alpha^i)}{T}}}{\sum\limits_{\alpha^i_j \in A_i} e^{\frac{EV(\alpha^i_j)}{T}}} \tag{16.1}$$

where $EV(\alpha^i)$ denotes the expected value of an action α^i and T is the temperature parameter that is controlled to diminish over time so that the exploitation probability is increased. Temperature determines the likelihood for an agent to explore other actions: when T is high, even when the $EV(\alpha^i)$ of an action is high, an agent may still choose an action that appears to be less desirable. This exploration strategy is especially important in stochastic environments like the one we are examining, where payoffs received for the same action combination may vary. For effective exploration, high temperature is used at the early stages of a task. The temperature is then decreased over time to favour exploitation, as the agent is more likely to have discovered the true values of different actions. The temperature T as a function of iterations is given by: $T(t) = 1 + T_{\max} \cdot e^{-st}$, where t denotes here the iteration number, s is the rate of decay and $T_{\max}$ is the initial temperature.

Now, let us elaborate on the definition of the expected value $EV(\alpha^i)$. The presence of multiple agents, each one learning simultaneously with others, is a potential impediment to the successful employment of Q-learning (and RL in general) in multi-agent settings like the one considered in this paper. When an agent a_i is learning the value of its actions in the presence of other agents, it is learning in a non-stationary environment. Thus, convergence of the Q-values is not guaranteed. What we need is each agent's policy to settle. This is a key issue and is discussed below. In general, there are two distinct ways in which Q-learning could be applied in a multi-agent system; the Independent and the Joint-Action Learner algorithm (Lauer and Riedmiller 2004; McGlohon and Sen 2004; Claus and Boutilier 1998). In an Independent Learner algorithm each agent learns its Q-values regardless of what the other agents are performing. This method is appropriate to be used when an agent has no reason to believe that other agents are acting strategically. Joint action learner algorithm is the one where the agents do not learn Q-values of their individual actions but the Q-values of their joint actions. This implies that each agent can observe the actions of other agents. Each agent in such a system maintains beliefs about the policies of other agents. So, an agent i assesses the expected value $EV(a^i)$ of its individual action a^i to be

$$EV(\alpha^i) = \sum_{\alpha^{-i} \in A_{-i}} \left\{ Q\left(s, \alpha^{-i} \cup \{\alpha^i\}\right) \cdot \prod_{\alpha_k^j \in \alpha^{-i}} \left[P^i(\alpha_k^j) \right] \right\} \qquad (16.2)$$

In the above equation, A_{-i} denotes the set of all possible joint actions that can performed by the group of agents that are considered "visible" by agent a_i, in the nested hierarchical sense (i.e., in our case, agents that are below in the hierarchy), $\alpha^{-i} \in A_{-i}$ denotes one such joint action performed by this group of agents, and $\alpha_k^j \in \alpha^{-i}$ is an individual action performed by a single agent a_j in this group. The reward that an agent receives at time instant t, after selecting certain action and moving to a new state, is defined by the reward function $R(t)$, which is formulated as follows:

$$\left\{ \begin{array}{l} \texttt{if } (Dist_{goal}(t) \leq Dist_{min}) \wedge (\Delta Dist_{goal}) \leq 0) \texttt{ then } R(t) = e^{-c \cdot (Dist_{goal}(t))} \\ \texttt{if } (Dist_{goal}(t) > Dist_{min}) \texttt{ then } R(t) = -2 \\ \texttt{if } (Dist_{goal}(t) < Dist_{min}) \wedge (\Delta Dist_{goal}) > 0) \texttt{ then } R(t) = -1 \end{array} \right\}$$

$$(16.3)$$

where $Dist_{goal(t)}$ is the distance from the goal at the iteration time t. $Dist_{min}$ is a threshold distance after which the agents starts receiving reward. $\Delta Dist_{goal}$ is the rate of change of distance from the goal.

16.4 Agent Architecture: The Case of Collaborative Mobile Robots

Following the same design principles as in the case of kinematic chains, the multi-agent architecture is here adapted in the case of two mobile robots pushing an object to a goal position. The challenge in this case is in representing the state of the system. Referring to the definition of an agent, let us examine the corresponding figure (Fig. 16.6). The agents are nested in a similar manner as in the case of the kinematic chain. What is different, though, now is the fact that the kinematic configuration of the system has no mounting point that could be used as a reference for the state-space representation (like it had before, in the case of a fixed-base kinematic chain). In other words, we have a system (the two robots and the object) that is moving. Keeping the same multi-agent design principles as before (i.e., hierarchy of agents, where the wheels -instead of the links- are the independent agents), we can see this system as a moving kinematic chain with both ends floating. This is in fact a nonholomonic multi-robot system, controlled only through the angular speed of

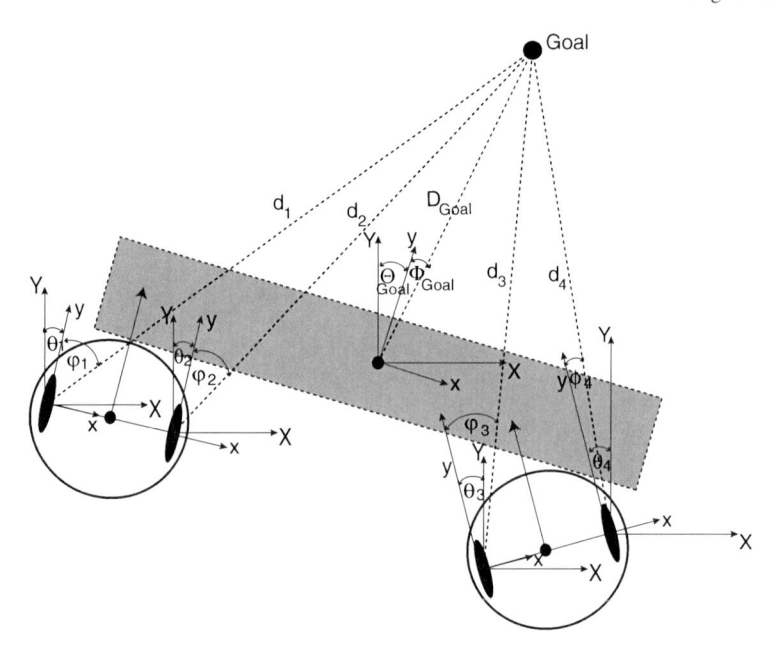

Fig. 16.6 Two mobile robots trying to collaboratively push the box to a goal position

Fig. 16.7 Agents are
assigned to degrees
of freedom (wheels)
of the mobile robots

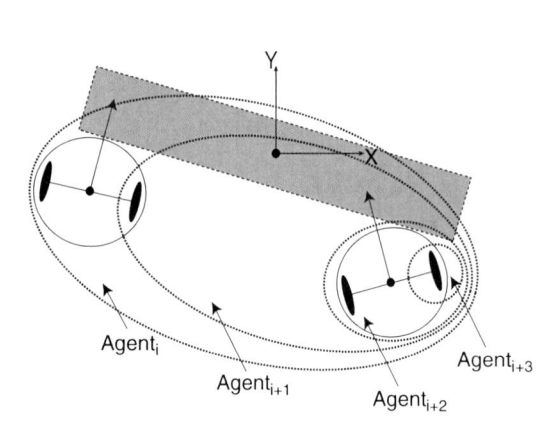

the wheels, where the goal is to coordinate their activity so that the object is pushed
to the goal position. Figure 16.7 depicts the hierarchical, nested arrangement of the
agents. As can be seen in this figure, the nested hierarchy starts with the agent corresponding to the left wheel of the left robot. Each agent sees only the agents below
in this hierarchical structure, meaning that $agent_i$ observes only $agent_{i+1}$, $agent_{i+1}$
sees only $agent_{i+2}$, and so on. What is interesting to point out here is the fact that
we are trying to fit a hierarchical relationship within physical entities that have no
direct physical connectivity.

For instance, $agent_{i+1}$ corresponding the right wheel of the left robot contains $agent_{i+2}$ which actually corresponds to the right robot. The state that describes the multi-agent system is now described as follows. For each $agent_i$, its state is defined as:

$$S_i = \langle \theta_i, \phi_i, d_i, \omega_i, \Theta_{\mathrm{goal}_i}, \Phi_{\mathrm{goal}_i}, D_{\mathrm{goal}_i} \rangle \tag{16.4}$$

where i is an index referring to an individual agent i, θ_i is the orientation of the agent with respect to global reference frame, ϕ_i is the orientation of the agent with respect to the goal, d_i is the distance of the agent from the goal, ω_i is the speed of the agent, Θ_{goal_i} is the orientation of the box with respect to the global reference frame, Φ_{goal_i} is the orientation of the box with respect to the goal, and D_{goal_i} is the distance of the box from the goal. Having defined the state of a single agent, the rest of the design principles of this multi-agent framework are exactly the same as already described before in Sect. 16.3.

16.4.1 Continuous Reinforcement Learning: Mobile Robots

The following subsections address the theoretical analysis of the learning mechanism as adapted for the case of collaborative mobile robots. The specific experimental setup poses significant challenges due to the fact that the robot skill evolves from the actions selected by its wheels. It is clear that we are dealing with a nonholonomic system, where the current setup is represented as an autonomous kinematic chain with both ends loose, moving inside the workspace and pushing the box. The state representation posed significant difficulties throughout the learning process, due to the fact that a minimum set of parameters is needed that uniquely defines the configuration of the system in a continuous space. Regarding the learning process, several stages of development have been followed and different approaches have been implemented, such as Q-leaning and TD(λ). As an initial approach, the typical Q-learning algorithm presented in Sect. 16.3.2.1 was used in the box-pushing problem. As expected, the enormous state-space led to over 400 million elements stored in the look-up table. Hence, the dimensionality of the box-pushing problem directed us toward the linear function approximation method. Combining another form of RL, namely TD(λ) Learning, with function approximation provided a more suitable approach, as described in the sequel.

16.4.1.1 TD(λ) Learning Method

Temporal difference (TD) learning is an approach to the problem of learning how to predict a quantity that depends on future values of a given signal. The name TD derives from its use of changes, or differences, in predictions over successive time steps to drive the learning process. The prediction at any given time step is updated to bring it closer to the prediction of the same quantity at the next time step. Here we use the TD algorithm to predict a measure of the total amount of reward expected over the future. More specifically, with the TD algorithm an agent computes an estimate about the future reward received for a transition from state s_t to state s_{t+1}. As the TD algorithm is a model-free method, we build a Q-model from experience without any reference to a dynamic model. Therefore, if we consider the value of the current state as $Q(s_t)$, the estimated value of the next state that we will reach will be $Q(s_{t+1})$ and the update for the value of $Q(s_t)$ will become:

$$Q(s_t) \leftarrow Q(s_t)(1 - \alpha) + \alpha[R(s_t) + \gamma Q(s_{t+1})] \qquad (16.5)$$

according to the stochastic version of Bellman backups. Similarly, for the next time instant $t + 1$ the update will be:

$$Q(s_{t+1}) \leftarrow Q(s_{t+1})(1 - \alpha) + \alpha[R(s_{t+1}) + \gamma Q(s_{t+2})] \qquad (16.6)$$

which takes into account the estimate for $Q(s_{t+2})$. Having now obtained a better estimate for $Q(s_{t+1})$, the TD(λ) algorithm goes back and, based on that new information, improves its updated estimate for the value of $Q(s_t)$:

$$Q(s_t) \leftarrow Q(s_t) + \alpha \gamma \lambda \delta_{t+1} \qquad (16.7)$$

where $\delta_{t+1} = R(s_{t+1}) + \gamma Q(s_{t+2}) - Q(s_{t+1})$, and λ is the factor that determines how heavily we weight the changes in Q values for s_{t+1}. In the same manner, we proceed to the time instant $t + 2$, where we obtain:

$$Q(s_{t+2}) \leftarrow Q(s_{t+2})(1 - \alpha) + \alpha[R(s_{t+2}) + \gamma Q(s_{t+3})] \qquad (16.8)$$

and moving backward to update the previous estimates we get:

$$Q(s_{t+1}) \leftarrow Q(s_{t+1}) + \alpha \gamma \lambda \delta_{t+2} \qquad (16.9)$$

and

$$Q(s_t) \leftarrow Q(s_t) + \alpha \gamma^2 \lambda^2 \delta_{t+2} \qquad (16.10)$$

The term $\gamma^2 \lambda^2$ is called eligibility vector $e(s_t)$. This vector is maintained as the agents gain experience and can be used to update the Q values accordingly.

TD learning can often be accelerated by the addition of eligibility traces. By the use of eligibility traces, the TD algorithm does not only update the table entry for the immediately preceding prediction but, since future predictions provide useful information for learning earlier predictions as well, one can extend TD learning so that at each step it updates a collection of many earlier predictions as well. Eligibility traces do this by providing a short-term memory of many previous input signals, so that each new observation can update the parameters related to these signals. Eligibility traces are usually implemented by an exponentially decaying memory trace, with decay parameter λ. This generates a family of TD algorithms TD(λ), $0 \leq \lambda \leq 1$, with TD(0) corresponding to updating only the immediately preceding prediction as described above, and TD(1) corresponding to equally updating all the preceding predictions. We summarize this approach in the algorithm presented in Fig. 16.8.

Fig. 16.8 TD(λ) algorithm implementation

TD(λ) - Look Up Table

Initialize Q(s) arbitrarily, Initialize e(s) to zero
Iterate for t =1... n
    Select action a_t and execute it
    Calculate $\delta_t \leftarrow$ R(s_t) + γQ(s_{t+1}) - Q(s_t)
    Calculate e(s) $\leftarrow$ e(s) + 1
    Iterate for all states s
        Q(s) $\leftarrow$ Q(s) + α e(s) δ_t
        e(s) $\leftarrow$ $\gamma \lambda$ e(s)
    end
end

16.4.1.2   TD(λ) Learning with Linear Function Approximation

State spaces may be extremely large and the look-up table approach may require excessive memory space. The TD(λ) algorithm can be generalized to use an approximate value function instead of keeping an explicit table of states. Let us define the approximate function: $Q(s) \simeq f_\theta(s)$, where f_θ is a function parameterized in θ, and s is the state vector. Now, instead of updating the $Q(s)$ directly, the value of θ can be updated instead. More formally, we seek to learn the parameter vector $\theta \in \mathbb{R}^n$ of an approximate value function $Q_\theta : S \to \mathbb{R}$, such that $Q_\theta(s) = \theta^T \phi_s$ (where $\phi_s \in \mathbb{R}^n$ is a feature vector characterizing state s) to minimize an objective function. There are multiple methods suitable for updating the parameter vector θ. One approach is the gradient method, where the updates to θ are proportional to the gradient of a suitable objective function with respect to θ. One natural choice might be the mean squared error (MSE) between the approximate value function Q_θ and the true value function Q. Hence, we define the objective function:

$$E = (1/2)(\hat{Q}(s_t) - f_\theta(s_t))^2 \tag{16.11}$$

By taking the gradient of function E, we simply get:

$$\nabla_\theta(E) = (\hat{Q}(s_t) - f_\theta(s_t))(0 - \nabla_\theta f_\theta(s_t)) \tag{16.12}$$

and by rearranging the terms we obtain:

$$-\nabla_\theta(E) = (\hat{Q}(s_t) - f_\theta(s_t))\nabla_\theta f_\theta(s_t) \tag{16.13}$$

From this, we can define the conventional linear TD algorithm of the following form:

$$\theta_{t+1} \leftarrow \theta_t + \alpha(\hat{Q}(s_t) - f_\theta(s_t))\nabla_\theta f_\theta(s_t) \tag{16.14}$$

Next we incorporate this θ update mechanism to our TD(λ) algorithm, which is summarized in Fig. 16.10. A fuzzy rule-base, as will be described in Sect. 16.4.1.4, is an instance of a linear parameterized function approximation architecture, where the weight of each rule i can be used as a feature $\phi_i(s)$. This is the learning mechanism that will be used in the case-study regarding collaborative mobile robots, presented later on in this chapter; it should be noted, however, that this approach does not always converge. Very recently, Sutton et al. (2009) presented a fast Gradient-Descent method for TD(λ) learning with linear approximation, which was proved to always converge. In the following subsection, a brief description of this approach is provided (full analysis, as well as proofs can be found in Sutton et al. 2009).

16.4.1.3 TD(λ) Learning Method with Gradient Correction

As has been described in the previous section, the objective function used was the MSE between the approximate value function Q_θ and the true value function Q. In this case, the mean square error is:

$$\text{MSE}(\theta) = ||Q_\theta - Q||^2 \tag{16.15}$$

The problem that may arise with this objective function concerns the fact that this approximated value function might not satisfy the Bellman equation. Most efforts made to develop gradient-descent algorithms have focused on finding a solution to that problem. In order to get around this problem, another objective function has been introduced, called Mean Square Bellman error (MSBE), which constitutes in fact an attempt to find a solution that satisfies the Bellman equation. More specifically, let $Q*$ be the optimal value function that satisfies the Bellman equation:

$$Q = R + \gamma PQ = TQ \tag{16.16}$$

where R is the reward vector and T is what is known as the Bellman operator. Under a stationary policy, the solution $Q*$ that satisfies the equation $Q* = TQ*$ is unique. Therefore, we can find an approximation Q_θ of that unique optimal solution for some θ's using the objective function:

$$\text{MSBE}(\theta) = ||Q_\theta - TQ_\theta||^2 \tag{16.17}$$

With this approach, the approximated value function will definitely satisfy the Bellman equation. However, it might be that this unique solution $Q*$ is not representable as Q_θ for any θ. Thus, what has been recently proposed by Sutton et al. in (2009) is based on the definition of what is called Mean Square Projected Bellman Error (MSPBE), which takes a projection operator Π that projects any value function u to the nearest value function representable as Q_θ by the function approximator:

$$\Pi u = Q_\theta \quad \text{where} \quad \theta = \text{argmin}_\theta ||Q_\theta - u||^2 \tag{16.18}$$

The approximate value function can now be defined as: $Q_\theta = \Pi T Q_\theta$, resulting to a new objective function defined as:

$$\text{MSPBE}(\theta) = ||Q_\theta - \Pi T Q_\theta||^2 \tag{16.19}$$

In their work (Sutton et al. 2009), Sutton et al. prove that this algorithm always converges, and that its complexity is $O(n)$.

16.4.1.4 Linear Function Approximation Using a Fuzzy Rule Base

Having described in the previous sections the $TD(\lambda)$ function approximation method, let us now discuss the specific function approximation architecture that we use in this work, which is tuned with the RL mechanism already discussed. The mechanism used is a fuzzy rule base (FRB), which can be defined as a function f that maps an input vector in a scalar output. If $f_\theta(\vec{s})$ is the function that we are trying to approximate, and $\vec{s}$ is a vector of state parameters, the input to the FRB function is this vector of state parameters. The next element that is required is a set of fuzzy rules. The form that these rules have is the following:

$$\text{Rule} - i : \text{ IF } (s_1 \in A_1^i) \text{ AND } (s_2 \in A_2^i) \text{ AND} \ldots (s_n \in A_n^i) \text{ THEN } (\text{output} = O^i)$$

where s_n is the nth parameter of the state parameters vector, A_n^i are fuzzy membership functions used in the ith rule, and O^i are tunable output parameters. Thus, the output of the FRB is a weighted average of O^i:

$$f_\theta(\vec{s}) = \theta_1 O^1 + \theta_2 O^2 + \theta_3 O^3 + \cdots + \theta_n O^n$$

$$\text{i.e.,: } f_\theta(s_1, s_2, s_3 \ldots s_n) = \theta_1 O^1 + \theta_2 O^2 + \theta_3 O^3 + \cdots + \theta_n O^n$$

It should be noted here that all values O^i are normalized.

In order to explain further the values obtained by O^i, let us consider an example where an input vector of three state variables (s_1, s_2, s_3) is considered, each variable having two membership functions as shown in Fig. 16.9. In this figure, it can be seen that each input state variable has two localized signal values (basically two sigmoids) (for the purpose of demonstration, let us assume these two values being HIGH and LOW). For this case of three state variables with two localized signal values each, eight rules need to be included in the FRB. All rules, from $1 \ldots 8$, will provide, respectively, eight outputs $O^1 \ldots O^8$.

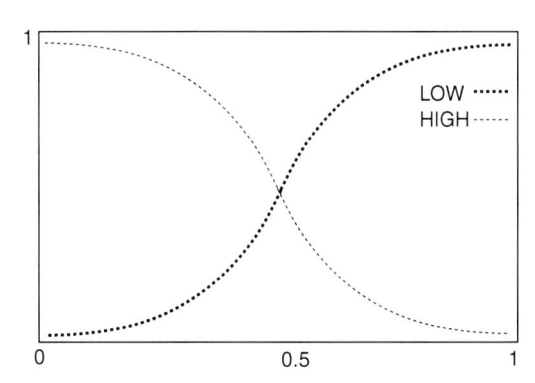

Fig. 16.9 Localized signal values describing every state variable

These outputs are then estimated based on the following set of probabilities: $P_{HIGH}^{s_1}, P_{LOW}^{s_1}, P_{HIGH}^{s_2}, P_{LOW}^{s_2}, P_{HIGH}^{s_3}, P_{LOW}^{s_3}$. The probability $P_{LOW}^{s_1}$ can be defined as:

$$P_{LOW}^{s_1} = \frac{1}{1 + c_i \, e^{-c_j s_1}} \tag{16.20}$$

and: $P_{HIGH}^{s_1} = 1 - P_{LOW}^{s_1}$.

The value of O^i can now be estimated as: $O^i = \prod_{j=1}^{k} P_{A_j^i}^{s_j}$, where k is the number of state parameters and i is the number of rules. So, in our example where $k = 3$ and $i = 8$, we have:

$$O^1 = P_{HIGH}^{s_1} \cdot P_{HIGH}^{s_2} \cdot P_{HIGH}^{s_3}, O^2 = P_{HIGH}^{s_1} \cdot P_{HIGH}^{s_2} \cdot P_{LOW}^{s_3}, \ldots, O^8 = P_{LOW}^{s_1} \cdot P_{LOW}^{s_2} \cdot P_{LOW}^{s_3}$$

All the outputs O^i will have to be normalized before computing the value for the function $f_\theta(s_1, s_2, s_3)$. Having calculated the FRB function f_θ as described above, its gradient $\nabla_\theta f_\theta(s_1, s_2, s_3)$ needs to be computed next, to update the eligibility traces. Thus, for every state variable the derivative of $f_\theta \vec{s}$ needs to be computed. For every θ_i, $\nabla_{\theta_i} f_\theta(s_1, s_2, s_3)$ equals O^i. The eligibility traces can then be calculated and, hence, the θ values can be updated. Having described the learning mechanism that is used in our specific problem setting, we will now go through the entire flow of operations as well as the action selection mechanism. Initially the agents (wheels of the mobile robots) have to establish a Joint Action that will be executed. Then some reward will be generated for all agents according to the evaluation (positive/negative) of their joint action. Same as in the case of the kinematic chain, the root agent selects an action by stochastically observing what the other agents have done in the past. More specifically, a table is maintained containing the last time instant that a specific action has been selected by an agent. By doing so, an "Action-Current-Tendencies" (ACT) table is created for all agents in the system. Every agent uses this ACT table to evaluate its beliefs over each specific action, before finalizing the action that it will apply to participate to the system's Joint Action. Thus, $agent_i$ will propose an $action j$. Then, possible Joint Actions are generated and evaluated, and according to this evaluation, the system estimates the corresponding possible states that will be obtained by performing those actions. From those possible states, the related state variables are extracted and subsequently inserted to the estimation of f_θ. For every possible joint action, a corresponding f_θ value is computed. Using Boltzman distribution, a decision over what Joint Action should be selected is then obtained. Finally, based on this decision, the action that each $agent_i$ will perform is finalized, and this information is passed to the next level (i.e., $agent_{i+1}$).

Following the proposed hierarchical architecture, the next-level agent receives as input the action that the above agent has decided to execute, and based on this information it goes through the same process (but this time, one action less has to be decided for the system's Joint Action). When this iterative process concludes, every agent locks its own action and, thus, a cumulative Joint Action is executed. At that instant, the ACT table is updated with the current time instant that every action is selected. Subsequently, the selected final Joint Action is executed and the system moves from its "previous state" to its "current state". A reward is then calculated. The state variables of the current state are used to compute $f_\theta(s_{\text{current}})$, while the state variables of the previous state are used to compute $f_\theta(s_{\text{previous}})$. By computing these two values along with the reward, we can calculate δ and subsequently the eligibility traces and the θ values.

Let us elaborate more on the update of the eligibility traces and θ values. In the specific example of three state variables and two localized signals, resulting to a total number of eight eligibility traces, we have:

$$\delta \leftarrow R(s_{\text{previous}}) + \gamma f_\theta(s_{\text{current}}) - f_\theta(s_{\text{previous}}) \tag{16.21}$$

One should not get confused with the names "previous state" and "current state" because TD(λ) is in fact an algorithm that is looking at the past, meaning that it is evaluating the state-transition by updating the variables starting from each state of the system and going backwards:

$$e(s_{\text{previous}}) \leftarrow \gamma \lambda \vec{e} + \nabla_\theta f_\theta(\vec{s}) \tag{16.22}$$

The above relationship states that all the eligibility traces of the previous state are updated, where $\vec{e}$ is the vector of the eligibility traces, ∇_θ is the gradient of function f_θ with respect to the parameter vector θ, and $\vec{s}$ is the vector of all state variables defining the state of the system. So, recalling that:

$$f_\theta(s_1, s_2, s_3 \ldots s_n) = \theta_1 O^1 + \theta_2 O^2 + \theta_3 O^3 + \cdots + \theta_n O^n \tag{16.23}$$

for the specific example of three state variables, with two localized signal values for each state variable, we have:

$$f_\theta(s_1, s_2, s_3) = \theta_1 O^1 + \theta_2 O^2 + \theta_3 O^3 + \theta_4 O^4 + \theta_5 O^5 + \theta_6 O^6 + \theta_7 O^7 + \theta_8 O^8 \tag{16.24}$$

The eligibility traces are then computed as:

$$e_i(s_{\text{previous}}) \leftarrow e_i(s_{\text{previous}}) \lambda \gamma + \nabla_{\theta_i} f_\theta(s_1, s_2, s_3) \quad (\text{for every } i = 1, \ldots 8) \tag{16.25}$$

which, if further elaborated, gives:

$$e_i(s_{\text{previous}}) \leftarrow e_i(s_{\text{previous}}) \lambda \gamma + O^i \quad (\text{for every } i = 1, \ldots 8) \tag{16.26}$$

Given the above updates for all the eligibility traces, the last step that remains is to update the parameter values of θ, as follows:

$$\theta_i \leftarrow \theta_i + \alpha \, \delta \, e_i(s_{\text{previous}}) \quad (\text{for every } i = 1, \ldots 8) \tag{16.27}$$

The overall algorithm is summarized in Fig. 16.10.

TD(λ) - Function Approximation

Initialize Q(s) arbitrarily, Initialize e(s) to zero
Iterate for t =1... n
    Select action a_t and execute it
    Calculate $\delta_t \leftarrow R(s_t) + \gamma f_\theta(s_{t+1}) - f(s_t)$
    Calculate $e(s_t) \leftarrow \gamma \lambda e(s_t) + \nabla_\theta f_\theta(s)$
    Calculate $\theta \leftarrow \theta + \alpha \delta_t e(s_t)$
end

Fig. 16.10 TD(λ) algorithm, with function approximation

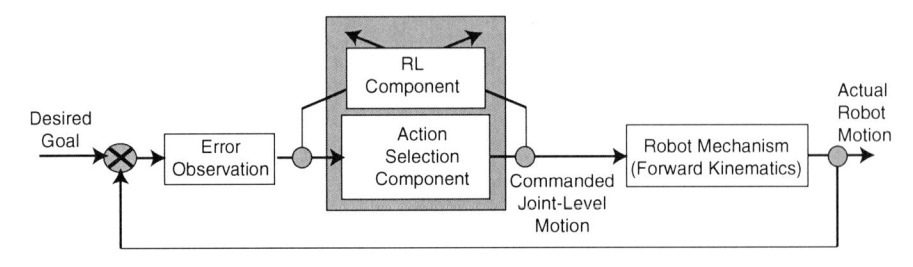

Fig. 16.11 The proposed RL-based robot control architecture

16.5 The RL-Based Robot Control Architecture

The robot control architecture used has three layers. The first layer is the error observation layer, the second one is the action selection layer, and the last one is the servo control layer. The first layer receives as input the desired goal for the multi-agent system along with the feedback from the last layer (as shown in Fig. 16.11). The error observation layer provides input to the next layer, which is the action selection layer. It can be seen that the action selection layer is coupled with the RL module. The RL module, augmenting the action selection mechanism, receives the error observation data, provides appropriate input to the action selection mechanism, for certain action(s) to be generated (without any initial prior knowledge regarding appropriate joint-level motion). Subsequently, the action selection mechanism generates the joint-level motion commands or the robot(s) servos, based on stochastic models, while providing information to the RL module regarding the probabilistic distribution that was assigned to the different action(s) by the action selection process. The commanded joint-level motion propagates to the third layer which corresponds to the robotic mechanism. This last layer generates the actual robot motion for the agent(s) involved in the system.

16.6 Numerical Experiments: Results and Discussion

To evaluate the performance of the proposed architecture, we performed a series of numerical experiments grouped into two categories, involving: (a) dexterous robot manipulation, and (b) cooperative mobile robots. For the first category, we initially consider a single (redundant) kinematic chain, like the one presented in Fig. 16.1, with four links using the proposed multi-agent architecture depicted in Fig. 16.4. We then consider, in Sect. 16.6.2, a simulated three-finger quasi-static grasp of a rectangular object, which is supposed to have infinite mass and specific (unknown) friction coefficient, where each finger of the simulated dexterous manipulator consists of an independent kinematic chain with four links. For the second category of numerical experiments, we consider two cooperating mobile robots performing a simulated box-pushing task, and we explore the application of the proposed learning architecture with respect to the development of collaborative robot control skills.

16.6.1 Single Kinematic Chain

The detailed parameters used in the simulation for the single kinematic chain are shown in Table 16.1. In order to achieve continuity over the joint space of the system, the space is fuzzified with several signals. In total, for every joint angle we have 20 signals with 15% overlap. Similarly, for the task space, we use nine signals each one spanning over a range of 40° (in total nine localized signals) to achieve continuity of the state definition. The overlap of these signals is 25%.

The objectives of this multi-agent system are twofold: (a) to respect the limitations that their physical interconnectivity imposes, and (b) to learn how to collaborate and communicate information among agents, to reach the goal position, without any previous experience or knowledge or task model. The first phase of the process is the training of the multi-agent system. The system is trained over 200 epochs, each epoch lasting 500 time units. This means that we allow the group of four agents to come up with a solution to the problem assigned within the time slot of 500 time units. In any case (whether a solution has been found or not), the knowledge obtained within the specific time interval is stored, and the agents are expected to further improve their behavior during the subsequent epochs. The behavior of the agents at the beginning of their collaborative activity is focusing

Table 16.1 Experimental parameters for the case of a single kinematic chain

	Link 1	Link 2	Link 3	Link 4
Length	3 cm	2 cm	4 cm	3 cm
Initial joint angle	0°	20°	30°	40°
Step: Joint angle	18°	18°	18°	18°
Step: Goal angle	40°	40°	40°	40°
Overlap of function f_1	15%	15%	15%	15%
Overlap of function f_2	25%	25%	25%	25%

on exploration; thus, to satisfy this requirement, the parameters in the equation: $T(t) = 1 + T_{\max} \times e^{-st}$ are defined in such a way that for the first 30 time units of each epoch we obtain high exploration, a behavior achieved with a decay factor set at the value of $s = 0.35$ (meaning that during this first period the agents might select actions that do not have the highest Q-values), while in the remaining time the agents select those actions with the higher Q-values. The maximum temperature for $t \in [1 \dots 500]$ is set as $T_{\max} = 100$, while the learning rate is set rather low (i.e., $\lambda = 0.1$), which means that we promote a slow learning approach. Furthermore, the agents interacting with the environment must receive appropriate rewards. In our simulation, we set $\text{Dist}_{\min} = 3$, meaning that the agents do not receive any positive reward if the end-effector is not closer than 3 distance units to the goal position.

Figure 16.12 summarizes the above learning results. This initial set of results depicts the state of the agents in four distinct epochs. During epoch 1, we see the agents exploring their workspace, a behavior that results in an error on position, with respect to the goal position that they are trying to reach. This error is not re-duced over the entire time duration of this epoch. We allow the process to repeat itself for another ten epochs. In epoch 10, a similar behavior can be observed with a slight improvement. In epoch 100 and 200, however, the evolution of the agents' collaborative behavior to reduce the error becomes apparent, since the actions with highest Q-values are now dominating the joint actions that the agents are selecting. In Fig. 16.13, we run again the same test with only one minor modification, the de-cay factor being now increased to $s = 0.75$. The results show that, already by epoch 10, the positioning error is now considerably reduced. In the subsequent epochs, we again see that the agents' behavior does settle to a sequence of actions that have the highest Q-values.

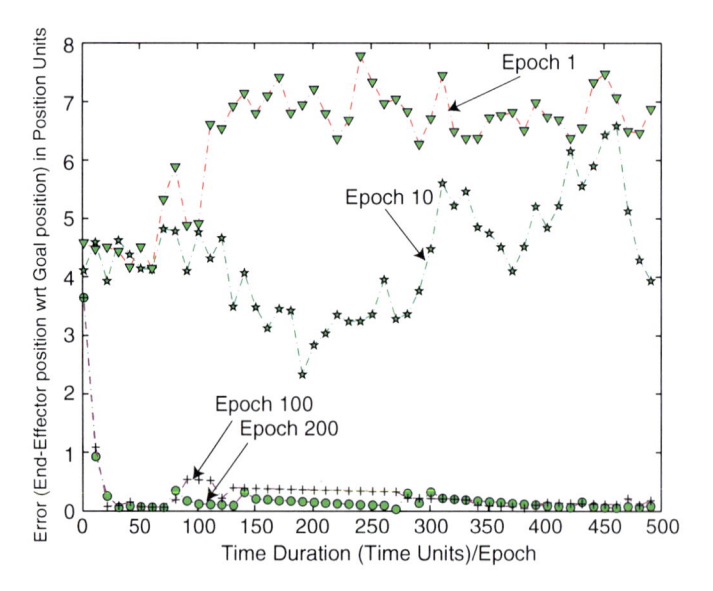

Fig. 16.12 Error over time over different epochs for factor $s = 0.35$

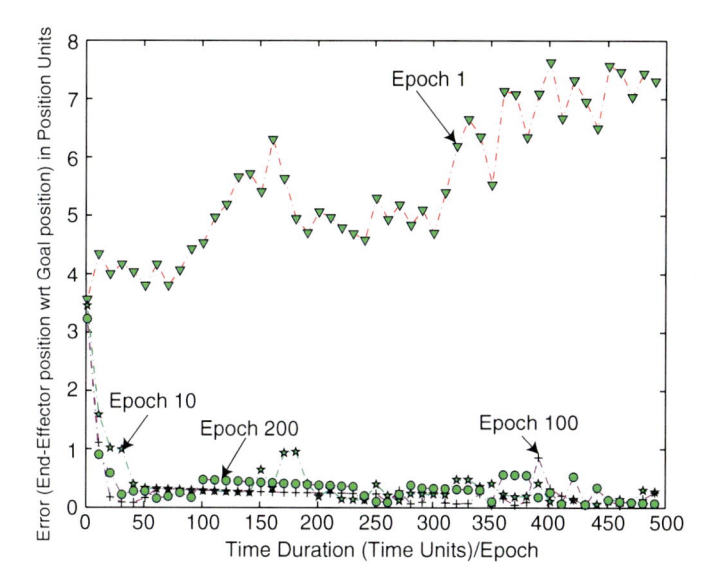

Fig. 16.13 Error over time over different epochs, for factor $s = 0.75$

Table 16.2 Different kinematic solutions derived by the proposed multi-agent architecture compared to the solution obtained using the pseudoinverse method ($J^\dagger$)

	$J^\dagger$	$s = 0.75$			$s = 0.35$			$s = 0.05$		
		c_1	c_2	c_3	c_1	c_2	c_3	c_1	c_2	c_3
Q_1	0	−2.0	−5.8	−11.8	−13.9	−39.0	29.4	−46.8	−46.9	−27.6
Q_2	24.1	63.9	41.9	57.8	39.1	137.3	22.6	120.8	119.0	93.2
Q_3	53.7	0.1	40.1	28.1	64.7	−33.3	−2.9	5.3	25.7	6.3
Q_4	59.8	84.1	59.8	55.9	26.1	27.2	120.3	10.4	−25.0	43.9

Besides the error convergence properties of the proposed approach, another important issue concerns evaluating the quality of the kinematic control solution(s) provided by this mechanism. To solve the inverse kinematic problem of the 4 DoF kinematic chain, an iterative method can be used, based on the computation of the pseudoinverse of the Jacobian matrix, $J^\dagger$. The Jacobian matrix J in our case is a 3×4 matrix consisting of four column vectors: $\vec{J}_1, \vec{J}_2, \vec{J}_3, \vec{J}_4$. Each $\vec{J}_i$ vector ($i = 1 \dots 4$) can be computed as the cross product of the vector representing the axis of rotation of the ith link against the vector expressing the distance between the corresponding ith joint and the end-effector. We can then write $\Delta\theta = J^\dagger \cdot \Delta p$, where $\Delta\theta$ is the increment of the joint angular displacement vector that causes the end-effector of the chain to move by Δp, where the pseudoinverse $J^\dagger$ is equal to: $J^T \cdot (J \cdot J^T)^{-1}$. Using this iterative method, we obtain an optimal set of angular displacements Q_i for $i = 1 \dots 4$ as shown in Table 16.2 (optimum in the least-square sense, leading to a minimum joint displacement for every link in the kinematic chain). An indicative set of corresponding solutions obtained with our proposed architecture are also shown in this table, varying according to the decay factor s selected.

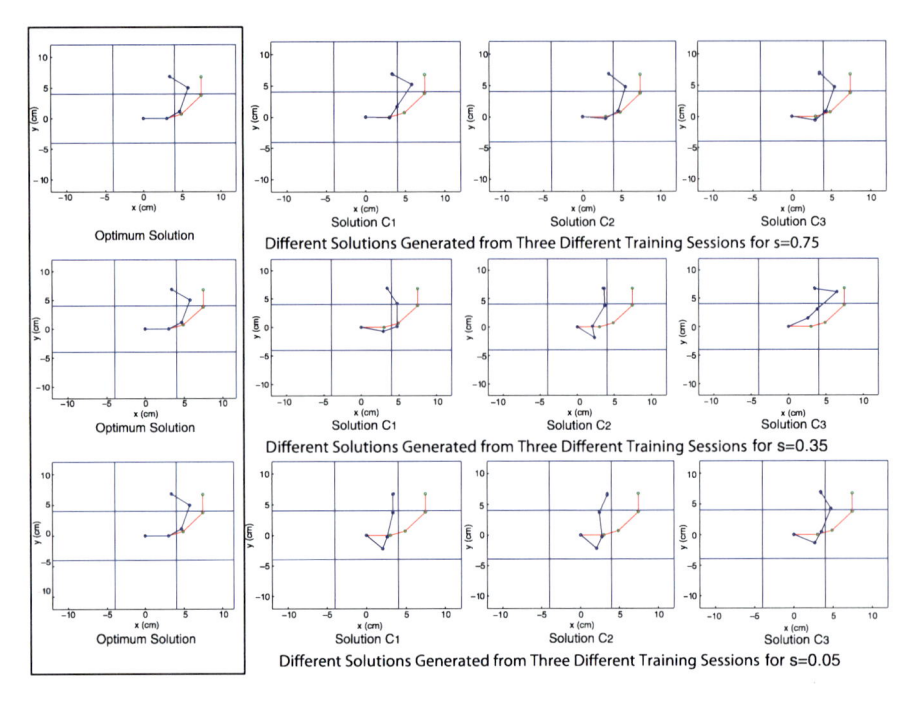

Fig. 16.14 Kinematic solutions generated for $s = 0.75$, $s = 0.35$ and $s = 0.05$

Figure 16.14 depicts graphically the results contained in Table 16.2, showing that for almost all different decay factors the multi-agent system proposed in this chapter generates quite natural solutions, while in certain cases close to the optimum one, as can be seen in the corresponding schematics (stick diagrams). Each solution provided in the corresponding figure has been generated after the completion of three different training sessions (each training session lasting 200 epochs), for three different decay factor, leading to a total of nine different solutions. To assess further the "optimality" of the emerging solutions, we performed a statistical analysis of the results obtained (error with respect to the theoretical LS optimal configuration) for a decay factor s varying in the interval $[0.005, \dots, 0.9995]$. The mean values and standard deviations (STD) of the error obtained are depicted in Figs. 16.15 and 16.16, at two different instances of the learning process, more specifically after 10 and after 100 learning epochs, respectively. These results demonstrate that, for a wide range of s values, the solutions stochastically generated by the system are near-optimal in the LS sense. Therefore, though it is not argued in any way that the proposed multi-agent learning system is theoretically guaranteed to provide such an optimal solution to this specific problem, it is shown that the system stochastically converges to a range of solutions that not only effectively resolve the kinematic redundancy but are also experimentally found (depending on the exploration-vs-exploitation strategy employed) to be indeed close to the theoretically optimal solutions in this sense, for a wide range of values of the exploration decay factor. One could then argue that the

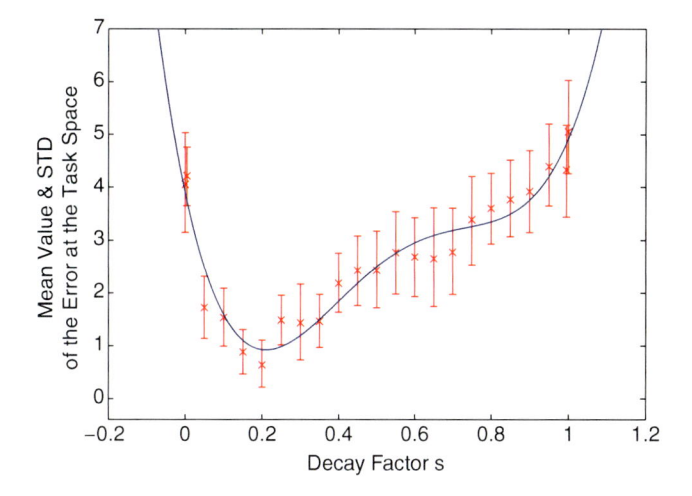

Fig. 16.15 Mean Values and Standard Deviation of the error (w.r.t to the LS optimal configuration), for a varying decay factor s, after 10 learning epochs

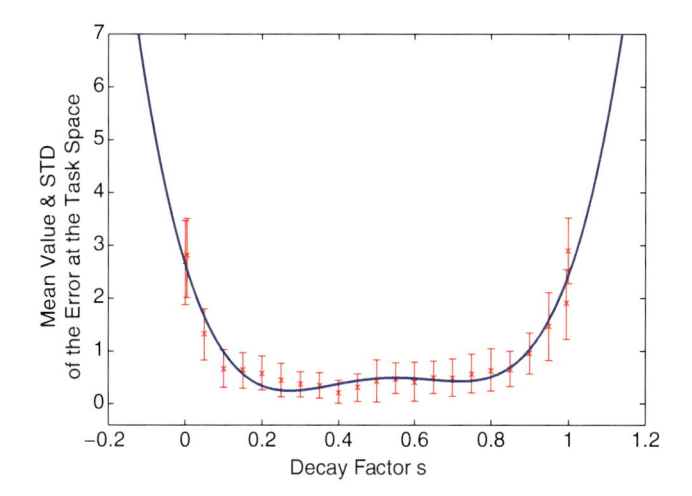

Fig. 16.16 Mean Values and Standard Deviation of the error (w.r.t to the LS optimal configuration), for a varying decay factor s, after 100 learning epochs

reinforcement approach adopted in this paper statistically and progressively rewards actions and converges towards behaviours that lead to faster (and more direct) distance error minimization, thus creating state-to-action maps that efficiently achieve the desired target and solve the redundancy by rewarding minimum-effort type of motions. This is an important experimental finding that supports the applicability of the proposed methodology in such a dexterous manipulation context.

When the learning process is completed and the agents have acquired knowledge about how to collaboratively reach their goal, the next step is to evaluate the generalization properties of the system, that is, to see how the agents use the knowledge

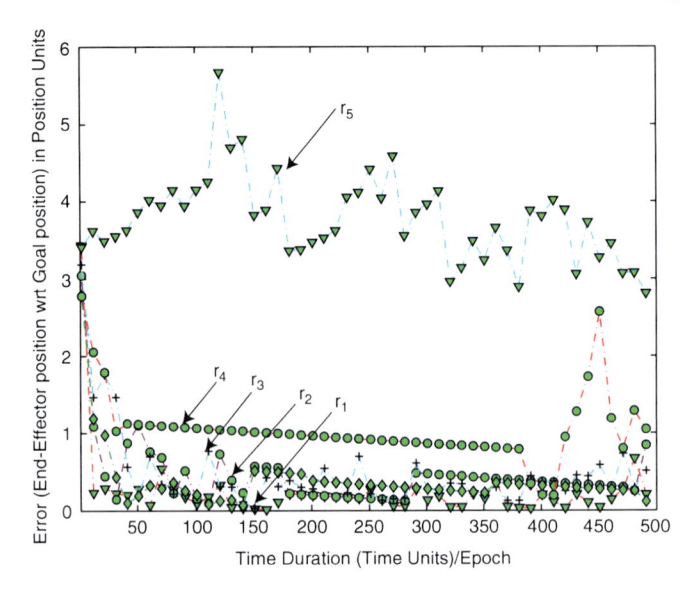

Fig. 16.17 Positioning error observed while reaching new goals in five different areas (of increasing radius $r_1, \ldots r_5$)

acquired and how, without any additional training, they can explore and reach targets that are related (but not identical) to the ones trained. The approach followed to test the behavior of our system is defined on the basis of selecting goals that are located within an area of a given radial distance measured from the initial (trained) goal position. We want to see whether the system is in a position to handle existing knowledge and exploit it further. So, having as a center of the potential exploitation area the initial goal position, we define five areas, each one having a different radius ($r_1 = 0.7, r_2 = 1.4, r_3 = 2.1, r_4 = 2.8, r_5 = 4.2$). For each of these areas, we assign five different new goals to the system, leading to a total of 25 new goals. We average the errors obtained per area and the results obtained are presented in Fig. 16.17. We can see that points in areas close to the initial goal position are indeed reached without any further exploration, thus illustrating the generalization capacity of the proposed multi-agent architecture.

16.6.2 Multi-Finger Grasp

Having seen how the proposed multi-agent architecture can be used on a single kinematic chain (composed of four independent agents that try to collaborate to reach a specific goal position without any prior knowledge or build-in behavior model), we are next attempting to evaluate the same architecture on a more complex task. We assume a three-finger dexterous manipulator, attempting to perform a quasi-static grasp for an object of supposedly infinite mass, as depicted in Fig. 16.18.

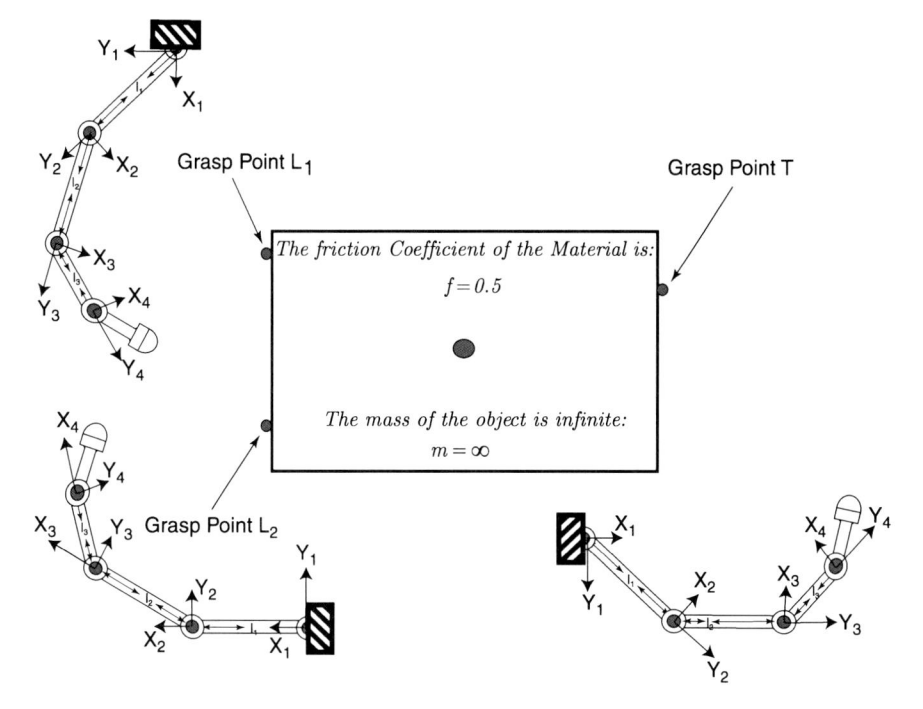

Fig. 16.18 Three-finger quasi-static grasp

The numerical experiment presented in the sequel attempts to evaluate the learning capacity of a multi-agent system comprised of 13 nested agents, performing a grasp without slipping on any fingertip of the manipulator, and with desired resultant force and torque equal to zero.

Before proceeding to the simulation of this specific setup, what is interesting to visualize is how the kinematic chains, depicted in Fig. 16.18, are mapped to the proposed multi-agent environment. This multi-agent representation is shown in Fig. 16.19, clearly illustrating both the nested and the hierarchical nature of the proposed architecture. The detailed parameters used in the simulation for the three-finger grasp are shown in Table 16.3. We can see the initial joint configuration of every kinematic chain, and the friction cone range at every grasp point, where each multi-agent finger will have to learn to operate within. We note that all contact points are simulated using a simple linear elasticity model, with specific spring and damping coefficients, as shown in Fig. 16.20, where k_t is the tangential and k_n is the normal spring coefficient, and c_n, c_t are the normal and tangential damping coefficients, respectively. The rest of the simulation parameters are the same as have been described in the previous section. The fuzzification parameters and the learning parameters have been kept the same as before. The system is trained over 200 epochs, each epoch lasting 500 time units. Within the corresponding time interval, the agents will have to learn simultaneously how to solve two problems: (a) first, to reach the grasp points (according to the previous experiment, their goal positions), and (b)

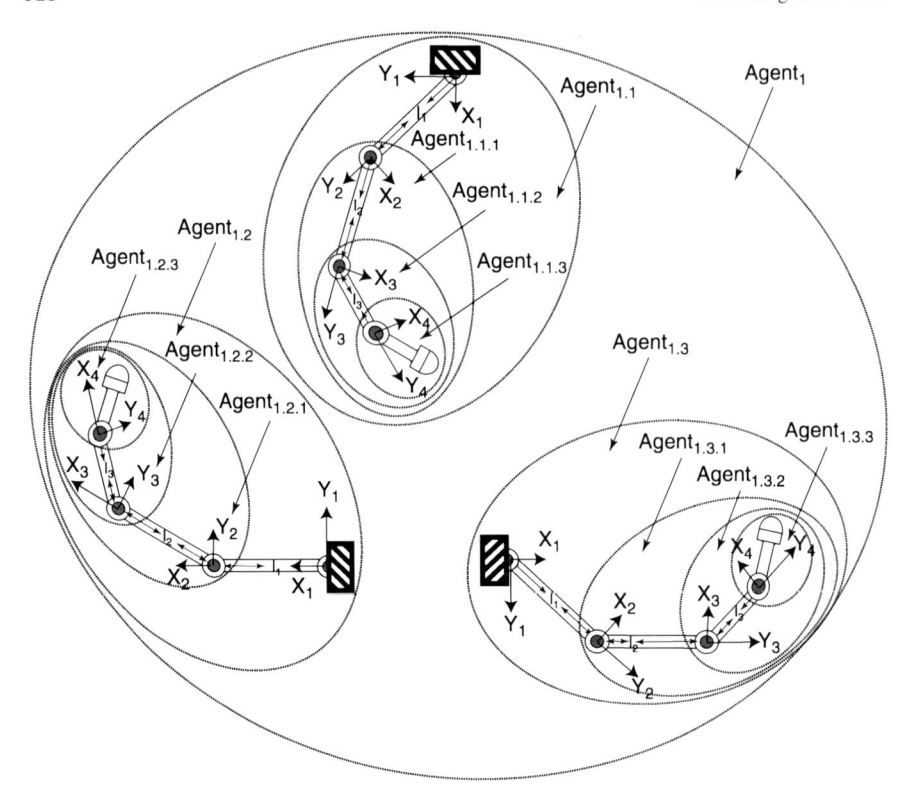

Fig. 16.19 Multi-agent representation of a three-finger manipulator

Table 16.3 Grasp parameters

	Finger L_1	Finger L_2	Finger T
Num of links (l_i)	4	4	4
Initial joint angle l_1	130°	180°	10°
Initial joint angle l_2	20°	340°	20°
Initial joint angle l_3	30°	330°	30°
Initial joint angle l_4	50°	310°	40°
Step: Joint angle	18°	18°	18°
Step: Goal angle	40°	40°	40°
Friction coefficient	0.5	0.5	0.5

second, to successfully coordinate their actions to apply appropriate contact forces at the corresponding grasp points, always maintained within the indicated friction cone to avoid slipping effects, and summing up to a desired net force and moment (in the case of the considered experiment, summing up to zero).

According to these requirements, the goal of the multi-agent system for this experiment has been augmented. Since the system is now attempting to learn, at a single step, two different goals, the reward function has to be modified appropriately to correctly guide the RL mechanism. The reward function that we saw in the previous section, which supports the task of reaching a goal position, is still valid

Fig. 16.20 Simulated contact point with tangential and normal spring-damping coefficients

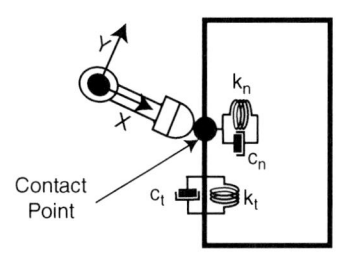

Contact
Point

but is now augmented by reward terms R_i that correspond to the constraints of three subtasks, namely: R_τ being the reward for satisfying the net moment constraint, R_f being the reward for satisfying the net force constraint, and R_s being the reward for satisfying the friction cone (no slip) constraint. Therefore, the reward function driving the current learning process takes the following form:

$$R(t) = \begin{cases} \text{if } (\text{Dist}_{\text{Goal}}(t) \leq \text{Dist}_{\min}) \wedge (\Delta \text{Dist}_{\text{Goal}}) \leq 0) \text{ then } R(t) = e^{-c \times \text{Dist}_{\text{Goal}}(t)} + \sum_i R_i \\ \text{else if } (\text{Dist}_{\text{Goal}}(t) > \text{Dist}_{\min}) \text{ then } R(t) = -2 + \sum_i R_i \\ \text{else if } (\text{Dist}_{\text{Goal}}(t) < \text{Dist}_{\min}) \wedge (\Delta \text{Dist}_{\text{Goal}}) > 0) \text{ then } R(t) = -1 + \sum_i R_i \end{cases}$$

$$(16.28)$$

where the reward terms R_i are defined as follows, for $i = f$ (force constraint), $i = \tau$ (torque constraint), and $i = s$ (no slip constraint):

$$R_f = e^{-c \times \text{Net}_{\text{force}}(t)}, \; R_\tau = e^{-c \times \text{Net}_{\text{torque}}(t)}, \; R_s = e^{-c \times \text{Friction}_{\text{Cone}}(t)}$$

with t indicating the specific trial. The experimental process is composed of three sets of experiments, each one with a different value for the decay factor s ($s = 0.05$, $s = 0.35$, and $s = 0.75$). The obtained results are presented in Figs. 16.21–16.26. In each figure, the results for four different epochs (1, 10, 100, and 200) are shown. These figures show the evolution over time of the applied net force (desired net force equal zero) and mean force error, illustrating the learning performance of the multi-agent grasping system and demonstrating the capacity of the proposed architecture to achieve simultaneous collaborative adaptation of the dexterous manipulator's fingertips actions. Finally, in Fig. 16.27, we see the mean square error over all epochs for three different decay factors.

Regarding the friction coefficient (which, we recall, has been set in this case to $\mu = 0.5$), this defines a friction cone limit within which all fingers must operate (in the sense that all fingertip contact forces must lie inside this limit). Figure 16.28 illustrates how the contact forces applied by each one of the three fingers (kinematic chains) progressively adapt (epochs 1, 10, 100, 200) to fit within these limits of the considered friction cone constraints. Finally, in Fig. 16.29, we see different sets of solutions that the system generates for different simulation parameters. The proposed multi-agent system proceeds using the knowledge acquired, without any additional training, by exploring and reaching new contact points that are related to the ones trained, in a manner similar to the one already described in the previous section for the case of a single kinematic chain.

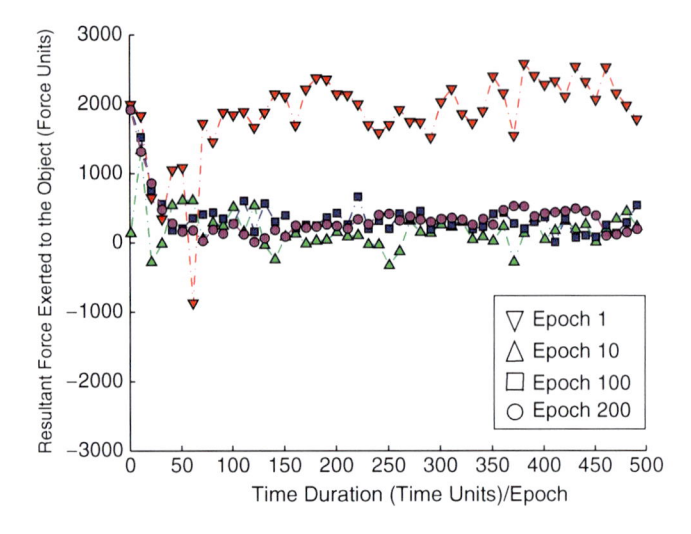

Fig. 16.21 Net force, over time, for different epochs (for $s = 0.05$)

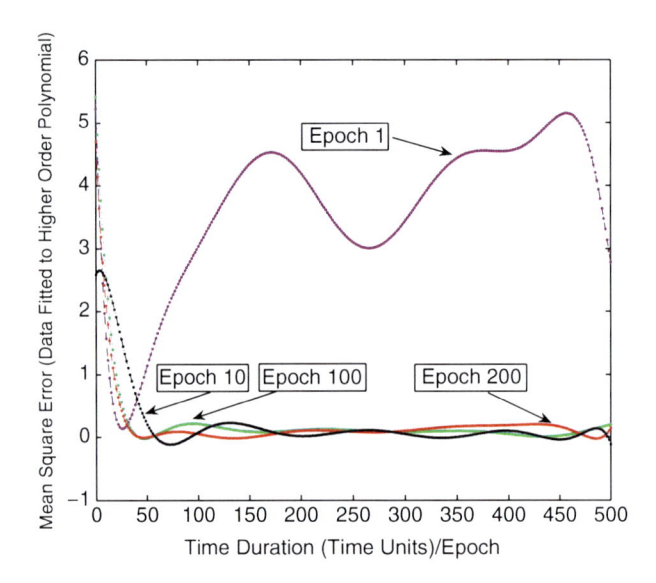

Fig. 16.22 Mean square error for different epochs (for $s = 0.05$)

16.6.3 Collaborative Mobile Robots: Box-Pushing Task

In this last experimental section, we perform an initial evaluation of the proposed multi-agent architecture and learning framework in the domain of mobile robots. The task assigned to two mobile robots is to build certain skills over time that consist of collaboratively pushing a box toward a specified goal position. The system in this

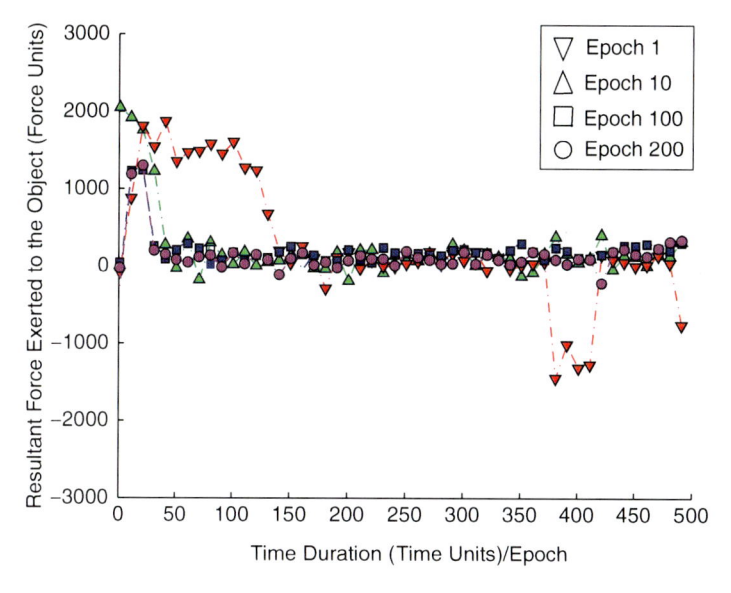

Fig. 16.23 Net force, over time, for different epochs (for $s = 0.35$)

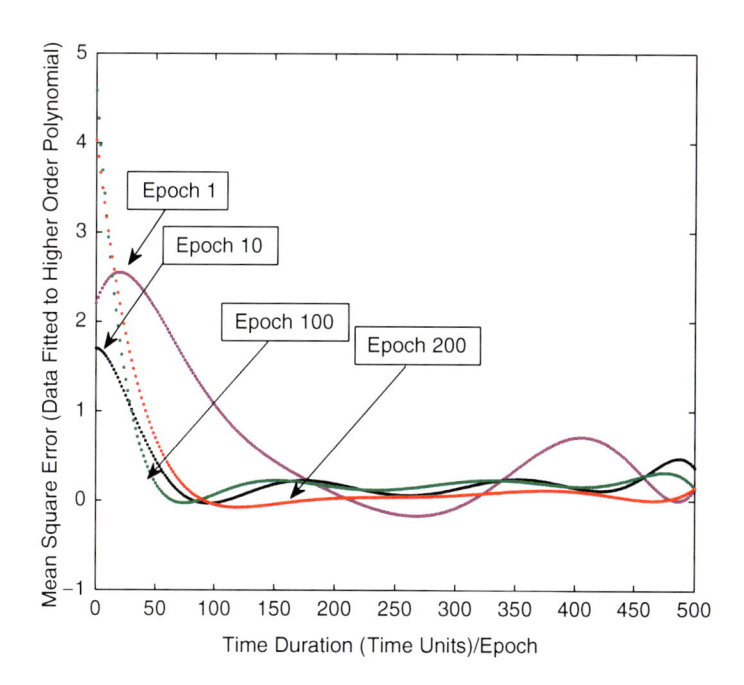

Fig. 16.24 Mean square error over different epochs (for $s = 0.35$)

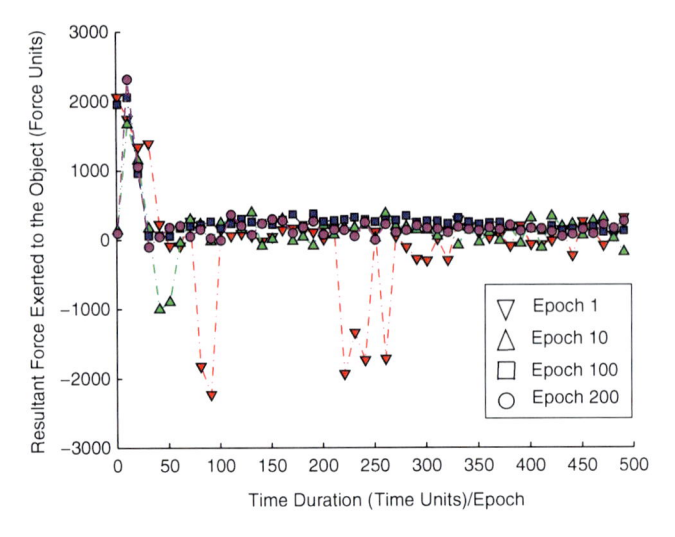

Fig. 16.25 Net force, over time, for different epochs ($s = 0.75$)

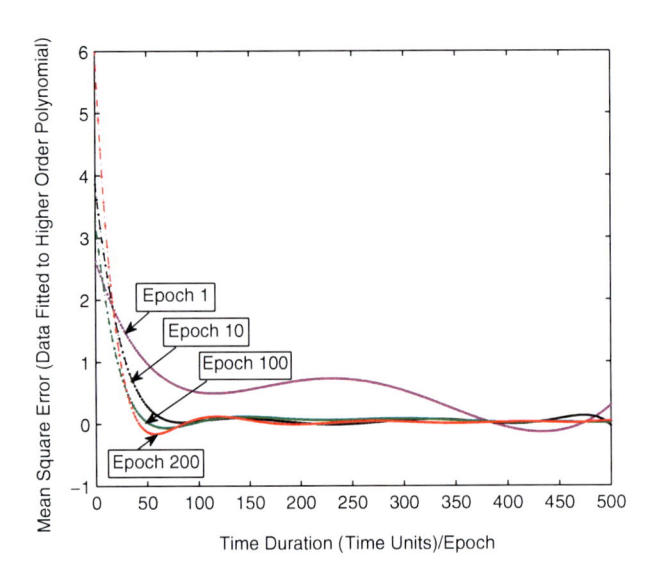

Fig. 16.26 Mean square error over different epochs (for $s = 0.75$)

comprises four distinct agents: Robot 1 – left wheel, Robot 1 – right wheel, Robot 2 – left wheel, and Robot 2 – right wheel. Thus, rephrasing the initial problem statement, four agents have to collaborate to push a box to the desired location. In addition, in this experiment we assume that there is no previous knowledge or build-in behavior model. The simulation environment that we use is Webots 6.1.5, where a

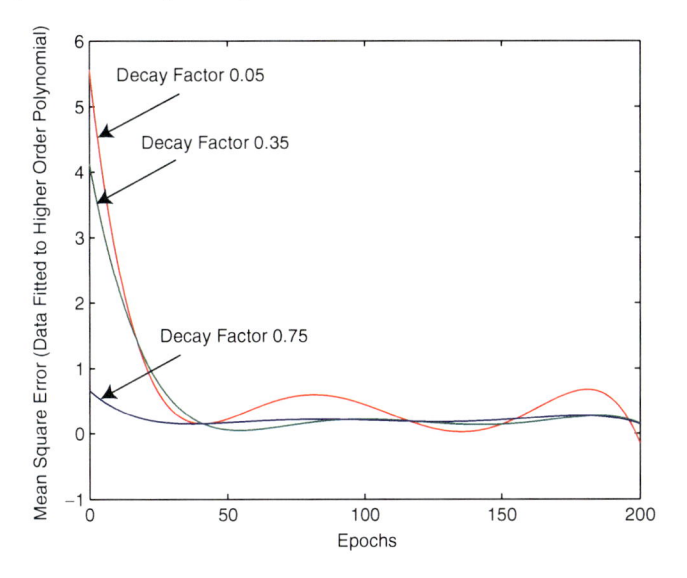

Fig. 16.27 Mean square error over all epochs for different decay factors s

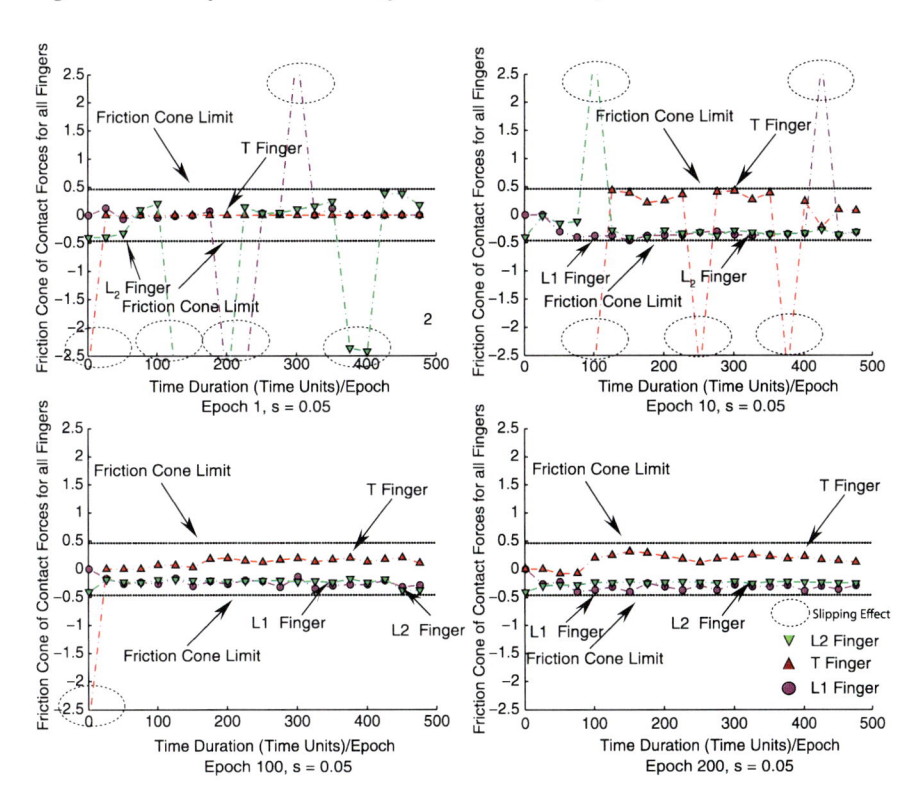

Fig. 16.28 Adaptation of fingertip force within the friction cone limit, $s = 0.05$

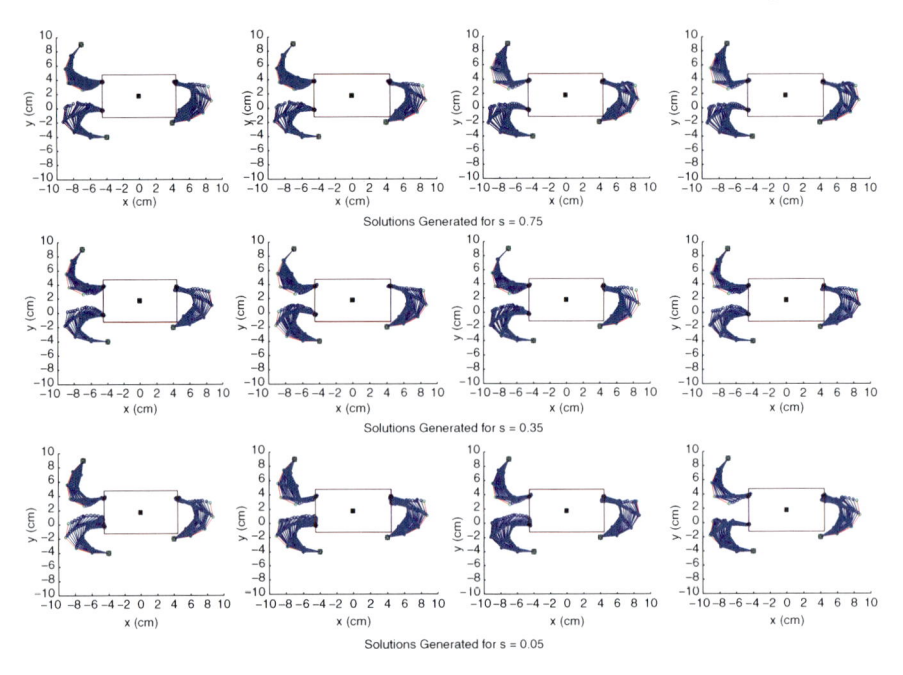

Fig. 16.29 Examples of generated grasps (for $s = 0.75, 0.35,$ and 0.05)

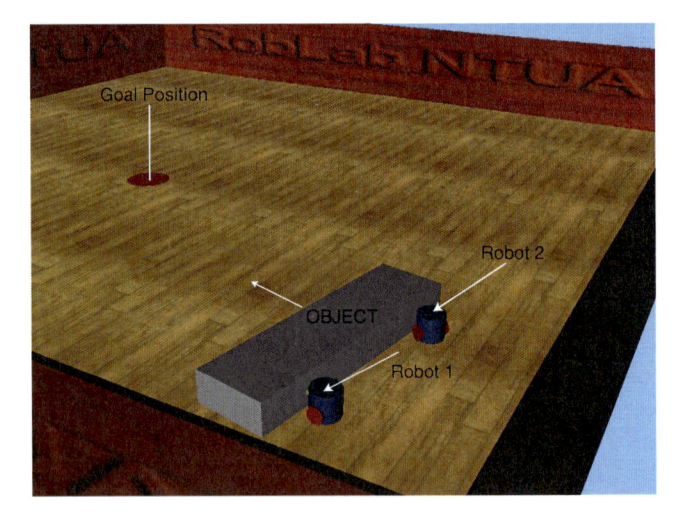

Fig. 16.30 Two mobile robots pushing a box to a specified goal location

synthetic environment has been designed, as can be seen in Fig. 16.30. The detailed parameters used in the simulation for the mobile robots are shown in Table 16.4.

The system is allowed to tune its θ parameters for a series of epochs, each one lasting for 900 trials. The learning parameters are: learning rate $\alpha = 0.0005$, the

Table 16.4 Experimental parameters

	Robot 1	Robot 2	Object
Left/Right wheel radius	0.025 m	0.025 m	N/A
Left/Right wheel coulomb friction	1	1	N/A
Distance between the wheels	0.09 m	0.09 m	N/A
Robot radius	0.045 m	0.045 m	N/A
Robot mass	0.5	0.5	N/A
Touch sensor	Yes	Yes	N/A
Object mass	N/A	N/A	0.5
Object inertia	N/A	N/A	0.02417
Object width	N/A	N/A	0.23 m
Object length	N/A	N/A	0.82 m
Object height	N/A	N/A	0.1 m
Object coulomb friction	N/A	N/A	1

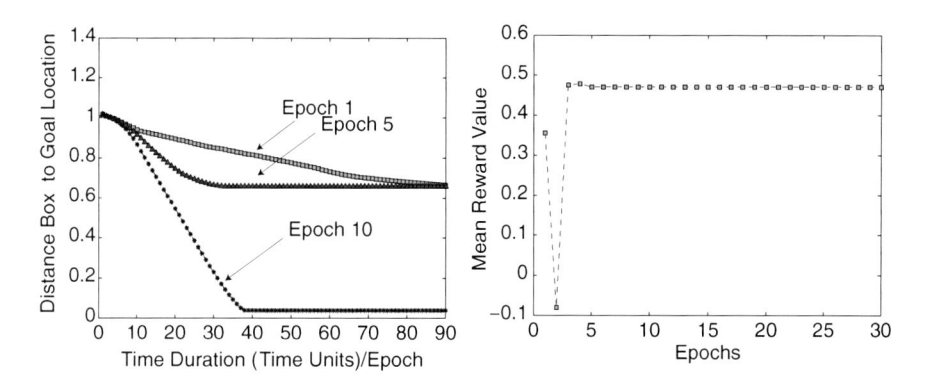

Fig. 16.31 The wheels decreasing the distance of the box from the goal position. The mean reward the multi-agent system receives over different epochs

discount factor $\gamma = 0.99$, $\lambda = 0.6$, while the decay factor $s = 0.005$. The results of the simulations are presented in Figs. 16.31 and 16.32. We can see in Fig. 16.31a that the agents (wheels) manage to build skills over time, so that they eventually collaborate to drive the robots in a way that the box is finally pushed to the goal location. As depicted in the corresponding plot, the goal is approached during the initial epochs (1-5), and the situation is significantly improved in the subsequent epochs, reaching epoch 10 when the box is successfully pushed by the cooperating robots to reach the desired goal location. Figure 16.31b shows the mean reward value that the system receives over different epochs. We note that the mean reward the system receives converges to a constant value after epoch 5 and maintains that constant value until the end. Figure 16.32 shows an instance of the simulation results obtained using Webots 6.1.5 software, for epoch 1, epoch 5, and epoch 10.

Fig. 16.32 Simulation results for epoch 1, epoch 5, and epoch 10

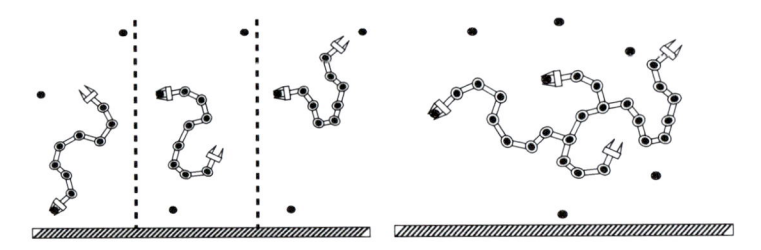

Fig. 16.33 Dexterous robotic chains performing hybrid locomotion/manipulation tasks (like climbing)

16.7 Conclusion and Future Work

By combining RL methods and traditional control approaches over a hierarchical multi-agent architecture, in a fuzzified state-space, we obtain a hybrid robot architecture, with respect to control topology, which is applied in a continuous state-space to perform a multi-finger quasi-static robotic grasp. The nested, self-evolving multi-agent architecture proposed in this chapter (which constitutes in fact an implementation of a recursive mechanism able to search for solution(s) in a specific problem-set) appears to be particularly suitable for robot control problems where increased degree of dexterity and cooperative skills are required. The basic advantage of such an approach is that no global task model (in the case-studies presented in this chapter, no robot inverse kinematic model) is needed. Moreover, the proposed multi-agent system, owing to its homogeneous characteristics (all agents obey the same structural/modular internal architecture), as well as to its hierarchical formation, facilitates scaling of the system to more complex structures.

In the last part of this work, we show the application of the proposed architecture to the field of mobile robotics, where the wheels of the robots become the independent agents that explore behaviors, and through them collaborative skills evolve. Figure 16.33 depicts another potential application of the proposed framework, where a rather more complicated multi-agent environment could be envisaged. Using the proposed framework in the domain of dexterous manipulation, we believe that challenging problems in this specific area can be tackled in a very elegant, interesting, and powerful way (for instance, in the sense of modularity, robustness, and extensibility). Similar (in some ways equivalent) problem settings, like grasp planning, locomotion control, or designing optimal climbing (and generally gaiting or locomotion) patterns, could also be approached within the same framework, leading to the notions of evolving cooperative learning and developmental robot control skills. Bringing together the areas of multi-agent architectures, machine learning and dexterous robotics will create new challenges, both theoretic and application-oriented, for all these domains of research.

References

M. N. Ahmadabadi and E. Nakano, "A "Constrain and Move" Approach to Distributed Object Manipulation", *Robotics and Automation, IEEE Transactions on*, 17(2), 157–172, 2001.

B. D. Argall et al., "A Survey of Robot Learning from Demonstration", *Robotics and Autonomous Systems*, 2008, doi:10.1016/j.robot.2008.10.024.

D. P. Bertsekas and J. N. Tsitsiklis, *Neuro-Dynamic Programming*, Athena Scientific, Belmont, MA, 1996.

R. A. Brooks, "A Robust Layered Control System for Mobile Robots", *IEEE Journal of Robotic Automation*, RA-2, 14–23, 1986.

G. W. Brown, "Iterative Solution of Games by Fictitious Play." In T. C. Koopmans editor, *Activity Analysis of Production and Allocation*, Wiley, New York, 1951.

Y. Cao, A. S. Fukunaga, A. Kahng, and F. Meng, "Cooperative Mobile Robots: Antecedents and Directions", In *Proceedings of the IEEE/RSJ International Conference on Intelligent Robots and Systems*, vol. 1, pp. 226–243, 1995.

C. Claus and C. Boutilier, "The Dynamics of Reinforcement Learning in Cooperative Multiagent Systems", *AAAI/IAAI*, pp. 746–752, 1998.

P. Dayan and L. F. Abbott, *Theoretical Neuroscience, Computational and Mathematical Modeling of Neural Systems*, MIT, Cambridge, MA, 2001.

K. Doya, "Temporal Difference Learning in Continuous Time and Space", *Advances in Neural Information Processing Systems 8*, MIT, Cambridge, MA, 1996.

D. R. Donald, J. Jennings, and D. Rus, "Information Invariant for Distributed Manipulation", *International Journal of Robotics Research*, 16(5), 673–702, 1997.

D. Fundenberg and D. M. Kreps, "Lectures on Learning and Equilibrium in Strategic Form Games", *CORE Foundation*, Louvain-La-Neuve, Belgium, 1992.

M. Iida, M. Sugisaka, and K. Shibata, "Application of Direct-Vision-Based Reinforcement Learning to a Real Mobile Robot", *Artificial Life and Robotics*, 7(3), 102–106, 2004.

L. P. Kaelbling, M. L. Littman, and A. W. Moore, Reinforcement Learning: A Survey, *Journal of Artificial Intelligence Research*, 4, 237–285, 1996.

O. Khatib et al., "Vehicle/Arm Coordination and Multiple Mobile Manipulator Decentralized Cooperation", In *Proceedings of the IEEE/RSJ International Conference on Intelligent Robotics and Systems*, vol. 2, Osaka, Japan, pp. 546–553, 1996.

J. R. Kok and N. Vlassis, "Sparse Tabular Multiagent Q-Learning", *Proceedings of Annual Machine Learning Conference of Benelearn* 2004.

T. Kondo and K. Ito, "A Reinforcement Learning using Adaptive State Space Construction Strategy for Real Autonomous Mobile Robots", *Robotics and Autonomous Systems*, vol. 46, no.2 pp. 111–124, Elsevier, 2004.

M. Lauer and M. Riedmiller, "Reinforcement Learning for Stochastic Cooperative Multi-Agent Systems," *aamas, pp. 1516–1517, Third International Joint Conference on Autonomous Agents and Multiagent Systems – Volume 3 (AAMAS'04)*, 2004.

J. Liu et al., "Reinforcement Learning for Autonomous Robotic Fish", *Studies in Computational Intelligence (SCI)*, 50, 121–135, 2007.

M. Lopes and J. Santos-Victor, "A Developmental Roadmap for Learning by Imitation in Robots", *Systems, Man, and Cybernetics Part B: Cybernetics, IEEE Transactions on*, 37(2), 2007.

M. Lungarella, G. Metta, R. Pfeifer, and G. Sandini, "Developmental Robotics: A Survey," *Connection Science*, 15(4), 151–190, 2003.

T. Matsui, T. Omata, and Y. Kaniyoshi, "Multi-Agent Architecture for Controlling a Multi-finger Robot", *Proceedings of the 1992 IEEE/RSJ International Conference on Intelligent Robots and Systems*, Raleigh, NC, 1992.

M. McGlohon and S. Sen, "Learning to Cooperate in Multi-Agent Systems by Combining Q-Learning and Evolutionary Strategy", *World Conference on Lateral Computing*, December 2004.

R. B. Myerson, *Game Theory: Analysis of Conflict*, Harvard University Press, Cambridge, 1991.

Y. Nakamura, Advanced Robotics: Redundancy and Optimization. Reading, MA, Addison-Wesley, 1990.

D. Rus, "Coordinated Manipulation of Objects", *Algorithmica*, 19(1), 129–147, 1997.

S. Schaal, "Is Imitation Learning the Route to Humanoid Robots", *Trends in Cognitive Sciences*, 3(6), 233–242, 1999.

K. Shibata, M. Sugisaka, and K. Ito, "Fast and Stable Learning in Direct-Vision-Based Reinforcement Learning", *Proceedings of International Symposium On Artificial Life and Robotics (AROB) 6th*, pp. 562–565, 2001.

K. Shibata and Y. Okabe, "Smoothing-Evaluation Method in Delayed Reinforcement Learning", 1995.

K. Shibata and Y. Okabe, "A Robot that Learns an Evaluation Function for Acquiring of Appropriate Motions" *World Congress on Neural Networks-San Diego, 1994 International Neural Network Society Annual Meeting*, Vol. 2., pp. II. 29-II34, 1994.

K. Shibata and K. Ito, "Effect of Force Load in Hand Reaching Movement Acquired by Reinforcement Learning", *ICONIP'02, Proceedings of the 9th International Conference on Neural Information Processing*, Computational Intelligence for the E-Age, 2002.

K. Shibata and K. Ito, "Hidden Representation After Reinforcement Learning of Hand Reaching Movement with Variable Link Length", *Proceedings of IJCNN(International Confernce on Neural Networks) 2003*, 1475–674, pp. 2619–2624, 2003.7.

Y. Shoham and M. Tennenholtz, "On the synthesis of useful social laws for artificial agent societies", *Proceedings AAAI-92*, pp. 276–281, San Jose, 1992.

R. S. Sutton and A. G. Barto, *Reinforcement Learning: An Introduction*, MIT, Cambridge, MA, 1998.

R. S. Sutton, H. R. Maei, D. Precup, S. Bhatnagar, D. Silver, C. Szepesvari, and E. Wiewiora, "Fast Gradient-Descent Methods for Temporal-Difference Learning with Linear Function Approximation", *26th International Conference on Machine Learning*, Montreal, Canada, 2009.

T. Takahashi, T. Tanaka, K. Nishida, and T. Kurita, "Self-Organization of Place Cells and Reward-Based Navigation for a Mobile Robot", *ICONIP* 2001.

C. Watkins, "Learning from Delayed Rewards", *PhD Thesis*, University of Cambidge, England, 1989.

Y. Yoshikawa and X. Zheng, "Coordinated Dynamic Hybrid Position/Force Control for Multiple Robot Manipulators Handling One Constrained Object", Int. J. Robot. Res., vol. 12, pp. 219–230, 1993.

J. Zamora, J. d. R. Millan, A. Murciano, "Specialization in Multi-Agent Systems Through Learning", *Biological Cybernetics*, vol. 76, pp. 375–382, Springer, Berlin, 1997.

Chapter 17
Actions and Imagined Actions
in Cognitive Robots

**Vishwanathan Mohan, Pietro Morasso, Giorgio Metta,
and Stathis Kasderidis**

Abstract Natural/Artificial systems that are capable of utilizing thoughts at the service of their actions are gifted with the profound opportunity to mentally manipulate the causal structure of their physical interactions with the environment. A cognitive robot can in this way virtually reason about how an unstructured world should "change," such that it becomes a little bit more conducive towards realization of its internal goals. In this article, we describe the various internal models for real/mental action generation developed in the GNOSYS Cognitive architecture and demonstrate how their coupled interactions can endow the GNOSYS robot with a preliminary ability to virtually manipulate neural activity in its mental space in order to initiate flexible goal-directed behavior in its physical space. Making things more interesting (and computationally challenging) is the fact that the environment in which the robot seeks to achieve its goals consists of specially crafted "stick and ball" versions of real experimental scenarios from animal reasoning (like tool use in chimps, novel tool construction in Caledonian crows, the classic trap tube paradigm, and their possible combinations). We specifically focus on the progressive creation of the following internal models in the behavioral repertoire of the robot: (a) a passive motion paradigm based forward inverse model for mental simulation/real execution of goal-directed arm (and arm + tool) movements; (b) a spatial mental map of the playground; and (c) an internal model representing the causality of pushing objects and further learning to push intelligently in order to avoid randomly placed traps in the trapping groove. After presenting the computational architecture for the internal models, we demonstrate how the robot can use them to mentally compose a sequence of "Push–Move–Reach" in order to Grasp (an otherwise unreachable) ball in its playground.

V. Mohan (✉)
Cognitive Humanoids Lab, Robotics Brain and Cognitive Sciences Department,
Italian Institute of Technology, Genoa, Italy
e-mail: vishwanathan.mohan@iit.it

V. Cutsuridis et al. (eds.), *Perception-Action Cycle: Models, Architectures,
and Hardware*, Springer Series in Cognitive and Neural Systems 1,
DOI 10.1007/978-1-4419-1452-1_17, © Springer Science+Business Media, LLC 2011

17.1 Introduction

The world we inhabit is an amalgamation of structure and chaos. There are regularities that could be exploited. Species biological or artificial, which do this best, have the greatest chances of survival. We may not have the power of an ox or the mobility of an antelope but still our species surpasses all the rest in our flair by inventing new ways to think, new ways to functionally couple our bodies with the structure afforded by our worlds. Simply stating, it is this ability to "explore, identify, internalize, and exploit" the possibilities afforded by the structure in one's immediate environment to counteract limitations "of perceptions, actions, and movements" imposed by one's embodied physical structure, and to do this in accordance with one's "internal goals," that forms the hallmark of any kind of cognitive behavior. In addition, natural/artificial systems that are capable of utilizing "thoughts" at the service of their "actions" are gifted with the profound opportunity to mentally manipulate the causal structure of their physical interactions with the environment. Complex bodies can in this way decouple behavior from direct control of the environment and react to situations that "do not really exist" but "could exist" as a result of their actions on the world. However, the computational basis of such cognitive processes has still remained elusive.

This is a difficult problem, but there are many pressures to provide a solution – from the intrinsic viewpoint of better understanding ourselves to creating artificial agents, robots, smart devices, and machines that can reason and deal autonomously with our needs and with the peculiarities of the environments we inhabit and construct. This has led researchers toward several important questions regarding the nature of the computational substrate that could drive an artificial agent to exhibit flexible, purposeful, and adaptive behavior in complex, novel, and sometimes hostile environments. How do goals, constraints, and choices "at multiple scales" meet dynamically to give rise to the seemingly infinite fabric of reason and action? Is there an internal world model (of situations, actions, forces, causality, abstract concepts)? If yes, "How" and "What" is modeled, represented, and connected? How are they invoked? What are the planning mechanisms? How are multiple internal models coordinated to generate (real/mental) sequences of behaviors "at appropriate times" so as to realize valued goals? How should a robot respond to novelty and how can a robot exhibit novelty? What kind of search spaces (physical and mental) is involved and how are they constrained? This chapter is in many ways an exploration into some of these questions expressed through the life of a moderately complex robot "GNOSYS" playing around in a moderately complex playground (which implicitly hosts artificially reconstructed scenarios inspired from animal cognition), trying to use its perceptions, actions, and imaginations "flexibly and resourcefully" so as to cater "rewarding" user goals.

In spite of extensive research in multiple fields, scattered across multiple scientific disciplines, it is fair to say that the present day artificial agents still lack much of the resourcefulness, purposefulness, flexibility, and adaptability that humans so effortlessly exhibit. Cognitive agent architectures are found in the current literature, ranging from purely reactive ones implementing the cycle of perception and action in a simplistic hardwired way to more advanced models of perception, state estimation, and action generation (Brooks 1986; Georgeff 1999; Toussaint 2006; Shanahan 2005; Gnadt and Grossberg 2008; Sun 2007, CLAR-ION architecture), architectures for analogy making (Hofstadter 1984; French 2006; Kokinov and Petrov 2001), causal learning (Pearl 1998; Geffner 1992), probabilistic/statistical inference (Yuille et al. 2006; Pearl 1988), and brain-based devices (Edelman et al. 2001, 2006, DARWIN BBDs). Even though symbols and symbol manipulation have been the main stay of cognitive sciences (Newell and Simon 1976) ever since the days of its early incarnations as AI, the disembodied nature of traditional symbolic systems, the need to presuppose explicit representations, symbol grounding and all other associated problems discussed in Sun (2000) have been troubling many cognitive scientists (Varela and Maturana 1974).

This led to the realization of the need for *experience* to precede *representation*, in other words the emergence of representational content as a consequence of sensory–motor interactions of the agent with its environment, a view that can be traced back to many different contributions spanning the previous decades, e.g., Wiener's Cybernetics (1948), Gibson's ecological psychology (1966), Maturana and Varela's autopoesis (1974), Beer's neuroethology (1990), and Clark's situatedness (1997). In this view, adaptive behavior can best be understood within the context of the (biomechanics of the) body, the (structure of the organism's) environment, and the continuous exchange of signals/energy between the nervous system, the body, and the environment. Hence the appropriate question to ask is not what the neural basis of adaptive behavior is, but what the contributions of all components of the coupled system to adaptive behavior and their mutual interactions are (Morasso 2006).

In other words, the ability to autonomously explore, identify, internalize, and exploit possibilities afforded by the structure in one's immediate environment is critical for an artificial agent to exercise intelligent behavior in a messy world of objects, choices, and relationships. Intelligent agents during the course of their lifetimes gradually master this ability of coherently integrating the information from the bottom (sensory, perceptual, conceptual) with the drives from the top (user goals, self goals, reward expectancy), thereby initiating actions that are maximally rewarding. A major part of this process of transformation takes place in the mental space (Holland and Goodman 2003), wherein the agent, with the help of an acquired internal model, executes virtual actions and simulates the usefulness of their consequences toward achieving the active goal. Hence, unlike a purely reactive system where the motor output is exclusively controlled by the actual sensory input, the idea that a cognitive system must be capable of mentally simulating action sequences aimed at achieving a goal has been gaining prominence in literature. This also resonates very well with emerging biological evidence in support of the simulation hypothesis toward generation of cognitive behavior, mainly *simulation of action*: we are able to activate motor structures of the brain in a way that resembles activity during a normal action but does not cause any overt movement (Metzinger and Gallese 2003; Grush 2004); *simulation of perception*: imagining perceiving something is actually similar to the perceiving it in reality, only difference being that, the perceptual activity is generated by the brain itself rather than by external stimuli (Grush 1995); *anticipation*: there exist associative mechanisms that enable both behavioral and perceptual activity to elicit other perceptual activity in the sensory areas of the brain. Most important, a simulated action can elicit perceptual activity that resembles the activity that would have occurred if the action had actually been performed (Hesslow 2002).

Computationally, this implies the need to have two different kinds of loops in the agent architecture: first, a situation–action–consequence loop or forward model that allows contemplated decision making (without actual execution of action) and second, a Situation–Goal–Action loop to solve the inverse problem of finding action sets which map the transformation from initial condition to active goal. That such forward models of the motor system occur in the brain has been demonstrated by numerous authors. For example, Shadmehr (1999) has shown how adaptation to novel force fields by humans is only explicable in terms of both an inverse controller and a learnable forward model. More recent work has proposed methods by which such forward models can be used in planning (where actual motor action is inhibited during the running of the forward model) or in developing a model of the actions of another person (Oztop et al. 2004). Engineering Control frameworks of attention, using modules of control theory (Taylor 2000) extended so as to be implemented using neural networks, have been extensively applied to modeling motor control in the brain (Morasso 1981; Wolpert, Ghahrmani and Jordan 1994; Imamizu 2000), with considerable explanatory success. Such planning has been analyzed in these and numerous other publications for motor control and actions but not for more general thinking, especially including reasoning. Nor has the increasingly extensive literature on imagining motor actions been appealed to: it is important to incorporate how motor actions are imagined as taking place on imagined objects, so as to "reason" what objects and actions are optimally rewarding. Others have also emphasized the need to combine working memory modules for imagining future events with forward models, for example, the process termed "prospection" in Emery and Clayton (2004). Guided by the experimental results from functional imaging and neuropsychology, computational architectures have recently begun to emerge in the literature for open-ended, goal-directed reasoning in artificial agents, most importantly incorporating the creation and use of internal models and motor imagery. A variety of computational architectures incorporating these ideas have been proposed recently, for example, an architecture that combines internal simulation with a global workspace (Shanahan 2005), Internal Agent Model (IAM) theory of consciousness (Holland 2003), learning a world model using interacting self-organizing maps (Toussaint 2004, 2006), and learning motor sequences using recurrent neural networks with parametric bias (Tani et al. 2007).

The idea of using internal models to aid generation of intelligent behavior also resonates very well with compelling evidence from several neuropsychological, electrophysiological, and functional imaging studies, which suggest that much of the same neural substrates underlying modality perception are also used in imagery; and imagery, in many ways, can "stand in" for (re-present, if you will) a perceptual stimulus or situation (Zattore et al. 2007; Behrmann 2000; Fuster 2003). Studies show that imagining a visual stimulus or performing a task that requires visualization is accompanied by increased activity in the primary visual cortex (Kosslyn et al. 1993; Klein et al. 2000). The same seems to be true for specialized secondary visual areas like fusiform gyrus, an area in the occipito-temporal cortex which is activated both when we see faces (Op de Beeck et al. 2008) and also when we imagine them (O'Craven and Kanwisher 2000). Lesions that include this area impair both face recognition and the ability to imagine faces. Brain imaging studies also illustrate heavy engagement of the motor system in mental imagery, i.e., we are able to activate motor structures of the brain in a way that resembles activity during a normal action but does not cause any overt movement (Parsons et al. 2005; Rizzolati et al. 2001; Grush 2004). EEG recordings on subjects performing mental rotation tasks have revealed activation of premotor and parietal cortical areas, indicating that they may be performing covert mental simulation of actions by engaging the same motor cortical areas that are used for real action execution. FMRI studies have similarly found activation of the supplementary motor area as well as of the parietal cortex during mental rotation (Cohen 1996). Similar results have also been obtained from experiments that involve auditory imagery of melodies that activates both the superior temporal gyrus (an area crucial for auditory perception) and the supplementary motor areas. Further, metallization also affects the autonomic nervous system, the emotional centers, and the body in same ways as actual perceptual experiences (Damasio 2000).

To summarize, the increasing complexity of our society and economy places great emphasis on developing artificial agents, robots, smart devices, and machines that can reason and deal autonomously with our needs and with the peculiarities of the environments we inhabit and construct. On the other hand, considerable progress in brain science, emergence of internal model-based theories of cognition, and experimental results from animal reasoning has resulted in tremendous interest of the scientific community toward investigation of higher level cognitive functions using autonomous robots as tools. Rapid increase in robots' computing capabilities, quality of their mechanical components, and subsequent development of several interesting (and complicated) robotic platforms, for example, Cog (Brooks 1997) with 21 Degrees of Freedom (DoFs), DB (Atkeson et al. 2000) with 30 DoFs, Asimo (Hirose and Ogawa 2007) with 34 DoFs, H7 (Nishiwaki et al. 2007) with 35 DoFs, iCub (Natale et al. 2007) with 53 DoFs, raise the challenge to propose concrete computational models for reasoning and action generation capable of driving these systems to exhibit purposeful, intelligent response and develop new skills for structural coupling with their environments.

The computational machinery driving the action generation system of the GNOSYS robot presented in this chapter contributes solutions to a number of issues that need to be solved to realize these competences:

(a) Account for forward/inverse functions of sensorimotor dependencies for a range of motor actions/action sequences
(b) Provide a proper neural representation to realize goal-directed planning, virtual experiments, and reward-related computations
(c) Capable of learning the state representations (sensory/motor) by exploration (and importantly without hand coded states, unrealistic assumptions in data acquisition)

(d) Models that are scalable (wrt dimensionality) and have an organized way to deal with novelty in state space
(e) Plastic and capable of representing/dealing with dynamic changes in the environment
(f) Capable of accommodating heterogeneous optimality criteria in a goal dependent fashion (and not being governed by a single predefined minimization principle to constrain solution space/resolve redundancy)
(g) Built in mechanisms for temporal synchrony and maintenance of continuity in perception, action, and time
(h) A clear framework for the integration of three important streams of information in any cognitive system: the top-down (simulated sensorimotor information), the bottom-up (real sensorimotor information) and the active goal
(i) Using the measure of coherence between these informational streams to alter behavior from normal dynamics to explorative dynamics, with a goal to maintain psychological consistency in the sensorimotor world
(j) Demonstrate the effectiveness of the architecture in a physical instantiation that allows active sensing/autonomous movement in ecologically realistic environments that permits comparisons to be made with experimental data acquired from animal nervous systems animal reasoning tasks

The rest of the chapter is organized as follows: Section 17.2 presents a general overview of the environmental set-up we constructed for training/validating the reasoning-action generation system of the GNOSYS robot, experiments from animal reasoning that inspired the design of the playground, and the intricacies involved in different scenarios that the environment implicitly affords to the robot during phases of user goal/curiosity driven explorative play. Section 17.3 presents a concise overview of the forward/inverse model for simulating/executing a range of goal-directed arm (and arm + tool) movements. Section 17.4 describes how a spatial map of the playground and an internal model for pushing objects is learnt by the GNOSYS robot with specific focus on acquisition, dynamics, generation of goal-directed motor behavior and dealing with dynamic changes in the world. How these internal models can operate unitedly in the context of an active goal is the major focus of Sect. 17.5. A discussion concludes.

17.2 The GNOSYS Playground

Emerging experimental studies from animal cognition reveal many interesting behaviors demonstrated by animals that have shades of manipulative tactics, mental swiftness, and social sophistication commonly attributed to humans. Such experiments generally focus on many open problems that are of great interest to the cognitive robotics community, mainly attention, categorization, memory, spatial cognition, tool use, problem solving, reasoning, language, and consciousness. Seeing a tool using chimp or a tool making corvid often falls short of astonishing us unless we question their computational basis or try to make robots do similar tasks that we often take as granted for humans. The advantages of creating a rich sensorimotor world for a cognitive robot are several: (a) facilitate exploration-driven development of different sensorimotor contingencies of the robot; (b) development of goal-dependent value systems; (c) allow realistic and experience driven internal representation of different cause effect relations, outcomes of interventions; (d) aid the designer to understand various computational mechanisms that may be in play (and should be incorporated in the cognitive architecture) based on the amazingly infinite ways by which goals may be realized in different scenarios; and (e) serve as a test bed to evaluate the performance of the system as a whole and make comparisons of the robot's behavior with that of real organisms, other cognitive architectures. Guided by experiments from animal reasoning, we constructed a playground for GNOSYS robot that implicitly hosts experimental scenarios of tasks related to physical cognition known to be solved by different species of primates, corvids, and children below 3 years. As seen in Fig. 17.1, The GNOSYS playground is a 3×3 m enclosure (every square approx. 1 m^2) with goal objects placed at arbitrary locations on the floor and on the centrally placed table. Objects like cylinders of various sizes, sticks of different lengths (possible tools to reach/push otherwise unreachable goal objects), and balls are generally placed randomly in the environment. Among the available sticks, the small red sticks are magnetized. Hence the robot can discover (through intervention) an additional affordance of making even longer sticks using them. Further, as seen in the Fig. 17.1, a horizontal groove is cut

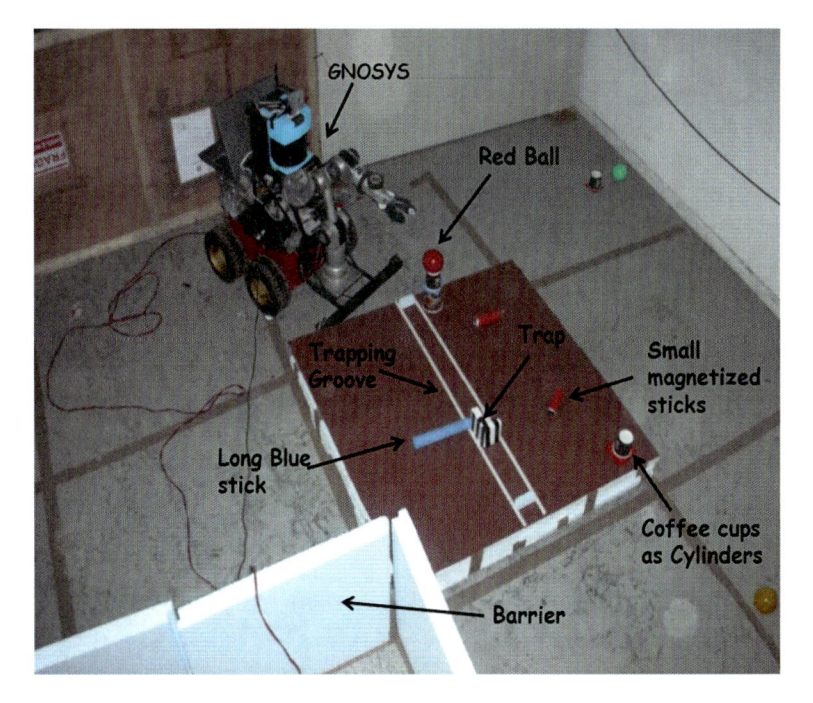

Fig. 17.1 3×3 m GNOSYS playground with different randomly placed goal objects and tools

and run across the table from one side to another, which enables the robot to slide sticks (a grasped tool) along the groove to push out a rewarding object to the edge of the table (this could eventually result in spatial activations that drive the robot to move to the edge of the table closest to the object). Moreover, traps could be placed all along the groove so as to prevent the reward from moving to the edges of the table when pushed by the robot (similar to the trap tube paradigm) hence blocking the action initiated by the robot and forcing it to change its strategy intelligently (and internalize the causal effect of traps).

The environment was designed to implicitly host three specific experiments from animal cognition studies (and their combinations):

(1) *The n-stick paradigm.* It is a slightly more complicated version of the task in which the animal reasons about using a nearby stick as a tool to reach a food reward that was not directly reachable with its end-effector (Visalberghi 1993; Visalberghi and Limongelli 1996) The two-sticks paradigm for example, involves two sorts of sticks: Stk1 (short) and Stk2 (long), one of each being present on a given trial, only the small one being immediately available, and the food reward only being reachable by means of the larger stick. We can easily see that a moderately complex sequence of actions involving tool use, pushing, reaching, and grasping is required to grasp a goal object under the two stick paradigm scenario. Both sticks and long cylinders could be opportunistically exploited by the robot as tools in different environmental scenarios.

(2) *Betty's hook shaping task.* If the previous task was about exploiting tools, this experiment relates to a primitive case of making a simple tool (based on past experience) to realize an otherwise unrealizable goal. This scenario is a "stick and ball" adaptation of an interesting case of novelty in behavior demonstrated by Betty, the Caledonian crow who lived and "performed" in Oxford under the discrete scrutiny of animal psychologists (Weir et al. 2002). She exploited her past experience of playing with flexible pipe cleaners to make a hook-shaped wire tool out of a straight wire in order to pull her food basket form a transparent vertical tube. The magnetized small sticks were introduced in the playground so that the robot could learn (accidentally) their special utility and use them creatively when nothing else works. Computationally, it implies making a cognitive architecture that enables a robotic artifact to reason about things that do not exist, but could exist as a result of its actions on the world.

(3) *Trap tube paradigm.* The trap tube task is an extremely interesting experimental paradigm that has been conducted on several species of monkeys and children (between 24 and 65 months), with an aim to investigate the level of understanding they have about the solution they employ to succeed in the task (Visalberghi and Tomasello 1997). Of course a robot that is capable of realizing goals under the previous two scenarios (i.e n sticks paradigm and betty's hook shaping task) is going to fail in the when traps are introduced in the trapping groove (as in figure 1) at least during the initial trials. This failure contradicts robot's earlier experiences of carrying out the same actions, for which it was actively rewarded. Can this contradiction at the level of reward values be used to trigger higher levels of reasoning and/or exploration activities in order to seek the

cause of failure? To achieve this computationally, the robot must have at least the following three capabilities:

(a) Achieving awareness that, for some reason, the physical world works differently from the mental (simulated) world
(b) Identifying the new variables in the environment that determine this inconsistency (in the trap tube case the robot should discover that the essential novelty are the holes/traps introduced by the experimenter)
(c) Initiating new actions that can block the effect of this new environmental variable (change the direction of pushing the ball, i.e., away from the hole/trap in the simplest case)

In this environmental layout, the robot is asked to pursue relatively simple high level user goals like reaching, grasping, stacking, pushing, and fetching different objects. The interesting fact is that even though the high level goals are simple, the complexity of the reasoning process (and subsequent action generation) needed to successfully realize these goals increases more than proportionately with the complexity of the environment in which the goal is attempted. Further, using a set of few sticks, balls, traps, and cylinders and combining/placing them in different ways, an enormous amount of complex environmental situations can be created, the only limitation being the imagination of the experimenter itself.

17.3 Forward/Inverse Model for Reaching: The Passive Motion Paradigm

The action of "reaching" is fundamental for any kind of goal-directed interaction between the body and the world. Tasks and goals are specified at a rather high, often symbolic level ("Stack 2 cylinders," "Grasp the red ball," etc.) but the motor system faces the daunting and under-specified task of eventually working out the problem at a much more detailed level in order to specify the activations which lead to joint rotations, movement trajectory in space, and interaction forces. In addition to dealing with kinematic redundancies, the generated action must be compatible with a multitude of constraints: internal, external, task specific, and their possible combinations. In this section, we describe the forward/inverse model for reaching that coordinates arm/tool movements in the GNOSYS robot during any kind of manual interaction with the environment.

The central theme behind the formulation of the forward inverse models is the observation that motor commands for any kind of motor action, for any configuration of limbs, and for any degree of redundancy can be obtained by an "internal simulation" of a "passive motion" induced by a "virtual force field" (Mussa Ivaldi et al. 1988) applied to a small number of task-relevant parts of the body. Here "internal simulation" identifies the relaxation to equilibrium of an internal model of limb (arm, leg, etc., according to the specific task); "passive motion" means that the joint rotation patterns are not specifically computed in order to accomplish a goal but are the indirect consequence of the interaction between the internal model of the limb and the force field generated by the target, i.e., the intended/attended goal. The model is based on nonlinear attractor dynamics where the attractor landscape is obtained by combining multiple force fields in different reference systems. The process of relaxation in the attractor landscape is similar to coordinating the movements of a puppet by means of attached strings, the strings in our case being the virtual force fields generated by the intended/attended goal and the other task dependent combinations of constraints involved in the execution of the task.

As shown in Fig. 17.2, the basic structure of the forward inverse models is composed of a fully connected network of nodes either representing forces or representing flows (displacements) in different motor spaces (end-effector space, joint space, muscle space, tool space, etc.). We also observe that a displacement and force node belonging each motor space is grouped as a work (force. displacement) unit (WU). There are only two kinds of connections (1) between a force and displacement node belonging to WU that describes the elastic causality of the coordinated system (determined by the stiffness and admittance matrices) and (2) between two different motor spaces that describes the geometric causality of the coordinated system (Jacobian matrix).

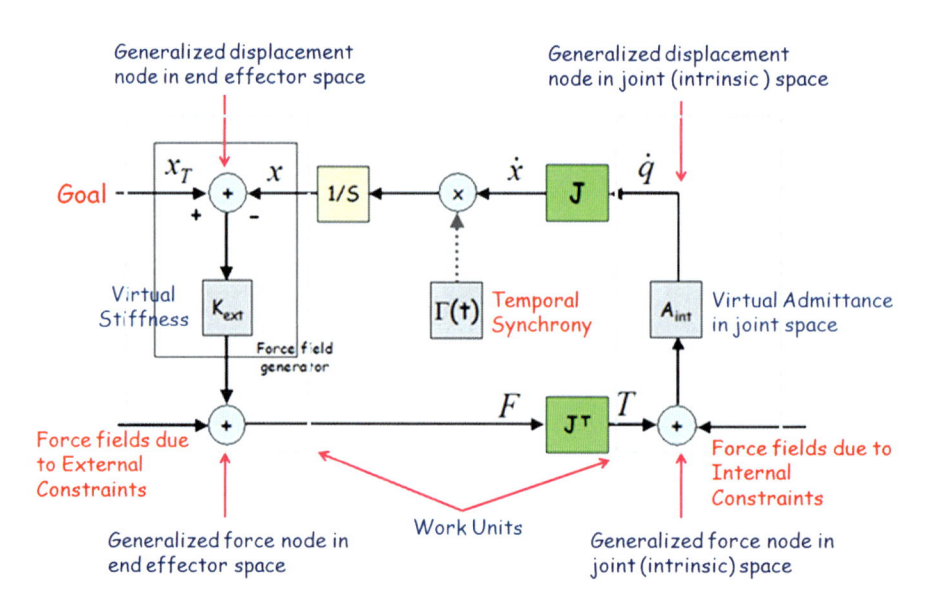

Fig. 17.2 Basic computational scheme of the PMP for a simple kinematic chain. x is the position/orientation of the end-effector, expressed in the extrinsic space; x_T is the corresponding target; q is the vector of joint angles in the intrinsic space; J is the Jacobian matrix of the kinematic transformation $x = f(q)$; K_{ext} is a virtual stiffness that determines the shape of the attractive force field to the target; "external constraints" are expressed as force fields in the extrinsic space; "internal constraints" are expressed as force fields in the intrinsic space; A_{int} is a virtual admittance that distributes the relaxation motion to equilibrium to the different joints; $\Gamma(t)$ is the time-varying gain that implements the terminal attractor dynamics

Let x be the vector that identifies the pose of the end-effector of a robot in the extrinsic workspace and q the vector that identifies the configuration of the robot in the intrinsic joint space: $x = k(q)$ is the kinematic transformation that can be expressed, for each time instant, as follows: where $J(q)$ is the Jacobian matrix of the transformation. The motor planner/controller, which expresses in computational terms the PMP, is defined by the following steps that are also represented graphically by the PMP network of Fig. 17.2:

1. Associate to the designated target x_T an attractive force field in the extrinsic space:

$$F = K_{ext}(x_T - x), \qquad (17.1)$$

   where K_{ext} is the virtual impedance matrix in the extrinsic space. The intensity of this force decreases monotonically as the end-effector approaches the target.
2. Map the force field into an equivalent torque field in the intrinsic space, according to the principle of virtual works:

$$T = J^T F. \qquad (17.2)$$

   Also the intensity of this torque vector decreases as the end-effector approaches the target.
3. Relax the arm configuration in the applied field:

$$\dot{q} = A_{int} \cdot T, \qquad (17.3)$$

   where A_{int} is the virtual admittance matrix in the intrinsic space: the implicit or explicit modulation of this matrix affects the relative contributions of the different joints to the reaching movement.
4. Map the arm movement into the extrinsic workspace:

$$\dot{x} = J \cdot \dot{q} \qquad (17.4)$$

5. Integrate over time until equilibrium:

$$x(t) = \int_{t_0}^{t} J\dot{q}\,\mathrm{d}\tau. \qquad (17.5)$$

Kinematic inversion is achieved through well posed direct computations, and no predefined cost functions are necessary to account for motor redundancy. While the forward model maps tentative trajectories in the joint space into the corresponding trajectories of the end-effector variables in the workspace, the inverse model maps desired trajectories of the end-effector into feasible trajectories in the joint space. The timing of the relaxation process can be controlled by using a TBG (Time Base

Generator) and the concept of terminal attractor dynamics (Zak 1988): this can be simply implemented by substituting the relaxation (17.4) with the following one:

$$\dot{q} = \Gamma(t) \cdot B \cdot T, \tag{17.6}$$

where a possible form of the TBG or time-varying gain that implements the terminal attractor dynamics is the following one (it uses a minimum-jerk generator with duration τ):

$$\Gamma(t) = \frac{\dot{\xi}}{1 - \xi}, \tag{17.7}$$

where

$$\xi(t) = 6(t/\tau)^5 - 15(t/\tau)^4 + 10(t/\tau)^3. \tag{17.8}$$

In general, a TBG can also be used as a computational tool for synchronizing multiple relaxations in composite PMP networks, coordinating relaxation of movements of two arms or even the movements of two robots. The algorithm always converges to an equilibrium state, in finite time (that is set using the TBG) under the following conditions:

(a) When the end-effector reaches the target, thus reducing to 0 the force field in the extrinsic space (17.1)
(b) When the force field in the intrinsic space becomes zero (17.2), although the force field in the extrinsic space is not null and this can happen in the neighborhood of kinematic singularities

Case (a) is the condition of success termination. But also in case (b), in which the target cannot be reached, for example, because it is outside the workspace, the final configuration has a functional meaning for the motion planner because it encodes geometric information valuable for replanning (breaking an action into a sequence of subactions like using a tool of appropriate length).

Multiple constraints can be concurrently imposed in a task-dependent fashion by building composite F/I models (in other words simply switching on/off different task relevant force field generators). In the composite F/I model of Fig. 17.3, there are three weighted, superimposed force fields that shape the spatio temporal behavior of the system.

1. To the end-effector (to reach the target)
2. To the wrist (for proper orientation)
3. A force field in joint space as internal constraints of Joint limits

The same TBG coordinates all the three relaxation processes. This compostie PMP network is effective in tasks like grasping a stick placed in the table, with a specific wrist orientation or an extended case of reaching a goal object (like a ball) with a specific tool orientation. In this case, the force field F1 of Fig. 17.2 is applied at the stick (tool) and field F2 applied at the end-effector. Figure 17.4 shows snapshots of the performance of the computational model of Fig. 17.3 on the GNOSYS robot during different manipulation scenarios.

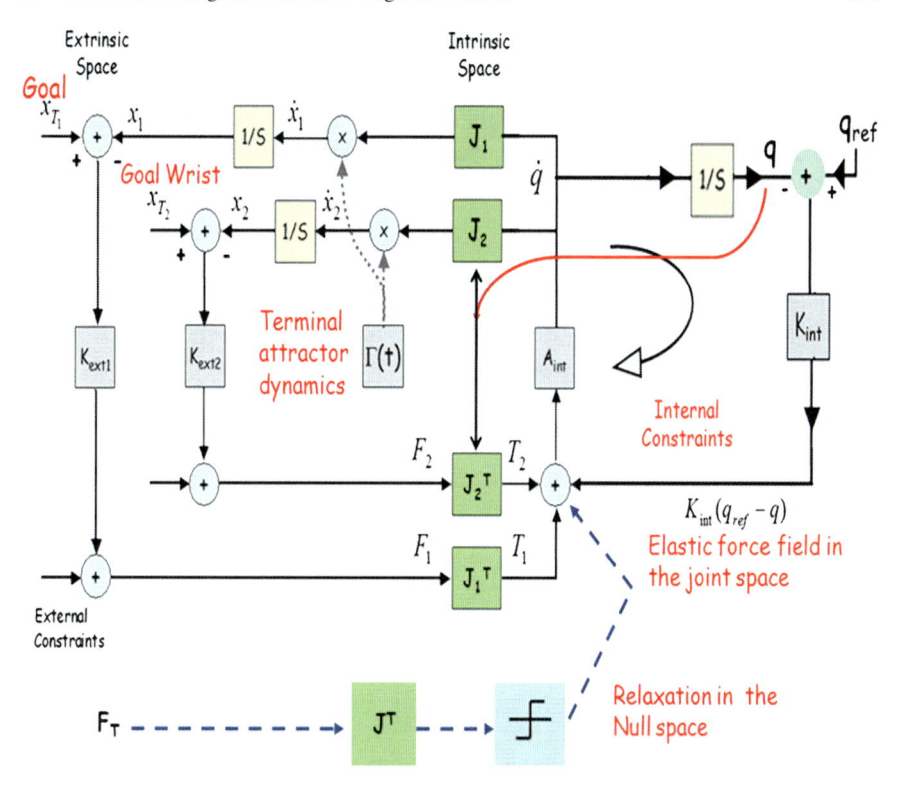

Fig. 17.3 Composite forward/inverse model with two attractive force fields applied to the arm, a field F1 that identifies the desired position of the hand/fingertip and a field that helps achieving a desired pose of the hand via an attractor applied to the wrist. Force fields representing other constraints like joint limits and net effort to be applied (scaled appropriately based on their relevance to the task) are also superimposed on the earlier fields F1 and F2. The time base generator takes care of the temporal aspects of the relaxation of the system to equilibrium. In this way, superimposed force fields representing the goals and task relevant mixtures of constraints can pull a network of task relevant parts of an internal model of the body to equilibrium in the mental space

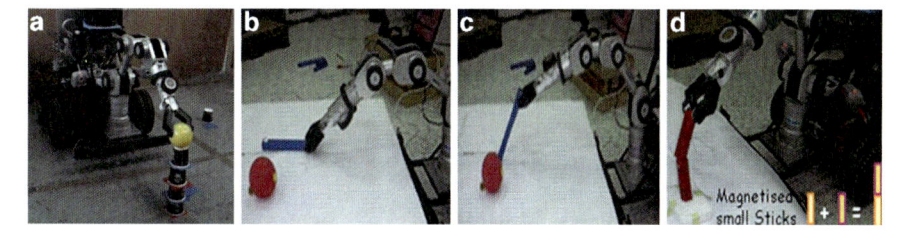

Fig. 17.4 Performance of the F/I model on GNOSYS. (**a**) Stacking task; (**b**) Reaching/Grasping a stick with specified wrist orientation; (**c**) Using a Stick as tool to reach a ball, adapting the kinematics with respect to the grasped tool; (**d**) Coupling two small red magnetized sticks (orienting the gripped first stick appropriately)

17.4 Spatial Map and Pushing Sensorimotor Space

A large body of neuroanatomical and behavioral data acquired from experiments conducted on mammals (primarily rodents) suggest involvement of a range of neural systems being involved in spatial memory and planning, like the head direction cells (Blair et al. 1998), spatial view cells (Georges-Francois et al. 1999), and hippocampal place cells (O'Keefe and Dostrovsky 1971) that exhibit a high rate of firing whenever an animal is in a specific location in the environment corresponding to the cell's "place field" and the recently found grid cells located in the entorhinal cortex in rats, known to constitute a mental map of the spatial environment. Like animals, the GNOSYS Robot also faces problems related to learning a mental map of the spatial topology of its environment and use it in coordination with the forward/inverse models for arm to realize goals in more complex scenarios. In addition, it also needs to learn the causality of pushing objects in the trapping groove using sticks. The spatial map and the pushing internal model essentially share the same computational substrate with the only difference being the sensorimotor variables that are at play in the two internal models. Hence we describe the two internal models jointly in this section. The computational architecture for the development of these internal models and the associated dynamics (that organize goal oriented behavior) is novel and brings together several interesting ideas from the theory of self organizing systems (Kohonen 1995), their extensions to growing maps (Fritzke 1995), neural field dynamics (Amari 1977), sensorimotor maps (Toussaint 2006), reinforcement learning (Sutton and Barto 1998), and temporal hebbian learning (Abbot and Sejnowski 1999). For reasons of space, we restrict ourselves to the following issues in this chapter:

(a) Learning the sensorimotor space (through self organization of sequences of randomly generated sensory motor data)
(b) Dynamics of the sensorimotor space (SMS): How activity moves bidirectionally between sensory and motor units
(c) Value Field Dynamics: How activity moves bidirectionally between sensory and motor units in a "goal-directed fashion"
(d) Dealing with dynamic changes in the world, cognitive dissonance (e.g., learning to nullify the effect of traps in the trapping groove).

17.4.1 Acquisition of the Sensorimotor Space

The sensorimotor variables for the spatial map are relatively straight forward, the sensory space composed of the global location of the robot in the playground ($x-y$ coordinates and orientation) coming from the localization system (Baltzakis 2004), the motor space is 2D, composed of translation commands appropriately converted into speed set commands communicated to low-level hardware. For the pushing internal model, sensory information coded is the location of the object (being pushed).

This information is derived after a visual scene analysis using the GNOSYS visual modules and reconstructed into 3D space coordinates using a motor babbling based algorithm (Mohan and Morasso 2007a). The function of the visual modules is out of scope for discussion in this article and interested reader may refer to GNOSYS documentation for further information on this issue. The motor space consists of the following variables (shown in Fig. 17.5):

(a) Location of the tool with respect to Goal
(b) The amount of force applied to the object. This is approximately proportional to the change in the DOF $\theta 1$ and $\theta 5$ of the KATANA arm of the robot

Figure 17.6 shows the general computational structure for the pushing and moving related internal models. The central element of the architecture is a growing intermediate neural layer common to both perception and action, called the sensorimotor space, henceforth SMS (Toussaint 2006).

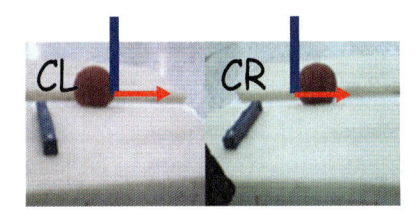

Fig. 17.5 Pushing to the right in the case of CL will not induce any motion on the ball. Pushing to the right in the case CR will displace the ball based on the amount of force applied (i.e., approximately equal the displacement of the stick in contact with the ball along the trapping groove)

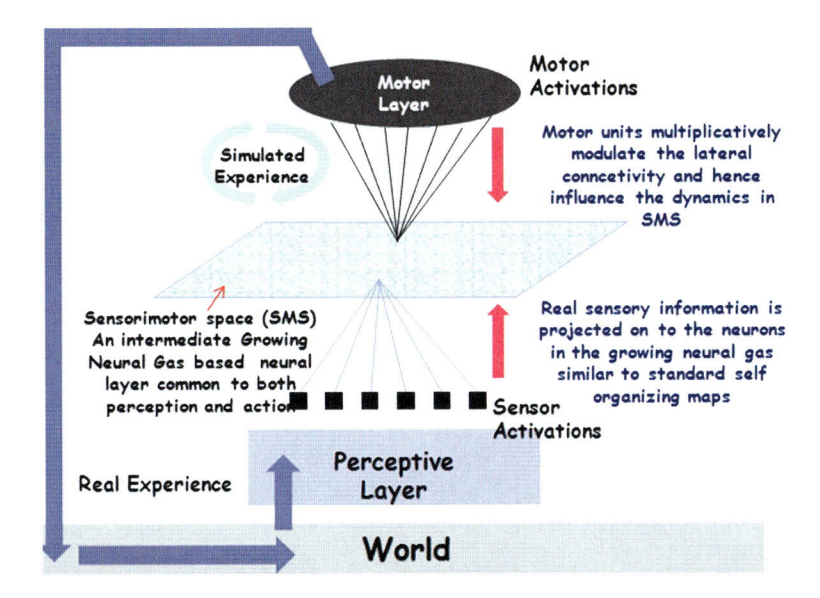

Fig. 17.6 General computational structure for the spatial map and pushing internal model

This neural layer not only self organizes sequences of sensorimotor data generated by the robot through random motor explorations (through the loop of real experience) but also sub symbolically represents the forward inverse functions of various sensorimotor dependencies (that is encoded in the connectivity structure). Further, it also serves as a proper computational substrate to realize goal-directed planning (using quasistationary value fields) and perform "what if" experiments in the mental space (through the loop of simulated experience shown in Fig. 17.6). During the process of learning the SMS, the simulated experience loop is turned off. In other words, the only loop active in the system is the loop of real experience. To learn the spatial mental map, the agent is allowed to move randomly in the play ground with a maximum translation of 14 cm and maximum rotation of 20° in one time step (in order to achieve the representational density necessary to perform motor tasks in future that require high precision). These movements generate the data, i.e., sequences of sensory and motor signals $S^{(t)}$ and $M^{(t)}$ using which the sensory weights, lateral connections between neurons, and the motor weights of the motor modulated lateral connections are learnt. Both the SMS and the complete lateral connectivity structure are learnt from zero using sequences of sensor and motor data generated by the robot through a standard growing neural gas algorithm and extended to encode motor information into the connectivity structure like the sensorimotor maps of Toussaint. Hence, in addition to incrementally self organizing the state space based on incoming sensorial information (like a standard GNG), the motor information is also fully integrated with the SMS at all times of operation. As seen in Fig. 17.5, motor units project to lateral connections in between the neurons in the SMS and influence their dynamics. This allows motor activity to multiplicatively modulate these lateral connections hence cause anticipatory shifts in neural activity in the SMS similar to that which would have occurred if the action was actually performed. Moreover, provided that the world is consistent, both mental simulation (top-down through motor modulated lateral connections) and real performance (bottom-up through self organizing competition) should activate the same neural population in the SMS, the coherence between them forming the basis for the stability of the sensorimotor world of GNOSYS robot. Figure 17.7 shows the

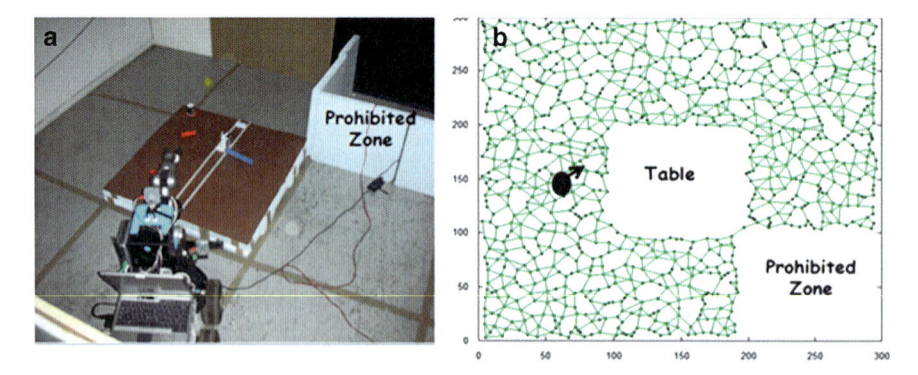

Fig. 17.7 Lateral topology of the spatial map after 23,350 iterations of self organization after which the map becomes almost stationary. Number of neurons $= 933$

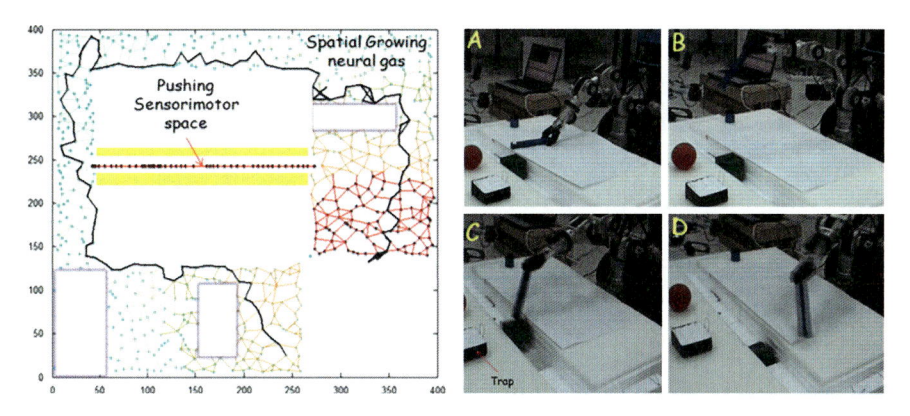

Fig. 17.8 Learnt lateral topology of the spatial map and the pushing SMS in the trapping groove. (A–D) A typical push sequence

lateral topology of the SMS of the spatial map learnt by the robot after this initial phase of self organization on sequences of sensory motor data.

Similar to the development of the SMS for spatial map, a growing SMS for pushing in the trapping groove was built using the data generated by repeated sequences of reaching a goal object with a stick (using the F/I Model pair for reaching), pushing in different directions (with different amount of force) and then tracking the new location of the ball. We simplify this scenario by considering pushing to be only functional along the horizontal axis. Figure 17.8 shows the internal spatial map of the Gnosys playground along with the SMS for pushing in the trapping groove. The other panels show a typical pushing sequence for data generation.

17.4.2   Dynamics of the Sensorimotor Space

After learning the SMS through self organization of sequences of sensory and motor data generated by the robot, we now focus on the dynamics of SMS that determines how activations move back and forth between the sensorimotor-action spaces and realize goal-directed behavior. A zoomed view of the interactions between two neurons in the scheme of Fig. 17.6 is shown in Fig. 17.9. The dynamical behavior of each neuron in the SMS is as follows: To every neuron "i" in the SMS we associate an activation x_i governed by the following dynamics:

$$\tau_x \dot{x_i} = -x_i + S_i + \beta_{\text{if}} \sum_{i,j} (M_{ij} W_{ij}) x_j. \tag{17.9}$$

We observe that the instantaneous activation of a neuron in SMS is a function of three different components. The first term induces an exponential relaxation to the dynamics (and is analogous to spatially homogenous neural fields of

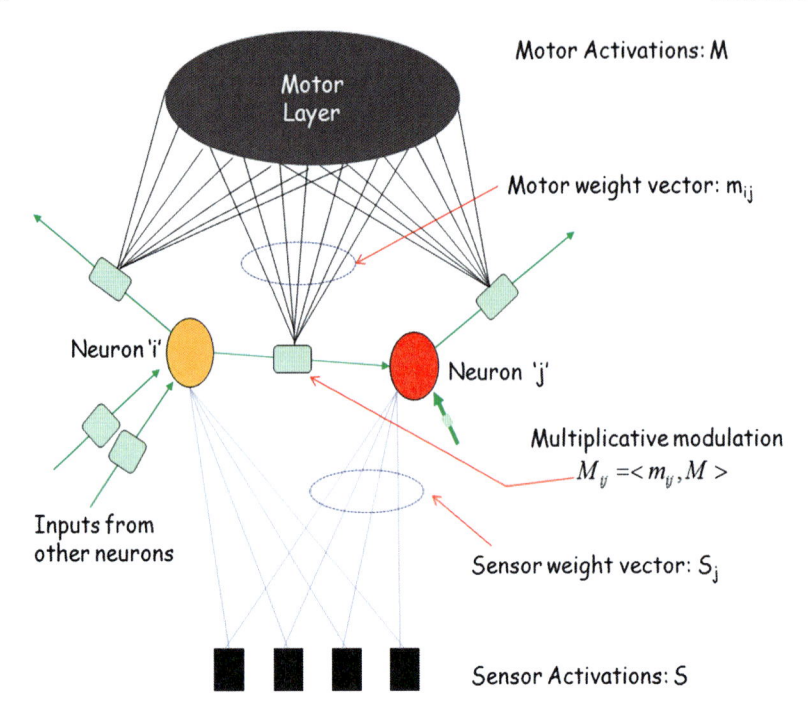

Fig. 17.9 Zoomed view of interactions between two neurons in the SMS, interactions between perceptive layer, motor layer, and the SMS

Amari 1977). The second term is the net feed forward (or alternatively bottom-up) input coming from the sensors at any time instant. The Gaussian kernel compares the sensory weight s_i of neuron i with current sensor activations S^t.

$$S_i = \frac{1}{\sqrt{2\pi}\sigma_s}\, \mathrm{e}^{\frac{-(S_i-S)^2}{2\sigma_S^2}}. \qquad (17.10)$$

Finally, the third term represents the lateral interactions between different neurons in the SMS, selectively modulated by the ongoing activations in the motor space. Hence, through this input the motor signals can couple with the dynamics of the SMS. If M is the current motor activity, and m_{ij} the motor weight encoded in the lateral connection between neuron i and j, the instantaneous motor modulated lateral connection M_{ij} between neurons i and j is defined as (and shown in Fig. 17.8):

$$M_{ij} = <m_{ij}, M>. \qquad (17.11)$$

The instantaneous value M_{ij}, i.e., the scalar product of motor weight vector m_{ij} with the ongoing motor activations M keeps changing with the activity in the action space and hence influences the dynamics of SMS. Due to this multiplicative coupling, a lateral connection contributes to lateral interaction between two neurons only when the current motor activity correlates with the motor weight vector of this connection. Inversely, by multiplicatively modulating lateral interactions between neurons in the SMS as a function of the motor activity in the action space, it is possible to predict the sensorial consequences of executing a motor action. Interaction between action space and SMS by virtue of motor modulated lateral connectivity thus embeds "Situation–Action–Consequence" loops or Forward Models into the architecture and offers a way of eliciting perceptual activity in the SMS, similar to that which would have occurred if the action was performed in reality.

The element β_{if} in (17.9) is called the bifurcation parameter and is defined as follows:

$$\beta_{if} = \frac{1}{\sqrt{2\pi}\sigma} e^{\frac{-(S_{Anticip}-S)2}{2\sigma^2}}. \tag{17.12}$$

This parameter basically estimates how closely the top-down (predicted) sensory consequence $S_{Anticip}$ of the virtual execution of any incremental motor action M correlates with the bottom-up (real sensory information) S. $S_{Anticip}$ can be easily computed by only considering the effect of top-down modulation in (17.4) and finding the neuron "k" in the SMS that shows maximum activation x_k among all neurons.

$$x_k = \sum_{k,j} (M_{kj} W_{kj}) x_j, \quad \text{for all } k, j \in (1, N). \tag{17.13}$$

Since sensory weights of every neuron is approximately tuned to the average sensory stimulus for which it was the best match, the anticipated sensory consequence $S_{Anticip}$ is nothing but the sensory weights of the neuron k that shows maximum activation under the effect of top-down modulation. The bifurcation parameter hence is a measure of the accuracy of the internal model at that point of time. $\beta_{if} \to 0$ implies that the internal model is locally inaccurate or there is a dynamic change in the real world, i.e., "the world is working differently in comparison to the way the robot thinks the world should be working."

What should the robot do when it detects the fact that the world is functioning in ways that are contrary to its anticipations? The best possible solution is to work on real sensory information and engage in a incremental cycle of exploration to adapt the SMS, learn some new lateral connections, grow new neurons, and eliminate few neurons (like the initial phase of acquiring the SMS). This flexibility is incorporated in the dynamics in the following fashion: As we can observe from (17.9) that as $\beta_{if} \to 0$, the top-down contribution to the dynamics also gradually decreases, in other words the system responds real sensory information only. Hence in this case only the real experience loop (of Fig. 17.5) is functional in the system. Now comes the next problem of how to trigger motor exploration dynamically, and this is the third important function of the bifurcation parameter. The bifurcation parameter controls the gradual switch between random exploration and planned behavior by controlling the amount of randomness (r) in motor signals in the dynamics of the action space as evident in (17.14).

$$\bar{a} = \beta_{if} \left(\sum_{i=1}^{N} x_i \overline{m_{k_i i}} \right) + \zeta(\bar{r}). \tag{17.14}$$

The second term in (17.14) triggers random explorative motor actions, where r is a vector of small random motor commands (in the respective motor DoFs) and $\zeta = 1 - \beta_{if}$. So under normal operations (when β_{if} is close to 1), the amount of randomness is very less and the motor signals are incrementally planned to achieve the goal at hand using the first term of (17.14). We will enter into details of this component after formulating the value field dynamics in the next section.

We also note that x_i in (17.9) are the time-dependent activations and the dot notation $\tau_x \dot{x}_i = F(x)$ is algorithmically implemented using an Euler integration step:

$$x(t) = x(t-1) + \frac{1}{\tau_x}(F(x(t-1))). \tag{17.15}$$

In sum, a consequence of the dynamics presented in this section is that at all times, information flows circularly between the SMS and the action space. While the current goal, connectivity structure, and the activity in the SMS project upwards to the action space and determine incremental motor excitations that are needed to realize the goal, motor signals from the action space influence top-down multiplicative modulations in the lateral connections of the SMS hence causing incremental shifts in the perceptual activity. In the next section, we will describe how the representational scheme described in the previous section and the dynamics described in this section serve as a general substrate to realize goal-directed planning (in simple terms, the problem of how goal couples with the internal model and influences the dynamics of the SMS).

17.4.3 Value Field Dynamics: How Goal Influences Activity in SMS

In addition to the activation dynamics presented in the previous section, there exists a second dynamic process that can be thought as an attractor in the SMS that performs the function of organizing goal-oriented behavior. The quasistationary value field V generated by the active goal together with the current (nonstationary) activations x_i (17.9) allows the system to incrementally generate motor excitations that lead toward the goal.

Value field dynamics acting on the SMS is defined as follows:

$$\tau_v \dot{v}_i = -v_i + R_i + \gamma(W_{ij}v_j)_{\max}, \tag{17.16}$$

$$R_i = DP + Q. \tag{17.17}$$

Let us assume that the dynamical system is given a goal G that corresponds to reaching a state s_G in the SMS. Just like the sensory signals couple with the neurons in the SMS through feed forward connections, the motor signals couple with the neurons in the SMS through motor modulated lateral connections, the goal G couples with the SMS by inducing reward/value excitations in all the neurons in the SMS. As seen in (17.16), the instantaneous value v_i of the ith neuron in the SMS at any time instance is a function of three factors (1) the instantaneous reward R_i, (2) the contribution of the expected future reward, where γ (approx 0.9) is the discount factor, and (3) the lateral connectivity structure of the SMS. Equation (17.17) shows the general structure of the instantaneous reward function we used in our computational model. The first term in the reward equation DP expresses the default plan if available (e.g., take the shortest or least energy path in the case of the spatial map). We will see in the later sections that it is in fact not really necessary to have a default plan in the reward structure and further there can be situations where new reward functions must be learnt by the system in order to initiate flexible behavior in the world. The second element in the reward function models these additional Goal dependent qualitative measures in the reward structure that are learnt through user/self penalization/rewards:

$$Q = Q_1 + Q_1 + \cdots + Q_n. \qquad (17.18)$$

Every component Q can be thought as a learnt additional value field (having a scalar value at each neuron of the SMS) and the net value field is a superposition of the Q component and the DP component. In this sense, the net attractor landscape is shaped by a task-specific superposition of value fields (similar to combinations of different force fields I the reaching F/I model), and behavior is nothing but an evolution of the system in this dynamically composed attractor landscapes. The Q components of the reward structure further play an important role in dealing with heterogeneous optimality, dealing with dynamic changes in the world, taking account of traps during pushing, etc. We will now present two examples to explain how different components in the model described by (17.9)–(17.18) interact under the presence of a goal.

17.4.4 Reaching Spatial Goals Using the Spatial Sensorimotor Space

Coming to the problem of reaching spatial goals using the spatial SMS, let us consider that the spatial goal induces a reward excitation to every neuron in the SMS (similar to Toussaint 2006) as given by (17.19), where s_i is the sensory codebook weight of the ith neuron, G is the spatial goal in the playground that has to be reached by the robot, and Z is chosen such that $\sum_i R_i = 1$,

$$R_i = \frac{1}{Z} e^{\frac{-(s_i-G)2}{2\sigma_R^2}}. \qquad (17.19)$$

Under the influence of this reward excitation, the value field on the spatial SMS will move quickly to its fixed point:

$$v_i^* = R_i + \gamma (W_{ij} v_j)_{\max}. \qquad (17.20)$$

The coupling between the value field and the dynamics of the SMS can now be understood by revisiting the expression for action selection (17.14). The element $m_{k_i i}$ represents the motor weights of a lateral connection between neuron i and its immediate neighbor k_i such that $k_i = \mathrm{argmax}_j(w_{ij} V_j)$. In simple terms, the value field influences the motor activity by determining the neighboring neuron (to the currently active neuron) that holds maximum value in the context of the currently active goal. In other words, it determines how valuable any motor excitation m_{hi} is with respect to the goal currently being realized. The motor action that is generated is hence the activation average of all the motor reference vectors $m_{k_i i}$ coded in the motor weights for all N neurons and at that time instance. In sum, the goal induces a value field that influences the computation of the incremental motor action to move toward the goal for the next time step; this motor activation in turn influences the dynamics of the SMS and causes a shift in activity; now the next step of valuable motor activation is computed, and this process progresses till the time the system achieves equilibrium. Hence, the information flows between the SMS and the motor system is in both ways: In the "tracking" process as given by (17.9), the information flows from the motor layer to the SMS: Motor signals activate the corresponding connections and cause lateral, predictive excitations. In the action selection process as given by (17.14), information moves from the SMS back to the motor layer to induce the motor activations that will enable the system to move closer to the goal. In sum, the output of this circular dynamics involving SMS, action space, and the goal induced value field is a trajectory: a trajectory of perceptions in the SMS and a trajectory of motor activations in the action space. Figure 17.10 shows the trajectories generated by the robot while moving to different spatial goals in the playground.

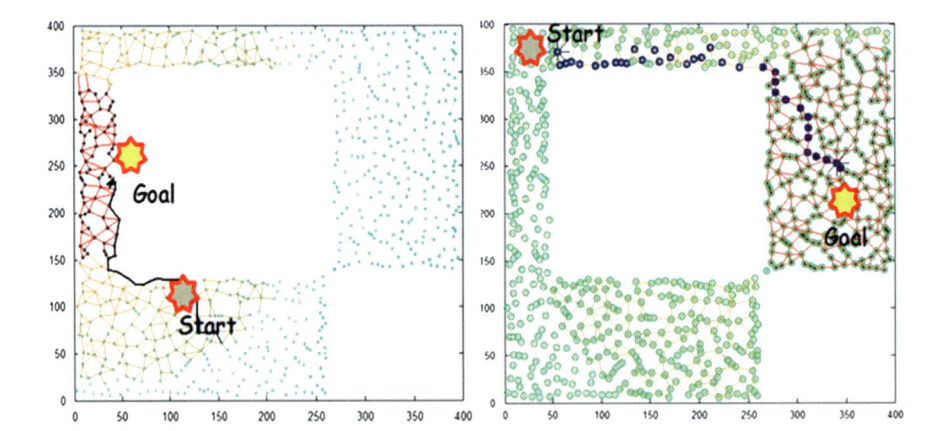

Fig. 17.10 Movements to different spatial goals in the GNOSYS play ground. The goal dependent value field (quasistationary) is shown superimposed on the spatial map. As seen in the figure, using the simple reward structure of (17.19) (i.e., only the DP component and no learnt value fields), neurons closer to the goal induce greater rewards

17.4.5 Learning the Reward Structure in "Pushing" Sensorimotor Space

In order to realize any high level goal that requires a pushing action to be initiated, it is not just important to be able to simulate the consequences of pushing, but also to be able to "push in ways that are rewarding." In other words, after learning the pushing SMS as described in Sect. 17.4.1, we now have the task of making the robot learn the reward structure involved in a pushing action, so that it can coordinate the pushing in a goal-directed fashion. In the set up of pushing in the trapping groove, we can estimate that pushing the goal to the either edges of the table should be maximally rewarding, since it ensures that the robot can move around the table and grasp the goal. We note here that no default plan (DP component) needs to be defined. Rather, the reward structure can be learnt directly by repeated trials of random explorative pushing of the goal in different directions along the groove, followed by an attempt to grasp the goal (by moving and pushing) after which the robot is presented by a reward by the user. These trials can also be done in the mental space by initiating virtual pushing commands, simulating the consequence, virtually evaluating the possibility of reaching the now displaced goal (using the GNG for spatial navigation and the forward/inverse model for reach/grasp), and finally self evaluating its success. Full reward is given to the neuron that fired last (that represents the location from where chances of reaching the goal are maximum) and gradually scaled versions, and the total reward is distributed to all the other neurons in the pushing SMS that were sequentially active during the trial. Energetic issues can also have their effects in the learnt reward structure, since there are multiple solutions to get the reward successfully by pushing in different directions. Influence of energetic issues in the reward field can be introduced by adding a decaying element in the promised net reward for achieving a goal successfully (17.21), which is a function of the amount of energy spent in the process of getting the goal (e.g., if the ball is pushed toward the right, more energy will be spent in navigation to achieve the goal of grasping the ball):

$$R_T = R_{net} \text{ if } \text{Dist}_{iter} < \delta,$$

$$R_T = R_{nst} e^{-(\text{Dist}_{iter}/125)} \text{ if } \text{Dist}_{iter} < \delta,$$

$$\delta = \frac{\text{Goal} - \text{Initpos}}{1.5}, \tag{17.21}$$

where R_T is the actual reward received in the end of the Tth trial in case of success, R_{net} is the net reward promised in each trial (we kept all promised rewards for success as 50), Dist_{iter} is an approximate calculation of the distance navigated by the robot to get the goal, which is estimated based on the number of neurons in the spatial SMS that were active in the trajectory from initial position of the robot to the goal. We must note that this distance travelled Dist_{iter} is a consequence of the pushing action that preceded navigation and not a result of the constraints in spatial navigation in the playground. In other words, if the robot pushed the goal to the right, it needs to navigate a much greater distance that it would have had to in case it had pushed the goal to the left. This is reflected in the number of neurons that are sequentially activated during the path from source to goal, i.e., $Dist_{iter}$. Since navigation has a high cost in terms of battery power consumed and since navigating greater distances than that was necessary directly implies spending more energy than that was necessary, the term $Dist_{iter}$ is one of the parameter that helps in distributing rewards based on the energetic efficiency of the solution. The other term δ is the ratio of the shortest distance between "the initial position (*Initpos*) of the robot from the final location of goal after pushing (Goal)" and "representational density of neurons covering the spatial SMS" that we conservatively approximated as 1.5. After every trial of pushing, the reward received by each neuron in the pushing SMS is added to its previous accumulated reward value. After about 50 trials, we averaged the rewards received by each neuron in each trial in order to generate the final reward structure for pushing. This reward structure can now be used to compute the value field which then drives the pushing SMS dynamics. This works exactly the same way as the spatial map dynamics, i.e., based on the value field, the next incremental motor action for pushing the goal object (a ball) is computed. This then modulates the lateral connections to cause a shift in activity that corresponds to the anticipated movement of the ball in the trapping groove. Now based on this new predicted location of the ball in the pushing SMS and the value field, the next incremental pushing action is computed and so on till the time the system attains equilibrium. The final anticipated spatial position of the ball after the pushing SMS dynamics is complete, in turn induces a quasistationary value field in the spatial map that triggers the spatial SMS dynamics so as to eventually pull the body toward it. Figure 17.11 shows a combined sequence of pushing and moving in the respective sensorimotor spaces.

We can observe from Fig. 17.11 that the pushing value field encourages the robot to push toward the left, since it is an energy efficient strategy and hence more rewarding. However, this may not always be true if there are dynamic changes in the world (like introduction of traps) during which always pushing the goal to the left may result in a failure to get the reward. Under such cases new experience based value fields need to be learnt [Q components in (17.17)] that dynamically shape the field structure appropriately taking into account these issues.

We now introduce these additional constraints on the pushing scenario by placing traps randomly at different locations along the trapping groove. Traps were indicated to the robot through visual markers so that their location in the groove can be estimated by reconstructing the information coming from the visual recognition system. When traps are introduced initially in the trapping groove, the behavior of the system is only governed by the previously learnt reward structure. Hence the robot follows the normal strategy as in the previous section. As seen in the three trials after introduction of traps shown in Fig. 17.12, the ball is pushed as a function of the value field learnt in the previous section (shown in pink on top of the trapping groove) and is always constant. This normal behavior continues till the time a contradiction is encountered between the anticipated position of the ball as a result of an incremental pushing action and the real location of the ball coming from the 3D reconstruction system. In other words, the ball is not really in the place where the

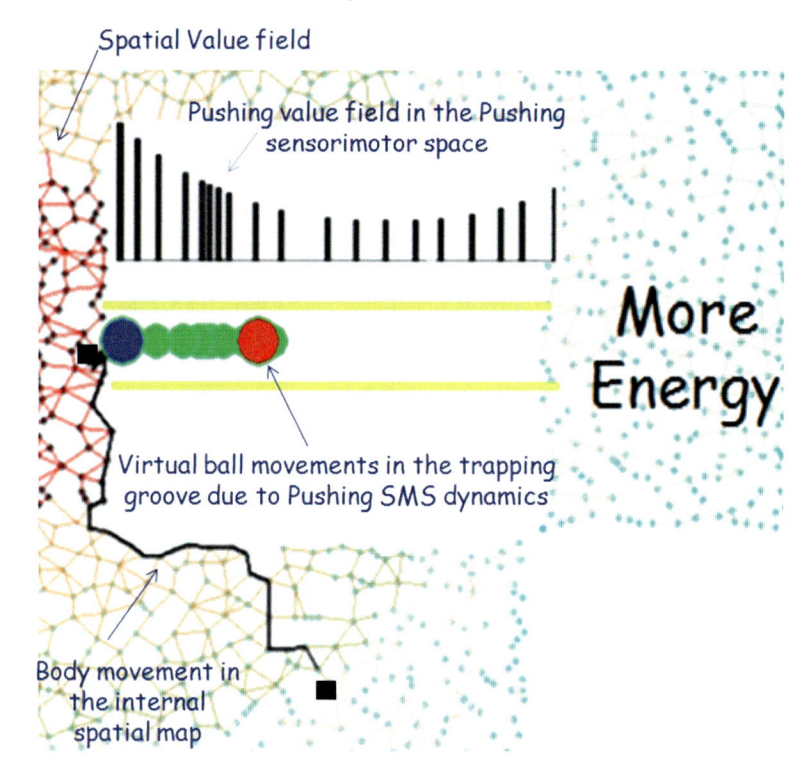

Fig. 17.11 Combined sequence of pushing and moving in the mental space. Note that the pushing reward structure encourages pushing to the left (that is more energy efficient). The final anticipated position of the ball once pushing SMS reaches its equilibrium is a spatial goal for the spatial SMS. This spatial goal induces a quasistaionary value field in the spatial SMS there by triggering the dynamics in the spatial map, hence pulling the body closer to the goal

robot thinks it should be as a result of the pushing action initiated by it. A contradiction automatically implies that there are new changes taking place in the world whose effects are not represented internally by the system. Such contradictions result in a phase of active exploration [since $\beta_{\text{if}} \to 0$, (17.9) and (17.14)] at least till the time the system is pulled back to the normal behavior by the already existing value field. Now the robot initiates incremental random pushing in different directions till the time the ball begins to move as anticipated, in which case pushing is once again governed by the preexisting plan. The path of the ball during random pushing and normal behavior is shown in Fig. 17.12 for four different cases. In the first case, since the initial location of the ball is close to the right end, following the normal behavior, the ball was pushed rightwards where it collides with the trap placed at around 220; this motion of the ball is shown in green with the white arrow. Now there is an active phase of random pushing for a while with the ball moving forwards and backwards, till the time it reaches a position from where the preexisting value field takes over. The motion of the ball due to explorative pushing is shown in blue with the direction indicated by the yellow arrow. Once it is on the other end of the table, it can be easily reached. Case 3 and case 4 are also similar to the first case, however, with a different environmental configurations. In the case 2, the trap was placed around 150, and the initial location of the ball shown is approximately 135. In this case, there was no exploration at all because the previously existing value field automatically causes the ball to be pushed to the left and the goal was achieved. In fact, the robot was blind about the existence of the trap in the sense that it was not the trap that caused the direction of pushing but the preexisting reward field it had developed earlier. This may also be a limitation of the approach because the knowledge is represented more in the form of associations of experiences (like the capuchins) rather than a still higher level of understanding of the real physical causality. Is there such a still higher level of understanding or is it just associative rules learnt by experience that are exploited intelligently is still an issue of debate, which we will not enter in this section.

What should the robot do with these sequences of new experiences, the experience of a new environment, a contradiction which it did not encounter before while solving similar goal, a phase of exploration to try to find an alternative solution that eventually results in success and rewards? We suggest that it should represent them as a memory and in the form of the q_i components in the reward structure given by (17.19). Further, the reward that was received on success needs to be distributed to the contributing neurons in the pushing SMS. This distribution is done as follows: in case of rewards, the most distal element receives the maximum reward and all contributing elements receive gradually scaled versions, circular solutions being actively penalized. The panels on the right show the new reward fields (qi) learnt after each trial. In case 1, for example (Fig. 17.12), the reward structure representing this experience reflects the fact that if the initial position of the ball is around 180 and the location of the trap is somewhere around 220, it is rewarding to push leftwards. For case 4, it reflects the fact that if the trap is somewhere around 60, and the initial position of the ball is around 150, it is more rewarding to push to the right. We also note that there is no need to predecide how many trials of such learning have to take place. Learning in the system takes place when it is needed, i.e., when there is a contradiction and things are not working as expected. After eight different single trap configurations, the behavior produced was intelligent enough that no further training was required.

The additionally learnt qi components of reward field also begin to influence the value field dynamics now and hence the value field structure is no longer constant like it was in Fig. 17.12. It changes based on the configuration of the problem. The net reward structure is a superposition of the default plan which was learnt previously in the absence of traps and the new experience related fields that were learnt after introduction of traps scaled appropriately based on their relevance to the currently active goal (Fig. 17.13):

$$R = R_{\text{default}} + \sum_{T=1}^{N} \sum_{E=1}^{m} R_E \cdot \frac{1}{\sqrt{2\pi}\sigma_T} \, e^{\frac{-(\text{Trap}_T - \text{Trap}_E)^2}{2\sigma_T^2}}. \tag{17.22}$$

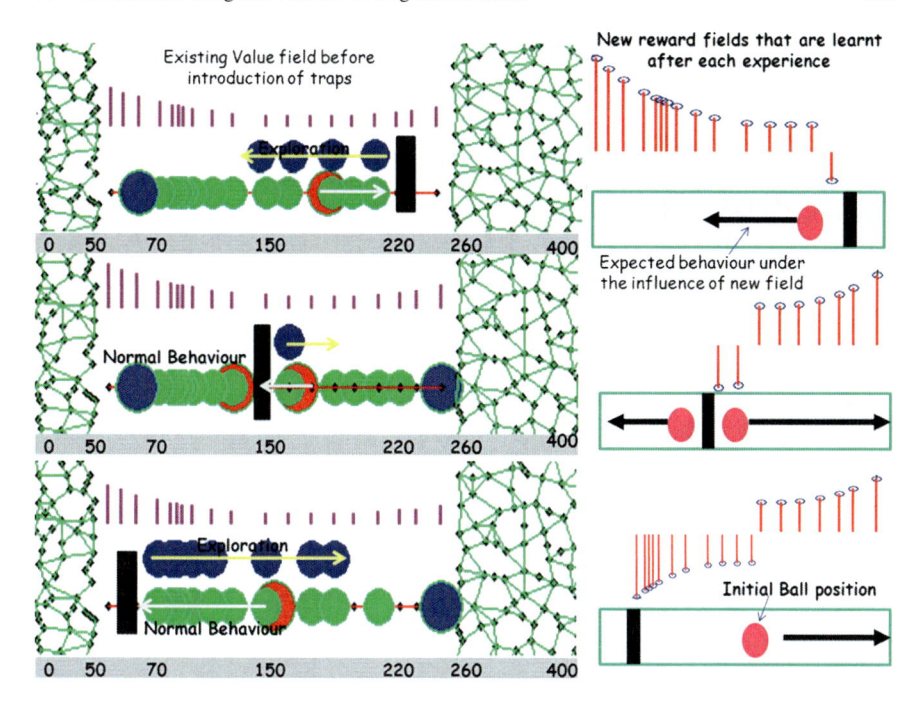

Fig. 17.12 Three trials of pushing under the influence traps placed at different locations along the groove are shown in the figure. The panels on the *right* show the new reward components q_i learnt after being rewarded due to successful realization of the goal partly because of random explorative pushing. In every trial, the robot has an experience, an experience of contradiction because of the trap, an experience of exploration which characterizes its attempt to nullify the effect of the trap so as to realize the goal and an experience of being rewarded by the user/self in case of success. This experience is represented in the form of a reward field in the pushing sensorimotor space. For example, in trial 3, what is represented is the simple fact that if the initial position of the ball is around 150 and the position of the trap is around 65, it is more rewarding to push toward the right and navigate all around the table to reach closer to the ball. These experiences, based on their relevance to the goal being attempted, will influence the behavior of the robot in the future

Here R_{default} is the pushing reward structure learnt in the previous section. T stands for number of traps. E stands for the number of experiences during which new reward fields were learnt (eight in our case). R_E is the Eth reward field. And the final term computes how relevant an experience E is with respect to the situation considering trap T present alone in the environment. Figure 17.13 shows examples of the pushing in the trapping groove for single trap configurations, after the learnt reward fields began contributing to the value field structure and hence actively influencing the behavior.

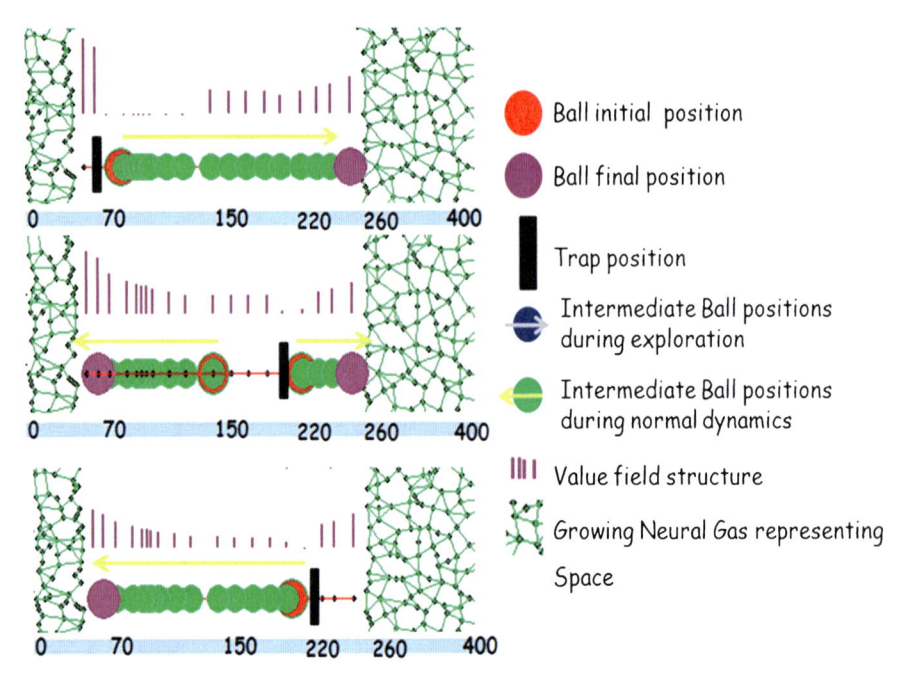

Fig. 17.13 Pushing in the presence of traps in the trapping groove. In the previous cases of pushing shown in Fig. 17.12, the value field superimposed on the pushing sensorimotor space was constant. In this figure, we can observe goal/trap specific changes in the value field. Experiences encountered in the past and represented in terms of fields are superimposed in a task relevant fashion to give rise to a net resultant field that drives the dynamics of the system. Also we see that in this case pushing direction is a function of both the relative position of the hole and the starting position of the reward/ball

17.5 A Goal-Directed, Mental Sequence of "Push–Move–Reach"

How can the internal models for reaching, spatial navigation, and pushing cooperate in simulating a sequence of actions leading toward the solution of a high level goal? Let us consider a scenario where the robot is issued a user goal to grasp a Green ball as shown in panel 1 of Fig. 17.14. In the initial environment, the ball is placed in the center of the trapping grove, unreachable from any direction. In addition, one trap is placed in the trapping groove as an additional constraint. It is quite a trivial task for even children to mentally figure out how to grasp the ball through a sequence of "push–move–reach," using the available blue stick as a tool and avoiding the trap. However, the amazing complexity of such seemingly easy tasks is only realized when we question the computational basis of these acts or make robots act in similar environmental scenarios. How can the robot use the internal action models presented in Sects. 17.3 and 17.4 to mentally figure out a plan to achieve its goal? Of course it can employ the F/I model for reaching to virtually evaluate

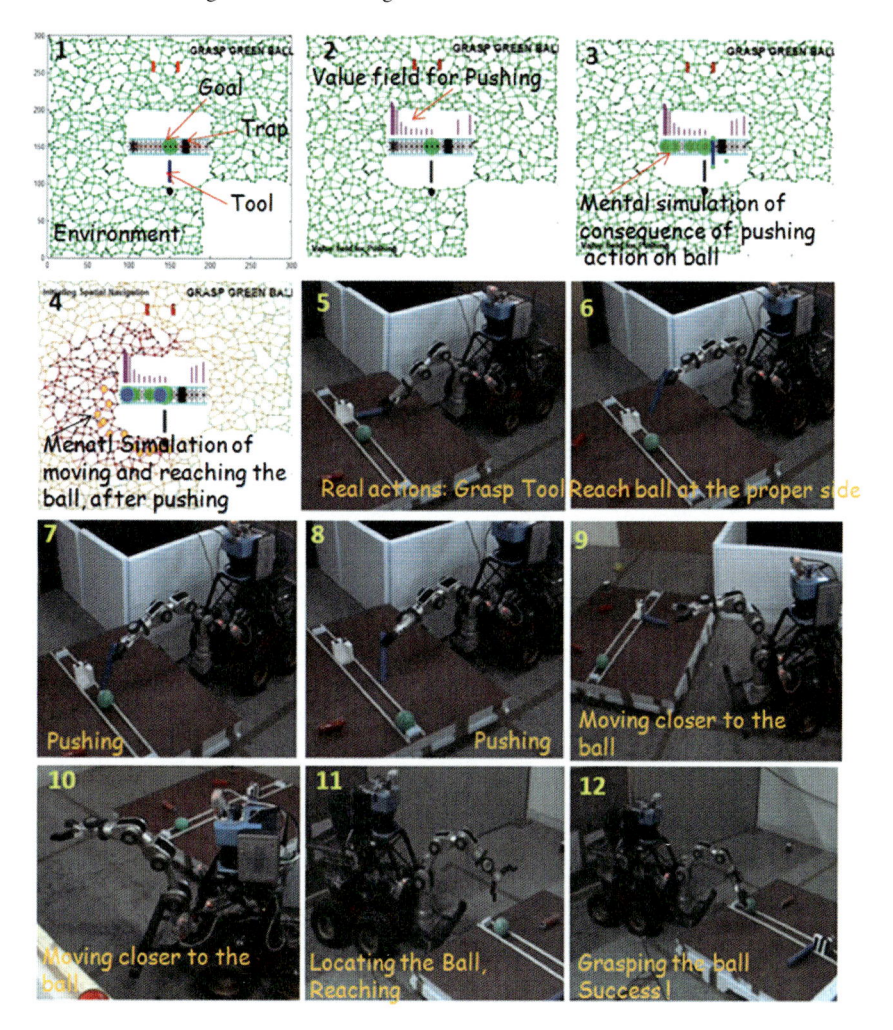

Fig. 17.14 Panels 1-4: Mental simulation of Virtual Push-Move-Reach actions to realize an otherwise impossible goal (Grasping the Green ball placed at the centre of the table unreachable directly to GNOSYS; there is a blue stick present in the environment, there is a trap placed along the trapping groove); Panel 5-12 initiation of real motor actions and successful realization of the goal.

the fact that the ball is not directly reachable with the end-effector, but reachable using the long blue stick (which is directly reachable to its end-effector). Using the pushing internal model, the robot can now perform a virtual experiment to evaluate the consequence of pushing the ball using the stick. The value field in the Pushing SMS (Panel B) incrementally generates actions that are needed to push the ball in the most rewarding way. We note that the pushing value field shown in panel B also includes trap specific adaptations, though a simple learnt pushing value field like the one shown in Fig. 17.8 is equally applicable when traps are not present. On the other hand, these motor activations modulate the lateral connectivity in the pushing

SMS and anticipate the position of the ball as the result of the virtual pushing. On reaching equilibrium, the output of the pushing internal model is a set of trajectories: the trajectory of the ball in the SMS and the trajectory of motor actions that is needed to push the ball in the action space. The anticipated final position of the ball in the trapping groove induces reward excitations on the neurons in the spatial sensorimotor space and triggers the spatial dynamics. The spatial dynamics functions exactly the same way moving in a dynamically generated value field in the internal spatial map, taking into account the set of constraints that are relevant to the task. The output of the spatial dynamics is once again a set of trajectories: the trajectory of the body in the spatial SMS and the trajectory of motor commands that needs to be executed in order to move the body closer to the spatial goal (i.e., the anticipated final position of the ball which was the output of the pushing internal model). Once the dynamics of spatial growing neural gas becomes stationary, Gnosys has the two crucial pieces of information needed to trigger passive motion paradigm (forward/inverse model for the arm): the location of the target (predicted by Pushing model) and the initial conditions (location of the body/end-effector predicted by the equilibrium configuration of the dynamics in the internal spatial map). As we saw in Sect. 17.2, the output of the forward inverse model is also a set of trajectories: the trajectory of the end-effector in the distal space and the trajectory of the joint angles in the proximal space. Starting from a mentally simulated initial body/end-effector position (coming from spatial sensorimotor map), the robot can now mentally simulate a reaching action directed toward a mentally simulated position of the goal target (coming from the pushing sensorimotor space), using the forward inverse model for reaching (passive motion paradigm). In sum, using the three internal models presented in this article, GNOSYS now has the seamless capability to mentally simulate sequences of actions (in different sensorimotor spaces) and evaluate their resulting perceptual consequences: "... since there is a trap there, it is advantageous to push in this direction; if I push in this direction, the ball may eventually go to that side of the table; in case I move my body closer to that edge, I may be in a position to grasp the ball ...."

17.6 Discussion

The functional role played by explorative sensorimotor experience acquired during play toward the overall cognitive development of an agent (natural/artificial) is now well appreciated by experts from diverse disciplines like child psychology, neuroscience, motor control, machine learning, linguistics, and cognitive robotics, among others. No wonder, playing is the most natural thing we do and there is much more to it than just having fun. In this article, we initially introduced the playground we designed for the GNOSYS robot and described the scenarios from animal reasoning that inspired its creation. Three internal models for action generation (reaching, spatial Map, and pushing) all critical for initiating intelligent motor behavior in the playground were presented. We further showed how using the acquired internal models, Gnosys can virtually simulate sequences of "actions and perceptions" in multiple sensory motor state spaces in order to realize a high level goal in a complicated environmental set up. The core action models like Pushing, Moving, Reaching, and Grasping form a closely connected network, predictions of one slowly driving the other (or providing enough information to make the other mental simulation possible). One key feature regarding various internal models (arm, spatial map, pushing, and abstract reasoning system) created in the GNOSYS architecture is the fact that all of them are structurally and functionally identical, use the same protocols for acquisition of information, same computational mechanisms for planning, adaptation, and prediction. The only difference is that they operate on different sensorimotor variables, move in the presence of different value fields toward different goals (local to their computational scope), using different resources of the body/environment. The output of the system ultimately is a set of temporally chunked trajectories (of end-effector, body, external object, etc.) all shaped due to combinations of superimposed fields applied to respective sensorimotor spaces.

While extending the architecture beyond the internal action models presented in this paper, we note that the computational complexity in the problem of realizing an user goal like "Reaching a Red Ball" in a complex environment results from the fact that before reaching the red ball itself with end-effector, there may be several intermediate sequences of real/virtual "Reaching," "Grasping," "Pushing," and "Moving," etc. directed at "potentially useful" environmental objects, information regarding which is not specified by the root goal itself (which was just "reach the red ball"). So before realizing the root goal, the robot has to "track down" and "realize" a set of useful subgoals that "transform" the world in ways that would then make the successful execution of the root goal possible. Hence, even though the high level goals are simple, the complexity of the reasoning process and actions needed to achieve them can increase more than proportionately with the complexity of the environment in which they need to be accomplished. So how can the robot reduce/distribute a high level goal into temporally chunked atomic goals for the different internal models? How can the robot do this flexibly for a large set of environmental configurations each having its own affordances and constraints? What happens if the constraints in some environments do not allow the goal to be realized (e.g., there are two traps in the trapping groove and the goal is placed in between them)? Can robot mentally evaluate the fact that it is in fact impossible to realize the goal in that scenario? Will it Quit without executing any physical action at all? If yes, does it have a reason to Quit? and Can we see the reasons that caused the Quitting by analysing the field structure? We are currently developing and evaluating the extended GNOSYS reasoning-action generation architecture to possibly attack some of these questions.

Acknowledgment This research was partly supported by the EU FP6 project GNOSYS and EU FP7 projects iTalk (Grant No: 214668) and HUMOR (Grant No: 231724).

References

Abbot, L. and Sejnowski, T.J. (1999). Neural codes and distributed representations. Cambridge, MA: MIT.

Amari, S. (1977). Dynamics of patterns formation in lateral-inhibition type neural fields. Biological Cybernetics, 27, 77–87.

Atkeson, C.G., Hale, J.G., Pollick, F. (2000). Using humanoid robots to study human behavior. IEEE Intelligent Systems, 15, 46–56.

Baltzakis, H. (2004). A hybrid framework for mobile robot navigation: modelling with switching state space networks. PhD Thesis, University of Crete.

Behrmann, M. (2000). The mind's eye mapped onto the brain's matter. Current Directions in Psychological Science. April 2000 9, 50–54, doi:10.1111/1467-8721.00059.

Blair, H.T., Cho, J., Sharp, P.E. (1998). Role of the lateral mammillary nucleus in the rat head direction circuit: a combined single unit recording and lesion study. Neuron, 21, 1387–1397.

Brooks, R.A. (1986). A robust layered control system for a mobile robot. IEEE Journal of Robotics and Automation, 2(1), 14–23.

Brooks, R.A. (1997). The Cog Project. T. Matsui (ed.), Special Issue (Mini) on Humanoid, Journal of the Robotics Society of Japan, vol. 15, No. 7.

Clark, A. (1997). Being there: putting brain, body and world together again. Cambridge, MA: MIT.

Cohen, M.S. (1996). Changes in cortical activity during mental rotation. A mapping study using functional MRI. Brain 119, 89–100.

Damasio, A.R. (2000). The feeling of what happens: body, emotion and the making of consciousness. New York: Vintage.

Edelman, G.M. (2006). Second nature: brain science and human knowledge. New Haven, London: Yale University Press.

Edelman, G.M. and Tononi, G. (2001). A universe of consciousness: how matter becomes imagination. New York: Basic Books.

Emery, N.J. and Clayton, N.S. (2004). The mentality of crows: convergent evolution of intelligence in corvids and apes. Science, 306, 1903–1907.

French, R.M. (2006). The dynamics of the computational modelling of analogy-making, CRC handbook of dynamic systems modelling. Fishwick, P. (ed.), Boca Raton, FL: CRC, LLC.

Fritzke, B. (1995). A growing neural gas network learns topologies. In Tesauro, G., Touretzky, D., Leen, T. (eds.), Advances in neural information processing systems, 7 (pp. 625–632). Cambridge, MA: MIT.

Fuster, J.M. (2003). Cortex and mind: unifying cognition. Oxford: Oxford University Press.

Geffner, H. (1992). Default reasoning: causal and conditional theories. MIT Press.

Georgeff, M.P. (1999). The belief-desire-intention model of agency. In Müller, J.P., Smith, M.P., Rao, A.S. (eds.), Intelligent agents, V LNAI. 1555, pp. 1–10. Berlin: Springer.

Georges-Francois, P., Rolls, E.T., Robertson, R.G. (1999). Spatial view cells in the primate hippocampus: allocentric view not head direction or eye position or place. Cerebral Cortex, 9(3), 197–212.

Gnadt, W. and Grossberg, S. (2008). SOVEREIGN: an autonomous neural system for incrementally learning planned action sequences to navigate towards a rewarded goal. Neural Networks, 21, 699–758.

GNOSYS project documentation: www.ics.forth.gr/gnosys.

Grush, R. (1995). Emulation and cognition, doctoral dissertation, University of California, San Diego.

Grush, R. (2004). The emulation theory of representation: motor control, imagery, and perception. Behavioral and Brain Sciences, 27, 377–396.

Hesslow, G. (2002). Conscious thought as a simulation of behavior and perception. Trends in Cognitive Sciences, 6(6), 242–247.

Hesslow, G. and Jirenhed, D.A. (2007). The inner world of a simple robot. Journal of Consciousness Studies, 14, 85–96.

Hirose, M. and Ogawa, K. (2007). Honda humanoid robots development. Philos Transact A Math Phys Eng Sci, 365, 11–19.

Hofstadter, D.R. (1984). The Copycat project: an experiment in nondeterminism and creative reasoning in intelligent systems. San Fransisco, CA: Morgan Kaufmann.

Holland, O. and Goodman, R. (2003). Robots with internal models: a route to machine consciousness? Journal of Consciousness Studies, Special Issue on Machine Consciousness, 10(4), 77–109.

Imamizu, N. (2000). Human cereballar activity reflecting an acquired internal model of a new tool. Nature, 403, 192–196.

Klein, I., Paradis, A.L., Poline, J.B., Kosslyn, S.M., Le Bihan, D. (2000) Transient activity in the human calcarine cortex during visual-mental imagery: an event-related fMRI study. Journal of Cognitive Neuroscience, 12 Suppl 2, 15–23.

Kohonen, T. (1995). Self-organizing maps. Berlin: Springer.

Kokinov, B.N. and Petrov, A. (2001). Integration of memory and reasoning in analogy-making: the AMBR model, the analogical mind: perspectives from cognitive science, Cambridge, MA: MIT.

Kosslyn, S.M. et al. (1993). Visual mental imagery activates topographically organized visual cortex: pet investigations. Journal of Cognitive Neuroscience, 5, 263–287.

Metzinger, T. and Gallese, V. (2003). Motor ontology: the representational reality of goals, actions and selves, Philosophical Psychology, 16, 365–388.

Mohan, V. and Morasso, P. (2007a). Towards reasoning and coordinating action in the mental space. International Journal of Neural Systems, 17(4), 1–13.

Mohan, V. and Morasso, P. (2007b). Neural network of a cognitive crow: an interacting map based architecture. Proceedings of IEEE international conference on self organizing and self adaptive systems, MIT Boston, MA, USA.

Mohan, V., Morasso, P., Metta, G., Sandini, G. (2009). A biomimetic, force-field based computational model for motion planning and bimanual coordination in humanoid robots. Autonomous Robots, 27(3), 291–301.

Morasso, P. (1981). Spatial control of arm movements. Experimental Brain Research, 42, 223–227.

Morasso, P. (2006). Consciousness as the emergent property of the interaction between brain body and environment: the crucial role of haptic perception, Artificial Consciousness, Exeter, UK: Imprint Academic.

Mussa Ivaldi, F.A, Morasso, P., Zaccaria, R. (1988). Kinematic networks. A distributed model for representing and regularizing motor redundancy. Biological Cybernetics, 60, 1–16.

Natale, L., Orabona, F., Metta, G., Sandini, G. (2007). Sensorimotor coordination in a "baby" robot: learning about objects through grasping. Prog Brain Res, 164, 403–424.

Newell, A. and Simon, H. (1976). Computer science as empirical enquiry: symbols and search, Communications of ACM, 19, 113–126.

Nishiwaki, K., Kuffner, J., Kagami, S., Inaba, M., Inoue, H. (2007). The experimental humanoid robot H7: a research platform for autonomous behaviour. Philos Transact A Math Phys Eng Sci, 365, 79–107.

O'Craven, K.M. and Kanwisher, N. (2000). Mental imagery of faces and places activates corresponding stimulus-specific brain regions. Journal of Cognitive Neuroscience, 12, 1013–1023.

O'Keefe, J. and Dostrovsky, J. (1971). The hippocampus as a spatial map. Preliminary evidence from unit activity in the freely-moving rat. Brain Research 34, 171–175.

Op de Beeck, H., Haushofer, J., Kanwisher, N. (2008). Interpreting fMRI data: maps, modules, and dimensions. Nature Reviews Neuroscience.

Oztop, E. Wolpert, D., Kawato, M. (2004). Mental state inference using visual control parameters. Cognitive Brain Research 158, 480–503.

Parsons, L.M., Sergent, J., Hodges, D.A., Fox, P.T. (2005). Cerebrally-lateralized mental representations of hand shape and movement. Journal of Neuroscience, 18, 6539–6548.

Pearl, J. (1988). Probabilistic analogies. AI Memo No. 755. Cambridge, MA: Massachusetts Institute of Technology.

Pearl, J (1998). Graphs, causality, and structural equation models, UCLA Cognitive Systems Laboratory, Technical Report R-253.

Rizzolatti, G., Fogassi, L., Gallese, V. (2001). Neurophysiological mechanisms underlying the understanding and imitation of action. Nature Reviews Neuroscience, 2, 661–670.

Shadmehr, R. (1999). Evidence for a forward dynamic model: human adaptive motor cotnrol. News in Physiological Sciences, 11, 3–9.

Shanahan, M.P. (2005). Perception as abduction: turning sensor data into meaningful representation. Cognitive Science, 29, 109–140.

Sun, R. (2000). Symbol grounding: a new look at an old idea. Philosophical Psychology, 13(2), 149–172.

Sun, R. (2007). The importance of cognitive architectures: An analysis based on CLARION. Journal of Experimental and Theoretical Artificial Intelligence, 19(2), 159–193.

Sutton, R. and Barto, A. (1998). Reinforcement learning. Cambridge, MA: MIT.

Tani, J., Yokoya, R., Ogata, T., Komatani, K., Okuno, H.G. (2007). Experience-based imitation using RNNPB Advanced Robotics, 21(12), 1351–1367.

Taylor, J.G. (2000). Attentional movement: the control basis for consciousness. Neuroscience Abstracts 26 (Part 2), 839(3), 2231.

Toussaint, M. (2004). Learning a world model and planning with a self-organizing dynamic neural system. Advances in neural information processing systems 16 (NIPS 2003), pp. 929–936, Cambridge: MIT.

Toussaint, M. (2006). A sensorimotor map: modulating lateral connections for anticipation and planning. Neural Computation, 18, 1132–1155.

Varela, F.J., Maturana, H.R., Uribe, R. (1974). Autopoiesis: the organization of living systems, its characterization and a model. Biosystems, 5, 187–196.

Visalberghi, E. (1993). Capuchin monkeys: a window into tool use activities by apes and humans. In Gibson, K. and Ingold, T. (eds.), Tool, Language and Cognition in Human Evolution (pp. 138–150). Cambridge: Cambridge University Press.

Visalberghi, E. and Limongelli, L. (1996). Action and understanding: tool use revisited through the mind of capuchin monkeys. In Russon, A., Bard, K., Parker, S. (eds.), Reaching into thought. The minds of the great apes pp. 57–79. Cambridge: Cambridge University Press.

Visalberghi, E. and Tomasello, M. (1997). Primate causal understanding in the physical and in the social domains. Behavioral Processes, 42, 189–203.

Wolpert, D.M., Ghahrmani, Z., Jordanm, M.I. (1994). An internal model for integration. Science, 269, 1880–1882.

Weir, A.A.S., Chappell, J., Kacelnik, A. (2002). Shaping of hooks in New Caledonian Crows. Science, 297, 981–983.

Yuille, A., Carter, N., Tenenbaum, J.B. (2006). Probabilistic models of cognition: conceptual foundations, Trends in Cognitive Sciences, 10(7), 287–291.

Zak, M. (1988). Terminal attractors for addressable memory in neural networks. Physics Letters. A, 133, 218–222.

Zatorre, R.J., Chen, J.L., Penhune, V.B. (2007). When the brain plays music. Auditory-motor interactions in music perception and production. Nature Reviews Neuroscience, 8, 547–558.

Chapter 18
Cognitive Algorithms and Systems: Reasoning and Knowledge Representation

Artur S. d'Avila Garcez and Luis C. Lamb

Abstract This chapter reviews recent advances in computational cognitive reasoning and their underlying algorithmic foundations. It summarises the neural-symbolic approach to cognition and computation. Neural-symbolic systems integrate two fundamental phenomena of intelligent behaviour: reasoning and the ability to learn from experience. The chapter illustrates how to represent, learn and compute several expressive forms of symbolic knowledge using neural networks. The goal is to provide computational models with integrated reasoning capabilities, where the neural networks offer the machinery for cognitive reasoning and learning while symbolic logic offers explanations to the neural models facilitating the necessary interaction with the world and other systems.

18.1 Introduction

Recent endeavors in cognitive science, artificial intelligence (AI) and evolutionary psychology have led to several hypotheses about the way cognitive models of reasoning, learning and language can be effected in or modelled by computational techniques (Pinker 2007; Pinker et al. 2008). Pinker has defended that the human mind is composed of computing constructions, or organs of computation (Pinker 2007). Furthermore, these models must cater for computation, specialisation and evolution. In computer science, recent efforts towards understanding and integrating learning, reasoning and action in artificial cognitive models have led to a number of developments, including approaches where learning and reasoning are modelled in a unified perspective (see, e.g. d'Avila Garcez et al. 2009; Valiant 2000). They have led also to the development of computational systems that are provably sound and have shown promise in a number of applications, including computational biology and fault diagnosis (d'Avila Garcez et al. 2002a).

A.S. d'Avila Garcez (✉)
Department of Computing, City University, London EC1V 0HB, UK
e-mail: aag@soi.city.ac.uk

V. Cutsuridis et al. (eds.), *Perception-Action Cycle: Models, Architectures, and Hardware*, Springer Series in Cognitive and Neural Systems 1, DOI 10.1007/978-1-4419-1452-1_18, © Springer Science+Business Media, LLC 2011

Three notable hallmarks of intelligent cognition are the ability to draw rational conclusions, the ability to make plausible assumptions and the ability to generalise from experience. In a logical setting, these abilities correspond to the processes of deduction, abduction, and induction, respectively. Although human cognition often involves the interaction of these three abilities, they are typically studied in isolation (a notable exception is Mooney and Ourston 1994). For example, in AI, symbolic (logic-based) approaches have been mainly concerned with deductive reasoning, while connectionist (neural networks-based) approaches have mainly focused on inductive learning.

Neural-symbolic computation seeks to integrate the processes of logical reasoning and learning within the neural-computation paradigm (d'Avila Garcez 2005; d'Avila Garcez et al. 2007a, 2009; d'Avila Garcez and Lamb 2005). When we think of neural networks, what springs to mind is their ability to learn from examples using efficient algorithms in a massively parallel fashion. In neural computation, induction is typically seen as the process of changing the weights of a network in ways that reflect the statistical properties of a dataset (set of examples), allowing for useful generalisations over unseen examples. When we think of symbolic logic, we recognise its rigour, semantic clarity and the availability of automated proof methods which can provide explanations to the reasoning process, e.g. through a proof history. In neural computation, deduction can be seen as the network computation of output values as a response to input values, given a particular set of weights. Standard feedforward and partially recurrent networks have been shown capable of deductive reasoning of various kinds depending on the network architecture, including nonmonotonic (d'Avila Garcez et al. 2002a), modal (d'Avila Garcez et al. 2007b, 2009), intuitionistic (d'Avila Garcez et al. 2006a, b), epistemic (d'Avila Garcez and Lamb 2006; d'Avila Garcez and Lamb 2004) and abductive reasoning (d'Avila Garcez et al. 2007a).

In what follows, we briefly review the work on the integration of a range of computer science logics and neural networks. These constitute the technical foundations of a rich model of cognitive computation. In particular, we consider how standard neural networks can represent modal logic and its variations such as temporal logic. The resulting neural-symbolic cognitive system is called *connectionist modal logic* (CML). We then investigate how different networks and their associated logics can be combined to give an expressive yet feasible model of computation. For example, a network encoding some nonmonotonic mechanism of vision processing may need to be combined with a network that uses a temporal database for planning. A methodology for combining systems called the fibring method is used for this (d'Avila Garcez and Gabbay 2004). The overall model consists of an ensemble of simple single-hidden-layer neural networks – each may represent the knowledge of an agent (or a possible world) at a particular time-point – with connections between networks representing the relationships between agents/possible worlds. Each ensemble may be at a different level of abstraction, so that networks at one level may be fibred onto networks at another level to form a structure combining met-alevel and object-level reasoning where high-level abstractions can be learned from low-level concepts. We claim that this structure offers the basis for an expressive yet computationally tractable cognitive model for integrated robust learning and expressive reasoning.

18.2 Neurons and Symbols

The modelling of behaviour is an important goal of psychology, cognitive science, computer science, neural computation, philosophy, communication and other areas. Among the most prominent tools in the modelling of behaviour are computational-logic systems (e.g. classical logic, nonmonotonic logic, modal and temporal logic) and connectionist models of cognition (e.g. feedforward and recurrent networks, deep networks, self-organising networks).

The goal of neural-symbolic computation is to provide a coherent, unifying view for logic and connectionist network reasoning, contributing to the modelling and understanding of cognitive behaviour, and producing better computational tools. Typically, translation algorithms from a symbolic to a connectionist representation and vice versa are used to provide (a) a neural implementation of a logic, (b) a logical characterisation of a neural system, or (c) a hybrid learning system that brings together features from connectionism and symbolic AI.

In what follows, we focus on nonclassical logics and their associated recurrent-network models. In particular, we consider modal and temporal logics. Modal logics are among the most successful applied-logic systems (Blackburn et al. 2006). Temporal logic and its combination with other modalities such as knowledge operators have been the subject of intensive investigation leading to some of the main logical systems used in computer science and AI (Fagin et al. 1995; Vardi 1997). Recurrent networks, in turn, have been widely studied within neural computation and cognitive science, and applied to temporal sequence learning problems such as time-series prediction (Elman 1990). Our goal is to produce a robust computational system for modal and temporal knowledge representation using logic and connectionist recurrent networks. We claim that nonclassical reasoning has a major role to play in computer science. In addition, we subscribe to the view that computational cognitive modelling can lead to valid theories of cognition and offer a better understanding of certain cognitive processes (d'Avila Garcez et al. 2009; Sun 2009). Finally, we argue that a purely symbolic approach would not be sufficient, as also argued by Valiant (Valian 2008), and that a hybrid connectionist-symbolic approach can accommodate robustness and produce a more effective model of cognitive computation.

Our methodology is to transfer principles and mechanisms between nonclassical logic computation and neural computation. In particular, we consider how principles of symbolic computation can be implemented by connectionist mechanisms. The reason for this is that we see connectionism as the hardware to build upon with the use of different levels of abstraction according to the needs of the application. This methodology, looking at principles, mechanisms and applications, has proven a fruitful way of progressing the research in the area (d'Avila Garcez et al. 2009) and abiding by Pinker's models of mind and cognition (Pinker 2007; Pinker et al. 2008). It has produced a connectionist system for nonclassical reasoning that strikes an adequate balance between complexity and expressiveness. In this system – known as a neural-symbolic system – neural networks provide the machinery for parallel computation and robust learning, while logic provides the necessary explanation to the network models, facilitating the necessary interaction with the world and other systems. In this integrated model, no conflict arises between a continuous and a discrete component of the system. Instead, a tightly coupled hybrid system exists that is continuous by nature (the neural network), but which has a clear discrete interpretation (its logic) at a different level of abstraction.

18.2.1 Abstraction

Growing attention has been given recently to deep network architectures where it is hoped that high-level abstract representations will emerge from low-level unprocessed datasets. The main example here are deep belief networks (Hinton et al. 2006), which use a sequence of restricted Boltzmann Machines to learn abstract representations from a grid of pixels, obtaining similar or better classification performance than support vector machines[1] (SVMs) (Shawe-Taylor and Cristianini 2004). This highlights the question of which representation is more appropriate in cognitive science: deep network models or shallow networks like SVMs. Neural-symbolic computation can help answer this question. It provides – almost as a side-effect – precise and useful expressiveness results for network models with respect to logic.

18.2.2 Modularity

Another key characteristic of neural-symbolic systems is modularity. In line with the ideas behind deep networks, neural-symbolic networks are modular and can be built through the careful engineering of network ensembles.[2] Modularity is of course important for comprehensibility and maintenance. On the other hand, massive integration of neural circuits is seen by many as a key feature of cognitive systems. There is a tension here between modularity and integration which emerges from

---

[1] It is worth noting that Hinton et al. (2006) is concerned mainly with unsupervised learning, while SVMs are supervised learning systems that require a certain amount of data preprocessing.

[2] Each network in the ensemble can be responsible for a specific task or logic, with the overall model being potentially very expressive. The methodology that we use to combine networks is that of fibring (Gabbay 1999) as discussed in some detail later.

computational concerns; it will probably fall on the field of cognitive computation to provide an alternative. Some strong hints as to the direction to follow can already be found in Taylor (2009).

18.2.3 Applications

Neural-symbolic systems have had important applications in many areas such as bioinformatics, simulation, robotics, fraud prevention and text processing. In such areas, a computational system is required to learn from experience and to reason about what has been learned (Browne and Sun 2001; Valiant 2003). For this process to be successful, the system must be robust (in the sense that the accumulation of errors resulting from the intrinsic uncertainty associated with the problem domain can be controlled). One such system that is already providing a contribution to problems in bioinformatics and engineering is the connectionist inductive learning and logic programming system (CILP) (d'Avila Garcez et al. 2002a). The merging of theory (known as background knowledge in machine learning) and data learning (i.e. learning from examples) in CILP networks have been shown more effective than purely symbolic or purely connectionist systems, especially in the case of noisy datasets (Towell and Shavlik 1994). Such results have contributed to the growing interest in developing neural-symbolic systems that are capable of learning from examples and background knowledge. It is important to consider the needs of the application. Complex applications can drive the research in this area further towards more effective systems.

18.2.4 Expressiveness

Until recently, neural-symbolic systems were not able to represent, compute and learn languages other than propositional logic and some fragments of first-order logic (Browne and Sun 2001; Cloete and Zurada 2000; Hölldobler and Kalinke 1994). In d'Avila Garcez et al. (2002b, 2003a, 2004c) and d'Avila Garcez and Lamb (2004), a new approach to knowledge representation and reasoning using neural-symbolic systems has been proposed, establishing a class of connectionist nonclassical logics, including connectionist modal, intuitionistic, temporal and epistemic logics (d'Avila Garcez et al. 2003b, 2004c; d'Avila Garcez and Lamb 2004). This new approach shows that a variety of nonclassical logics can be effectively represented by neural network ensembles. More recently, it has been shown that argumentation frameworks can also be represented by the same network ensemble models, offering an integrated approach to learning and reasoning of arguments, including non-standard forms of argumentation (d'Avila Garcez et al. 2004a, b, 2005) and of analogy (Borger et al. 2008).

As claimed in Browne and Sun (2001), if connectionism is a possible paradigm to cognitive science and AI, neural networks must be able to compute symbolic reasoning in an efficient and effective way. Moreover, in hybrid learning systems, usually the connectionist component is fault-tolerant, while the symbolic component may be "brittle and rigid." By integrating connectionist systems and sound nonclassical logics, we tackle this problem and offer a principled way to effectively compute, represent and learn various nonclassical logics within connectionist models.

18.2.5 Representation

A historical criticism of neural networks has been raised by John McCarthy already back in 1988 (McCarthy 1988). McCarthy referred to neural networks as having a "propositional fixation", in the sense that they were not able to represent first-order logic. This per se has remained a challenge for a decade, but several approaches have now dealt with first-order reasoning in neural networks, see, e.g. Browne and Sun (2001).

Perhaps in an attempt to address McCarthy's criticism, many researchers in the area have focused their attention only on first-order logic. More recently, it has been shown that nonclassical, practical reasoning can be used in a number of applications in neural-symbolic systems (d'Avila Garcez and Lamb 2004; d'Avila Garcez et al. 2003b, 2004a, b, c, 2005). Nonclassical logics have been shown adequate in expressing several reasoning features, allowing for the representation of temporal, epistemic and probabilistic abstractions in computer science and AI (Fagin et al. 1995; Gabbay et al. 2003; Halpern 2003). Some applications of nonclassical logics include the characterisation of timing analysis in combinatorial circuits (Mendler 2000) and in spatial reasoning (Bennett 1994), with possible use in geographical information systems. For instance, Bennett's propositional intuitionistic approach provided for tractable yet expressive reasoning about topological and spatial relations. Thus, a connectionist nonclassical logic can offer a richer cognitive model of computation, more realistic for modelling the many dimensions of an autonomous agent.

18.2.6 Nonclassical Reasoning

In summary, we believe that for neural computation to achieve its promise, connectionist models must be able to cater for nonclassical reasoning. We believe that the different communities cannot ignore the achievements and impact that nonclassical logics have had in computer science. Temporal logic, for instance, has had a large impact on both academia and industry (Gabbay et al. 1994; Pnueli 1977).

Modal logics, in turn, have become a lingua franca for the specification and analysis of knowledge and communication in multi-agent and distributed systems (Fagin et al. 1995; Wooldridge 2001). Epistemic logics have found a large number of applications, notably in game theory and in models of knowledge and interaction in multi-agent systems (Fagin et al. 1995; Gabbay et al. 1994; Pnueli 1977). Nonmonotonic reasoning has dominated the research on logic and AI in the 1980s and 1990s, and intuitionistic logic is considered by many as providing not only an adequate logical foundation for several core areas of theoretical computer science, including type theory and functional programming (van Dalen 2002), but also a solid basis for constructive reasoning (d'Avila Garcez et al. 2006a, b, 2009).

Notwithstanding all this evidence, little attention has been given to nonclassical reasoning and their integration with neural networks. If neural networks are to represent rich models of reasoning, it is undeniable that nonclassical logic should be at the core of this enterprise.

In the long run, neural-symbolic computation seeks to achieve a characterisation of a rich semantics for cognitive computation. This has been identified as a major challenge for computer science (Valiant 2003). We are proposing a methodology for the representation of several forms of nonclassical reasoning in artificial neural networks. Such expressive logics have been successfully used in computer science. Connectionist approaches should consider them by means of adequate computational models catering for integrated reasoning, knowledge representation and learning in cognitive science.

18.3 Neural-Symbolic Learning Systems

For neural-symbolic integration to be effective as a model of computation, we need to investigate how to represent, reason and learn expressive logics in neural networks. We also need to find effective ways of expressing the knowledge encoded in a trained network in a comprehensible symbolic form. There are at least two lines of action. The first is to take standard neural networks and try and find out which logics they can represent. The other is to take well-established logics and concepts (e.g. recursion) and try and encode those in a neural network architecture. Both lines require a principled approach, so that whenever we show that a particular logic can be represented by a particular neural network, we need to show that the network and the logic are in fact equivalent (a way of doing this is to prove that the network computes a formal semantics of the underlying logic). Similarly, if we develop a knowledge extraction algorithm, we need to make sure that it is correct (sound) in the sense that it produces rules that are encoded by the network, and that it is *quasi-*complete in the sense that the extracted rules increasingly approximate the exact behaviour of the network.

During the past 20 years, a number of models for neural-symbolic integration have been proposed [mainly in response to John McCarthy's note *Epistemological challenges for connectionism*[3] (McCarthy 1988), itself a response to Paul Smolensky's *On the proper treatment of connectionism* (Smolensky 1988)]. Broadly speaking, researchers have made contributions in three main areas, providing (a) a logical characterisation of a connectionist system, (b) a connectionist implementation of a logic, or (c) a hybrid system bringing together features from connectionist systems and symbolic AI (Hitzler et al. 2004). Early relevant contributions include Hölldobler and Kalinke (1994), Shastri (1999) and Sun (1995) on knowledge representation, d'Avila Garcez and Zaverucha (1999) and Towell and Shavlik (1994) on learning with background knowledge, and Bologna (2004), d'Avila Garcez et al. (2001), Jacobsson (2005), Setiono (1997) and Thrun (1994) on knowledge extraction. The reader is referred to d'Avila Garcez et al. (2002a) for a detailed presentation of neural-symbolic learning systems and applications.

Neural-symbolic learning systems contain six main phases: (1) *background knowledge insertion,* (2) *inductive learning from examples,* (3) *massively parallel deduction,* (4) *theory fine-tuning,* (5) *symbolic knowledge extraction* and (6) *feedback* (see Fig. 18.1). In phase (1), symbolic knowledge is translated into the initial architecture of a neural network with the use of a *translation algorithm.* In phase (2), the neural network is trained with examples by a neural learning algorithm, which revises the theory given in phase (1) as *background knowledge.* In phase (3), the network can be used as a massively parallel system to compute the logical consequences of the theory encoded in it. In phase (4), information obtained from the computation carried out in phase (3) may be used to help fine-tune the network to better represent the problem domain. This mechanism can be used, for example, to resolve inconsistencies between the background knowledge and the training examples. In phase (5), the result of training is explained by the extraction of revised symbolic knowledge. As with the insertion of rules, the *extraction algorithm* must

---

[3] McCarthy (1988) identifies four knowledge representation problems for neural networks: the problem of *elaboration tolerance* (the ability of a representation to be elaborated to take additional phenomena into account), the *propositional fixation* of neural networks (based on the assumption that neural networks cannot represent relational knowledge), the problem of how to make use of any available *background knowledge* as part of learning and the problem of how to obtain domain *descriptions* from trained networks as opposed to mere discriminations. Neural-symbolic integration can address each of the above challenges. In a nutshell, the problem of elaboration tolerance can be resolved by having networks that are fibred forming a modular hierarchy, similar to the idea of using self-organising maps (Gärdenfors 2000; Haykin 1999) for language processing, where the lower levels of abstraction are used for the formation of concepts that are then used at the higher levels of the hierarchy. CML (d'Avila Garcez et al. 2007b) deals with the so-called propositional fixation of neural networks by allowing them to encode relational knowledge in the form of accessibility relations; a number of other formalisms have also tackled this issue as early as 1990 (Bader et al. 2005, 2007; Hölldobler 1993; Shastri and Ajjanagadde 1990), the key question being how to have simple representations that promote effective learning. Learning with background knowledge can be achieved by the usual translation of symbolic rules into neural networks. Problem descriptions can be obtained by rule extraction; a number of such translation and extraction algorithms have been proposed (e.g. Bologna 2004; d'Avila Garcez et al. 2001; d'Avila Garcez and Zaverucha 1999; Lozowski and Zurada 2000; Hitzler et al. 2004; Jacobsson 2005; Nunez et al. 2006; Setiono 1997; Sun 1995).

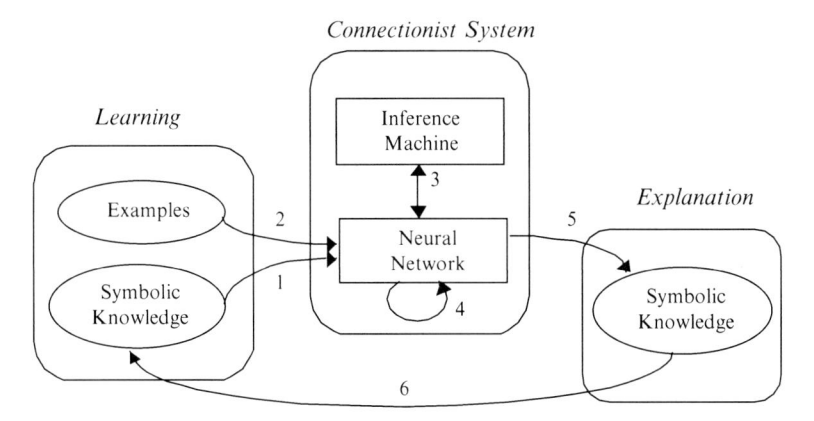

Fig. 18.1 Neural-symbolic learning systems

be provably correct, so that each rule extracted is guaranteed to be a rule of the network. Finally, in phase (6), the knowledge extracted may be analysed by an expert to decide whether it should feed the system again, closing the learning and reasoning cycle.

Our neural network models consist of feedforward and partially recurrent networks, as opposed to the symmetric networks investigated, e.g. in Smolensky and Legendre (2006). It uses a localist rather than a distributed representation,[4] and it works with backpropagation, the most successful neural learning algorithm used in industrial-strength applications (Rumelhart et al. 1986).

18.4 Technical Background

In this section, we introduce some technical aspects of neural-network and neural-symbolic computation. The reader can skip this section if interested mainly in the overall cognitive model architecture.

18.4.1 Neural Networks and Neural-Symbolic Systems

An artificial neural network is a directed graph. A unit (or neuron) in this graph is characterised, at time t, by its input vector $I_i(t)$, its input potential $U_i(t)$, its

---

[4] We depart from distributed representations for two main reasons: localist representations can be associated with highly effective learning algorithms such as backpropagation, and in our view localist networks are at an appropriate level of abstraction for symbolic knowledge representation. As advocated in Page (2000), we believe one should be able to achieve the goals of distributed representations by properly changing the levels of abstraction of localist networks, while some of the desirable properties of localist models cannot be exhibited by fully distributed ones.

activation state $A_i(t)$ and its output $O_i(t)$. The units of the network are interconnected via a set of directed and weighted connections. If there is a connection from unit i to unit j, then $W_{ji} \in \mathbb{R}$ denotes the weight of this connection. The input potential of neuron i at time t $(U_i(t))$ is obtained by computing a weighted sum for neuron i such that $U_i(t) = \sum_j W_{ij} I_i(t)$. The activation state of a neuron i at time t $(A_i(t))$ is a bounded real or integer number. $A_i(t)$ is given by the neuron's *activation rule* h_i, which is a function of the neuron's input potential, i.e. $A_i(t) = h_i(U_i(t))$. Typically, h_i is either a linear, a non-linear or a sigmoid activation function, e.g. $tanh(x)$. In addition, θ_i is known as the threshold of neuron i. We say that neuron i is *activated* at time t if $A_i(t) > \theta_i$. Finally, the neuron's output value $O_i(t)$ is given by $f_i(A_i(t))$; usually, f_i is the identity function. The units of a neural network can be organised in layers. An n-layer feedforward network is an acyclic graph containing one input layer, $n - 2$ hidden layers and one output layer. It computes a function $\varphi : \mathbb{R}^r \to \mathbb{R}^s$, where r and s denote the number of units occurring in the input and output layers, respectively. Most neural models also have a *learning rule*, responsible for changing the weights of the network so that it learns to approximate φ given a number of *training examples*, e.g. input vectors and their respective target output vectors.

The CILP (d'Avila Garcez et al. 2002a) is a computational model based on neural networks that integrates inductive learning from examples and background knowledge with deductive learning using logic programming. In CILP, a translation algorithm maps a logic program $\mathcal{P}$ into a single hidden layer neural network $\mathcal{N}$ such that $\mathcal{N}$ computes the fixed-point operator $T_\mathcal{P}$ of $\mathcal{P}$. This provides a massively parallel model for computing the stable model semantics of $\mathcal{P}$. In addition, $\mathcal{N}$ can be trained with examples using a neural learning algorithm, having $\mathcal{P}$ as background knowledge. The knowledge acquired by training can then be extracted, closing the learning cycle (d'Avila Garcez et al. 2002a).

Let us exemplify how CILP translation algorithm works. Each rule (r_l) of $\mathcal{P}$ is mapped from the input layer to the output layer of $\mathcal{N}$ through one neuron (N_l) in the single hidden layer of $\mathcal{N}$. Intuitively, the translation algorithm from $\mathcal{P}$ to $\mathcal{N}$ has to implement the following conditions: (c_1). The input potential of a hidden neuron (N_l) can only exceed N_l's threshold (θ_l), activating N_l, when all the positive antecedents of r_l are assigned truth-value *true* while all the negative antecedents of r_l are assigned *false*; and (c_2). The input potential of an output neuron (A) can only exceed A's threshold (θ_A), activating A, when at least one hidden neuron N_l that is connected to A is activated.

Example 1. (CILP) Consider the logic program $\mathcal{P} = \{r_1 : B \wedge C \wedge \neg D \to A, r_2 : E \wedge F \to A, r_3 : B\}$, where $\neg$ stands for *default negation*. The translation algorithm derives the network $\mathcal{N}$ of Fig. 18.2, setting weights (W) and thresholds (θ) in such a way that conditions (c_1) and (c_2) above are satisfied. Note that if $\mathcal{N}$ ought to be fully connected, any other link (not shown in Fig. 18.2) should receive weight zero initially. Each input and output neuron of $\mathcal{N}$ is associated with an atom of $\mathcal{P}$. As a result, each input and output vector of $\mathcal{N}$ can be associated with an interpretation for $\mathcal{P}$, so that an atom (e.g. A) is true if its corresponding neuron (neuron A) is activated. Note also that each hidden neuron N_l corresponds

Fig. 18.2 Neural network
for logic programming

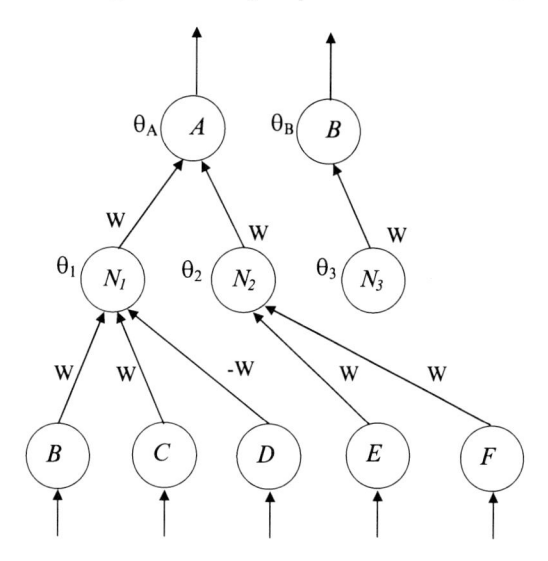

to a rule r_l of $\mathcal{P}$. In order to compute a stable model, output neuron B should feed input neuron B such that $\mathcal{N}$ is used to iterate the fixed-point operator $T_{\mathcal{P}}$ of $\mathcal{P}$ (d'Avila Garcez et al. 2002a). This is done by transforming $\mathcal{N}$ into a recurrent network $\mathcal{N}_r$, containing feedback connections from the output to the input layer of $\mathcal{N}$, all with fixed weights $W_r = 1$. In the case of $\mathcal{P}$ above, given any initial activation to the input layer of $\mathcal{N}_r$, it always converges to the following stable state: $A = false$, $B = true$, $C = false$, $D = false$, $E = false$, and $F = false$, that represents the unique fixed-point of $\mathcal{P}$.

18.4.2 The Language of Connectionist Modal Logic

In CML, the CILP system is extended to the language of modal logic programming (Orgun and Ma 1994) further extended to allow modalities such as necessity ($\square$) and possibility ($\lozenge$) to occur also in the head of clauses. The modalities shall allow us to represent several modes of reasoning, as we will illustrate in the coming sections. A modal translation algorithm then sets up an ensemble of CILP neural networks (d'Avila Garcez et al. 2002a), each network representing a possible world that can be trained by examples just like CILP networks. The ensemble computes a (fixed-point) semantics of modal theories, thus working as a massively parallel system for modal logic (d'Avila Garcez et al. 2004c). Since each network can be trained efficiently by a neural learning algorithm (e.g. backpropagation Rumelhart et al. 1986), one can adapt the ensemble by performing inductive learning.

A main feature of modal logics is the use of Kripke's *possible world semantics*. Under this interpretation, we say that a proposition is necessary in a world if it is true in all worlds which are possible in relation to that world, whereas it is possible in a world if it is true in at least one world which is possible in relation to that same world. This is expressed in the semantics formalisation by a (binary) relation between possible worlds. In modal logic programming, a *modal atom* is of the form MA, where $M \in \{\Box, \Diamond\}$ and A is an atom. A *modal literal* is of the form ML, where L is a literal. A *modal program* is a finite set of clauses of the form $MA_1, \ldots, MA_n \rightarrow A$. We define *extended modal programs* as modal programs extended with modalities $\Box$ and $\Diamond$ also in the head of clauses and default negation in the body of clauses. In addition, each clause is labelled by the possible world in which it holds, similarly to Gabbay's labelled deductive systems (Broda et al. 2004). Thus, an *extended modal program* is a finite set of clauses C of the form $\omega_i : ML_1, \ldots, ML_n \rightarrow MA$, where ω_i is a label representing a world in which the associated clause holds, and a finite set of relations $\mathcal{R}(\omega_i, \omega_j)$ between worlds ω_i and ω_j in C.

A (Kripke) model M is a tuple $M = (\mathcal{W}, \mathcal{R}, \pi)$, where (a) $\mathcal{W}$ is a set of possible worlds; (b) $\mathcal{R}$ is a binary accessibility relation over worlds; and (c) π is a mapping associating worlds to formulas. We write $(M, \omega) \models \alpha$ if α is true at ω in M. Formally:

$(M, \omega) \models$ p iff $\omega \in \pi(p)$ for a propositional letter p
$(M, \omega) \models \neg \alpha$ iff $(M, \omega) \not\models \alpha$
$(M, \omega) \models \alpha \wedge \beta$ iff $(M, \omega) \models \alpha$ and $(M, \omega) \models \beta$
$(M, \omega) \models \Box \alpha$ iff $\forall \omega' \in \mathcal{W}$, if $\mathcal{R}(\omega, \omega')$ then $(M, \omega') \models \alpha$
$(M, \omega) \models \Diamond \alpha$ iff $\exists \omega'$ such that $\mathcal{R}(\omega, \omega')$ and $(M, \omega') \models \alpha$

When computing the semantics of a modal program, we consider what is computed in individual worlds, and the fixed-point of the program as a whole. When computing the fixed-point at each world, we consider the consequences derived locally and the consequences derived from the interaction between worlds. Locally, fixed-points are computed as before, by simply renaming each modal literal ML_i by a new literal L_j not in the language, and computing stable models. When considering interacting worlds, there are two cases to address, according to the $\Box$ and $\Diamond$ modalities and the accessibility relation $\mathcal{R}$, which might render additional consequences in each world.

Briefly, whenever $\Box A$ is true in a world (i.e. a neuron labelled $\Box A$ is activated in the corresponding neural network), A must be true in every world related to that (i.e. connections in the ensemble of networks must be established so that the firing of neuron $\Box A$ activates all neurons A in the related networks). Whenever $\Diamond A$ is true in a world (neuron $\Diamond A$ is activated), A must be true in one world related to that (connections must be established so that the firing of $\Diamond A$ activates A in one related world). The choice of the world in which to have A activated is arbitrary, reflecting the semantics of the $\Diamond$ modality. The following example illustrates this.

Example 2. (CML) Let $\mathcal{P} = \{\omega_1 : r \rightarrow \Box q, \omega_1 : \Diamond s \rightarrow r, \omega_2 : s, \omega_3 : q \rightarrow \Diamond p, \mathcal{R}(\omega_1, \omega_2), \mathcal{R}(\omega_1, \omega_3)\}$. We start by creating three CILP neural networks to represent the worlds ω_1, ω_2 and ω_3 (see Fig. 18.3). Then, we interconnect networks according to the meaning of $\Box$ and $\Diamond$. Hidden neurons labelled M, $\vee$ *and* $\wedge$ are

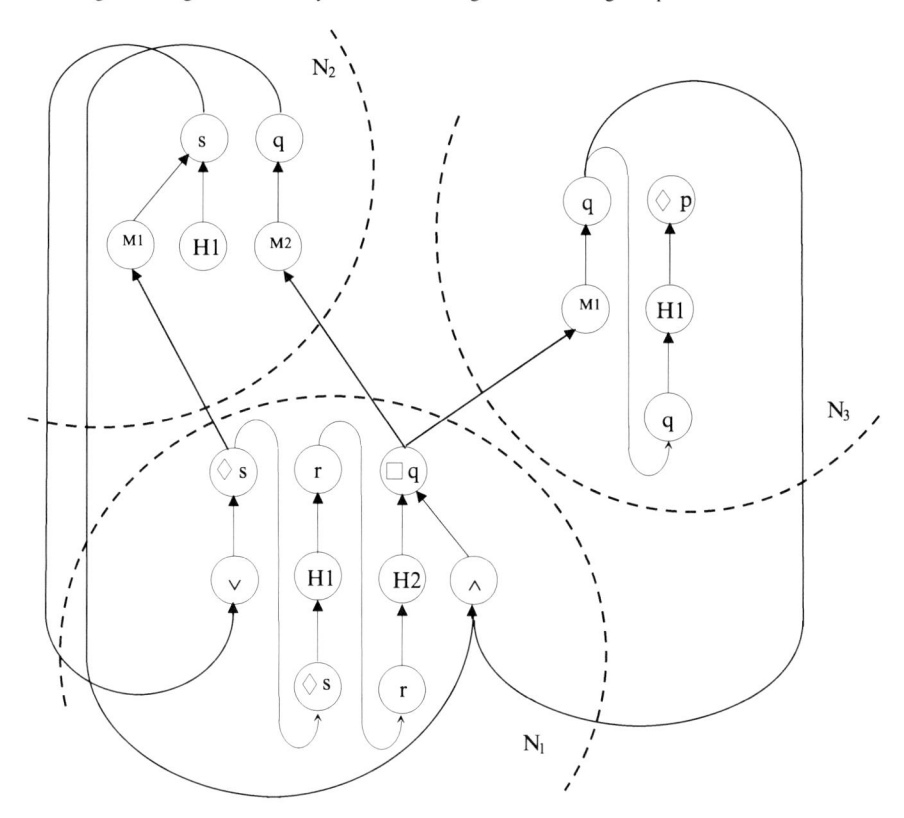

Fig. 18.3 Neural network ensemble for modal reasoning

created to do so. The remaining neurons are all created by CILP. For example, whenever neuron $\Box q$ is activated in ω_1, neuron q should be activated in both ω_2 and ω_3; whenever neuron $\Diamond s$ is activated in ω_1, neuron s should be activated in ω_2. This is implemented by using the hidden neurons labelled as M in the network. Dually, whenever q is activated in both ω_2 and ω_3, $\Box q$ should be activated in ω_1; whenever s is activated in ω_2, $\Diamond s$ should be activated in ω_1. This is implemented by using neurons labelled as $\land$ and $\lor$, respectively.

18.4.3   Reasoning About Time and Knowledge

In order to reason about the truth of sentences in time and represent knowledge evolution through time, we need to add temporal operators to the language of CML, as described below. We consider a temporal logic of knowledge that combines knowledge and time operators (see Fagin et al. 1995 for complete axiomatisations of such logics). The language of CML is extended with a set of agents $\mathcal{A} \subseteq \mathbb{N}$, a set of

Fig. 18.4 Network ensemble
for temporal reasoning

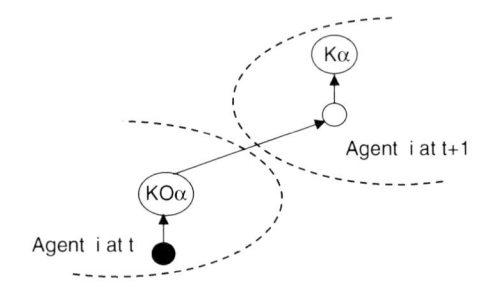

Agent i at t+1

Agent i at t

unary connectives: K_i, $i \in \mathcal{A}$, where $K_i p$ reads "agent i knows p", and the temporal operator $\bigcirc$ (next time). A temporal translation algorithm then is responsible for converting temporal rules of the form $t : K_{[\mathcal{A}]} L_1, ..., K_{[\mathcal{A}]} L_k \rightarrow \bigcirc K_{[\mathcal{A}]} L_{k+1}$, into neural network ensembles, where $[\mathcal{A}]$ denotes an element selected from $\mathcal{A}$ for each literal L_j, $1 \leq j \leq k + 1$, $1 \leq t \leq n$; $k, n \in \mathbb{N}$ (d'Avila Garcez and Lamb 2004).

To each time-point, we associate the set of formulas holding at that point and extend the definition of a model M as follows: $M = (T, \mathcal{R}_1, ..., \mathcal{R}_n, \pi)$, where (a) T is a set of (linearly) ordered points; (b) $\mathcal{R}_i$ ($i \in A$) is an agent accessibility relation over points; (c) π is a mapping associating time points to formulas. We write $(M, t) \models \alpha$ if α is true at point t in M. Formally:

$(M, t) \models \bigcirc \alpha$ iff $(M, t + 1) \models \alpha$
$(M, t) \models K_i \alpha$ iff $\forall u \in T$, if $R_i(t, u)$ then $(M, u) \models \alpha$

It is worth noting that whenever a rule's consequent is preceded by $\bigcirc$, a forward connection from t to $t + 1$ and a feedback connection from $t + 1$ to t need to be added to the ensemble. For example, if $t : a \rightarrow \bigcirc b$ is a rule in $\mathcal{P}$, then not only must the activation of neuron a at t activate neuron b at $t + 1$, but the activation of neuron b at $t + 1$ must also activate neuron $\bigcirc b$ at t.

Example 3. One of the typical axioms of temporal logics of knowledge is $K_i \bigcirc \alpha \rightarrow \bigcirc K_i \alpha$ (Fagin et al. 1995), which means that an agent does not forget tomorrow what he knew today. This can be represented in an ensemble of CILP networks by connecting output neuron $K \bigcirc \alpha$ of agent i at time t to a hidden neuron that connects to output neuron $K \alpha$ of agent i at time $t + 1$. In Fig. 18.4, the black circle denotes a neuron that is always activated (*true*), and the activation value of output neuron $K \bigcirc \alpha$ at time t propagates to output neuron $K \alpha$ at time $t + 1$ via a hidden neuron. Weights must be such that $K \alpha$ at $t + 1$ is also activated (*true*).

18.5   Connectionist Nonclassical Reasoning

As discussed in Sect. 18.2, we believe that for neural (and cognitive) computation to achieve their promise, their models must be able to cater for nonclassical reasoning. Nonclassical logics have had a great impact in philosophy and AI (d'Avila Garcez and Lamb 2005; Fagin et al. 1995). For instance, nonmonotonic reasoning has dominated the research on logic and AI in the 1980s and 1990s, temporal logic has had a large impact on both academia and industry, and modal logics have become a lingua franca for the specification and analysis of knowledge and communication in multi-agent and distributed systems (Fagin et al. 1995). In this section, we consider modal and temporal reasoning as key representatives of nonclassical reasoning.

It is well known that modal logics correspond, in terms of expressive power, to the two-variable fragment of first-order logic (van Benthem 1984). Furthermore, as the two-variable fragment of first-order logic is decidable, this explains why modal logics are so "robustly decidable" and amenable to applications (Vardi 1997). Both AI and computer science have made extensive use of decidable modal logics, including in the analysis and model checking of distributed and multi-agent systems, program verification and specification, and hardware model checking. More recently, description logics, whose models are similar to modal logic (Kripke) models, have been instrumental in the study of the semantic web (Baader et al. 2003).

The basic idea behind connectionist nonclassical reasoning and CML is simple. Instead of having a single network as in the case of CILP, we now consider a set of CILP networks, and we label them, say, ω_1, ω_2, etc. Then, we can talk about a concept L holding at ω_1 and the same concept L holding at ω_2 separately. In this way, we can see ω_1 as a possible world and ω_2 as another, and this allows us to represent modalities such as necessity and possibility, time, arguments (d'Avila Garcez et al. 2005), epistemic states (d'Avila Garcez and Lamb 2006; d'Avila Garcez et al. 2003a, 2004c) and intuitionistic reasoning (d'Avila Garcez et al. 2006a, b, 2009). It is useful noting that this avenue of research is of interest in connection with McCarthy's conjecture on the propositional fixation of neural networks (McCarthy 1988), as discussed in Sect. 18.3, because of the correspondence between propositional modal logic and the above-mentioned two-variable fragment of first-order logic. In other words, CML shows that relatively simple neural-symbolic systems may go beyond classical propositional logic in terms of expressive power.

18.5.1 Connectionist Modal Reasoning

Modal logic deals with the analysis of concepts such as *necessity* (represented by $\Box L$, read "box L" and meaning that L is *necessarily true*), and *possibility* (represented by $\Diamond L$, read "diamond L" and meaning that L is *possibly true*). A key aspect of modal logic is the use of *possible worlds* and a binary (accessibility) relation $\mathcal{R}(\omega_i, \omega_j)$ between worlds ω_i and ω_j. In possible world semantics, a proposition is necessary in a world if it is true in all worlds which are possible in relation to that world, whereas it is possible in a world if it is true in at least one world which is possible in relation to that same world.

CML uses ensembles of neural networks (instead of single networks) to represent the language of modal logic programming (Orgun and Ma 1994). The theories are now sets of modal clauses each of the form $\omega_i : ML_1, \ldots, ML_n \to MA$, where

ω_i is a label representing a world in which the associated clause holds and $M \in \{\Box, \Diamond\}$, together with a finite set of relations $\mathcal{R}(\omega_i, \omega_j)$ between worlds ω_i and ω_j. Such theories are implemented in a network ensemble, each network representing a possible world, with the use of labels in the ensembles allowing the representation of the accessibility relations.

In CML, each network in the ensemble is a simple single-hidden-layer CILP network to which standard neural learning algorithms can be applied. Learning, in this setting, can be seen as learning the concepts that hold in each possible world independently, with the accessibility relation providing the information on how the networks should interact. For example, take three networks all related to each other. If neuron $\Diamond a$ is activated in one of these networks, then neuron a must be activated in at least one of the networks. If neuron $\Box a$ is activated in one network, then neuron a must be activated in all the networks. This implements in a connectionist setting the possible-world semantics mentioned above. This is achieved by defining the connections and the weights of the network ensemble, following a translation algorithm.

Figure 18.3 is an example: it shows an ensemble of three neural networks labelled N_1, N_2 and N_3, which might *communicate* in different ways. We look at N_1, N_2 and N_3 as *possible worlds*. Input and output neurons may now represent $\Box L, \Diamond L$ or L, where L is a literal. $\Box A$ will be *true* in a world ω_i if A is *true* in all worlds ω_j to which ω_i is related. Similarly, $\Diamond A$ will be *true* in a world ω_i if A is *true* in some world ω_j to which ω_i is related. As a result, if neuron $\Box A$ is activated in network N_1, denoted by $\omega_1 : \Box A$, and world ω_1 is related to worlds ω_2 and ω_3, then neuron A must be activated in networks N_2 and N_3. Similarly, if neuron $\Diamond A$ is activated in N_1, then a neuron A must be activated in an arbitrary network that is related to N_1.

It is also possible to make use of CML to compute the fact that $\Box A$ holds at a possible world, say ω_i, whenever A holds at *all* possible worlds related to ω_i, by connecting the output neurons of the related networks to a hidden neuron in ω_i which connects to an output neuron labelled as $\Box A$. Dually for $\Diamond A$, whenever A holds at *some* possible world related to ω_i, we connect the output neuron representing A to a hidden neuron in ω_i which connects to an output neuron labelled as $\Diamond A$. Due to the simplicity of each network in the ensemble, when it comes to learning, we can still use backpropagation on each network to learn the local knowledge inside each possible world.

18.5.2 Connectionist Temporal Reasoning

The representation of temporal dimensions and symbolic temporal variables in cognitive science remains a relevant research field, with a number of implications not only in psychology but also in computer science and AI. Developments in this area demand, therefore, sound symbolic temporal inference systems underlying the neural and cognitive machinery. Existing approaches have now only started to make some headway towards sound representation of time in cognitive computational processes (d'Avila Garcez and Lamb 2006; Shastri 2007).

By extending the CML framework, we allow reasoning and learning about temporal, epistemic and probabilistic knowledge dealing with different reasoning dimensions of an idealised agent. Learning is achieved by training each individual network in the ensemble, which in turn corresponds to the current knowledge of an agent within a possible world. Such a form of learning aims to attend to the need of learning mechanisms in multi-agent systems in which modal logics are an essential feature to represent several kinds of knowledge dimensions an agent is typically endowed with (Wooldridge 2001).

The *Connectionist Temporal Logic of Knowledge* (CTLK) is an extension of CML, which considers temporal and epistemic knowledge (d'Avila Garcez and Lamb 2006). Generally speaking, the idea is to allow, instead of a single ensemble, a number n of ensembles, each representing the knowledge held by a number of agents at a given time-point t. Figure 18.5 illustrates how this dynamic feature can be combined with the symbolic features of the knowledge represented in each network, allowing not only for the analysis of the current state (possible world or time-point) but also for the analysis of how knowledge changes through time.

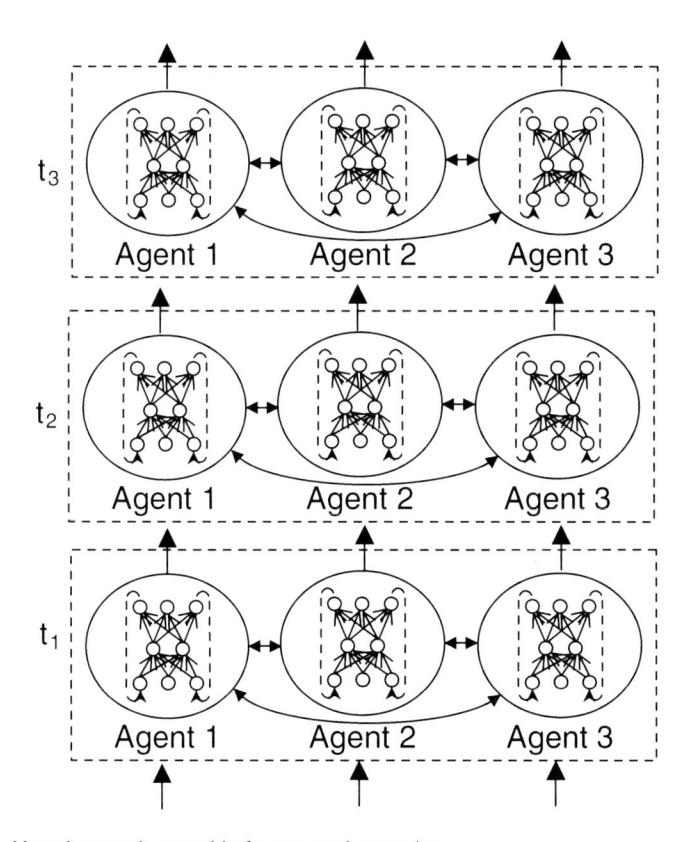

Fig. 18.5   Neural network ensemble for temporal reasoning

CML deals with time implicitly in snapshots; CTLK deals with time explicitly allowing the model to reason about time and knowledge. Different applications may be better suited to one or the other: the snapshot solution is simpler computationally, while the explicit model is richer. The *muddy children* case study, which we will consider in Sect. 18.5.3, will illustrate this.

The number of ensembles n that is necessary to solve a given problem will depend on the problem domain, in particular on the number of time-points needed for reasoning about the problem. For example, in the case of the *muddy children puzzle* (described below) (Fagin et al. 1995), which is a distributed knowledge representation problem, it is sufficient to use as many ensembles as the number of children that are muddy. The choice of n in a different domain might not be as straightforward, possibly requiring a fine-tuning process similar to that performed by learning, but with a varying network architecture. Other considerations around CTLK include the need for more extensive evaluations of the model with respect to learning, and the question of the trade-off between space and time complexity in such bounded networks. The fact, however, that the model is sufficient to deal with a variety of reasoning tasks is encouraging. Recently, CTLK was applied effectively on multi-process synchronisation and learning in concurrent programming (Lamb et al. 2007).

18.5.3 Case Study

In this section, we apply CTLK to the muddy children puzzle, a classic example of reasoning in multi-agent environments. In contrast to the also well-known wise men puzzle (Fagin et al. 1995; Huth and Ryan 2000), in which the reasoning process is sequential, in the muddy children puzzle, reasoning is distributed and simultaneous. There is a group of n children playing in a garden. A certain number of children k $(k \leq n)$ has mud on their faces. Each child can see if the others are muddy, but cannot see if they themselves are muddy.[5]

A caretaker announces that at least one child is muddy $(k \geq 1)$ and asks "do you know if you have mud on your face?" To help in the understanding of the puzzle, let us consider the cases where $k = 1, k = 2$ and $k = 3$.

If $k = 1$ (only one child is muddy), the muddy child answers *yes* at the first instance since she cannot see any other muddy child. All the other children answer *no* at the first instance.

If $k = 2$, suppose children 1 and 2 are muddy. In the first instance, all children can only answer *no*. This allows 1 to reason as follows: "if 2 had said *yes* the first time round, she would have been the only muddy child. Since 2 said *no*, she must be

---

[5] We follow the muddy children problem description presented in Fagin et al. (1995). We must also assume that all the agents involved in the situation are truthful and intelligent.

seeing someone else muddy, and since I cannot see anyone else muddy apart from 2, I myself must be muddy!" Child 2 can reason analogously and also answers *yes* at the second time round.

If $k = 3$, suppose children 1, 2 and 3 are muddy. Every child can only answer *no* at the first two time rounds. Again, this allows 1 to reason as follows: "if 2 or 3 had said *yes* at the second time round, they would have been the only two muddy children. Thus, there must be a third person with mud. Since I can see only 2 and 3 with mud, this third person must be me!" Children 2 and 3 can reason analogously to conclude as well that *yes*, they are muddy.

The puzzle illustrates the need to distinguish between an agent's individual knowledge and *common knowledge* about the world in each situation. For example, when $k = 2$, after everybody says *no* in the first round, it becomes common knowledge that at least two children are muddy. Similarly, when $k = 3$, after everybody says *no* twice, it becomes common knowledge that at least three children are muddy, and so on. In other words, when it is common knowledge that there are at least $k - 1$ muddy children, after the announcement that nobody knows if they are muddy or not, then it becomes common knowledge that there are at least k muddy children, for if there were $k - 1$ muddy children all of them would have known that they had mud on their faces. Note that this reasoning process can only start once it is common knowledge that at least one child is muddy, as announced by the caretaker.[6]

The snapshot version of the muddy children puzzle, where time is implicit, can be solved by CML. The interested reader is referred to d'Avila Garcez et al. (2009) for the details of the CML implementation. A full solution to the puzzle, however, can only be obtained with the use of CTLK. The addition of an explicit temporal variable to the puzzle allows one to reason about knowledge acquired after each time round. For example, assume as above that there are three muddy children playing in the garden. First, they all answer *no* when asked if they know whether they are muddy or not. Moreover, as each muddy child can see the other children, they will reason as previously described and answer *no* again at the second time round, reaching the correct conclusion at time round three. This solution requires, at each round, that the CML networks be extended with the knowledge acquired from reasoning about what is seen and what is heard by each agent. This clearly requires each agent to reason about time. There are alternative ways of modelling this, but one possible representation is as follows (below, a rule of the form $t_1 : \neg K_1 p_1 \wedge \neg K_2 p_2 \wedge \neg K_3 p_3 \rightarrow \bigcirc K_1 q_2$ states that if at time-point t_1, child 1 does not know that she is muddy, denoted by $\neg K_1 p_1$, and neither do children 2 and 3, then at the next

---

[6] The representation of common knowledge in neural networks throws some interesting questions. In CML, common knowledge is represented implicitly by connecting neurons appropriately as reasoning progresses (e.g. as it becomes known at round two that at least two children should be muddy). The representation of common knowledge explicitly in the object level would require the use of neurons that are activated when "everybody knows" something (implementing in a finite domain the common knowledge axioms of Fagin et al. 1995), but this would complicate the formalisation of the puzzle given in this chapter. This explicit form of representation and its ramifications are worth investigating though and can be treated in their own right in future work.

time-point t_2, denoted by $\bigcirc$, child 1 knows that at least two children must be muddy, denoted by $K_1 q_2$):

Temporal rules for agent(child) 1:
$t_1 : \neg K_1 p_1 \wedge \neg K_2 p_2 \wedge \neg K_3 p_3 \rightarrow \bigcirc K_1 q_2$
$t_2 : \neg K_1 p_1 \wedge \neg K_2 p_2 \wedge \neg K_3 p_3 \rightarrow \bigcirc K_1 q_3$
Temporal rules for agent(child) 2:
$t_1 : \neg K_1 p_1 \wedge \neg K_2 p_2 \wedge \neg K_3 p_3 \rightarrow \bigcirc K_2 q_2$
$t_2 : \neg K_1 p_1 \wedge \neg K_2 p_2 \wedge \neg K_3 p_3 \rightarrow \bigcirc K_2 q_3$
Temporal rules for agent(child) 3:
$t_1 : \neg K_1 p_1 \wedge \neg K_2 p_2 \wedge \neg K_3 p_3 \rightarrow \bigcirc K_3 q_2$
$t_2 : \neg K_1 p_1 \wedge \neg K_2 p_2 \wedge \neg K_3 p_3 \rightarrow \bigcirc K_3 q_3$

The temporal rules above can be translated into a network structure similar to that of Fig. 18.5. Each network in the ensemble can be trained from examples using standard backpropagation. We have trained two groups of CTLK network ensembles to compute a solution to the muddy children puzzle. To one of them, we have added a temporal rule $t_1 : \neg K_1 p_1 \wedge \neg K_2 p_2 \wedge \neg K_3 p_3 \rightarrow \bigcirc K_1 q_2$ as background knowledge. To the other, we did not add any background knowledge. We then compared average test-set performances. We have considered the case in which Agent 1 is to decide whether or not she is muddy at time t_2. Each training example expresses the knowledge held by Agent 1 at t_2, according to the truth-values of atoms $K_1 \neg p_2$, $K_1 \neg p_3$, $K_1 q_1$, $K_1 q_2$, and $K_1 q_3$. As a result, there are 32 examples, i.e. all possible combinations of truth-values for input neurons $K_1 \neg p_2$, $K_1 \neg p_3$, $K_1 q_1$, $K_1 q_2$, $K_1 q_3$, with input value 1 denoting truth-value *true*, and input -1 denoting *false*. For each example, we are concerned about whether Agent 1 will know that she is muddy or not, i.e. whether output neuron $K_1 p_1$ is active.

From the description of the muddy children puzzle, we know that at t_2, $K_1 q_2$ should be *true* (i.e. $K_1 q_2$ is a *fact*). This information can be derived from the temporal rule given as background knowledge above, but not from the training examples. Although the background knowledge can be changed by the training examples, it places a bias on certain combinations (in this case, the examples in which $K_1 q_2$ is *true*), and this may produce a better performance, in particular when the background knowledge is correct. This effect has been observed, for instance, in Towell and Shavlik (1994) on experiments on DNA sequence analysis, in which background knowledge is expressed by *production rules*. In Towell and Shavlik (1994) (and also d'Avila Garcez et al. 2002a), the set of examples is noisy and the background knowledge counteracts the noise and reduces the chances of data overfitting.

We have evaluated the CTLK model using an eightfold *cross-validation* methodology, whereby eight CTLK network ensembles were created, each having been trained with 28 of the 32 available examples, with four examples being left out for testing at each time. This process was then repeated for the CTLK networks with background knowledge. For each training round, the training set was presented to the networks for 10,000 epochs. All the networks have reached a training-set error of 0.01. The ensembles containing no background knowledge achieved an average test-set accuracy of 81.25%. The ensembles to which the temporal rule was added as background knowledge achieved an average test-set accuracy of 87.5%, a noticeable difference in performance. The result corroborates the importance of using background knowledge. In both cases, the same training parameters were used: learning rate $\eta = 0.2$, term of momentum $\alpha = 0.1$ and activation function $\tanh(x)$.

18.6 Fibring Neural Networks

In certain applications, different CTLK neural networks may need to be combined. In complex systems, it is frequently useful to create components that can be analysed independently and combined as appropriate in a modular way. A methodology for combining systems called *fibring* can be used for this (d'Avila Garcez and Gabbay 2004). Fibring promotes structured learning by combining networks at different levels of abstraction. Networks at one level may be fibred onto networks at another level to form a structure combining metalevel and object-level reasoning where high-level abstractions can be learned from low-level concepts.

In Bader et al. (2005), the idea of fibring was used to encode first-order logic programs in neural-network ensembles: a neural network is used to iterate a global counter and this counter is combined (fibred) with another neural network. This allows logic programs with an infinite number of ground instances to be translated into a finite neural network structure (e.g. $\neg even(x) \rightarrow even(s(x))$ for $x \in \mathbb{N}, s(x) = x + 1$). The translation is made possible because fibring implements a key feature of symbolic computation in neural networks, namely, *recursion*, as described below.

The idea of fibring neural networks is simple. Fibred networks may be composed not only of interconnected neurons but also of other networks, forming a recursive architecture. A fibring function then defines how this architecture behaves by defining how the networks in the ensemble should relate to each other. Typically, the fibring function will allow the activation of neurons in one network to influence the change of weights in another network. Intuitively, this can be seen as a master network being responsible for training a slave network. Interestingly, albeit being a combination of simple and standard neural networks, fibred networks are very expressive and can approximate any polynomial function in an unbounded domain, thus being more expressive than standard networks.

Figure 18.6 exemplifies how a network (B) can be fibred onto another network (A). Of course, the idea of fibring is not just to organise networks as a number of subnetworks; in Fig. 18.6, the output neuron of A is expected to be a neural network in its own right. The input, weights and output of B may depend on the activation state of A's output neuron, according to the fibring function φ. Fibred networks can be trained by examples in the same way as standard networks. For instance, networks A and B could have been trained separately before having been fibred. Networks can be fibred in a number of different ways as far as their architectures are concerned: network B could have been fibred onto a hidden neuron of network A.

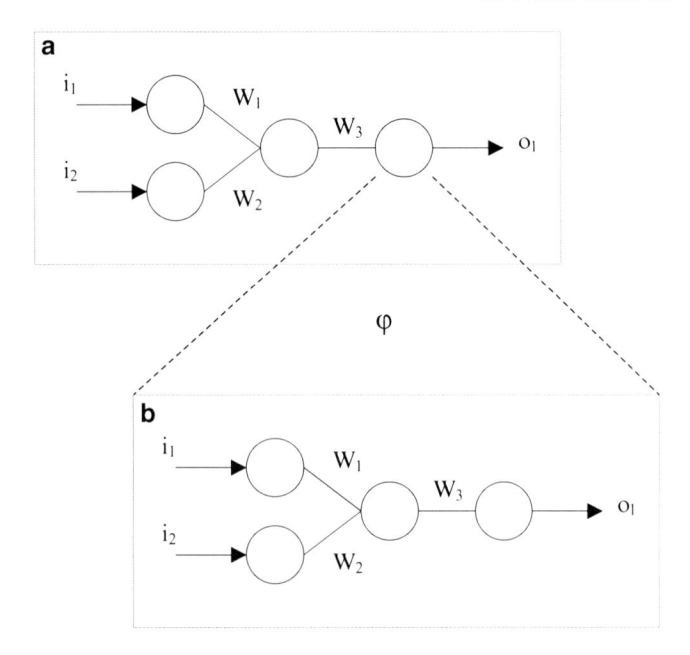

Fig. 18.6 Fibring neural networks

More generally, fibring offers a methodology for combining systems. As an example, network A could have been trained with a robot's visual system, while network B would have been trained with its planning system, and fibring would serve to perform the composition of the two systems (Gabbay 1999). Fibring can be very powerful. It offers the extra expressiveness required by complex applications at low computational cost (that of computing fibring function φ). Of course, we would like to keep φ as simple as possible so that it can be implemented itself by simple neurons in a fully connectionist model. Interesting work remains to be done in this area, particularly in what regards the question of emergence, modularity versus integration, and attentional focus (Taylor 2009), and the more practical question of how one should go about fibring networks in real applications. With respect to cognitive models, one can envisage fibring of several abilities: for instance, fibring of temporal and spatial reasoning with actions in an arbitrary environment can provide a useful framework to areas such as cognitive robotics.

18.7 Concluding Remarks

The need for rich, logic-based knowledge representation formalisms and algorithms in computational cognitive systems has been argued for a long time (d'Avila Garcez et al. 2009; Pinker 2007; Smolensky 1988; Stenning and van Lambalgen 2008; Valiant 1984). The foundational approach proposed in this chapter aims to attend to such a need. The integration of other modes of reasoning, including conditional (Broda et al. 2002; Leitgeb 2007) and BDI logics (Rao and Georgeff 1998) would also contribute to the foundational model and corresponding experimental developments in neural-symbolic computation. The use of neural-symbolic systems also facilitates knowledge evolution and adaptability through learning. It would be interesting to apply the formalism to belief revision in the context of distributed, multi-agent systems.

CML and its variations offer an illustration of how the area of neural computation may contribute to the area of cognitive reasoning, while fibring is an example of how logic can bring insight into neural and cognitive computation. CML offers parallel models of computation for modal logics that can be integrated with an efficient learning algorithm. Fibring is an example of how concepts from symbolic computation, in this case *recursion*, can help in the development of new neural models. This is not necessarily conflicting with the ambition of biological plausibility, e.g. fibring functions can be understood as a model of *presynaptic weights*, which play an important role in biological neural networks.

Connectionist nonclassical reasoning and network fibring are the cornerstones of our overall cognitive model, which we may call *fibred network ensembles*. In this model, a network ensemble A (representing, e.g. a temporal theory) may be combined with another network ensemble B (representing, e.g. an intuitionistic theory d'Avila Garcez et al. 2006a). Higher level concepts (say, in A) may be combined and brought into the object-level (say, in B) without blurring the distinction between the two levels and maintaining modularity. One may reason in the metalevel and use that information in the object-level, a typical example being (metalevel) reasoning about actions in (object-level) databases containing inconsistencies (Gabbay and Hunter 1993). Relations between networks/concepts in the object-level may be represented and learned in the metalevel. If two networks denote, for example, concepts $P(X, Y)$ and $Q(Z)$, a meta-network can learn to map a representation of concepts P and Q onto a third network, denoting say $R(X, Y, Z)$, such that for example $P(X, Y) \land Q(Z) \rightarrow R(X, Y, Z)$. The interested reader is referred to d'Avila Garcez et al. (2009) for details.

Figure 18.7 illustrates the fibred network ensembles. The overall model takes the most general knowledge representation ensemble of Fig. 18.5 and allows a number of such ensembles to be combined at different levels of abstraction through fibring. Relations between concepts at level n may be generalised to level $n + 1$ with the use of metalevel networks. Abstract concepts at level n may be specialised (or instantiated) at level $n - 1$ with the use of a fibring function. Knowledge evolution through time occurs at each level. Alternative outcomes, possible worlds and the nonmonotonic reasoning process of multiple interacting agents can be modelled at each level. Learning can take place inside each modular network or across networks in the ensemble.

The question of how the human mind integrates reasoning and learning capacities is only starting to be answered (Gabbay and Woods 2005; Stenning and van Lambalgen 2008). We argue that the prospects are better if we investigate the connectionist processes of the brain together with the logical processes of symbolic

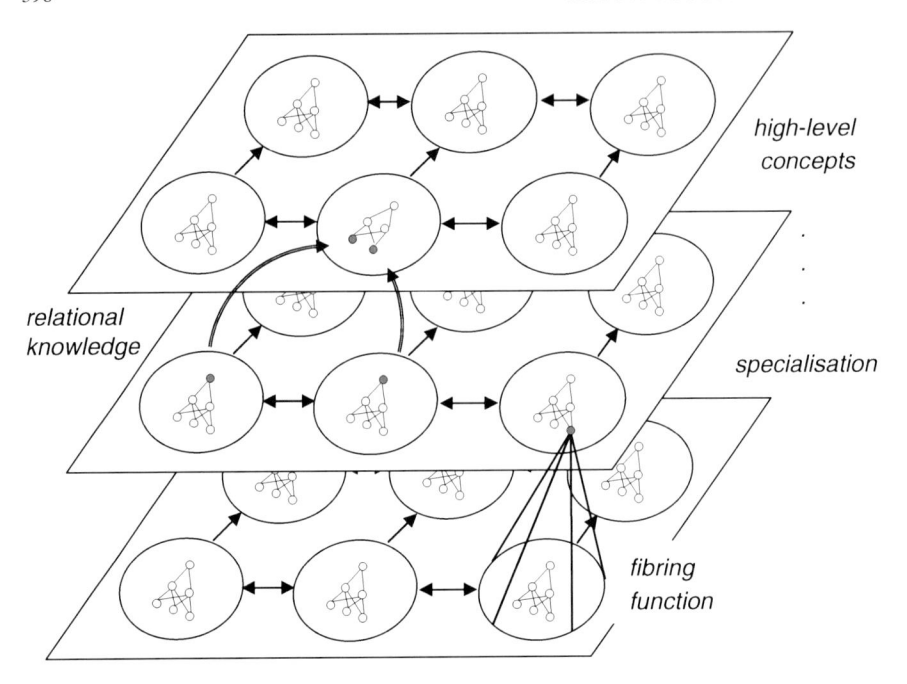

Fig. 18.7 Fibred network ensembles: structured learning and knowledge representation

computation, and not as two isolated paradigms. The framework of *fibred network ensembles* is expressive and tractable to address most current applications. Further development of the framework includes testing in controlled cognitive tasks.

The challenges for neural-symbolic cognitive computation today emerge from the goal of effective integration, expressive reasoning and robust learning. While adding reasoning capabilities to neural models, one cannot afford to lose learning performance. This means that one cannot depart from the key idea that neural networks are composed of simple processing units organised in a massively parallel architecture (i.e. one should not allow some clever neuron to perform complex symbol processing). It seems that learning is most effective at the propositional level, while some higher-level reasoning is useful at the first-order level. Computationally, the challenges are associated with the more practical aspects of the application of neural-symbolic cognitive systems in areas such as fault diagnosis, robotics, simulation and the semantic web. These challenges include the effective computation of logical models, the efficient extraction of comprehensible knowledge and, ultimately, the striking of the right balance between tractability and expressiveness. The reader is referred to d'Avila Garcez and Hitzler (2009) for a number of recent papers dealing with some of these challenges, including some real applications on multi-modal processing, simulation and defense.

In summary, by paying attention to the developments on either side of the division between the symbolic and the sub-symbolic paradigms, we are getting closer to a unifying theory of cognition, or at least promoting a faster and principled development of the field of AI. This chapter described a family of connectionist nonclassical reasoning systems and hinted at how they may be combined at different levels of abstraction by fibring. We hope it serves not only as a stepping stone towards such a theory to reconcile the symbolic and connectionist approaches, but also as a foundational model for effective reasoning in cognitive computational systems.

Human beings are quite extraordinary at performing practical reasoning as they go about their daily business. *There are cases where the human computer, slow as it is, is faster than AI systems. Why are we faster? Is it the way we perceive knowledge as opposed to the way we represent it? Do we know immediately which rules to select and apply? We must look for the correct representation in the sense that it mirrors the way we perceive and apply the rules* (Gabbay 1998). Ultimately, neural-symbolic cognitive computation is about asking and trying to answer these questions, and about the associated provision of neural-symbolic systems with integrated expressive reasoning and robust learning capabilities.

References

F. Baader, D. Calvanese, D. L. McGuinness, D. Nardi, and P. F. Patel-Schneider, editors. *The Description Logic Handbook: Theory, Implementation, and Applications*. Cambridge University Press, Cambridge, 2003.

S. Bader, A. d'Avila Garcez, and P. Hitzler. Computing first-order logic programs by fibring artificial neural networks. In *Proceedings of the AAAI International FLAIRS Conference*, pages 314–319, 2005.

S. Bader, P. Hitzler, S. Holldobler, and A. Witzel. A fully connectionist model generator for covered first-order logic programs. In *Proceedings of the International Joint Conference on Artificial Intelligence IJCAI-07*, pages 666–671, Hyderabad, India, 2007. AAAI.

B. Bennett. Spatial reasoning with propositional logics. In *Proceedings of the Fourth International Conference on Principles of Knowledge Representation and Reasoning KR-94*, pages 51–62, 1994.

P. Blackburn, J. van Benthem, and F. Wolter, editors. *Handbook of Modal Logic*. Studies in Logic and Practical Reasoning. Elsevier, Amsterdam, 2006.

G. Bologna. Is it worth generating rules from neural network ensembles? *Journal of Applied Logic*, 2(3):325–348, 2004.

R. V. Borger, A. d'Avila Garcez, and L. Lamb. A neural-symbolic perspective on analogy. *Behavioral and Brain Sciences*, 31(4):379–380, 2008.

K. Broda, D. Gabbay, L. Lamb, and A. Russo. Labelled natural deduction for conditional logics of normality. *Logic Journal of the IGPL*, 10(2):123–163, 2002.

K. Broda, D. Gabbay, L. Lamb, and A. Russo. *Compiled Labelled Deductive Systems: A Uniform Presentation of Non-classical Logics*. Studies in Logic and Computation. Research Studies Press/Institute of Physics Publishing, Baldock, UK, Philadelphia, PA, 2004.

A. Browne and R. Sun. Connectionist inference models. *Neural Networks*, 14:1331–1355, 2001.

I. Cloete and J. Zurada, editors. *Knowledge-Based Neurocomputing*. MIT, Cambridge, MA, 2000.

A. d'Avila Garcez. Fewer epistemological challenges for connectionism. In S. B. Cooper, B. Lowe, and L. Torenvliet, editors, *Proceedings of Computability in Europe, CiE 2005*, volume LNCS 3526, pages 139–149, Amsterdam, The Netherlands, June 2005. Springer, Berlin.

A. d'Avila Garcez and D. Gabbay. Fibring neural networks. In *Proceedings of 19th National Conference on Artificial Intelligence (AAAI-04)*, pages 342–347, San Jose, CA, 2004.

A. d'Avila Garcez and P. Hitzler, editors. *Proceedings of IJCAI International Workshop on Neural-Symbolic Learning and Reasoning NeSy09*, Pasadena, California, USA, 2009.

A. d'Avila Garcez and L. Lamb. Reasoning about time and knowledge in neural-symbolic learning systems. In S. Thrun, L. Saul, and B. Schoelkopf, editors, *Advances in Neural Information Processing Systems 16*, Proceedings of NIPS 2003, pages 921–928. MIT, Cambridge, MA, 2004.

A. d'Avila Garcez and L. Lamb. Neural-symbolic systems and the case for non-classical reasoning. In S. Artëmov, H. Barringer, A. d'Avila Garcez, L. Lamb, and J. Woods, editors, *We Will Show Them! Essays in Honour of Dov Gabbay*, pages 469–488. College Publications, International Federation for Computational Logic, UK, 2005.

A. d'Avila Garcez and L. Lamb. A connectionist computational model for epistemic and temporal reasoning. *Neural Computation*, 18(7):1711–1738, 2006.

A. d'Avila Garcez and G. Zaverucha. The connectionist inductive learning and logic programming system. *Applied Intelligence Journal, Special Issue on Neural Networks and Structured Knowledge*, 11(1):59–77, 1999.

A. d'Avila Garcez, K. Broda, and D. Gabbay. Symbolic knowledge extraction from trained neural networks: A sound approach. *Artificial Intelligence*, 125:155–207, 2001.

A. d'Avila Garcez, K. Broda, and D. Gabbay. *Neural-Symbolic Learning Systems: Foundations and Applications*. Perspectives in Neural Computing. Springer, Berlin, 2002a.

A. d'Avila Garcez, L. Lamb, and D. Gabbay. A connectionist inductive learning system for modal logic programming. In *Proceedings of the 9th International Conference on Neural Information Processing ICONIP'02*, pages 1992–1997, Singapore, 2002b. IEEE.

A. d'Avila Garcez, L. Lamb, K. Broda, and D. Gabbay. Distributed knowledge representation in neural-symbolic learning systems: A case study. In *Proceedings of AAAI International FLAIRS Conference*, pages 271–275, St. Augustine, FL, 2003a. AAAI.

A. d'Avila Garcez, L. Lamb, and D. Gabbay. Neural-symbolic intuitionistic reasoning. Frontiers in Artificial Intelligence and Applications Vol. 104, pages 399–408. IOS, 2003b.

A. d'Avila Garcez, D. Gabbay, and L. Lamb. Argumentation neural networks. In *Proceedings of the 11th International Conference on Neural Information Processing, ICONIP'04*, volume 3316 of *Lecture Notes in Computer Science*, pages 606–612. Springer, New York, 2004a.

A. d'Avila Garcez, D. Gabbay, and L. Lamb. Towards a connectionist argumentation framework. In *Proceedings of the 16th European Conference on Artificial Intelligence, ECAI 2004, including Prestigious Applicants of Intelligent Systems, PAIS 2004, Valencia, Spain, August 22–27, 2004*, pages 987–988, 2004b.

A. d'Avila Garcez, L. Lamb, K. Broda, and D. Gabbay. Applying connectionist modal logics to distributed knowledge representation problems. *International Journal on Artificial Intelligence Tools*, 13(1):115–139, 2004c.

A. d'Avila Garcez, D. Gabbay, and L. Lamb. Value-based argumentation frameworks as neural-symbolic learning systems. *Journal of Logic and Computation*, 15(6):1041–1058, 2005.

A. d'Avila Garcez, L. Lamb, and D. Gabbay. Connectionist computations of intuitionistic reasoning. *Theoretical Computer Science*, 358(1):34–55, 2006a.

A. d'Avila Garcez, L. Lamb, and D. Gabbay. A connectionist model for constructive modal reasoning. In *Advances in Neural Information Processing Systems 18*, Proceedings of NIPS 2005, pages 403–410. MIT, 2006b.

A. d'Avila Garcez, D. M. Gabbay, O. Ray, and J. Woods. Abductive reasoning in neural-symbolic systems. *TOPOI: An International Review of Philosophy*, 26:37–49, 2007a.

A. d'Avila Garcez, L. Lamb, and D. Gabbay. Connectionist modal logic: Representing modalities in neural networks. *Theoretical Computer Science*, 371(1–2):34–53, 2007b.

A. d'Avila Garcez, L. Lamb, and D. Gabbay. *Neural-Symbolic Cognitive Reasoning*. Cognitive Technologies. Springer, Berlin, 2009.

J. Elman. Finding structure in time. *Cognitive Science*, 14(2):179–211, 1990.

R. Fagin, J. Halpern, Y. Moses, and M. Vardi. *Reasoning About Knowledge*. MIT, Cambridge, MA, 1995.

D. Gabbay. *Elementary Logics: a Procedural Perspective*. Prentice Hall, London, 1998.

D. Gabbay. *Fibring Logics*. Oxford University Press, Oxford, 1999. Oxford Logic Guides, Vol. 38.

D. M. Gabbay and A. Hunter. Making inconsistency respectable: Part 2 – meta-level handling of inconsistency. In *Symbolic and Quantitative Approaches to Reasoning and Uncertainty ECSQARU'93*, volume LNCS 747, pages 129–136. Springer, Berlin, 1993.

D. Gabbay and J. Woods. *A Practical Logic of Cognitive Systems, Volume 2: The reach of abduction: Insight and trial*. Elsevier, New York, 2005.

D. Gabbay, I. Hodkinson, and M. Reynolds. *Temporal logic: mathematical foundations and computational aspects*, volume 1. Oxford University Press, Oxford, 1994. Oxford Logic Guides, Vol. 28.

D. Gabbay, A. Kurucz, F. Wolter, and M. Zakharyaschev. *Many-dimensional Modal Logics: Theory and Applications*, volume 148 of *Studies in Logic and the Foundations of Mathematics*. Elsevier Science, Amsterdam, The Netherlands, 2003.

P. Gärdenfors. *Conceptual Spaces: The Geometry of Thought*. MIT, Cambridge, MA, 2000.

J. Halpern. *Reasoning About Uncertainty*. MIT, Cambridge, MA, 2003.

S. Haykin. *Neural Networks: A Comprehensive Foundation*. Prentice Hall, New Jersey, 1999.

G. E. Hinton, S. Osindero, and Y. Teh. A fast learning algorithm for deep belief nets. *Neural Computation*, 18:1527–1554, 2006.

P. Hitzler, S. Holldobler, and A. K. Seda. Logic programs and connectionist networks. *Journal of Applied Logic*, 2(3):245–272, 2004. Special Issue on Neural-Symbolic Systems.

S. Hölldobler. Automated inferencing and connectionist models. Postdoctoral Thesis, Intellektik, Informatik, TH Darmstadt, 1993.

S. Hölldobler and Y. Kalinke. Toward a new massively parallel computational model for logic programming. In *Proceedings of the Workshop on Combining Symbolic and Connectionist Processing, ECAI 1994*, pages 68–77, 1994.

M. Huth and M. Ryan. *Logic in Computer Science: Modelling and Reasoning About Systems*. Cambridge University Press, Cambridge, 2000.

H. Jacobsson. Rule extraction from recurrent neural networks: A taxonomy and review. *Neural Computation*, 17(6):1223–1263, 2005.

L. Lamb, R. Borges, and A. d'Avila Garcez. A connectionist cognitive model for temporal synchronisation and learning. In *Proceedings of the Twenty-Second AAAI Conference on Artificial Intelligence AAAI 2007*, pages 827–832. AAAI, 2007.

H. Leitgeb. Neural network models of conditionals: an introduction. In X. Arrazola and J. M. Larrazabal et al., editors, *Proceedings of ILCLI International Workshop on Logic and Philosophy of Knowledge, Communication and Action*, pages 191–223, Bilbao, 2007.

A. Lozowski and J. Zurada. Extraction of linguistic rules from data via neural networks and fuzzy approximation. In I. Cloete and J. Zurada, editors, *Knowledge-Based Neurocomputing*, pages 403–417. MIT, Cambridge, 2000.

J. McCarthy. Epistemological challenges for connectionism. *Behavioral and Brain Sciences*, 11(1):44, 1988.

M. Mendler. Characterising combinatorial timing analysis in intuitionistic modal logic. *Logic Journal of the IGPL*, 8(6):821–852, 2000.

R. Mooney and D. Ourston. A multistrategy approach to theory refinement. In R. Michalski and G. Teccuci, editors, *Machine Learning: A Multistrategy Approach*, volume 4, pages 141–164. Morgan Kaufmann, San Mateo, CA, 1994.

H. Nunez, C. Angulo, and A. Catala. Rule based learning systems for support vector machines. *Neural Processing Letters*, 24(1):1–18, 2006.

M. Orgun and W. Ma. An overview of temporal and modal logic programming. In *Proceedings of the International Conference on Temporal Logic ICTL'94*, volume 827 of *Lecture Notes in Artificial Intelligence*, pages 445–479. Springer, Berlin, 1994.

M. Page. Connectionist modelling in psychology: A localist manifesto. *Behavioral and Brain Sciences*, 23:443–467, 2000.

S. Pinker. *The Stuff of Thought: Language as a Window into Human Nature*. Viking, New York, 2007.

S. Pinker, M. A. Nowak, and J. J. Lee. The logic of indirect speech. *Proceedings of the National Academy of Sciences USA*, 105(3):833–838, 2008.

A. Pnueli. The temporal logic of programs. In *Proceedings of 18th IEEE Annual Symposium on Foundations of Computer Science*, pages 46–57, 1977.

A. Rao and M. Georgeff. Decision procedures for BDI logics. *Journal of Logic and Computation*, 8(3):293–343, 1998.

D. Rumelhart, G. Hinton, and R. Williams. Learning internal representations by error propagation. In D. Rumelhart and J. McClelland, editors, *Parallel Distributed Processing: Explorations in the Microstructure of Cognition*, volume 1, pages 318–362. MIT, Cambridge, 1986.

R. Setiono. Extracting rules from neural networks by pruning and hidden-unit splitting. *Neural Computation*, 9:205–225, 1997.

L. Shastri. Advances in SHRUTI: a neurally motivated model of relational knowledge representation and rapid inference using temporal synchrony. *Applied Intelligence Journal, Special Issue on Neural Networks and Structured Knowledge*, 11:79–108, 1999.

L. Shastri. Shruti: A neurally motivated architecture for rapid, scalable inference. In B. Hammer and P. Hitzler, editors, *Perspectives of Neural-Symbolic Integration*, pages 183–203. Springer, Heidelberg, 2007.

L. Shastri and V. Ajjanagadde. From simple associations to semantic reasoning: A connectionist representation of rules, variables and dynamic binding. Technical report, University of Pennsylvania, 1990.

J. Shawe-Taylor and N. Cristianini. *Kernel Methods for Pattern Analysis*. Cambridge University Press, Cambridge, 2004.

P. Smolensky. On the proper treatment of connectionism. *Behavioral and Brain Sciences*, 44:1–74, 1988.

P. Smolensky and G. Legendre. *The Harmonic Mind: From Neural Computation to Optimality-Theoretic Grammar*. MIT, Cambridge, MA, 2006.

K. Stenning and M. van Lambalgen. Human reasoning and cognitive science, 2008.

R. Sun. Robust reasoning: integrating rule-based and similarity-based reasoning. *Artificial Intelligence*, 75(2):241–296, 1995.

R. Sun. Theoretical status of computational cognitive modeling. *Cognitive Systems Research*, 10(2):124–140, 2009.

J. Taylor. Cognitive computation. *Cognitive Computation*, 1(1):4–16, 2009.

S. Thrun. Extracting provably correct rules from artificial neural networks. Technical report, Institut für Informatik, Universität Bonn, 1994.

G. Towell and J. Shavlik. Knowledge-based artificial neural networks. *Artificial Intelligence*, 70(1):119–165, 1994.

L. Valiant. A theory of the learnable. *Communications of the ACM*, 27(11):1134–1142, 1984.

L. Valiant. A neuroidal architecture for cognitive computation. *Journal of the ACM*, 47(5):854–882, 2000.

L. Valiant. Three problems in computer science. *Journal of the ACM*, 50(1):96–99, 2003.

L. Valiant. Knowledge infusion: In pursuit of robustness in artificial intelligence. In *Proceedings of the 28th Conference on Foundations of Software Technology and Theoretical Computer Science*, pages 415–422, Bangalore, India, 2008.

J. van Benthem. Correspondence theory. In D. Gabbay and F. Guenthner, editors, *Handbook of Philosophical Logic*, chapter II.4, pages 167–247. D. Reidel Publishing Company, Dordrecht, 1984.

D. van Dalen. Intuitionistic logic. In D. Gabbay and F. Guenthner, editors, *Handbook of Philosophical Logic*, volume 5. Kluwer, Dordrecht, 2nd edition, 2002.

M. Vardi. Why is modal logic so robustly decidable. In N. Immerman and P. Kolaitis, editors, *Descriptive Complexity and Finite Models*, volume 31 of *Discrete Mathematics and Theoretical Computer Science*, pages 149–184. DIMACS, 1997.

M. Wooldridge. *Introduction to Multi-agent Systems*. Wiley, New York, 2001.

Chapter 19
Information Theory of Decisions and Actions

Naftali Tishby and Daniel Polani

Abstract The perception–action cycle is often defined as "the circular flow of *information* between an organism and its environment in the course of a sensory guided sequence of actions towards a goal" (Fuster, Neuron 30:319–333, 2001; International Journal of Psychophysiology 60(2):125–132, 2006). The question we address in this chapter is in what sense this "flow of information" can be described by Shannon's measures of information introduced in his mathematical theory of communication. We provide an affirmative answer to this question using an intriguing analogy between Shannon's classical model of communication and the perception–action cycle. In particular, decision and action sequences turn out to be directly analogous to codes in communication, and their complexity – the minimal number of (binary) decisions required for reaching a goal – directly bounded by information measures, as in communication. This analogy allows us to extend the standard reinforcement learning framework. The latter considers the future expected reward in the course of a behaviour sequence towards a goal (value-to-go). Here, we additionally incorporate a measure of information associated with this sequence: the cumulated information processing cost or bandwidth required to specify the future decision and action sequence (information-to-go).

Using a graphical model, we derive a recursive Bellman optimality equation for information measures, in analogy to reinforcement learning; from this, we obtain new algorithms for calculating the optimal trade-off between the value-to-go and the required information-to-go, unifying the ideas behind the Bellman and the Blahut–Arimoto iterations. This trade-off between value-to-go and information-to-go provides a complete analogy with the compression–distortion trade-off in source coding. The present new formulation connects seemingly unrelated optimization problems. The algorithm is demonstrated on grid world examples.

N. Tishby (✉)
School of Engineering and Computer Science, Interdisciplinary Center for Neural
Computation, The Suadrsky Center for Computational Biology,
Hebrew University Jerusalem, Jerusalem, Israel
e-mail: tishby@cs.huji.ac.il

V. Cutsuridis et al. (eds.), *Perception-Action Cycle: Models, Architectures,*
and Hardware, Springer Series in Cognitive and Neural Systems 1,
DOI 10.1007/978-1-4419-1452-1_19, © Springer Science+Business Media, LLC 2011

19.1 Introduction

To better understand intelligent behaviour in organisms and to develop such behaviour for artificial agents, the principles of perception, of intelligent information processing and of actuation undergo significant study. Perception, information processing and actuation per se are often considered as individual, separate input–output processes. Much effort is devoted to understand and study each of these processes individually. Conceptually, the treatment of such input–output models is straightforward, even if its details are complex.

Compared to that, combining perception, information processing and actuation together introduces a feedback cycle that considerably changes the "rule of the game". Since actuation changes the world, perception ceases to be passive and will, in future states, generally depend on actions selected earlier by the organism. In other words, the organism controls to some extent not only which states it wishes to visit, but consequently also which sensoric inputs it will experience in the future.

The "cycle" view has intricate consequences and creates additional complexities. It is, however, conceptually more satisfying. Furthermore, it can help identifying biases, incentives and constraints for the self-organized formation of intelligent processing in living organisms – it is no surprise that the *embodied intelligence* perspective is adopted by many AI researchers Pfeifer and Bongard (2007) which is intimately related to the perception–action cycle perspective. It has been seen as a path towards understanding where biological intelligence may have risen from in evolution and how intelligent dynamics may be coaxed out of AI systems.

A challenge for the quantitative treatment of the perception–action cycle is that there are many ways of modeling it which are difficult to compare. Much depends on the choice of architecture, the selected representation and other aspects of the concrete model. To alleviate this unsatisfactory situation, recent work has begun to study the perception–action cycle in the context of a (Shannonian) information-theoretic treatment (Bialek et al. 2001; Touchette and Lloyd 2000, 2004; Klyubin et al. 2004, 2007), reviving early efforts by Ashby (1956).

The information-theoretic picture is universal, general, conceptually transparent and can be post hoc imbued with the specific constraints of particular models. On the informational level, scenarios with differing computational models can be directly compared with each other. At the same time, the informational treatment allows one to incorporate limits in the information processing capacity that are fundamental properties of a particular agent–environment system.

This is especially attractive in that it seems to apply to biologically relevant scenarios to some extent; details of this view are increasingly discussed (Taylor et al. 2007; Polani 2009). Under this perspective, in essence, one considers, e.g. the informational cost of handling a given task. Vice versa, one can ask how well one can actually handle a task if constraints on the computational power are imposed (here in the form of limited informational bandwidth).

On the other side, there is the established framework of *Markovian Decision Problems* (MDPs) which is used to study how to find *policies* (i.e. agent strategies) that perform a task well, where the quality of the performance is measured via some cumulative reward value which depends on the policy of the agent. The MDP framework is concerned with describing the task and with solving the problem of finding the optimal policy. It is not, however, concerned with the actual processing cost that is involved with carrying out the given (possibly optimal) policies. Thus, an optimal policy for the MDP may be found which maximizes the reward achieved by an agent, but which does not heed possible computational costs or limits imposed on an agent – this is in contrast to simple organisms which cannot afford a large informational bandwidth or minimal robots for which a suboptimal, but informationally cheap performance would be sufficient.

It is therefore the goal of this chapter to marry the MDP formalism with an information-theoretic treatment of the processing cost required by the agent (and the environment) to attain a given level of performance (in terms of rewards). To combine these disparate frameworks, we need to introduce notions from both information theory and the theory of MDPs.

To limit the complexity of the present exposition, this chapter will concentrate only on modelling action rewards; we will not address here its symmetric counterpart, namely (non-informational) costs of sensoric data acquisition. Furthermore, we will skim only the surface of the quite intricate relation of the present formalism with the framework of predictive information (e.g. in Sect. 19.5.3.2).

19.2 Rationale

An important remark about this chapter is that its core intention is the development of a general and expressive framework, and not a particularly efficient algorithm. In adopting the principled approach of information theory, we are here not concerned with producing an algorithm which would compare performance-wise with competitive state-of-the-art MDP learners (such as, e.g. Engel et al. 2003; Jung and Polani 2007); and neither are we concerned with proposing or investigating particular flavours of the perception–action architecture. Instead, the idea behind the application of information theory to the perception–action cycle is to open the path towards a new way of looking at the perception–action cycle with the hope that this will lead to new concepts, new insights and, possibly, new types of questions.

In particular, because of its universality, and because the framework of information theory has deep ramifications into many fields, including physics and statistical learning theory, it makes different architectures, models and scenarios comparable under a common language. This allows information theory to be applied across various and quite disparate domains of interest (such as, e.g. robotics, language or speech); in addition, it opens up a natural approach to bridge the gap between the study of artificial and of biological systems. Information theory gives rise to a rich and diverse set of theorems and results, far beyond its original application to communication theory. These results include the formulation of fundamental bounds for computational effort and/or power. Furthermore, information-optimal solutions exhibit a significant array of desirable properties, among other being least biased or committed, or being maximally stable (see Sect. 19.8.2).

The particular slant that this chapter takes is essentially to consider the issue of optimal control in the guise of MDPs and expressing it in the language of information. This is not a mere equivalent re-expression of the control task. Rather, it adds a rich structural and conceptual layer to it.

In a recent work, the classical task of optimal control has found an elegant reformulation in probabilistic and information-theoretic language: the control task is formulated as a probabilistic perturbation of a system by an agent, and the Bellman equation governing the optimal control solution can be expressed instead as a Kullback–Leibler divergence minimization problem Todorov (2009); by identifying optimal control problems with Bayesian inference problems Strens (2000), the efficient methods for graphical model inference become available for the computation of the optimal policies Kappen et al. (2009). This can be generalized to consider directly the desired distribution of outcomes of the agent's action (Friston et al. 2006; Friston 2009).

In fact, this interpretation invites an even more general picture of the control problem: instead of specifying the reward structure externally, one can consider the intrinsic informational dynamics of an agent interacting with its environment. On the one hand, the study of the information flows in such a system gives important insights into the operation of the agent–environment system (Ay and Wennekers 2003; Lungarella and Sporns 2006; Ay and Polani 2008). On the other hand, one can obtain natural, intrinsically driven ("self-motivated", "reward-less") agent behaviours by optimizing information flows (Lungarella and Sporns 2005; Sporns and Lungarella 2006), predictive information Ay et al. (2008) or the agent-external channel capacity of the perception–action cycle ("empowerment", Klyubin et al. 2005a, 2008). Because of its cyclical character, the interplay between agent-external and agent-internal information processing has been likened to an informational "Carnot Cycle" Fry (2008), whose optimal thermal efficiency would find an informational analog in the concept of *Dual Matching*, which is essentially the *joint source-channel coding* proposed in Gastpar et al. (2003).

Here, we will revisit the MDP problem mentioned earlier and again we will use an informational view. However, here, unlike in Todorov (2009); Kappen et al. (2009), the informational picture will not be used to implement a Bayesian inference mechanism that realizes a solution to the Bellman equation – quite the opposite: we will, in fact, stick to the classical decision-theoretic Bellman approach to MDP. We will, however, combine this approach with a perspective that elevates the information used in the agent's decision process to the "first class object" of our discourse. Adopting this philosophy, we will see how the ramifications of the information approach will project into a wide array of disparate issues, ranging from the analogy between the perception–action cycle and Shannon's communication channel to the relation between the Dual Matching condition mentioned earlier ("Carnot optimality") and the perfectly adapted environment in Sects. 19.5.3.2 and 19.7.2.

19.3 Notation

19.3.1 Probabilistic Quantities

We will use uppercase characters for random variables $X, Y, Z \ldots$, lowercase characters for concrete values $x, y, z \ldots$ that they assume and curved characters $\mathcal{X}, \mathcal{Y}, \mathcal{Z}$ for their respective domains. For simplicity, we will assume that the domains are finite.

The probability that a random variable X assumes a value $x \in \mathcal{X}$ is denoted by $\Pr(X = x)$. However, to avoid unwieldy expressions, we will, by abuse of notation, write $p(x)$ with x being the lowercase of the random variable X in question. Occasionally, we will need to associate two different distributions with the same random variable domain $\mathcal{X}$; these cases will be clearly indicated, and the corresponding distributions will be denoted by the notation $p(x), \hat{p}(x), q(x) \ldots$ and similar. Where a random variable is subscripted, such as in X_t, a different index t will denote different variables.

19.3.2 Entropy and Information

We review some of the basic concepts of Shannon's information theory and notational conventions relevant for this chapter. For a more complete discussion, see e.g. Cover and Thomas (1991).

Define the *entropy* $H(X)$ of a random variable X as:

$$H(X) := - \sum_{x \in \mathcal{X}} p(x) \log p(x) \tag{19.1}$$

where the result is expressed in the unit of *bits* if the logarithm is taken with respect to base two; we assume the identification $0 \log 0 = 0$. Furthermore, in the following we will drop the summation domain $\mathcal{X}$ when the domain is obvious from the context. Also, we will write $H[p(x)]$ instead of $H(X)$ if we intend to emphasize the distribution p of X.

The entropy is a measure of uncertainty about the outcome of the random variable X before it has been measured or seen, and is a natural choice for this Shannon (1949). The entropy is always non-negative. It vanishes for a deterministic X (i.e. if X is completely determined) and it is easy to see (e.g. using the Jensen inequality) that $H[p(x)]$ is a convex functional over the simplex of probability distributions which attains its maximum $\log |\mathcal{X}|$ for the uniform distribution, reflecting the state of maximal uncertainty.

If a second random variable Y is given which is jointly distributed with X according to the distribution $p(x, y)$, then one can define the joint entropy $H(X, Y)$

trivially as the entropy of the joint random variable (X, Y). Furthermore, one can now define the *conditional entropy* as:

$$H(X|Y) \stackrel{\Delta}{=} \sum_y p(y) H(X|Y = y) := - \sum_y p(y) \sum_x p(x|y) \log p(x|y).$$

The conditional entropy measures the remaining uncertainty about X if Y is known. If X is fully determined on knowing Y, it vanishes. With the conditional entropy, one obtains the basic additivity property, or *chain rule* for the entropy:

$$H(X, Y) = H(Y) + H(X|Y) = H(X) + H(Y|X). \tag{19.2}$$

This additivity of the entropy for (conditionally) independent variables can, in fact, be taken as the defining property of Shannon's entropy, as it uniquely determines it under mild technical assumptions.

Instead of the uncertainty that remains in a variable X, once a jointly distributed variable Y is known, one can ask the converse question: how much uncertainty in X is resolved if Y is observed, or stated differently, how much *information* does Y convey about X. This gives rise to the highly important notion of *mutual information* between X and Y, which is expressed in the following equivalent ways:

$$I(X; Y) = H(X) - H(X|Y) = H(Y) - H(Y|X) = H(X) + H(Y) - H(X, Y). \tag{19.3}$$

It is non-negative and symmetric, and vanishes if and only if X and Y are independent. The mutual information turns out to play a major role in Shannon's source and channel coding theorems and other important ramifications of information theory.

A closely related, technically convenient and theoretically important quantity is the *relative entropy*, or *Kullback–Leibler divergence*: assume two distributions over the same domain $\mathcal{X}$, $p(x)$ and $q(x)$, where p is absolutely continuous with respect to q (i.e. $q(x) = 0 \Rightarrow p(x) = 0$). Then define the relative entropy of p and q as:

$$D_{\mathrm{KL}}[p||q] \stackrel{\Delta}{=} \sum_x p(x) \log \frac{p(x)}{q(x)}, \tag{19.4}$$

with the convention $0 \log \frac{0}{0} \stackrel{\Delta}{=} 0$. The relative entropy is a measure how much "compression" (or prediction, both in bits) could be gained if instead of an hypothesized distribution q of X, a concrete distribution p is used. It is the mean code length difference if $q(x)$ is assumed for the prior distribution of X but $p(x)$ is the actual distribution.

We mention several important properties of the relative entropy needed for the rest of the paper (for details, see e.g. Cover and Thomas 1991). First, one has $D_{\mathrm{KL}}[p||q] \geq 0$ with equality if and only if $p = q$ everywhere (almost everywhere in the case of continuous domains $\mathcal{X}$). In other words, one cannot do better than actually assuming the "correct" q. Second, the relative entropy can become infinite if for an outcome x that can occur with non-zero probability $p(x)$ one assumes a probability $q(x) = 0$. Third, the mutual information between two variables X and Y can be expressed as:

$$I(X;Y) = D_{\mathrm{KL}}[p(x, y)\|p(x)p(y)] \tag{19.5}$$

where we write $p(x)p(y)$ for the product of the marginals by abuse of notation. Basically, this interprets mutual information as how many bits about Y can be extracted from X if X and Y are not independent, but jointly distributed. If they are indeed independent, that value vanishes.

Furthermore, we would like to mention the following property of the relative entropy:

Proposition 1. $D_{KL}[p\|q]$, is convex in the pair (p, q).

The first important corollary deriving from this is the strong concavity of the entropy functional, i.e. that there is a unique distribution with maximum entropy in any (compact) convex subset of the simplex. Since the mutual information can be written as:

$$I(X;Y) = D_{\mathrm{KL}}[p(x, y)\|p(x)p(y)] = \mathbb{E}_y D_{\mathrm{KL}}[p(x|y)\|p(x)],$$

one can assert that $I(X;Y)$ is a concave function of $p(x)$ for a fixed $p(y|x)$, and a convex function of $p(y|x)$ given $p(x)$. This is relevant because it guarantees the unique solution of the two fundamental optimization problems of information theory. The first is $\min_{p(y|x)} I(X;Y)$, given $p(x)$ and subject to other convex constraints, i.e. the source coding or rate-distortion function. The second is $\max_{p(x)} I(X;Y)$, given $p(y|x)$ and possibly other concave constraints, i.e. the channel capacity problem.

19.4 Markov Decision Processes

19.4.1 MDP: Definition

The second major building block for the treatment of perception–action cycles in this chapter is the framework of Markov Decision Processes (MDPs). It is a basic model for the interaction of an organism (or an artificial agent) with a stochastic environment. We note that, as discussed in Sect. 19.8.2, the Markovicity of the model is not a limitation; while our exposition of the formalism used below will indeed assume access to the full state of the system for the purposes of computation, it will fully retain the ability to model the agent's subjective view.

In the MDP definition, we follow the notation from Sutton and Barto (1998). Given a state set $\mathscr{S}$, and for each state $s \in \mathscr{S}$ an action set $\mathscr{A}(s)$, an MDP is specified by the tuple $(\mathbf{P}_{s,a}^{s'}, \mathbf{R}_{s,a}^{s'})$, defined for all $s, s' \in \mathscr{S}$ and $a \in \mathscr{A}(s)$, where

$\mathbf{P}^{s'}_{s,a}$ defines the probability that performing an action a in a state s will move the agent to state s' (hence

$$\sum_{s'} \mathbf{P}^{s'}_{s,a} = 1 \tag{19.6}$$

holds) and $\mathbf{R}^{s'}_{s,a}$ is the expected reward for this particular transition (note that $\mathbf{R}^{s'}_{s,a}$ depends not only on starting state s and action a but also on the actually achieved final state s').

19.4.2 The Value Function of an MDP and Its Optimization

The MDP defined in Sect. 19.4.1 defines a transition structure plus a reward. A *policy* specifies an explicit probability $\pi(a|s)$ to select action $a \in \mathscr{A}(s)$ if the agent in a state $s \in \mathscr{S}$. The policy π has the character of a conditional distribution, i.e. $\sum_{a \in \mathscr{A}(s)} \pi(a|s) = 1$.

Given such a policy π, one can now consider the cumulated reward for an MDP, starting at time t at state s_t, and selecting actions according to $\pi(a|s_t)$. This action gives a reward r_t and the agent will find itself in a new state s_{t+1}. Iterating this forever, one obtains the total reward for the agent following policy π in the MDP by cumulating the rewards over the future time steps:

$$\mathfrak{R}_t = \sum_{t'=t}^{\infty} r_{t'}, \tag{19.7}$$

where $\mathfrak{R}_t$ is the total reward accumulated[1] into the future starting at time step t.

Due to the Markovian nature of the model, this cumulated reward will depend only on the starting state s_t and, of course, the policy π of the agent, but not time, so that the resulting future expected cumulative reward value can be written as $V^\pi(s)$, where we dropped the index t from the state s.

Usually in the reinforcement learning literature, a distinction is made between *episodic* and *non-episodic* MDPs. Strictly spoken, (19.7) applies to the non-episodic case. For the episodic case, the sum in (19.7) is not continued to infinity, but stops at reaching so-called goal states. However, to unify the treatment, we will stick to (19.7) and rather suitably adapt ($\mathbf{P}^{s'}_{s,a}$, $\mathbf{R}^{s'}_{s,a}$).

A central result from the theory of dynamic programming and reinforcement learning is the so-called Bellman recursion. Even while the function V^π would be, in principle, computed by averaging over all possible paths generated through the

---

[1] To simplify the derivations, we will always assume convergence of the rewards and not make use of the usual MDP discount factor; in particular, we assume either episodic tasks or non-episodic (continuing) tasks for which the reward converges. For further discussion, see also the remark in Sect. 19.8.2 on *soft policies*.

policy π, it can be expressed through the *Bellman Equation* which is a recursive equation for V^π:

$$V^\pi(s) = \sum_{a \in \mathcal{A}(s)} \pi(a|s) \cdot \sum_{s' \in \mathcal{S}(s)} \mathbf{P}^{s'}_{s,a} \cdot \left[\mathbf{R}^{s'}_{s,a} + V^\pi(s') \right] \qquad (19.8)$$

where the following assumptions hold:

1. a sums over all actions $\mathcal{A}(s)$ possible in s
2. s' sums over all successor states $\mathcal{S}(s)$ of s
3. For an episodic MDP, $V^\pi(s)$ is defined to be 0 if s is a goal state

In the reinforcement learning literature, it is also common to consider the value function V being expanded per-action into the so-called Q function as to reflect which action a in a state s achieves which reward:

$$Q^\pi(s,a) = \sum_{s' \in \mathcal{S}(s)} \mathbf{P}^{s'}_{s,a} \cdot \left[\mathbf{R}^{s'}_{s,a} + V^\pi(s') \right] . \qquad (19.9)$$

In turn, $V^\pi(s)$ can be obtained from $Q^\pi(s,a)$ by averaging out the actions a with respect to the policy π.

To streamline and simplify the notation in the following, we assume without loss of generality that, in principle, all actions $\mathcal{A} \overset{\Delta}{=} \bigcup_{s \in \mathcal{S}} \mathcal{A}(s)$ are available at each state s and all states $\mathcal{S}$ could be potential successor states. For this, we will modify the MDP in a suitable way: actions $a \in \mathcal{A}$ that are not in the currently legal action set $\mathcal{A}(s)$ will be excluded from the policy, either by imposing the constraint $\pi(a|s) \overset{!}{=} 0$ or, equivalently, by setting $\mathbf{R}^{s'}_{s,a} \overset{!}{=} -\infty$. Similarly, transition probabilities $\mathbf{P}^{s'}_{s,a}$ into non-successor states s' are assumed to be 0. Furthermore, for an episodic task, a goal state s is assumed to be absorbing (i.e. $\mathbf{P}^{s'}_{s,a} = \delta_{s,s'}$ where the latter is the Kronecker delta) with transition reward $\mathbf{R}^{s'}_{s,a} = 0$.

The importance of the Bellman Equation (19.8) is that it provides a fixed-point equation for V^π. The value function that fulfils the Bellman Equation is unique and provides the cumulated reward for an agent starting at a given state and following a given policy. Where a value function V which is not a fixed point is plugged into the right side of (19.8), the equation provides a contractive map which converges towards the fixed point of the equation and thus computes the value of V^π for a given policy π. This procedure is called *value iteration*.

Finally, for an optimization of the policy – the main task of reinforcement learning – one then greedifies the policy, obtaining π', inserts it back into the Bellman Equation, recomputes $V^{\pi'}$ and continues the double iteration until convergence Sutton and Barto (1998). This two-level iteration forms a standard approach for MDP optimization. In this chapter, we will develop an analogous iteration scheme which, however, will also cater for the informational cost that acting out an MDP policy entails. How to do this will be the topic of the coming sections.

19.5   Coupling Information with Decisions and Actions

The treatment of an MDP, fully specified by the tuple $(\mathbf{P}_{s,a}^{s'}, \mathbf{R}_{s,a}^{s'})$, is usually considered complete on finding an optimal policy. However, recent arguments concerning biological plausibility indicate that any hypothesized attempt to seek optimal behaviour by an organism needs to be balanced by the considerable cost of information processing which increasingly emerges as a central resource for organisms (Laughlin 2001; Brenner et al. 2000; Taylor et al. 2007; Polani 2009). This view has recently led to a number of investigations studying the optimization of informational quantities to model agent behaviour (Klyubin et al. 2005a; Prokopenko et al. 2006; Ay et al. 2008). The latter work creates a connection between *homeokinetic* dynamics and the formalism of predictive information (Der et al. 1999; Bialek et al. 2001).

In view of the biological ramifications, the consideration of an *explicit* reward becomes particularly relevant. The question then becomes to find the optimal rewards that an organism can accumulate under given constraints on its informational bandwidth, or, more generally, the best possible trade-off between how much reward the organism can accumulate vs. how much informational bandwidth it needs for that purpose.

In this context, it is useful to contemplate the notion of *relevant information.* The concept of relevant information stems from the information bottleneck formalism Tishby et al. (1999). To do so, one interprets an agents' actions as the relevance indicator variables of the bottleneck formalism. This provides an informational treatment of single-decision sequential (Markovian) Decision Problems Polani et al. (2001, 2006), quantifying how much informational processing power is needed to realize a policy that achieves a particular value/reward level. Since we determine the relevance of information by the value it allows the agent to achieve, we speak in this context also of *valuable information.*[2] A related, but less biologically and resource-motivated approach is found in Saerens et al. (2009). The topic of this chapter is a significant generalization of these considerations to the full-fledged perception–action cycle, and their treatment in the context of a generalized Bellman-type recursion.

19.5.1   Information and the Perception–Action Cycle

To apply the formalism of information theory, the quantities involved are best represented as random variables. Specifically in the context of an MDP, or more

---

[2] Valuable information is not to be confused with the *value of information* introduced in Howard (1966). Serving similar purposes, it is conceptually different, as it measures the value difference attainable in a decision knowing vs. not knowing the outcome of a given random variable. Stated informally, it could be seen as "non information-theoretic conjugate" of valuable information.

specifically of an agent acting in an MDP, one can make use of the formalism of (Causal) Bayesian Networks (Pearl 2000; Klyubin et al. 2004, 2007). Here, we introduce for completeness the general Bayesian perception–action cycle formalism, before specializing it to the particular case studied in this chapter.

19.5.1.1 Causal Bayesian Networks

We briefly recapitulate the definition of *(Causal) Bayesian Networks*, also called *probabilistic graphical models* (see Pearl 2000). Causal Bayesian Networks provide a way to describe compactly and transparently the joint distribution of a collection of random variables.

A Bayesian network G over a set of random variables $\mathbf{X} \equiv \{X_1, \ldots, X_n\}$ is a directed acyclic graph (DAG) G in which each vertex is annotated by (or identified by) names of the random variables X_i. For each such variable X_i, we denote by $\mathsf{Pa}[X_i]$ the set of parents of X_i in G and $\mathsf{Pa}[x_i]$ their values. We say that a distribution $p(\mathbf{x})$ is *consistent* with G, written $p \sim G$, if it can be factored in the form:

$$p(x_1, \ldots, x_n) = \prod_{i=1}^{n} p(x_i \mid \mathsf{Pa}[x_i]).$$

(To simplify notation, we adopt the canonical convention that a conditional distribution conditioned against an empty set of variables is simply the unconditional distribution).

19.5.1.2 Bayesian Network for a Reactive Agent

To see how Bayesian Networks can be applied to model an agent operating in a world, consider first a minimal model of a reactive agent. The agent carries out actions A after observing the state W of the world to which it could, in principle, have full access (this assumption will be revisited later). Such an action, in turn, transforms the old state of the world into a new world state.

In different states, typically different actions will be taken, and if one wishes to model the behaviour of the agent through time, one needs to unroll the cycle over time. This leads to the Causal Bayesian Network which is shown in (19.10); here the random variables are indexed by the different time steps $\ldots t - 3, t - 2, \ldots, t, t + 1 \ldots$. Each arrow indicates a conditional dependency (which can also be interpreted as causal in the sense of Pearl). For instance, at time t, the state of the world is W_t which defines the (possibly probabilistic) choice of action A_t immediately (no memory); this action, together with the current state of the world, determines the next state of the world W_{t+1}, etc.

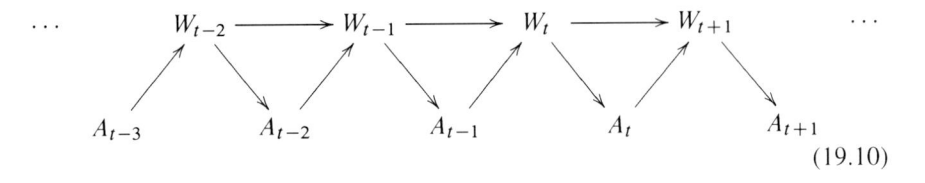

$$(19.10)$$

Note that in this model there is a priori no limitation on the arrow from W_t to A_t, i.e. on $p(a_t|w_t)$, and the agent could theoretically have full access to the state W_t.

19.5.1.3 Bayesian Network for a General Agent

After the simplest of the cases, consider a significantly more general case, the perception–action cycle of an agent with sensors and memory, as in (19.11).

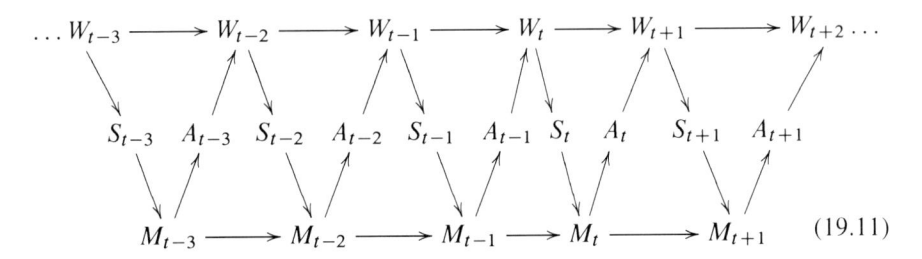

$$(19.11)$$

Here, W_t represents again the state of the world at time t, A_t the actions; additionally, one has S_t as the sensor and a memory variable M_t. The sensor variable S_t models the limited access of the agent to the environment, and the memory variable allows the agent to construct an internal model of the external state that can depend on earlier sensoric observations.

A few notes concerning this model:

1. It is formally completely symmetric with respect to an exchange of agent and environment – the state of the environment corresponds to the memory of the agent; the interface between environment and agent is formed by the sensors and actuators (the only practical difference is that the arrows inside the agent are amenable to adaptation and change, while the environmental arrows are slow, difficult and/or expensive to modify, but this is not reflected in the model skeleton).

2. From the agent's point of view, the sensor variable essentially transforms the problem into a POMDP; however, in the Bayesian Network formalism, there is nothing special about the limited view of the agent as opposed to an all-knowing observer. Where necessary, the eagle's eye perspective of an all-knowing observer can be assumed. In particular, in conjunction with informational optimization, this allows one to select the level of informational transparency in the network between various nodes, in particular, to study various forms and degrees of access to world information that the agent could enjoy, such as, e.g. in sensor evolution scenarios Klyubin et al. (2005b).

3. In the following, we will consider only agents which, in principle, could have full access to the world state. In other words, we will collapse W and S into the same variable (we postpone the discussion of the full system to a future study).

4. We note furthermore that modeling a memory implies that one wishes to separately consider the informational bandwidth of the environment and that of the agent; else the environment itself could be considered as part of the agent's memory. In this chapter, we are indeed interested in the information processed in the complete agent–environment system and not the individual components. Here, we will therefore consider a reactive agent without a memory. However, this is not a limitation in principle, as the approach presented is expected to carry over naturally to models of agents with memory.

With these comments, for this chapter, we finally end up at the following diagram which will form the basis for the rest of the chapter:

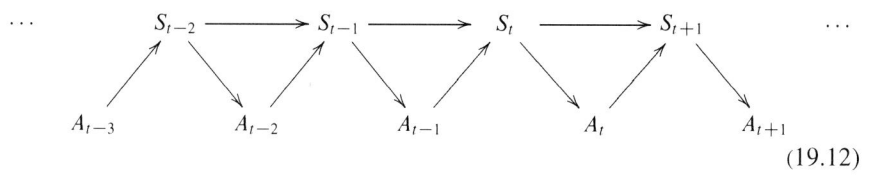

$$(19.12)$$

which is essentially diagram (19.10), but where environmental state and sensor state have been collapsed into the same random variable S.

19.5.2 Actions as Coding

To obtain an intuition how actions can be interpreted in the language of coding, represent a given MDP as a graph[3] with transition edges between a state s and its successor state s'. In this graph, the edges are labeled by the actions a available in the given states (as well as the transition probabilities and rewards). Multiple edges can connect a given state s and its successor state s' as long as they differ by their action label, in other words differing actions may lead to the same transition.

From this picture, it is easy to see that any MDP can be modified without loss of generality in such a way that the actions have the same cardinality in every state. In the simplest case, we can always make this cardinality 2 and all the actions/decisions binary. To see this, consider the standard example of a grid-world, or a maze. As explained above, describe such a world by a (directed) graph with vertices of various degrees. Replace now every junction (which could have an arbitrary number of possible action choices) by a roundabout. In a roundabout, every decision becomes binary: continue on the roundabout or take the turn. Thereby, the decisions/actions become binary (left/right), without changing anything else in the problem.

---

[3] This is a transition graph and should not be confused with the Bayesian Network Graph.

More generally, this can be done by transforming any complex decision into a binary decision tree. Thus, long decision/action sequences are encoded as long bit sequences. It is important to note, though, that each such bit can have entirely different semantic context, depending on its location on the graph.

Assume now for this section that our MDP is made of only binary decisions and/or actions, denoted by $a_j \in \{0, 1\}$, and treat decision sequences as finite binary strings $a_1 a_2 \ldots a_k \equiv \mathbf{a}$. Denote furthermore by l the standard string length function operating on these strings: $l(\mathbf{a}) \equiv l(a_1 a_2 \ldots a_k) \overset{\Delta}{=} k$.

We can now draw simple direct analogies between binary decision sequences and binary codes in communication. We assume that there exist finite sequences of actions (at least one) which deterministically connect any state s with any other state s' (the MDP is simply connected, and deterministic). Denote by $\mathbf{a}_{s,s'}$ such a (binary) action sequence, and by $l(\mathbf{a}_{s,s'})$, its length. The following properties either follow immediately or are easy to verify:

- Unique decoding: a binary decision sequence uniquely determine the end-state s', given the starting state s.
- Concatenation: for any three states s_1, s_2, s_3, the concatenation of $\mathbf{a}_{s_1,s_2} \circ \mathbf{a}_{s_2,s_3}$ is a sequence that connects s_1 to s_3.
- Minimal length: for any two states s, s', we can define $l_{s \rightarrow s'} \overset{\Delta}{=} \min_{\mathbf{a}_{s,s'}} l(\mathbf{a}_{s,s'})$. We will call this minimal length between s and s' the *decision complexity* of the path $s \rightarrow s'$ in the MDP.

Now, in general, while decision/action sequences have similarities to codes as descriptions of trajectories on a graph, there are also important differences. First, in general it is not always the case that the goal of an MDP is reaching a particular target state (or a set of target states), as in the case of the maze. Sometimes the optimal solution for an MDP is distributed and more diffused, or it involves an extended behaviour of potentially infinite duration as is required in optimal control, wealth accumulation or survival. In particular, the required action trajectory may turn out not to be deterministic, or not finite, or neither. Moreover, if the MDP is noisy (i.e. $\mathbf{P}_{s,a}^{s'}$ is not deterministic), a given action sequence does not necessarily determine a unique trajectory of states, but rather a distribution of state sequences.

This poses no problem though, since, as in standard information theory, above notion of decision complexity can be smoothly generalized to still have meaning in the probabilistic or non-finite action sequence: for this purpose, one defines it to be the *expected* number of binary decisions, or bits of information, required by the agent in the future to achieve a certain expected cumulated reward value $V^\pi(s_t)$, if one starts out at a particular state s_t at a given time t in the MDP. The decision complexity is essentially a generalization of the notion of relevant information (mentioned earlier in Sect. 19.5) from individual actions to complete action sequences starting at the current time and extending into the future.

Furthermore, one can extend the information processed by the agent alone to the information processed by the whole system, encompassing both agent and environment. Hereby, one moves from decision complexity to *process information*. This quantity is computed for the whole agent–environment system, beginning with the current state (and agent action) and extended towards the open-ended future of the system. Strictly spoken, we only consider the process information towards the future, not the past. In analogy to the term "cost-to-go", we will speak of *process information-to-go*, or simply *information-to-go* $\mathfrak{I}^{\pi}(s_t, a_t)$ which is computed specifying a given starting state s_t and an initial action a_t and accumulating information-to-go into the open-ended future.

The information-to-go is the information needed (or missing) to specify the future states and actions relative to some prior knowledge about the future. One component of this information is due to uncertainty in the decisions; the other is the information that is "given" or "processed" by the environment in the state transitions. For discrete states and actions, we can think of it as the future state-action entropy, conditioned on the current state-action pair. More generally, one would use the Kullback–Leibler divergence D_{KL} of the actual distribution relative to the prior distribution. This actually implements an "informational regret", which generalizes the simple conditional entropy (and is better-behaved in the case of continuous states and actions). We define the notion of information-to-go formally in Sect. 19.6.2 (19.20).

Let us emphasize again that the philosophy of the present approach is intimately related to the concept of relevant information (Tishby et al. 1999; Polani et al. 2001, 2006) which quantifies the minimal informational cost for individual decisions an agent needs to take to achieve a given future expected cumulated reward (i.e. value). A difference between relevant information and the present study is that here we consider the information processed by whole system, not just the agent. However, by far the most important distinction is that we generalize this concept to include the information to be processed by the system not just for one time step, but over the *whole* period of the run and thus through multiple cycles of the perception–action cycle projected into the future.

The intimate relation between the theories of information and of coding would suggest an interpretation of the formalism in terms of coding. However, statements in coding theory are typically of asymptotic nature. Now, in general, a run of an agent through an MDP does not need to be of infinite length, not even on average: finite-length runs are perfectly possible. To reconcile this conflict, consider the following two interpretations which we expect to be able to recover the asymptotics required for a coding-theoretic interpretation:

1. Runs are restarted after completion, infinitely often, thereby extending the MDP into an infinite future.
2. Assume that the total available information processing power of the agent is pooled and shared by a large number of separate decision processes; a particular decision process will use a certain amount of information processing bandwidth at a given time, and the information-theoretic formalism then describes the usage averaged over all processes (weighted with their probability of occurring). This second view introduces an ensemble interpretation of the formalism.

We believe that both the infinite length run and the ensemble interpretation allow a connection of the present formalism to coding theory and provide tight bounds in the asymptotic case. This question will be pursued further in the future, but is outside the remit of this chapter.

19.5.3 Information-To-Go

To prepare the ground for the later technical developments, in the present section, we first give a high-level outline of the coming discussions.

19.5.3.1 A Bellman Picture

The information-to-go, $\mathfrak{I}^\pi(s_t, a_t)$, is an information-theoretic quantity that is associated with every state-action pair, in analogy to the $Q(s_t, a_t)$ function which can be considered as the *value- (or reward-)to-go* from reinforcement learning. The information-to-go quantifies how many bits one needs on average to specify the future state-action sequence in an MDP (or its informational regret) relative to a prior.

In Sect. 19.6, we shall suggest a principled way of generating $\mathfrak{I}^\pi$-optimal softmax policies. Strikingly, these solutions obey a Bellman-type recursion equation, analogous to reinforcement learning; they involve a locally accumulated quantity which can be interpreted as *(local) information gain*, the information provided by the environment associated with each state transition and action, in analogy to the local reward in MDPs.

The local information gain will be shown to be composed of two natural terms. The first is the information measuring the environmental response to an action, i.e. the information processed in the transition as the system moves to a new state; it is determined by the MDP probabilities $\mathbf{P}_{s,a}^{s'}$. The second term is the information *required* by the agent to select the valuable actions and is determined by the policy $\pi(a|s)$.

The combination of the value and information Bellman equations gives a new Bellman-like equation for the linear (Lagrangian) combination of the two, the *free energy* of the MDP, denoted $F^\pi(s_t, a_t, \beta)$, which reflects the optimal balance between the information-to-go and value-to-go achieved. For given transition probabilities, $\mathbf{P}_{s,a}^{s'}$, and given policy $\pi(a|s)$, we can calculate the free energy for every state/action pair by solving the Bellman equation, for any given trade-off between information-to-go and value-to-go. This essentially implements a novel variant of the rate-distortion formalism which applies to the value-information trade-off of MDP sequences.

19.5.3.2 Perfectly Adapted Environments

This formulation gives rise to a second new insight, namely the characterization of the *perfectly adapted environment* by further minimizing the free energy with respect to the MDP probabilities, $\mathbf{P}_{s,a}^{s'}$. The MDP transition probabilities that minimize the free energy are shown to be exponential in the reward, $\mathbf{R}_{s,a}^{s'}$. In that particular case, all the information about the future is valuable and the optimal policy turns out to be also the one that minimizes statistical surprises.

Perfectly adapted environments form another family of scenarios (besides the classical model of Kelly gambling, Kelly 1956) with the property that the maximization of information about the future is equivalent to the maximization of the value of the expected reward. In general, this is not the case: rather, the current state of the system will provide valuable (relevant) as well as non-valuable information about the future. Non-valuable (irrelevant) information about the future is the information that can be safely ignored (in a bottleneck sense) without affecting the future expected reward. It is the valuable (relevant) information only which affects the future reward. In the special cases of both Kelly gambling as well as our case of the perfectly adapted environment, however, all information is valuable – maximizing the future information is equivalent to maximizing the expected future reward. When the bandwidth for future information is limited, this leads to a suboptimal trade-off with respect to achievable future expected reward.

The interest in studying perfectly adapted environments stems from the fact that they provide the key for linking predictive information with task-relevant information. It has been hypothesized that living organisms maximize the predictive information in their sensorimotor cycle, and this hypothesis allows to derive a number of universal properties of the perception–action cycle (Bialek et al. 2001; Ay et al. 2008), and, if this hypothesis has merit, this would imply an interpretation of organismic behaviour in terms of (Kelly) gambling on the outcome of actions. On the other hand, actual organismic rewards could in principle be structured in such a way that much of the predictive information available in the system would turn out to be irrelevant to select the optimal action; the distinction between relevant and irrelevant information provides a characterization of an organism's niche in its information ecology.

However, under the assumption that information acquisition and processing is costly for an organism (Laughlin 2001; Polani 2009), one would indeed expect a selection pressure for the formation of sensors that capture just the value-relevant component of the predictive information, but no more. A step further goes the hypothesis that, over evolutionary times, selective pressure would even end up realigning the reward and the informational structures towards perfectly adapted environments. Although here we are concentrating only on their theoretical implications, it should be mentioned that all these hypotheses imply quantitative and ultimately experimentally testable predictions.

19.5.3.3 Predictive Information

We assume in (19.12) that the agent has full sensoric access to the state of the world. This is a special case of the more general case where the information-to-go is the information that the agent at the current time has on the future of the system and which is extracted from past observations. One implication of this assumption is that the future information of the organism is bounded by the *predictive information* (defined, e.g. in Shalizi and Crutchfield 2002; Bialek et al. 2001) of the environment Bialek et al. (2007). In Fig. 3 from Bialek et al. (2007), information about the past, the future, adaptive value and resources are put in relation to each other, and the predictive information (i.e. the information which the past of the system carries about the future) corresponds to its third quadrant. Note that this is not the full predictive information, but only the valuable (relevant) component of the predictive information, in the sense that it identifies the information necessary to achieve a given value in a given reward structure. Its supremum is the total valuable information that the environment carries about the future. The organism cannot have more future valuable information than is present in the environment and will, usually, have less since it is bounded by metabolic, memory or computational resources. As was shown by Bialek et al. (2001), for stationary environments the predictive information grows sub-linearly with the future horizon window (it is sub-extensive). On the other hand, for stationary environments and ergodic MDPs, the information-to-go grows linearly with the horizon (future window size), for large enough windows.

19.5.3.4 Symmetry

We wish to attract attention to a further observation: we commented already in Sect. 19.5.1.3 that the Bayesian Network is symmetric with respect to environment and agent – but, in addition, the Bayesian Network is also structurally symmetric with respect to an interchange of past and future, whereby the role of sensing and acting is switched. This structural symmetry is reflected in the essential interchangeability of the past and future axes in the third quadrant of Fig. 3 in Bialek et al. (2007). To complete the symmetry, we would need to additionally introduce a sensoric cost in analogy to the actuatoric reward which is already implemented in this chapter. Of course, the symmetry is only structural; in the framework, past and future are of course asymmetric, since we compress the past and predict the future – i.e. we minimize the information about the past and maximize the information about the future (and see, e.g. also Ellison et al. 2009). Likewise, the symmetry between environment and agent is only structural, but the flexibility and the characteristic dynamics will in general differ strongly between the environment and the agent.

19.5.4 The Balance of Information

The relevance of the informational treatment of the perception–action cycle arises to some degree from the fact that information, while not a conserved quantity, observes a number of consistency and bookkeeping laws. Other such laws are incarnated as informational lower or upper bounds. Always implicit to such considerations is of course the fact that, in the Shannon view, the source data can – in principle – be coded, transmitted through the channel in question and then suitably decoded to achieve the given bounds.

We note that the multi-staged treatment of communication channels separating source and channel coding as espoused in the classical presentation by Shannon is not necessarily the most biologically relevant scenario. For instance, it was noticed by Berger (2003) that biology might be using non-separable information in the sense of using joint source-channel coding Csiszár and Körner (1986). This view is of particular interest due to the discovery of the general existence of optimally matched channels Gastpar et al. (2003). The simplicity and directness they afford suggest that such channels may have relevance in biology. Specifically, with these advantages, it is conceivable that biological perception–action cycles would profit from co-evolving all their components towards optimally matched channels; this is particularly true in biology since biological channels are likely to have had sufficient time and degrees of freedom to evolve optimally matched channels.

If this hypothesis is valid, biological channels and perception–action cycles will not only strive to be informationally optimal, but also fulfil the additional constraints imposed by the optimally matched channel condition. In a metaphorical way, this hypothesis corresponds to an "impedance match" or a balance criterion for information flowing between the organism and the environment in the cycle.

The optimal match hypothesis, for one, contributes to the plausibility of the informational treatment for the understanding of biological information processing; in addition, it provides a foundation for predictive statements, both quantitative and structural. It is beyond the scope of this chapter to dwell on these ramifications in detail. However, it should be kept in mind that these are an important factor behind the relevance of informational bookkeeping principles for the studies of the perception–action cycle.

With these preliminaries in place, the present section will now review some elementary information-theoretical bookkeeping principles. The reader is already acquainted with them and is invited to only lightly skim this section for reference.

19.5.4.1 The Data Processing Inequality and Chain Rules for Information

Consider a simple linear *Markov Chain*, a special case of a Bayesian Network, consisting of three random variables: $U \to X \to Y$. Then, the *Data Processing Theorem* states that Y cannot contain more information about U than X, formally

$$I(X;U) \geq I(U;Y).$$

In other words, Y can at most reflect the amount of information about U that it acquires from X, but no more than that. While information cannot grow, it can be lost in such a linear chain. However, to reacquire lost information, it would need to feed in from another source. The insight gained by the data processing inequality can furthermore be refined by not just quantifying, but actually identifying the information from a source variable that can be extracted downstream in a Markov Chain. One method to do so is, for instance, the Information Bottleneck Tishby et al. (1999).

As a more general case, consider general finite sequences of random variables, $X^n \equiv (X_1, X_2, \ldots, X_n)$. From (19.2), it is easy to see that one has

$$H(X_1, X_2, \ldots, X_n) = H(X_1) + H(X_2|X_1) + H(X_3|X_2, X_1) + \ldots + H(X_n|X^{n-1}). \tag{19.13}$$

In the case of finite – say m-th order – Markov chains, this simplifies drastically. Here, a variable X_k is screened by the m previous variables $X_{k-m}, \ldots, X_{k-1}$ from any preceding variable, that is, a conditional entropy simplifies according to

$$H(X_k|X^{k-1}) \equiv H(X_k|X_1, \ldots, X_{k-1}) = H(X_k|X_{k-m}, \ldots, X_{k-1})$$

(without loss of generality assume $k > m$, or else pad by empty random variables X_i for $i \leq 0$).

Similarly to (19.13), one has for the mutual information with any additional variable Y the relation

$$I(X_1, X_2, \ldots, X_n; Y) = I(X_1; Y) + I(X_2; Y|X_1) + I(X_3; Y|X_2, X_1) + \ldots + I(Y; X_n|X^{n-1}), \tag{19.14}$$

where the *conditional mutual information* is naturally defined as $I(X; Y|Z) \equiv H(X|Z) - H(X|Y, Z)$. The conditional mutual information can be interpreted as the information shared by X and Y once Z is known.

We have seen in (19.5) that mutual information can be expressed in terms of the Kullback–Leibler divergence. Thus, the chain rule of information is, in fact, a special case of the chain rule of the Kullback–Leibler divergence (see also Cover and Thomas 1991):

$$D_{\text{KL}}[p(x_1, \ldots, x_n)||q(x_1, \ldots, x_n)] = D_{\text{KL}}[p(x_1)||q(x_1)] + D_{\text{KL}}[p(x_2|x_1)||q(x_2|x_1)] + \ldots$$
$$+ D_{\text{KL}}[p(x_n|x^{n-1})||q(x_n|x^{n-1})] \tag{19.15}$$

with the conditional Kullback–Leibler divergence defined as

$$D_{\text{KL}}[p(y|x)||q(y|x)] \stackrel{\Delta}{=} \sum_x p(x) \sum_y p(y|x) \log \frac{p(y|x)}{q(y|x)}. \tag{19.16}$$

19.5.4.2 Multi-Information and Information in Directed Acyclic Graphs

A multivariate generalization of the mutual information is the multi-information. It is defined as

$$\mathbf{I}[p(X)] = \mathbf{I}(X_1, X_2, \ldots, X_n) = D_{\text{KL}}[p(x_1, x_2, \ldots, x_n)||p(x_1)p(x_2)\ldots p(x_n)]. \tag{19.17}$$

There are various interpretations for the multi-information. The most immediate one derives from the Kullback–Leibler representation used above: in this view, the multi-information measures by how much more one could compress the random variable $(X_1, X_2, \ldots, X_n)$ if one treated it as a joint random variable as opposed to a collection of independent random variables $X_1, X_2, \ldots, X_n$. In other words, this is a measure for the overall dependency of these variables that could be "squeezed out" by joint compression. The multi-information has proven useful in a variety of fields, such as the analysis of graphical models (see, e.g. Slonim et al. 2006).

Proposition 2. *Let* $\mathbf{X} = \{X_1, \ldots, X_n\} \sim p(\mathbf{x})$, *and let* G *be a Bayesian network structure over* $\mathbf{X}$ *such that* $p \sim G$. *Then*

$$\mathbf{I}[p(\mathbf{x})] \equiv \mathbf{I}(\mathbf{X}) = \sum_i I(X_i; \mathsf{Pa}[X_i]).$$

That is, the total multi-information is the sum of "local" mutual information terms between each variable and its parents (related additivity criteria can be formulated for other informational quantities, see also Ay and Wennekers 2003; Wennekers and Ay 2005).

An important property of the multi-information is that it is the cross-entropy between a model multivariate distribution and the "most agnostic", completely independent prior over the variables; therefore, it can be used to obtain finite sample generalization bounds using the PAC-Bayesian framework (McAllester 1999; Seldin and Tishby 2009).

19.6   Bellman Recursion for Sequential Information Processing

The language and the formalisms needed to formulate the central result of this chapter are now in place. Recall that we were interested in considering an MDP not only in terms of maximized rewards but also in terms of information-to-go.

We consider complete decision sequences and compute their corresponding information-to-go during the whole course of a sequence (as opposed to the information processed in each single decision, as in Polani et al. 2001, 2006). We will combine the Bayesian Network formalism with the MDP picture to derive trade-offs between the reward achieved in the MDP and the informational effort or cost required to achieve this reward. More precisely, unlike in the conventional picture of MDP where one essentially seeks to maximize the reward, no matter what the cost of the decision process, we will put an informational constraint on the cost of the decision process and ask what the best reward is which can be achieved under this processing constraint.

It turns out that the resulting formalism resembles closely the Bellman recursion which is used to solve regular MDP problems, but it applies instead to informational quantities. This is in particular interesting since informational costs are not extensive as MDP rewards are Bialek et al. (2001).

Before we proceed to introduce the algorithm, note that the reward is only associated with the agent's choice of actions and the ensuing transitions. Thus, only that information about the future is relevant here which affects the rewards. In turn, the component of entropy of the future which is not going to affect the reward can be ignored. Basically, this is a "rate-distortion" version of the concept of statistical sufficiency: we are going to ignore the variability of the world which does not affect the reward.

19.6.1 Introductory Remarks

Consider now the stochastic process of state-action pairs

$$S_t, A_t, S_{t+1} A_{t+1}, S_{t+2} A_{t+2}, \ldots, S_{t+n} A_{t+n}, \ldots$$

where the state-action pairs derive from an MDP whose Bayesian Network corresponds to (19.12), beginning with the current state S_t and action A_t. The setup is similar to Klyubin et al. (2004), Still (2009).

We reiterate the argument from Sect. 19.5.1.3, point 2 and emphasize once more that for our purposes, it is no limitation to assume that the agent has potentially unlimited access to the world state, and we therefore can exclusively consider MDPs instead of POMDPs. The information/value trade-off will simply find the best possible pattern of utilization of information.

For a finite informational constraint, this still implicitly defines a POMDP, however one that is not defined by a particular "sensor" (i.e. partial observation) structure, but rather by the quantitative limits of the informational bandwidth. The formalism achieves the best value for a given informational bandwidth in the sense that no other transformation of the MDP into a POMDP using the same information processing bandwidth will exceed the optimal trade-off solution with respect to value. In the following, we will thus consider the system from an eagle's eye perspective where for the purposes of the computation we have – in principle – access to all states of the system, even if the agent itself (due to its information bandwidth constraints) may not.

To impose an explicitly designed POMDP structure (e.g. incorporating physical, engineering or other constraints), one could resort to the extended model from (19.11) instead. The latter incorporates sensors (i.e. explicit limitations to what the agent can access from the environment) as well as memory. Considering the latter turns the perception–action cycle into a full formal analogy of Shannon's communication channel. In this case, however, one typically needs to include also the agent memory into the picture. For such a scenario, preliminary results indicate that informational optimality criteria have the potential to characterize general properties of information-processing architectures in a principled way van Dijk et al. (2009).

Concludingly, the formalism introduced here is not limited to reactive agents, and in future work we will extend it to memory-equipped agents. Here, however, we limit ourselves to reactive agents, as these already represent an important subset of the systems of interest and provide a transparent demonstration of the central ideas of our approach.

19.6.2 Decision Complexity

Assume that, at time t, the current state and action are given: $S_t = s_t$, $A_t = a_t$. The distribution of successor states and actions in the following time steps $t+1, t+2, \ldots$ is given by $p(s_{t+1}, a_{t+1}, s_{t+2}, a_{t+2}, \ldots | s_t, a_t)$. We assume now a fixed prior on the distribution of successive states and actions: $\hat{p}(s_{t+1}, a_{t+1}, s_{t+2}, a_{t+2}, \ldots)$.

Define now the process complexity as the Kullback–Leibler divergence between the actual distribution of states and actions after t and the one assumed in the prior:

$$\Im^{\pi}(s_t, a_t) \triangleq \mathbb{E}_{p(s_{t+1}, a_{t+1}, s_{t+2}, a_{t+2}, \ldots | s_t, a_t)} \log \frac{p(s_{t+1}, a_{t+1}, s_{t+2}, a_{t+2}, \ldots | s_t, a_t)}{\hat{p}(s_{t+1}, a_{t+1}, s_{t+2}, a_{t+2}, \ldots)}.$$

$$(19.18)$$

$\Im^{\pi}(s_t, a_t)$ measures the informational regret of a particular sequence relative to a prior probability for the sequence. The prior encodes all information known about the process which can range from a state of complete ignorance up to a full model of the process (in which case $\Im^{\pi}(s_t, a_t)$ would vanish).

However, we want to consider priors which are simpler than the full MDP model. Of particular interest are those where the components of process $S_{t+1} A_{t+1}, S_{t+2} A_{t+2}, \ldots$ are independent, i.e. where the prior has the form

$$\hat{p}(s_{t+1}, a_{t+1}, s_{t+2}, a_{t+2}, \ldots) = \hat{p}(s_{t+1}) \hat{\pi}(a_{t+1}) \hat{p}(s_{t+2}) \hat{\pi}(a_{t+2}) \ldots$$

where all "hatted" distributions are the individual priors on the respective random variables (we denote the priors for the actions by $\hat{\pi}$ instead of $\hat{p}$ for reasons that will become clearer below, see e.g. (19.19)). With such a choice of the prior, $\Im^{\pi}(s_t, a_t)$ becomes a measure for the interaction between the different steps in the decision cascade.[4]

Selecting the priors $\hat{p}(s_{t+1})$, $\hat{\pi}(a_{t+1})$, $\hat{p}(s_{t+2})$, $\hat{\pi}(a_{t+2}), \ldots$ beforehand and independently from the MDP corresponds to the most agnostic assumption. Another specialization is the *stationarity* assumption that the random variables $S_{t+1}, S_{t+2}, \ldots$ and $A_{t+1} A_{t+2}, \ldots$ are i.i.d. and share the same state distributions $\hat{p}(s_{t+1})$, $\hat{p}(s_{t+2}), \ldots$ and action distributions $\hat{\pi}(a_{t+1})$, $\hat{\pi}(a_{t+2}), \ldots$.

For our purposes, it is useful to mention the criterium of *consistency*. Consistency can be total or partial. *Total consistency* means that $\hat{p}(s_{t+1})$, $\hat{\pi}(a_{t+1})$, $\hat{p}(s_{t+2})$, $\hat{\pi}(a_{t+2}), \ldots$ result from the marginalization of the *total* original distribution which itself is consistent with the Bayesian Network (19.12). In the case of total consistency, $\Im^{\pi}(s_t, a_t)$ becomes the multi-information between the state/action variables $S_{t+1} A_{t+1}, S_{t+2} A_{t+2}, \ldots$ throughout the sequence.

---

[4] Note that, unless stated otherwise, we always imply that the distributions $\hat{p}(s_{t+1})$, $\hat{p}(s_{t+2}), \ldots$ as well as $\hat{\pi}(a_{t+1})$, $\hat{\pi}(a_{t+2}), \ldots$ can be different for different t.

On the other hand, *partial consistency* means that only parts of the relations in the Bayesian Network are respected in forming the factorization.

In this chapter, we will use close to minimal assumptions: we assume stationarity with partial consistency, where the state distributions $\hat{p}(s_{t+1})$, $\hat{p}(s_{t+2})$, ... are the same for all times, and the action distributions are consistent with them via the policy π which we assume constant over time for all t:

$$\hat{\pi}(a_t) = \sum_{s_t \in \mathscr{S}} \pi(a_t|s_t) \cdot \hat{p}(s_t). \tag{19.19}$$

The prior $\hat{p}(s_t)$ is chosen as uniform distribution over the states for all t.

Stronger consistency assumptions, such as requiring the $\hat{p}(s_{t+1})$, $\hat{p}(s_{t+2})$, ... to respect the transition probabilities $p(s_{t+1}|s_t, a_t)$ in the Bayesian Network (we call this *ergodic stationarity* in the special case of $\hat{p}(s_{t+1})$, $\hat{p}(s_{t+2})$, ... being identical distributions) will be considered in the future, but are outside of the remit of this chapter.

With above comments, the information-to-go will be defined in the following as the Kullback–Leibler divergence of the future sequence of states and actions, starting from s_t, a_t, with respect to stationary prior state distributions over the state sequence $\hat{p}(s_{t+1})$, $\hat{p}(s_{t+2})$, ... and policy-consistent (19.19) action distributions $\hat{\pi}(a_{t+1})$, $\hat{\pi}(a_{t+2})$, ... :

$$\Im^\pi(s_t, a_t) \overset{\Delta}{=} \mathbb{E}_{p(s_{t+1}, a_{t+1}, s_{t+2}, a_{t+2}, \ldots | s_t, a_t)}$$

$$\log \frac{p(s_{t+1}, a_{t+1}, s_{t+2}, a_{t+2}, \ldots | s_t, a_t)}{\hat{p}(s_{t+1})\hat{\pi}(a_{t+1})\hat{p}(s_{t+2})\hat{\pi}(a_{t+2}) \ldots}. \tag{19.20}$$

The interpretation of this quantity is as follows: $\Im^\pi(s_t, a_t)$ measures the informational cost for the system to carry out the policy π, starting at time t into the indefinite future with respect to the prior. In general, this quantity will grow with the length of the future. It measures how much information is processed by the whole agent–environment system in pursuing the given policy π. This quantity can also be interpreted as the number of bits that all the states and actions share over the extent of the process as opposed to the prior. One motive for studying this quantity is that it provides important insights about how minimalistic agents can solve external tasks under limited informational resources.

The central result of this chapter is that the optimization of V^π under constrained information-to-go $\Im^\pi(s_t, a_t)$, although encompassing the whole future of the current agent, can be computed through a one-step-lookahead recursion relation; moreover, this recursion relation closely mirrors the Bellman recursion used in the value iteration algorithms of conventional reinforcement learning.

19.6.3 Recursion Equation for the MDP Information-To-Go

We obtain a recursion relation for this function by separating the first expectation from the rest. With Proposition 2 in the context of (19.12), it is easy to see that one has

$$\mathfrak{I}^{\pi}(s_t, a_t) = \mathbb{E}_{p(s_{t+1}, a_{t+1}|s_t, a_t)} \left[\log \frac{p(s_{t+1}|s_t, a_t)}{\hat{p}(s_{t+1})} + \log \frac{\pi(a_{t+1}|s_{t+1})}{\hat{\pi}(a_{t+1})} \right.$$
$$\left. + \mathfrak{I}^{\pi}(s_{t+1}, a_{t+1}) \right], \tag{19.21}$$

with $p(s_{t+1}, a_{t+1}|s_t, a_t) = p(s_{t+1}|s_t, a_t)\pi(a_{t+1}|s_{t+1})$ (for more general Bayesian graphs, more general statements can be derived).

The terms associated with the first state-action-transition,

$$\Delta I_{s_t, a_t}^{s_{t+1}} = \log \frac{p(s_{t+1}|s_t, a_t)}{\hat{p}(s_{t+1})} + \log \frac{\pi(a_{t+1}|s_{t+1})}{\hat{\pi}(a_{t+1})} \tag{19.22}$$

can be interpreted as the *information gain* associated with this transition. In this recursion, the information gain takes on the role of the local reward, in complete analogy with the quantity $\mathbf{R}_{s,a}^{s'}$ from reinforcement learning.

Information gain-type quantities appear as natural Lyapunov functions for master equations and in the informational formulation of exploration and learning problems (Haken 1983; Vergassola et al. 2007). Note that the quantities in (19.22) can be both positive and negative (even if the prior is marginal, e.g. Lizier et al. 2007).[5] Only by averaging, the familiar non-negativity property of informational quantities is obtained.

19.6.3.1 The Environmental Response Term

The information gain (19.22) consists of two terms which we discuss in turn. The first term quantifies the statistical surprise in the transition due to our action (relative to the prior). It can be seen as the *environmental response information* as it measures the response of the world to the agent's control action. It is also interpretable as the information gained if one can *observe* the next state (in a fully observed MDP), or as the information *processed* by the environment in this state transition. In Sect. 19.7.2, this term combined together with the MDP reward will give rise to the concept of the perfectly adapted environment which reflects the perception–action cycle version of the notion of optimally matched channels by Gastpar et al. (2003).

---

[5] The interpretation of a negative information gain is that under the presence/observation of a particular condition, the subsequent distributions are blurred. One caricature example would be that, to solve a crime, one would have a probability distribution sharply concentrated on a particular crime suspect. If now additional evidence would exclude that suspect from consideration and reset the distribution to cover all suspects equally, this would be an example for negative information gain.

The environmental response information can be considered an information-theoretic generalization or a soft version of the control-theoretic concept of *controllability* (in this context, see also Ashby 1956; Touchette and Lloyd 2000, 2004; Klyubin et al. 2005b; Todorov 2009). As here we do not limit the agent's access to the world state and also do not model the sensoric cost, the information gain term does not contain an analogous information-theoretic term corresponding to observability, but this is only a restriction of the current scenario, not of the model in general.

Strictly spoken, when we talk above about controllability/observability, we refer only to the actual level of control (and observation) exerted, not about controllability (and observability) in the sense of the maximally achievable control/observation. For an information-theoretic treatment of *combined* controllability and observability in the latter (i.e. maximality) sense, see e.g. Klyubin et al. (2008).

19.6.3.2   The Decision Complexity Term

We now turn briefly to the second term in (19.22); the second term reflects the decision complexity, i.e. the informational effort that the agent has to invest in the subsequent decision at time $t + 1$. The average of this term according to (19.21) measures the information required for the selection of the agent's action at time $t + 1$. Importantly, note that this value for the decision complexity at time $t + 1$ as calculated from the recursive (19.21) and (19.22) is always conditional on the state s_t and the action a_t at the current time t.

These two components make clear that the information processing exhibited by the agent–environment system decomposes into two parts, one that captures the environmental information processing and one that reflects the agent's decision. This decomposition is related to that known from compositional Markov chains Wennekers and Ay (2005) and provides an elegant and transparent way of distinguishing which part of a system is responsible for which aspect of information processing.

19.7   Trading Information and Value

We can now calculate the minimal information-to-go (i.e. environmental information processing cost plus decision complexity) that is required to achieve a given level of value-to-go.

19.7.1   The "Free-Energy" Functional

At this point, we remind the reader of (19.9) which is used in the reinforcement learning literature to characterize the value- or reward-to-go in terms of state-action pairs instead of states only:

$$Q^{\pi}(s_t, a_t) = \sum_{s_{t+1}} \mathbf{P}_{s_t, a_t}^{s_{t+1}} \cdot \left[\mathbf{R}_{s_t, a_t}^{s_{t+1}} + V^{\pi}(s_{t+1}) \right]. \tag{19.23}$$

As V^{π} quantifies the future expected cumulative reward when starting in state s_t and then following the policy π, the function Q^{π} separates out also the initial action a_t, in addition to the initial state s_t.

The constrained optimization problem of finding the minimal information-to-go at a given level of value-to-go can be turned into an unconstrained one using the Lagrange method; for this the quantity to minimize (the information-to-go) is complemented by the constraint (the value-to-go) multiplied by a Langrange multiplier β:

$$F^{\pi}(s_t, a_t, \beta) \overset{\triangle}{=} \Im^{\pi}(s_t, a_t) - \beta Q^{\pi}(s_t, a_t). \tag{19.24}$$

This Lagrangian builds a link to the free energy formalism known from statistical physics: the Langrange multiplier β corresponds to the inverse temperature, and the information-to-go $\Im^{\pi}$ corresponds to the physical entropy. The value-to-go (here expressed as Q^{π}) corresponds to the energy of a system, and F^{π}/β corresponds to the free energy from statistical physics. However, for simplicity we will apply the notion of *free energy* to F^{π} itself. The analogy with the free energy from statistical physics provides an additional justification for the minimization of the information-to-go under value-to-go constraints: the minimization of F^{π} identifies the least committed policy in the sense that the future is the least informative, i.e. the least constrained and thus the most robust.

This philosophy is closely related to the minimum information principle Globerson et al. (2009): if one has an input–output relationship, one selects a model that processes the least information that is consistent with the observations. This corresponds again to the least committed solution that covers the observations (and is, in general, not identical to the maximum entropy solution with which it coincides only in certain cases, see Globerson et al. 2009).

For the later purposes, it is useful to expand the free energy as follows:

$$\Im^{\pi}(s_t, a_t) - \beta Q^{\pi}(s_t, a_t)$$

$$= \mathbb{E}_{p(s_{t+1}|s_t,a_t)\pi(a_{t+1}|s_{t+1})} \left[\log \frac{p(s_{t+1}|s_t,a_t)}{\hat{p}(s_{t+1})} + \log \frac{\pi(a_{t+1}|s_{t+1})}{\hat{\pi}(a_{t+1})} \right.$$

$$\left. - \beta \mathbf{R}_{s_t a_t}^{s_{t+1}} + \Im^{\pi}(s_{t+1}, a_{t+1}) - \beta V^{\pi}(s_{t+1}) \right]$$

$$= \mathbb{E}_{p(s_{t+1}|s_t,a_t)} \left[\log \frac{p(s_{t+1}|s_t,a_t)}{\hat{p}(s_{t+1})} - \beta \mathbf{R}_{s_t a_t}^{s_{t+1}} \right.$$

$$+ \mathbb{E}_{\pi(a_{t+1}|s_{t+1})} \left[\log \frac{\pi(a_{t+1}|s_{t+1})}{\hat{\pi}(a_{t+1})} + \Im^{\pi}(s_{t+1}, a_{t+1}) \right.$$

$$\left. \left. - \beta Q^{\pi}(s_{t+1}, a_{t+1}) \right] \right] \tag{19.25}$$

where the last equality follows from

$$V^\pi(s_{t+1}) = \mathbb{E}_{\pi(a_{t+1}|s_{t+1})}[Q^\pi(s_{t+1}, a_{t+1})].$$

This leads to the following recursive relation for the free energy:

$$F^\pi(s_t, a_t, \beta) = \mathbb{E}_{p(s_{t+1}|s_t,a_t)}\left[\log \frac{p(s_{t+1}|s_t,a_t)}{\hat{p}(s_{t+1})} - \beta \mathbf{R}_{s_t a_t}^{s_{t+1}}\right.$$
$$\left. + \mathbb{E}_{\pi(a_{t+1}|s_{t+1})}\left[\log \frac{\pi(a_{t+1}|s_{t+1})}{\hat{\pi}(a_{t+1})} + F^\pi(s_{t+1}, a_{t+1}, \beta)\right]\right].$$
$$(19.26)$$

The task of finding the optimal policy, i.e. the one minimizing its information-to-go under a constraint on the attained value-to-go is solved by the unconstrained minimization of the corresponding Lagrangian, i.e. the free energy functional F^π:

$$\operatorname*{argmin}_\pi F^\pi(s_t, a_t, \beta) = \operatorname*{argmin}_\pi [\Im^\pi(s_t, a_t) - \beta Q^\pi(s_t, a_t)]$$

where the minimization ranges over all policies. A particular constraint on the value-to-go is imposed by selecting the respective "inverse temperature" Lagrange multiplier β.

Extending (19.26) by the Lagrange term for the normalization of π and taking the gradient with respect to π, and then setting the gradient of F^π to 0 (both for the entire term as well as inside the brackets) provides us with a Bellman-type recursion for the free energy functional as follows: an optimal policy π satisfies the recursive (19.26) as well as the relations

$$\pi(a|s) = \frac{\hat{\pi}(a)}{Z^\pi(s,\beta)} \exp(-F^\pi(s, a, \beta)) \qquad (19.27)$$

$$Z^\pi(s,\beta) = \sum_a \hat{\pi}(a) \exp(-F^\pi(s, a, \beta)) \qquad (19.28)$$

$$\hat{\pi}(a) = \sum_s \pi(a|s)\hat{p}(s). \qquad (19.29)$$

in a self-consistent fashion. In turn, iterating the system of self-consistent (19.26)–(19.29) till convergence for every state will produce an optimal policy. This system of equations essentially unifies the Bellman equation and the Blahut–Arimoto algorithm from rate-distortion theory.

Note that as result of the algorithm, we obtain a non-trivial soft-max policy for every finite value of β. Furthermore, if the optimal policy is unique, the equations will recover it as a deterministic policy for the limit $\beta \to \infty$. The compound iterations converged to a unique policy for any finite value of β. While we believe that a convergence proof (possibly without uniqueness guarantees) could be developed along the lines of the usual convergence proofs for the Blahut–Arimoto algorithm, we defer this to a future paper.

It should be mentioned at this point that the reward structure determining the form of Q is an externally defined part of the system description. In this chapter, the reward can be of any type. However, the reward could be realized as a more specific quantity, e.g. an informational measure, such as, for example, a predictive information gain in which case the formalism would reduce to a particular form.

19.7.2 Perfectly Adapted Environments

An intriguing aspect of the free-energy formalism is that we can consider the optimality not only of the agent's policy but also of the environment. This is particularly relevant for the characterization of a "best match" between the organism's action space and the responses of the environment, which realizes a *perfectly adapted environment*. We already commented earlier on the close conceptual analogy between the concept of the perfectly adapted environment and what we suggest to be its information-theoretic counterpart: the environment being considered as the channel, and the agent's actions as the source in an optimally matched channel Gastpar et al. (2003).

In the following paragraphs, we characterize the optimal (i.e. perfectly adapted) environment using the language of our formalism, as the MDP that minimizes the free–energy. Define the notation

$$q_\beta(s'|s,a) = q_{s,a}^{s'} \overset{\Delta}{=} \frac{p(s')}{Z(\beta,s,a)} \exp\left(\beta \mathbf{R}_{s,a}^{s'}\right). \tag{19.30}$$

Then, with the free energy functional $F^\pi(s_t, a_t, \beta) = \Im^\pi(s_t, a_t) - \beta Q^\pi(s_t, a_t)$, the Bellman equation can be rewritten as:

$$F^\pi(s_t, a_t, \beta) = \mathbb{E}_{p(s_{t+1}|s_t,a_t)} \left[\log \frac{p(s_{t+1}|s_t, a_t)}{q_\beta(s_{t+1}|s_t, a_t)} - \log Z(\beta, s_t, a_t) \right.$$

$$\left. + \mathbb{E}_{\pi(a_{t+1}|s_{t+1})} \left[\log \frac{\pi(a_{t+1}|s_{t+1})}{\hat{\pi}(a_{t+1})} + F^\pi(s_{t+1}, a_{t+1}, \beta) \right] \right].$$

Note that in this form the first term averages to the Kullback–Leibler divergence between the actual probability $p(s_{t+1}|s_t, a_t)$ and the "optimal distribution" $q_\beta(s_{t+1}|s_t, a_t)$ of the next state s_{t+1}, for fixed current state s_t and action a_t.

The first term in F is minimized[6] with respect to the environment transition probabilities precisely when the MDP is fully adapted to the reward, namely, when

---

[6] Alternatively, one could minimize F^π by setting the gradient of F^π with respect to $p(s_{t+1}|s_t, a_t)$ to 0 similar to the derivation of (19.27)–(19.29) under the assumption that π is already optimized. This implements the assumption that the adaptation of the environmental channel is "slow", corresponding to the adaptation of the agent policy.

$$p(s_{t+1}|s_t, a_t) = q_\beta(s_{t+1}|s_t, a_t).\tag{19.31}$$

In this case, the Kullback–Leibler divergence vanishes. Plugging in the optimal policy π (satisfying (19.26)–(19.29)), and using the special relationship between the state transitions and the rewards ((19.30) and (19.31)), the accumulated term reduces to the sum $-\log Z(\beta, s_t, a_t) - \log Z^\pi(s_{t+1}, \beta)$, i.e. the local free energy purely of the current step which itself consists of the environmental and the agent component.

In the particular case of perfectly adapted environments, all the future information is indeed valuable. In other words, minimizing the statistical surprise or maximizing the predictive information is equivalent to maximizing the reward. This can be interpreted as a generalization of the classical Kelly gambling scenario Kelly (1956). Note that this is not the case for general reward structures $\mathbf{R}_{s,a}^{s'}$. In view of the hypothesized central role of information for biological systems, it will be of significant interest for future research to establish to which extent the environments of living organisms are indeed perfectly adapted.

19.8 Experiments and Discussion

19.8.1 Information-Value Trade-Off in a Maze

The recursion developed in Sect. 19.7.1 can be applied to various scenarios, of which we study one specific, but instructive case. We consider a simple maze (inset of Fig. 19.1) where an agent starts out at the bright spot in the lower left corner of

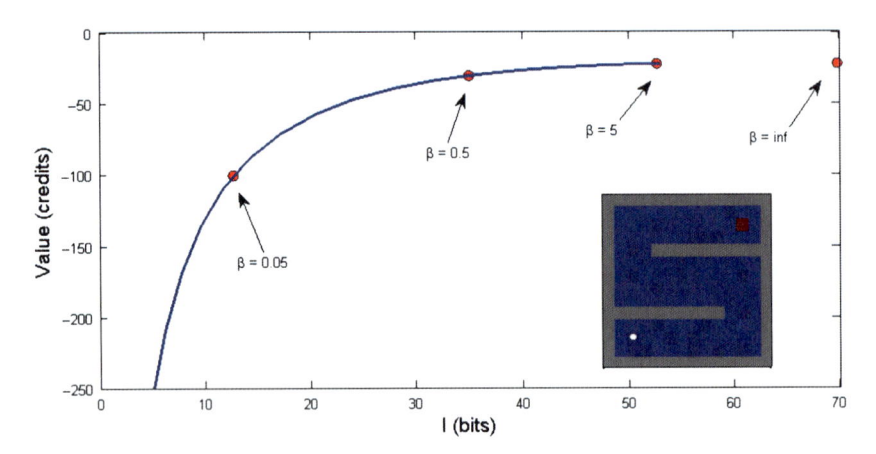

Fig. 19.1 Trade-off curve between value-to-go and information-to-go. This is in full analogy to the rate-distortion plots, if we consider (negative) distortion replaced by value-to-go and rate by information-to-go

the grid world and needs to reach the target in the right upper corner, marked by the red dot. The task is modeled through a usual reinforcement learning reward, by giving each step a "reward" (i.e. a penalty) of -1 until the target is reached. The target cell is an absorbing state, and once the agent reaches it, any subsequent step receives a 0 reward, realizing an episodic task in the non-episodic framework of the Bellman-recursion from Sect. 19.6.

Figure 19.1 shows how as one permits increasing amounts of information-to-go, the future expected cumulated reward achieved also increases (it is the negative value of the length of the route – i.e. as one is ready to invest more information bandwidth, one can shorten the route). Note that when β vanishes, this attempts to save on information-to-go while being indifferent to the achievement of a high value-to-go. As opposed to that, letting $\beta \to \infty$ aims for a policy that is indeed optimal in its value-to-go.

Now, in the case of $\beta \to \infty$, similar to Polani et al. (2006), the informational Bellman recursion will find a policy which is optimal for the reinforcement learning task. However, unlike the conventional policy or value iteration algorithms, the algorithm will not be "satisfied" with a value-optimal solution, but select a policy among the optimal policies which at the same time minimizes the information-to-go.

19.8.2 Soft vs. Sharp Policies

Figure 19.2 shows actual policies resulting for various values of β. For small β, the policy is almost a random walk. Such a walk will ultimately end up in the (absorbing) goal state, at little informational cost, but at quite negative reward values, since it takes a long time to find the goal.

As one increases β and thus increases the available information capacity, sharper, more refined and accurate policies emerge. Note that, in general, the policies we obtain by the informational trade-off algorithm from Sects. 19.6 and 19.7 will be soft policies for finite β, and an agent following them will produce state trajectories which increasingly expand and branch out over time which will typically reduce the future expected cumulated reward. This may allow additional scenarios to those mentioned in Footnote 1 (Sect. 19.4.2) to exhibit converging rewards without having

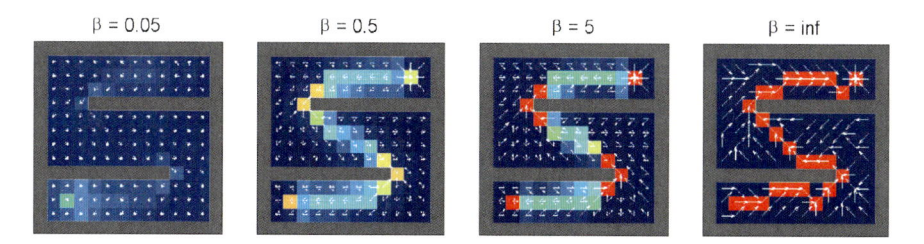

Fig. 19.2 Policies resulting from the trade-off between value-to-go and information-to-go

to use the usual discount factor, as long as β is finite. In these cases, if the cumulated rewards diverge for $\beta \to \infty$ (zero temperature, i.e. optimal policy), this would only constitute a "pathological" boundary case.

Under the PAC-Bayes perspective McAllester (1999), our free energy is composed of the cumulated Kullback–Leibler distances between posterior and prior distribution of S_{t+1} and A_{t+1} (see (19.21) and (19.22)). This gives rise to another interesting interpretation of the soft policies obtained by the above formalism: namely, the policies minimizing the respective Kullback–Leibler expressions in these equations provide a bound on the variation of the accumulated reward over different episodes of the agent's run; in fact, those policies provide *stable* results in the sense that the bound on variations from run to run is the tightestRubin et al. (2010).

The soft policies we obtained above are similar to the "softened paths" suggested in Saerens et al. (2009), derived from entropic quantities used as regularization term for the reinforcement learning task. In this chapter, however, as in Polani et al. (2006), we use Shannon information not just as a regularization quantity, but with a specific interpretation as an information processing cost: the minimal informational cost of the decision process that the agent (or, in the present paper, the agent–environment) system has to undergo to achieve a particular reward.

In the light of this interpretation, the study of information has immediate repercussions for the biological picture of information processing, i.e. the "information metabolism" of organisms. If one adopts the view that organisms tend to implement an information parsimony principle (Laughlin 2001; Polani 2009), then this implies that biological systems will exhibit a tendency to achieve a given level of performance at the lowest informational cost possible (or perform as well as possible under a given informational bandwidth). In our formalism, this would correspond to operating close to the optimal reward/information (strictly spoken, decision complexity) trade-off curve, always assuming that a suitable reward function can be formulated (Taylor et al. 2007; Bialek et al. 2007).

In this chapter, we demonstrated how the trade-off curve between value-to-go and information-to-go can be computed for the agent–environment system over whole sequence histories using a Bellman-type recursive backup rule. In the future, we will apply these techniques introduced here to other variants of the problem. One is the calculation of the decision complexity (i.e. the relevant information) only, the minimal amount of information that needs to acquired and processed by the agent itself, but not by the environment, to achieve a certain reward. In Polani et al. (2006), the relevant information was computed only for a single-step action sequence. With the Information-Bellman backup rule introduced here, we will be able in the future to generalize the relevant information calculation to multi-step action sequences. To quantify parsimonious information acquisition in the multi-step case, we will use Massey's concept of *directed information* Massey (1990).

At this point, some comments are in place concerning the Markovicity of the world state in our models. The Markovicity condition seems, at first sight, a comparatively strong assumption which might seem to limit the applicability of the formalism for modeling the subjective knowledge of an organism or agent. However, note that, while we compute various quantities from a "eagle's eye perspective" under knowledge of the full state, in the model the agent itself is not assumed to have full access to the state. Rather, the information bandwidth but not its precise form is constrained in this chapter. Finally, using the full formalism from (19.11), more complex structural constraints on the information acquisition can easily be incorporated in the form of sensors.

Let us here emphasize another final point: the presented perception–action cycle formalism implements an information-theoretic analogy for the classical treatment of optimal control problems. However, as we propose to consider information not merely as an auxiliary quantity, but in fact as a "first class" quantity in its own right, the present treatment aims to go beyond just an equivalent restatement of stochastic optimal control, rather, to provide a conceptually enriched framework, in which the informational view gives rise to a refined set of notions, insights, tools and, ultimately, research questions.

19.9 Conclusions

In this chapter, we have treated the reward-driven decision process in the perception–action cycle of an agent in a consistently information-theoretic framework. This was motivated by increasing biological evidence for the importance of (Shannon) information as resource and by the universality that the language of information is able to provide.

We consider a particular incarnation of this problem, namely an agent situated in an MDP defining a concrete task; this task is encoded as a cumulated reward which the agent needs to maximize. The information-theoretic view transforms this problem into a trade-off between the reward achieved at a given informational cost. This extends classic rate-distortion theory into the context of a full-fledged perception–action cycle. At the same time, the methodology gives a precise quantitative meaning to J.M. Fuster's above-quoted intuition about the perception–action cycle being the "circular flow of information between an organism and its environment".

The chapter shows that not only it is possible and natural to reframe the treatment of perception–action cycles in this way, but that MDP formalisms such as the Bellman recursion can be readily extended to provide a unified Blahut–Arimoto/Value Iteration hybrid that computes the quantities of interest. In this chapter, we illustrated this idea in a simple setting. More comprehensive settings which are of significant interest for both biology and artificial intelligence can be readily incorporated due to the flexibility of the formalism and will be treated in future work.

We hypothesize that the ability to trade off the value and the informational cost of whole behaviours lies at the core of any understanding of organismic behaviours. The hypothesis is that organisms attempt to realize valuable behaviours at the lowest possible informational cost, and that they will seek slightly suboptimal solutions if these solutions can be afforded at a significantly lower informational cost. Thus, the informational treatment of the perception–action cycle promises to open a quantitative and predictive path to understand the structure of behaviours and information processing in living organisms. At the same time, it can provide a systematic handle on how to develop AI systems according to principles which are both biologically plausible and relevant.

Acknowledgements The authors would like to thank Jonathan Rubin for carrying out the simulations and the preparation of the corresponding diagrams.

References

Ashby, W. R., (1956). *An Introduction to Cybernetics*. London: Chapman & Hall Ltd.

Ay, N., Bertschinger, N., Der, R., Güttler, F., and Olbrich, E., (2008). Predictive Information and Explorative Behavior of Autonomous Robots. *European Journal of Physics B*, 63:329–339.

Ay, N., and Polani, D., (2008). Information Flows in Causal Networks. *Advances in Complex Systems*, 11(1):17–41.

Ay, N., and Wennekers, T., (2003). Dynamical Properties of Strongly Interacting Markov Chains. *Neural Networks*, 16(10):1483–1497.

Berger, T., (2003). Living Information Theory – The 2002 Shannon Lecture. *IEEE Information Theory Society Newsletter*, 53(1):1,6–19.

Bialek, W., de Ruyter van Steveninck, R. R., and Tishby, N., (2007). Efficient representation as a design principle for neural coding and computation. arXiv.org:0712.4381 [q-bio.NC].

Bialek, W., Nemenman, I., and Tishby, N., (2001). Predictability, complexity and learning. *Neural Computation*, 13:2409–2463.

Brenner, N., Bialek, W., and de Ruyter van Steveninck, R., (2000). Adaptive rescaling optimizes information transmission. *Neuron*, 26:695–702.

Cover, T. M., and Thomas, J. A., (1991). *Elements of Information Theory*. New York: Wiley.

Csiszár, I., and Körner, J., (1986). *Information Theory: Coding Theorems for Discrete Memoryless Systems*. Budapest: Academiai Kiado.

Der, R., Steinmetz, U., and Pasemann, F., (1999). Homeokinesis – A new principle to back up evolution with learning. In Mohammadian, M., editor, *Computational Intelligence for Modelling, Control, and Automation*, vol. 55 of *Concurrent Systems Engineering Series*, 43–47. Amsterdam: IOS.

Ellison, C., Mahoney, J., and Crutchfield, J., (2009). Prediction, Retrodiction, and the Amount of Information Stored in the Present. *Journal of Statistical Physics*, 136(6):1005–1034.

Engel, Y., Mannor, S., and Meir, R., (2003). Bayes meets Bellman: The Gaussian Process Approach to Temporal Difference Learning. In *Proceedings of ICML 20*, 154–161.

Friston, K., (2009). The free-energy principle: a rough guide to the brain? *Trends in Cognitive Sciences*, 13(7):293–301.

Friston, K., Kilner, J., and Harrison, L., (2006). A free energy principle for the brain. *Journal of Physiology-Paris*, 100:70–87.

Fry, R. L., (2008). Computation by Neural and Cortical Systems. Presentation at the Workshop at CNS*2008, Portland, OR: Methods of Information Theory in Computational Neuroscience.

Fuster, J. M., (2001). The Prefrontal Cortex – An Update: Time Is of the Essence. *Neuron*, 30:319–333.

Fuster, J. M., (2006). The cognit: A network model of cortical representation. *International Journal of Psychophysiology*, 60(2):125–132.

Gastpar, M., Rimoldi, B., and Vetterli, M., (2003). To Code, or Not to Code: Lossy Source-Channel Communication Revisited. *IEEE Transactions on Information Theory*, 49(5):1147–1158.

Globerson, A., Stark, E., Vaadia, E., and Tishby, N., (2009). The Minimum Information principle and its application to neural code analysis. *PNAS*, 106(9):3490–3495.

Haken, H., (1983). *Advanced synergetics*. Berlin: Springer.

Howard, R. A., (1966). Information value theory. *IEEE Transactions on Systems Science and Cybernetics*, SSC-2:22–26.

Jung, T., and Polani, D., (2007). Kernelizing LSPE(λ). In *Proceedings of the 2007 IEEE International Symposium on Approximate Dynamic Programming and Reinforcement Learning, April 1–5, Hawaii*, 338–345. `

Kappen, B., Gomez, V., and Opper, M., (2009). Optimal control as a graphical model inference problem. arXiv:0901.0633v2 [cs.AI].

Kelly, J. L., (1956). A New Interpretation of Information Rate. *Bell System Technical Journal*, 35:917–926.

Klyubin, A., Polani, D., and Nehaniv, C., (2007). Representations of Space and Time in the Maximization of Information Flow in the Perception-Action Loop. *Neural Computation*, 19(9):2387–2432.

Klyubin, A. S., Polani, D., and Nehaniv, C. L., (2004). Organization of the Information Flow in the Perception-Action Loop of Evolved Agents. In *Proceedings of 2004 NASA/DoD Conference on Evolvable Hardware*, 177–180. IEEE Computer Society.

Klyubin, A. S., Polani, D., and Nehaniv, C. L., (2005a). All Else Being Equal Be Empowered. In *Advances in Artificial Life, European Conference on Artificial Life (ECAL 2005)*, vol. 3630 of *LNAI*, 744–753. Berlin: Springer.

Klyubin, A. S., Polani, D., and Nehaniv, C. L., (2005b). Empowerment: A Universal Agent-Centric Measure of Control. In *Proceedings of the IEEE Congress on Evolutionary Computation, 2–5 September 2005, Edinburgh, Scotland (CEC 2005)*, 128–135. IEEE.

Klyubin, A. S., Polani, D., and Nehaniv, C. L., (2008). Keep Your Options Open: An Information-Based Driving Principle for Sensorimotor Systems. *PLoS ONE*, 3(12):e4018. **URL:** *http://dx.doi.org/10.1371/journal.pone.0004018*, Dec 2008.

Laughlin, S. B., (2001). Energy as a constraint on the coding and processing of sensory information. *Current Opinion in Neurobiology*, 11:475–480.

Lizier, J., Prokopenko, M., and Zomaya, A., (2007). Detecting non-trivial computation in complex dynamics. In Almeida e Costa, F., Rocha, L. M., Costa, E., Harvey, I., and Coutinho, A., editors, *Advances in Artificial Life (Proceedings of the ECAL 2007, Lisbon)*, vol. 4648 of *LNCS*, 895–904. Berlin: Springer.

Lungarella, M., and Sporns, O., (2005). Information Self-Structuring: Key Principle for Learning and Development. In *Proceedings of 4th IEEE International Conference on Development and Learning*, 25–30. IEEE.

Lungarella, M., and Sporns, O., (2006). Mapping Information Flow in Sensorimotor Networks. *PLoS Computational Biology*, 2(10):e144.

Massey, J., (1990). Causality, feedback and directed information. In *Proceedings of the International Symposium on Information Theory and its Applications (ISITA-90)*, 303–305.

McAllester, D. A., (1999). PAC-Bayesian model averaging. In *Proceedings of the Twelfth Annual Conference on Computational Learning Theory, Santa Cruz, CA*, 164–170. New York: ACM.

Pearl, J., (2000). *Causality: Models, Reasoning and Inference*. Cambridge, UK: Cambridge University Press.

Pfeifer, R., and Bongard, J., (2007). *How the Body Shapes the Way We think: A New View of Intelligence*. Bradford Books.

Polani, D., (2009). Information: Currency of Life?. *HFSP Journal*, 3(5):307–316. **URL:** *http://link.aip.org/link/?HFS/3/307/1*, Nov 2009.

Polani, D., Martinetz, T., and Kim, J., (2001). An Information-Theoretic Approach for the Quantification of Relevance. In Kelemen, J., and Sosik, P., editors, *Advances in Artificial Life (Proceedings of the 6th European Conference on Artificial Life)*, vol. 2159 of *LNAI*, 704–713. Berlin: Springer.

Polani, D., Nehaniv, C., Martinetz, T., and Kim, J. T., (2006). Relevant Information in Optimized Persistence vs. Progeny Strategies. In Rocha, L. M., Bedau, M., Floreano, D., Goldstone, R., Vespignani, A., and Yaeger, L., editors, *Proceedings of Artificial Life X*, 337–343.

Prokopenko, M., Gerasimov, V., and Tanev, I., (2006). Evolving Spatiotemporal Coordination in a Modular Robotic System. In Nolfi, S., Baldassarre, G., Calabretta, R., Hallam, J. C. T., Marocco, D., Meyer, J.-A., Miglino, O., and Parisi, D., editors, *From Animals to Animats 9: 9th International Conference on the Simulation of Adaptive Behavior (SAB 2006), Rome, Italy,* vol. 4095 of *Lecture Notes in Computer Science,* 558–569. Berlin: Springer.

Rubin, J., Shamir, O., and Tishby, N., (2010). A PAC-Bayesian Analysis of Reinforcement Learning. In Proceedings of AISTAT 2010.

Saerens, M., Achbany, Y., Fuss, F., and Yen, L., (2009). Randomized Shortest-Path Problems: Two Related Models. *Neural Computation,* 21:2363–2404.

Seldin, Y., and Tishby, N., (2009). PAC-Bayesian Generalization Bound for Density Estimation with Application to Co-clustering. In *Proceedings of the 12th International Conference on Artificial Intelligence and Statistics (AIStats 2009),* vol. 5 of *JMLR Workshop and Conference Proceedings.*

Shalizi, C. R., and Crutchfield, J. P., (2002). Information Bottlenecks, Causal States, and Statistical Relevance Bases: How to Represent Relevant Information in Memoryless Transduction. *Advances in Complex Systems,* 5:1–5.

Shannon, C. E., (1949). The Mathematical Theory of Communication. In Shannon, C. E., and Weaver, W., editors, *The Mathematical Theory of Communication.* Urbana: The University of Illinois Press.

Slonim, N., Friedman, N., and Tishby, N., (2006). Multivariate Information Bottleneck. *Neural Computation,* 18(8):1739–1789.

Sporns, O., and Lungarella, M., (2006). Evolving coordinated behavior by maximizing information structure. In Rocha, L. M., Bedau, M., Floreano, D., Goldstone, R., Vespignani, A., and Yaeger, L., editors, *Proceedings of Artificial Life X,* 323–329.

Still, S., (2009). Information-theoretic approach to interactive learning. *EPL (Europhysics Letters),* 85(2):28005–28010.

Strens, M., (2000). A Bayesian Framework for Reinforcement Learning. In Langley, P., editor, *Proceedings of the 17th International Conference on Machine Learning (ICML 2000), Stanford University, Stanford, CA, USA, June 29 – July 2, 2000.* Morgan Kaufmann.

Sutton, R. S., and Barto, A. G., (1998). *Reinforcement Learning.* Cambridge, Mass.: MIT.

Taylor, S. F., Tishby, N., and Bialek, W., (2007). Information and Fitness. arXiv.org:0712.4382 [q-bio.PE].

Tishby, N., Pereira, F. C., and Bialek, W., (1999). The Information Bottleneck Method. In *Proceedings of the 37th Annual Allerton Conference on Communication, Control and Computing, Illinois.* Urbana-Champaign.

Todorov, E., (2009). Efficient computation of optimal actions. *PNAS,* 106(28):11478–11483.

Touchette, H., and Lloyd, S., (2000). Information-Theoretic Limits of Control. *Physical Review Letters,* 84:1156.

Touchette, H., and Lloyd, S., (2004). Information-theoretic approach to the study of control systems. *Physica A,* 331:140–172.

van Dijk, S. G., Polani, D., and Nehaniv, C. L., (2009). Hierarchical Behaviours: Getting the Most Bang for your Bit. In Kampis, G., and Szathmáry, E., editors, *Proceedings of the European Conference on Artificial Life 2009, Budapest.* Springer.

Vergassola, M., Villermaux, E., and Shraiman, B. I., (2007). 'Infotaxis' as a strategy for searching without gradients. *Nature,* 445:406–409.

Wennekers, T., and Ay, N., (2005). Finite State Automata Resulting From Temporal Information Maximization. *Neural Computation,* 17(10):2258–2290.

Chapter 20
Artificial Consciousness

Antonio Chella and Riccardo Manzotti

Abstract "Artificial" or "machine" consciousness is the attempt to model and implement aspects of human cognition that are identified with the elusive and controversial phenomenon of consciousness. The chapter reviews the main trends and goals of artificial consciousness research, as environmental coupling, autonomy and resilience, phenomenal experience, semantics or intentionality of the first and second type, information integration, attention. The chapter also proposes a design for a general "consciousness oriented" architecture that addresses many of the discussed research goals. Comparisons with competing approaches are then presented.

20.1 Introduction

Artificial consciousness, sometimes labeled as *machine consciousness*, is the attempt to model and implement those aspects of human cognition which are identified with the often elusive and controversial phenomenon of consciousness (Aleksander 2008; Chella and Manzotti 2009). It does not necessarily try to reproduce human consciousness as such, insofar as human consciousness could be unique due to a complex series of cultural, social, and biological conditions. However, many authors have suggested one or more specific aspects and functions of consciousness that could, at least in principle, be replicated in a machine.

At the beginning of the information era (in the 1950s), there was no clear-cut separation between intelligence and consciousness. Both were considered vaguely overlapping terms referring to what the mind was capable of. For instance, the famous Turing Test was formulated in such a way so as to avoid any commitment about any distinction between a human-like intelligence machine and a human-like conscious machine. As a result, the main counterargument to the Turing Test was raised by a philosopher of the mind (Searle 1980). There was no boundary between

A. Chella (✉)
Department of Computer Engineering, University of Palermo,
Viale delle Scienze – Building 6, 90128 Palermo, Italy
e-mail: chella@unipa.it

V. Cutsuridis et al. (eds.), *Perception-Action Cycle: Models, Architectures,*
and Hardware, Springer Series in Cognitive and Neural Systems 1,
DOI 10.1007/978-1-4419-1452-1_20, © Springer Science+Business Media, LLC 2011

intelligence and consciousness. Similarly, most of the cybernetic theory explicitly dealt with the mind as a whole. It is not by chance that the first mention of the term artificial consciousness appears in those years in to a book of cybernetics by Tihamér Nemes, *Kibernetikai gépek* (Nemes 1962), later translated in English (Nemes 1969).

In the following years, in the aftermath of the cybernetic decline, the idea of designing a conscious machine was seldom mentioned because the very notion of consciousness was considered highly suspicious. The reasons for this long lasting scientific banishment are articulated at length in many excellent books (Searle 1992; Chalmers 1996). However, since the beginning of the 1990s, a new scientific interest for consciousness arose (Crick 1994; Hameroff et al. 1996; Miller 2005) leading to current widespread approaches in neuroscience (Jennings 2000; Koch 2004; Adami 2006). Such increased acceptance of the topic allowed many researches in Robotics and AI to reconsider the possibility of modeling and implementing a conscious machine (Buttazzo 2001; Holland 2003, 2004; Adami 2006; Chella and Manzotti 2007; Aleksander 2008; Aleksander et al. 2008; Buttazzo 2008; Chrisley 2008; Manzotti and Tagliasco 2008; Chella and Manzotti 2009).

Before outlining the details of some critical yet promising aspects of artificial consciousness, a preliminary caveat is useful. As we have mentioned, artificial consciousness assumes that there is some aspect of the mind (no ontological commitments here, we could have used the word *cognition* had it not been associated with a conscious-hostile view of the mind) that has not yet been adequately addressed. Therefore, scholars in the field of artificial consciousness suspect that there could be something more going on in the mind than what is currently under scrutiny in field such as artificial intelligence, cognitive science, and computer science. Of course, most AI researchers would agree that there is still a lot of work to do: better algorithms, more data, more complex, and faster learning structures. However, it could be doubted whether this improvement in AI would ever lead to an artificial agent equivalent to a biological mind or it would rather miss some necessary aspect. In the field of artificial consciousness, scholars suspect that AI missed something important.

There are two main reasons that support this suspicion and that encourage an upsurge of interest in artificial consciousness: (1) the gap between artificial and biological agents; (2) the unsatisfactory explanatory power of cognitive science as to certain aspects of the mind such as phenomenal experience and intentionality.

The former problem is how to bridge the still huge chasm dividing biological intelligent and conscious agents from artificial ones – most notably in terms of autonomy, semantic capabilities, intentionality, self-motivations, resilience, and information integration. These are the main problems still waiting to have a solution in engineering terms and it is, at the same time, encouraging and disparaging that conscious agents seem to deal so seamlessly with them.

The latter problem is more theoretical, but endorses many practical issues as well. Not only artificial agents are a poor replica of biological intelligent agents, but – more worryingly – the models derived from artificial intelligence and cognitive science did not succeed in explaining the human and the animal mind. What is the missing ingredient? Could it be either consciousness or something closely linked to it?

For instance, why are not we able to design a working semantic search engine? Yes, we are aware that there is a great deal of interest as to the semantic web and other related project that herald their grasp of semantic aspects of information. Unfortunately, most of the work in the area, apart from its technical brilliance, does not uncover a lot of ground as to what semantics is. For instance, most of semantic engines are simply multiple layered syntactic engines. Instead of storing just the word "Obama," you store also other tags like "president," "USA," and such. While this could provide the impression that the system knows something about who Obama is, there is no semantics inside the system as to what such strings mean.

It is not a secret that the problem of consciousness is hindered by many deep scientific and philosophical conundrums. It is not altogether obvious whether anything like artificial consciousness is possible. After all there could be some physical constraints unbeknownst to us that would prevent a machine without an evolutionary history and a DNA-based constitution to ever express consciousness. Yet, for lack of a strong reason to believe in the a priori impossibility of artificial consciousness, it seems more fruitful to approach it head on.

Before getting into the details of the various aspects of artificial consciousness, it is useful to tackle with a few rather broad question that need to be answered, at least temporarily, in order to be able to flesh out the general structure of the problem.

The first question to consider is whether consciousness is real or not. For many years, the problem has been either simply dismissed or declared to be ill-conceived. Such approach is no longer acceptable mainly because of the wide acceptance that consciousness studies had in the neurosciences (Atkinson et al. 2000; Jennings 2000; Crick and Koch 2003; Miller 2005).

The second question is whether there is any theoretical constraint preventing humans to build a device with a mind comparable to that of a human being. Once more we are confronted more with our prejudices than with either any real empirical evidence or theoretical reason. In 2001, at the Cold Spring Harbour Laboratories (CSHL), a distinguished set of scholars in neuroscience and cognitive science answered negatively to this question by agreeing that "There is no known law of nature that forbids the existence of subjective feelings in artifacts designed or evolved by humans" (quoted in Aleksander 2008) – thereby opening the road to serious attempts at designing and implementing a conscious machine. After all, our brain and body allow the occurrence of consciousness (we have been careful not to say that the brain creates consciousness, we do not want to fall in the mereological fallacy, see Bennett and Hacker 2003).

The third question addresses the issue of the physical or functional nature of consciousness. In other words, is consciousness the result of a particular functional organization or does it emerge out of specific physical phenomena? It could be that the principle of multiple realizability simply does not hold for consciousness: a specific physical process could be required. As simulated water is not wet, a functionally equivalent agent could be consciousness-free whether realized without certain critical components. However, there is no evidence as to what such critical phenomena could be. Up to now all suggested physical phenomena (e.g., the infamous microtubule) did not live up to empirical investigation. On the other hand, if consciousness stems out of functional structures what they could be.

The fourth and final issue is a more specific version of the previous one – if consciousness emerges out of functional states, is it a kind of computation or something else? Is consciousness an algorithmic phenomenon of computational nature? After all, since David Marr, the dominant view is that cognition is computation, if not, what else? It is a view that has been sometimes labeled as symbolic *computationalism*. Famously, Newell stated that "...although a small chance exists that we will see a new paradigm emerge for mind, it seems unlikely to me. Basically, there do not seem to be any viable alternatives [to computationalism]" (Newell 1990, p. 5). Although many argued strongly against this view, there is no consensus as to what the next step could be (Bringsjord 1994; Chrisley 1994; Harnad 1994; Van Gelder 1995): Could artificial consciousness bridge the gap? Time will tell.

Luckily, artificial consciousness does not need to be committed to a particular theoretical view since it can appeal to a real implementation that would hopefully overcome any theoretical limits. This is the invoked strength of an engineering discipline that does not always need to be constrained by our epistemic limits (Tagliasco 2007). So basically, a researcher in the field of artificial consciousness should answer optimistically to all of the previous three questions: (1) consciousness is a real physical phenomenon; (2) consciousness could be replicated by an artificial system designed and implemented by humans; and (3) consciousness is either a computational phenomenon or something more, either way it can be both understood and replicated.

20.2   Goals of Artificial Consciousness

Here we would not outline an implementable model of consciousness since, frankly, such a model is still to be conceived. Besides there are strong doubts as to what be necessary for consciousness. Recently, Giulio Tononi and Cristof Koch argued that consciousness does not require many of the skills that roboticists are trying to implement: "Remarkably, consciousness does not seem to require many of the things we associate most deeply with being human emotions, memory, self-reflection, language, sensing the world, and acting in it." (Koch and Tononi 2008, p. 50). Rather we will focus on what an artificial conscious machine should achieve. Too often cognitive scientists, roboticists, and AI researchers present their architecture labeling their boxes with intriguing and suggestive names: "emotional module," "memory," "pain center," "neural network," and so on. Unfortunately, labels on boxes in a architecture model constitute empirical and theoretical claims that must be justified elsewhere – at best, they are "explanatory debts that have yet to be discharged" (Dennett 1978).

Roughly speaking, machine consciousness lies in the middle between the two extremes of biological chauvinism (only brains are conscious) and liberal functionalism (any functional systems behaviorally equivalent is conscious). Its proponents maintain that biological chauvinism could be too narrow and yet they concede that some kind of physical constraints could be unavoidable (no multiple realizability).

Recently, many authors emphasized the alleged behavioral role of consciousness (Baars 1998; Aleksander and Dunmall 2003; Sanz 2005; Shanahan 2005) in an attempt to avoid the problem of phenomenal experiences. Owen Holland suggested that it is possible to distinguish Weak Artificial Consciousness from Strong Artificial Consciousness (Holland 2003). The former approach deals with agents which behave as if they were conscious, at least in some respects. Such view does not need any commitment to the hard problem of consciousness thereby suggesting a somehow smoother path to the final target (Seth 2009). On the contrary, strong artificial consciousness deals squarely with the possibility of designing and implementing agents capable of real conscious feelings.

Although the distinction between weak and strong artificial consciousness could set a useful temporary working ground, it could also suggests a misleading view. Setting aside the crucial feature of the human mind – namely phenomenal consciousness – could miss something indispensable for the understanding of the cognitive structure of a conscious machine. Skipping the so-called hard problem could not be a viable option in the business of making conscious machines.

The distinction between weak and strong artificial consciousness is questionable since it should be matched by a mirror dichotomy between true conscious agents and "as if" conscious agents. Yet, human beings are conscious, and there is evidence that most animals exhibiting behavioral signs of consciousness are phenomenally conscious. It is a fact that human beings have phenomenal consciousness. They have phenomenal experiences of pains, pleasures, colors, shapes, sounds, and many more other phenomena. They feel emotions, feelings of various sort, bodily, and visceral sensations. Arguably, they also have phenomenal experiences of thoughts and of some cognitive processes. Finally, they experience being a self with a certain degree of unity. Human consciousness entails phenomenal consciousness at all levels. In sum, it would be very bizarre whether natural selection had gone at such great length to provide us with consciousness if there was a way to get all the advantages of a conscious being without it actually being phenomenally so. Could we really hope to be smarter than natural selection in this respect, sidestepping the issue of phenomenal consciousness? Thus we cannot but wonder whether it could be possible to design a conscious machine without dealing squarely with the hard problem of phenomenal consciousness. If natural selection went for it, we strongly doubt that engineers could avoid doing the same. Hence it is possible that the dichotomy between phenomenal and access consciousness – and symmetrically the separation between weak and strong artificial consciousness – is eventually fictitious.

While some authors adopted an open approach that does not rule out the possibility of actual phenomenal states in current or future artificial agents (Chella and Manzotti 2007; Aleksander et al. 2008), other authors (Manzotti 2007; Koch and Tononi 2008) maintained that a conscious machine is necessarily a phenomenally conscious machine. Yet, whether having phenomenal consciousness is a requisite or an effect of a unique kind of cognitive architecture is still a shot in the dark.

On the basis of artificial consciousness, a research working program can be outlined. One way to do it is focusing on the main goals that artificial consciousness should achieve: autonomy and resilience, information integration, semantic capabilities, intentionality, and self-motivations. There could be some skepticism as to the criteria to select such goals. There are two main criteria: one is contingent and the other is more theoretical: (1) the relevance in the artificial consciousness literature; (2) the incapability of traditional cognitive studies to outline a convincing solution.

It is significant that artificial consciousness is not the only game in town challenging the same issues. Other recent approaches were suggested in order to overcome more or less the same problems. An incomplete list includes topics like artificial life, situated cognition, dynamic systems, quantum computation, epigenetic robotics, and many others. It is highly probable that there is a grain of truth in each of these proposals, and it is also meaningful that they share most of their ends if not their means.

Let us consider the main goal of artificial consciousness one by one starting with a rather overarching issue: does consciousness requires a tight coupling with the environment or is it an internal feature of a cognitive system?

20.2.1 Environment Coupling

Under the label of environment coupling we refer to a supposedly common feature of conscious agents: they act in a world they refer to. Their goals are related to facts of the world and their mental states refer to world events. It is perhaps paradoxical that the modern notion of the mind, originated with the metaphysically unextended and immaterial Cartesian soul, is now a fully embodied situated goal-oriented concept.

As we will see, there are various degrees of environment coupling, at least in the theoretical landscape: from embodied cognition to situations, from semantic externalism to radical externalism.

Yet – as much reasonable as the idea that the mind is the result of a causal coupling with the world may seem – it was not the most popular neither for the layman nor for the scientist. The most common doctrine as to the nature of mental phenomena is that consciousness stems out of the brain as an intrinsic and internal property of it. A few years ago, John Searle unabashedly wrote that "Mental phenomena are caused by neurophysiological processes in the brain and are themselves features of the brain" (Searle 1992, p. 1). Some years later, on *Nature*, George Miller stated that "Different aspects of consciousness are probably generated in different brain regions." (Miller 2005, p. 79). Others added that if the problem of consciousness had to shift from a philosophical question to a scientific one resurrecting "a field that is plagued by more philosophical than scientifically sound controversies" (Changeux 2004, p. 603), the only conceivable option is to look inside the nervous system – as if physical reality were restricted to neural activity alone. However, such premises are still empirically undemonstrated and could get overturned by future experimental results. Koch provides a clear statement summing up the core of these widespread beliefs (Koch 2004, p. 16, italics in the original): "The goal is to discover *the minimal set of neuronal events and mechanisms jointly sufficient for a specific conscious percept.*"

This goal is based on the premise that there must be a set of neural events *sufficient* for a specific conscious percept (Crick and Koch 1990). Yet, such a premise has never been empirically demonstrated so far. Just to be clear, what is at stake is not whether neural activity is necessary but whether neural activity is either *sufficient for or identical with* phenomenal experience. Is consciousness produced *inside* the brain?

By and large, this view does not conflict with the obvious fact that human brains need to develop in a real environment and are the result of their individual history. The historical dependency on development holds for most biological subsystems: for instance, muscles and bones need gravity and exercise in order to develop properly. But, once developed, they are sufficient to deliver their output, so to speak. They need gravity and weight in order to develop, but when ready, muscles are sufficient to produce a variable strength as a result of the contraction of myofibrils. Alternatively, consider the immune system. It needs a contact with the *Varicella Zoster* virus (chicken pox) in order to develop the corresponding antigens. Yet, subsequently, the immune system is sufficient to produce such an output. In short, historical dependence during development is compatible with sufficiency once the system is developed. In the case of consciousness, for most neuroscientists, the environment is necessary for development of neural areas, but it is not constitutive of conscious experience when it occurs in a grown-up normally developed human being. Many believe that consciousness is produced by the nervous system like strength is produced by muscles – the neural activity and consciousness having the same relation as myofibrils and strength.

What has to be stressed is that, although many scientists boldly claim that there is plenty of evidence showing that "the entire brain is clearly sufficient to give rise to consciousness" (Koch 2004, p. 16), actually there is none. The "central plank of modern materialism – the supposition that consciousness supervenes on the brain" (Prinz 2000, p. 425) is surprisingly poorly supported by experimental evidence, up to now.

If the thesis that consciousness is only the result of what is going on inside the brain (or a computer), we must then explain how the external world can contribute to it and thus adopt a more ecological oriented view. The most notable example is offered either by situated or embodied cognition – i.e., the idea that cognition is not only a symbolic crunching performed inside a system, but rather a complex and extended network of causal relation between the agent, its body, and its environment (Varela et al. 1991/1993; Clark 1997; Ziemke and Sharkey 2001; Pfeifer and Bongard 2006; Clark 2008). Although a view very popular, it gained its share of detractors too (for instance, Adams and Aizawa 2008, 2009). It is a view that has been developed in various degrees.

The more conservative option is perhaps simple embodiment – namely, the hypothesis that the body is considered as an integral part of the cognitive activity. Thus, cognitive activity is not constrained to the dedicated information processing hardware, but rather it comprehends the perception–action loops inside the agent body. Such an approach is considered to be advantageous since the body takes into account many physical aspects of cognition that would be difficult to emulate and simulate internally. These approaches were triggered by the seminal work of Rodney Brooks in the 1990s (Brooks 1990, 1991; Brooks et al. 1998, 1999;

Collins et al. 2001; Metta and Fitzpatrick 2003; Paul et al. 2006). However, further efforts have not provided the expected results. In other words, the morphology of the agent seems a successful cognitive structure only when the task is relatively physical and related to bodily action. In short, embodiment is great to take charge of the computational charge of walking, grasping, running, but it is unclear whether it can be useful to develop higher cognitive abilities (Prinz 2009). Sometimes situated cognition is nothing but a more emphasized version of embodiment although it usually stresses the fact that all knowledge is situated in activity bound to social, cultural, and physical contexts (Gallagher 2009).

A more literal version of environmental coupling is developed under the shorthand of externalism in its various versions (Rowlands 2003; Hurley 2006). By and large these positions assume that the perceived environment is not only necessary or useful, but literally constitutive of what the mind is. In other words, these views go beyond both functionalism and embodiment/situatedness. They maintain that the cognitive agent's boundaries have to be extended so to comprehend a relevant part of the environment. This goal can be addressed at various degrees of commitment to the general thesis. To cut a long story short, we can distinguish between cognitive externalism and phenomenal externalism. The former holds that, though cognition is extended in the sense that the agent take advantage of structures which are external to the body of the agent, phenomenal experience as such is still a property of processes occurring inside the agent – the most notable example is perhaps the extended mind model by Clark and Chalmers (1999) though others have defended similar views (Robbins and Aydede 2009). The latter view, labeled as phenomenal externalism, is more daring and takes in consideration the possibility that the vehicles of phenomenal experience are physically larger than the agent body. The conscious mind would thus literally comprehend part of the environment in a strong physical sense (Rockwell 2005; Honderich 2006; Manzotti 2006).

20.2.2   Autonomy and Resilience

A conscious agent is a highly autonomous agent. It is capable of self development, learning, and self-observation. Is the opposite true?

According to Sanz, there are three motivations to pursue artificial consciousness (Sanz 2005): (1) implementing and designing machines resembling human beings (cognitive robotics); (2) understanding the nature of consciousness (cognitive science); and (3) implementing and designing more efficient control systems. The third goal is strongly overlapped with the issue of autonomy. A conscious system is expected to be able to take choices in total autonomy as to its survival and the achievements of its goals. Many authors believe that consciousness endorses a more robust autonomy, a higher resilience, a more general problem-solving capability, reflexivity, and self-awareness.

It is unquestionable that a conscious agent seems to have a higher adaptability to unpredictable situations. This conflates against the so-called epiphenomenal view of consciousness according to which consciousness would have no real positive effects on behavior (Libet et al. 1983; Pockett 2004). Yet this view fails when considered the appearance of conscious agents as a result of natural selection – whatever consciousness is, it consumes resources, thus it must have same advantage. Since many evidence does show that most if not all repetitive mental skills can be performed more efficiently in the absence of consciousness (from playing videogames to more complex sensorimotor tasks), it makes sense to suppose that consciousness has a role in keeping together all cognitive processes so as to pursue a common goal.

A conscious agent is thus characterized by a strong autonomy that often leads also to resilience to an often huge range of disturbances and unexpected stimuli. Many authors addressed these aspects trying to focus on the importance of consciousness as a control system. Taylor stressed the relation between attention and consciousness (Taylor 2002, 2007, 2009) that will be sketched at greater length below. Sanz et al. aim to develop a full-fledged functional account of consciousness (Sanz 2005; Sanz et al. 2007; Hernandez et al. 2009). According to their view, consciousness necessarily emerges from certain, not excessively complex, circumstances in the dwelling of cognitive agents. Finally, it must be quoted Bongard who is trying to implement resilient machines able to recreate their internal model of themselves (Bongard et al. 2006). Though he does not stress the link with consciousness, it has been observed that a self-modeling artificial agent has many common traits with a self-conscious mind (Adami 2006).

20.2.3   *Phenomenal Experience*

This is the most controversial and yet the more specific aspect of consciousness. It is so difficult that in the 1990 had been labeled as the "hard problem" meaning that it entails some very troublesome and deep aspect of reality. Most authors agree that there is no way to sidestep it, and yet there does not seem to be any way to deal with it.

In short, the hard problem is the following. The scientific description of the world seems devoid of any quality of which a conscious agent (a human being) has an experience of. So, in scientific terms, a pain is nothing but a certain configuration of spikes through nerves. However, from the point of view of the agent, the pain is *felt* and not as a configuration of spikes but rather with a very excruciating and unpleasant quality. What is this quality? How it is produced? Where does it take place? Why is it there? "Consciousness is feeling, and the problem of consciousness is the problem of explaining how and why some of the functions underlying some of our performance capacities are felt rather than just 'functed'." (Harnad and Scherzer 2008). Either there is a deep mystery or our epistemic categories are fundamentally flawed. The difficulty we have in tackling with the hard problem of consciousness could indeed be the sign that we need to upgrade our scientific worldview.

Whether the mental world is a special construct concocted by some irreproducible feature of most mammals is still an open question. There is neither empirical evidence nor theoretical arguments supporting such a view. In the lack of a better theory, we wonder whether it would be wiser to take into consideration the rather surprising idea that the physical world comprehends also those features that we usually attribute to the mental domain (Skrbina 2009).

In the case of machines, how is it possible to take over the so-called *functing vs. feeling* divide (Lycan 1981; Harnad and Scherzer 2008)? As far as we know, a machine is nothing but a collection of interconnected modules functioning in a certain way. Why the functional activity of a machine should transfigure in the feeling of a conscious experience? However, the same question could be asked about the activity of neurons. Each neuron, taken by itself, does not score a lot better than a software module or a silicon chip as to the emergence of feelings. Nevertheless, we must admit that we could discount a too simplistic view of the physical world.

It is thus possible that a successful approach to the phenomenal aspect of artificial consciousness will not only be a model of a machine, but rather a more general and overarching theory that will deal with the structure of reality.

Given the difficulties of this problem, it is perhaps surprising that many scholars tried to tackle it. There are two approaches, apparently very different: the first approach tries to mimic the functional structure of a phenomenal space (usually vision). The advantage is that it is possible to build robots that exploit the phenomenal space of human beings. Although there in neither explicit nor implicit solution of the hard problem, these attempts highlight that many so-called ineffable features of phenomenal experiences are nothing but functional properties of perceptual space. This is not to say that phenomenal qualities are reducible to functional properties of perceptual spaces. However, these approaches could help to narrow down the difference. For instance, Chrisley is heralding the notion of synthetic phenomenology as an attempt "either to characterize the phenomenal states possessed, or modeled by, an artifact (such as a robot); or 2) any attempt to use an artifact to help specify phenomenal states" (Chrisley 2009, p. 53). Admittedly, Chrisley does not challenge the hard problem. Rather his theory focuses on the sensorimotor structure of phenomenology. Not so differently, Igor Alexander defended various versions of depictive phenomenology (Aleksander and Dunmall 2003; Aleksander and Morton 2007) that suggest the possibility to tackle from a functional point of view the space of qualia.

Another interesting and related approach is that pursued by one of the authors (Chella) who developed a series of robots aiming to exploit sensorimotor contingencies and externalist inspired frameworks (Chella et al. 2001, 2008). An interesting architectural feature is the implementation of a generalized closed loop based on the perceptual space as a whole. In other words, in classic feedback only a few parameters are used to control robot behavior (position, speed, etc.). The idea behind the robot is to match a global prediction of the future perceptual state (for instance by a rendering of the visual image) with the incoming data. The goal is to achieve a tight coupling between robot and environment. According to these models and implementations, the physical correlate of robot phenomenology would not lie in the images internally generated but rather in the causal processes engaged between the robot and the environment (Chella and Manzotti 2009).

Not all authors would consider such approaches satisfying. Most philosophers would argue that unless there is a way to naturalize phenomenal experience it is useless to try to implement it. Although it is a feasible view, it is perhaps worth considering that phenomenal experience in human being is a result of natural selection and that, as such, was not selected as the answer to a theory, but as a solution to one or more practical problems. In this sense, keep trying to model phenomenal experience in robots, albeit with all well-known limitations, may unwittingly help us to understand what phenomenal experience is.

20.2.4 Semantics or Intentionality of the First Type

Semantics is the holy grail of AI symbolic manipulation. It is the missing ingredient of AI. Without semantics it is even doubtful, whether it is meaningful talking of symbols. For, what is a symbol without a meaning? A gear is not a symbol, although is part of a causal network and although it has causal properties that could be expressed in syntactical terms. In fact, all implementations of symbols are causal networks isomorphic to a syntactical space.

Although, since the 1950s, AI and computer science have tried to ground semantics in syntax, it seems conceivable that it ought to be the other way round: syntax could be grounded on semantics.

As we have mentioned, semantics can be seen as a synonym of intentionality of the first type – i.e., intentionality as it was defined by the philosopher Franz Brentano in his seminal work on psychology (Brentano 1874/1973). According to him, intentionality is the arrow of semantics: what links a symbol (or a mental state) to its content whether it be a physical event, a state of affair in the real world, or a mental content. It is not by chance that intentionality of this kind was suggested as the hallmark of the mental.

Brentano's suggestion triggered a long controversy that is still lasting. A famous episode what John Searle's much quoted paper on the Chinese room (Searle 1980). In that paper, he challenged the view that information, syntax, and computation endorse the meaning (the intentionality) of symbols. More recently, the problem of semantics was reframed as the "symbol grounding problem" by Harnad (Harnad 1990).

Not surprisingly, one of the main argument against machine consciousness has been the apparent lack of semantics of artificial systems whose intentionality is allegedly derived from that of their users and designers (Harnad 2003). Human beings (and perhaps animals) have intrinsic intentionality that gives meaning to their brain states. On the contrary, artificial systems like computers or stacks of cards do not have intrinsic intentionality. A variable x stored in my computer is not about my bank account balance. It is the use that I make of that variable that allows me to attribute it a meaning. However, by itself, that variable has no intrinsic intentionality, no semantics, and no meaning. Or so it goes the classic argument.

Yet, it remains vague both what intrinsic intentionality is and how a human being is able to achieve it. For lack of a better theory, we could envisage to turn the argument upside down. First, we could observe that there are physical systems (human beings) capable of intentionality of the first kind. These systems are conscious agents. Then, we could suspect that, as Brentano suggested, being conscious and having intentionality could be two aspects of the same phenomenon. If this were true, there would be only one way to achieve semantics – namely, designing a conscious agent.

20.2.5 Self-Motivations or Intentionality of the Second Type

Usually artificial agents do not develop their own motivations. Their goals and hierarchy of subgoals are carefully imposed at design time. This is desirable since the goal of artificial agents is to fulfill their designers' goals, not to wander around purchasing unbeknownst goals. Yet, conscious agents like humans seem capable to develop their own goals and, indeed, it seems a mandatory feature of a conscious being. A human being incapable of developing his/her own agenda would indeed seem curiously lacking something of important.

The capability of developing new goals is deeply intertwined with intentionality according to the Dennett's definition of it: having an intention of pursuing a certain state of affairs (Dennett 1987, 1991). A conscious human being is definitely an agent that is both capable of having intentions and capable of developing new ones. While the latter condition has been frequently implemented since Watt's governor, the latter one is still a rather unexplored feature of artificial agents. We could distinguish between teleologically fixed and teleologically open agents. A teleologically fixed agent is an agent whose goals (and possibly rules to define new ones) are fixed at design time. For instance, a robot whose goal is to pick up as many coke cans as it can is teleologically fixed – similarly, a software agent aiming at winning chess games or a controller trying to control the agent's movements in the smoothest possible way is teleologically fixed too. Learning usually does not affect goals. On the contrary, often learning is driven by existing goals.

On the contrary a human teenager is developing new kinds of goals all the time and most of them are only loosely connected with the phylogenetic instincts. Culture is a powerful example of phenomenon building on top of itself continuously swallowing new and unexpected goals. Is this a feature that is intrinsic of consciousness or it is only a contingent correlation between two otherwise separate phenomena?

It is a fact that artificial agents score very poorly in such area, and thus it is fair to suspect that there is a strong link between being conscious and developing new goals. Up to now there was a lot of interest as to how to learn achieving a goal in the best possible way, but not too much interest as to how to develop a new goal. For instance, in their seminal book on neural network learning processes, Richard S. Sutton and Andrew G. Barto stress that they design agent in order to "learn what to do – how to map situations to actions – so as to maximize a numerical reward signal [the goal] [...] All learning agents have explicit goals" (Sutton and Barto 1998, p. 3–5). In other words, learning deals with situations in which the agent seeks "how" to achieve a goal despite uncertainty about its environment. Yet the goal is fixed at design time. Nevertheless, there are many situations in which it could be extremely useful to allow the agent to look for "what" has to be achieved – namely, choosing new goals and developing corresponding new motivations. In most robots, goals are defined elsewhere at design time (McFarland and Bosser 1993; Arkin 1998), but, at least, behavior changes according to the interaction with the environment.

Interestingly enough, in recent years various researchers tried to design agents capable of developing new motivations and new goals (Manzotti and Tagliasco 2005; Bongard et al. 2006; Pfeifer et al. 2007), and their efforts were often related with machine consciousness.

It is interesting to stress the strong analogy between the two types of intentionality. Both are considered strongly correlated with conscious experience. Both are seen as intrinsic features of biological beings. Both play a role in linking mental states with external state of things. In short, it is fair to suspect that a conscious architecture could be the key to address both kind of intentionality and, indeed, it could be the case that the two types of intentionality are nothing but two aspects of consciousness.

20.2.6 Information Integration

Consciousness seems to be deeply related with the notion of unity. But what is unity in an artificial agent? The closest notion is that of information integration – namely, the necessity of unifying somewhere the many streams of otherwise separate and scattered flows of information coming from a multiplicity of heterogeneous sources.

In neuroscience, there is an analogous problem called the "binding problem" (Revonsuo and Newman 1999; Engel and Singer 2001; Bayne and Chalmers 2003). Such problem is usually called the "binding problem," and it has received no clear solution. Many binding mechanisms have been proposed ranging from temporal synchronization to hierarchical mechanism. So far, no one was satisfying.

Perhaps the most striking everyday demonstration of the binding problem is offered by our field of view. Watching at a normal scene with both eyes, we perceive a unified and continuous field of view with no gap between the left and the right side of it. And yet the visual processing is taking place in two separate areas, each in one of the two cerebral hemispheres. So to speak, if we watch someone walking from the right to the left of our field of view, the corresponding visual activities are going to shift from the left to the right hemisphere of our head. How can we explain the fact that we do not perceive any gap in the visual field, ever?

This notion is dangerously close to the easily discredited idea of an internal locus of control – an artificial version of the *homunculus* – a sort of Cartesian theatre where the information is finally presented to the light of consciousness. There are all sorts of objections to this view both from neuroscience and from AI. To start, possibly the view conflates together attention and consciousness, and this is not necessarily a good move as shown by all cases where the two takes place independently of one another (Koch and Tsuchiya 2006; Mole 2007; Kentridge et al. 2008).

So the question as to what gives unity to a collection of separate streams of information remains largely unanswered notwithstanding the everyday familiar experience of consciously perceiving the world a one big chunk of qualities. Yet what does it give unity to a collection of parts, being them events, parts, processes, computations, instructions? The ontological analysis has not gone very far (Simons 1987; Merrick 2001), and the neuroscience wonders at the mystery of neural integration (Revonsuo 1999; Hurley 2003). Machine consciousness has to face the same issue. Would it be enough to provide a robot with a series of capabilities for the emergence of a unified agent? Should we consider the necessity of a central locus of processing or the unity would stem out of some other completely unexpected aspect?

Classic theories of consciousness are often vague as to what gives unity. For instance, would the Pandemonium-like community of software demons championed by Dennett (1991) gain a unity eventually? Does a computer program have unity out of its programmer's head? Would embodiment and situatedness be helpful?

A possible and novel approach to this problem is the notion of integrated information introduced by Tononi (2004). According to him, certain ways of processing information are intrinsically integrated because they are going to be implemented in such a way that the corresponding causal processes get entangled together. Although still in its final stage, Tononi's approach could cast a new light on the notion of unity in an agent. Tononi suggested that the kind of information integration necessary to exhibit the kind of behavioral unity and autonomy of a conscious being is also associated with certain intrinsic causal and computational properties which could be responsible for having phenomenal experience (Tononi 2004).

20.2.7 Attention

If consciousness has to play a role in controlling the behavior of an agent, a mechanism which cannot be overlooked is attention control. Attention seems to play a crucial role in singling out to which part of the world to attend. However, it is yet unclear: what is the exact relation between attention and consciousness? Though it seems that there cannot be consciousness without attention (Mack and Rock 1998; Simons 2000), there is not sufficient evidence to support the thesis of the sufficiency of attention to bestow consciousness. However, implementing a model of attention is fruitful since it introduces many aspects from control theory that could help in figuring out what are the functional advantages of consciousness. This is of the utmost importance since any explanation of consciousness should be tied down to suitable functional ground truth. A satisfying attention control mechanism could satisfy many of the abovementioned goals of consciousness such as autonomy, information integration, perhaps intentionality.

A promising available model of attention is the CODAM neural network control model of consciousness whose main is to provide a functional account (Taylor and Rogers 2002; Taylor 2003, 2007). Such model has several advantages since it suggests various ways to speed up the response and the accuracy of the agent.

The model is capable of making several predictions as to how to achieve better performances in dealing with incoming stimuli. Furthermore, it explains other features shared by the brain such as the competition between bottom-up signals from unexpected and strong inputs and top-down control aimed at goals. It also provides a mechanism to justify the inhibition between incoming signals. Other aspects addressed by the model are the Posner benefit effect in vision (Taylor and Fragopanagos 2005), some evolutionary and developmental aspects of the control systems, some features of the attentional blink, and some aspects of subliminal processing (Taylor and Fragopanagos 2007).

Another advantage of the CODAM neural network control model is that it provides suggestions as to how the brain could implement it. The central idea is that the functional role of the attention copy signal is endorsed by the corollary discharge of attention movement (which is the reason of the name of the model). The possible neural basis of the CODAM has been addressed at length by Taylor (Taylor and Rogers 2002; Taylor 2000, 2003, 2007).

It is worthwhile to describe briefly the structure of the CODAM model since it shows several common features with the model that will be described in the last part of this chapter. As shown in Fig. 20.1, the input enters through an incoming module generically labeled as input and is then sent both to a goal dedicated module and to an object map. In the object module is stored a set of higher level representation of

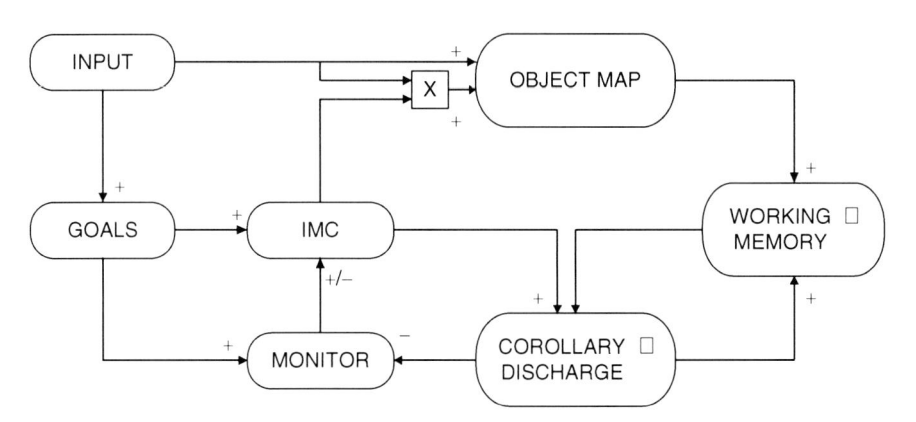

Fig. 20.1 The CODAM model architecture and the functional relations between its modules. Adapted from Taylor 2007, p. 987

previous relevant input objects. In this way, the incoming stimulus can be matched against the previously stored knowledge. At the same time, the input signal reaches the goal module, thus activating both an inverse module controller (IMC) and an internal monitor module. The role of the inverse controller is that of re-modulating the strength and thus the importance of the input signal as to the object module. On the other hand, the monitor module computes the difference between the desired goal state and the estimated state of the system.

20.3   A Consciousness-Oriented Architecture

In this paragraph, a tentative consciousness-oriented architecture is summarized. The architecture does not pretend to be either conclusive or experimentally satisfying. However, it is a cognitive architecture, so far only partially implemented (Manzotti 2003; Manzotti and Tagliasco 2005), whose design aims at coping with issues such as intentionality (of both kinds), phenomenal experience, autonomy, resilience, environment coupling, and information integration that are mandatory in order to address machine consciousness (Chella and Manzotti 2007, 2009). Or, at least, it ought to suggest how such issues are to be addressed. In many ways, it capitalizes on previous attempts and hopefully is making some predictions on what phenomenal content could be.

Before getting into the details of the suggested architecture, we would like to point out some other general principles that were used as guidelines in this work. These principles are derived both from theoretical considerations and from empirical facts about the brain.

- *Antirepresentationalist stance*. It is fair to say that the representationalist/ antirepresentationalist debate has not been solved yet. Do conscious subjects perceive the world or their representations? Are internal representations the object of our mental life or are they just epistemic tools referring to slices of the causal chain that gets us acquainted with the environment? Should we prefer indirect or direct model of perception? In AI, representations were long held to be real. However, for many reasons outlined elsewhere (Manzotti 2006), a strong antirepresentationalist stance is here adopted: representations have a fictitious existence. Like centers of mass, they do not really exist. They are concepts with a high explanatory efficacy in describing the behavior of agents. Hence, in designing a conscious-oriented architecture, we better get rid of them and conceive our architecture as a part of the environment from the very start.
- *One architectural principle "to rule them all"*. The highly complex structure of the brain cannot be completely specified by the genetic code. Indeed, it could even be undesirable, since it would reduce neural adaptability. Further, many studies showed that most areas of the brain can emulate the neural circuitry of any other one (Sur et al. 1988; Roe et al. 1993; Sharma et al. 2000; Kahn and Krubitzer 2002; Hummel et al. 2004; Ptito et al. 2005). The blue print of the brain contained inside genes seems rather coarse. Probably, the highly complex final

structure of the brain, as it has been studied in neuroscience (Krubitzer 1995), is not the result of complex design specifications but rather of a few architectural principles capable of producing the final results (Aboitz et al. 2003). In this spirit, complex models are suspicious, simple rules able to generate complexity in response to environmental challenges are to be preferred.

- *Memory vs. speed.* It is not uncommon to utilize processor capable of Giga or even Teraflops. In terms of pure speed, the sheer brute force of artificial processors is no match for brain-ware. However, a somehow underestimated feature of the brain is the apparent identity between neural structures for memory and neural structures for processing. The rigid separation between the incredibly fast processor and the sequentially accessed memory of computers has no equivalent in the brain. Thus, it could make sense to consider architectures that would trade speed for some kind of memory integration, albeit implemented on Neumannesque hardware.
- *Embodied and situated cognition.* This is an issue strongly related with our antirepresentationalist stance. Representations are a trick devised to carry a replica of the world inside the agent. It is doubtful whether it is a successful strategy. Thus, from the start, the cognitive architecture will be taken as an element of a larger causal network made of both the body and of a relevant part of the environment too.
- *Phenomenal externalism.* This is, of course, the most provocative hypothesis – namely that the vehicles of phenomenal content are to be taken as the physical processes engaged between the agent and the environment and not as the properties of the symbolic processes taking place inside the agent. Bizarre as it may seem, the presented model has a twofold advantage. On one hand, it does not dwell either on any kind of dualism or on any ontologically diaphanous "inner reality." On the other hand, it makes predictions as to when phenomenal content occurs as a result of situated cognition. After all, the model suggest that phenomenal content is nothing but certain kind of physical processes.

The above mentioned are hunches that can be derived from various evidences. Yet they are not criteria to measure the success of such an architecture, whose goal is different from executing a task as in classic AI. A good check list could be provided by the topics considered in the previous paragraphs – i.e., intentionality of both kinds (Dennett 1987; Harnad 1995), phenomenal experience (Manzotti 2006), autonomy (Clark et al. 1999; Chella et al. 2001; Di Paolo and Iizuka 2008), resilience (Bongard et al. 2006), environment coupling (Robbins and Aydede 2009), and information integration (Tononi 2004; Manzotti 2009).

The architecture presented here is based on a causal structure that can be replicated again and again at different levels of complexity. The architecture will span three levels: the unit level, the module level, and the architecture level. A rather simple structural principle should be able to generate a complete architecture managing an unlimited amount of incoming information. The generated architecture is going to use all the memory at its disposal and it should not make an explicit distinction between data and processing.

20.3.1 The Elementary Intentional Unit

The basic element of our architecture is rather simple. It is a unit receiving an input (whether it be as simple as a bit or as complex as any data structure you could envisage) and producing an output (a scalar, an integer or a logical value). From many to one, so to speak. As we will see, the unit has also a control input, but this is a detail that will be specified below.

The main goal of the unit is getting matched with an arbitrary stimulus and, after such a matching, having the task of being causally related with such original stimulus. It could be implemented in many different ways. Formally, it can be expressed by an undefined function waiting its first input before being matched to it forever. Basically, it is like having a selective gate that is going to be burned on its first input. After being "burned," the unit has a significant output only if the current input resembles the first input. If a continuous similarity function and a continuous output are used, the gate tunes the passage of information rather than blocking/allowing it, yet the general principle remains the same.

The final detail is the control input signal. Due to the irreversible nature of the matching, it could make sense to have a way to signal when the first input is received. Since the unit could be activated only at a certain moment, it is useful preserving its potential until certain conditions are obtained. Thus the need for an external control signal that will activate the unit signaling that the first input is on its way.

Due to its role rather than to its elementary behavior, we label the unit as the *intentional unit*. It is important that, up to now, the unit is not committed to any particular kind of data or internal implementation. The unit is potentially very general. It can be adapted to any kind of inputs: characters, words, numbers, vectors, and images.

A more formal description will help getting the gist of the unit (Fig. 20.2). Any kind of input domain C can be defined. The output domain is more conveniently defined as a real number ranging from 0 to 1. Finally, a similarity function has to be

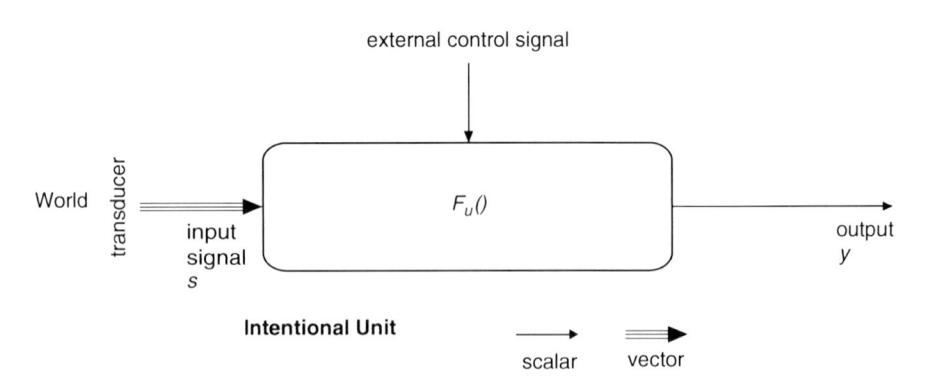

Fig. 20.2 The intentional unit

chosen – at worst, the identity function could be used. The similarity function fs : $C \times C$ a $[0, 1]$ must be such that standard conditions such as $fs(c, c) = 1, \forall c \in C$ & $fs(c_1, c_2) = fs(c_2, c_1), \forall c_1, c_2 \in C$ & $fs(c_1, c_2) \leq fs(c, c), \forall c_1, c_2, c \in C$ apply.

The similarity function is used to implement the intentional unit internal function $F_u : C$ a$[0, 1]$ – the function that will change its behavior forever after its first input. F is defined as follows:

$$F_u(s_t) = \begin{cases} 0 & t < t_0 \\ 1 & t = t_0 \\ fs_t(s_t, s_{t_0}) & t > t_0 \end{cases} .$$

It must be stressed that t_0 is an arbitrary chosen instant marked by the external control signal mentioned above. In short, the above function can be rewritten as:

$$F_u(r, s_t) = v.$$

F_u is a function waiting for something to happen before adopting its final and fixed way of working. After the input $s_{t_0} \in C$ occurring simultaneously with the first time that $r = 1$, the output $y \in [0, 1]$ becomes the output of the similarity between the incoming input $s_t \in C$ and the original input s_{t_0}. The similarity function is as simple as the identity function or as complex as the designer likes. The most important requirement is that its maximum output value must hold for any value like the first received input.

Consider a few examples. Suppose that the incoming domain C is constituted by alphabetic characters, that the similarity function $fs : C \times C$ a$[0, 1]$ is the identity function, and that the output domain is the binary set $\{0, 1\}$. The function F_u is thus complete and the intentional unit can be implemented. The function F has no predictable behavior until it receives the first input. After that it will output 1 only when a character identical to the one received at its beginning is received. To conclude the example, imagine that a possible input is the following sequence of characters: "S," "T," "R," "E," "S," "S." The output will then be 1, 0, 0, 0, 1, 1.

A simple variation on the similarity function would permit a slightly fuzzier notion of similarity: two characters are similar (output equal to 1) either if they are the same or if they are next to each other in the alphabetical order. Given such a similarity function and the same input, the output would now be 1, 1, 1, 0, 1, 1. To make things more complex, a continuous output domain such as $[0, 1]$ is admitted and the similarity function is changed to $fs (c) = 1 - (AD)/(TC)$, whereas AD is the alphabetical distance and TC is the total number of alphabetical character. With the above input, the output would then be: 1, 0.96, 0.96, 0.65, 1, 1. Clearly, everything depends on the first input. ,

A useful formalization of the unit is the one having vectors as its input domain. Given two vectors $\vec{v}, \vec{w} \in \mathfrak{R}^n$, a simple way to implement the similarity function is using a normalized version of a distance function between vectors $d (\vec{v}, \vec{w})$, $d : (\mathfrak{R}^n \times \mathfrak{R}^n) \mapsto \mathfrak{R}$. Suitable candidates are the Minkowski function or the Tanimoto distance (Duda et al. 2001). Using grey images as input vectors, elsewhere, a simple correlation function proved useful enough (Manzotti and Tagliasco 2005):

$$fs\left(\vec{v}, \vec{w}\right) = \frac{1}{2}\left[1 - C\left(\vec{v}, \vec{w}\right)\right] = \frac{1}{2}\left[1 - \frac{\sum (v_i - \mu_v)(w_i - \mu_w)}{\sqrt{\sum (v_i - \mu_v)^2 \cdot \sum (w_i - \mu_w)^2}}\right].$$

These are all a few of the examples that could be made. It is important to stress that the unit is very adaptable and open to man different domains and implementation. Furthermore, the unit has a few features that are worth being stressed.

First, the unit does not distinguish between data and processing. In some sense, it is a unit of memory since its internal function is carved on a certain input thereby keeping a trace of it. Yet, there is no explicit memory, since there is not a stored value, but rather a variation in its behavior by means of its internal function. Indeed, the function can be implemented storing somewhere the value of the first input, but there is no need to have any explicit memory.

Another interesting aspect is that the unit shows a behavior which is the result of the coupling with the environment. When the unit is "burned," it is also forever causally and historically matched to a certain aspect of the environment (the one that produced that input).

Furthermore, there is no way to predict the unit future behavior since it is the result of the contingent interaction with the environment. If the input is unknown, the unit behavior is unknown too. If the input comes from the environment, there is no way to predict the unit behavior.

Finally, the unit seems to mirror, to a certain extent, some aspect of its own environment without having to replicate it. Slightly more philosophically, the unit allows to a pattern in the environment to exist by allowing it to produce effects through itself (more detailed considerations on this issue are outlined in Manzotti 2009).

By itself the intentional unit could seem pretty useless. However, things get more interesting once a large number of them are assembled together.

20.3.2 The Intentional Module

Suppose to have the capability of implementing and packing many intentional units into the same physical or logical package. The result could be a slightly more interesting structure here labeled as *intentional module*. This module has already been put to test in a very simplified robotic setup aiming at developing new motivations and controlling the gaze of a camera toward unexpected classes of visual stimuli (Manzotti and Tagliasco 2005; Manzotti 2007).

The simplest way to step from the intentional unit to the module is to pack together a huge number of intentional units all receiving the same input source. To avoid them behaving exactly the same, some mechanisms that prevents them from being shaped by the same input at the same time must be added. There are various ways to do it. A simple way is to number them and then to enforce that the units can be burned sequentially. This could be obtained by means of the external control signal each intentional unit has. Because of its importance the external control signal is labeled *relevant signal*. The name expresses that such a signal is relevant in the life of each intentional unit since it controls to which input value the unit is matched forever.

Since the burning of each intentional unit will surely have a cost in terms of resources, it could make sense to add some more stringent conditions before switching on a relevant signal (and thus coupling forever an intentional unit to a certain input). Which conditions? To a certain extent they could be hard-wired and thus derived from some a priori knowledge the designer wants to inject into the system. Or, if the system is the result of some earlier generations of similar systems, it could be derived from the past history of previous generations. But it is definitely interesting whether the system could develop its own criteria to assign resources to further incoming inputs. By and large, the size of the input domain is surely much larger than that can be mapped by the available intentional units. In short, a feasible solution is having two explicitly divided sets of criteria working concurrently and then adding their outputs together so to have a relevant signal to be sent to the intentional units. The first set could be a set of hardwired functions trying to pin down external conditions that, for one reason or another, could be relevant for the system. The second set should be somehow derived from the growing set of matched intentional units themselves.

On the basis of what information these two sets of criteria should operate. A simple solution is on the basis of the incoming information with an important difference. The hardwired criteria could operate straight on the incoming data since they are hardwired and thus apply off-the-shelf rules. On the other hand, the derived set of criteria could use the output of the intentional units thereby using a historically selected subset of the incoming signals.

Shifting from the logical structure to the level of implementation, the above mentioned elements are packed into three submodules (Fig. 20.3). First the intentional units are packed into a huge array. Second, the hard-wired criteria are grouped together in such a way that they receive the signals jointly with the array of intentional units. Third, there is a third submodule grouping together the criteria derived by the past activity of the intentional units.

As it is clear from the structure, another small detail has been added: the module receives an external relevant signal that flanks the two internally generated ones. The reason for this will get clearer in the following paragraph.

A few more observations are needed. The module receives a vector (the incoming signal) and, possibly, an external relevant signal. It has two outputs: the internal relevant signal (the sum of hard-wired criteria, historically derived criteria, and possibly the external relevant signal) and a vector. For simplicity, all real values (both scalar signals and values of the vector elements) are normalized.

The output relevant signal is extremely important since it compresses all the past history of the system with respect to the current input signal.

The vector output is also a result both of the history and of the hard-wired criteria. Each element of this vector is the output of an intentional unit. Therefore, the output

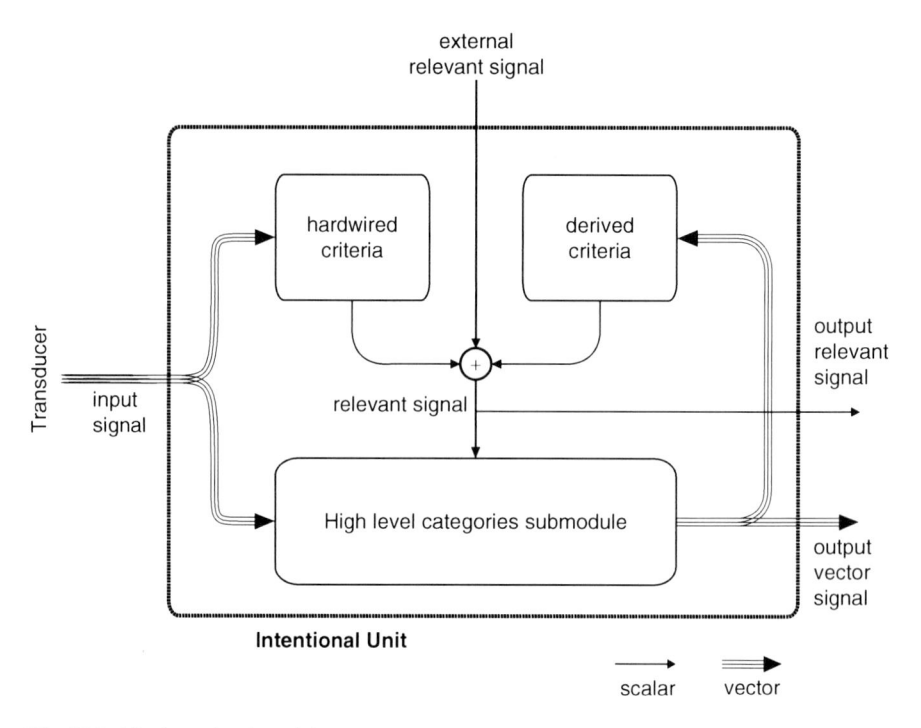

Fig. 20.3 The intentional module

vector has as many elements as there are intentional units. The value of each element expresses how much the corresponding intentional unit is activated by the current input signal, which in turn means how much the current input signal is similar to a given past input.

Formally, the intentional module implements a function:

$$F_{\mathrm{m}}\left(r, \vec{v}\right) = \begin{pmatrix} r_n \\ \vec{v}_n \end{pmatrix},$$

whereas r is the external control signal, $\vec{v}$ is the input vector, rn is the output relevant signal, and $\vec{v}_n$ is the output vector signal. Given an array of N intentional units, the output vector is:

$$\vec{v}_n = \begin{pmatrix} F_{\mathrm{u}}^1\left(r, \vec{v}\right) \\ \cdots \\ F_{\mathrm{u}}^N\left(r, \vec{v}\right) \end{pmatrix}.$$

It is interesting to note that each intentional unit (F_{u}^i) has a different starting time t_0^i, and thus it is matched to a different input vector $\vec{v}_i$.

20.3.3 The Intentional Architecture

The above intentional module is only slightly more interesting than the intentional unit, although it is already sufficient to implement classic conditioning, attentive behavior, and a rough self-generation of new goals (Manzotti 2009). There are a few features of the described intentional modules that are worth to be stressed once more.

First, the module can process any kind of data. It does not have to know in advance whether the incoming data is originated by images, sounds, texts, or whatever. In principle, at least, it is very general. Second, the module receives a vector and a scalar, and it outputs a vector and a scalar as well. There is ground to use the intentional module as the building block of a much larger architecture.

Second, the module uses vectors to send and receive data and scalars to send and receive controls.

Third, the module embeds its history in its structure. The module is unpredictable since its behavior is the result of a close coupling with the environment.

Now we will outline how to exploit these three features in order to design a more complex architecture.

Consider the case of a robot moving in an environment such as our own. In a real environment, there are multiple sources of information as well as multiple ways to extract different channels out of the same data source. Consider a visual color channel. It can be subdivided into a grey scale video, a color video, a filtered gray scale video (edges), and many other interesting channels. Besides, there are many more sources of information about the environment such as sound, tactile information, proprioception, and so on.

Consider having the capability of implementing a huge number of intentional modules. Consider having M incoming sources of information corresponding to as many vectors. For instance, vision could give rise to many different source of information. Sound capability could add a few more sources (different bandwidths, temporal vectors, spectral vectors), and so on. Suppose having M intentional modulestaking care of each of these sources. Suppose that each intentional module has a reasonably large number of intentional units inside. At this point, a few fixed rules will suffice to build dynamically an architecture made of intentional modules. Out of uniformity, we label it as *intentional architecture*. The rules are the followings:

1. Assign to each source of data a separate intentional module. Whether the capacity of the module is saturated (all intentional units are assigned), assign other intentional modules as needed. These modules make the first level of the architecture.
2. When the first level is complete, use the output of the first level modules as inputs for further levels of intentional modules.
3. The further level intentional modules are assigned to every possible earlier level modules output. However, due to many factors (the richness of the original external source of data, the implemented similarity function inside the intentional units, the incoming data, and so on), the output vector sizes are going to diminish

increasing the level. When this happens the intentional module will recruit a smaller and smaller number of intentional units. In that case, its output will get merged with that of other intentional modules with similarly smaller output.

4. In a while, the previous conditions will obtain for all intentional modules of the higher levels, thereby pushing toward a convergence.

5. All of the above applies for input and output vectors. As to the control signals, the rule is the opposite: backward connecting the higher level intentional modules with the lower level ones. In this way, the relevant signal produced by the highest possible intentional units will orient the activity of the lowest level intentional unit modules.

6. An example of a tentative final result is shown in Fig. 20.4. It has four sources of information (one of which is split onto two intentional units, thus providing five input signals). The end result is an architecture with four levels and twelve intentional modules (five modules in the first, four in the second, two in the third, and one in the fourth). Formally, the whole architecture behave like a giant module and its behavior could be expressed by the formula:

$$F_a\left(r, \vec{v}_1 \ldots \vec{v}_M\right) = \begin{pmatrix} r_a \\ \vec{v}_a \end{pmatrix},$$

whereas r_a is the total relevant signal, $\vec{v}_a$ is the final output vector, and $\vec{v}_1 \ldots \vec{v}_M$ is a series of input signals. And r? Although it is not represented in Fig. 20.4, it cannot be excluded that that the architecture would admit a global relevant signal for all its modules. If this were the case, r_a would be the global relevant signal alerting all of the modules that something relevant is indeed arriving.

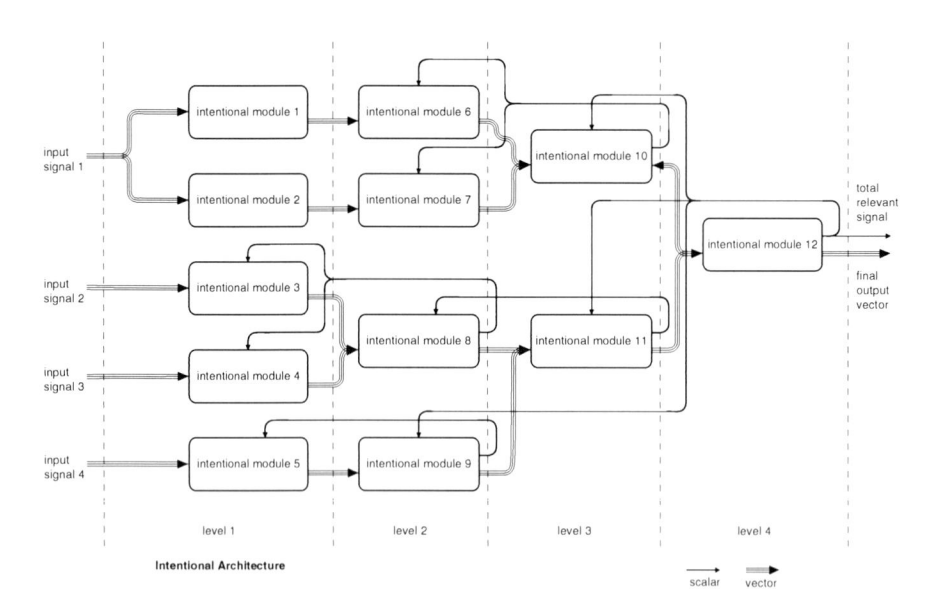

Fig. 20.4 The intentional architecture

20.3.4 Check List for Consciousness-Oriented Architectures

It is time to use the checklist for consciousness-oriented architectures suggested above. Consider the suggested intentional architecture as a test bed to check the issues mentioned above.

Bottom-up vs. top-down. The architecture is neither bottom-up nor top-down. On one hand, it is built from scratch by the incoming data. On the other hand, at every level it sends backward the relevant control signal thereby tuning the way in which the information recruits new intentional units and new intentional modules. In some sense, the final choice depends on the highest level relevant signal. However, such a signal is the result of all the bottom-up activity taking place before.

Information integration. All the incoming information flows through the system producing two final outputs (a vector and a scalar) that, in different ways, summarize all that happened before. It could be interesting to apply Tononi's measures to the information flow of it. Further, the integration happens at every step (both at the unit level, at the module level, and at the architecture level). The existence of backward control signals ensures that the structure is not a simple feed-forward one, but rather a complex recurrent one.

Environment coupling. All the information received in the past gets embedded into the structure of the architecture. The history of the architecture is embedded into its structure. Each intentional units is matched to a past input signal and thus to a past event. The architecture is carved out by the environment step by step and, in a deep causal sense, it mirrors the structure of the world to a certain extent.

Intentionality of the first kind. Here deep philosophical ground has to be addressed. Intentionality as semantics is not going be an easy game. Yet, the proposed architecture suggests some useful approach. The external events, which are the source of the input, are not causally inactive (for the architecture, at least) before being matched whether to an intentional unit or to a cluster of intentional units. After the matching, the event has a new causal role. Consider a pattern. Before it being seen by someone, does it really exist? Similarly a pattern in the world gets coupled with the activity inside the architecture by means of the causal relation endorsed by the intentional unit. Such a causal relation is the same one that allows that very pattern to be causally active. The outlined causal structure could be a way to tackle with the problem of semantics. It could be the way to naturalize it.

Intentionality of the second kind. The system has both hardwired fixed criteria and newly developed criteria. Each intentional module can develop new criteria that depend only on its past history. In a similar way, the architecture is combining all the relevant signals into a global relevant signal that could be the foundation for a global motivational system. It is a teleologically open system.

One architectural principle to rule them all. This requirement has been satisfied rather well. The architecture is the repetition of the same causal structure again and again. From the intentional unit up to the architecture itself, the causal structure is always following the same dictum: receive inputs and change accordingly as to become causally sensitive to those very inputs. Match yourself with the environment! This is rather evident if we compare the three functions describing the behavior of the unit level, the module level, and the architecture level:

$$F_u(r, s_t) = v,$$

$$F_m\left(r, \vec{v}\right) = \begin{pmatrix} r_n \\ \vec{v}_n \end{pmatrix},$$

$$F_a\left(r, \vec{v}_1 \ldots \vec{v}_M\right) = \begin{pmatrix} r_a \\ \vec{v}_a \end{pmatrix}.$$

The only significant difference is the increase in the complexity of the incoming and outgoing information. It will not be difficult to envisage an even higher level made of whole architectures combining together.

Antirepresentationalist stance. There are neither explicit representations nor variable stored anywhere. Every complex incoming cluster of events is distributed in all the architecture as a whole. There is no way to extract a particular representation from the values stored inside the architecture. However, when exposed to a certain stimulus, the architecture will recognize it. The architecture has a meaning only if considered as a whole with the environment. It is both embodied and situated.

Autonomy and resilience. First, it is interesting to note that the architecture could suffer substantial damage both in the intentional units and even in the intentional modules without being totally destroyed. For instance, the loss of a whole module would imply the loss of the capability to deal with the corresponding information source. Yet, the system will continue to behave normally with respect to all other information sources. As to the other sources, the system will show no sign of reduced performances. Similarly, new information sources could be added anytime, although if they are present at the very beginning, the architecture will incorporate them more seamlessly. As a result, the architecture resilience seems pretty good. As to the autonomy, it can be stressed that the system develops both epistemic categories (by means of intentional units of higher levels) and goals (by means of the backward control signals). In this way, the system is going to be rather unpredictable and strongly coupled with its environment thereby implementing a strong decisional and epistemic autonomy.

Phenomenal experience. It would be unfair to pretend a final proof of the occurrence of phenomenal experience since its final solution will probably require some ontological breakthrough whose scope is definitely a lot more far reaching than cognitive science and artificial intelligence could envisage. Yet, the present model is flanked by the externalist view that so far has neither being validated nor rejected. We think it is possible to foresight that this kind of architecture (or a much improved one in the same spirit) would start to address phenomenal experience since the kind of engagement between such an architecture and the environment could lead to the continuity required to have phenomenal experience. Consider that all the events occurring inside the architecture are the result of environmental phenomena and that the emergence of top-down signal is the eventual outcome of stimuli originating externally to the architecture. The architecture is significantly shaped by the environment.

20.3.5 A Comparison with Other Approaches

It is useful to compare other attempts at modeling a conscious machine. The following is by no means neither an exhaustive list nor a throughout description of the architectures mentioned. However, for those already aware of other approaches a quick glance could help to gain a better grasp of the gist of the architecture presented here. One more caveat is that we will focus on differences rather than on the commonalities.

The most obvious candidate is Taylor's CODAM model of attention. We would like to point out at the similarity between the intentional module and the CODAM model previously quoted and sketched. Compare the two diagrams in Figs. 20.1 and 20.3. In both cases, a similar interplay between control and memory occurs. Both models have separate and parallel control signals going from the input either to the memory or to the controller. A few models, and their connections too, can be rather convincingly matched against each other. The CODAM object module is similar in role and in internal structure to the category array of the intentional module. Both store high level representations of the stimuli received. Both have a control signal which is the result of the combination of a goal system and the bottom-up contributions of the match between the incoming stimuli and the learning achieved. In the CODAM model, this is achieved by means of the cooperation between the goal module, the IMC, and the working memory. In the presented intentional module, this is achieved thanks to the sum of two signals: the output of the hardwired criteria submodule, which is analogous to the goal module, and the output of the acquired criteria submodule, which receives from the category submodule (as the working memory too is fed by the object map in the CODAM model).

Of course, there are also many relevant differences. The CODAM model is closer to the brain neural architecture, and it is particularly suited to match many attentional processes in the visual system. On the other hand, the intentional architecture does not try to explain either attentional or the cognitive features of biological systems, rather it capitalizes on them in order to achieve information integration, environmental coupling, autonomy, and intentionality. Whether it is successful, it is matter of further experimental research.

Another candidate for comparison is Stan Franklin's IDA whose goal is to mimic many high-level behaviors (mostly cognitive in the symbolic sense) gathering together several functional modules. In IDA's top-down architecture, high-level cognitive functions are explicitly modeled (Franklin 1995, 2003). They aim at a full functional integration between competing software agencies. The biggest difference with our approach is that we make explicit hypothesis as to how information has to be processed in order to reproduce the same properties of a conscious brain. While Franklin's approach is fully compatible with the functionalistic tenet of multiple realizability, our approach isn't. IDA is essentially a functionalist effort. We maintain that consciousness is something more than information processing – it involves embodiment, situatedness, and physical continuity with the environment in a proper causal entanglement.

Consider now Baars' Global Workspace as it has been implemented by Shanahan (Shanahan and Baars 2005; Shanahan 2006). Shanahan's model addresses explicitly several aspects of conscious experience such as imagination and emotion. Moreover, it addresses the issue of sensory integration and the problem of how information is processed in a centralized workspace. It is an approach that, on the one hand, suggests a specific way to deal with information and, on the other hand, endorses internalism to the extent that consciousness is seen as the result of internal organization. Consciousness, in short, is a style of information processing (the bidirectional transfer of information from/to the global workspace) achieved through different means – "conscious information processing is cognitively efficacious because it integrates the results of the brain's massively parallel computational resources" (Shanahan 2006, p. 434). He focuses on implementing a hybrid architecture that mixes together the more classic cognitive structure of global workspace with largely not symbolic neural networks.

Both Shanahan's and our approach exploit a tight coupling between the environment and the agent. Yet, our architecture does not aim at modeling explicitly either a common workspace or a predesigned organization between various cognitive functions. Rather, the "conscious" bottleneck, so to speak, ought to stem out of the bottom-up vs. top-down interplay described above. Another important difference is that, at least conceptually, Shanahan's approach assumes that there is a separation between the inside and the outside of the agent. Inside the agent, it is meaningful to speak of information, computation, broadcasting, and representations. Outside, there are just physical events. Shanahan's architecture matches such theoretical chasm. On the contrary, since we stress the concrete possibility that consciousness could be spread in the environment, our architecture is designed as to become seamlessly causally connected with its historical and individual surroundings.

Similar considerations hold for Hesslow's emphasis on emulation (Hesslow and Jirenhed 2007). Since he is apparently committed to the view that consciousness inheres on the existence of an inner world, Hesslow defends the so-called simulation hypothesis which does not rely on any assumptions about the nature of imagery or perception except that activity in sensory cortex can be elicited internally. He conceives emulation as an internally generated feedback loop. On the contrary, in the presented architecture, everything can always be traced back to something occurred externally. To a certain extent, our architecture could be criticized for being almost completely environment-driven. Yet this is not a shortcoming but rather an unavoidable condition shared by agents once they are seen as a part of the environment rather than as a separate domain.

As for the broad category of approaches explicitly referring to embodiment and situatedness, we can quote both Holland's work on *Cronos* and Bongard on resilient autonomous agents (Holland 2004; Bongard et al. 2006; Pfeifer and Bongard 2006; Pfeifer et al. 2007). Both cases highlight the importance of embodiment. However, although they suggest the importance of taking advantage of morphology and embodiment to achieve superior performance, they still seem to consider that models and representations are instantiated inside the cognitive domain of the agent. So the world acts either as an external memory or as an external collection of functional modules or, finally, as a simplifier of otherwise too complex computational problems. We do pursue a different approach. Our architecture tries to exploit the causal structure of the environment to develop accordingly. However, it remains incomplete whether not coupled with the causal history it triggered its growth. Yet, the spirit of our architecture is definitely closer to such embodied approach rather than to the pure functionalist approach of classic AI.

There are many more authors we could only quote such as Chrisley's synthetic phenomenology, Aleksander's axioms, Haikonen's conscious machines, and many others (Haikonen 2003; Aleksander et al. 2008; Aleksander and Morton 2008; Chrisley 2009). Yet it is perhaps more useful to outline two critical dimensions along which most attempts lie. The first dimension is the dichotomy between internalist approaches and externalist approaches passing through embodiment and situatedness. The second dimension corresponds to the dichotomy between pure multiple realizability vs. critical style of information processing. The former view assumes that as long as two agents perform to share the same functional behavior, they are the same. The latter view assumes that, at some level, it matters the way in which a certain cognitive task is performed. We can thus sketch in Fig. 20.5 a recapping map of the various views and available architectures.

A final remark. Most of the aforementioned differences are rooted in the different conceptual frameworks adopted by authors. Thus, it is possible that, once expressed using a completely neutral jargon, two randomly picked up architectures would end up being much closer than they would appear when disguised under their authors' theoretical commitments. Yet this is not enough to set aside all differences. At this stage of research, an architecture is not simply either a functional or a mathematical diagram. An architecture is a bundle with the viewpoint it endorses. To make an example from the recent history of AI, consider neural networks and not linear mathematical approximators. At a certain descriptive levels, they are the same, but if represented in their theoretical framework they look different. Neural networks represented a cognitive tool whose efficacy went much beyond its mathematical properties for better or for worse.

20.4 Conclusion

Although AI achieved impressive results (Russell and Norvig 2003), it is always astonishing the degree of overvaluation that many nonexperts seem to stick to. In 1985 (!), addressing the Americal Philosophical Association, Fred Drestke was sure that "even the simple robots designed for home amusement talk, see, remember

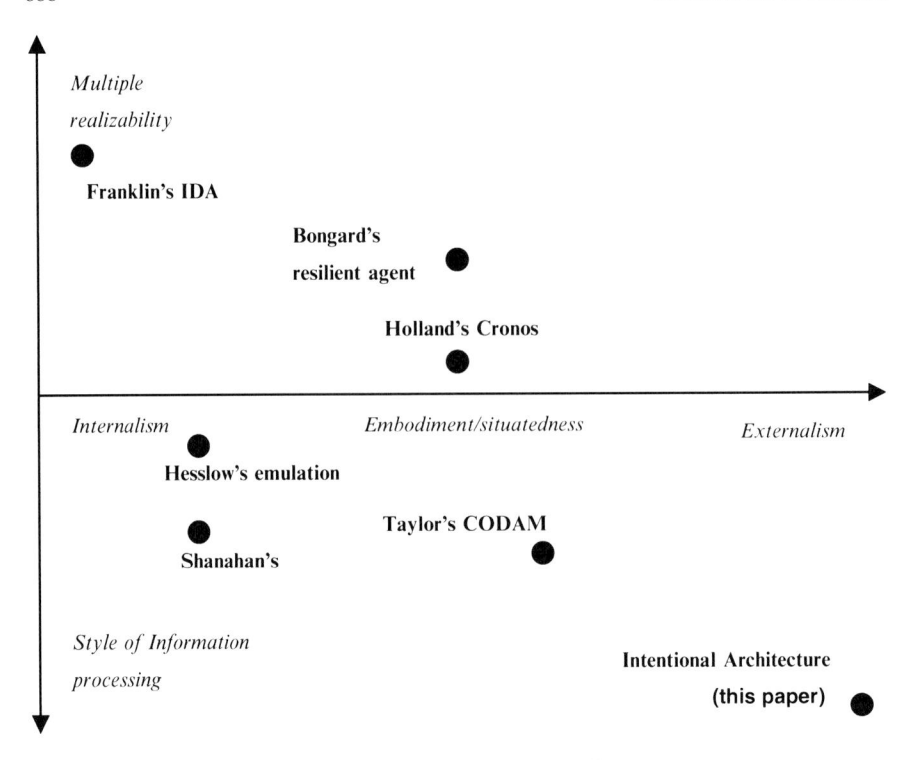

Fig. 20.5 A map of current attempt at achieving machine consciousness

and learn" (Dretske 1985, p. 23). It is not unusual to hear that robots are capable of feeling emotions or taking autonomous and even moral choices (Wallach and Allen 2009). It is a questionable habit that survives and that conveys false hopes about the current status of AI research. Occasionally, even skilled scholars slip into this habit quoting implausible Legobots used in first-year undergraduate robot instruction as agents capable of developing new motivations (Aleksander et al. 2008, pp. 102–103).

Such approximate misevaluations of the real status of AI hinder new researchers from addressing objectives allegedly but mistakenly assumed as already achieved. Due to various motivations, not all of strict scientific nature, in the past, many AI researchers made bold claims about their achievements so to endorse a false feeling about the effective level of AI research.

Studying consciousness and suggesting artificial model of it could help addressing those aspects of the mint that have remained so far elusive. The available results encourage to keep designing conscious machines or, at least, machines that reproduce some aspects of conscious experience. Although a complete theory of consciousness will very probably require some ground-breaking conceptual revisions, the deliberate attempt at mimicking consciousness will pave the way to a future scientific paradigm to understand the mind.

References

Aboitz, F., D. Morales, et al., (2003), "The evolutionary origin of the mammalian isocortex: Towards an integrated developmental and functional approach." in *Behavioral and Brain Sciences*, 26: 535–586.

Adami, C., (2006), "What Do Robots Dreams Of?" in *Science*, 314(58): 1093–1094.

Adams, D., K. Aizawa, (2008), *The Bounds of Cognition*, Singapore, Blackwell.

Adams, F., K. Aizawa, (2009), "Why the Mind is Still in the Head" in P. Robbins and M. Aydede, Eds, *The Cambridge Handbook of Situated Cognition*, Cambridge, Cambridge University Press: 78–95.

Aleksander, I., (2008), "Machine consciousness." in *Scholarpedia*, 3(2): 4162.

Aleksander, I., U. Awret, et al., (2008), "Assessing Artificial Consciousness." in *Journal of Consciousness Studies*, 15(7): 95–110.

Aleksander, I., B. Dunmall, (2003), "Axioms and Tests for the Presence of Minimal Consciousness in Agents." in *Journal of Consciousness Studies*, 10: 7–18.

Aleksander, I., H. Morton, (2007), "Depictive Architectures for Synthetic Phenomenology" in A. Chella and R. Manzotti, Eds, *Artificial Consciousness*, Exeter, Imprint Academic(30–45).

Aleksander, I., H. Morton, (2008), "Depictive Architectures for Synthetic Phenomenology" in A. Chella and R. Manzotti, Eds, *Artificial Consciousness*, Exeter, Imprint Academic(30–45).

Arkin, R. C., (1998), *Behavior-Based Robotics*, Cambridge (Mass), MIT.

Atkinson, A. P., M. S. C. Thomas, et al., (2000), "Consciousness: mapping the theoretical landscape." in *TRENDS in Cognitive Sciences*, 4(10): 372–382.

Baars, B. J., (1988), *A Cognitive Theory of Consciousness*, Cambridge, Cambridge University Press.

Bayne, T., D. Chalmers, (2003), "What is the Unity of Consciousness?" in A. Cleeremans, Ed., *The Unity of Consciousness: Binding, Integration, and Dissociation*, Oxford, Oxford University Press: 23–58.

Bennett, M. R., P. M. S. Hacker, (2003), *Philosophical Foundations of Neuroscience*, Malden (Mass), Blackwell.

Bongard, J., v. Zykov, et al., (2006), "Resilient Machines Through Continuous Self-Modeling." in *Science*, 314(5802): 1118–1121.

Brentano, F., (1874/1973), *Psychology From an Empirical Standpoint*, London, Routledge & Kegan Paul.

Bringsjord, S., (1994), "Computation, among other things, is beneath us." in *Minds and Machines*, 4(4): 469–488.

Brooks, R. A., (1990), "Elephants Don't Play Chess." in *Robotics and Autonomous Systems*, 6: 3–15.

Brooks, R. A., (1991), "New Approaches to Robotics." in *Science*, 253: 1227–1232.

Brooks, R. A., C. Breazeal, et al., (1998), "Alternate Essences of Intelligence", in *AAAI 98*.

Brooks, R. A., C. Breazeal, et al., (1999), "The Cog Project: Building a Humanoid Robot" in C. Nehaniv, Ed., *Computation for Metaphors, Analogy, and Agents*, Berlin, Springer(1562): 52–87.

Buttazzo, G., (2001), "Artificial Consciousness: Utopia or Real Possibility." in *Spectrum IEEE Computer*, 34(7): 24–30.

Buttazzo, G., (2008), "Artificial Consciousness: Hazardous Questions." in *Journal of Artificial Intelligence and Medicine*(Special Issue on Artificial Consciousness).

Chalmers, D. J., (1996), *The Conscious Mind: In Search of a Fundamental Theory*, New York, Oxford University Press.

Changeux, J. P., (2004), "Clarifying Consciousness." in *Nature*, 428: 603–604.

Chella, A., M. Frixione, et al., (2008), "A Cognitive Architecture for Robot Self-Consciousness." in *Artificial Intelligence in Medicine*(Special Issue of Artificial Consciousness).

Chella, A., S. Gaglio, et al., (2001), "Conceptual representations of actions for autonomous robots." in *Robotics and Autonomous Systems*, 34(4): 251–264.

Chella, A., R. Manzotti, Eds, (2007), *Artificial Consciousness*, Exeter (UK), Imprint Academic.

Chella, A., R. Manzotti, (2009), "Machine Consciousness: A Manifesto for Robotics." in *International Journal of Machine Consciousness*, 1(1): 33–51.

Chrisley, R., (1994), "Why Everything Doesn't Realize Every Computation." in *Minds and Machines*, 4(4): 403–420.

Chrisley, R., (2008), "The philosophical foundations of Artificial Consciousness." in *Journal of Artificial Intelligence and Medicine*(Special Issue on Artificial Consciousness).

Chrisley, R., (2009), "Synthetic Phenomenology." in *International Journal of Machine Consciousness*, 1(1): 53–70.

Clark, A., (1997), Being there: putting brain, body and world together again, Cambridge (Mass), MIT.

Clark, A., (2008), *Supersizing the Mind*, Oxford, Oxford University Press.

Clark, A., D. Chalmers, (1999), "The Extended Mind." in *Analysis*, 58(1): 10–23.

Clark, O. G., R. Kok, et al., (1999), "Mind and autonomy in engineered biosystems." in *Engineering Applications of Artificial Intelligence*, 12: 389–399.

Collins, S., M. Wisse, et al., (2001), "A Three-dimensional Passive-dynamic Walking Robot with Two Legs and Knees." in *The International Journal of Robotics Research*, 20(7): 607–615.

Crick, F., (1994), *The Astonishing Hypothesis: the Scientific Search for the Soul*, New York, Touchstone.

Crick, F., C. Koch, (1990), "Toward a Neurobiological Theory of Consciousness." in *Seminars in Neuroscience*, 2: 263–295.

Crick, F., C. Koch, (2003), "A framework for consciousness." in *Nature Neuroscience*, 6(2): 119–126.

Dennett, D. C., (1978), *Brainstorms: philosophical essays on mind and psychology*, Montgomery, Bradford Books.

Dennett, D. C., (1987), *The intentional stance*, Cambridge (Mass), MIT.

Dennett, D. C., (1991), *Consciousness explained*, Boston, Little Brown and Co.

Di Paolo, E. A., H. Iizuka, (2008), "How (not) to model autonomous behaviour." in *BioSystems*, 91: 409–423.

Dretske, F., (1985), "Machines and the Mental." in *Proceedings and Addresses of the American Philosophical Association* 59(1): 23–33.

Duda, R. O., P. E. Hart, et al., (2001), *Pattern classification*, New York, Wiley.

Engel, A. K., W. Singer, (2001), "Temporal binding and the neural correlates of sensory awareness." in *TRENDS in Cognitive Sciences*, 5: 16–25.

Franklin, S., (1995), *Artificial Minds*, Cambridge (Mass), MIT.

Franklin, S., (2003), "IDA: A Conscious Artefact?" in *Journal of Consciousness Studies*, 10: 47–66.

Gallagher, S., (2009), "Philosophical Antecedents of Situated Cognition" in P. Robbins and M. Aydede, Eds, *The Cambridge Handbook of Situated Cognition*, Cambridge, Cambridge University Press.

Haikonen, P. O., (2003), *The Cognitive Approach to Conscious Machine*, London, Imprint Academic.

Hameroff, S. R., A. W. Kaszniak, et al., (1996), *Toward a science of consciousness: the first Tucson discussions and debates*, Cambridge (Mass), MIT.

Harnad, S., (1990), "The Symbol Grounding Problem." in *Physica*, D(42): 335–346.

Harnad, S., (1994), "Computation is just interpretable symbol manipulation; cognition isn't " in *Minds and Machines*, 4(4): 379–390.

Harnad, S., (1995), "Grounding symbolic capacity in robotic capacity" in L. Steels and R. A. Brooks, Eds, *"Artificial Route" to "Artificial Intelligence": Building Situated Embodied Agents*, New York, Erlbaum.

Harnad, S., (2003), "Can a machine be conscious? How?" in *Journal of Consciousness Studies*.

Harnad, S., P. Scherzer, (2008), "First, Scale Up to the Robotic Turing Test, Then Worry About Feem." in *Journal of Artificial Intelligence and Medicine* (Special Issue on Artificial Consciousness).

Hernandez, C., I. Lopez, et al., (2009), "The Operative mind: A Functional, Computational and Modeling Approach to Machine Consciousness." in *International Journal of Machine Consciousness*, 1(1): 83–98.

Hesslow, G., D.-A. Jirenhed, (2007), "Must Machines be Zombies? Internal Simulation as a Mechanism for Machine Consciousness", in *AAAI Symposium*, Washington DC, 8–11 November 2007.

Holland, O., Ed. (2003), *Machine consciousness*, New York, Imprint Academic.

Holland, O., (2004), "The Future of Embodied Artificial Intelligence: Machine Consciousness?" in F. Iida, Ed., *Embodied Artificial Intelligence*, Berlin, Springer: 37–53.

Honderich, T., (2006), "Radical Externalism." in *Journal of Consciousness Studies*, 13(7–8): 3–13.

Hummel, F., C. Gerloff, et al., (2004), "Cross-modal plasticity and deafferentiation." in *Cognitive Processes*, 5: 152–158.

Hurley, S. L., (2003), "Action, the Unity of Consciousness, and Vehicle Externalism" in A. Cleeremans, Ed., *The Unity of Consciousness: Binding, Integration, and Dissociation*, Oxford, Oxford University Press.

Hurley, S. L., (2006), "Varieties of externalism" in R. Menary, Ed., *The extended mind*, Aldershot, Ashgate publishing.

Jennings, C., (2000), "In Search of Consciousness." in *Nature Neuroscience*, 3(8): 1.

Kahn, D. M., L. Krubitzer, (2002), "Massive cross-modal cortical plasticity and the emergence of a new cortical area in developmentally blind mammals." in *Proceedings of National Academy of Science*, 99(17): 11429–11434.

Kentridge, R. W., T. C. W. Nijboer, et al., (2008), "Attended but unseen: Visual attention is not sufficient for visual awareness." in *Neuropsychologia*, 46: 864–869.

Koch, C., (2004), *The Quest for Consciousness: A Neurobiological Approach*, Englewood (Col), Roberts & Company Publishers.

Koch, C., G. Tononi, (2008), "Can Machines be Conscious?" in *IEEE Spectrum*: 47–51.

Koch, C., N. Tsuchiya, (2006), "Attention and consciousness: two distinct brain processes." in *TRENDS in Cognitive Sciences*, 11(1): 16–22.

Krubitzer, L., (1995), "The organization of neocortex in mammals: are species differences really so different?" in *Trends in Neurosciences*, 18: 408–417.

Libet, B., C. A. Gleason, et al., (1983), "Time of conscious intention to act in relation to onset of cerebral activity. The unconscious initiation of a freely voluntary act." in *Brain*, 106(3): 623–642.

Lycan, W. G., (1981), "Form, Function, and Feel." in *The Journal of Philosophy*, 78(1): 24–50.

Mack, A., I. Rock, (1998), *Inattentional Blindness*, Cambridge (Mass), MIT.

Manzotti, R., (2003), "A process based architecture for an artificial conscious being" in J. Seibt, Ed., *Process Theories: Crossdisciplinary studies in dynamic categories*, Dordrecht, Kluwer: 285–312.

Manzotti, R., (2006), "An alternative process view of conscious perception." in *Journal of Consciousness Studies*, 13(6): 45–79.

Manzotti, R., (2006), "Consciousness and existence as a process." in *Mind and Matter*, 4(1): 7–43.

Manzotti, R., (2007), "Towards Artificial Consciousness." in *APA Newsletter on Philosophy and Computers*, 07(1): 12–15.

Manzotti, R., (2009), "No Time, No Wholes: A Temporal and Causal-Oriented Approach to the Ontology of Wholes." in *Axiomathes*, 19: 193–214.

Manzotti, R., V. Tagliasco, (2005), "From "behaviour-based" robots to "motivations-based" robots." in *Robotics and Autonomous Systems*, 51(2–3): 175–190.

Manzotti, R., V. Tagliasco, (2008), "Artificial Consciousness: A Discipline Between Technological and Theoretical Obstacles." in *Artificial Intelligence in Medicine*, 44(2): 105–118.

McFarland, D., T. Bosser, (1993), *Intelligent Behavior in Animals and Robots*, Cambridge (Mass), MIT.

Merrick, T., (2001), *Objects and Persons*, Oxford, Oxford Clarendon Press.

Metta, G., P. Fitzpatrick, (2003), "Early integration of vision and manupulation." in *Adaptive Behavior*, 11(2): 109–128.

Miller, G., (2005), "What is the Biological Basis of Consciousness?" in *Science*, 309: 79.

Mole, C., (2007), "Attention in the absence of consciousness." in *TRENDS in Cognitive Sciences*, 12(2): 44–45.

Nemes, T., (1962), *Kibernetic Gépek*, Budapest, Akadémiai Kiadò.

Nemes, T., (1969), *Cybernetic machines*, Budapest, Iliffe Books and Akademiai Kiadò.

Newell, A., (1990), *Unified Theories of Cognition*, Cambridge (Mass), Harvard University Press.

Paul, C., F. J. Valero-Cuevas, et al., (2006), "Design and Control of tensegrity Robots." in *IEEE Transactions on Robotics*, 22(5): 944–957.

Pfeifer, R., J. Bongard, (2006), *How the Body Shapes the Way We Think: A New View of Intelligence (Bradford Books)* New York, Bradford Books.

Pfeifer, R., M. Lungarella, et al., (2007), "Self-Organization, Embodiment, and Biologically Inspired Robotics." in *Science*, 5853(318): 1088 - 1093.

Pockett, S., (2004), "Does consciousness cause behaviour?" in *Journal of Consciousness Studies*, 11(2): 23–40.

Prinz, J., (2000), "The Ins and Outs of Consciousness." in *Brain and Mind*, 1: 245–256.

Prinz, J., (2009), "Is Consciousness Embodied?" in P. Robbins and M. Aydede, Eds, *The Cambridge Handbook of Situated Cognition*, Cambridge, Cambridge University Press: 419–436.

Ptito, M., S. M. Moesgaard, et al., (2005), "Cross-modal plasticity revealed by electrotactile stimulation of the tongue in the congenitally blind." in *Brain*, 128: 606–614.

Revonsuo, A., (1999), "Binding and the phenomenal unity of consciousness." in *Consciousness and Cognition*, 8: 173–85.

Revonsuo, A., J. Newman, (1999), "Binding and consciousness." in *Consciousness and Cognition*, 8: 127–127.

Robbins, P., M. Aydede, Eds, (2009), *The Cambridge Handbook of Situated Cognition*, Cambridge, Cambridge University Press.

Rockwell, T., (2005), *Neither ghost nor brain*, Cambridge (Mass), MIT.

Roe, A. W., P. E. Garraghty, et al., (1993), "Experimentally induced visual projections to the auditory thalamus in ferrets: Evidence for a W cell pathway." in *Journal of Computational Neuroscience*, 334: 263–280.

Rowlands, M., (2003), *Externalism. Putting Mind and World Back Together Again*, Chesham, Acumen Publishing Limited.

Russell, S., P. Norvig, (2003), *Artificial Intelligence. A Modern Approach*, New York, Prentice Hall.

Sanz, R., (2005), "Design and Implementation of an Artificial Conscious Machine", in *IWAC2005*, Agrigento.

Sanz, R., I. Lopez, et al., (2007), "Principles for consciousness in integrated cognitive control." in *Neural Networks*, 20: 938–946.

Searle, J. R., (1980), "Minds, Brains, and Programs." in *Behavioral and Brain Sciences*, 1: 417–424.

Searle, J. R., (1992), *The Rediscovery of the Mind*, Cambridge (Mass), MIT.

Seth, A., (2009), "The Strength of Weak Artificial Consciousness." in *International Journal of Machine Consciousness*, 1(1): 71–82.

Shanahan, M., (2006), "A Cognitive Architecture that Combines Internal Simulation with a Global Workspace." in *Consciousness and Cognition*, 15: 433–449.

Shanahan, M., B. J. Baars, (2005), "Applying Global Workspace Theory to the Frame Problem." in *Cognition*, 98(2): 157–176.

Shanahan, M. P., (2005), "Global Access, Embodiment, and the Conscious Subject." in *Journal of Consciousness Studies*, 12(12): 46–66.

Sharma, J., A. Angelucci, et al., (2000), "Visual behaviour mediated by retinal projections directed to the auditory pathway." in *Nature*, 303: 841–847.

Simons, D. J., (2000), "Attentional capture and inattentional blindness." in *TRENDS in Cognitive Sciences*, 4: 147–155.

Simons, P. M., (1987), *Parts. A Study in Ontology*, Oxford, Clarendon.

Skrbina, D., Ed. (2009), *Mind that abides. Panpsychism in the new millennium*, Amsterdam, John Benjamins Pub.

Sur, M., P. E. Garraghty, et al., (1988), "Experimentally induced visual projections into auditory thalamus and cortex." in *Science*, 242: 1437–1441.

Sutton, R. S. and A. G. Barto, (1998), *Reinforcement Learning*, Cambridge (Mass), MIT.

Tagliasco, V., (2007), "Artificial Consciousness. A Technological Discipline" in A. Chella and R. Manzotti, Eds, *Artificial Consciousness*, Exeter, Imprint Academic: 12–24.

Taylor, J. G., (2000), "Attentional movement: The control basis for consciousness." in *Society for Neuroscience Abstracts*, 26: 2231#839.3.

Taylor, J. G., (2002), "Paying attention to consciousness." in *TRENDS in Cognitive Sciences*, 6(5): 206–210.

Taylor, J. G., (2003), "Neural Models of Consciousness" in M. A. Arbib, Ed., *The Handbook of Brain Theory and Neural Networks*, Cambridge (Mass), MIT: 263–267.

Taylor, J. G., (2007), "CODAM: A neural network model of consciousness." in *Neural Networks*, 20: 983–992.

Taylor, J. G., (2009), "Beyond Consciousness?" in *International Journal of Machine Consciousness*, 1(1): 11–22.

Taylor, J. G. and N. Fragopanagos, (2005), "The interaction of attention and emotion." in *Neural Networks*, 18(4): 353–369.

Taylor, J. G. and N. Fragopanagos, (2007), "Resolving some confusions over attention and consciousness." in *Neural Networks*, 20(9): 993–1003.

Taylor, J. G. and M. Rogers, (2002), "A control model of the movement of attention." in *Neural Networks*, 15: 309–326.

Tononi, G., (2004), "An information integration theory of consciousness." in *BMC Neuroscience*, 5(42): 1–22.

Van Gelder, T., (1995), "What Might Cognition Be, If Not Computation?" in *The Journal of Philosophy*, 92(7): 345–381.

Varela, F. J., E. Thompson, et al., (1991/1993), *The Embodied Mind: Cognitive Science and Human Experience*, Cambridge (Mass), MIT.

Wallach, W. and C. Allen, (2009), *Moral Machines. Teaching Robots Right from Wrong*, New York, Oxford University Press.

Ziemke, T. and N. Sharkey, (2001), "A stroll through the worlds of robots and animals: applying Jakob von Uexküll's theory of meaning to adaptive robots and artificial life." in *Semiotica*, 134(1/4): 701/46.

Part III
Hardware Implementations

Vassilis Cutsuridis, Amir Hussain, and John G. Taylor

In this part, leading engineers present hardware implementations of the various components of the perception–action cycle, namely sensation, multisensory fusion, planning and action. These systems make use of knowledge from the areas of computer networking, machine learning, artificial neural networks and neuromorphic engineering.

In the chapter entitled "Smart sensor networks", Lim describes the current wireless sensor network technology requirements and issues in designing and implementing smart sensor networks. The motivations for this work comes from a wide spectrum of applications, ranging from distributed multi-robot perception, navigation and manipulation applications to distributed sense-and-response systems to dynamic situation awareness and decision support systems. He describes the systems for implementing smart sensor network applications through the use of key components, such as appropriate sensor network protocols, distributed services and reconfigurable smart nodes. He provides two proofs of concepts for using cyclic perception–reason–action modules in multi-robot control applications and real-time target tracking applications. He presents preliminary results from these two types of applications that show the effectiveness in using these methods. He concludes with future work and extensions that will improve these approaches.

In the chapter entitled "Multisensor fusion for low-power wireless microsystems", Tang and Murray examine the use of artificial neural networks for multisensor fusion in low-power wireless microsystems. They present an integrated multisensor microsystem named *Lab-in-a-Pill* and discuss the requirements of its high-dimensional sensor fusion. They present a neural algorithm named "Continuous Restricted Boltzmann Machine" (CRBM) as a form of multisensor fusion for the Lab-in-a-Pill. The hardware implementation of the CRBM, its practical issues such as stochastic noise, time-dependent drift and biofouling, and its performance in clustering and classification of high-dimensional sensory signals are then described. They conclude their chapter with several future research directions.

In the final chapter entitled "Bio-inspired mechatronics and control interfaces", Artemiadis and Kyriakopoulos present a methodology for estimating human arm motion using a decoding method of electromyographic (EMG) signals from muscles of the upper limb and a probabilistic filtering technique for arm motion. Their methodology is assessed through real-time experiments in controlling a remote robot arm in random 3D movements using only EMG signals recorded from able-bodied subjects. Their proposed methodology can found important applications in human–robot control interfaces, such as teleoperated robot arms, arm exoskeletons and prosthetic devices.

Chapter 21
Smart Sensor Networks

Alvin S. Lim

Abstract The concepts of cyclic perception–reason–action behavior are useful for implementing smart sensor networks in a wide spectrum of applications, ranging from distributed multi-robot perception, navigation, and manipulation applications to distributed sense-and-response systems to dynamic situation awareness and decision support systems. Although there are many challenging problems, current wireless sensor network technology can be augmented to meet the requirements and address the issues in designing and implementing smart sensor network applications. The key components for supporting this are low-powered microdevices for sensors and actuators, efficient sensor network protocols, reconfigurable smart nodes, and distributed services, such as distributed lookup, composition, and adaptation services. These components simplify the implementation of multiple and independent distributed perception–reason–action modules that are executed in parallel but may communicate with each other to achieve an integrated system-level goal. Two proofs of concepts for using cyclic perception–reason–action modules are demonstrated in multi-robot control applications and real-time target tracking applications. Preliminary results from these two types of applications show the effectiveness in using these methods. In summary, the key principles of perception–reason–action cycles are instrumental in implementing many real-time cyber-physical applications. This approach can be extended in future work for improving these system components and real-time applications.

21.1 Overview

Smart sensor networks that emulate the cyclic perception–reason–action behavior are useful to many applications, such as industrial systems, business enterprises, battlefield situation awareness, healthcare facilities, smart offices, structural health, intelligent vehicles, and smart homes. These networks are involved in gathering

A.S. Lim (✉)
Department of Computer Sciences and Engineering, Auburn University, Auburn,
AL 36849, USA
e-mail: lim@eng.auburn.edu

V. Cutsuridis et al. (eds.), *Perception-Action Cycle: Models, Architectures,*
and Hardware, Springer Series in Cognitive and Neural Systems 1,
DOI 10.1007/978-1-4419-1452-1_21, © Springer Science+Business Media, LLC 2011

and disseminating real-time sensor information, processing the information through collaborative tasks in the distributed sensors, and propagating control signals to actuators for controlling the behavior of the physical systems. Smart sensor networks may consist of very large number of highly mobile sensor data sources, and actuators and users may be scattered over a wide area with little or no fixed network infrastructure support. These smart sensor networks must be self-organizing in order to adapt rapidly to dynamic changes in sensor nodes configuration and distributed sensor tasks For instance, dynamic query processing and target tracking through an unstructured sensor network of surveillance information sources and users must use the appropriate distributed services and network protocols to solve the problems of mobility, dispersion, weak and intermittent disconnection, dynamic reconfiguration, and limited power availability. Three main distributed services useful for supporting self-organization in spite of these sensor network problems are: lookup service, composition service, and dynamic adaptation service. Through a distributed implementation of these services, new application-specific network and system services can be deployed spontaneously in the sensor network. They also enable dynamic adaptation of these services to incremental addition and removal of sensor nodes, device failure and degradation, migration of sensor nodes, and changing requirements in tasks and networks. When placed together impromptu, sensor nodes should immediately know about the capabilities and functions of other smart nodes and work together as a community system to perform coordinated tasks and networking functionalities.

To provide an understanding of what design and implementation are feasible, the remainder of Sect. 21.1 will describe the current wireless sensor network technology, requirements and issues in designing smart sensor networks, and implementation issues. The engineering justification for perception–reason–action system is described in Sect. 21.2, where motivations for this work comes from a wide spectrum of applications, ranging from distributed multi-robot perception, navigation, and manipulation applications to distributed sense-and-response systems to dynamic situation awareness and decision support systems. It also reviews the challenges of these applications and previous engineering technology. Section 21.3 describes the systems for implementing smart sensor network applications through the use of key components, such as appropriate sensor network protocols, distributed services, and reconfigurable smart nodes. These components simplify the implementation of various distributed perception–reason–action modules. It provides two proofs of concepts for using cyclic perception–reason–action modules in multi-robot control applications and real-time target tracking applications. Preliminary results from these two types of applications show the effectiveness in using these methods. Section 21.4 summarizes the key principles and describes future work and extensions that will improve on these approaches.

21.1.1 Wireless Sensor Networks Technology

Smart sensor networks allow many different types of wireless sensor nodes (or actuators), and mobile devices may be assembled impromptu and reconfigured dynamically in a mobile and ad hoc wireless network. Each wireless sensor node contains battery power source, wireless communications, multiple sensing modality, computation unit and limited memory. Dual processors may be included for enhancing computation and real-time sensor processing. Different sensing modalities are supported for different types of applications, e.g., acoustic sensors, seismic sensors, piezo-electric contact sensors, ultrasonic, infrared, tactile, humidity, temperature, light intensity sensors, water quality sensors, and motion sensors using infrared imagers. Wireless transceivers in the nodes provide communications between nodes, using various transmission techniques, such as TDMA, CDMA, and OFDMA. Each node may contain a global positioning system (GPS) receiver that allows the node to determine its current location and time. GPS uses triangulation method with signals received from three or four satellites to calculate the location of the node with the accuracy of 1 m. However, without clear line-of-sight to the satellites as in urban canyons, tunnels, and indoor, GPS cannot be used. To overcome this problem, other localization devices may also be attached. Message routing and query processing often uses location information.

Wireless sensor nodes can be constructed from either commercial off the shelf (COTS) devices or specially fabricated components. A large variety of commercial devices are available, depending on the requirements of applications, such as size, battery life, robustness, wireless range, computing power, and memory needs. The following are commonly used wireless sensor devices.

The Mica and Rene motes are originally developed at the University of California, Berkeley, in the late 1990s. They are commercially available through the company Crossbow, in different versions, Mica2, Mica2dot, Imote2, and TelosB, depending on their size and capabilities. A Mica2 mote contains 868/916 MHz multi-channel radio transceiver, 38.4 kbps data rate radio, communication range of 500 ft, 512 KB flash memory, and two AA batteries. Extension boards are also available, including light, temperature, RH, barometric pressure, acceleration/seismic, acoustic, magnetic, and other sensor boards. These Mica motes typically run the open-source TinyOS operating system. Other software may also be used with TinyOS, such as Mate and MoteWorks, to enhance programmability and provide reliable, ad hoc mesh networking, development tools, and enterprise middleware. The more capable TelosB motes contain IEEE 802.15.4/ZigBee compliant 2.4 GHz wireless communication, 250 kbps data rate radio, 1 MB flash memory, and 8 MHz TI MSP430 microcontroller with 10 kB RAM.

Wireless sensor nodes can also be constructed from COTS PC104 boards and other related peripheral boards with different sensor attachments. Based on the application requirements, different devices can be attached. The following are typical devices that may be attached: IEEE 802.11 wireless PC card, 11 Mbps data rate, communication range of 100 m, 4 GB flash memory, and SLA battery. It runs the Linux operating systems, which enable smart sensor applications to be rapidly developed. Its main disadvantage is its larger size and power consumption which exceeds that of the Mica2 and TelosB motes.

Other experimental wireless sensor devices have been designed to provide extremely small form factor and low power. For example, Smart Dust mote has been designed at the University of California, Berkeley, to support these features, particularly its small 2-mm diameter size. Integrated into the single package are MEMS sensors, a semiconductor laser diode, a MEMS corner-cube retro-reflector (CCR), an optical receiver, signal processing and control circuitry, and a power source based on thick-film batteries and solar cells. This small package is self-powered to sense the physical environments and transmit its sensor information. Ultra-low-power wireless communication is provided through active optical transceivers. To support even lower power data transmission, the Smart Dust includes a CCR with electrostatic actuators that can deflect one of its mirrors at kilohertz rate, allowing the CCR to transmit data back at kilobits per second when illuminated by an external light source. The communication range is 150 m using a 5-mW illuminating laser.

21.1.2 *Design Requirements and Issues*

The selection of the smart sensor network technology to be deployed will depend on the requirements of the applications. We will first discuss some common tasks of smart sensor applications and then discuss the features of smart sensor networks required for supporting these tasks. In general, most smart sensor networks have some or all of the following requirements.

1. Determining and retrieving basic sensor values, e.g., humidity, light intensity, temperature, or barometric pressure, at a particular location or region and at a particular time or time interval. The sensor values must be disseminated to the user requesting the information possibly through multiple wireless sensor nodes configured as a mobile and ad hoc network.
2. Detecting an occurrence of an event and determining its related parameters. For example, when a collision occurs in the highway, vehicles approaching the collision must be alerted and adaptive cruise control system must be notified immediately to avoid further collisions. The exact location of the collision and time can be determined and forwarded to the approaching cars, so they can avoid the lanes with the collided vehicles.
3. Determining a physical object by evaluating multi-modal sensor data, e.g., determining the types of vehicles in the roadway by determining and fusing the characteristic of the signals from acoustic, seismic, and infrared sensors, possibly attached to a single wireless sensor node.
4. Determining the time sequence behavior of physical objects, e.g., tracking the movement of a vehicle or flow of materials in a plant. This type of tasks is more complex and often requires collaborative computation and communication from multiple wireless sensor nodes. A complex distributed algorithm will fuse sensor values from different sensor nodes to determine the object movement, e.g., velocity, direction, and location.
5. Analyzing the sensors' inputs to determine the optimal course of action through distributed and parallel simulation, multi-level simulation, and intelligent methods, such as genetic algorithms. The best set of actions will be selected to control

the desired behavior of the enterprise. If available, more powerful backend processors may be used for these analysis and simulations. Otherwise they are performed in the networked sensor nodes.

6. Determining the specific actuators and correcting parameters values and the routes for propagating the control signals to implement the selected sequence of actions for controlling the behavior of the physical enterprise.

In order to support the above tasks, smart sensor networks are required to support the following capabilities.

1. Large number of (possible mobile) wireless sensor nodes. Typical sensor deployment may consist of 10,000 or 100,000 sensors, which could be mobile, e.g., vehicle sensor networks. The networks and distributed algorithms must be scalable.

2. Low-energy consumption and long network lifetime. The network protocols and distributed algorithms must be efficiently designed to conserve energy and extend the lifetime of the sensor network in order to avoid the need for regular maintenance after nodes are deployed in remote locations.

3. Self-organization of the sensor network. When deployed in hostile environments, sensor nodes may malfunction or be destroyed, their batteries may be depleted, subnetworks could be disconnected, and new nodes may join the network. Despite these adverse conditions, the sensor network, task structures, and the algorithms must be capable of self-organization to provide uninterrupted perception, reason, and action functions as though there is no change in the network.

4. Real-time communication. Some critical sensor information, e.g., collision occurrence, must be analyzed rapidly, and the information must be transmitted to approaching vehicles within a certain time constraint. The action control signal must be propagated to the adaptive cruise control system by a certain deadline to avoid further collisions. The network and system must handle real-time communication and collaborative processing.

5. Collaborative signal and information processing. The algorithms for fusing sensor information and determining the physical object and its behavior could be distributed across multiple sensor nodes to speed up the reasoning process through parallel processing techniques.

6. Distributed query processing. Information from sensor values, event occurrences, object detection, and tracking behavior may be stored in distributed sensor nodes. Sensor query and analytical information can be efficiently processed through enhanced distributed querying techniques and optimized retrieval method from these distributed storage nodes.

21.1.3 Implementation Issues

Based on the requirements for smart sensor nodes described above, current COTS wireless sensor network node technology can support most of the underlying functionality, although some of the more advanced capabilities still need to be developed.

Most of these sensor nodes allow new sensors to be easily attached, including imaging, seismic, and acoustic sensors. Each sensor node also have computing capabilities and self-contained energy sources to process the sensor information and participate in collaborative processing. The devices have wireless communication capability, primarily using RF transceivers for exchanging sensor, analytical, and control information between one another to produce a specific global behavior for the entire enterprise. The RF transmission range depends on many factors, including power of the energy sources, terrain, obstructing objects, and interferences. Depending on the communication and sensing range, designers must derive the optimal placement of the sensors to provide the maximum coverage of a region of interest with the minimum number of sensor nodes. However, redundant sensor nodes may be deployed to increase reliability of the smart sensor network. Redundant sensor nodes also increase the lifetime of the sensor network when some sensor nodes are placed in staggered sleep state to conserve energy.

21.2   Engineering Technology Justifications

While wireless sensor network technologies are commercially available, more needs to be developed to fully support all the capabilities required for implementing the cyclic perception, reason, and action behavior. For instance, efficient self-organizing sensor networks components will support these distributed and dynamic cyclic behaviors. Though some capabilities have been developed, more new technologies still need to be developed to simplify and enhance the implementation of smart sensor applications with these behaviors.

21.2.1   *Application of Perception–Reason–Action Sensor Networks*

Networks of perception–reason–action sensor and actuator nodes are useful for implementing many complex distributed control applications, such as distributed multi-robot control, distributed sense-and-response systems, and dynamic situation awareness. These applications involve dynamic real-time interaction between the smart sensor network and the physical environment which is being controlled. The typical cycle of perception to actions progresses as follows (Fennema and Hanson 1992; Fuster 2004). The smart sensor network senses the appropriate physical energy quantity from the environment and converts the sensed quantity into its perception of the state of the environment, which is fed into some internal and possibly distributed reasoning technique to derive the best action to take. The action triggers the energy that activates the actuator and moves the device (e.g., robot) or items in the environment. MEMS devices, such as millimeter-scale silicon accelerometer, silicon pressure sensors, gyros, and flow sensors have become commonly used and effective sensors.

21.2.1.1 Distributed Multi-robot Perception, Navigation, and Manipulation

Much of the advancement in implementing perception–reason–action cyclic behavior has been in the domain of multi-robotic sensing, navigation, and manipulation (Estrin et al. 2002). However, reliable and human-like robots still require more research advances in the areas of sensing and perception, planning-based control (Rusu et al. 2009), reactive control, perception-based navigation and manipulation (Lee 2009), reinforcement learning and behavior-based robotic control (Goldberg and Mataric 2002).

Sensing and perception are required in robotics for recognizing, modeling, and understanding its environment for the purpose of reliable navigation and manipulation performance (Lee 2009). An array of robot sensors, such as camera, ultrasonic, infrared, and tactile, enhance the reliability in recognizing and understanding the environment (Goldberg and Mataric 2002). Information from these sensors will be integrated through appropriate data fusion algorithms to generate a coherent model of the environment (Lee 1997). Since each sensor type creates its own representation of the real-world object and environment, data fusion of multiple sensors' information may require integration of artificial intelligence representation and reasoning techniques (Cassimatis et al. 2004; Datteri et al. 2003). For example, two different sensors that detect an object must combine their representation of the object with real-world knowledge about the dynamics and properties of the object to determine if they are the same object.

Based on its perception of the environment and its goals, the reasoning process selects the appropriate action and sends outputs to its actuators, such as wheels, gripers, arms, and speech, for initiating the correct navigation or manipulation operations. The navigation and manipulation performance of robots depend on their capability in recognizing, modeling, and understanding its environment (Shell and Mataric 2005). Perception-guided manipulation is a more complex problem than perception-guided navigation because it requires higher level of intelligence and more precise and comprehensive modeling of its heavily cluttered 3D workspace, more direct interaction between its multiple sensor modalities (e.g., vision, force, and tactile) and its environment, and more adaption to unexpected local variations.

How a robot reacts to its perception of the environment and its goals, i.e., its reasoning process, depends on its control strategies. There is a spectrum of control strategies ranging from planner-based to purely reactive strategies (Matarić 1992). Planner-based strategies use centralized models for verifying and processing sensory information from multiple sensors and integrating them with its real-world model used by the planner to generate a sequence of actions, i.e., plan. The disadvantage of planner-based algorithms is that they rely on logic and model of the robots and their environments and is thus unresponsive and slow to adapt to dynamic environments. On the other end of the spectrum, purely reactive strategies use a collection of condition–action pairs, i.e., reactive rules. These strategies perform merely lookups of a matching set of sensor readings to generate appropriate actions. This tight coupling between the sensors and actions provides a fast and responsive feedback loop but lacks generality and the ability to store knowledge representations. The system maintains minimal internal state and no internal model and search. Hybrid strategies combine both purely reactive and planner-based strategies by using the reactive approach for low-level control and the planner-based approach for higher level decision making. Although the two control systems are independent, they may communicate through an intermediate software layer for coordinating their actions. Typically, the low-level reactive process takes care of the immediate safety of the robot, while the high-level planner selects the action sequences.

Another approach that provides more powerful and flexible reasoning capabilities is using behavior-based strategies that generalize reactive systems by introducing the notion of behavior as an encapsulated, time-extended sequence of actions. Behavior-based strategies combine tightly coupled perception–action components (as in reactive systems) with behavior representation and adaptation. Behavior-based systems are distributed and may consist of parallel and concurrently executing behaviors without centralized arbiter or reasoning process. They may use different forms of internal representations and perform computation on them to determine the actions to take. The decision is made based not only on the different sensor information but also on the input from other behaviors in the system. Thus, unlike reactive systems, behavior-based systems have greater expressive and learning capabilities. While reactive systems require every tasks and states for achieving goals to be defined at design time, behavior-based system allows interaction of the various system components and behavior with task-specific modules to produce an emergent behavior for achieving the desired global goals.

Determining the correct action to take will involve not only conditions that can be immediately sensed but also the environment in the past and future, occluded, spatially distant and hypothetical situations that cannot be immediately sensed. To achieve this will require the integration of multiple representation and reasoning techniques. Multi-agent system architectures are useful for composing multiple reasoning and planning techniques and integrating them continuously with multiple representations spanning from low-level perception and action to higher level knowledge representation schemes. Multiple sequences of simulations of the different reasoning and planning schemes can be interleaved and executed in parallel.

21.2.1.2 Distributed Sense-and-Response Systems

In general, sense-and-respond (S&R) systems simply sense (and perceive) the environment and respond (reason and act) appropriately. In their simpler forms, S&R systems (Chandy 2005) consist of rules defined using when–then semantics. The when clause corresponds to the detection of an event, and the then clause describes the execution of the response. The condition complexity may vary. For instance, the when clause may be based on a single event, a history of events, a single stream, or the fusion of multiple streams. The response may be human centered or automated. The cost of an S&R system can be broken into three parts: the cost of false positives, the cost of false negatives, and the incremental costs of running the system.

There are several interesting applications of S&R systems.

1. A sensor-based emergency medical response system can be developed in the field of healthcare (Hashmi et al. 2005). At the lowest level, sensors worn by patients transmit vital signs and location information to local command centers (e.g., ambulances) via IEEE 802.15.4 networks. Real-time communication capability is important since information about vital signs is being communicated and timely response of the system is critical to the care of patients. The goal of the system is to provide greater situational awareness about patient condition and arrival time so that doctors can make more informed decisions.

2. Many business applications (Cohen et al. 2005; Haeckel 1999; Kapoor et al. 2005) can benefit from S&R systems. For instance, a unified event stream processing system can be used to monitor, analyze, and detect critical business events. This provides an accurate perception of the situations. The objective is to improve efficiency, reduce cost, and enable the enterprise to react quickly. The event stream processor provides the reasoning capability by calculating metrics based on event messages received from a variety of sources such as complaint databases or real-time production statistics. A domain expert creates rules based on the metrics to provide alerts for important business situations. Real-time communication mechanism is well suited to the time-critical nature of many phenomena that lead to immediate response with the appropriate action command.

3. S&R systems have also been developed for hazardous weather detection and response, such as tornados, hurricanes, and tsunamis. For instance, NetRad (Zink et al. 2005) is a distributed adaptive collaborative sensing (DACS) system for early tornado detection, with the goal of detecting a tornado within 60 s of formation and track its centroid within a 60-s temporal region. NetRad is composed of a dense network of low-powered radars that collaborate to accurately predict tornado formation. The radars report atmospheric readings to the System Operations and Control Center (SOCC) where a merged version of the data is analyzed to detect meteorological features. Unlike other S&R systems, the NetRad system uses a high-bandwidth, highly structured, wired network to connect the sensors. Other smart sensor networks also emphasize fast response time on the order of subseconds, as opposed to the 30-s turnaround time of NetRad.

An interesting application of distributed and large-scale S&R systems is a smart sensor network for detecting tsunamis and mitigating their effects. The current tsunami warning system is composed of ten buoys in the Pacific and five in the Atlantic/Caribbean. Each buoy contains a submerged tsunameter for detecting subtle pressure changes which indicate tsunami waves. This system is maintained by the National Oceanic and Atmospheric Administration (Gonzelez et al. 2005; Meinig et al. 2005a, b) through the Deep-Ocean Assessment and Reporting of Tsunamis (DART) project.

An alternative system was proposed (Casey et al. 2006, 2008) for more accurate detection and response to tsunamis, particularly in coastal areas with valuable assets (e.g., nuclear reactors). The smart sensor network consists of a large wireless ad hoc network of about 80 sensor nodes, each equipped with submerged tsunameter to collect underwater pressure reading across a large coastal area. The sensing data from these networked sensors are exchanged between the sensor nodes and an analysis algorithm to provide the accurate perception of detection of tsunamis and prediction of the path of the tsunami waves. The reasoning is provided by an analysis algorithm (Casey et al. 2006) that uses a general regression neural network (GRNN) (Specht 1991) to analyze the pressure data from the tsunameters and predict the set of actuators that should be fired at a subset of four to ten inflatable barriers for impeding the tsunami waves. The system includes a real-time response mechanism with real-time network protocols to ensure that these action commands are delivered to the barriers within the short deadline for impeding the tsunami waves.

21.2.1.3 Dynamic Situation Awareness and Decision Support Systems

In many automated rapid response systems, decisions for action must be made quickly as the system develops accurate awareness of the current situation and the perception of the dynamic environments (Yilmaz et al. 2007). Sensors on the ground, such as cameras, acoustics, and seismic sensors, may detect some cache of weaponry and a hostile crowd formation. Since it is impossible to foresee all moves in a conflict, decision makers must use a reasoning method that considers several contingency models on demand to explore and determine plausible consequences of their courses of actions. Because the environment will dynamically evolve in real time, the sensors must continuously generate accurate perception of the observed state that will in turn help context decision makers to identify and bring one or more new family of models. For instance, the decision maker may reason that the original plan may result in civilian casualty within the new observed context. Hence, the reasoning process must experiment with several contingency models in real time on demand. Using this decision support reasoning method, the system can be applied to unstructured problems with the characteristics of (1) deep uncertainty, (2) dynamic environments, (3) shifting, ill-defined, and competing goals, (4) action/feedback loops, and (5) time stress.

21.2.2 Review of Application Challenges

Building self-organizing smart sensor networks and implementing perception–reason–action sensor networks are challenging problems for the following main reasons.

1. Many different types of sensors with a range of capabilities may be deployed with different specialized network protocols and application requirements. Data-centric network protocols are becoming common in sensor network (Esler 1999; Estrin 1999). With many mixed types of sensors and applications, sensor networks may need to support several data-centric network protocols simultaneously.

2. These mixed types of sensor nodes may be deployed incrementally and spontaneously with little or no pre-planning. The networks must be extensible to new types of sensor nodes and services. They must be deployed spontaneously to form efficient ad hoc networks using sensors with limited computational, storage, GPS and short-range wireless communication capabilities. They rapidly coordinate with each other for detecting, tracking, reporting activities and disseminating the information efficiently through the impromptu network of sensors.

3. Wireless mobile sensor nodes need to form temporary networks in lieu of any established infrastructure or centralized network administrator. Runtime facilities for information processing and communication must be capable of adapting to the problems of amorphous networks.

4. The sensor network must react rapidly to changes in the sensors composition, task or network requirements, device failure and degradation, and mobility of sensor nodes. Sensor devices may be deployed in very harsh environment and subject to destruction and dynamically changing conditions. The configuration of the network will frequently change due to constant changes in sensor position, reachability, power availability and task requirements. The network protocols must be survivable in spite of device failure and frequent real-time changes. Sensor network must be secure in the face of this open and dynamic environment.

5. No fixed network infrastructure exists for the large number of sensors in the location, such as hostile battlefield environments. Networking protocols cannot rely on the existence of fixed networks such as the current version of Mobile IP. Mobility of sensor nodes and RF relay nodes affects distributed algorithms that rely on the topology of the network. Capabilities must also be included for locating, identifying and tunneling information to and from mobile sensor nodes.

6. Sensing and actuation involve real-time energy flow between the sensor network and the physical environment which is problematic because of uncertainty in the estimation of the state of the physical system. Since sensors and actuators are physical devices with accuracy and precision limitations, the measured sensor data are only approximation to the actual values. Accuracy and fault tolerance must be enhanced using techniques such as redundancy with overlapping sensor field of view, sensor fusion algorithms, and placement strategies. Another problem is the stochastic latency in actuation which may cause instability and unreliability in closed loop control.

21.2.3 Review of Previous Engineering Technology Systems

Other alternate technology systems are also available for implementing smart sensor networks, such as the cellular phone network system and the internet with fixed network and mobility enhanced capabilities (e.g., Mobile IP and cellular broadband network). There are advantages in using the cellular phone communication system. The base station infrastructures are pervasive, access to the wireless network is easily established, and area of coverage is extensive in most places. The main disadvantage in using cellular network is the long delay in setting up connections and also the long end-to-end delay for sensor-to-sensor communication. Also, there is a high cost involved in subscribing to the service provider and for setting up connection, particularly for large number of sensor nodes that are deployed in the field. Furthermore, in remote areas or hostile environments, e.g., battlefield, cellular infrastructure is nonexistent or cannot be depended. Another viable technology option for implementing smart sensor networks is using IEEE 802.11 or WiFi networks that provide the last hop wireless communication service for connecting mobile and unthethered wireless sensor nodes to a fixed internet networking system. Since the IEEE 802.11 communication range is about 100 m, the fixed internet network routers or access points must be within 100 m of all the sensor nodes in the network. In remote areas, placement of wireless IEEE 802.11 access points may not be feasible. Another disadvantage of using this approach is that the power consumption of IEEE 802.11 wireless cards is very high, partly due to its frequent scanning algorithm.

21.3 The System

The required basic functionalities of smart sensor networks can be supported through three reconfigurable and mobility-aware subsystem layers: sensor signal and information processing systems, reconfigurable distributed services, and the sensor network and physical device layer. This architecture allows efficient implementation of the functionalities and avoids duplication of functionalities in the different layers It promotes efficient coordination between them. The sensor signal and information processing layer contains mobility-aware mediators and adaptive sensor query processing. The runtime reconfigurable distributed system contains distributed services for supporting mobile smart sensor applications. The network and physical layer contains data-centric network routing protocols, physical wireless transmission modules and sensors that generate the raw data. The architecture consists of three key system layers (Fig. 21.1):

1. *Application systems*, e.g., sensor information processing layer and collaborative signal processing
2. *Reconfigurable distributed systems* that provide distributed self-organizing services to the application systems
3. *Sensor networking and physical device layer* that routes messages through the ad hoc sensor network.

At the physical device layer, different physical sensor and mobile devices may be assembled impromptu and self-organize dynamically in an ad hoc wireless network. Each sensor node contains battery power source, wireless communications, multiple sensing modality, computation unit and limited memory. Dual processors may be included for computation-intensive tasks and real-time sensor signal processing. Many sensing modality may be supported, including acoustic

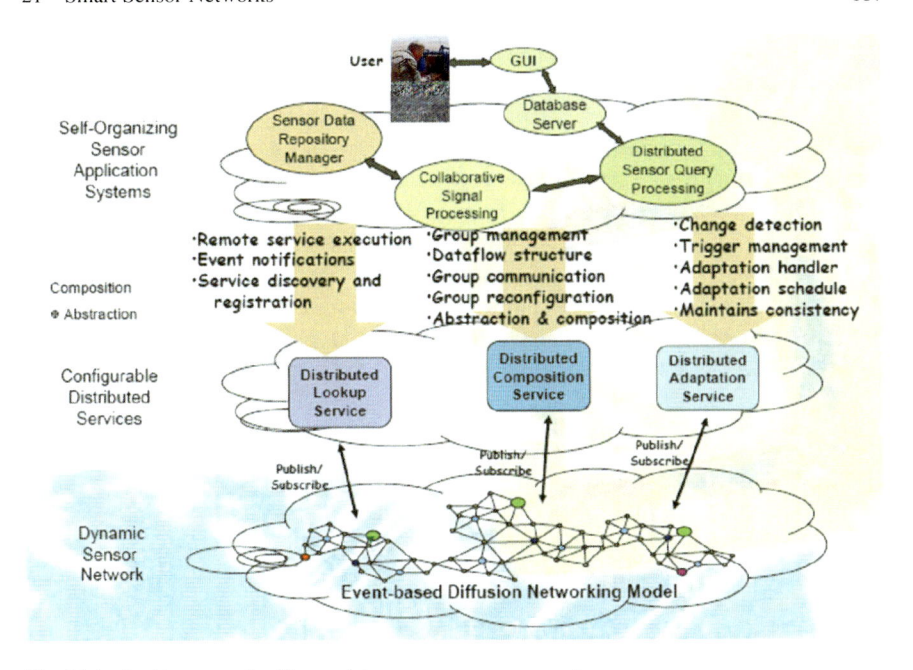

Fig. 21.1 Architecture of self-organizing smart sensor networks

sensing using high-sensitivity and noise-canceling microphones, seismic vibration using geophones, and motion detection using two-pixel infrared imagers. Wireless transceivers in the nodes provide wireless communications between nodes. Hierarchical communication between clusters of neighboring nodes can be dynamically established through appropriate selection of wireless channels by the cluster head nodes. Each node contains a GPS receiver or other localization devices that allow nodes to determine their current location and time.

At the networking layer, ad hoc routing protocols allow messages to be forwarded through multiple physical clusters of sensor nodes. Directed diffusion routing is often used because of its ability to dynamically adapt to changes in sensor network topology and its energyefficient localized algorithms. To retrieve sensor information, a node will set up an interest gradient through all the intermediate nodes to the data source. Upon detecting an interest for its data, the source node will transmit its data at the requested rate.

The reconfigurable distributed system uses the diffusion network protocol to route its messages in spite of dynamic changes in the sensor network. These distributed services will support applications systems, such as distributed query processing, collaborative signal processing, and other applications. The advantage of using these services is that application and system programs may use simpler communication interfaces and abstraction than the raw network communication interface and metaphor (e.g. subscribe/publish used in diffusion routing). Furthermore, these distributed services may enhance the overall performance, such as throughput and delay. These services will be implemented on top of the directed diffusion protocol which can still be used by applications concurrently with these distributed services. Directed diffusion can simultaneously support retrieval of sensor data and also be used by some of the distributed reconfigurable services. On the other hand, distributed services provide other forms of communication – such as reconfigurable inter-process communication and impromptu establishment of community of smart sensor services – as required by smart sensor applications.

At the application system layer, distributed query processing and collaborative signal processing modules could communicate with each other to support smart sensor applications, such as surveillance and tracking functions of the enterprise. In smart sensor systems, the cooperation between mobility-aware mediators, sensor agents and collaborative signal processing modules provide efficient access to diverse heterogeneous sensor data, surveillance and tracking information through the sensor network. The mobile sensor information layer is supported by three major components: interoperable mobile object, dynamic query processing and mobile transactions. In the interoperable mobile object model, cooperative network of mobility-aware mediators and sensor agents will be configured to support interfaces to remote sensor data sources through multihop wireless network protocols.

In applications that are based on perception, reasoning, and action cycles, each smart sensor node implements its own behavior supported by its sensors that allows it to perceive its environment, its actuators to control its movement and manipulation of the environment, and its local reasoning process which may take inputs not only from its sensors but also from other smart sensor nodes that implement other behaviors. These distributed behaviors communicate with each other through the wireless sensor network supported by directed diffusion. As more task-specific behaviors are added to the application, its capabilities will be more complex as more complex properties could emerge from the interactions of these various behaviors and system components (Matarić 1992). While the smart sensor nodes may communicate through wireless links, they may also communicate indirectly by sensing the current state of the environment and goals that could be affected by actuators of other mobile sensor nodes. Information can thus be exchanged between these nodes at higher levels of consistency and correctness.

21.3.1 Components

21.3.1.1 Data-Centric Sensor Network Protocols

Data-centric networking, such as directed diffusion protocol (Intanagonwiwat et al. 2003), is useful not only for sensor data retrieval but also for implementing all the distributed reconfigurable services over dynamically changing ad hoc sensor networks. Diffusion routing converges quickly to network topological changes, conserves mobile sensor energy, and reduces the network bandwidth overhead since routing information is not periodically advertised. Routing is based on the data contain in sensor nodes rather than unique identification. Directed diffusion is a type of reactive routing protocols which only update routing information on demand. In contrast, proactive routing protocols, such as links state routing, frequently exchange routing information. For sensor networks that experience greater dynamic changes, reactive routing algorithms are more appropriate, whereas for those that are more static and experience infrequent topological change, proactive routing algorithms are more efficient.

Directed diffusion is a data-centric protocol, i.e., nodes are not addressed by IP addresses but by the data they generate. Data generated by a node is named by the *attribute–value* pairs. A sink node requests for a certain data by broadcasting an *interest* for the named data in the sensor network. The interest and gradient is established at intermediate nodes for this request throughout the sensor network. When a source node has a data that matches the interest, the data will be "drawn" down toward that sink node using this interest gradient that was established. Intermediate nodes may cache, transform data, or direct interests based on previously cached data. The sink node can determine if a neighbor node is in the shortest path whenever it received a new data earliest from that node. The sink node will reinforce this shortest path by sending a reinforcement packet with a higher data rate to this neighbor node which forwards it to all the nodes in the shortest path. Other nonoptimal paths may be negatively reinforced so that they do not forward data at all or at a lower rate.

Distributed services and applications use *publish and subscribe* API provided by directed diffusion. Through the `subscribe()` function, an application declares an interest that consists of a list of attribute–value pairs. The subscription is then diffused through the sensor network. A source node may indicate the type of data it offers through the `publish()` function. It then sends the actual data through the handle returned from the `publish()` function. The sink node then receives the data that has propagated through the sensor network using a `recv()` function call with the handle returned from the `subscribe()` call.

21.3.1.2 Distributed Services

Self-organizing smart sensor networks may be built from reconfigurable smart sensor nodes that may be developed independently but may interact with other smart sensor nodes through well-defined interfaces that encapsulate networking and interaction functionalities. Some smart sensor nodes may execute autonomously to provide networking and system services or control various information retrieval and dissemination in the dynamically changing sensor network (Lim 1999). They may simultaneously be service providers for other sensor nodes and be clients of services that other smart nodes provide. To enhance the ability to reconfigure their networking, configuration, and adaptation functionalities, smart sensor nodes may make use of three main classes of distributed services: Lookup service, composition service, and adaptation service (Fig. 21.1).

The lookup service enables new application, system, and network services to be registered and made available to other sensor nodes. Methods for calling the services remotely are also provided by the lookup service. The composition service allows clusters of sensor nodes to be formed and managed reliably. The adaptation service allows sensors nodes and clusters to reconfigure dynamically as a result of sensor node mobility, failure, and spontaneous deployment. These servers enable sensor nodes to form spontaneous communities in ad hoc sensor networks that may be dynamically reconfigured and hierarchically composed to adapt to real-time information changes and events.

These distributed servers may be replicated for higher availability, efficiency, and robustness. Distributed servers coordinate with each other to perform decentralized services, e.g., distributed lookup servers may work together to discover the location of a particular remote service requested by a node.

Reconfigurable Smart Nodes

By exploiting these distributed services, sensor nodes can be enabled to be self-aware, self-reconfigurable, and autonomous. These sensor nodes, known as reconfigurable smart nodes, can be used to build scalable and self-organizing sensor networks. (We will refer to reconfigurable smart sensor nodes as smart nodes or sensor nodes.) Smart nodes may represent sensor nodes, actuator nodes, other types of mobile nodes, fixed nodes, or cluster of these nodes. They may simultaneously be service providers for other smart nodes and clients of services that other smart nodes provide. Smart nodes may be dynamically composed into impromptu networked clusters forming clustered smart nodes that work together to provide abstract services for the agile sensor network. They may also adapt rapidly to abrupt changes in the sensors capabilities, events, and new real-time information. Very large networks with hundreds of thousands of sensors nodes can be deployed by hierarchically composing reconfigurable smart nodes.

Smart sensor nodes may consist of hardware devices and software for interacting with the real-world systems. The hardware may contain computational, memory, wireless communication, and sensing devices. Smart nodes may contain control software for monitoring information from real-world devices such as simple sensors, engaging in distributed signal processing and generating appropriate control signals to produce a desired result in the real-world system. The control software takes advantage of the functionalities provided by the networking and system software.

Smart nodes interact with other smart nodes through well-defined interfaces (for networking and systems operations) which also maintain interaction states to allow nodes to be dynamically reconfigured. These explicit interaction states and behavior information allow localized algorithms with the adaptation servers to maintain consistency when autonomous nodes and clusters are reconfigured dynamically, move around, or recover from failure. The implementation and data of nodes (software and hardware) are encapsulated (hidden) from other nodes.

Different designers may independently develop smart nodes and their network and system services using different methods. For example, one designer may use a network protocol that is suited for a particular sensor application with its set of network requirements, such as low latency, power conservation, GPS capability, high error rate, and disconnection. In order to ensure consistency during dynamic reconfiguration and failure recovery of sensor nodes, the protocols may be analyzed by the adaptation servers based on their specification model.

When new smart nodes are added to the sensor network, they register their services with a lookup server. Other nodes that require a service will discover the services available in a cluster through the lookup servers that return the location and interface of the service nodes. This is similar to Jini (Arnold 1999) that manages system-level services based on Java code executing in IP-based networks. On the other hand, reconfigurable smart nodes may provide lower level networking services using generic mobile codes executing in data-centric sensor networks. Client nodes then interact directly with the service node. Smart nodes are self-aware of their own location, configuration, and services that they perform.

Lookup Server

New network and system services may be introduced (or deleted) by a sensor node to other nodes in a self-organizing sensor network. A sensor node that provides a service is called a service provider and a node that uses the service is called a service client. Since service providers may be introduced or removed from the sensor network at any time, a lookup server is needed to keep track of the availability of these services. A sensor node may register a resource that it maintains or service that it can perform with a lookup server. Each smart node has a home lookup server which keeps track of the location of the node when it moves. A lookup server may contain information on services or resources at multiple clusters. Other nodes that require the service may request the service through a lookup server. If the service is recorded in the lookup server, it will return the location of that service to the requesting node. Otherwise, if the service is not recorded in the lookup server of the region, a discovery protocol is used to locate the service through other lookup servers. A request message is propagated to all the lookup servers and the server that contains the service registration information will return the reply with the service location. It may also return the cluster name of that service. The lookup server that made the request will then cache that service location and cluster name information in its local registration cache. At regular frequency, service and resource registration information may be disseminated from one lookup server to other lookup servers in the agile sensor network.

Lookup servers support mobility of sensor nodes. When a sensor node moves to a different cluster at another location, it notifies, whenever possible, the previous lookup server that it is moving. When it arrives at another cluster in a new location, it will register with the new lookup server which will notify the previous lookup server. The new lookup server will propagate the change in the nodes' location to other lookup servers. The lookup server responsible for a sensor node that is interacting with the mobile node will notify the sensor node of the service location change. Existing interactions between the mobile node and other nodes will thus be handed over to the new location. Adaptation servers may be involved in the handover operation to preserve global consistency during the handoff, as discussed below, resulting in uninterrupted use of the service.

Compositional Server

The compositional server manages various smart nodes that may be added to (or removed from) clusters in the agile sensor network. It also manages network abstractions (or group behavior) of clusters and hierarchical composition of clusters.

The compositional server simplifies dynamic reconfiguration of services provided by each smart node or cluster. It also simplifies the development of large self-organizing sensor network by allowing individual node and cluster to be specified and designed independently, while the compositional behavior and constraints on a cluster of components may be separately specified.

Compositional servers enhance compositionality and clustering abstraction of sensor networks. To enhance adaptive sensor networks, each node can be designed independently and the networking requirements, interaction, and synchronization operations with other nodes may be specified separately. This decoupling of autonomous smart nodes from their networking requirements and interaction behavior enables smart nodes to be easily adapted, replaced, and reconfigured when triggered by dynamic events in the sensor network.

Clusters of smart sensor nodes may be formed under the management of a compositional server. Hierarchical clusters are also possible for larger scale sensor networks. Sensors may cooperate in providing services and resources as a group. A cluster of sensors may also provide distributed services by coordinating the tasks among the sensors, such as aggregating summary information. Clustered smart nodes encapsulate the abstract group behavior, networking and system capabilities provided cooperatively by the group of smart nodes. The abstract behavior may map to the lower level behavior of the individual adaptable components. There will be a head smart node in the cluster that is responsible for the control of the cluster and inter-cluster communications and networking functions. Group communication to nodes in a cluster can be efficiently implemented by sending a message first to the cluster head which then multicasts it to the member nodes. Member nodes will elect a cluster head from the set of nodes with most powerful networking and system capabilities. Smart nodes in a cluster may cooperate to perform the networking and system functions for the cluster. Synchronization constraints associated with network protocols and system services among smart nodes may be specified in clustered smart nodes. Clients of the services may work cooperatively together to perform common objectives in the abstract services. The formalism used for defining separate component behavior and interaction behavior simplifies the mechanisms for replaceable components and alterable interaction patterns. The capability to specify hierarchical composite clusters enables designers to build large and complex sensor networks by clustering together smaller network-enabled sensor devices at each level.

Adaptation Server

Adaptation servers utilize information from the compositional server, lookup server, and analytical tools to control smart nodes during dynamic reconfiguration and failure recovery (Lim and Friedberg 1992) of the agile sensor network. Each smart node may execute autonomously to control different network operations in the sensor network and may interact and coordinate independently with other smart nodes to perform collaborative networking and control operations (Lim 1995). For example, in a flexible manufacturing cell, a component may control a milling machine and may communicate with components that control the robot and workpiece to coordinate the task of removing a milled workpiece and transporting it to the next workstation.

Adaptation servers monitor clusters of smart nodes during normal execution either through the spontaneous signal from the sensors, probing of the smart nodes, or explicit network management directives for reconfiguration and failure recovery. When a runtime reconfiguration is requested or triggered, the adaptation server will generate the appropriate schedule of reconfiguration operations that will ensure the reconfigured and affected sensor nodes are globally consistent. Servers are not customized for particular network protocols or system services. Instead, they store and analyze the component composition and analytical results for coordination, adaptation, and failure recovery. To ensure correct adaptation and maintain consistency, the adaptation server makes use of analytical tools for dependency analysis and relevant information from compositional servers and lookup servers. When smart nodes are added or removed from the agile sensor network, a suite of analytical tools may be utilized to ensure that the sensor network still maintains its safety and liveness properties (Lim 1995). Smart nodes (or clusters of smart nodes) may be specified and analyzed independently. The server may take the smart node and network specifications and produce intermediate results to a diagnostic checker that checks for syntactic correctness, overall reachability of all states, and feasibility of each abstract behavior of the networking functions. The suite of analytical tools automatically verifies the potential for deadlock, livelock, and starvation. While deadlocks are usually easily removed by specifying additional constraints, livelock and starvation are probabilistic properties which are not easily removed by modifying the specification. Starvation and livelock are better handled by dynamically detecting them and recovering affected processes. The analytical tools also analyze dependencies for preserving global consistency during dynamic adaptation and failure recovery.

21.3.1.3 Distributed Perception–Reason–Action Modules

Reconfigurable smart nodes are useful for implementing distributed perception–reason–action modules that may communicate with one another to perform sequences of actions that will achieve common global tasks. They enable the system to be easily modified where new task-specific modules may be added at runtime

(Gerkey and Matarić 2001). Existing community of behaviors (modules) can discover newly introduced behaviors and communicate with each other to perform more complex tasks by combining their capabilities and sharing more extensive and reliable information about the states of the environment.

For maximum flexibility, each task for perception (including sensing), reasoning, and action (including actuator) may be implemented by independent but communicating reconfigurable smart node. For efficiency and practical reasons, some of the tasks (or even all tasks) may be combined in some reconfigurable smart nodes, e.g., perception and reasoning in a smart node.

A behavior-based controller is a network of these reconfigurable smart nodes that can be grouped under three categories: perception, reason, and action. In a typical cycle, information flow from the perception to the reason components, from reason to other reason components, and from reason to action components (Fuster 2004; Klyubin et al. 2004). Based on the current state of the environment and behaviors, the most appropriate action is selected. After the action is performed, the resulting changes in the state of the environment will be perceived through a collection of sensors. This may trigger the activation of another perception, reason, and action cycle. This process will repeat continuously until the eventual goal is incrementally achieved.

Perception

The perception components consist of various types of sensors that produce sensor inputs from the physical world into the perception system and algorithms for determining the perceived state of the environment. Examples of sensors are cameras, piezo-electric bumpers, infrared sensors, tactile sensors, MEMS devices, seismic sensors, acoustic sensors, and positioning system. For example, images from the cameras can be processed by an image understanding software that will determine the existence of an obstacle. In robotic manipulation system, it is important to provide visual perception with 3D recognition and modeling of the unstructured environment which may be cluttered with various daily life objects. Recognition of objects and their poses in 3D must be accurately and reliably determined in real time using advanced vision techniques, such as the particle filter-based probabilistic method (Lee et al. 2008). The accuracy and dependability of recognition and modeling of the environment will determine the overall reliability of a robot in performing its navigation and manipulation tasks.

Reliability in perception can be achieved using techniques, such as "behavior perception" where a robot collects and selects an optimal set from multiple evidences (visual inputs) from the environment and then matches them against different models stored in the robot database. The matching algorithm for recognizing an object and estimating its pose is based on sequence of images, in which particles are used to represent the probability distribution of object pose in 3D space. Different techniques are used, such as SIFT (scale invariant feature transform) (Lowe 1999), color, 3D line, 2D square (Lee et al. 2006), etc., to generate multiple evidences of the environment, such as photometric and geometric features. Using consecutive observation in a sequence of images, the particles are converged into a single pose. The process of matching evidences with models is repeated multiple times until a sufficiently high level of certainty in recognition and pose estimation is reached. The integration of these evidences enables perception to be reliable even in noisy and unfriendly visual environments. The main challenge is to develop a method that will perform well in real-world environments, when the 3D data is not accurate, texture condition of the object is poor, object is a distance away (scaling), rotation or the illumination is low.

There are several techniques for object recognition using statistical approaches. One approach uses maximum a posteriori (MAP) estimation (Boykov and Huttenlocher 1999) to find the match between object and scene by using a Markov random field (MRF) model (Boykov and Huttenlocher 1999) as a probabilistic model for capturing dependencies between features of the object model. Another approach uses multidimensional receptive field histogram (Schiele and Crowley 1996). The particle filtering approach described above (Lee et al. 2008) has the advantage that it can also recognize textureless objects since it can deal with various features such as photometric features (SIFT; Lowe 1999, color) and geometric features (line, square) (Lee et al. 2006). This probabilistic method is based on a sequence of images to recognize an object and estimate its pose through the following procedures. First, it calculates some information on the environment, such as texture density, illumination, and distance of expected object pose, from the input image and the 3D point cloud of in situ monitoring. Next, it selects the valid features in an input image through the cognitive perception engine (CPE), which automatically perceives the environment from in situ monitoring information and stores evidences of all objects to be recognized. Then, these features are used for generating multiple poses which are in turn used for determining observation likelihood. The object pose is represented by particles that are propagated from the previous state using the motion information. Based on weights assigned to predicted particles, it then re-samples the particle. The above procedures are repeated until the particles converge to a single pose. Through this method, 3D objects can be recognized even in unstructured environments that are cluttered with various everyday objects.

Reason

The reasoning system consists of multiple modules implemented in independent but communicating reconfigurable smart nodes. Each module receives information from either the raw sensor input or the perception of the states of the environment from some recognition or modeling components. In behavior-based control, reasoning system consists of multiple units called behavior. Each behavior receives input from sensors and/or other behaviors, processes the inputs, and sends outputs to actuators or other behaviors. Each behavior generally performs some independent tasks, such as homing, puck detection, and puck grabbing, based on the information it received from the sensors, from the perceived state of the environment, or from other behaviors. All behaviors in a controller execute in parallel, simultaneously receiving inputs and generating outputs. When multiple outputs are transmitted simultaneously to an actuator or behavior, an action selection mechanism will resolve the conflict and generate the most appropriate output for the actuator or behaviors from the multiple outputs (Pirjanian 1998).

A collection of behaviors make decision on the most appropriate next action to take based on the input information they collectively receive. As each behavior receives input information, it individually makes a decision to generate an actuator command to perform an action that it reasons is the most appropriate reaction to the input sensor and behavior information. However, multiple actuator commands may be generated simultaneously by multiple behaviors. The decision on which of these actuator commands is the best next action is determined by control signals sent between behaviors. The control signal mechanism is used in the reasoning system for three main purposes. First, a behavior may also transmit signal to temporarily inhibit another behavior from executing or permanently inhibits it until it lifts the inhibition at a later time. Second, a behavior may transmit a signal to another behavior to perform a certain action. Third, the control signals may be used to arbitrate between actuation commands from different behaviors to decide on the correct actions. More details of the action selection mechanism are described in the next subsection.

Action

Action components receive actuator commands from the behavior and the action selection mechanism to activate their actuator devices. The actuator commands are converted to signal for controlling the actuators, e.g., robotic arm, in order to activate a specific trajectory based on the dynamics of the arm.

Sensing and action are closely related in cyclic loops. Each module is task specific and implements a behavior by specifying the sensing thresholds in combination with a reasoning process to trigger specific actions. A collection of these task-specific modules in collaboration will result in the desired global behavior (Jones and Matarić 2005). The advantage of this approach is that it enables properties to emerge from the interaction of the various system modules and components (Maravall and de Lope 2003). Information flow between modules is done indirectly by the modules themselves from sensing of the environment and using the robot's current state and goal, rather than through direct communication and dependence. This minimizes the complexity of inter-module communication and synchronization. In the absence of a centralized controller, parallel execution of task-specific modules that interact with each other will achieve an integrated system-level goal.

Because of their effect on the overall goal, it is important to design task-specific modules correctly. Some modules implement reactive behavior, such as guaranteeing safety and survival in a dynamic and unstructured environment, while other modules involve more complex behavior that includes reasoning for sequences of incremental tasks that will achieve the system-level goal. Reactive modules implement reflexes for safety and survival and contain collection of simple rules which receive sensor inputs and send command directly to actuators. These bottom-up processes also operate under top-down constraints to ensure that the purpose of the system is achieved.

Task-specific modules should be carefully designed using the following strategies. First, the behavior of the modules should be defined qualitatively in the observer space. Second, the modules should be specified in terms of actions in the observer space, for achieving the subgoal that will reduce the difference toward the main goal. Third, the actions should be specified in terms of the robot's actuators with as fine grained a precision as possible, so that small and incremental motion is made by each action instead of a large trajectory to the final position. This allows reactive control to be combined with behavior-based control where constant sensing enables probabilistic convergences toward the system goal. This incremental process allows errors to be corrected dynamically rather than using a single large action where the environment may change during execution of the large action.

In planner-based control systems, all sensory information are combined to derive the global world view using different sensor fusion techniques, ranging from simple potential field computation (Brooks 1986) to Kalman filters (Ayache and Faugeras 1988; Crowley 1985). These techniques for global integration are complex with more stringent requirements on the sensing resources. On the other hand, behavior-based control systems are decentralized and do not attempt to merge all sensory information into a single model. Instead, subsets of the sensory information are combined into task-specific behavior that generates small incremental actions. For example, hand content and arm position can determine if the robot can head home (Connell 1990). Since there are multiple such behaviors with small incremental action executing in parallel, the hard problem is arbitrating among potentially many actions and selecting the most appropriate action.

Since multiple actions generated simultaneously by multiple behaviors may be directed to the same actuator, the controller must select only one most appropriate action. This ensures that only one behavior has control over the actuator at any time. This is achieved in behavior-based systems using a fixed control hierarchy that defines a priority ordering on the behaviors. Only the action of the highest priority behavior in a given context is selected. Two types of decision mechanisms are used for deriving the correct actions: (1) behavior selection, where only one actuator command is selected out of multiple possible commands from different behaviors and (2) subsumption-style priority scheme, where a priority ordering of the actuator commands will determine which action will take precedence when merging the actions. In addition to fixed priority schemes above, other more flexible approaches are also available. Most common methods use a multivariable function embedded in each behavior to determine an activation level for each behavior. The thresholds are tuned or determined through a learning process to trigger the most appropriate action (Brooks 1991; Matarić 2001). Other approaches use voting schemes to arbitrate among behaviors.

The overall behavior of the system is defined by the sequence of actions that are triggered by different modules. This action sequence will emerge over time based on the dynamic changes in the environment and the robot states and goals. Because of the importance of ensuring correct sequencing, behavior-based modules must be designed with anticipation of the appropriate sequences that will contribute toward the eventual goals. In contrast, planner-based control system computes the sequence of actions at runtime based on the conditions at the beginning of the time period.

21.3.2 Proof of Concept

As proof of concepts to demonstrate the usefulness of the systems and techniques, we will describe two different applications: multi-robot control and target tracking applications. Both of these applications involve distributed sensors and actuators where distributed sensing data generates the appropriate perception and reasoning processes which trigger the appropriate sequence of actions toward the eventual goals.

21.3.2.1 Multi-robot Control Applications

The behavior-based perception–reason–action technique is useful for implementing an integrated control for a group of mobile robots which perform navigation, landmark detection, map learning, and planning without the support of a centralized controller (Goldberg and Mataric 2002). The behavior modules can be easily implemented using reconfigurable smart sensor nodes that support reconfiguration, discovery of behaviors, and communication.

The implementation of behavior-based multi-robot control is described in Goldberg and Mataric (2002). The main tasks demonstrated with these multi-robot control applications are distributed versions of de-mining, toxic waste cleanup, and terrain mapping. Multiple robots execute in parallel to cooperate in performing the global tasks while avoiding interference with each other. Although their implementation is not based on reconfigurable smart nodes, it could be more efficiently and flexibly built using reconfigurable smart components.

The main task of the multiple mobile robots is to search and collect certain objects in a search region and bring them to a goal region using some form of navigation. The entire area, assumed to be contiguous and bounded, can be divided into three nonoverlapping types of regions: search regions, goal regions, and empty regions that do not contain any object and are not goal regions. The experiment was performed with four mobile robots in a 11 × 14 ft rectangular enclosure with the region configuration as shown in Fig. 21.2. The search region is 126 sq. ft where 27 small metal cylinders (pucks) are evenly distributed. The goal region, called *Home*, is a 90° quadrant with 2 ft radius. The empty region is divided into a boundary and a buffer region.

Each of the four IS Robotics R2e robots is equipped with a two-fingered gripper and a differentially steered base powered by two drive motors. The sensors attached to the robots include piezo-electric contact (bump) sensors around the base and in the gripper, five infrared (IR) sensors around the chasis and one on each finger for

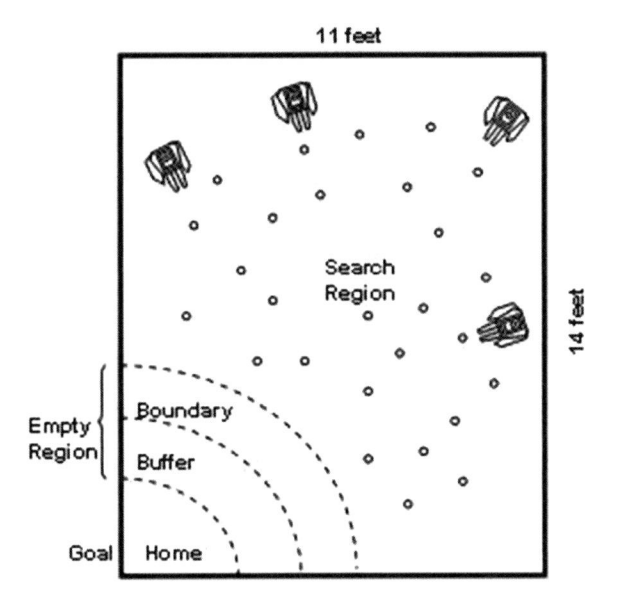

Fig. 21.2 Configuration for the collection task

proximity detection, a color sensor in the gripper, a radio transceiver for communication and data gathering, and an ultrasonic/radio positioning system (using triangulation). Programming the robots is done using Behavior Language (Brooks 1990) which supports a parallel, asynchronous, behavior-based programming language based on the subsumption architecture (Brooks 1986). A single Motorola 68332 16-bit 16 MHz microcontroller provides the computational need of each robot.

The main purpose of the experiments is to demonstrate the effectiveness in using behavior-based control paradigms in single-robot and distributed multi-robot control. The behavior-based control consists of a collection of modules, called behaviors, which implement the perception–reason–action cyclic processes. As they execute in parallel, each of these behaviors receives inputs from the sensors and/or other behaviors, processes the input to perceive the state of the environment, reasons about the most appropriate action based on its perception, selects and transmits the action to the actuators and/or other behaviors. Each of these behaviors performs an independent function, such as avoiding obstacles or moving toward the goal region. The collection of behaviors in each robot can either be identical (i.e., homogeneous controller) or be different (i.e., heterogeneous controller). Heterogeneous controllers may use different behavior sets in different robots to perform different types of task, e.g., search and deposit puck in goal region.

In homogeneous controller, each robot executes concurrently and independently as controlled by the same collection of behaviors. Each behavior performs a specific function, such as puck detection, puck grabbing, homing, reverse homing, wandering, avoiding, creeping, and exiting. Each behavior receives a subset of the sensor inputs and control signals from other behaviors. The reasoning process takes the input information and generates a control signal to other behavior and/or action commands to actuators. Some behaviors will generate only the control signals, while others generate action commands to actuators. Two actuators attached to the robots are gripper motor that grabs the puck and drive motor that moves the robot. When multiple behaviors generate different actions to an actuator, only one action will be selected through two mechanisms, behavior selection and subsumption-style priority scheme. The action command can be generated to either the gripper motor to grab a puck or the drive motor to move the robot toward the goal region or to avoid collision with another robot. Eventually a sequence of these incremental actions will cause the robots to search, grab, and deposit all the pucks from the search region to the goal region.

In heterogeneous controllers, robots may be divided into two or more groups, whereby each group of robot is controlled by a different collection of behaviors. One group may stay only in the search region, where they contain behaviors that will search for pucks and then bring them to the other group of robots that bring the pucks all the way to the goal region. The robots in the search region contain the following behaviors: puck detection, puck grabber, wandering, homing, reverse homing, avoiding, and exiting, while the goal robot contains the following behaviors: creeping, sweeping, and exiting. Although they have different subgoals, the two groups of robots cooperate to achieve the overall goal of searching and moving all the pucks to the goal region.

21.3.2.2 Real-Time Target Tracking Applications

In real-time target tracking applications, we have successfully developed a smart sensor network for localizing and tracking moving targets through the sensor field (Yang et al. 2010). A cluster of smart sensors use an enhanced closest point of approach (ECPA) algorithm, where our modification allows the algorithm to correctly compute the bearing of the target trajectory, the relative position between the sensors and the trajectory, and the velocity of the target (Yang et al. 2010). To validate ECPA, we designed and implemented the algorithm over a data-centric sensor network. This ECPA software also communicates over a collaborative mixed wireless sensor network with control software for controlling video sensor nodes that capture real-time images or video of the target at its predicted location. Our experimental results show that we can achieve our goals of detecting the target and predicting its location, velocity, and direction of travel with reasonable accuracy. In addition, results from the target detection algorithm can be used to predict the future target location so that a camera can capture video of the moving target for identification purposes.

An example smart sensor network for target tracking may consist of wireless sensor nodes, each of which is an x86-based PC104-based sensor node. The sensor nodes including the cluster head are PC104 nodes with a high-sensitivity noise cancellation microphone, 128 MB RAM, 4 GB flash memory, and an IEEE 802.11b wireless PC card. The gateway node and camera control node are also PC104-based sensor nodes. Video capture is performed on a Pentium 4-based Dell Latitude computer using a Diamond VC5000 One Touch Video Capture USB 2.0 which converts the camera's RCA output to digital video and interfaces with the Windows XP machine using USB 2.0. The camera is a Sony EVID30. The interface to the controller is RS-232C, 9,600 bps, serial port.

The OS used for the sensor nodes is Slax 6.0.1. A network time protocol daemon was downloaded from http://www.slax.org and incorporated into Slax 6.0.1 for use in the time synchronization process. Windows XP SP2 is installed on the video capture computer for compatibility with the Diamond VC5000 One Touch Video Capture.

To evaluate the performance of our systems, we set up the target detection and tracking sensor networks at two field locations: a vacant parking lot at Auburn University and AU's National Center for Asphalt Technologies (NCAT) test tracks. We experimented with several wireless ad hoc sensor network configurations: (1) a basic sensor network and camera network configuration, (2) a network with three additional wireless network hops from the cluster head to a remote camera control node, and (3) a network that enables more directions (orientations) for targets to move.

Our results show that we can achieve our goals of detecting the target and predicting its location, velocity, and direction of travel with reasonable accuracy. The results from the algorithm that computes the target location and velocity are shown below. Our second set of results shows that the results of target detection algorithm can be used by a camera to take pictures or video of the target for identification purposes.

21.3.3 Preliminary Results

The results of preliminary experiments demonstrate the usefulness of the perception–reason–action cycles for many applications, such as multi-robot control and target tracking applications.

21.3.3.1 Performance of Multi-robot Control Applications

The results show that the robots successfully completed collection of 14 out of the 27 pucks from the search region and deposited them at the goal region. For the homogeneous controller, the results show high level of physical interference among the robots near the goal region because multiple robots running the same set of behaviors simultaneously attempt to deliver pucks to the goal region. This interference is reduced in the design of the heterogeneous controller where only one robot is at the goal region and the surrounding empty area. The other robots deliver pucks to near the goal region and the goal robot asynchronously picks up those pucks and delivers them to the goal region. The results of the heterogeneous controller also show that the robots successfully collected 14 out of the 27 pucks. However, the interference near the home is much reduced.

In terms of the total time for completing the task, the homogeneous controller took 549 s, while the heterogeneous controller took 1,081 s, about twice as long. The homogeneous controller spent 143 s in avoiding interference between robots, while the heterogeneous controller spent 442 s in avoiding interference. This shows that although the heterogeneous controller avoids interference near the goal region, there was more interference in the search region. In calculating the total distance traveled, similar results show that in the homogeneous controller, the total distance traveled by all the robots is 475 ft, whereas in the heterogeneous controller, the total distance is 1,227 ft. The reason for this is the same as in the interference case.

Robustness is an important property of practical multi-robot systems. During the experiments, the R2e robots are prone to failure from many causes in the physical world, such as static electricity that corrupts the memory system and causes the robot's computer to crash. The experimental results show that despite these failures the multi-robot controllers were still able to perform the tasks satisfactorily. Because the controllers are behavior based, the tight coupling between sensing and action makes the system less susceptible to noise and uncertainty in the sensors and actuators. The homogeneous controller was robust because the robots are identical and independent. As such, partial or complete failure of a group of robots will cause the entire system to fail, as long as there is one robot, the task can still be accomplished. The heterogeneous controller is also behavior based and similarly less susceptible to noise and uncertainty in sensing and action. In addition, its robustness is a result of the asynchronous interaction between the two groups of robot as they pass the pucks asynchronously.

21.3.3.2 Performance of Target Tracking Applications

In the experiments, it was demonstrated that the ECPA algorithm computes the target location, velocity, and direction of travel based on the CPA time reported from the five sensors. The results show that the target location and velocity can be accurately computed (Yang et al. 2010).

Figure 21.3 shows the predicted position and actual path of the vehicle for the eight runs in our experiments (Yang et al. 2010). Note that the computed target speed and trajectory almost exactly match the actual values. The shorter dashed lines show the results from two runs in the AC direction and two runs in the CA directions. These trajectories are very close to the actual one shown by the solid AC path. The longer dash lines show the results of two runs in the BD direction and two runs in the DB direction. These trajectories (with one exception) are very close to the actual trajectory.

The camera first points in an initial position (in our case, west). The target then moves through the sensor field and quickly passes through the camera's field of view. The networked sensors then detect the target and calculate the target location and velocity. The predicted position of target is calculated based on the velocity and

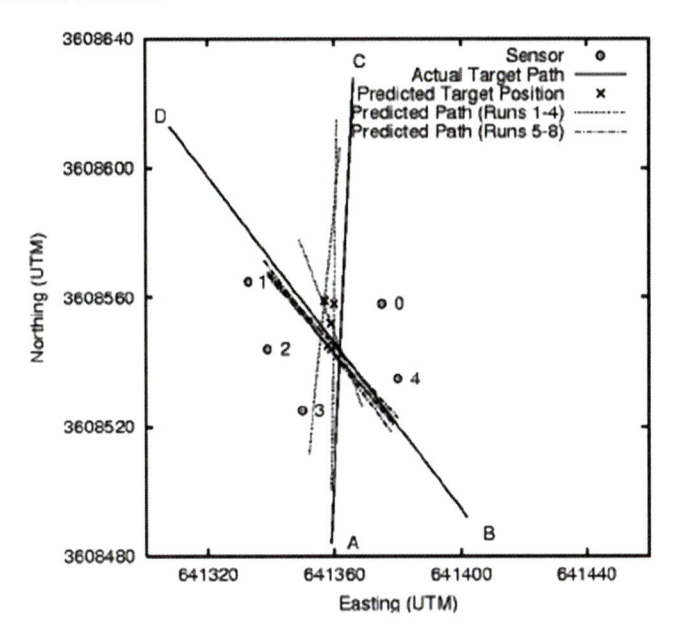

Fig. 21.3 Plot of target tracking results showing computed target location and velocity match the actual value

Fig. 21.4 The video camera automatically tracks a moving target. (**a**) The target is first detected by the sensor field. (**b**) The target position and velocity are predicted and the target is tracked by the video camera

transmission delay from sensors to the camera controller, whose action command causes the camera to pan and zoom toward the predicted position to capture the target on video. The images from the video capture can be used to identify the target more accurately. Figure 21.4 shows the video clips of this sequence of events where Fig. 21.4a shows the target first appearing in the sensor network field and Fig. 21.4b shows target at the predicted position.

21.4 Future Work

In summary, we have shown that cyclic perception–reason–action modules are useful for implementing smart sensor network applications such as distributed multi-robot perception, navigation, and manipulation and real-time target detection and tracking. Since a large network of these modules execute in parallel and interact in complex ways, their implementation can be simplified using reconfigurable smart nodes which include sensors and actuators, efficient sensor network protocols, and distributed services support, such as lookup, composition, and adaptation services. Reconfigurable smart nodes enable multiple independent distributed perception–reason–action modules to be executed in parallel, evolve dynamically and still communicate and collaborate with each other to achieve an integrated system-level goal. We have shown that cyclic perception–reason–action modules can be used in multi-robot control applications and real-time target tracking applications, where preliminary results show the effectiveness in using these methods.

While we have demonstrated the feasibility of implementing smart sensor networks for useful sensor and actuator applications, such as multi-robot control and target detection and tracking, much work remains to be done to implement even more complete distributed sensor services to simplify the development of large variety of smart sensor applications. The implementation of these distributed services requires careful consideration so that there is a balance between including the required functionalities and design simplicity for ensuring efficiency, energy conservation, and small software footprints.

21.4.1 Future Extensions

Future self-organizing smart sensor networks would be more easily implemented if there are full implementations of the distributed reconfigurable sensor services and perception–reason–action modules as described in this chapter.

We will focus in the future on the implementation of efficient perception–reason–action modules and distributed services – lookup, composition, and adaptation – that manage changes in the available services in the sensor network, migration of sensor nodes, and changes in task and network requirements. The distributed lookup service is implemented using directed diffusion network. The distributed lookup service provides more efficient and large-scale support for these changes that would not be possible using directed diffusion directly. Through a distributed implementation of the lookup services, other application-specific network and system services, such as new perception–reason–action modules, can be defined spontaneously in the sensor network. The lookup service is capable of adapting to changes in the ad hoc sensor network topology.

Discovery of lookup servers must first be provided in the ad hoc sensor network. When a node (that may implement a lookup server, a service provider, or a service client) moves into a new area or is rebooted, it first sends out a discovery interest to search for a nearby lookup server. A perception–reason–action module may act as a service client, service provider, or both. If there is a lookup server available, the lookup server will send an acknowledgment with the information of itself, such as its name, ID, location, and other information, to the node by publishing this information. If several lookup servers reply, the node will accept the first one that replied and ignore the others. The node stores the lookup server information in a table called "lookup table" for future reference. To reduce bandwidth usage, the service client will only broadcast its discovery interest to within a limited region in its vicinity. The size of the region will depend on number of lookup servers in the sensor network. Each intermediate node runs an application-specific routing filter program that checks if the node is outside the region for the discovery interest. If it is, the node will not propagate those interests any further. Otherwise, it will broadcast the discovery interest to its other neighbors.

The service provider registration function is developed as follows. When a service provider, such as a new perception–reason–action module, moves into a new area or restarts a service, it first tries to find a new lookup server and registers its service with the lookup server. This is done by sending out the registration interest to the lookup server that includes the information on the service provider itself, such as its location, ID, name, and other information on the service and interface. It then waits for acknowledge from the lookup server. The service lookup function can be implemented as follows. When a sensor node needs to find information on a specific service and the relevant service provider, it calls its local service lookup function that sends a lookup interest to the lookup server to fetch the updated service information. After the lookup server received the lookup interest, it searches its local service table first. If it finds the service record, it will return a data packet to the node by publishing the data. Otherwise, it sends the lookup request interest to all other lookup servers. If the service still cannot be found, it returns a failure message to the requesting node.

Another useful function provided by the lookup service is enabling of remote service execution. When a node wants to execute a service from a service provider, it searches its local service table to get the information on the service and the service provider. If it cannot find the relevant information, it will call service lookup function to update its local service table. Three types of service interfaces may be specified by the service provider including: location or address of the service provider, interface definition of the service, or mobile code for the interface protocol.

Development of the distributed composition service will be useful for implementation of smart sensor networks, with large variety of perception–reason–action modules. Most applications in sensor networks involve groups of sensor nodes and behavior modules that coordinate with each other to perform a global task. Distributed composition service enables applications to compose sensors and behavior modules to form a task group and maintains the group of sensors and behaviors on the fly to achieve higher task availability. When the development of the composition service is complete, we can show how the process of re-organization can be automated via an adaptation server in collaboration with the composition service.

We can also develop a RCSP (refinery constraint satisfaction problem) algorithm that enables self-adaptation of groups of sensor nodes and behavior modules. Further testing and simulation can validate the framework of the composition service and performance improvement in terms of service availability, load balancing, and energy preserving as enabled by the adaptation algorithm, which is implemented through the composition service.

Implementing control systems with multiple, parallel, and independent perception–reason–action modules is complex and challenging. Composing these modules with complex interactions to achieve global objectives may be simplified with hierarchical levels of controls, such as action-level, plan-level, and goal-level control. Although these hierarchical structuring of control levels is important, there are still many other challenging problems. First, there need to be capabilities for composing many disparate perception–reason–action algorithms. Since multiple types of perception–reason–action modules may be executing in parallel, the system must combine multiple knowledge representation and inference techniques (Cassimatis et al. 2004), including data fusion, symbol grounding, and merging of reasoning, planning, perception, and action. Second, more thorough studies are needed to investigate the effects of uncertainty in the estimation of the state of the physical system on the behavior of the controlled system. We need to develop better techniques for enhancing accuracy and precision limitations of sensors and actuators using techniques such as redundancy with overlapping sensor field of view, sensor fusion algorithms, and placement strategies. Finally, the stochastic latency in actuation may cause instability and unreliability in closed loop control. Improved techniques must be developed to predict instability and enhance reliability in these real-time control systems.

References

K. Arnold, et al., *"The Jini Specification,"* Addison Wesley, Reading, 1999.

N. Ayache, O. Faugeras, "Building, Registering and Fusing Noisy Visual Maps," *International Journal of Robotics Research*, Vol. 7, Issue 6, December 1988, pp. 45–65.

Y. Boykov, D. Huttenlocher, "A New Bayesian Framework for Object Recognition," *IEEE Conference on Computer Vision and Pattern Recognition*, 1999, pp. 517–523.

R. Brooks, "A Robust Layered Control System for a Mobile Robot," *IEEE Journal of Robotics and Automation*, April 1986, pp. 14–23.

R. Brooks, *"The Behavior Language: User's Guide, Technical Report AIM-1227,"* MIT AI Lab, Cambridge, 1990.

R. Brooks, "Intelligence Without Reason," *Proceedings of IJCAI*, 1991.

K. Casey, A. Lim, G. Dozier, "Data-Centric Network Services for Tsunami Detection and Response," *International Journal on Distributed Sensor Networks*, Vol. 4, Issue 1, January 2008, pp. 28–43.

K. Casey, A. Lim, G. Dozier, "A Sensor Network Architecture for Tsunami Detection and Response," *Innovations and Real-Time Applications of Distributed Sensor Networks (DSN) Symposium*, October 16–17, Washington DC, 2006.

K. Casey, A. Lim, G. Dozier, "Evolving General Regression Neural Networks for Tsunami Detection and Response," *Proceedings of the International Congress on Evolutionary Computation (CEC), IEEE*, July 2006.

N. Cassimatis, J. Trafton, M. Bugajska, A. Schultz, "Integrating Congition, Perception and Action Through Mental Simulation in Robots," *Robotics and Autonomous Systems*, Vol. 49, 2004, pp. 13–23.

K. Chandy, "Sense and Respond Systems," *Thirty-First Annual International Conference of the Association of System Performance Professionals*, December 2005.

M. Cohen, J. Sairamesh, M. Chen, "Reducing Business Surprises Through Proactive, Real-Time Sensing and Alert Management," *EESR '05: Proceedings of the 2005 Workshop on End-to-End, Sense-and-Respond Systems, Applications and Services*, USENIX Association, Berkeley, CA, USA, 2005, pp. 43–48.

J. Connell, "A Colony Architecture for an Artificial Creature," MIT A.I. Lab Technical Report 1151, June 1990.

J. Crowley, "Dynamic World Modeling for an Intelligent Mobile Robot Using a Rotating Ultra-Sonic Ranging Device," *IEEE Conference on Robotics and Automation*, St. Louis, MO, March 1985.

E. Datteri, et al., "Expeccted Perception: An Anticipation-Based Perception-Action Scheme in Robots," *IEEE International Conference on Intelligent Robots and Systems*, Las Vegas, Nevada, October 2003, pp. 934–939.

M. Esler, et al., "Next Century Challenges: Data-Centric Networking for Invisible Computing," *ACM Mobicom*, 1999.

D. Estrin, et al., "Next Century Challenges: Scalable Coordination in Sensor Networks," *ACM Mobicom*, 1999.

D. Estrin, D. Culler, K. Pister, G. Sukhatme, "Connecting the Physical World with Pervasive Networks," *IEEE Pervasive Computing*, Vol. 1, Issue 1, January 2002, pp. 59–69.

C. Fennema, A. Hanson, "Interweaving Reason, Action, and Perception," *Intelligent Robots and Computer Vision XI: Algorithms, Techniques, and Active Vision*, D. Casasent, Eds, Vol. 1825, SPIE, Bellingham, 1992, pp. 144–158.

J. Fuster, "Upper Processing Stages of the Perception-Action Cycle," *Trends in Cognitive Sciences*, Vol. 8, Issue 4, April 2004, pp. 143–145.

B. Gerkey, M. Matarć, "Principled Communication for Dynamic Multi-Robot Task Allocation," *Experimental Robotics VII, LNCIS 271*, D. Rus, S. Singh, Eds, Springer-Verlaag, Berlin, 2001, pp. 253–362.

D. Goldberg, M. Mataric, "Design and Evaluation of Robust Behavior-Based Controllers for Distributed Multi-Robot Collection Tasks," *Robot Teams: From Diversity to Polymorphism*, T. Balch, L.E. Parker, Eds, A.K. Peters, Ltd., Natick, April 2002.

F.I. Gonzelez, E.N. Bernard, C. Meifg, M. Eble, H.O. Mofjeld, S. Stalin, "The Nthmp Tsunameter Network," *National Hazards*, Vol. 35, Issue 1, 2005, pp. 25–39.

S. Haeckel, "Adaptive Enterprise: Creating and Leading Sense-and-Respond Organizations," Harvard Business School Press, Cambridge, 1999.

N. Hashmi, D. Myung, M. Gaynor, S. Moulton, "A Sensor-Based, Web Service-Enabled, Emergency Medical Response System," *EESR '05: Proceedings of the 2005 Workshop on End-to-End, Sense-and-Respond Systems, Applications and Services*, USENIX Association, Berkeley, CA, USA, 2005, pp. 25–29.

C. Intanagonwiwat, R. Govinda, D. Estrin, J. Heidermann, F. Silva, "Directed Diffusion for Wireless Sensor Networking," *IEEE/ACM Transactions on Networking*, Vol. 11, Issue 1, February 2003, pp. 2–16.

C. Jones, M. Matarć, "Behavior-Based Coordination in Multi-Robot Systems," *Autonomous Mobile Robots: Sensing, Control, Decision-Making, and Applications*, S. Ge, F. Lewis, Eds, Marcel Dekker, Inc., New York, 2005.

S. Kapoor, K. Bhattacharya, S. Buckley, P. Chowdhary, M. Ettl, K. Katircioglu, E. Mauch, L. Phillips, "A Technical Framework for Sense-and-Respond Business Management," *IBM Systems Journal*, Vol. 44, Issue 1, 2005, pp. 5–24.

A. Klyubin, D. Polani, C. Nehaniv, "Organization of the Information Flow in the Perception-Action Loop of Evolved Agents," *NASA/DoD Conference on Evolvable Hardware (EH'04)*, 2004, pp. 177–184.

J. Lee, S. Baek, C. Choi, S. Lee, "Particle Filter Based Robust Recognition and Pose Estimation of 3D Objects in a Sequence of Images," *Recent Progress in Robotics: Viable Robotic Service to Human: An Edition of the Selected Papers from the 13th International Conference on Advanced Robotics (ICAR 2007)*, Springer, New York, 2008.

S. Lee, *"Summary of Perception Guided Navigation and Manipulation,"* Springer, Berlin, 2009.

S. Lee, E. Kim, Y. Park, "3D Object Recognition Using Multiple Features for Robot Manipulation," *Proceeding of IEEE International Conference on Robots and Automation (ICRA2006)*, Orlando, Florida, USA, May 15–19, 2006.

S. Lee, "Sensor Fusion and Planning with Perception-Action Network," *Journal of Intelligent and Robotic Systems*, Vol. 19, 1997, pp. 271–298.

A. Lim, "Architecture for Autonomous Decentralized Control of Large Adaptive Enterprises," *DARPA-JFACC Symposium on Advances in Enterprise Control*, San Diego, California, November 1999.

A. Lim, S.A. Friedberg, "A State Machine Approach to Reliable Distributed Systems," *Proceedings of the 11th IEEE Symposium on Reliable Distributed Systems*, Houston, Texas, October 1992, pp. 204–212.

A. Lim, "A Uniform Software Architecture for Cooperation, Reliability and Reconfiguration of Autonomous Decentralized Systems," *Second International Symposium on Decentralized Systems, IEEE*, Phoenix, Arizona, April 25–27, 1995.

A. Lim, "Automatic Analytical Tools for Reliability and Dynamic Adaptation of Complex Distributed Systems," *IEEE ICECCS*, November 1995.

D. Lowe, "Object Recognition from Local Scale Invariant Features," *Proceedings of 17th International Conference on Computer Vision (ICCV'99)*, Kerkyra, Greece, September 1999, pp. 1150–1157.

D. Maravall, J. de Lope, "Emergent Reasoning from Coordination of Perception and Action: An Example Taken from Robotics," *EUROCAST 2003*, Spain, 2003, pp. 436–447.

M. Matarić, "Learning in Behavior-Based Multi-Robot Systems: Policies, Models, and Other Agents," *Cognitive Systems Research, Special Issue on Multi-Disciplinary Studies of Multi-Agent Learning*, R. Sun, Ed, Vol. 2, No. 1, April 2001, pp. 81–93.

M. Matarić, "Behavior-Based Systems: Main Properties and Implications," *IEEE International Conference on Robotics and Automation, Workshop on Architectures for Intelligent Control Systems*, Nice, France, May 1992, pp. 46–54.

C. Meinig, S.E. Stalin, A.I. Nakamura, F. Gonzelez, H.G. Milburn, "Technology Developments in Real-Time Tsunami Measuring, Monitoring and Forecasting," *Oceans 2005 MTS/IEEE*, Washington DC, September 2005a.

C. Meinig, S.E. Stalin, A.I. Nakamura, H.B. Milburn, "Real-Time Deep-Ocean Tsunami Measuring, Monitoring, and Reporting System: The NOAA DART II Description and Disclosure," Technical Report, NOAA, Washington DC, 2005b.

P. Pirjanian, "Multiple Objective Action Selection and Behavior Fusion Using Voting," Doctoral Dissertation, Institute of Electronic Systems, Alborg University, Denmark, 1998.

R. Rusu, I. Sucan, B. Gerkey, S. Chitta, M. Beetz, L. Kavraki, "Real-Time Perception-Guided Motion Planning for a Personal Robot," *Proceedings of the 22nd IEEE/RSJ International Conference on Intelligent Robots and Systems (IROS)*, St. Louis, MO, USA, October 11–15, 2009.

B. Schiele, J. Crowley, "Probabilistic Object Recognition using Multidimensional Receptive Field Histograms," *International Conference on Pattern Recognition*, August 1996.

D. Shell, M. Mataric, "Behavior-Based Methods for Modeling and Structuring Control of Social Robots," *Cognition and Multi-Agent Interaction from Cognitive Modeling to Social Simulation*, R. Son, Ed, Cambridge University Press, Cambridge, 2005, pp. 279–306.

D. Specht, "A General Regression Neural Network," *IEEE Transactions on Neural Networks*, Vol. 2, No. 6, 1991, pp. 568–576.

L. Yilmaz, A. Lim, S. Bowen, T. Ören, "Requirements and Design Principles for Multisimulation with Multiresolution, Multistage Models," *2007 Winter Simulation Conference*, Washington, DC, December 2007.

Q. Yang, A. Lim, K. Casey, R. Neelisetti, "An Empirical Study on Real-Time Target Tracking with Enhanced CPA Algorithm in Wireless Sensor Networks," *Ad Hoc and Sensor Wireless Networks: An International Journal*, Vol. 7, No. 3–4, 2009, pp. 225–249.

Q. Yang, A. Lim, K. Casey, R. Neelisetti, "An Enhanced CPA Algorithm for Real-Time Target Tracking in Wireless Sensor Networks," *International Journal on Distributed Sensor Networks*, Vol. 5, Issue 5, 2010, pp. 619–643.

M. Zink, D. Westbrook, S. Abdallah, B. Horling, V. Lakamraju, E. Lyons, V. Manfredi, J. Kurose, K. Hondl, "Meteorological Command and Control: An End-to-End Architecture for a Hazardous Weather Detection Sensor Network," *EESR '05: Proceedings of the 2005 Workshop on End-to-end, Sense-and-Respond Systems, Applications and Services*, USENIX Association, Berkeley, CA, USA, 2005, pp. 37–42.

Chapter 22
Multisensor Fusion for Low-Power Wireless Microsystems

Tong Boon Tang and Alan F. Murray

Abstract This chapter addresses the use of artificial neural network (ANN) as a form of multisensor fusion for low-power microsystems in wireless sensor networks. The ANN is configured to perform local preprocessing and early clustering/classification of high-dimensional sensory signals. This chapter reviews the use of ANNs applied to fuse electrochemical sensory data, and the status of state-of-the-art VLSI neural hardware is presented. The hardware-amenability of these neural algorithms creates an opportunity to integrate multiple sensors and their data fusion within a single silicon chip, thus miniaturizing the physical size of microsystems and improving the signal integrity of measurements. Besides the operation of early classification, several other practical issues (i.e., stochastic noise, time-dependent drift, and biofouling) of electrochemical sensors are also discussed. Subsequently, a multisensor microsystem named *Lab-in-a-Pill* is used as a case study. We demonstrate how to implement an ANN to perform early classification and thus to autocalibrate an array of electrochemical sensors online. The chapter concludes with some discussion and future research directions.

22.1 Introduction

Since the late 1970s, silicon has been identified as a promising sensor substrate because it permits the integration of the sensing element and the signal processing circuit within a single chip, and allows the cost-effective batch-fabrication process. The types of transduction available in silicon include radiant, mechanical, thermal, magnetic, and chemical signal conversion from a wide range of measurands (as summarized in Table 22.1). While silicon displays sufficient physical effects to most types of measurands, some modification to the silicon surface is required to enhance

T.B. Tang (✉)
School of Engineering, The University of Edinburgh, King's Buildings, Mayfield Road, Edinburgh EH9 3JL, UK
e-mail: Tong-Boon.Tang@ed.ac.uk

V. Cutsuridis et al. (eds.), *Perception-Action Cycle: Models, Architectures, and Hardware*, Springer Series in Cognitive and Neural Systems 1, DOI 10.1007/978-1-4419-1452-1_22, © Springer Science+Business Media, LLC 2011

Table 22.1 List of signal domains with their examples and physical effects in silicon (Middelhoek and Hoogerwerf 1985)

Signal domains	Example measurands	Physical effects
Radiant	Light intensity, wavelength, polarization, phase	Photoconductivity, photovoltaic, photoelectric and photomagneto-electric effects
Mechanical	Force, pressure, vacuum, flow, tilt, thickness	Piezoresistivity, lateral photoelectric and lateral photovoltaic effects
Thermal	Temperature, temperature gradient, heat, entropy	Seebeck effect, temperature dependence of conductivity and junction, Nernst effect
Magnetic	Field intensity, flux density, permeability	Magnetoresistance, Hall and Suhl effects
Chemical	Concentration, toxicity, pH, reduction potential	Ion-sensitive field effect

the sensor sensitivity to detect other measurands. For example, to make a magnetic field sensor, a thin Nickel-iron layer must be deposited on top of the silicon substrate; to create a pH sensor, the metal gate of a silicon field-effect transistor has to be replaced by an ion-sensitive membrane which has affinity for protons.

Typically, the transduction has limited sensitivity, and sensor outputs are in analogue format, making them particularly susceptive to noise when transmitted through long wires. The signal-to-noise (SNR) can be improved considerably when the sensor and the signal conditioning circuit are integrated onto the same substrate. Such a sensor is better known as an "integrated sensor." It achieves better SNR using local amplification of the raw sensory signal. With shorter interconnects, integrated sensor also suffers less parasitic capacitance/resistance effects. From the manufacturing point of view, it is more cost-effective to be fabricated and packaged than those which use a separate signal conditioning chip.

Moreover, direct on-chip signal processing can offer the following advantages:

- *Improved sensor characteristics*
  Most sensors demonstrate some level of nonlinearity. Using an on-chip feedback system or a look-up table can improve linearity. Besides, most sensors have an undesirable sensitivity to other measurands (e.g., temperature and interfering ions). Accurately predicting these side effects by incorporating the relevant sensors and circuits on the same chip can counteract such cross-sensitivity problem. Another common sensing issue is "offset," i.e., the sensor does not give a zero output when the readout should be zero. The offset can usually be compensated using a bias circuit with either laser-trimming resistors or a digital-to-analogue converter (DAC) for recalibration at the input stage of signal amplification. A more critical issue is "drift" in parameters because it can lead to a change in sensitivity, offset, linearity, etc. Normally, continuous autocalibration is required to ensure the reliability of measurements. Finally, the operating bandwidth of sensor can also be improved by on-chip signal processing through a feedback system.
- *Signal conditioning and formatting*
  Nearly all sensors generate analogue signals. An on-chip analogue-to-digital convertor (ADC) can translate the signals immediately into pulse streams which

have better noise immunity. Besides, most sensors have high internal impedance, and thus their measurements are very susceptive to output load. This can be resolved by simply adding a buffer at each sensor output node. Another use of on-chip signal processing is to convert all measurements into one single format (voltage/current), facilitating multisensor integration. It can also be used to pre-process measurements from multiple redundant sensors via a median filter, for instance, to minimize the required bandwidth for data communications.

- *High-level signal processing*
  Integrating more sophisticated signal processing functions on the transducer chip is in parallel to the trend of very large-scale integration (VLSI) technology (Ko and Fung 1982). Functions such as binary classification, pattern recognition, and sensor fault detection can be implemented into a microprocessor or a digital signal processor (DSP). A power-saving scheme can also be implemented to improve electrical energy efficiency.

In the present days, it is common to integrate multiple sensors into the same chip to produce an *integrated multisensor microsystem*. Using a huge number of identical/disparate sensors, as inspired by biological systems, can improve the robustness of the microsystem and provide information on different attributes of a target entity. In addition to the sensors, the microsystem incorporates some low-power circuits and, in some cases, wireless communications for a broad range of applications (Wise 1999). For instance, the wireless integrated network sensors (WINS) (Pottie and Kaiser 2000) are deployed for surveillance applications in place of the more expensive hard-wired systems. Limited to a narrow bandwidth, sensory signals are thus processed locally. Only events of interest can provoke an alarm to a nearby basestation for further decision making. Local data fusion (typically classification) uses an artificial neural network (ANN) because it allows data fusion at all levels (signal, pixel, feature, and symbol) (Luo et al. 2002).

Another two similar works on ubiquitous low-cost, low-energy sensor arrays are the *SMARTDUST* (Warneke et al. 2002) and the *picoNode* (Rabaey et al. 2000). Each Dust is built from off-the-shelve components (namely the sensors, the microprocessor, and the battery) and packaged into 1 cm^3 and powered by an on-chip solar cell or combustion fuel. Dusts communicate to each other via laser or radio frequency (at 916 MHz) signals, and the transmission range can be over 20 m. Likewise, each picoNode has a dimension of 1 cm^3 and weights less than 100 g. To be deployed in many hundreds, each picoNode uses ultra-low power ($\leq 100 \mu$W) to avoid frequent battery replacement. Its communication protocol stack is specially optimized to minimize power consumption by making a trade-off between the amount of wireless communication and local data computation.

As mentioned, integrated multisensor microsystems today are more sophisticated than ever and naturally, with increasing amount of measurement data, there is a commensurate demand for a more efficient, intelligent sensor fusion technique. Ideally, the technique should be hardware-amenable and compute at microwatt level. Additionally, it should be robust against stochastic noise, sensor drift, and occasional sensor failures, which all can lead to inaccurate measurements.

This chapter sets out to examine the use of ANNs for multisensor fusion in low-power wireless microsystems. As this is a very broad topic, we focus only on the cases where ANNs are used to auto-compensate sensor drift, a practical problem experienced by most, if not all, electrochemical sensors. The remaining of this chapter is thus organized as follows. Section 22.2 reviews the use of ANNs in fusing electrochemical sensory signals and Sect. 22.3 reports the status of state-of-the-art neural VLSI hardware. In Sect. 22.4, current analytical techniques used to overcome sensor drift are presented. Subsequently, an integrated multisensor microsystem named *Lab-in-a-Pill* is introduced in Sect. 22.5, and the requirements of its sensor fusion are detailed. Section 22.6 presents a neural algorithm named "continuous restricted Boltzmann Machine" (CRBM) as a form of multisensor fusion for the Lab-in-a-Pill. The hardware implementation of the CRBM and the experimental results are described in Sect. 22.7. We further identify several future research directions in Sect. 22.8 before concluding the chapter with a summary.

22.2 ANNs in Electrochemical Sensor Fusion

To date, the applications of ANNs are vast in variety, primarily in the fields of control and pattern recognition. We, however, only consider applications using ANNs to fuse electrochemical measurements in this section. A list of example applications is shown in Table 22.2.

There are two major sensing systems under this topic: the electronic nose and the electronic tongue. An electronic nose is defined as an instrument which comprises a sampling system, an array of chemical gas sensors with differing selectivity, and a computer with an appropriate pattern-classification algorithm, capable of analyzing simple or complex gases, vapors, or odors qualitatively (Stetter and Penrose 2001). A similar definition can be applied to an electronic tongue, except it analyzes *liquid* samples. The primary aim of these two systems is to achieve similar sensing capability, if not better, to human's olfactory and taste systems. Inspired by nature, these systems are constructed by an increasingly large array of sensors; thus local data fusion becomes more attractive and of necessity.

In the literature, principal component analysis (PCA) has been widely used to preprocess arrays of sensory signals (Kermani et al. 1999; Lindquist and Wide 2001; Rodriguez-Mendez et al. 2004; Sarry and Lumbreras 2000; Shin et al. 2000; Sundic et al. 2000; Wide et al. 1998). PCA is a linear transformation technique that select a new coordinate system for the input dataset such that the greatest variance by any projection of the dataset comes to lie on the first axis (known as the first principal component, PC1), the second greatest variance on the second axis, and so on. Because sensor signals are often not differential in a straightforward way, PCA is used to extract the key features in the dataset. PCA is also used to reduce the input data dimension when a large array of sensors (either identical or disparate) are used, providing a more compact perspective of an entity. There is a similar statistical technique called "discriminant function analysis (DFA)." Unlike PCA, however,

Table 22.2 List of example applications using ANN to fuse electrochemical sensor signals

No.	Application	Sensor	Algorithm	H/S[a]	Drift	Remark
1	Contaminant identification (Keller et al. 1994)	Metal-oxide sensors	MLP	S	No	Use complimentary sensors (each with associative memory/MLP)
2	Chemical sensing (Natale et al. 1995)	Not given	Kohonen's SOM	S	Yes	Adaptive model. Identify sampling rate to ensure trackability of drift
3	Juice classification (Wide et al. 1998)	Metal-oxide and conductivity sensors	MLP	S	No	PCA to reduce input dimension
4	Gas identification (Marco et al. 1998)	Metal-oxide sensors	SOM	S	Yes	Adaptive model
5	Prediction of health of dairy cattle (Gardner et al. 1999)	Metal-oxide sensors	MLP	S	No	Time-dependent MLP
6	Breath alcohol detection (Roppel et al. 1999)	Metal-oxide sensors	MLP/RBF	S	No	Preprocessed with PCNN or rank-order-filter
7	Odorant classification (Kermani et al. 1999)	Conducting polymer sensors	GA-supervised MLP	S	No	Dimension reduction by PCA
8	Gas discrimination in an air-conditioned system (Sarry and Lumbreras 2000)	Metal-oxide sensors	PCA/DFA	S	No	Better result with DFA
9	Potato chips and cream classification (Sundic et al. 2000)	Conductivity and metal-oxide sensors	MLP/RBF	S	No	Use PCA to reduce input dimension
10	Classification of the strain and growth phase of cyanobacteria (Shin et al. 2000)	Metal-oxide sensors	PCA/MLP/LVQ/ Fuzzy-ARTMAP	S	Yes	Sensor signals are pre-processed Off-line data fusion – not autocalibration
11	Water quality test (Lindquist and Wide 2001)	Conductivity sensors	PCA	S	No	
12	Detection of toxic gases (Sayago et al. 2002)	Metal-oxide sensors	MLP	S	No	
13	Online water pollution monitoring (Bermejo et al. 2002; Bedoya et al. 2004)	ISFETs	Blind source separation	S	No	Plan to implement in MCU/DSP
14	Characterization of red wine (Rodriguez-Mendez et al. 2004)	Gas, optical and electro-chemical sensors	PCA	S	No	Fusion of three sensory systems

[a] H/S refers to Hardware/Software implementation

DFA allows nonorthogonal axial feature extraction, and hence provides a better discrimination between clusters in classification problems (Sarry and Lumbreras 2000).

One of the most popular ANNs is multilayer perceptron (MLP) mainly because of its strength in classification and function approximation. It is in essence a static network, yet it can be adapted to process dynamic data by adding a tapped delay line, forming the time delay neural network (Lang et al. 1990). MLP has been applied successfully in many real-world problems, ranging from contaminant identification (Keller et al. 1994) to juice classification (Wide et al. 1998), from dairy cattle's health prediction (Gardner et al. 1999) to breath alcohol detection (Roppel et al. 1999), and many more. Obviously, there are times when other types of ANN are preferred over the MLP. One such example is radial basis function (RBF) (Roppel et al. 1999; Sundic et al. 2000) which is known to be more efficient for data interpolation and generalization than the MLP, when the input data dimension is small. Another example is Kohonen's self-organizing map (SOM) (Natale et al. 1995) which can learn without a teacher. This is particularly important when there is no predefined *priori* for clustering problems. Other applied ANNs include learning vector quantization (LVQ) (Shin et al. 2000), Fuzzy-ARTMAP (Shin et al. 2000), and blind source separation (Bedoya et al. 2004; Bermejo et al. 2002; Rodriguez-Mendez et al. 2004).

Thus far, these electrochemical sensor fusions have merely been applied in software as it is a fairly straightforward and rapid development task. Many commercial and noncommercial ANN simulators are now widely available, and most are accompanied with a good collection of established ANN algorithms. Even nonspecialists in the ANN field can easily apply the neural algorithms for data fusion. However, such a software-based signal preprocessing scheme is often slow in computing and is thus performed off-line. This is not ideal when a real-time signal analysis is required on site. More importantly, it does not capitalize the parallelism in neural architecture, which, in nature, can offer high-speed computation in pattern recognition applications.

With the emergence of new applications such as the distributed sensor networks, there is a growing interest to translate neural algorithms to hardware [microcontroller unit (MCU) or digital signal processor (DSP)] (Bedoya et al. 2004; Bermejo et al. 2002). Example implementation can be found commonly in robotic research works but are yet to be seen in electrochemical sensing. Translation to MCU and DSP is comparatively straightforward due to their rich hardware resources. However, it also means that the hardware is not power- and silicon area-efficient. Therefore, researchers have started building custom hardwares for the neural algorithms using analogue/mixed-signal VLSI technology. Examples of such implementation will be reviewed in the next section.

Interestingly, the issue of drift has been raised on several occasions but is hardly been addressed (refer to Table 22.2). In 2000, Shin et al. used off-line processing to determine the growth phase of cyanobacteria in water for up to 40 days. Inevitably, sensor drift occurred but in this case it was just treated as a noise. No recalibration against drift was performed. The only attempts recorded in the literature came from

Marco et al. (1998) and Natale et al. (1995), where adaptive Kohonen's SOM was used to track a linear drift. Their simulation results highlighted the importance of annealing parameter K_σ to ensure that data from two different classes were distinguishable by a trained SOM, despite the imposed drift. If K_σ was too small, the learning of SOM would cease at an early stage, probably even before the second class was presented. On the other hand, if K_σ was too large, the annealing would be too slow and the continuous learning might suffer from catastrophic interference (CI).[1] The way they reduced the effects of CI was by feeding data from the two classes alternately into the SOM. Undoubtedly, this drift tracking task would be much tougher if only a single class of data was available over a long period of time. Since the aforementioned scenario is quite possible, if not typical, it will be considered in the drift tracking simulations in Sect. 22.6.

22.3 Neural Hardware in VLSI Technology

In recent years, many ANNs have been realized in dedicated hardware to take advantage of the parallelism in neural architecture. This parallelism is recognized as the enabling feature of our biological systems which allow us to perform pattern recognition more effectively and faster, in spite of all the slow biological neurons, than a conventional von Neumann processor. This motivates electronic engineers to develop neural hardware which can extract useful information from multidimensional data in a similar manner as our biological systems.

Generally, there are two types of hardware implementation. The first type tries to mimic biological structures with very simple circuits – usually based on analogue VLSI technology. Meanwhile, the second type aims to support ANN as a mathematical tool and is commonly implemented in digital platforms [an architectural survey is available at Ienne et al. (1996)]. Analogue neural hardware often has fixed functions but offers a simple, direct interface with the real world. Moreover, it can be achieved in compact layout with a low-power requirement (Card et al. 1998; Woodburn and Murray 1997). On the other hand, digital neural hardware is more programmable, easier to communicate with other systems (especially for robust long-term storage and retrieval of ANN parameters in memory), and exhibits better immunity to noise and interference. More recently, hybrid techniques have also been exploited (Coggins et al. 1995; Leong and Jabri 1993) to gain the benefits from both technologies – analogue VLSI for data computation and digital VLSI for data communication and storage. In the following sections, we review neural hardwares according to their learning methods – with or without a teacher.

---

[1] CI can be defined as the phenomenon that occurs when later training disrupts results of previous training and is characterized by the inability to incrementally learn sets of training patterns. CI is readily observed in studies of backpropagation. This phenomenon is also referred to as sequential learning and sometimes lifelong learning (Sarkaria 2004).

22.3.1 Supervised ANN-Based Hardware

Several successful implementations of supervised neural algorithms, mainly of the MLP, have been reported in the literature (Asanovic et al. 1991; Coggins et al. 1995; Holt and Hwang 1993; Jabri and Flower 1992; Leong and Jabri 1993; Mayes et al. 1996). The implementations vary by their weight adaptation techniques. The three available techniques (Moerland and Fiesler 1996) are as follows:

1. *Off-chip learning*, where the neural hardware is not involved in the training process. The ANN parameters (known as weights) are trained using a computer with high precision. The values of trained weights are subsequently downloaded to the neural hardware.
2. *Chip-in-the-loop learning*, where the neural hardware is involved in feedforward propagation only. The calculation of new weights is done off-chip on a computer and is downloaded to the neural hardware after each training cycle.
3. *On-chip learning*, where the entire training of the ANN is performed on-chip offering the possibility to adapt any nonidealities of hardware (e.g., device-mismatch and nonlinear response) and continuous training.

Early experiments (Asanovic et al. 1991; Holt and Hwang 1993) show that the training of MLP requires at least 16 bits weight accuracy (which translates into a huge chip area). MLP with lower than 16 bits weight accuracy will not learn because the weight updates are often less than the quantization step which prevents the weights from changing. One proposed solution is to use a "weight perturbation (Jabri and Flower 1992)" algorithm where weights are constantly updated at a predetermined step size. Leong and Jabri (1993) reported that 6 bits weight accuracy was sufficient for the training of a MLP – a major improvement from previous MLP implementations, at the expense of using a more complex *backpropagation* (Rumelhart et al. 1986) as their weight adaptation algorithm.

22.3.2 Unsupervised ANN-Based Hardware

While the implementation of the MLP is well established, the algorithm itself (MLP) is not adequate for applications where no *priori* classes are being predefined. A more suitable solution is to use an unsupervised ANN, e.g., PCA, SOM, associative memory (AM), cellular neural network (CNN), and Boltzmann Machine. Examples of their implementation can be found in Alspector et al. (1989), Chua and Roska (1993), Higuchi et al. (1991), Hsu et al. (2002), Macq et al. (1993), and Nishizawa and Hirai (1998). Among all the algorithms, CNN possesses the characteristics that suit analogue neural hardware implementation the most. It consists of simple generic circuitries as basic building blocks, which can be replicated many times. Moreover, it uses a local connectivity strategy which eases the complexity of wiring and also permits very high speed operations (due to distributed computing).

Despite all these advantages, CNN like many others (e.g., PCA, SOM, and AM) can only perform direct transformation (of possibly complex inputs to more "manageable" outputs) as in a look-up table. Each input is treated as a discrete variable. The underlying relationship (i.e., correlation) between the variables is not exploited as a feature in assisting the detection of dichotomy between two classes of data, for instance. This type of ANN is commonly known as a *deterministic* network. Its counterpart is a *probabilistic* network, also known as Bayesian belief network. It can model the joint probability distribution of all variables represented by each sensor node. Such capability makes the network more robust against incomplete input data entries (possibly caused by temporary or permanent occlusion of sensing surface), if compared with those deterministic ANNs. Other advantages offered uniquely by Bayesian belief networks are detailed in Heckerman (1999).

The exact inference methods in Bayesian belief networks can become intractable easily because their complexity grows exponentially with the number of variables (Heckerman 1999). Therefore, several approximate inference methods have been developed. Among them, Boltzmann Machine is one which is hardware-amenable. Example works that contribute to the VLSI implementation of Boltzmann Machine are reported in Alspector (1989, 1991), where the networks are formed by stochastic binary neurons. The stochasticity is achieved by injecting artificially generated noise (i.e., amplified on-chip thermal noise) into each neuron. These Boltzmann Machines have been tested successfully in the classic "XOR" problem.

This type of Boltzmann Machine is, however, very restricted to binary problems. In 2003, Chen and Murray demonstrated that binary neurons were less suitable for modeling signals in continuous-valued format, a format in which all real-world signals appear in nature. In particular, the binary neurons were poor in modeling asymmetrical data distributions. For a binary neuron, the possible states are "0" and "1." Assume that a data distribution should be encoded by two hidden neurons with states $\{0,0\}$, $\{0,1\}$, and $\{1,0\}$. However, simulation result revealed that there could also be data generated by state $\{1,1\}$, albeit at a lower probability (Chen and Murray 2003).

Furthermore, the *Gibbs sampling* method in the network optimization process is time-consuming. It needs to be run iteratively until the Boltzmann Machine settles at an equilibrium state. Only thereafter, it can update the weight, which is a function of the Kullback–Leibler divergence for the data distributions of the initial and the equilibrium states. This leads to the proposal of minimizing contrastive divergence (MCD) by Hinton (2002). With the MCD learning rule, merely one-step Gibbs sampling is required for weight update. This accelerates the training process.

In search for a Boltzmann Machine that can process real-world signals directly, a new variant named "continuous restricted Boltzmann Machine (CRBM)" has been developed (Chen and Murray 2003). The MCD learning rule is adopted to ensure a faster convergence in the training process. Additionally, there is no interconnect between neurons in the same layer (visible/hidden), thus reducing the architecture into a simpler form. More importantly, the training rules are hardware amenable. Only simple mathematical operations, i.e., subtraction and multiplication, are used, which can be easily translated into hardware. The MCD learning rule has been implemented using analogue VLSI technology, and the neural hardware has demonstrated successfully its ability to perform on-chip learning to model simple distributions (Chen et al. 2003; Fleury et al. 2004). Therefore, CRBM is the natural choice for our study in Sect. 22.6.

22.4 Analytical Techniques for Counteracting Drift

Drift can be caused by several different dynamic processes within a sensing system, e.g., poisoning and aging of the sensor, or environmental changes, e.g., temperature and pressure variations. When sensor drift occurs, any subsequent data fusion will be impaired, if not become impossible. Thus, the study of drift has been important to the sensor community. In this section, four general techniques to counteract drift are reviewed.

22.4.1 Recalibration

For each measurement, some form of uncertainty is encompassed inherently. There are two types of uncertainty, namely systematic errors (bias) and randomness. Repeating measurements can eliminate the latter but not the former. Change in bias while the environment remains constant is known as drift. One of the most popular techniques to counteract drift is to recalibrate either at scheduled time intervals or when the measurements exceed a predefined limit (Chen and You 2008; Clarke 1999; Steinhage and Winkel 2000). In a more complex system such as Tsai et al. (1999), slow drift is distinguished from a stochastic noise, sudden failure, and genuine change of operating conditions by measuring not only the uncertainty of measurements but also the sum of uncertainty and the change of uncertainty in repeated measurements. Table 22.3 illustrates the characteristics of different sensor errors.

Traditionally, recalibrating electrochemical sensors is achieved using a reference buffer (Bris et al. 1997). But, such technique is not suitable for integrated multisensor microsystems because it is very difficult to scale-down the physical size of the buffer while ensuring its seal is tight and no chemical reaction can occur (Smith and Scott 1986). An alternative recalibration technique is to apply an error correction algorithm. The sensors are continuously modeled using hidden variable

Table 22.3 Characteristics of different sensor errors

Type of sensor error	Uncertainty	Sum of uncertainty	Change of uncertainty
Drift	Not large	Not small	Small
Noise	Small	About zero	Large
Sudden failure	Any	Large	Large
Change of conditions	Any	Medium	Medium

models (Holmberg et al. 1997; Jamasb et al. 1997) or Kohonen's SOM (Marco et al. 1998) for example. When the estimation error exceeds the permitted tolerance, the model parameters will be updated to compensate for the drift.

22.4.2   Data Filtering

Another counteracting drift technique is to select the best estimation (i.e., one with the least drift) from an array of redundant sensors – i.e., a winner-takes-all approach (Yen and Feng 2000). Estimation of sensor signals is achieved with Kalman filters. The array of sensors are divided into groups with overlaps. Each group is trained/modelled by an MLP. Output from the MLP which gives the least standard deviation (i.e., drift) is then presented as the best estimation. The standard deviation is additionally used as an indicator of the confidence in the estimations and hence the health of the sensing system. No drift-correction is attempted in this technique. Hence, the performance of the sensing system will degrade since sensor drift is inevitable in all electrochemical sensors.

22.4.3   Drift Insensitivity

The third technique is to make the output signal of sensing system insensitive to drift. The most straightforward way in achieving this is to measure in a *differential mode* (Errachid et al. 1999; Hendrikse et al. 1998). For instance, a pH-sensing system uses one ion-sensitive FET (ISFET) and one ion-insensitive FET (REFET). The REFET uses buffered hydrogel or parylene as an ion-blocking membrane replacing the usual metal gate. Since both ISFET and REFET use a common reference electrode, they suffer the same amount of drift. The differential output has the drift effect canceled out, and thus is independent of the drift. However, the two sensors suffer from "crosstalk" and the differential sensitivity is usually small.

Another way to make the output signal insensitive to drift is to model drift based on historical data. If the cause of drift is known, it is possible to develop mathematical models to predict the drift effect (Grattarola et al. 1992; Jamasb 2004; Jamasb et al. 1998). However, the cause of drift and its influence are normally unknown for electrochemical sensors. To resolve this, Sachenko et al. (2001) suggested using a recurrent neural network (RNN) to predict drift based on *short-term* historical data. Meanwhile, Platonov used a general mathematical model to fit the drift pattern and optimized it by minimizing the mean-square error (MSE) recursively (Platonov et al. 1998).

Drift effect can also be eliminated by carefully removing it as a feature during data fusion. Artursson and Holmin (Artursson et al. 2000; Holmin et al. 2001) used PCA to extract key features from historical data, in which drift was first identified as one of the key principal components. This component was then eliminated from contributing to the subsequent stage of data fusion. Similarly, Lazzerini and Marcelloni (2000) adopted a k-nearest neighbor (KNN) algorithm to extract features and used a supervised fuzzy isodata (SFI) algorithm to select features. Assumed that a rich historical dataset is available, such technique can permanently immunize the sensing system from any drift effect.

22.4.4 Fault Isolation

The fourth technique to counteract drift is to isolate drifting sensors from contributing to data fusion. Isolation is applied to a sensor when the standard deviation of repeated measurements exceeds the permitted tolerance. Drifting sensors are identified by comparing the actual sensor output with the estimated output of trained models such as autoassociative memory (Chung and Merat 1996), fuzzy logic (Park and Lee 1993), MLP (Guo and Nurre 1991), and PCA (Seiter and DeGrandpre 2001). Those with excessive estimation errors will be considered as faulty sensors and hence will be isolated.

22.5 Lab-in-a-Pill

Lab-in-a-Pill is an example of integrated multisensor microsystems. Both Lab-on-a-Chip[2] and System-on-Chip[3] (SoC) technologies were exploited to build the pill as a generic measurement and monitoring tool for environmental and biomedical applications (Tang et al. 2002). Main efforts were on system-level integration and miniaturization. Wireless communication capability was also included for its use in distributed wireless networks. A photograph of the LIAP is shown in Fig. 22.1. This chapter will use it as a platform to examine how an ANN can be used to fuse multiple sensory signals.

Four types of silicon sensors were incorporated in the pill. The temperature sensor measures the ambient temperature and provides supplementary information for the temperature-dependent sensors. The pH sensor measures the acidity and the dissolved oxygen sensor detects the activity of aerobic bacteria in the environment of interest. The conductivity sensor gives an estimation of local content by measuring the total level of dissolved solids (TDS). The sensors were fabricated on two chips for prototyping purpose. Details can be found in Tang et al. (2002).

---

[2] Lab-on-a-Chip is a technology that integrates complex laboratory sensors and sample-handling capabilities onto a glass or silicon plate (Figeys and Pinto 2000).

[3] SoC is an electronics design methodology that integrates data analysis, instrumentation, and communication capabilities onto a single piece of silicon (Martin and Chang 2001). The reuse of pre-developed electronic circuit modules is an essential component of the SoC concept.

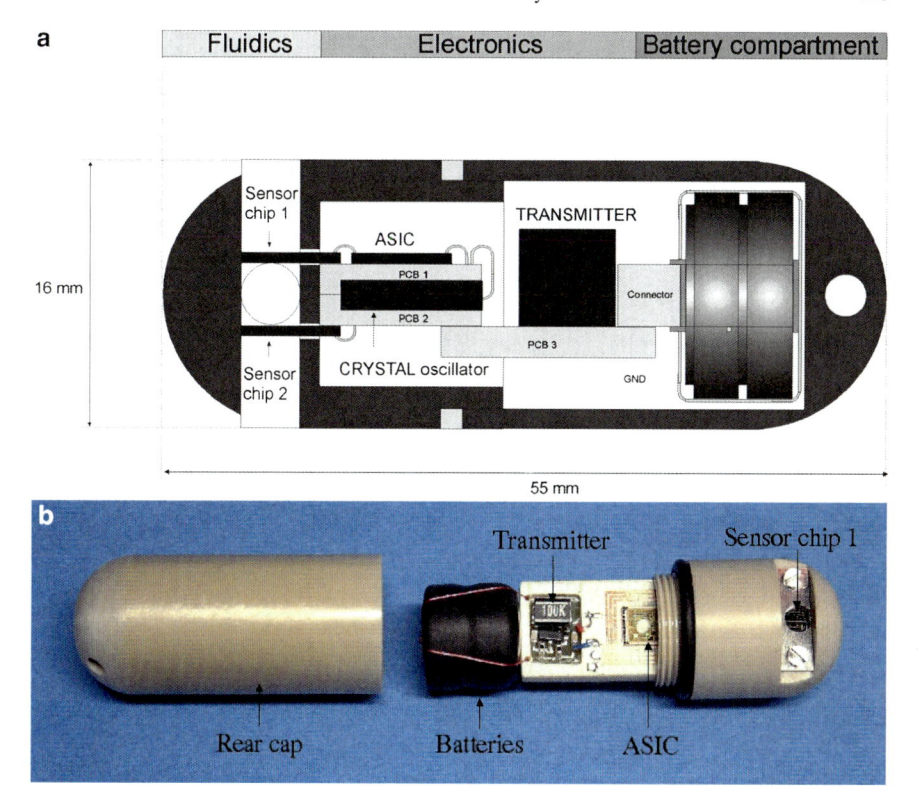

Fig. 22.1 Lab-in-a-Pill in (**a**) schematic diagram and (**b**) photograph with its rear cap opened for the inner view of the microsystem

Besides the sensor chips, an application-specific integrated circuit (ASIC) chip has been designed and fabricated. It consists of sensor interface circuits, a 10-b analogue-to-digital converter (ADC), a digital microcontroller (modified Motorola 6805), a clock divider, a direct-sequence spread spectrum (DS-SS) encoder, and some radio frequency (RF) circuit. The DS-SS encoder not only can improve the bit error rate (BER) of the wireless data transmission, but also enable the use of multiple pills with a single base-station since each encoder has a unique code sequence (Aydin et al. 2005).

Together with a 40 MHz transmitter and two SR44 Ag_2O batteries, the chips were packaged into a biocompatible capsule which has a solid chemical resistant polyether-terketone (PEEK) coating. Its dimensions are 16×55 mm and it weighs 13.5 g. The microsystem consumes 12.1 mW of power in continuous mode (the worst case scenario) and the batteries last for 40 h (Johannessen et al. 2004). From the test bench results, both pH and dissolved oxygen sensors were found to be less stable and reliable due to corrosion on the encapsulation layer of the miniaturized on-chip reference electrodes. When extended over time, the encapsulation layer will break and the sensors will malfunction (Fig. 22.2).

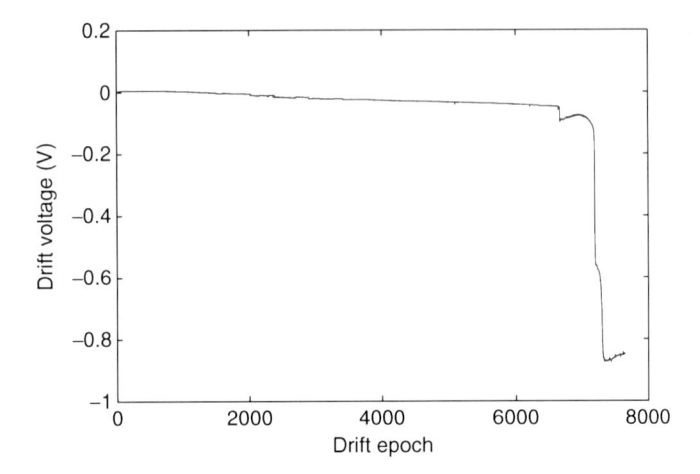

Fig. 22.2 Example drift data recorded with a physiological electrolyte (0.9% saline) solution of pH 7. The ambient temperature was maintained at 36.5°C throughout the experiment using a water bath and having the assay solution continuously stirred and recirculated by a peristaltic pump. One unit of drift epoch is equivalent to 10 s

Moreover, there is a significant variation in some parameters of the sensors due to fabrication process. For instance, the threshold voltage of pH sensors from one single batch of fabrication can vary from 0.28 to 1.88 V. Such magnitude of variation means that it is difficult to use a generic signal conditioning circuit for all (functionally identical) sensors without compromising the sensor sensitivity, given a limited voltage supply. To achieve the maximum sensitivity, all sensors need to be calibrated individually, but this is an expensive practice in terms of time and labor.

Another problem in this particular application is *biofouling*. Biofouling refers to the gradual accumulation of waterborne organisms (such as bacteria and protozoa) on the surfaces of engineering structures in water, which contribute to a decrease in the sensitivity of the sensor. A compelling solution is to incorporate more redundant sensors on chip to improve the robustness of the measurements. This, however, means a broader databus bandwidth will be required. To avoid it, an effective solution is to introduce local data fusion. Therefore, we explore the option of having an ANN to perform binary classification to identify events of interest.

Given the aforementioned practical issues of electrochemical sensors, the specifications for the ANN can be summarized as follows:

1. Perform binary classification as a form of data fusion
2. Ameliorate the imperfection due to intrinsic device variation
3. Be robust against stochastic noise
4. Autocalibrate against sensor drift
5. Operate with analogue signals directly
6. Consume low electric energy for data computation
7. Require simple computation only and be hardware-amenable
8. Have a small silicon footprint

22.6 The "Neural" Solution: Adaptive Stochastic Classifier

As discussed in Sect. 22.3.2, the CRBM is a suitable candidate for the specified task – binary classification of multiple sensory signals in the presence of stochastic noise and sensor drift. In this section, we provide a formal introduction to the CRBM and present some application-specific modification to its training methodology. The learning ability and adaptivity to drifting data, of the CRBM, will subsequently be examined.

22.6.1 Continuous Restricted Boltzmann Machine

The CRBM is a generative model that is able to perform autonomous feature extraction and is based on Hinton's Products-of-Experts architecture (Hinton 1999). It has one visible and one hidden layer with interlayer connection only. The visible and hidden neurons are connected by weight vectors {W}. Chen and Murray develop the model such that it can interface with analogue world directly. Continuous-valued neurons are used in contrary to the binary neurons in RBM (Smolensky 1986). Figure 22.3 shows a CRBM with three visible, four hidden, and two (permanently "on") bias neurons, v_0 and h_0.

22.6.1.1 Continuous Stochastic Neuron

Let s_i and s_j represent the states of a visible neuron i and a hidden neuron j, respectively, and $w_{ij} = w_{ji}$ the bidirectional weights. The state of the neuron j is defined as:

$$s_j = \varphi_j \left(\sum_i w_{ij} s_i + \sigma \cdot N_j(0, 1) \right) \qquad (22.1)$$

$$\text{with} \quad \varphi_j(x_j) = \theta_L + (\theta_H - \theta_L) \cdot \frac{1}{1 + \exp(-a_j x_j)} \qquad (22.2)$$

Fig. 22.3 A CRBM with three visible, four hidden, and two bias neurons. Each neuron except the bias neurons has a sigmoidal activation function where its state s_i is confined by $-1 \leq s_i \leq +1$. The visible layer is connected to the hidden layer via symmetrical weights, i.e., $w_{ij} = w_{ji}$

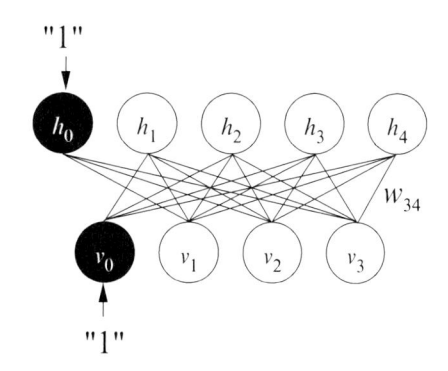

where σ is a noise scaling constant, $N_j(0, 1)$ is a Gaussian random variable with zero mean and unit variance, and a_j is a noise control parameter.

The product of σ and $N_j(0, 1)$ forms a noise input component in the neuron, providing it a sense of stochasticity. This is required in searching for the optimum model through the *Gibbs sampling* method. The $\varphi_j(x)$ is a sigmoidal function with asymptotes at θ_L and θ_H. Commonly, it is set to be a *tanh* function (i.e., the asymptotes θ_L and θ_H are -1 and $+1$, respectively).

The parameter a_j controls the slope of the sigmoidal function, and thus the nature and the extent of the stochastic neuron behavior. A small value of a_j leads to an almost-linear sigmoidal function, renders input noise[4] negligible, and results a near-deterministic neuron. On the other hand, a large value of a_j leads toward a step function and results in an approximately binary stochastic neuron.

22.6.1.2 CRBM Learning Rule

The CRBM uses the concept of "Minimizing Contrastive Divergence (MCD)" (Hinton 2002) in the learning rules of its weight, $\{w_{ij}\}$ and noise control parameter, $\{a_j\}$, which are defined as:

$$\Delta w_{ij} = \eta_w \left(\langle s_i s_j \rangle_0 - \langle \hat{s}_i \hat{s}_j \rangle_1 \right) \tag{22.3}$$

$$\Delta a_j = \frac{\eta_a}{a_j^2} \left(\langle s_j^2 \rangle_0 - \langle \hat{s}_j^2 \rangle_1 \right) \tag{22.4}$$

where $\langle x \rangle_0$ is the expectation value of x at time zero in a Markov chain, $\langle x \rangle_1$ is the expectation value of x after one-step Gibb sampling, η_w is the learning rate for weights, w_{ij}, and η_a is the learning rate for noise control parameter a_j. With no intralayer connections, the states of neurons can be updated in parallel. This further simplifies the hardware implementation.

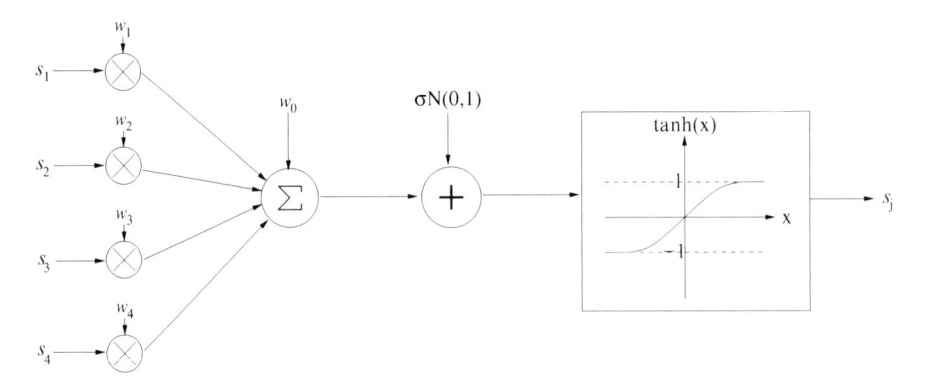

Fig. 22.4 A continuous stochastic neuron with four inputs

---

[4] This input noise is external, not the purposely injected N_j.

In this work, the CRBM is used to model the sensing characteristics of the electrochemical sensors. To quantify the modeling performance, we use a single-layer-perceptron (SLP) (Widrow and Hoff 1960) as a linear classifier for the activities of the CRBM hidden neurons with respect to two different classes of data. The success of sensor modeling determines the accuracy of this adaptive stochastic classifier (CRBM + SLP).

22.6.2 Training Methodology

First of all, the CRBM is trained without supervision in batch. The noise scaling constant σ is set to a value which avoids both over-fitting and complete loss of information (due to excessive noise). The learning rate must also be considered carefully. The "rule of thumb" drawn from several CRBM-modeling projects is to have the visible noise control parameters' learning rate η_v ten times that of hidden neurons and the weights. This encourages autonomous annealing through adaptation of the visible layer's noise control parameters a_v on a shorter timescale than that for adaptation of the weights w_{ij} and hidden noise control parameters a_h to model the details of training data distribution.

Upon completion of training the CRBM, the SLP is supervised and trained in batch. It takes the corresponding activities (extracted features) of the CRBM hidden neurons $\{h_j\}$ as inputs, and outputs a label $\{0, 1\}$ with respect to the inputs at the visible neurons $\{v_i\}$. The mean square error (MSE), which is the average of the square of the difference between the desired output $d(t)$ and the actual output $y(t)$, is used as a performance indicator for the training. When the targeted MSE is achieved, the first phase of the training of the entire adaptive stochastic classifier is completed.

In the second phase where the adaptive stochastic classifier operates in a dynamic environment, a novel methodology is proposed to allow the classifier to counteract sensor drift and the potentially long absence of a complete representation of both classes of data. To do so, we need to:

1. Allow the CRBM to be adaptive to the latest data distribution (i.e., the original distribution plus drift).
2. Ensure that the CRBM continues to present *consistent* features to the SLP (linear classifier) to maintain high accuracy in data classification.

To accomplish **Task #1**, the weight of the hidden bias neuron w_{i0} needs to be adaptive because it encodes the base line for the input data distribution. The learning rule for w_{i0} is as in (22.3). The **Task #2**, on the other hand, demands the weight of the visible bias neuron w_{0j} to be updated because it encodes the thresholds for the activities of the hidden neurons, which in turn affects the classification by the SLP. Simply by updating the thresholds, consistent features can be presented to the SLP which is set to be nonadaptive in the dynamic environment.

From (22.1), it is obvious that an input drift (Δs_i) will result in a corresponding change in the hidden neuron state (s_j), moving away from the known features to the

SLP. Let $\Delta w_{0j}(t)$ be the amount of update required, and $\Delta s_i(t)$ denotes the average sensor drift in visible neuron v_i at time t. To maintain consistent input feature sets to the SLP:

$$\Delta w_{0j}(t) = -\left(\sum_{i=1}^{m-1} w_{ij}^* \Delta s_i(t)\right) \tag{22.5}$$

where w_{ij}^* denotes the weight after the initial CRBM training session in static environment. Assume that the CRBM manages to trace the latest distribution and hence succeeds in **Task #1**, the drift can be expressed as:

$$\Delta s_i(t) = w_{i0}^* - w_{i0}(t). \tag{22.6}$$

Substituting (22.6) into (22.5), we have the learning rule for w_{0j} as:

$$\Delta w_{0j}(t) = -\eta_{on}\left(\sum_{i=1}^{m-1} w_{ij}^* [w_{i0}^* - w_{i0}(t)]\right). \tag{22.7}$$

In (22.7), a new scaling term η_{on} has been introduced to prevent the weights of the model from getting into saturation, in the event of an instantaneous and huge noise spike, and $0 < \eta_{on} \leq 1$. The complete methodology flow can be summarized by Fig. 22.5.

22.6.3 Simulation Results

The adaptive stochastic classifier, together with its learning rules, was implemented in MATLAB. In this section, we first demonstrate the learning of the classifier with two simple but multidimensional clusters of simulated data. The clusters are overlapping each other due to limited sensitivity and stochastic noise. Subsequently, we examine the accuracy of the classifier at different pH levels, i.e., with the two

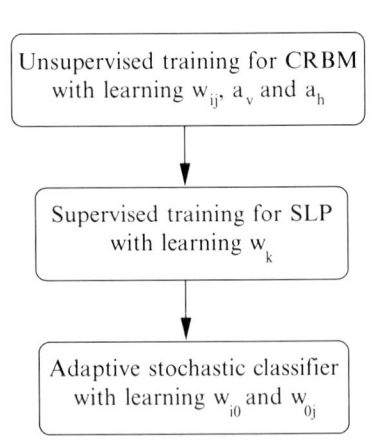

Fig. 22.5 Methodology flow of training the adaptive stochastic classifier (CRBM + SLP)

clusters separated by different distances graphically. Then, we extend the work to modeling 2D non-Gaussian meshed clusters. This is an interesting, nontrivial task because the CRBM is in essence formed by a set of Gaussian experts. Finally, we impose real drift upon the inputs to the classifier and examine whether the classifier can track the drift autonomously and maintain high accuracy in data classification.

22.6.3.1   With Simple, Multidimensional Overlapping Clusters

This simulation involved one temperature sensor (S1) and ten pH sensors (S2-11). Each pH sensor had a unique intrinsic threshold voltage, i.e., $V_{th} = \{0.50, 1.20, 0.28, 0.90, 1.60, 0.40, 1.10, 0.70, 1.88, 0.60\}$. Six hidden neurons were used to capture the features in the training data clusters which were generated using the sensor models described in Tang et al. (2004). Class A data was defined by the conditions whereby the temperature was 25°C and pH value was 4; class B data was defined whereby the temperature was 40°C and pH value was 10. The learning rates for the CRBM weight w_{ij} and the other parameters were 1.0 and 0.1, respectively. The classifier was trained with 200 samples from each class, according to the methodology detailed in previous section. The numbers of training epoch were 30,000 and 5,000 for the CRBM and the SLP, respectively.

Figure 22.6a–c depicts the training data clusters in 2D, while Fig. 22.6d–e shows their corresponding 20-step Gibbs sampling reconstruction[5] data distributions.

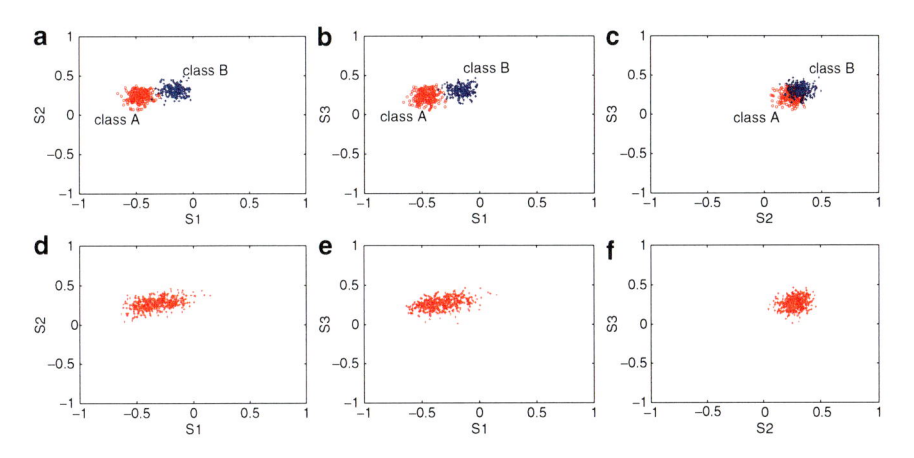

Fig. 22.6   Training data for the 11-dimensional clusters simulation: (**a**) temperature sensor (S1) and first pH sensor (S2); (**b**) temperature sensor (S1) and second pH sensor (S3); (**c**) first pH sensor (S2) and second pH sensor (S3). (**d**)–(**f**) are their corresponding 20-step Gibbs sampling reconstruction data

---

[5] In this context, "reconstruction" refers to the ability of the CRBM to converge and regenerate the distribution of training data at the visible layer with its trained weights after several Gibbs-sampling steps, disregard to what the initial states of the visible neurons were.

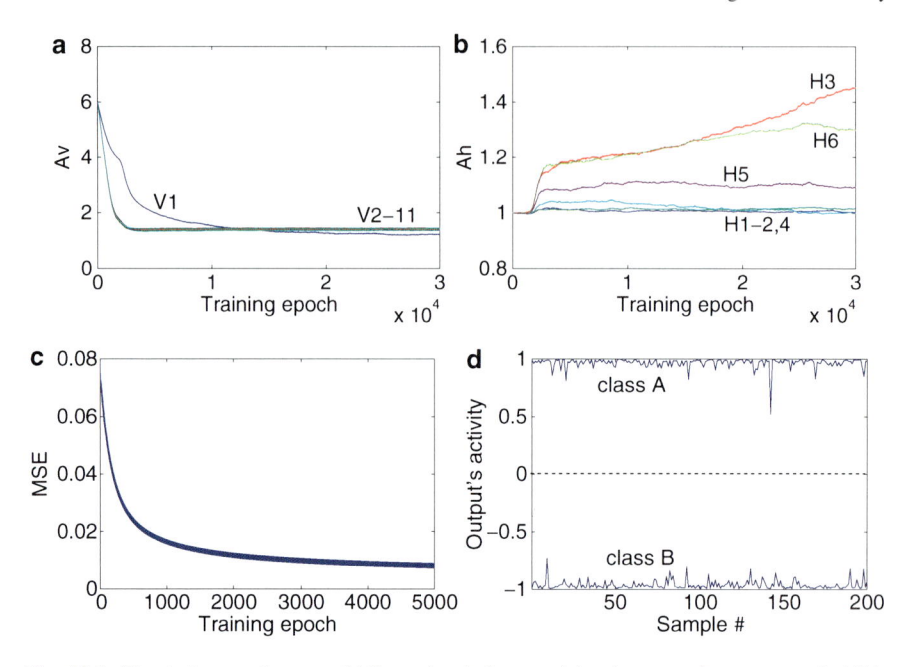

Fig. 22.7 Simulation results on multidimensional clusters: (**a**) noise control parameter of visible layer A_v; (**b**) noise control parameter of hidden layer A_h; (**c**) MSE during the SLP training stage; and (**d**) the SLP activity to 400 test samples

Table 22.4 Simulation results on the final noise control parameter A_h and SLP weight w_k

| Parameter | Hidden neuron | | | | | |
	H1	H2	H3	H4	H5	H6
Final A_h	1.0024	1.0134	1.4519	0.9995	1.0923	1.2986
Final w_k	1.2154	−0.7434	7.9178	1.6897	3.3714	6.8491

As illustrated, the CRBM has encoded the training data clusters reasonably well in its weight and noise control parameters after 30,000 training epochs. Further evidence was found in the noise control parameter A_v recorded over the entire training period (Fig. 22.7a). The initial high value (i.e., six) of A_v was autonomously annealed to less than 1.46. In contrast, the noise control parameters A_h for hidden units H3 and H6 were increased to relatively more significant values 1.4519 and 1.2986, making them more sensitive to the embedded information which could tell which class an input data was coming from (Fig. 22.7b and Table 22.4). The dichotomy between the two data classes was further augmented by the SLP weight w_k (Table 22.4). The trained classifier was subsequently tested with a new set of data, 200 samples for each class. Figure 22.7d shows the SLP activity with respect to the test data. Thresholding at zero, the adaptive stochastic classifier (CRBM + SLP) achieved 100% accuracy in data classification after this initial phase of training.

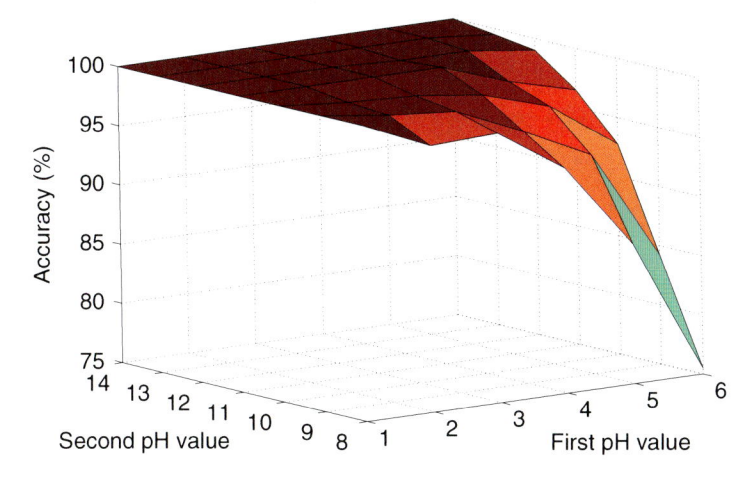

Fig. 22.8 Simulation results on different pairs (in pH value) of data clusters in terms of classification accuracy

Additionally, a series of simulations were run to examine the accuracy of the classifier on sensing different pairs (in pH value) of training data clusters. The temperature was fixed at a common point (37°C) for both clusters. The classifiers trained with each pair of data distributions are the same in terms of architecture size, noise scaling constants, and learning rates. The test result is plotted in Fig. 22.8. As expected, the closer the two training data clusters were, the lower the accuracy it achieved. The accuracy was at 75.5% in the worst case where class A represents pH 6 and class B represents pH 8.

22.6.3.2   With 2D Non-Gaussian Meshed Clusters

Noise in sensory data does not always obey a Gaussian distribution. Unlike in the previous simulations, modeling non-Gaussian clusters require the CRBM to learn not only the mean value and the dispersion but also the shape of the data cluster. We found that the noise scaling parameter A_v in particular needed to be set to a low value (e.g., 0.1) to suppress the noise component in the neuron and thus to permit the fine features of the training data cluster to be modeled (Tang and Murray 2007). With a larger A_v value, the reconstruction result was poor. Merely the mean value was learnt but not the general shape of the cluster.

Two simulations were performed to examine whether the adaptive stochastic classifier could classify 2D meshed data clusters. Both were linearly inseparable classification problems. In the first simulation, a CRBM with seven hidden neurons was trained for 5,000 epochs to encode the training data clusters with 400 samples for each class, as illustrated in Fig. 22.9a. The noise scaling constants were fixed at 0.4. Noise control parameter A_v at the visible layer was held constant at 0.1, while the noise control parameter A_h at the hidden layer was allowed to evolve into feature detectors. All the learning rates were set to 0.2.

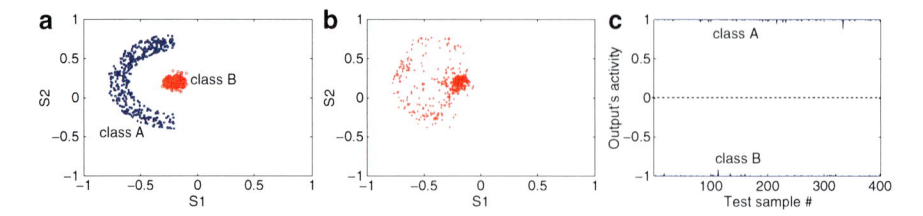

Fig. 22.9 Simulation 1 on meshed clusters: (**a**) training data; (**b**) reconstruction data after 5,000 epochs; and (**c**) the SLP activity with respect to a set of test data

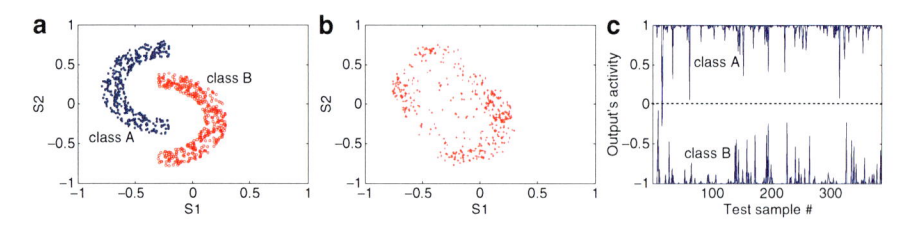

Fig. 22.10 Simulation two on meshed clusters: (**a**) training data; (**b**) reconstruction data after 20,000 epochs; and (**c**) the SLP activity with respect to a set of test data

Such data clusters were considerably complex to be modeled, especially with a common set of noise control parameters and noise scaling constants, because the two clusters had different dispersions and shapes. Figure 22.9b depicts the reconstruction data clusters by the CRBM after the initial training stage, which exhibited some resemblance to the training data clusters. To quantify the performance, an SLP was trained for 5,000 epochs to linearly classify the response of the CRBM, more precisely the activity of the hidden units, corresponding to the class of input data: as before, class A data yielded an output of "+1"; class B data gave an output of "−1". Figure 22.9c shows the SLP response to a set of test data, 400 samples for each class. Thresholding the response at zero, the trained adaptive stochastic classifier showed 100% accurate.

To further validate the capability of the classifier, a second set of meshed clusters were used. This time both training data clusters were non-Gaussian (refer to Fig. 22.10a). Due to the increase in complexity, the CRBM was trained for a longer period of time, 20,000 learning epochs to be exact. Upon completion, a 20-step Gibbs sampling reconstruction was generated, as depicted in Fig. 22.10b. Repeated as before, an SLP was used to evaluate the performance of the nonlinear data modeling by the CRBM. As expected, the dichotomy between the two clusters was more difficult to be modeled, and hence the level of accuracy degraded a little but remained reasonably high, i.e., 99.74%. The result demonstrated once again the strength of the CRBM in continuous-valued data modeling despite of its relatively simple and small architecture.

22.6.3.3   With Real Drifting Data

The pH ISFET sensor used in the Lab-in-a-Pill has a sensitivity of $-23\,\text{mV/pH}$. Therefore, a drift of $-163.8\,\text{mV}$, for instance, will result in a measurement error of $+7$ pH units. In this drift simulation, a CRBM with seven hidden neurons was first trained to model the data clusters as shown in Fig. 22.10a. Subsequently, the trained CRBM was fed with drifting data, in which the measurements from sensor $S1$ drifted toward the upper limit $(+1)$ and those from sensor $S2$ in the opposite direction. Figure 22.11a shows the drift data.

In addition, let us assume that only data from class A was available over the entire drift period (76,440s or 7,644 drift epochs). The absence of a complete set of data from all classes over a length of time is common in most real-world applications. In this particular case, we are interested to examine the stability and the plasticity of the learning CRBM. During the drift period, the CRBM's weight was allowed to adapt, at a learning rate of 0.1, to the sensor drift but with a constrained configuration described in Sect. 22.6.2. The SLP was, however, not subjected to any further training.

Figure 22.11b depicts the SLP's response/activity to the drifting data. As only class A data were fed, the SLP's activity should always be at '+1'. With the

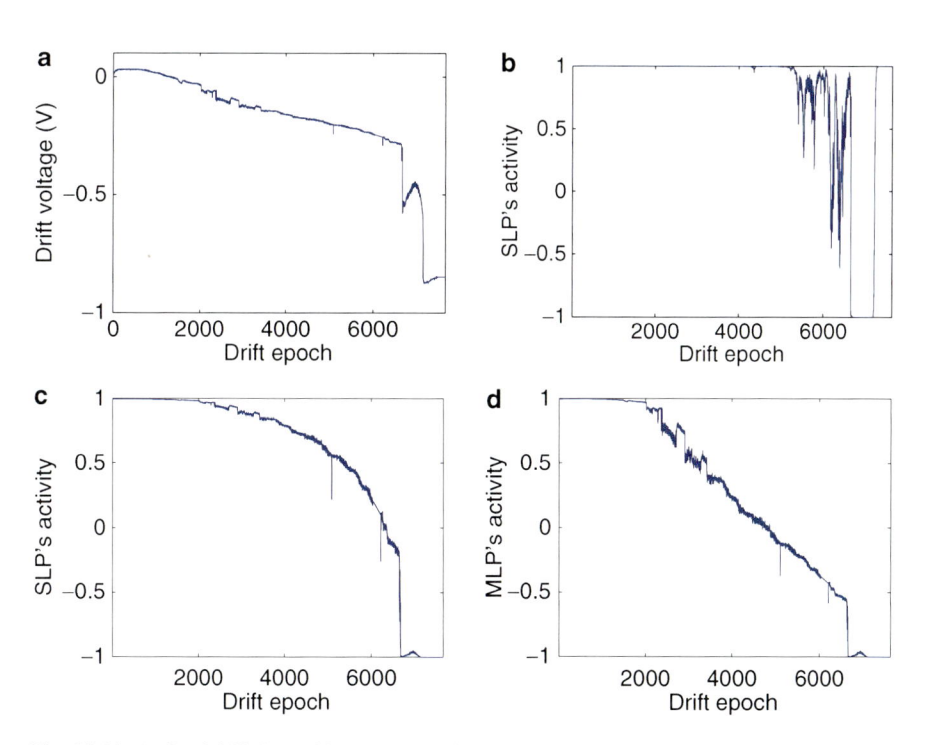

Fig. 22.11 (a) Real drift data with constant pH level and (**b–d**) the activity of output unit for each algorithm over the 7,644 drift epochs

proposed configuration, the CRBM managed to track the drift autonomously for at least first 5,000 drift epochs (Fig. 22.11b). However, as the drift was further increased, the CRBM was unable to compensate. At around 6,500th drift epoch, the abrupt drift step (caused by a malfunction in the reference electrode) resulted the CRBM to fail completely as expected. Sets of 400 test samples for each class were also used to evaluate the autocalibration by the CRBM at different drift epochs. Classification accuracy degraded from 100% (at start of the simulation), to 96.28% (after 4,000 epochs), and to 89.83% (after further 2,000 epochs).

To highlight the importance of the online adaptability feature, two trained but subsequently nonadaptive neural classifiers, namely another SLP and an MLPs network, were used as benchmarks. Both classifiers were trained and fed with the same datasets as the combined CRBM and SLP. After 20,000 training epochs, the SLP achieved an MSE of 1.37×10^{-1} and an accuracy of 95.55%, as this is not a linearly separable task. As predicted, the SLP lost track of the drift over time (Fig. 22.11c). Its accuracy dropped to 86.26% and to 79.26% at 40,000th and 6,000th epochs, respectively.

A two-layer MLP with seven hidden neurons was used. The MSE was 7.02×10^{-4} and the classification accuracy was 100% after 20,000 training epochs. Figure 22.11d depicts the response of the MLP's output neuron to the drifting data. A sharper fall in the response was observed, and this could be explained using higher-order (hence better-defined) hyperplanes by the MLP, as compared with the SLP. When tested with datasets at different drift epochs, the MLP's accuracy dropped to 85.12% and to 72.72% at 40,000th and 6,000th epochs, respectively, again falling more drastically than the SLP.

Alternatively, we can compensate sensor drift using a recurrent MLP (Brdys and Kulawski 1999; Parlos et al. 1994; Zimmermann et al. 2002). Drift is treated as a temporal feature which is then eliminated from the measurements. However, in most real applications like ours, historical data that are needed to train the MLP are not always available. The other approach is to recalibrate the MLP at scheduled time intervals and/or when a predetermined error tolerance is violated. As a demonstration, previous simulation was repeated, but this time the MLP was recalibrated at 2,500th and 5,000th drift epochs. When tested with datasets at 4,000th and 6,000th drift epochs, the accuracy of the adaptive MLP remained high, 99.88% and 99.26%, respectively. Figure 22.12a shows the MLP's activity during the drift simulation. As clearly illustrated, the recalibration provided the MLP a chance to update its weights to track the latest clusters and thus eliminate any previously accumulated drift error.

While the hardware implementation of MLP is well established, the scheduled recalibration requires storing a large set of measurements for the batch training. This can be expensive in terms of hardware. Moreover, unlike the CRBM, the MLP can only perform direct transformation. The underlaying relationship (i.e., correlation) among the input data is not exploited, thus making the classifier less robust against incomplete input data entries. As such, probabilistic models such as the CRBM are the preferred choices.

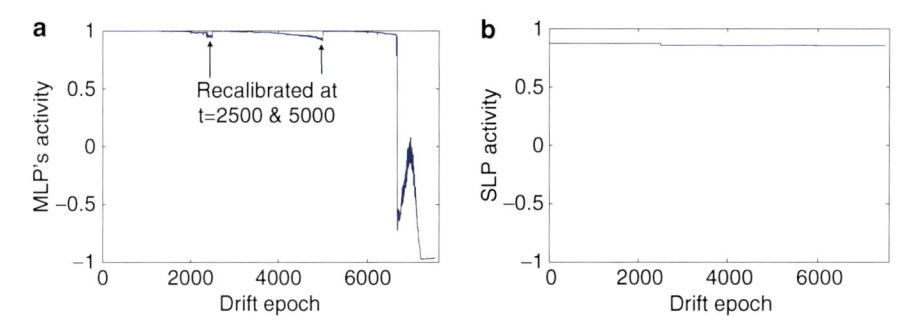

Fig. 22.12   The activity of output unit for each algorithm over the 7,644 drift epochs: (**a**) adaptive MLP; and (**b**) combined Kohonen's SOM and SLP

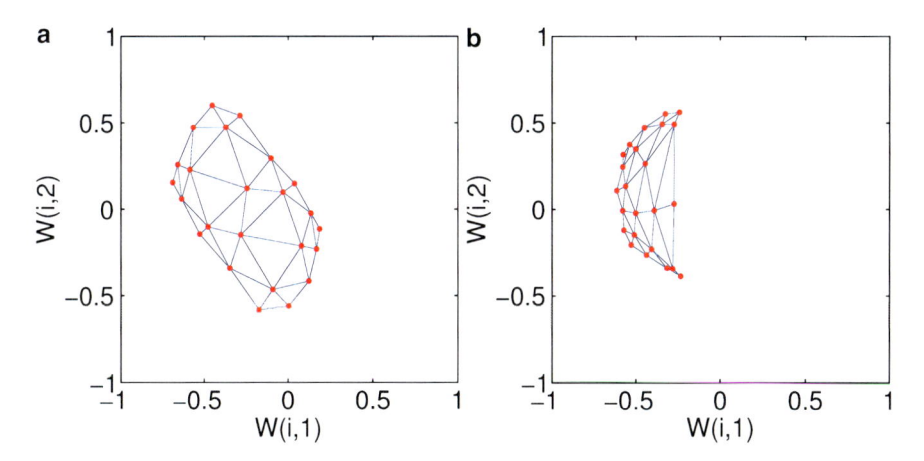

Fig. 22.13   The weight vectors of the SOM at (**a**) $t = 0$ and (**b**) $t = 2,500$th drift epoch

The CRBM is an associative memory-based model. One of the main challenges in using such models is catastrophic interference (CI). To illustrate the problem due to CI, we used a Kohonen's SOM (Haykin 1998), another example of associative memory-based models, to encode the training data clusters (Fig. 22.10a). The SOM comprised an array of 4×6 neurons and was trained for 2,000 epochs. To perform binary classification, a SLP was connected to the SOM. After an initial training, the MSE of the SLP was 1.92×10^{-2}, and the combined SOM and SLP could classify test data with 100% accurate. During the drift simulation, the SOM was allowed to adapt to sensor drift, but not the SLP.

Figure 22.12b shows the activity of the SLP over the 7,644 drift epochs. While the SLP was outputting correctly, the test data classification results were poor when sampled at 4,000th and 6,000th drift epochs. The levels of accuracy were merely 61.04% and 62.40%. When we examined the weight vectors of the SOM (Fig. 22.13), it became clear that the SOM suffered from the CI during the drift simulation as a result of its online learning. Figure 22.13a shows that the SOM

Table 22.5 Simulation results on the test data classification accuracy at different drift epochs by various algorithms

Algorithm	Accuracy (%) at drift epoch #			
	Start	4,000th	6,000th	End
CRBM + SLP	100.00	96.28	89.83	50.37
Fixed SLP	95.55	86.26	79.26	50.00
Fixed MLP	100.00	85.12	72.72	50.00
Adaptive MLP	100.00	99.88	99.26	52.22
SOM + SLP	100.00	61.04	62.40	50.00

have arranged its weight vectors to represent the two training clusters after the initial 2,000 training epochs. Figure 22.13b, on the other hand, shows the updated weight vectors after 2,500 drift epochs. They were arranged in a pattern similar to the drifted cluster A only. This is, as expected, because the SOM was fed with class A data merely during the drift simulation. This proved that CI has a major impact in our application, and an associative memory-based models could fail quite quickly. Our proposed training methodology (as mentioned in Sect. 22.6.2) can prevent such incident. The evidence is depicted in Fig. 22.11b and Table 22.5, which also lists the test results for other aforementioned algorithms at various drift epochs.

22.7 CRBM Hardware and Experimental Results

The CRBM has been implemented as a separate chip. It consists of two visible and four hidden neurons as well as their MCD training circuits. The following sections describe the implementation and the chip measurements. Using the *chip-in-the-loop learning* approach, it is demonstrated that the CRBM hardware can model data with three different types of distribution, namely symmetrical, nonsymmetrical, and doughnut-shaped.

22.7.1 Chip Implementation

First of all, the learning rules for the weights and the noise control parameters (22.3 & 22.4) were simplified into the following equations:

$$\Delta w_{ij} = \eta_w \text{sign}\left(\langle s_i s_j \rangle_4 - \langle \hat{s}_i \hat{s}_j \rangle_4\right) \qquad (22.8)$$

$$\Delta a_j = \eta_a \text{sign}\left(\langle s_j^2 \rangle_4 - \langle \hat{s}_j^2 \rangle_4\right) \qquad (22.9)$$

where $\langle \cdot \rangle_4$ denotes averaging over four training data. Although their original forms are intrinsically hardware-amenable, these simplified rules can further reduce the

Table 22.6 Mapping of parameter values between software simulations and hardware implementation

Parameter	MATLAB	VLSI (V)	Mapping ratio
s_i	$[-1.0, 1.0]$	$[1.5, 3.5]$	1:1
w_{ij}	$[-2.5, 2.5]$	$[0.0, 5.0]$	1:1
a_i	$[0.5, 9.0]$	$[1.0, 3.0]$	Logarithmic

complexity of the hardware and hence keep the silicon footprint to minimum. They take only the signs of contrastive divergence to update the weights and the noise control parameters at fixed step sizes, η_w and η_a. As shown in Chen et al. (2003) and Fleury et al. (2004), this increases the training time ($1.0\,s \rightarrow 7.2\,s$ to model data with a nonsymmetrical distribution) but has little or no effect on modeling ability.

A full CRBM VLSI chip with two visible and four hidden neurons has been designed with Cadence® Virtuoso® custom design platform. Table 22.6 summarizes the mapping of all parameters between software simulations and hardware implementation, with $+2.5\,V$ defined as the reference zero. This look-up table can be used to predict the chip measurements thus allowing us to study the training of the CRBM. The chip design was subsequently implemented with *austriamicrosystems* $0.6\,\mu m$ 2P3M CMOS process. Figure 22.14 shows a micrograph of the fabricated chip. It operates at $100\,kHz$ and the total power consumption (excluding the buffers that are only used for testing) is $7.164\,mA$, with a power supply voltage of $+5\,V$. For each neuron, a Linear Feedback Shift Register (LFSR) (Alspector et al. 1991a) was used as a random noise source. More details of the chip design can be found in Chen and Murray (2006).

22.7.2   Learning in Hardware

Early chip measurements revealed that the results from *on-chip learning* were not ideal due to some intrinsic offsets in certain training circuits. These offsets (as high as $200\,mV$) became an issue when they were larger than the contrastive divergence, $\Delta_D = (\langle s_i s_j \rangle_4 - \langle \hat{s}_i \hat{s}_j \rangle_4)$. They would dominate the training rules and therefore take control of the updating directions for weight and noise control parameters, often leading them to saturate at either the power rail or the ground. For those with less significant offsets, however, the weight and noise control parameter circuits did show evidential signs of on-chip MCD learning – only to be limited by those with huge offsets.

Further investigations unveiled that these offsets were primarily attributed to device mismatches and clock-feedthrough errors in the accumulator module of the neuron circuit design. They were approximately an average value of 7% of the voltage range (e.g., $140\,mV$ of a 2 V). To achieve a better on-chip learning, the training circuits must have an offset smaller than 1% (Chen and Murray 2006). One potential

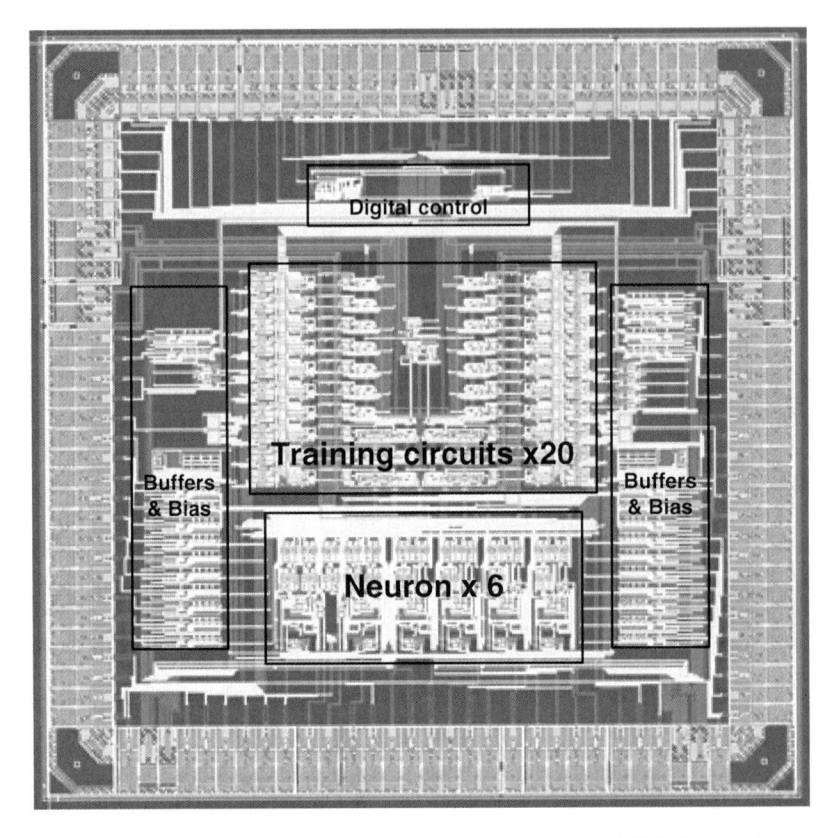

Fig. 22.14 Micrograph of the CRBM VLSI chip (Chen and Murray 2006). It comprises a total of six neurons (two visible and four hidden) and their MCD training circuits. Additional modules namely digital control for parameter training, buffers for testing, and bias circuits are also included. The total silicon area is $3.197 \times 2.928\,\mathrm{cm}$

solution is to redesign the accumulators with the dynamic current mirrors proposed in Wegmann and Tsividis (1989). The dynamic current mirror removes device mismatches using the same set of transistors to sample and to output a current. It has been reported that its output error can be as small as 0.05% (Wegmann and Tsividis 1989), $20\times$ smaller than the tolerable offset in a CRBM hardware.

In this particular case, a better approach to demonstrate the MCD in hardware is via *chip-in-the-loop learning*. First of all, offsets at all important circuit nodes were derived from chip measurements and added into our simulation model. With a better description of the hardware, the weights and the noise control parameters could be calculated more accurately off-the-chip on a computer. The CRBM hardware was involved in the feedforward propagation only. The simulation results were then downloaded to the CRBM hardware at the end of each training cycle. The subsequent sections report the experimental results.

22.7.3 Regenerating Data With a Symmetric Distribution

The CRBM hardware was trained to regenerate data with a symmetric distribution as in Fig. 22.15a. After 30,000 training epochs, the parameter values were mapped from the software simulations to the hardware implementation:

$$\{w_{ij}\} = \begin{bmatrix} \times\ 3.203\ 2.598\ 2.400\ 1.319 \\ 2.862\ 2.280\ 2.180\ 4.061\ 2.589 \\ 2.616\ 2.365\ 2.471\ 1.811\ 2.565 \end{bmatrix} (V)$$

$$\{a_v\} = \begin{bmatrix} 2.458\ 1.651 \end{bmatrix} (V)$$

$$\{a_h\} = \begin{bmatrix} 1.503\ 2.516\ 1.286\ 1.551 \end{bmatrix} (V). \tag{22.10}$$

The learning rates for parameters were $\eta_w = \eta_{ah} = 0.003$ and $\eta_{av} = 0.03$. Figure 22.15b illustrates a 20-step reconstruction generated by the CRBM hardware. As depicted, the CRBM hardware was able to model the two-cluster data after the offsets were compensated in software. Moreover, the voltage ranges in (22.10) indicate that the weights w_{ij} and the noise control parameters $a_{v/h}$ have not saturated. In other words, the chip design operates as intended and can provide sufficiently wide voltage ranges for parameters to be trained.

To demonstrate the probabilistic behavior of the CRBM hardware, two different sets of measurements on the visible neurons recording the movement of a test data to one of the two clusters were collected. The visible neurons were initialized to $V(0) = (2.5, 2.5)V$, and their states were subsequently Gibbs-sampled 20× consecutively. Figure 22.16 shows two distinct traces of sampled visible states. Despite their common start point, each moved to different cluster within the first few steps and subsequently "oscillated" mildly within the region. This clearly demonstrates the probabilistic behavior of the CRBM as a generative model and the role of the injected stochastic noise.

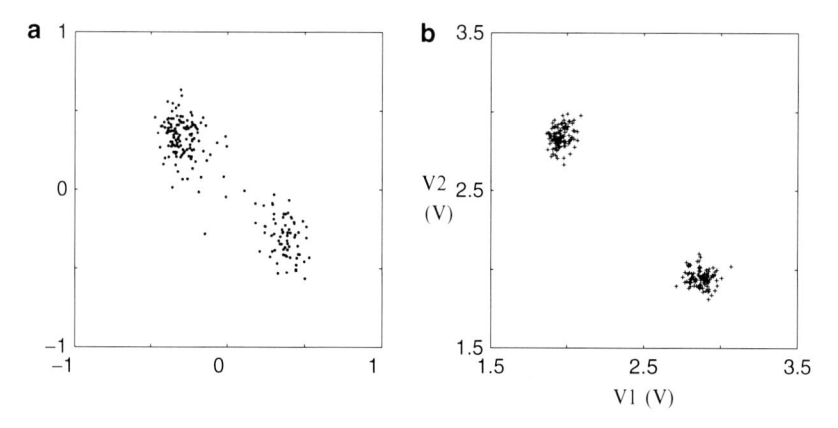

Fig. 22.15 (a) The training data and (b) the 20-step reconstruction generated by the CRBM hardware with parameters refreshed to the levels in (22.10)

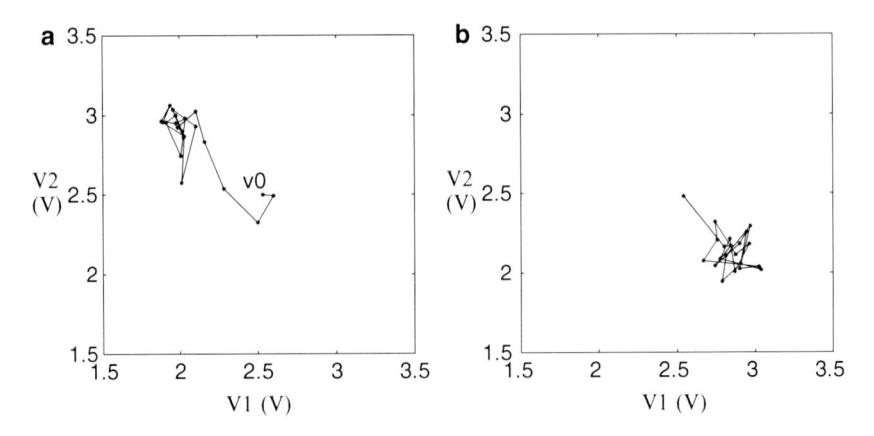

Fig. 22.16 Two distinct sets of measurements on the visible neurons during 20 steps of Gibbs sampling in the CRBM hardware. The trace in (**a**) moves toward the upper-left cluster, while the trace in (**b**) moves toward the lower-right cluster

22.7.4 Regenerating Data with a Nonsymmetric Distribution

The experiment was repeated to regenerate data with a nonsymmetric distribution, as shown in Fig. 22.17a. The upper-left cluster is formed by data points sampled from an elliptic Gaussian, while the bottom-right cluster is from a circular Gaussian. The aim of this experiment is to demonstrate that unlike the binary RBM, the CRBM can model nonsymmetric data distributions. As before, the CRBM hardware was trained via the chip-in-the-loop learning approach for 30,000 epochs with $\eta_w = \eta_{ah} = 0.003$ and $\eta_{av} = 0.03$. Figure 22.17b shows the 20-step reconstruction by a CRBM model in software simulation with a corresponding set of parameters as below:

$$\{w_{ij}\} = \begin{bmatrix} \times & 2.662 & 2.702 & 1.927 & 2.425 \\ 2.571 & 4.544 & 1.661 & 2.446 & 2.748 \\ 2.186 & 3.732 & 3.875 & 2.451 & 2.593 \end{bmatrix} (V)$$

$$\{a_v\} = \begin{bmatrix} 2.011 & 2.151 \end{bmatrix} (V)$$

$$\{a_h\} = \begin{bmatrix} 1.8287 & 1.2083 & 2.0883 & 2.4835 \end{bmatrix} (V). \tag{22.11}$$

By setting the parameters to (22.11), the CRBM hardware regenerated data as illustrated in Fig. 22.17c. The cluster at the bottom-right corresponds to the circular cluster in training data, while the elliptical cluster is now represented by two well-separated circular clusters. Their clear separation is caused by excessively large values of the weights and noise control parameters. Simply adjusting a_{v1} to 2.1, V, a_{h1} to 2.274 V and w_{10} to 2.45 V, a better reconstruction of the training data

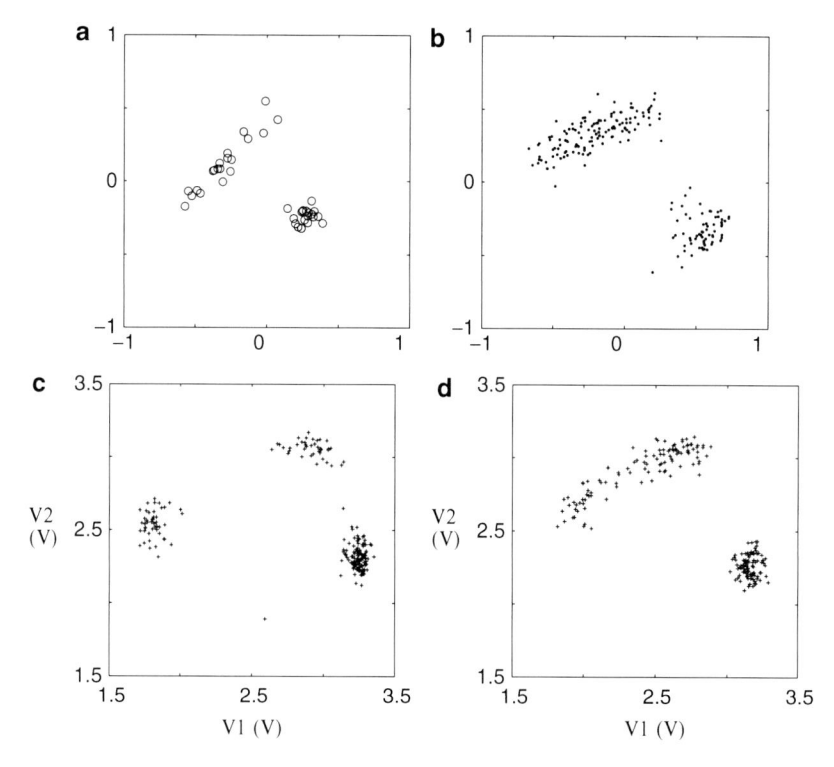

Fig. 22.17 (a) Training data sampled from one elliptic and one circular Gaussian. (b) Twenty-step reconstruction generated by a CRBM model in software after being trained for 30,000 epochs. (c) Twenty-step reconstruction generated by the CRBM hardware with parameters set to the levels in (22.11). (d) Twenty-step reconstruction by the CRBM hardware after some minor fine-tuning

can be obtained (Fig. 22.17d). This fine-tuning was required because not all the imperfections in hardware were taken into the consideration during the training. The CRBM hardware was only involved in the feedforward propagation.

22.7.5 Regenerating Data with a Doughnut-Shaped Distribution

To further prove that a CRBM could model non-Gaussian data distributions as well, the CRBM hardware was used to regenerate data with a doughnut-shaped distribution, as depicted in Fig. 22.18a. In order to accomplish the task, the CRBM must be able to capture correlations between probabilities in both dimensions.

The CRBM hardware was once again trained via the chip-in-the-loop approach for 20,000 epochs with $\eta_w = 0.00075$, $\eta_{av} = \eta_{ah} = 0.0075$. Figure 22.18b shows the 20-step reconstruction generated by a trained CRBM model in software. The reconstruction forms a circular band, albeit with a slightly uneven distribution.

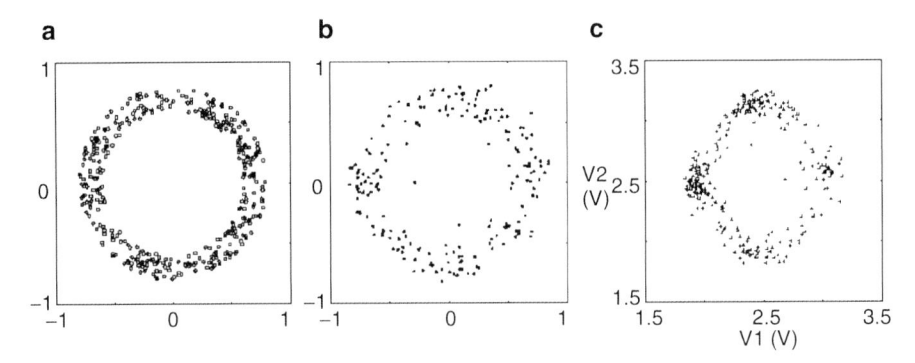

Fig. 22.18 (**a**) Training data with a doughnut-shaped distribution. (**b**) Twenty-step reconstruction generated by a CRBM model in software after being trained for 20,000 epochs. (**c**) Twenty-step reconstruction generated by the CRBM hardware with the parameters set to the levels in (22.12) plus some fine-tuning

This is due to the use of merely four hidden neurons to represent this fairly complex distribution. The parameters of this model were then mapped into the hardware implementation as:

$$\{w_{ij}\} = \begin{bmatrix} \times & 2.503 & 2.548 & 2.451 & 2.498 \\ 2.497 & 2.964 & 2.514 & 2.526 & 2.025 \\ 2.498 & 2.973 & 2.503 & 2.434 & 2.975 \end{bmatrix} (V)$$

$$\{a_v\} = \begin{bmatrix} 1.287 & 1.294 \end{bmatrix} (V)$$

$$\{a_h\} = \begin{bmatrix} 1.154 & 1.835 & 1.898 & 1.173 \end{bmatrix} (V). \tag{22.12}$$

By downloading (22.12) with slight adjustments on a_{v1} and a_{h1}, the CRBM hardware regenerated a 20-step reconstruction as shown in Fig. 22.18c. As in the software, the distribution of the reconstruction is not perfectly even, but it still forms a roughly circular band. The shape of the distribution is distorted due to the nonlinearity in the multipliers. Multiple Gibbs sampling amplifies the problem. To resolve it, *on-chip learning* is required to adapt the nonlinearity in the multipliers. Nevertheless, this experiment demonstrates that the CRBM hardware is able to model a complex distribution with a limited number of neurons.

22.8 Discussion and Future Works

In Sect. 22.6, we demonstrated how a neural network such as the CRBM could perform data fusion for the multisensor microsystem *Lab-in-a-Pill* in software. Then, in Sect. 22.7 we illustrated an example implementation of the CRBM and showed its continuous-valued probabilistic behavior in silicon, modeling different types of data distributions. The immediate next step is to integrate the CRBM with the sensors on a single substrate. Importantly, the accumulators need to be redesigned with dynamic current mirrors to permit on-chip learning.

On the neural algorithm itself, the simulation results in Sect. 22.6.3.3 suggested that the use of a *fixed* learning rate η_w is less than ideal, because drift occurs at different rates at different time instances, in most real-world applications. If $\eta_w \ll \eta_{op}$ where η_{op} is the optimum learning rate, the model will not be able to keep up with the sensor drift. This is a major issue in long-term measurements. On the other hand, if $\eta_w \gg \eta_{op}$, large weight update can cause the CRBM to be very susceptive to stochastic noise and may destabilize the CRBM away from the optimum solution. In Tang and Murray (2007), a series of simulations was carried out, and the results showed that, unsurprisingly, different drift rates have different optimal values of η_{op}. These manifest a demand for an adaptive learning rate for the classifier to improve its reliability in dynamic environment. One possible solution is the general adaptive learning rule proposed by Murata et al. (1996). It will require careful selection of the algorithm parameters to ensure stability in the online learning.

In this chapter, we have constrained the study to electrochemical sensors only. However, the use of ANN as a form of data fusion mechanism, as described in this chapter, is equally applicable to other types of silicon sensor. For instance, the piezoelectric active-sensors (Park et al. 2006) used to monitor the health of building structures can be equipped with an ANN for several purposes: (a) to track drift/degradation in the mechanical/electrical properties of the sensors themselves, (b) to eliminate temperature-dependent variation in measurements, and (c) to identify any damage in the structures, hence provide an early warning. Similarly, the vision chips (Lichtsteiner et al. 2008; Philipp et al. 2007; Shi 2002) which consist of an array of photodetectors, with an ANN, can be distributed at multiple locations to detect any abnormality in factory machinery. More recently, a cantilever wind sensor (Argyrakis et al. 2007) based on the microelectromechanical system (MEMS) technology has been built. It comprises an array of piezoresistors to detect the motion of wind. Using an ANN, the response from each piezoresistor can be fused to infer the direction of the wind, for example.

The introduction of ANN to low-power multisensor microsystems not only allows local data fusion, but the ability of ANN to *learn* provides a means to ameliorate device mismatch and time-dependent degradation in the VLSI circuits. Both problems (device mismatch and time-dependent degradation) have been identified as key challenges for future semiconductor technology and design (Itrs 2008). Early study on exploiting the plasticity of ANN to minimize the effects of device variations, within the context of a depth-from-motion algorithm, has recently been reported in Cameron and Murray (2008). Test chip results indicated that the ANN was able to compensate the device mismatch and provide a more consistent output. With the statistical variability of future semiconductor devices undoubtedly to be more significant, having a robust computing paradigm such as the ANNs is likely to be the way forward to facilitate more sensors and algorithmic functions to be implemented effectively in future integrated multisensor microsystems.

22.9 Summary

In this chapter, various types of silicon sensors are introduced, and the motivations for on-chip signal processing are explained. The use of ANNs to fuse electrochemical sensory data is then reviewed. It is found that while sensor drift is widely acknowledged as a major issue, there is yet a viable solution, especially one which can be implemented with low-power circuitry. This leads to a review of state-of-the-art neural hardware and another review of current analytical techniques for counteracting drift. Subsequently, we show how an ANN can be used as a form of multisensor fusion mechanism in an example integrated multisensor microsystem, named *Lab-in-a-Pill*. Like in most wireless sensor networks, this application requires the ANN to be robust against stochastic noise and sensor drift, as well as be hardware-amenable within a small silicon footprint. One possible solution is the generative model – CRBM. Its architecture and training methodology which is modified to enable online drift-tracking are later detailed. Simulation results show that CRBM not only can model non-Gaussian data clusters, but can also autocalibrate the sensors against stochastic drift, if configured. It is also identified that adaptive learning rate that corresponds to time-dependent drift rate can further improve the results. The hardware implementation of the CRBM and the chip measurements are also reported. Experimental results demonstrate the continuously valued probabilistic behaviors of the CRBM in silicon. With the VLSI technology that is set to progress further, we anticipate that the sensor array will get larger and more diversified, and local data fusion will have a major role in future wireless microsystems.

References

Alspector, J., Allen, R.B., Jayakumar, A., Zeppenfeld, T., Meir, R.: Relaxation networks for large supervised learning problems. In: Advances in Neural Processing Systems, **4**, 1015–1026 (1991)

Alspector, J., Gannett, J.W., Haber, S., Parker, M.B., Chu, R.: A VLSI-efficient technique for generating multiple uncorrelated noise sources and its application to stochastic neural networks. IEEE Transactions on Circuits and Systems **38**(1), 109–123 (1991)

Alspector, J., amd R. B. Allen, B.G.: Performance of a stochastic learning microchip. In: Advances in Neural Information Processing Systems, **1**, 748–760 (1989)

Argyrakis, P., Hamilton, A., Webb, B., Zhang, Y., Gonos, T., Cheung, R.: Fabrication and characterization of a wind sensor for integration with a neuron circuit. Microelectronic Engineering **84**(5–8), 1749–1753 (2007)

Artursson, T., Eklov, T., Lundstrom, I., Martensson, P., Sjostrom, M., Holmberg, M.: Drift correction for gas sensors using multivariate methods. Journal of Chemometrics **14**, 711–723 (2000)

Asanovic, K., Morgan, N.: Experimental determination of precision requirements for backpropagation training of artificial neural networks. In: Proceedings of International Conference on Microelectronics for Neural Network, pp. 9–15. Munich, Germany (1991)

Aydin, N., Arslan, T., Cumming, D.R.S.: A direct-sequence spread-spectrum communication system for integrated sensor microsystems. IEEE Transactions on Information Technology in Biomedicine **9**(1), 4–12 (2005)

Bedoya, G., Jutten, C., Bermejo, S., Cabestany, J.: Improving semiconductor-based chemical sensor arrays using advanced algorithms for blind source separation. In: Proceedings of the IEEE Sensors for Industry Conference, pp. 149–154. New Orleans, Louisiana, USA (2004)

Bermejo, S., Bedoya, G., Parisi, V., Cabestany, J.: An on-line water monitoring system using a smart ISFET array. In: Proceedings of the IEEE Conference on Industrial Electronics Society, pp. 2797–2802 (2002)

Brdys, M.A., Kulawski, G.J.: Dynamic neural controllers for induction motor. IEEE Transactions on Neural Networks 10(2), 340–355 (1999)

Bris, N.L., Birot, D.: Automated pH-ISFET measurements under hydrostatic pressure for marine monitoring application. Analytica Chimica Acta 356, 205–215 (1997)

Cameron, K.L., Murray, A.F.: Minimizing the effect of process mismatch in a neuromorphic system using spike-timing-dependent adaptation. IEEE Transactions on Neural Networks 19(5), 899–913 (2008)

Card, H.C., McNeill, D.K., Schneider, C.R.: Analog VLSI circuits for competitive learning networks. Analog Integrated Circuits and Signal Processing 15, 291–314 (1998)

Chen, H., Fleury, P., Murray, A.F.: Minimizing Contrastive Divergence in noisy, mixed-mode VLSI neurons. In: Advances in Neural Information Processing Systems, vol. 16 (2003)

Chen, H., Murray, A.F.: A Continuous Restricted Boltzmann Machine with an implementable training algorithm. IEE Proceedings on Vision, Image and Signal Processing 150(3), 153–158 (2003)

Chen, H., Murray, A.F.: Continuous-valued probabilistic behaviour in a vlsi generative model. IEEE Transactions on Neural Networks 17(3), 755–770 (2006)

Chen, T.L., You, R.Z.: A novel fault-tolerant sensor system for sensor drift compensation. Sensors and Actuators A: Physical 147(2), 623–632 (2008)

Chua, L.O., Roska, T.: The CNN paradigm. IEEE Transactions on Circuits and Systems-I: Fundamental Theory and Applications 40(3), 147–156 (1993)

Chung, D., Merat, F.L.: Neural network based sensor array signal processing. In: Proceedings of the IEEE International Conference on Multisensor Fusion and Integration for Intelligent Systems, pp. 757–764. Washington, DC, USA (1996)

Clarke, D.W.: Sensor, actuator and plant validation. IEE Colloquium on Intelligent and Self-Validating Sensors pp. 1–8 (1999)

Coggins, R., Jabri, M., Flower, B., Pickard, S.: A hybrid analog and digital VLSI neural network for intracardiacmorphology classification. IEEE Journal of Solid-States Circuits 30(5), 542–550 (1995)

Errachid, A., Bausells, J., Jaffrezic-Renault, N.: A simple REFET for pH detection in differential mode. Sensors and Actuators B 60, 43–48 (1999)

Figeys, D., Pinto, D.: Lab-on-a-chip: A revolution in biological and medical sciences. Analytical Chemistry 72(9), 330A–335A (2000)

Fleury, P., Chen, H., Murray, A.F.: On-chip Contrastive Divergence learning in analogue VLSI. In: Proceedings of the International Joint Conference on Neural Networks, pp. 1723–1728. Budapest, Hungary (2004)

Gardner, J.W., Hines, E.L., Molinier, F., Bartlett, P.N., Mottram, T.T.: Prediction of health of dairy cattle from breath samples using neural network with parametric model of dynamic response of array of semiconducting gas sensors. IEE Proceedings on Sci. Meas. Technology 146(2), 102–106 (1999)

Grattarola, M., Massobrio, G., Martinoia, S.: Modelling H^+-Sensitive FET's with SPICE. IEEE Transactions on Electron Devices 39(4), 813–819 (1992)

Guo, T.H., Nurre, J.: Sensor failure detection and recovery by neural networks. In: Proceedings of IJCNN, vol. 1, pp. 221–226. Seattle, WA, USA (1991)

Haykin, S.: Neural Networks: A Comprehensive Foundation. Prentice Hall (1998)

Heckerman, D.: Learning in graphical models, chap. A tutorial on learning with Bayesian networks, pp. 301–354. MIT, Cambridge, MA, USA (1999)

Hendrikse, J., Olthuis, W., Bergveld, P.: A method of reducing oxygen induced drift in iridium oxide pH sensors. Sensors and Actuators B 53, 97–103 (1998)

Higuchi, T., Furuya, T., Handa, K., Takahashi, N., Nishiyama, H., Kokubu, A.: IXM2: A parallel associative processor. In: Proceedings of the international symposium on Computer architecture, pp. 22–31. Toronto, Ontario, Canada (1991)

Hinton, G.E.: Products of experts. In: Proceedings of the 9th International Conference on Artificial Neural Networks, pp. 1–6. Edinburgh, Scotland (1999)

Hinton, G.E.: Training Products of Experts by Minimizing Contrastive Divergence. Neural Computation **14**, 1771–1800 (2002)

Holmberg, M., Davide, F.A.M., Natale, C.D., D'Amico, A., Winquist, F., Lundstrom, I.: Drift counteraction in odour recognition applications: lifelong calibration method. Sensors and Actuators B **42**, 185–194 (1997)

Holmin, S., Krantz-Rulcker, C., Lundstrom, I., Winquist, F.: Drift correction of electronic tongue responses. Institute of Physics Measurement Science Technology **12**, 1348–1354 (2001)

Holt, J.L., Hwang, J.N.: Finite precision error analysis of neural network hardware implementations. IEEE Transactions on Computers **42**(3), 281–290 (1993)

Hsu, D., Figueroa, M., Diorio, C.: Competitive learning with floating-gate circuits. IEEE Transactions on Neural Networks **13**(3), 732–744 (2002)

Ienne, P., Cornu, T., Kuhn, G.: Special-purpose digital hardware for neural networks: An architectural survey. Journal of VLSI Signal Processing Systems **13**, 5–25 (1996)

ITRS: International technology roadmap for semiconductors update. Technical report (2008)

Jabri, M., Flower, B.: Weight perturbation: An optimal architecture and learning technique for analog VLSI feedforward and recurrent multilayer networks. IEEE Transactions on Neural Networks **3**(1), 154–157 (1992)

Jamasb, S.: An analytical technique for counteracting drift in ion-selective field effect transistor (ISFETs). IEEE Sensors Journal (2004)

Jamasb, S., Collins, S.D., Smith, R.L.: Correction of instability in Ion-selective Field Effect Transistors for accurate continuous monitoring of pH. In: Proceedings of IEEE International Conference of EMBS, pp. 2337–2340. Chicago, IL, USA (1997)

Jamasb, S., Collins, S.D., Smith, R.L.: A physical model for threshold voltage instability in Si_3N_4-Gate H^+-Sensitive FET's (pH-ISFET's). IEEE Transactions on Electron Devices **45**(6), 1239–1245 (1998)

Johannessen, E.A., Wang, L., Cui, L., Tang, T.B., Ahmadian, M., Astaras, A., Reid, S.W., Yam, S., Murray, A.F., Flynn, B.W., Beaumont, S.P., Cumming, D.R.S., Cooper, J.M.: Implementation of multichannel sensors for remote biomedical measurements in a microsystems format. IEEE Transactions on Biomedical Engineering **51**(3), 525–535 (2004)

Keller, P.E., Kouzes, R.T., Kangas, L.J.: Three neural network based sensor systems for environmental monitoring. In: Proceedings of the IEEE Electro, pp. 378–382. Boston, MA, USA (1994)

Kermani, B.G., Schiffman, S.S., Nagle, H.T.: Using neural networks and genetic algorithms to enhance performance in an electronic nose. IEEE Transactions on Biomedical Engineering **46**(4), 429–439 (1999)

Ko, W.H., Fung, C.D.: VLSI and intelligent transducers. Sensors and Actuators **2**, 239–250 (1982)

Lang, K.J., Waibel, A.H., Hinton, G.E.: A time-delay neural network architecture for isolated word recognition. Neural Networks **3**(1), 23–43 (1990)

Lazzerini, B., Marcelloni, F.: Counteracting drift of olfactory sensors by appropriately selecting features. IEE Electronics Letters **36**(6), 509–510 (2000)

Leong, P.H.W., Jabri, M.A.: A low power trainable analogue neural network classifier chip. In: Proceedings of the IEEE Custom Integrated Circuits Conference, pp. 451–454. San Diego, CA, USA (1993)

Lichtsteiner, P., Posch, C., Delbruck, T.: A 128x128 120db 15µs latency asynchronous temporal contrast vision sensor. IEEE Journal of Solid-State Circuits **43**(2), 566–576 (2008)

Lindquist, M., Wide, P.: Virtual water quality tests with an electronic tongue. In: Proceedings of the IEEE IMTC, vol. 2, pp. 1320–1324 (2001)

Luo, R.C., Yih, C.C., Su, K.L.: Multisensor fusion and integration: Approachs, applications, and future research directions. IEEE Sensors Journal **2**(2), 107–119 (2002)

Macq, D., Verleysen, M., Jespers, P., Legat, J.D.: Analog implementation of a kohonen map with on-chip learning. IEEE Transactions on Neural Networks **4**(3), 456–461 (1993)

Marco, S., Ortega, A., Pardo, A., Samitier, J.: Gas identification with tin oxide sensor array and self-organizing maps: Adaptive correction of sensor drifts. IEEE Transactions on Instrumentation and Measurement **47**(1), 316–321 (1998)

Martin, G., Chang, H.: System-on-chip design. In: Proceedings of International Conference on ASIC, pp. 12–17. Shanghai, China (2001)

Mayes, D.J., Hamilton, A., Murray, A.F., Reekie, H.M.: A pulsed VLSI radial basis function chip. In: Proceedings of the IEEE International Symposium on Circuits and Systems, vol. 3, pp. 297–300. Atlanta, GA, USA (1996)

Middelhoek, S., Hoogerwerf, A.C.: Smart Sensors: When and Where? Sensors and Actuators **8**, 39–48 (1985)

Moerland, P., Fiesler, E.: Handbook of Neural Computation, chap. Chapter E1.2: Neural Network Adaptations to Hardware Implementations. Institute of Physics Publishing and Oxford University Publishing, New York, USA (1996)

Murata, N., Muller, K., Ziehe, A., Amari, S.: Adaptive on-line learning in changing environments. In: Advance in Neural Information Processing Systems, vol. 9, pp. 599–605 (1996)

Natale, C.D., Davide, F.A.M., D'Amico, A.: A self-organizing system for pattern classification: time varying statistics and sensor drift effects. Sensors and Actuators B **26-27**, 237–241 (1995)

Nishizawa, K., Hirai, Y.: Hardware implementation of PCA neural network. In: Proceedings of ICONIP, pp. 85–88. Kitakyushu, Japan (1998)

Park, G., Farrar, C.R., Rutherford, A.C., Robertson, A.N.: Piezoelectric active sensor self-diagnostics using electrical admittance measurements. Journal of Vibration and Acoustics **128**(4), 469–476 (2006)

Park, S., Lee, C.S.G.: Fusion-based sensor fault detection. In: Proceedings of IEEE International Symposium on Intelligent Control, pp. 156–161. Chicago, IL, USA (1993)

Parlos, A.G., Chong, K.T., Atiya, A.F.: Application of the recurrent multilayer perceptron in modelling complex process dynamics. IEEE Transactions on Neural Networks **5**(2), 255–266 (1994)

Philipp, R.M., Orr, D., Gruev, V., van der Spiegel, J., Etienne-Cummings, R.: Linear current-mode active pixel sensor. IEEE Journal of Solid-State Circuits **42**(11), 2482–2491 (2007)

Platonov, A.A., Szabatin, J., Jedrzejewski, K.: Optimal synthesis of smart measurement systems with adaptive correction of drifts and setting errors of the sensor's working point. IEEE Transactions on Intrumentation and Measurement **47**(3), 659–665 (1998)

Pottie, G.J., Kaiser, W.J.: Wireless integrated network sensors. Communications of the ACM **43**(5), 51–58 (2000)

Rabaey, J.M., Ammer, M.J., da Silva Jr., J.L., Patel, D., Roundy, S.: PicoRadio supports ad hoc ultra-low power wireless networking. Computer **33**(7), 42–48 (2000)

Rodriguez-Mendez, M.L., Arrieta, A.A., Parra, V., Bernal, A., Vegas, A., Villanueva, S., Gutierrez-Osuna, R., de Saja, J.A.: Fusion of three sensory modalities for the multimodal characterization of red wines. IEEE Sensors Journal **4**(3), 348–354 (2004)

Roppel, T., Wilson, D., Dunman, K., Becanovic, V., Padgett, M.L.: Design of a low-power, portable sensor system using embedded neural networks and hardware preprocessing. In: Proceedings of the IEEE International Joint Conference on Neural Networks, pp. 142–145 (1999)

Rumelhart, D.E., Hinton, G.E., Williams, R.J.: Learning Internal Representations by Error Propagation, *Computational models of cognition and perception*, vol. 1, chap. 8, pp. 319–362. MIT, Cambridge, MA, USA (1986)

Sachenko, A., Kochan, V., Turchenko, V., Tsahouridis, K., Laopoulos, T.: Error compensation in an intelligent sensing instrumentation system. In: Proceedings of IEEE Instrumnetation and Measurement Technology Conference, pp. 869–874. Budapest, Hungary (2001)

Sarkaria, S.: Catastrophic interference (2004). Http://www.ee.ubc.ca/elec592/PDFfiles/Catastrophic_Learning.pdf

Sarry, F., Lumbreras, M.: Gas discrimination in an air-conditioned system. IEEE Transactions on Instrumentation and Measurement **49**(4), 809–812 (2000)

Sayago, I., d. C. Horrillo, M., Baluk, S., Aleixandre, M., Fernandez, M.J., Ares, L., Garcia, M., Santos, J.P., Gutierrez, J.: Detection of toxic gases by a tin oxide multisensor. IEEE Sensors Journal 2(5), 387–393 (2002)

Seiter, J.C., DeGrandpre, M.D.: Redundant chemical sensors for calibration-impossible applications. Talanta pp. 99–106 (2001)

Shi, B.E.: A low power orientation selective vision sensor. IEEE Transactions on Circuits and Systems-II: Analog and Digital Signal Processing 47(5), 435–440 (2002)

Shin, H.W., Llober, E., Gardner, J.W., Hines, E.L., Dow, C.S.: Classification of the strain and growth phase of Cyanobacteria in potable water using an electronic nose system. IEE Proceedings on Science, Measurement and Technology 147(4), 158–164 (2000)

Smith, R.L., Scott, D.C.: An integrated sensor for electrochemical measurements. IEEE Transactions on Biomedical Engineering 33(2), 83–90 (1986)

Smolensky, P.: Parallel Distributed Processing: Explorations in Microstructure of Cognition, vol. 1, chap. Information processing in dynamical systems: Foundations of harmony theory, pp. 195–281. MIT (1986)

Steinhage, A., Winkel, C.: A robust self-calibrating data fusion architecture. In: Proceedings of IEEE National Geoscience and Remote Sensing Symposium, pp. 963–965. Honolulu, HI, USA (2000)

Stetter, J.R., Penrose, W.R.: The electrochemical nose. http://electrochem.cwru.edu/ed/encycl/art-n01-nose.htm (2001)

Sundic, T., Marco, S., Samitier, J., Wide, P.: Electronic tongue and electronic nose data fusion in classification with neural networks and fuzzy logic based models. In: Proceedings of the IEEE IMTC, vol. 3, pp. 1474–1479 (2000)

Tang, T.B., Chen, H., Murray, A.F.: Adaptive, integrated sensor processing to compensate for drift and uncertainty: a stochastic 'neural' approach. IEE Proceedings on Nanobiotechnology 151(1), 28–34 (2004)

Tang, T.B., Johannessen, E., Wang, L., Astaras, A., Ahmadian, M., Murray, A.F., Cooper, J.M., Beaumont, S.P., Flynn, B.W., Cumming, D.R.S.: Toward a miniature wireless integrated multisensor microsystem for industrial and biomedical applications. IEEE Sensors Journal: Special Issue on Integrated Multisensor Systems and Signal Processing 2(6), 628–635 (2002)

Tang, T.B., Murray, A.F.: Adaptive sensor modelling and classification using a continuous restricted boltzmann machine (crbm). Neurocomputing 70(7-9), 1198–1206 (2007)

Tsai, C.S., Tong, C.C., Oh, L.E.: Sensor data correction with neural network incorporating fuzzy logic. In: Proceedings of IEEE International Fuzzy Systems Conference, pp. 66–71. Seoul, Korea (1999)

Warneke, B.A., Scott, M.D., Leibowitz, B.S., Zhou, L., Bellew, C.L., Chediak, J.A., Kahn, J.M., Boser, B.E., Pister, K.S.J.: An autonomous 16mm^3 solar-powered node for distributed wireless sensor networks. In: Proceedings of IEEE Sensors, pp. 1510–1515. Orlando, FL, USA (2002)

Wegmann, G., Tsividis, Y.: Very accurate dynamic current mirrors. Electronics Letters 25(10), 644–646 (1989)

Wide, P., Winquist, F., Bergsten, P., Petriu, E.M.: The human-based multisensor fusion method for artificial nose and tongue sensor data. IEEE Transactions on Instrumentation and Measurement 47(5), 1072–1077 (1998)

Widrow, B., Hoff, M.E.: Adaptive switching circuits. IRE WESCON Convention Record pp. 96–104 (1960)

Wise, K.D.: Integrated microsystems: Merging MEMS, micropower electronics, and wireless commnunications. In: Proceedings of IEEE ASIC/SoC Conference, pp. xxiii–xxix (1999)

Woodburn, R., Murray, A.F.: Implementing artificial neural networks in analogue VLSI. In: Proceedings of the International Conference on Neural Information Processing, pp. 658–661. Dunedin, New Zealand (1997)

Yen, G.G., Feng, W.: Winner take all experts network for sensor validation. In: Proceedings of the IEEE International Conference on Control Applications, pp. 92–97. Anchorage, Alaska, USA (2000)

Zimmermann, H.G., Tietz, C., Grothmann, R.: Yield curve forecasting by error correction neural networks and partial learning. In: ESANN Proceedings, pp. 407–412. Bruges, Belgium (2002)

Chapter 23
Bio-Inspired Mechatronics and Control Interfaces

Panagiotis K. Artemiadis and Kostas J. Kyriakopoulos

Abstract There is great effort during the last decades toward building control interfaces for robots that are based on signals measured directly from the human body. In particular, electromyographic (EMG) signals from skeletal muscles have proved to be very informative regarding human motion, and therefore they are usually incorporated in control interfaces for robots that are either remotely operated or being worn by humans, i.e., arm exoskeletons. This chapter presents a methodology for estimating human arm motion using EMG signals from muscles of the upper limb, using a decoding method and an additional bio-inspired filtering technique based on a probabilistic model for arm motion. The method results in a robust human–robot control interface that can be used in many different kinds of robots (i.e., teleoperated robot arms, arm exoskeletons, prosthetic devices). The proposed methodology is assessed through real-time experiments in controlling a remote robot arm in random 3D movements using only EMG signals recorded from able-bodied subjects.

23.1 Overview

Robots that are controlled through natural human interfaces have gained increased attention during the last years. This is getting more obvious since a substantial percentage of robotic mechanisms being built during the last decade is *worn* or closely located to the human body. Prosthetic and orthotic mechanisms are increasingly used for aiding the people in need, as well as for augmenting human capabilities (i.e., power exoskeletons) or for providing a more direct control interface for remotely operated machines. The control interfaces for robots that are so close to humans should be safe and natural for the user, as well as able to mimic human motion characteristic or patterns, to be perceived as *extensions* of the subject's body.

P.K. Artemiadis (✉)
PostDoctoral Associate, Massachusetts Institute of Technology,
77 Massachusetts Avenue, Cambridge, MA 02139, USA
e-mail: partem@mit.edu

V. Cutsuridis et al. (eds.), *Perception-Action Cycle: Models, Architectures, and Hardware*, Springer Series in Cognitive and Neural Systems 1, DOI 10.1007/978-1-4419-1452-1_23, © Springer Science+Business Media, LLC 2011

In this chapter, a robotic platform driven by a novel control interface that is based on human body-oriented signals is analyzed. More specifically, a robot arm is controlled in real-time for performing everyday life tasks including but not limited to reaching objects in the three-dimensional (3D) space. The robot arm motion is based on decoding human upper limb muscle activity, represented by surface electromyographic (EMG) signals collected in a noninvasive and absolutely safe way from the subject's arm. Surface EMG activity is recorded from muscles that are responsible for actuating the upper limb in performing motion in the 3D arm workspace. Then, a decoding algorithm is used to decode the processed EMG activity to arm motion. This algorithm has been previously trained using EMG recordings and corresponding arm motion during a training session. The resulted decoded motion is then inserted to a bio-inspired filter that essentially corrects any non-anthropomorphic behavior. Finally, the resulted arm motion is mapped to the robot joints and sent to a robot controller. The controller then communicates with the robot servo-controller to drive the robot arm according to the subject's arm motion in real-time. The subject is actually *in the loop* since he or she can control the robot arm in real-time and correct its behavior by means of natural physical limb motions.

The overall architecture combines multisensory information processing, algorithmic processes, human motion statistics, and probabilistic modeling, as well as real-time control implementation. The combination of intelligent software and hardware realizes a solid intelligent system that embodies the perception–action behavior in the control of a robotic device in real-time.

One of the main aspects discussed in this chapter is how the natural information processing can result to the decoding of the performed motion. This is not implemented by going through the complicated biomechanics of the human upper limb, but instead, the relationship between the human nervous system control signals is mapped onto the performed motion using a trainable, intelligent system. This system is trained to decode the human control signals to arm motion using dimensionality reduction techniques and system identification algorithms. Therefore, the problem of the high complexity of human motor control and the musculoskeletal dynamics is indirectly tackled through a learning architecture.

Moreover, the anthropomorphism of human motions is also analyzed and modeled within this chapter. A probabilistic model, in comparison to absolute cost functions used in previous works, is used for describing multi-joint arm configuration and coordination for everyday life tasks. In this way, the hardware (robot arm) is trained also to *act like a human*, completing tasks in a more safe and efficient way.

Finally, all the systems are embodied in a solid architecture that combines intelligent algorithms and advanced hardware, to formulate an advanced human–robot control platform, that can be used in the majority of tasks including a robot being controlled by a human subject, either the robot is remotely operated or it is attached on the subject's body.

In Sect. 23.2, a literature review is presented on the human–machine control interfaces based on natural information processing and on describing anthropomorphism at human motions. The total system architecture is presented in Sect. 23.3, where all software and hardware pieces used are analyzed. Experimental results using the proposed architecture in a real platform are presented in Sect. 23.4, while Sect. 23.5 concludes the chapter and proposes future extensions of the presented system.

23.2 Previous Work

Control interfaces for robots that are based on signals measured directly from the human body have received increased attention during the last decade. The necessity of those kinds of interface stems from the fact that most conventional control interfaces require complex mechanisms or systems of sensors, while in most of the cases the user should be trained to map his or her action (i.e., three-dimensional (3D) motion of a joystick or a haptic device) to the resulted motion of the robot. Some examples can be found in Woo-Keun et al. (2004) and Park and Khatib (2006). In this chapter, a control interface is proposed, which uses EMG signals from muscles of the upper limb to control a remote robot arm. In particular, a decoding model that translates the recorded muscle activity to arm motion, in conjunction with a bio-inspired filtering technique based on the modeling of the arm movement, provides the motion commands to the robot arm, in real-time.

Till now, many researchers have investigated EMG signals as control interface for robots. The studies in this field can be classified into three groups: the control of prosthetic arms and hands; the application to orthoses mainly for rehabilitation purposes; and the control of remotely operated robots. As examples of the first, Smagt et al. (Bitzer and van der Smagt 2006) used EMG signals from ten muscles of the forearm to control a four-fingered robot hand. Myoelectric signals were also used for the control of a complex robot hand in Farry et al. (1996). A hand prosthesis using pattern recognition of myoelectric signals can be also found in Kato et al. (1967). Quite recently, Thakor et al. (2007) achieved to identify 12 individuated flexion and extension movements of the fingers using EMG signals from muscles of the forearm of an able-bodied subject, while a hydraulically driven multifunction prosthetic hand was driven by EMG signals in Schulz et al. (2005). A review on controlling prosthetic hands using EMG signals can be found in Zecca et al. (2002) and Scott and Parker (1988). The Boston Arm (Jerard et al. 1974) and the Utah Arm (Jacobson et al. 1982) were driven by EMG signals in the past too.

Regarding orthotic devices, a lot of robotic mechanisms intended for either rehabilitation or extension of human ability have been developed during the last decades. As examples of the latter, Kazerooni proposed a new class of robot manipulator worn by humans in Kazerooni (1990). In these devices, the human in physical contact with the robotic manipulator exchange power and information signals. Similar orthotic devices for the upper limb have been developed and presented in Cavallaro et al. (2005). In this case, the exoskeleton robot serving as an assistive device worn by the human functions as a human amplifier, while EMG signals are used as the main control signal providing high intuitiveness. For rehabilitation purposes, Krebs et al. (Dipietro et al. 2005) developed a training system for the upper limb movements of stroke patients, which incorporates EMG signals, introducing the EMG-triggered robot-assisted therapy. In this setup, the onset of the patient's attempt to move is detected by monitoring EMG signals in selected muscles, whereupon the robot assists the user to perform point-to-point movements in the horizontal plane.

Robotic manipulators have been often remotely operated using EMG signals. Fukuda et al. (2003) proposed a human-assisting manipulator teleoperated by EMG signals and arm motions. A position tracking system was used to track arm motions, while EMG signals recorded from the forearm muscles were used to estimate hand and wrist motion. A two-dimensional myoelectric control of a robot arm was realized recently in Celani et al. (2007).

Since most previous studies focused on EMG signal discrimination, a variety of algorithms have been proposed for this scope. A statistical log-linearized Gaussian mixture neural network has been proposed in Fukuda et al. (2003) to discriminate EMG patterns for wrist motions. Neural networks, since they can acquire the nonlinear mapping of learning data, are widely used in many EMG-based control applications, especially in the field of robotic hands. A standard back propagation network was used in Hiraiwa et al. (1989) for estimating five finger motion, while the same network was used in Huang and Chen (1999) for classifying eight motions based on signal features extracted (e.g., zero-crossing and variance). Farry et al. (1996) used the frequency spectrum of EMG signals to classify motions of the human hand to remotely operate a robot hand, while support vector machines were used for the classification of hand postures using EMG signals in Bitzer and van der Smagt (2006).

A few researchers have tried to build continuous models to decode arm motion from EMG signals. The Hill-based muscle model (Hill 1938), whose mathematical formulation can be found in Zajac (1986), is more frequently used in the literature (Cavallaro et al. 2005; Artemiadis and Kyriakopoulos 2005). However, only a few DoFs were analyzed (i.e., 1 or 2), since the nonlinearity of the model equations and the large numbers of unknown parameters for each muscle make this analysis rather difficult. Similarly, musculoskeletal models have been analyzed in the past (Potvin et al. 1996; Lloyd and Besier 2003; Manal et al. 2002), focusing on a small number of muscles and actuated DoFs.

In most of the studies mentioned earlier, EMG signals were used to decide the desired posture (in the cases of robot hands) or direction of motion (in the cases of prosthetic and remotely operated robot arms). However, a continuous representation of the user's motion would be essential for an effective EMG-driven robotic device. Particularly in the case of robot arm control, a control interface able to output smooth profiles of motion, rather than a set of binary or generally discrete variables, is of utmost importance. The authors have used EMG signals in the past to control in a continuous way a single DoF of a robot arm in Artemiadis and Kyriakopoulos (2005), as well as 2 DoFs during planar catching tasks in Artemiadis and Kyriakopoulos (2006). However, for an EMG-driven robot arm to be more effective, a set of more than 2 DoFs should be included, enabling the arm to be used for tasks in the 3D space. This would also entail recording from a large number of muscles contributing to those DoFs. A more recent work of the authors (Artemiadis and Kyriakopoulos 2010a) is decoding EMG signals for 3D arm motion, but without using algorithms for filtering the decoded motion using the bio-inspired architecture presented in this chapter.

Most of the previous works in the field do not incorporate the fact that the resulted EMG-based estimates for motion should be human-like. In other words, kinematic and dynamic characteristics that govern human arm movements can be identified, modeled, and finally incorporated into an EMG-based control interface. This would certainly improve the system accuracy, while it would increase system robustness to unforeseen cases, since muscles activation patterns never seen during training could cause a decoding model to fail. In such cases, a model of the human arm movement could filter erroneous EMG-based motion estimates and consequently improves system performance. In other words, a bio-inspired model of arm movements could deal with the uncertainty of the decoded motion and finally produce more accurate arm motion estimates, in real-time, improving the overall system efficiency.

23.3 System Architecture

As noted before, the proposed methodology needs a training session, where all the mathematical models and probabilistic relationships were structured and estimated using EMG and corresponding motion data. Therefore, the first phase would be the *training* phase. In the following paragraph, the problem definition is given while the assumptions and simplification are also defined.

23.3.1 Background and Problem Definition

There is no doubt that the musculoskeletal system of humans is quite efficient, while very complex. Narrowing our interest down to the upper limb and not considering finger motion, approximately 30 muscles actuate 7 DoFs. In this study, we are focusing on the principal joints of the upper limb, i.e., the shoulder and the elbow. The wrist motion is not included in the analysis for simplicity. Three rotational DoFs were used to model the shoulder joint and one rotational DoF for the elbow joint. Hence, 4 DoFs will be analyzed from a kinematic point of view.

23.3.2 System Training Phase

The training phase lasts about 3 min, during which arm motion and EMG data are recorded. The processing of the data is done off-line, for each subsystem included in the total architecture.

23.3.2.1 Recording Arm Motion

For the training of the proposed system, the motion of the upper limb should be recorded and joint trajectories should be extracted. For this scope, a magnetic position tracking system was used, equipped with two position trackers and a reference system, with respect to which the 3D position of the trackers is provided. In order to compute the four joint angles, one position tracker is placed at the user's elbow joint and the other one at the wrist joint. The reference system is placed on the user's shoulder. The setup as well as the four modeled DoFs are shown in Fig. 23.1. Let $\mathbf{T_1} = \begin{bmatrix} x_1 & y_1 & z_1 \end{bmatrix}^T$, $\mathbf{T_2} = \begin{bmatrix} x_2 & y_2 & z_2 \end{bmatrix}^T$ be the position of the trackers with respect to the tracker reference system. Let q_1, q_2, q_3, and q_4 be the four joint angles modeled as shown in Fig. 23.1. Finally, by solving the inverse kinematic equations (see Artemiadis and Kyriakopoulos (2008) for more details) the joint angles are given by:

$$q_1 = \arctan 2(\pm y_1, x_1)$$
$$q_2 = \arctan 2\left(\pm\sqrt{x_1^2 + y_1^2}, z_1\right)$$
$$q_3 = \arctan 2(\pm B_3, B_1)$$
$$q_4 = \arctan 2\left(\pm\sqrt{B_1^2 + B_3^2}, -B_2 - L_1\right)$$

(23.1)

where

$$B_1 = x_2 \cos(q_1)\cos(q_2) + y_2 \sin(q_1)\cos(q_2) - z_2 \sin(q_2)$$
$$B_2 = -x_2 \cos(q_1)\sin(q_2) - y_2 \sin(q_1)\sin(q_2) - z_2 \cos(q_2)$$
$$B_3 = -x_2 \sin(q_1) + y_2 \cos(q_1)$$

(23.2)

Fig. 23.1 The user moves his arm in the 3D space. Two position tracker measurements are used for computing the four joint angles. The tracker base reference system is placed on the shoulder. q_1 and q_2 jointly correspond to shoulder flexion–extension and adduction–abduction, q_3 corresponds to shoulder internal–external rotation, while q_4 corresponds to elbow flexion–extension

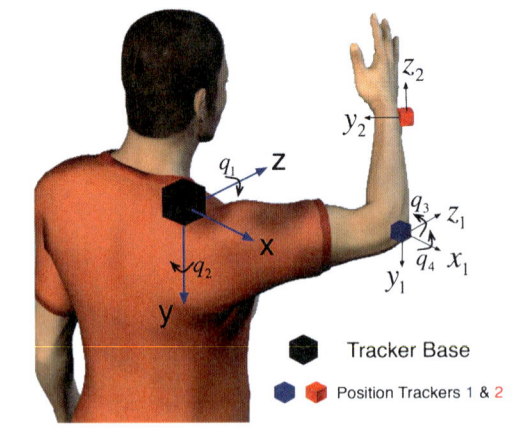

where L_1 is the length of the upper arm. The length of the upper arm can be computed from the distance of the first position tracker from the base reference system:

$$L_1 = \|\mathbf{T_1}\| = \sqrt{x_1^2 + y_1^2 + z_1^2} \qquad (23.3)$$

Likewise, the length of the forearm L_2 can be computed from the distance between the two position trackers, i.e.,

$$L_2 = \sqrt{(x_2 - x_1)^2 + (y_2 - y_1)^2 + (z_2 - z_1)^2} \qquad (23.4)$$

It must be noted that since the position trackers are placed on the skin and not in the center of the modeled joints, the lengths L_1, L_2 may vary as the user moves the arm. However, it was found that the variance during a 4-min experiment was less than 1 cm (i.e., approximately 3% of the mean values for the lengths L_1, L_2). Therefore, the mean values of L_1 and L_2 for a 4-min experiment were used for the following analysis.

The position tracking system provides the position vectors $\mathbf{T_1}$, $\mathbf{T_2}$ at the frequency of 60 Hz. Using an anti-aliasing FIR filter, these measurements are re-sampled at the frequency of 1 kHz, to be consistent with the muscle activations sampling frequency.

23.3.2.2 Recording Muscle Activity

Based on the biomechanics literature (Cram and Kasman 1998), a group of nine muscles mainly responsible for the studied motion is recorded: deltoid (anterior), deltoid (posterior), deltoid (middle), pectoralis major, pectoralis major (clavicular head), trapezius, biceps brachii, brachioradialis, and triceps brachii. A smaller number of muscles could have been recorded (e.g., focusing on one pair of agonist–antagonist muscles for each joint). However, to investigate a wider arm motion variability, where less significant muscles could play an important role in specific arm configurations, a group of nine muscles were selected. Surface bipolar EMG electrodes used for recording are placed on the user's skin following the directions given in Cram and Kasman (1998). Raw EMG signals after amplification are digitized at the sampling frequency of 1 kHz. Then a full wave rectification takes place, and then the signals are low-pass filtered using a fourth order Butterworth filter, with a cutoff frequency of 4 Hz. Finally, the signals from each muscle are normalized to their maximum voluntary isometric contraction value (Zajac 1986). Three able-bodied subjects were used (three males of 27 ± 3 years old), while during the experiment the subjects were standing close to the robot arm, with their neck positioned looking at front. All experimental procedures were conducted under a protocol approved by the National Technical University of Athens Institutional Review Board.

The training phase lasts about 3 min, during which motion and EMG data are recorded. Based on those data, the decoding model and the model of the human arm movements will be trained.

23.3.3 Data Representation

Since the number of muscles recorded is quite large (i.e., 9), a low-dimensional (low-D) representation of muscle activations will be used instead of individual activations. The problem of dimension reduction is introduced as an efficient way to overcome the curse of the dimensionality when dealing with vector data in high-dimensional spaces and as a modeling tool for such data. It is generally defined as the search for a low-dimensional manifold that embeds the high-dimensional data. The goal of dimension reduction is to find a representation of that manifold (i.e., a coordinate system) that will allow to project the original data vectors on it and obtain a low-dimensional, compact representation of them. In our case, muscle activations and joint angles are the high-dimensional data, which will be embedded into two manifolds of lower dimension. This should extract muscle synergies and motion primitives, which could be represented into the two new low-dimensional manifolds using the new coordinate systems of them. Later, having decreased the dimensionality of both the muscle activations and joint angles, the mapping between these two sets will be achievable.

The most widely used dimension reduction technique is principal component analysis (PCA). It is widely used due to its conceptual simplicity and the fact that relatively efficient algorithms exist for its computation. The central idea of PCA, since it is a dimension reduction method, is to reduce the dimensionality of a data set consisting of a large number of interrelated variables, while retaining as much as possible of the variation present in the data set. This is achieved by transforming to a new set of variables, the principal components (PCs), which are uncorrelated, and which are ordered so that the first few retain most of the variation present in all of the original variables. In this study, the PCA algorithm will be implemented twice: once for finding the new representation of the muscle activation data, and then once more for the representation of joint angles. For details about the method, the reader should refer to Jackson (1991) and Jolliffe (2002).

Let

$$\mathbf{U} = \begin{bmatrix} \mathbf{u}^{(1)} \ \mathbf{u}^{(2)} \ \cdots \ \mathbf{u}^{(9)} \end{bmatrix}^{T} \tag{23.5}$$

be a $9 \times m$ matrix containing the m samples of the muscle activations from each of the nine muscles recorded, i.e.,

$$\mathbf{u}^{(i)} = \begin{bmatrix} u_1^{(i)} \ u_2^{(i)} \ \cdots \ u_m^{(i)} \end{bmatrix}^{T}, i = 1, \ldots, 9, m \in \mathbb{N} \tag{23.6}$$

Likewise, let

$$\mathbf{Y} = \begin{bmatrix} \mathbf{y}^{(1)} \ \mathbf{y}^{(2)} \ \mathbf{y}^{(3)} \ \mathbf{y}^{(4)} \end{bmatrix}^{T} \tag{23.7}$$

be a $4 \times m$ matrix containing m samples of the four joint angles, i.e.,

$$\begin{aligned}
\mathbf{y}^{(1)} &= \begin{bmatrix} q_{11} \ q_{12} \ \cdots \ q_{1m} \end{bmatrix}^{T} \\
\mathbf{y}^{(2)} &= \begin{bmatrix} q_{21} \ q_{22} \ \cdots \ q_{2m} \end{bmatrix}^{T} \\
\mathbf{y}^{(3)} &= \begin{bmatrix} q_{31} \ q_{32} \ \cdots \ q_{3m} \end{bmatrix}^{T} \\
\mathbf{y}^{(4)} &= \begin{bmatrix} q_{41} \ q_{42} \ \cdots \ q_{4m} \end{bmatrix}^{T}
\end{aligned} \tag{23.8}$$

where $q_{1_k}, q_{2_k}, q_{3_k}, q_{4_k}, k = 1, \ldots, m$ denote the k-th measurement of the joint angle q_1, q_2, q_3, and q_4, respectively.

The computation of the PCA method entails the singular value decomposition (SVD) of the zero-meaned data covariance matrix. Hence, regarding muscle activations, mean values of each row of the matrix $\mathbf{U}$ in (23.5) are subtracted from the corresponding row. Then, the covariance matrix $\mathbf{F}_{9\times9}$ is computed. The singular value decomposition of the latter results to:

$$\mathbf{F} = \mathbf{GJG}^T \tag{23.9}$$

where $\mathbf{J}$ is the 9×9 diagonal matrix with the eigenvalues of $\mathbf{F}$, and $\mathbf{G}$ is the 9×9 matrix with columns the eigenvectors of $\mathbf{F}$. The principal component transformation of the muscle activation data can then be defined as:

$$\mathbf{M} = \mathbf{G}^T \mathbf{K} \tag{23.10}$$

where $\mathbf{K}$ is the $9 \times m$ matrix computed from $\mathbf{U}$ by subtracting the mean value of each muscle across the m measurements. The principal component transformation of the joint angles can be computed likewise. If $\mathbf{L}$ is the $4 \times m$ matrix computed from $\mathbf{Y}$ by subtracting the mean values of each joint angle across the m measurements, and $\mathbf{H}$ is the 4×4 matrix with columns the eigenvectors of the covariance matrix, then the principal component transformation of the joint angles is defined as:

$$\mathbf{Q} = \mathbf{H}^T \mathbf{L} \tag{23.11}$$

The PCA algorithm results to a representation of the original data to a new coordinate system. The axes of that system are the eigenvectors computed as analyzed before. However, the data do not appear to have the same variance across those axes. In fact, in most cases only a small set of the eigenvectors is enough to describe most of the original data variance. Here is where the dimension reduction comes into. Let p the total variables at the original data, i.e., the number of the eigenvectors calculated by the PCA algorithm. It can be proved that only the *first* r, $r \ll p$ eigenvectors can describe most of the original data variance. The *first* eigenvectors are the first eigenvectors we have if they are ranked to a descending order with respect to their eigenvalues. Therefore, the eigenvectors with the highest eigenvalues describe most of the data variance (Jolliffe 2002). Describing the p original variables with fewer dimensions r is finally the goal of the proposed method. There have been many proposed criteria for choosing the right number of the principal components to keep, to retain most of the original data variance. The reader should refer to Jolliffe (2002) for a complete review of those methods. Perhaps the most obvious criterion for choosing r is to select a cumulative percentage of total variation which one desires that the selected PCs contribute, i.e., 80% or 90%. The required number of the PCs is then the smallest value of r, for which this chosen percentage is exceeded. Since the eigenvalue of each eigenvector coincides with its contribution to the total variance, then the smallest value of r, r_- can be found by the following inequality:

$$\frac{\sum_{i=1}^{r_-} \lambda_i}{\sum_{n=1}^{p} \lambda_n} \geq P \ \%, \quad r_-, p \in \mathbb{N} \tag{23.12}$$

where $P \ \%$ is the total variance which one desires that the selected PCs contribute, i.e., 90%, and r_- the minimum integer number for which (23.12) is satisfied.

Using the above criterion, we can select the minimum number of eigenvectors of muscles activation and joint angles, capable of retaining most of the original data variance. In particular, for muscle recordings, the first two principal components were capable of describing 95% of the total variance. Regarding joint angles, the first two principal components described 93% of the total variance too. Therefore, the low-dimensional representation of the nine muscles activation during 3D motion of the arm is defined by:

$$\xi = \mathbf{V}^T \mathbf{K} \tag{23.13}$$

where $\mathbf{V}$ is a 9×2 matrix, whose columns are the two first eigenvectors of $\mathbf{G}$ in (23.10), resulting to the size of $2 \times m$ for matrix ξ. Likewise, the low-dimensional representation of joint angles during 3D motion of the arm is defined by:

$$\emptyset = \mathbf{W}^T \mathbf{Ł} \tag{23.14}$$

where $\mathbf{W}$ is a 4×2 matrix, whose columns are the two first eigenvectors of $\mathbf{H}$ in (23.11), resulting to the size of $2 \times m$ for matrix $\emptyset$.

Using the above dimension reduction technique, the high-dimensional data of muscle activations and corresponding joint angles were represented into two manifolds of fewer dimensions. In particular for joint angles, using two instead of four variables to describe arm movement suggests motor primitives, which is a general conception that has been extensively analyzed in the literature (d'Avella et al. 2006; Mussa-Ivaldi and Bizzi 2000).

23.3.4 Decoding Arm Motion from EMG Signals

Raw EMG signals from nine muscles are preprocessed and represented into a low-dimensional space resulting to only two variables (i.e., low-dimensional EMG embeddings). In addition, four joint angles are embedded into a two-dimensional space. Having those variables, we can define the problem of decoding as follows: find a function f that can map muscle activations to arm motion in real-time, being able to be identified using training data. Generally, we can define it by:

$$\mathbb{Y} = f (\mathbb{U}) \tag{23.15}$$

where $\mathbb{Y}$ denotes human arm kinematics embeddings and $\mathbb{U}$ muscle activation embeddings. As noted in the introduction, the scope of this study, unlike related past works, is to result to a continuous representation of motion using EMG signals. From a physiological point of view, a model that would describe the function of skeletal muscles actuating the human joints would be generally a complex one. This would entail highly nonlinear musculoskeletal models with a great number of parameters that should be identified (Artemiadis and Kyriakopoulos 2005). For this reason, we can adopt a more flexible decoding model in which we introduce "hidden", or "latent" variables we call $\mathbf{x}$. These hidden variables can model the unobserved, intrinsic system states and thus facilitate the correlation between the observed muscle activation $\mathbb{U}$ and arm kinematics $\mathbb{Y}$. Therefore, (23.15) can be rewritten as shown below:

$$\mathbb{Y} = f(\mathbb{X}, \mathbb{U}) \tag{23.16}$$

where $\mathbb{X}$ denotes the set of the "hidden" states. Regarding function f that describes the relation between the input and the hidden states with the output of the model, a selection of a linear one can be made to facilitate the use of well-known algorithms for training. The latter selection results to the following state space model:

$$
\begin{aligned}
\mathbf{x}_{k+1} &= \mathbf{A}\mathbf{x}_k + \mathbf{B}\xi_k + \mathbf{w}_k \\
\mathbf{z}_k &= \mathbf{C}\mathbf{x}_k + \mathbf{v}_k
\end{aligned}
\tag{23.17}
$$

where $\mathbf{x}_k \in \mathbb{R}^d$ is the hidden state vector at time instance kT, $k = 1, 2, \ldots, T$ the sampling period, d the dimension of this vector, $\xi \in \mathbb{R}^2$ is the vector of the low-dimensional muscle activations, and $\mathbf{z}_k \in \mathbb{R}^2$ is the vector of the low-dimensional joint kinematics. The matrix $\mathbf{A}$ determines the dynamic behavior of the hidden state vector $\mathbf{x}$, $\mathbf{B}$ is the matrix that relates muscle activations ξ to the state vector $\mathbf{x}$, while $\mathbf{C}$ is the matrix that represents the relationship between the joint kinematics $\mathbf{z}_k$ and the state vector $\mathbf{x}$. $\mathbf{w}_k$ and $\mathbf{v}_k$ represent zero-mean Gaussian noise in the process and observation equations, respectively, i.e., $\mathbf{w}_k \sim N(\mathbf{0}, \mathbf{W})$, $\mathbf{v}_k \sim N(\mathbf{0}, \mathbf{Z})$, where $\mathbf{W} \in \mathbb{R}^{d \times d}$, $\mathbf{Z} \in \mathbb{R}^{2 \times 2}$ are the covariance matrices of $\mathbf{w}_k$, $\mathbf{v}_k$, respectively.

Model training entails the estimation of the matrices $\mathbf{A}$, $\mathbf{B}$, $\mathbf{C}$, $\mathbf{W}$, and $\mathbf{Z}$. Given a training set of length m, including the low-dimensional embeddings of the muscle activations and joint angles, the model parameters can be found using an iterative prediction-error minimization (i.e., maximum likelihood) algorithm (Ljung 1999). The dimension d of the state vector should also be selected for model fitting. This is done in parallel with the fitting procedure, by deciding the number of states, greater of which any additional states do not contribute more to the model input–output behavior. The way to conclude to this is by observing the singular values of the covariance matrix of the system output (Ljung 1999).

23.3.5 *Modeling Human Arm Movement*

The modeling of human arm movement has received increased attention during the last decades, especially in the field of robotics (Billard and Mataric 2001) and graphics. This is because there is a great interest in modeling and understanding underlying laws and motion dependencies among the DoFs of the arm, to incorporate them into robot control schemes. Most of the previous works in this area focus on the definition of motor primitives (Fod et al. 2002) or objective functions that are minimized during arm motion. These models, however, lack the ability to describe dependencies among the DoFs of the arm, and consequently cannot be used for filtering noisy measurements of motion, or even generating a probability distribution over the possible arm configurations. In this work, to model the dependencies among the DoFs of the arm during random 3D movements, we are going to use graphical models.

23.3.5.1 Graphical Models

Graphical models are a combination between probability theory and graph theory. They provide a tool for dealing with two characteristics; the uncertainty and the complexity of random variables. Given a set $\mathbb{F} = \{ f_1 \ldots f_N \}$ of random variables with joint probability distribution $p(f_1, \ldots, f_N)$, a graphical model attempts to capture the conditional dependency structure inherent in this distribution, essentially by expressing how the distribution factors as a product of *local functions* (e.g., conditional probabilities) involve in various subsets of $\mathbf{F}$. Directed graphical models is a category of graphical models, also known as Bayesian Networks. A directed acyclic graph is a graphical model where there are no graph cycles when the edge directions are followed. Given a directed graph $G = (V, E)$, where V the set of vertices (or nodes) representing the variables $f_1, \ldots, f_N$, and E the set of directed edges between those vertices, the joint probability distribution can be written as follows:

$$p(f_1, \ldots, f_N) = \prod_{i=1}^{N} p(f_i | a(f_i)) \qquad (23.18)$$

where $a(f_i)$ the *parents* (or direct ancestors) of node f_i. If $a(f_i) = \emptyset$ (i.e., f_i has no parents), then $p(f_i | \emptyset) = p(f_i)$, and the node i is called the *root* node.

An advantage of graphical models is that the directed edges can be used to model causality explicitly. By inspecting the arrows in such models, it is easy to determine which variables directly influence others. Moreover, using (23.18) we can compute the joint probability for a set of variables to have a value, taking into account their dependencies learned from the training data. However, (23.18) requires the parents of each variable; therefore, the structure of the graphical model is required. This can be learned from the training data, using the algorithm presented below.

23.3.5.2 Building the Model

A version of a directed graphical model is a tree model. Its restriction is that each node has only one parent. The optimal tree for a set of variables is given by the Chow–Liu algorithm (Chow and Liu 1968). Briefly, the algorithm constructs the maximum spanning tree of the complete mutual information graph, in which the vertices correspond to variables of the model and the weight of each directed edge $f_i \rightarrow f_j$ is equal to the mutual information $I\left(f_i, f_j\right)$, given by:

$$I\left(f_i, f_j\right) = \sum_{f_i, f_j} p\left(f_i, f_j\right) \log \frac{p\left(f_i, f_j\right)}{p\left(f_i\right) p\left(f_j\right)} \qquad (23.19)$$

where $p\left(f_i, f_j\right)$ the joint probability distribution function for f_i, f_j, and $p\left(f_i\right)$, $p\left(f_j\right)$ the marginal distribution probability functions for f_i, f_j, respectively. Mutual information is a unit that measures the mutual dependence of two variables. The most common unit of measurement of mutual information is the bit, when logarithms to the base 2 are used. It must be noted that the variables $\{f_1 \ldots f_N\}$ are considered discrete in the definition of (23.19). Details about the algorithm of the maximum spanning tree construction can be found in Chow and Liu (1968).

In our case, we have a set of four discrete variables $\{q_1, q_2, q_3, q_4\}$. These variables correspond to the joint angles of the four modeled DoFs of the arm, in degrees, rounded to the nearest integer value. Discrete variables are used for simplifying subsequent algorithm for training and inference using the directed graphical model, without losing much information from the data, since the maximum error imposed by discretizing joint angle values to the nearest integer is 0.5 deg. Using joint angle data recorded during the training phase, we can build the tree model. The resulted tree structure is shown in Fig. 23.2. Therefore, using (23.18), we can now define the joint probability of the four variables representing joint angles by:

$$p\left(q_1, q_2, q_3, q_4\right) = \prod_{i=1}^{4} p\left(q_i \mid a\left(q_i\right)\right) \qquad (23.20)$$

where $a\left(q_i\right)$ the parents of variable q_i. Therefore, from the tree structure (see Fig. 23.2), we have

$$p\left(q_1, q_2, q_3, q_4\right) = p\left(q_1\right) p\left(q_4 \mid q_1\right) p\left(q_3 \mid q_1\right) p\left(q_2 \mid q_4\right) \qquad (23.21)$$

where $p\left(q_i \mid q_j\right)$, $i, j = 1, 2, 3, 4$, the conditional probability distribution of q_i, given its parent q_j. These conditional probabilities can be initially described by 2D histograms, constructed using the training data. That is for each value of the parent q_j, construct a histogram of the values taken by child q_i. In this way, we come up with conditional 2D histograms for each pair of parent–child.

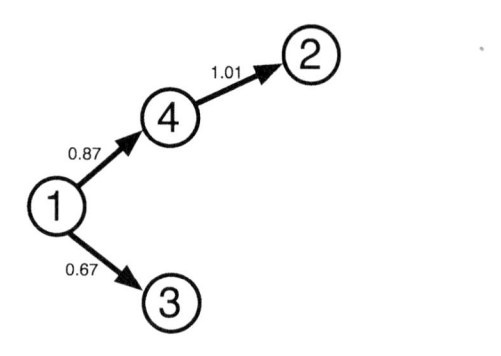

Fig. 23.2 The directed graphical model (tree) representing nodes (i.e., joint angles) dependencies. Node i corresponds to q_i. $i \rightarrow j$ means that node i is the parent of node j, where $i, j = 1, 2, 3, 4$. The mutual information $I(i, j)$ is shown at each directed edge connecting i to j. The value of the mutual information quantifies the information gained if we describe two variables through their dependency, instead of considering them as independent. Its value is in bits

Since training data are finite, and there can be cases where a specific value for a joint angle was not observed, while values around this were observed, we are going to move one step further toward a continuous model by fitting a Gaussian Mixture Model (GMM) to the conditional distribution of a variable and its parent. For the root variable of the tree, i.e., q_1, a GMM will also be fitted. Therefore, the marginal probability distribution function of the root is given by:

$$p(q_1) = \sum_{c=1}^{K_1} p_c \mathcal{N}(q_1; \mu_{1c}, \sigma_{1c}) \tag{23.22}$$

where K_1 the number of the Gaussian mixture components, p_c the mixing coefficients of the components, and $\mathcal{N}(q_1; \mu_{1c}, \sigma_{1c})$ a Gaussian with mean μ_{1c} and variance σ_{1c}. The conditional probability distribution of a child variable q_i given its parent variable q_j is given by:

$$p(q_i | q_j) = \sum_{c=1}^{K_j} p_c N(q_i; \mu_{ic} + \omega_{ic} q_j, \sigma_{ic}) \tag{23.23}$$

where ω_{ic} weights for the parent variable q_j calculated by the fitting procedure. Details about the GMMs and their fitting procedure [Expectation Maximization (EM)] can be found in McLachlan and Peel (2000). Having obtained the continuous estimates of the 2D histograms, we can then revert to the tabular (discrete) representation of the conditional distribution, using (23.23) for each parent–child pair value, and (23.22) for obtaining the one-dimensional histogram for the root node. The resulted 2D conditional histograms are shown in Fig. 23.3. Observing the 2D histograms, one can see that the conditional distributions are not Gaussian-like.

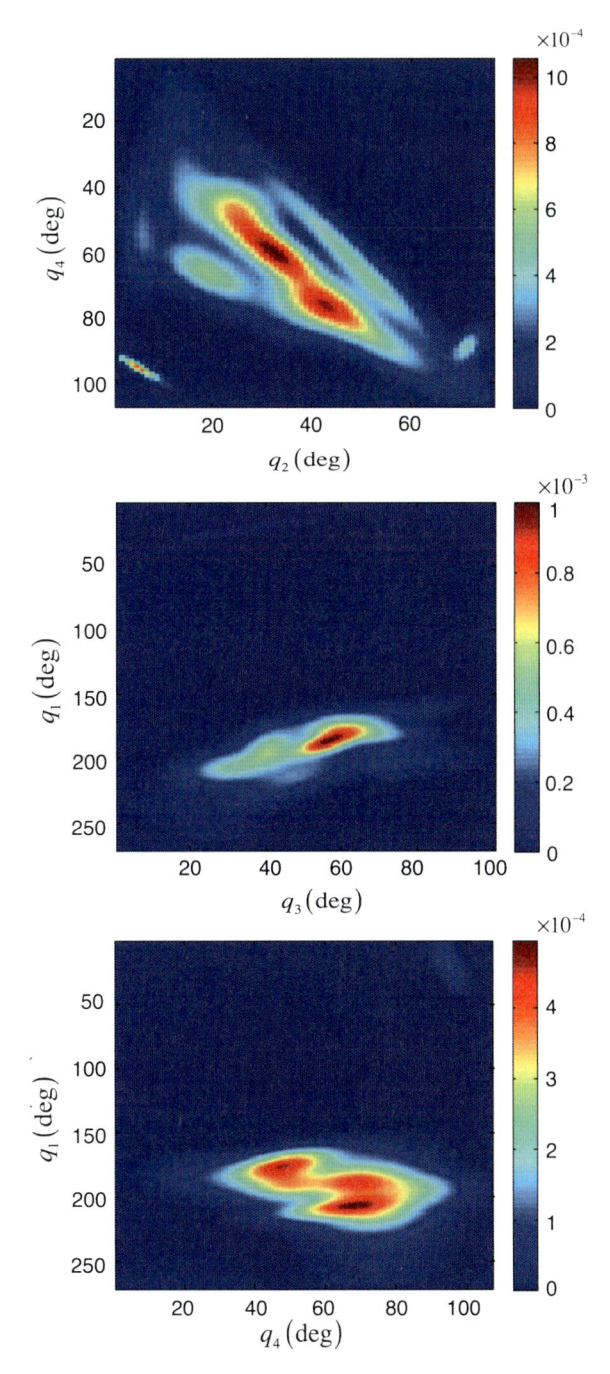

Fig. 23.3 Conditional probabilities represented in 2D histograms for each pair of the tree. Observed values for parents and children are shown in columns and rows, respectively

Therefore, techniques like PCA or Kalman Filter would definitely fail in describing arm movement and being a successful filter for our problem. However, a graphical model can incorporate these dependencies; therefore, it can be used for modeling the complexity of arm movements.

23.3.5.3 Inference Using the Graphical Model

Inference in probabilistic models in general, and in Bayesian Networks in our case, is the estimation of values of *hidden* nodes in a graph, given the values of the *observed* ones, where *hidden* nodes are the nodes that are not known either due to lack of measurement method or due to some missing measurements, and *observed* nodes are the nodes that are measured. In general, if a directed graph has a set of nodes $\mathbb{Q}$, we can partition the set into three disjoint subsets such as $\mathbb{Q}_F$, $\mathbb{Q}_E$, and $\mathbb{Q}_R$, where $\mathbb{Q}_F$ is the set of the hidden nodes, $\mathbb{Q}_E$ the set of the observed (or *evidence*) nodes, and $\mathbb{Q}_R$ the marginal nodes, which are the nodes that are neither observed nor hidden, since we do not care about their values, but they are intermediate nodes in the tree path connecting the observed nodes with the hidden ones. Then, the probability density function of $\mathbb{Q}_F$, given the $\mathbb{Q}_E$, is given by:

$$p\left(\mathbb{Q}_F \mid \mathbb{Q}_E\right) = \frac{\sum\limits_{\mathbb{Q}_R} p\left(\mathbb{Q}_E, \mathbb{Q}_F, \mathbb{Q}_R\right)}{\sum\limits_{\mathbb{Q}_F, \mathbb{Q}_R} p\left(\mathbb{Q}_E, \mathbb{Q}_F, \mathbb{Q}_R\right)} \tag{23.24}$$

which is based on the Bayes rule (Duda et al. 2001). In our case, where the conditional distributions are represented by the 2D histograms, the probabilistic inference method for a hidden node, given the observed ones, is realized through the following method: select the appropriate indices in the columns corresponding to $\mathbb{Q}_E$ based on the values, sum over the columns corresponding to $\mathbb{Q}_R$, and normalize the resulting table over $\mathbb{Q}_F$ (Xiang 2002). For this reason, a message-passing algorithm, called *junction tree*, is used. For details, the reader should refer to Xiang (2002).

Sometimes a node is not observed, but we have some distribution over its possible values; this is often called "soft" or "virtual" evidence. Using the *junction tree algorithm*, one can infer a hidden node, given a probability distribution over the possible values of the other, "observed" nodes. In this case, the *junction tree algorithm* finds the posterior marginal distributions for all the nodes in the tree. Since all nodes interact to each other according to the tree structure, there can be cases where the prior distribution for a node deriving from its "virtual" evidence is different from the posterior distribution after inference. Therefore, there can be cases where we have a prior probability distribution over the possible values of all the nodes of the tree, and by exploiting the tree connectivity (i.e., nodes dependency), find a *better* posterior probability distribution for the values of the nodes. This is the characteristic that will be used in our case for filtering the EMG-based estimates, using the tree structure built from the training data.

23.3.6   Filtering Motion Estimates Using the Graphical Model

Using model equation (23.17), at every time instance,[1] motion estimates are provided through the vector $\mathbf{y} = \begin{bmatrix} \hat{q}_1 & \hat{q}_2 & \hat{q}_3 & \hat{q}_4 \end{bmatrix}^T$.[2] However, from the definition of the model (23.17), these *prior* motion estimates belong to a Gaussian distribution given by:

$$\hat{q}_i \sim \mathcal{N}\left(\mathbf{C}_i \mathbf{x}, \sigma_i^2\right), \quad i = 1, 2, 3, 4 \tag{23.25}$$

where $\mathbf{C}_{i_{1 \times d}}$ is the ith row of the matrix $\mathbf{C}$ of the model, i.e.

$$\mathbf{C} = \begin{bmatrix} \mathbf{C}_1 & \mathbf{C}_2 & \mathbf{C}_3 & \mathbf{C}_4 \end{bmatrix}^T \tag{23.26}$$

and σ_i^2 the variance of each prior Gaussian distribution.[3] Therefore, at every time instance, the EMG-based decoding model outputs a prior distribution for every joint angle of the human arm. These distributions are considered as "soft" evidences for the nodes of the tree shown in Fig. 23.2. Then, using the junction tree algorithm, we get the posterior distributions for each tree node, which correspond to our final estimates for the human arm joint angles.

In this way, we use the graphical model as a filter for the prior estimates of joint angles, given by the EMG-based decoder. The human-like characteristics are embodied in our architecture through the graphical model, since it describes the dependencies between the joint angles during the arm motion. By filtering the EMG-based estimates using the graphical model, we avoid drifting phenomena in our motion estimates, while providing a safety net in cases where our EMG-based estimates are erroneous. The total architecture is depicted in Fig. 23.4. The way this filtering technique improves the overall system accuracy is assessed through various experiments, analyzed in the Sect. 23.4.

23.3.7   Robot Control

Having computed the final estimates for the human joints angles

$$\mathbf{q_H} = \begin{bmatrix} q_1 & q_2 & q_3 & q_4 \end{bmatrix}^T \tag{23.27}$$

---

[1] The EMG-based decoding model outputs motion estimates $\mathbf{y}$ at the frequency of the EMG acquisition, i.e., 1 kHz.

[2] Subscripts t denoting time instances are omitted for simplicity.

[3] Covariance matrix $\mathbf{Q}$ is defined as diagonal matrix during model fitting.

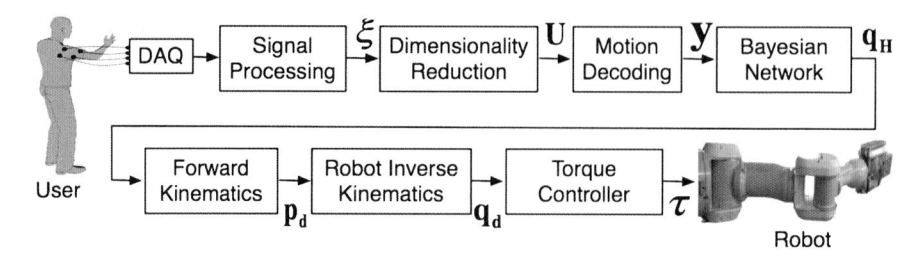

Fig. 23.4 The total system architecture. EMG signals after preprocessing ξ are represented in low-dimensional space, and then used by the decoding model to output the prior estimates for motion **y**. The final motion estimates in joint space $\mathbf{q_H}$ are produced using the Bayesian Network. The human hand position is obtained through forward kinematics, and then the robot joint angles are computed using the human hand position and the robot inverse kinematics. The final robot joint angles $\mathbf{q_d}$ are used in the torque control to actuate the robot arm

we can then command the robot arm. However, since robot and user's links have different length, the direct control in joint space would lead the robot end-effector in different position in the 3D space than that desired by the user. Consequently, the user's hand position should be computed using the estimated joint angles, and then we can command the robot to drive its end-effector at this point in space. This is realized using the forward kinematics of the human arm to compute the user's hand position and then solving the inverse kinematics for the robot arm to drive its end-effector to the same position in the 3D space. Hence, the final command to the robot arm is in joint space. Therefore, the subsequent robot controller analysis assumes that a final vector $\mathbf{q_d} = \begin{bmatrix} q_{1d} & q_{2d} & q_{3d} & q_{4d} \end{bmatrix}^T$ containing the four desired robot joint angles is provided, where these joint angles are computed through the robot inverse kinematics as described above.

A 7 DoF anthropomorphic robot arm (PA-10, Mitsubishi Heavy Industries) is used. Only four DoFs of the robot are actuated (joints of the shoulder and elbow), while the others are kept fixed at zero position via electromechanical brakes. The arm is horizontally mounted to mimic the human arm. The robot arm along with the actuated DoFs is depicted in Fig. 23.5. The robot motors are controlled in torque. In order to control the robot arm using the desired joint angle vector $\mathbf{q}_d$, an inverse dynamic controller is used, defined by:

$$\tau = \mathbf{I}(\mathbf{q_r})(\ddot{\mathbf{q}}_\mathbf{d} + \mathbf{K_v}\dot{\mathbf{e}} + \mathbf{K_p}\mathbf{e}) + \mathbb{G}(\mathbf{q_r}) + \mathbb{C}(\mathbf{q_r}, \dot{\mathbf{q}}_\mathbf{r})\dot{\mathbf{q}}_\mathbf{r} + \mathbf{F_{fr}}(\dot{\mathbf{q}}_\mathbf{r}) \qquad (23.28)$$

where $\tau = \begin{bmatrix} \tau_1 & \tau_2 & \tau_3 & \tau_4 \end{bmatrix}^T$ is the vector of robot joint torques, $\mathbf{q_r} = \begin{bmatrix} q_{1r} & q_{2r} & q_{3r} & q_{4r} \end{bmatrix}^T$ the robot joint angles, $\mathbf{K_v}$ and $\mathbf{K_p}$ gain matrices, and **e** the error vector between the desired and the robot joint angles, i.e.

$$\mathbf{e} = \begin{bmatrix} q_{1d} - q_{1r} & q_{2d} - q_{2r} & q_{3d} - q_{3r} & q_{4d} - q_{4r} \end{bmatrix}^T \qquad (23.29)$$

$\mathbf{I}$, $\mathbb{G}$, $\mathbb{C}$, and $\mathbf{F_{fr}}$ are the inertia tensor, the gravity vector, the Coriolis-centrifugal matrix, and the joint friction vector of the four actuated robot links and joints,

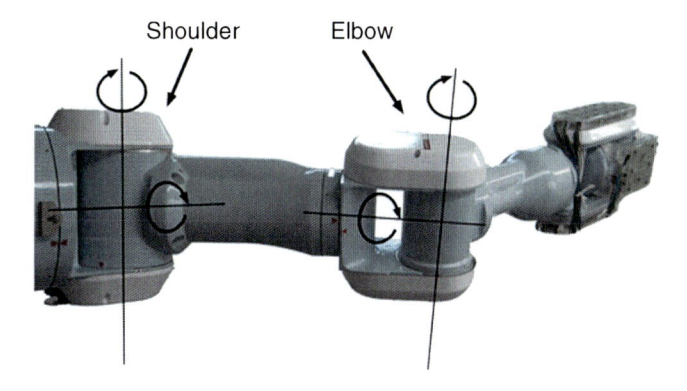

Fig. 23.5 The controlled robot arm is equipped with two rotational DoFs at the shoulder and two at the elbow

respectively, identified in Mpompos et al. (2007). The vector $\ddot{\mathbf{q}}_d$ corresponds to desired angular acceleration vector that is computed through simple differentiation of the desired joint angle vector $\mathbf{q}_d$ using a necessary low-pass filter to cut off high frequencies.

23.4   Experimental Results

23.4.1   Hardware and Experiment Design

The proposed architecture is assessed through a remote robot arm teleoperation scenario. The robot arm used is a 7 DoF anthropomorphic manipulator (PA-10, Mitsubishi Heavy Industries). Two personal computers (PCs) are used, running Linux operating system. One of the PCs communicates with the robot servo-controller through the ARCNET protocol in the frequency of 500 Hz, while the other acquires the EMG signals and the position tracker measurements (during training). The two PCs are connected through serial communication (RS-232) interface for synchronization purposes. EMG signals are acquired using a signal acquisition board (NI-DAQ 6036E, National Instruments) connected to an EMG system (Bagnoli-16, Delsys Inc.). Single differential surface EMG electrodes (DE-2.1, Delsys Inc.) are used. The position tracking system (Isotrak II, Polhemus Inc.) used during the training phase is connected with the PC through serial communication interface (RS-232). The size of the position sensors is 2.83(W) 2.29(L) 1.51(H) cm.

   The user is initially instructed to move his or her arm randomly in 3D space as shown in Fig. 23.1. During this phase, EMG signals and position trackers measurements are collected for not more than 3 min. These data are enough to train the EMG-based decoding model and the graphical model analyzed earlier. The models' computation time is less than 1 min. As soon as the models are estimated, the real-time operation phase takes place. The user is instructed to move the arm in 3D space, having visual contact with the robot arm. The position trackers measurements are not used during this phase. Using the proposed method, estimations about the human motion are computed using only the recorded EMG signals from the nine muscles mentioned earlier. However, the position trackers are kept in place (i.e., on the human's arm) for off-line validation reasons. The estimated hand trajectories versus the real ones during real-time operation phase are shown in Fig. 23.6. Moreover, the corresponding trajectories with and without the proposed bio-inspired filtering technique are illustrated, to prove the method efficiency. The proposed system was tested by three subjects in total with similar results.

23.4.2 Efficiency Assessment

Two criteria will be used for assessing the accuracy of the reconstruction of human motion using the proposed methodology. These are the root-mean-squared error (RMSE) and the correlation coefficient (CC). The latter describes essentially the similarity between the reconstructed and the true motion profiles and constitutes the most common means of reconstruction assessment for decoding purposes. The mathematical definitions of the criteria can be found in Artemiadis and Kyriakopoulos (2007b). Real and estimated motion data were recorded for 40 s during the real-time operation phase. Using the hand forward kinematics, the criteria values are computed in Cartesian space and listed in Table 23.1.

In general, the proposed methodology was used very efficiently for controlling the robot arm. As it can be seen from the results, the bio-inspired filtering method improves significantly the decoding performance, while increases the system robustness in cases where the decoding method could not track the human arm motion. From the teleoperation point of view, the users were able to quite easily control the robot arm, in real-time, having visual contact with it. All the three users that tested the system found it quite comfortable and efficient.

Another worth-assessing parameter is the type of the model used (i.e., linear model with hidden states and the bio-inspired filter). The authors feel that a comparison with a well-known and of similar complexity algorithm is rational. Thus, a comparison with the linear filter method was done. The linear filter method is a widely used method for decoding arm motions (especially when using neural signals), which has achieved exciting results so far (Carmena et al. 2003). Briefly, if $\mathcal{Q}_k$, the kinematics variables decoded (i.e., joint angles) at time $t_k = kT$ and $\mathcal{E}_{i,k-j}$ the muscle activation of muscle i at time t_{k-j}, the computation of the linear-filter entails finding a set of coefficients $\vartheta = \left[a \; \lambda_{1,1} \; \ldots \; \lambda_{i,j} \right]^T$, so that

$$\mathcal{Q}_k = a + \sum_{i=1}^{\zeta} \sum_{j=0}^{\mathfrak{s}} \lambda_{i,j} \mathcal{E}_{i,k-j} \qquad (23.30)$$

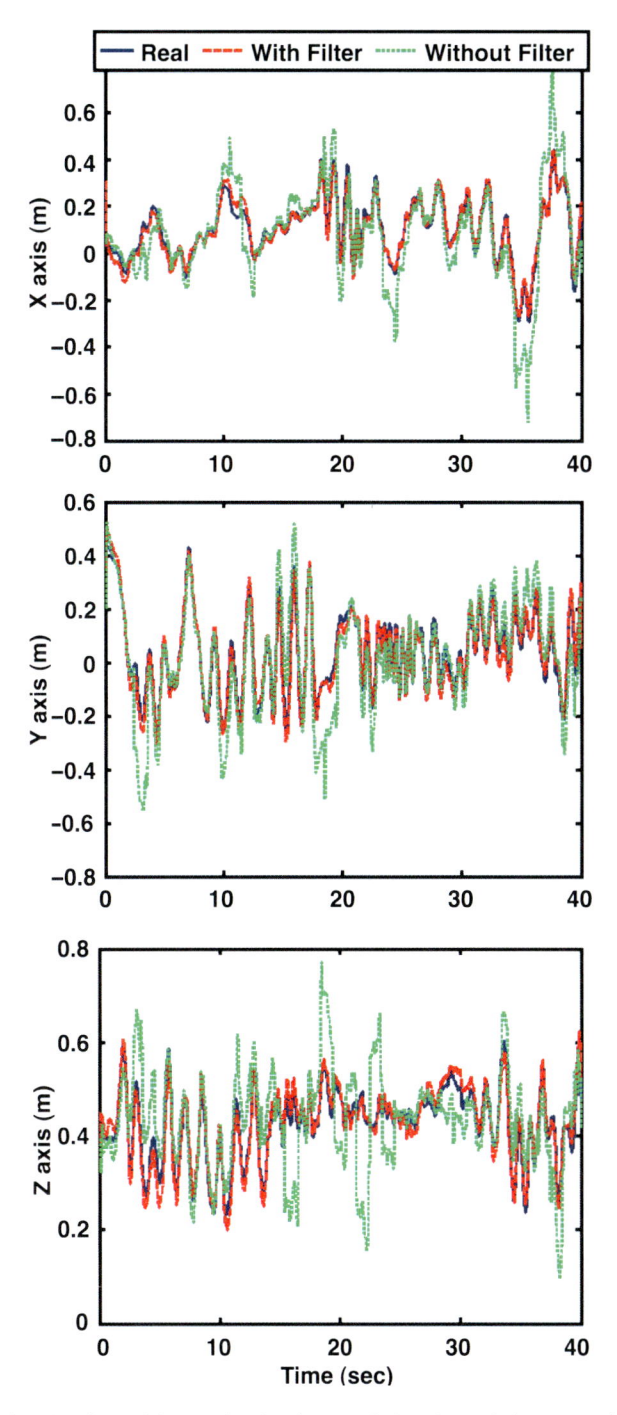

Fig. 23.6 Real and estimated human hand trajectory during the real-time operation phase. The estimated trajectory without the Bayesian Network filtering approach is also depicted

Table 23.1 Comparison between the proposed methodology using the bio-inspired filtering approach and a single decoding model without filtering, in Cartesian space

Method	CC_x	CC_y	CC_z	$RMSE_x$ (cm)	$RMSE_y$ (cm)	$RMSE_z$ (cm)
Decoding with bio-inspired filter	0.97	0.96	0.96	1.56	1.79	1.89
Single decoding	0.85	0.79	0.76	5.89	9.14	5.78

Table 23.2 Comparison between the proposed architecture and a linear filter in decoding motion in joint space

Decoding model	CC_1	CC_2	CC_3	CC_4	$RMSE_1$ (rad)	$RMSE_2$ (rad)	$RMSE_3$ (rad)	$RMSE_4$ (rad)
Proposed model	0.98	0.97	0.98	0.99	0.03	0.04	0.03	0.02
Linear filter	0.77	0.75	0.79	0.76	0.16	0.22	0.19	0.17

where a a constant offset, $\lambda_{i,j}$ the filter coefficients, ζ the number of muscles recorded, and the parameter $\mathcal{N}$ specifies the number of time bins used. A typical value of the latter is 100 ms, thus $\mathcal{N} = 100$, for a sampling period of 1 ms (Carmena et al. 2003). The coefficients can be learned from training data using simple least-squares regression. In our case, for the sake of comparison, the same training data were used for both the switching model and the linear filter, and after training, both models were tested using the same testing data as before. Values for RMSE and CC for these two methods are reported in Table 23.2 for joint angles. It must be noted that the low-dimensional representation for muscle activations and kinematics was used, since linear filter behaved better using this kind of data rather than using the high-dimensional data. Regarding complexity, our proposed model is slightly more demanding than the linear filter method. More specifically, the training of the proposed model lasted 1 min more than the linear filter, while the online computations needed were of equal computational complexity.

However, a comparison of the proposed architecture with a position tracking system is worth-analyzing. In other words, why the proposed EMG-based method for controlling robots is more efficient than a position tracking system that can track the performed user's arm motion?

One of the main reasons for which an EMG-based system is more efficient is the high frequency and the smoothness of the motion estimates that it provides, which is very crucial when a robot arm is to be controlled. The EMG-based system provides motion estimates at the frequency of 1 kHz,[4] compared to the 30 Hz frequency of the mostly used position tracking systems. Moreover, the proposed methodology guarantees smoothness in the computed estimates because of the linear dynamics of the hidden states used in the model. In order to depict the necessity for smooth and high-frequency estimates from the robot control point of view, a robot arm teleoperation test was realized using the EMG-based proposed architecture and a position

---

[4] Equal to the acquisition frequency of the EMG signal, which is usually high.

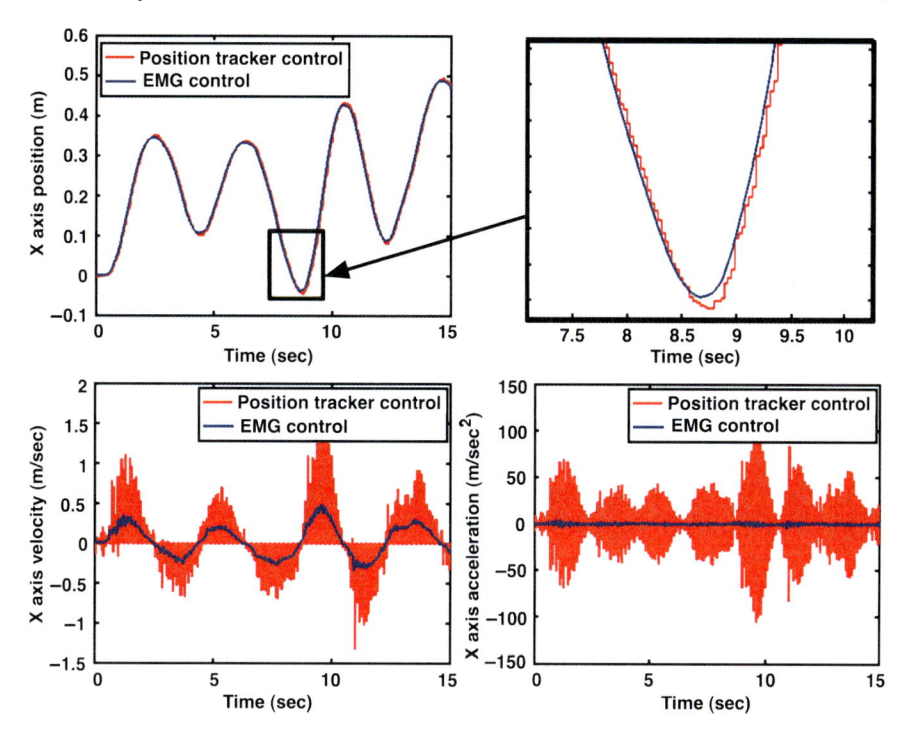

Fig. 23.7 Comparison of robot performance using EMG-based control and a position tracker. *Top left figure*: the robot end-effector position at the X robot axis. *Top right figure*: a detailed view of the robot end-effector position. Control using a position tracker proved to be very jerky, resulting in high robot velocity (*bottom left figure*) and high acceleration (*bottom right figure*), in comparison with EMG-based control that resulted to smooth robot trajectory. Robot end-effector along the X axis is only depicted, since similar behavior was noticed at the other axes

tracking system providing motion data at the frequency of 30 Hz. The resulted robot arm motion in the 3D space is shown in Fig. 23.7. As it can be seen, the resulted robot motion was very smooth when EMG-based estimates were used and very jerky when the position tracker measurements were used instead. It must be noted that filtering the position tracker measurements with an averaging filter was also attempted; however, the delays that the filter imposed to the desired motion made the user difficult to teleoperate the robot arm.

23.5 Conclusion and Future Extensions

In this chapter, a robotic platform driven by a novel control interface that is based on human body-oriented signals was analyzed. An EMG-based decoding model provided estimates of arm motion in 3D, in real-time. Then, a bio-inspired filter, realized through a Bayesian Network approach, improved the accuracy of the estimates, using human motion characteristics learned during model training, mathematically formed using joint angle dependencies for 3D arm movements. The final motion estimates were used to control a robot arm in 3D space, in real-time.

The novelty of the method proposed here can be centered around two main issues. First, human arm movements are modeled using a Bayesian Network, revealing significant joint angle dependencies, and finally constructing a probabilistic model for arm movements that can be used in many research fields (i.e., robotics, graphics, biomechanics). Second, using this Bayesian Network to filter the motion estimates of the EMG-based decoder, we achieved to improve the overall system performance, incorporating the human-like characteristics of the arm motion. In this way, the system is more robust in cases where the EMG-based decoding model could not track the human arm movements. This result enables the dexterous control of robotic devices, as presented in this study.

The necessity of introducing a natural human–robot interface was shown throughout this chapter. The subjects were easily acquainted with the device and the mapping between their arm motion and robot motion. Moreover, the simplicity and directness of the proposed interfaces allow its application to cases of robots being worn by human subjects, either as orthotic devices (exoskeletons) or as prosthetic devices.

With the use of EMG signals and robotic devices in the area of rehabilitation receiving increased attention during the last years, our method could be proved beneficial in this area. Moreover, the proposed method can be used in a variety of applications, where an efficient human–robot interface is required. For example, prosthetic or orthotic robotic devices mainly driven by user-generated signals can be benefited by the proposed method, concluding to a user-friendly and effective control interface. A comparison of the proposed interface with a conventional position tracking system was analyzed, and the advantages of the former were shown experimentally.

Another reason for which an EMG-based system is more efficient than a conventional position tracking system can be found if one realizes that EMG signals represent muscle activity, which is responsible not only for motion but also for force. Therefore, EMG signals can be used for the estimation of exerted force, when the user wants to teleoperate a robot in exerting force to the environment, or for the control of human–robot coupled system, i.e., arm exoskeletons. The authors have used EMG signals to decode arm exerted force and control human–robot coupled systems in the past (Artemiadis and Kyriakopoulos 2007a, 2009). Since conventional motion tracking systems cannot provide information about exerted force, EMG signals constitute a more convenient, inexpensive, and efficient interface for the control of robotic systems.

Future extensions of the proposed setup would entail the online learning of the models and algorithms used in the architecture. Changes in the neural signals should be detected and accommodated in the decoding model. Some ideas on how to detect changes at the EMG activity and compensate for them in the decoding process can be found in Artemiadis and Kyriakopoulos (2010b). However, algorithms that should be trainable online will certainly be beneficial for the field.

These algorithms would entail real-time tracking of EMG signal features that would essentially provide information necessary for online retraining the decoding model. Adaptation algorithms would be used to vary the decoding model parameters to different situations, according to the accuracy of the motion estimates.

Moreover, muscle activity or generally signals coming from the human nervous system carry much more information, not limited only to limb motion. For example, muscles not only actuate the limbs but also make the limb exert force at the environment, while control the impedance of the arm in cases of external disturbances. Therefore, the decoding of force and impedance information from nervous signals is also a very important extension of the presented work, which would especially benefit the control of orthotic devices, in which the problem of human–environment interaction through the worn robot is still an issue.

The way that human subjects control or generally interact with robots is undoubtedly very critical for ensuring efficiency and safety. Interfaces based on natural human *processes* can increase the directness of the human–robot interface, along with the degree of robustness for the robot control. Integrating the human subject in the loop and using as control signals information that is included in the human nervous system can benefit the field of robot control and mechatronics, since it augments the applications of robots in everyday life.

References

Artemiadis PK, Kyriakopoulos KJ (2005) Teleoperation of a robot manipulator using EMG signals and a position tracker. Proc of IEEE/RSJ Int Conf Intelligent Robots and Systems pp 1003–1008

Artemiadis PK, Kyriakopoulos KJ (2006) EMG-based teleoperation of a robot arm in planar catching movements using armax model and trajectory monitoring techniques. Proc of IEEE Int Conf on Robotics and Automation pp 3244–3249

Artemiadis PK, Kyriakopoulos KJ (2007a) EMG-based position and force control of a robot arm: Application to teleoperation and orthosis. Proc of IEEE/ASME International Conference on Advanced Intelligent Mechatronics, Switzerland

Artemiadis PK, Kyriakopoulos KJ (2007b) EMG-based teleoperation of a robot arm using low-dimensional representation. Proc of IEEE/RSJ Int Conf Intelligent Robots and Systems pp 489–495

Artemiadis PK, Kyriakopoulos KJ (2008) Assessment of muscle fatigue using a probabilistic framework for an EMG-based robot control scenario. Proc of IEEE Int Conf Bioinformatics and Bioengineering

Artemiadis PK, Kyriakopoulos KJ (2009) EMG-based position and force control of coupled human-robot systems: Towards EMG-controlled exoskeletons. In Experimental Robotics, Springer Berlin/Heidelberg pp 241–250

Artemiadis PK, Kyriakopoulos KJ (2010a) EMG-based control of a robot arm using low-dimensional embeddings. IEEE Transactions on Robotics 26(2):393–398

Artemiadis PK, Kyriakopoulos KJ (2010b) An EMG-based robot control scheme robust to time-varying emg signal features. IEEE Transactions on Information Technology in Biomedicine 14(3):582–588

Billard A, Mataric MJ (2001) Learning human arm movements by imitation: Evaluation of a biologically inspired connectionist architecture. Robotics and Autonomous Systems 37:2–3:145–160

Bitzer S, van der Smagt P (2006) Learning EMG control of a robotic hand: towards active prostheses. Proc of IEEE Int Conf on Robotics and Automation pp 2819–2823

Carmena JM, Lebedev MA, Crist RE, O'Doherty JE, Santucci DM, Dimitrov DF, Patil PG, C S Henriquez CS, Nicolelis MAL (2003) Learning to control a brain-machine interface for reaching and grasping by primates. PLoS, Biology 1:001–016

Cavallaro E, Rosen J, Perry JC, Burns S, Hannaford B (2005) Hill-based model as a myoprocessor for a neural controlled powered exoskeleton arm- parameters optimization. Proc of IEEE Int Conf on Robotics and Automation pp 4514–4519

Celani NML, Soria CM, Orosco EC, di Sciascio FA, Valentinuzzi ME (2007) Two-dimensional myoelectric control of a robotic arm for upper limb amputees. Journal of Physics: Conference Series 90

Chow CK, Liu CN (1968) Approximating discrete probability distributions with dependence trees. IEEE Transactions on Information Theory 14(3):462–467

Cram JR, Kasman GS (1998) Introduction to Surface Electromyography. Aspen Publishers, Gaithersburg, Maryland

d'Avella A, Portone A, Fernandez L, Lacquaniti F (2006) Control of fast-reaching movements by muscle synergy combinations. The Journal of Neuroscience 25(30):7791–7810

Dipietro L, Ferraro M, Palazzolo JJ, Krebs HI, Volpe BT, Hogan N (2005) Customized interactive robotic treatment for stroke: EMG-triggered therapy. IEEE Transactions on Neural Systems and Rehabilitation Engineering 13(3):325–334

Duda RO, Hart PE, Stork DG (2001) Pattern classification. Wiley, New York

Farry KA, Walker ID, Baraniuk RG (1996) Myoelectric teleoperation of a complex robotic hand. IEEE Transactions on Robotics and Automation 12(5):775–788

Fod A, Mataric MJ, Jenkins OC (2002) Automated derivation of primitives for movement classification. Autonomous Robots 12(1):39–54

Fukuda O, Tsuji T, Kaneko M, Otsuka A (2003) A human-assisting manipulator teleoperated by EMG signals and arm motions. IEEE Transactions on Robotics and Automation 19(2):210–222

Hill AV (1938) The heat of shortening and the dynamic constants of muscle. Proc R Soc Lond Biol pp 136–195

Hiraiwa A, Shimohara K, Tokunaga Y (1989) EMG pattern analysis and classification by neural network. Proc of IEEE Int Conf Systems, Man, Cybernetics pp 1113–1115

Huang HP, Chen CY (1999) Development of a myoelectric discrimination system for a multidegree prosthetic hand. Proc of IEEE Int Conf on Robotics and Automation pp 2392–2397

Jackson JE (1991) A user's guide to principal components. Wiley, New York, London, Sydney

Jacobson SC, Knutti DF, Johnson RT, Sears HH (1982) Development of the utah artificial arm. IEEE Transactions on Biomedical Engineering 29(4):249–269

Jerard RB, Williams TW, Ohlenbusch CW (1974) Practical design of an EMG controlled above elbow prosthesis. Proc of Conf Engineering Devices for Rehabilitation pp 73–73

Jolliffe IT (2002) Principal component analysis. Springer, New York, Berlin, Heidelberg

Kato I, Okazaki E, Kikuchi H, Iwanami K (1967) Electropneumatically controlled hand prosthesis using pattern recognition of myo-electric signals. In Dig 7th ICMBE pp 367–367

Kazerooni H (1990) Human-robot interaction via the transfer of power and information signals. IEEE Transactions on Systems, Man, and Cybernetics 20(2):450–463

Ljung L (1999) System identification: Theory for the user. Upper Saddle River, NJ: Prentice-Hall

Lloyd DG, Besier TF (2003) An EMG-driven musculoskeletal model to estimate muscle forces and knee joint moments in vivo. Journal of Biomechanics 36:765–776

Manal K, Buchanan TS, Shen X, Lloyd DG, Gonzalez RV (2002) Design of a real-time EMG driven virtual arm. Computers in Biology and Medicine 32:25–36

McLachlan G, Peel D (2000) Finite mixture models. Wiley, New York

Mpompos NA, Artemiadis PK, Oikonomopoulos AS, Kyriakopoulos KJ (2007) Modeling, full identification and control of the mitsubishi PA-10 robot arm. Proc of IEEE/ASME International Conference on Advanced Intelligent Mechatronics, Switzerland

Mussa-Ivaldi FA, Bizzi E (2000) Motor learning: the combination of primitives. Philosophical Transactions of the Royal Society of London B 355:1755–1769

Park J, Khatib O (2006) A haptic teleoperation approach based on contact force control. International Journal of Robotics Research 25(5-6):575–591

Potvin J, Norman R, McGill S (1996) Mechanically corrected EMG for the continuous estimation of erector spine muscle loading during repetitive lifting. European Journal of Applied Physiology 74:119–132

Schulz S, Pylatiuk C, Reischl M, Martin J, Mikut R, Bretthauer G (2005) A hydraulically driven multifunctional prosthetic hand. Robotica 23:293–299

Scott RN, Parker PA (1988) Myoelectric prostheses: state of the art. Journal of Medical Engineering and Technology 12(4):143–151

Tenore F, Ramos A, Fahmy A, Acharya S, Etienne-Cummings R, Thakor N (2007) Towards the control of individual fingers of a prosthetic hand using surface EMG signals. Proc 29th Annual International Conference of the IEEE EMBS pp 6145–6148

Woo-Keun Y, Goshozono T, Kawabe H, Kinami M, Tsumaki Y, Uchiyama M, Oda M, Doi T (2004) Model-based space robot teleoperation of ets-vii manipulator. IEEE Transactions on Robotics and Automation 20(3):602–612

Xiang Y (2002) Probabilistic reasoning in multiagent systems: A graphical models approach. Cambridge University Press, Gaithersburg, Maryland

Zajac FE (1986) Muscle and tendon: Properties, models, scaling, and application to biomechanics and motor control. Bourne, J R ed CRC Critical Reviews in Biomedical Engineering 17:359–411

Zecca M, Micera S, Carrozza MC, Dario P (2002) Control of multifunctional prosthetic hands by processing the electromyographic signal. Critical Reviews in Biomedical Engineering 30(4–6):459–485

Index

V. Cutsuridis et al. (eds.), *Perception-Action Cycle: Models, Architectures,*
and Hardware, Springer Series in Cognitive and Neural Systems 1,
DOI 10.1007/978-1-4419-1452-1, © Springer Science+Business Media, LLC 2011